The Science of Cooking

The Science of Cooking

Understanding the Biology and Chemistry Behind Food and Cooking

2nd Edition

Joseph J. Provost
Keri L. Colabroy
Brenda S. Kelly
Ashley L. Corrigan Steffey
Mark A. Wallert

Library of Congress Cataloging-in-Publication Data

Names: Provost, Joseph J., author. | Colabroy, Keri L., author. | Kelly,
  Brenda S., author. | Steffey, Ashley L. Corrigan, author. | Wallert,
  Mark A., author.
Title: The science of cooking : understanding the biology and chemistry
  behind food and cooking / Joseph J. Provost, Keri L. Colabroy, Brenda S.
  Kelly, Ashley L. Corrigan Steffey, Mark A. Wallert.
Description: 2nd edition. | Hoboken, New Jersey : Wiley, [2025] | Includes
  bibliographical references and index.
Identifiers: LCCN 2025001247 | ISBN 9781394158218 (paperback) | ISBN
  9781394158232 (epub) | ISBN 9781394158225 (adobe pdf)
Subjects: LCSH: Food–Analysis. | Biochemistry. | Food–Composition. |
  Food–Biotechnology.
Classification: LCC TX545 .P76 2025 | DDC 664/.07–dc23/eng/20250208
LC record available at https://lccn.loc.gov/2025001247

Cover Design: Wiley
Cover Images: © vitec/Shutterstock, © Olga Cranga/Shutterstock, © Hayati Kayhan/Shutterstock, © MaraZe/Shutterstock, © twomeows/Getty Images, © dem10/Getty Images, Courtesy of Joseph J. Provost

SKY10100670_032125

Contents

Preface

We are excited to present the second edition of "*The Science of Cooking. Understanding the Biology and Chemistry Behind Food and Cooking, Second Edition*" a comprehensive guide designed to enhance the learning experience for both students and instructors. With the addition of Ashley Corrigan Steffey as co-author, drawing from her extensive experience in teaching the course over the past several years, we bring exciting new ideas and perspectives to this edition.

In response to valuable feedback from both faculty and students, this edition introduces several key features that will help deepen understanding of the scientific principles behind food and cooking. Each chapter now includes clearly defined learning objectives and key concepts, which will enable students to focus on foundational scientific principles that are critical to understanding the material. Additionally, we have incorporated practice problems for each chapter to reinforce these principles and assist instructors in guiding students through challenging concepts.

One exciting new addition is the "Science for the Chef" section, which offers a diverse set of recipes with detailed explanations of the scientific processes involved in preparing each dish. This section bridges the gap between theory and practice, showing how the chemistry and biology discussed in the text manifest in the kitchen. By grounding scientific concepts in everyday cooking experiences, we aim to make the material more accessible and engaging for non-science majors.

In addition to these new pedagogical tools, we have restructured the content to make it more approachable and user-friendly. The first chapter, which previously combined both basic chemistry and biochemical concepts, has been divided into two distinct sections. The new introductory chemistry chapter focuses on essential topics such as bonding, molecular forces, and macromolecules—key elements necessary for understanding the rest of the book. The more advanced biochemical material has been relocated to its own chapter, ensuring that students are well-equipped with foundational knowledge before diving into more complex subjects. This reorganization allows for a more logical progression of ideas and a clearer presentation of material.

We are also pleased to announce the inclusion of updated and expanded inquiry-based cooking labs, now integrated directly into the text. These labs provide hands-on, experiential learning opportunities that encourage students to design experiments, test hypotheses, and explore the science behind food preparation. To further support this inquiry-driven approach, we have made the guided inquiry labs freely available online for instructors who adopt the text. These resources are designed to foster collaboration and critical thinking in the classroom, ensuring that students not only understand the scientific content but also engage with it actively.

The second edition of "*The Science of Cooking. Understanding the Biology and Chemistry Behind Food and Cooking, Second Edition*" retains the core philosophy of the first: to provide a robust scientific foundation for understanding food and cooking while offering engaging, practical applications of that knowledge. As we continue to refine and expand upon the material, our goal remains the same—to create a resource that helps instructors teach, students learn, and everyone enjoy the process of discovering the science behind the food we eat.

We sincerely hope that this new edition will inspire students and educators alike to explore the fascinating world of food science and that the tools and resources provided will help make teaching and learning about the science of food an enjoyable and enriching experience.

Inquire, Learn, Investigate, and Eat Well!

Acknowledgments

We acknowledge all the faculty and students that have motivated and inspired us to rethink and revise our content for this second edition. We also gratefully acknowledge the critically helpful collaboration of our student research assistant Dhivya Shepherd (Muhlenberg College, Class of 2026), whose diligent and persistent efforts brought this second edition to completion.

About the Authors

Dr. Joseph J. Provost is a professor of chemistry and biochemistry at the University of San Diego. He has helped create and teach a science of cooking class and taught to small and large classes. Provost has served on educational and professional development committees for the American Society for Biochemistry and Molecular Biology and the American Chemical Society while teaching biochemistry, biotechnology, introductory chemistry laboratories, and lectures. In addition to teaching, he is one of the leaders to integrate research into the classroom using malate dehydrogenase (MDH CURES) and studying the role of intracellular pH and metabolism in disease. When not in the lab or class, Provost can be found making wine, cheese, grilling, and then playing or coaching hockey.

Dr. Keri L. Colabroy is a professor of chemistry at Muhlenberg College in Allentown, Pennsylvania, where she created and teaches a course on Kitchen Chemistry for non-science majors. When she isn't evangelizing non-science majors with her love of chemistry, Colabroy is teaching organic chemistry, biochemistry courses, and a first-year writing course on coffee while also serving as director of the biochemistry program. Her scholarly research is in the area of bacterial antibiotic biosynthesis with a focus on metalloenzymes and actively involves undergraduates. Colabroy serves as coordinator for undergraduate research at the college and participates on the Council for Undergraduate Research in the Division of Chemistry. When not in the lab or class, Colabroy can be found driving her kids around to their activities or singing at church.

Dr. Brenda S. Kelly is an associate professor of biology and chemistry at Gustavus Adolphus College in St. Peter, Minnesota. Kelly's immersion into teaching about science and cooking began when she co-taught The Chemistry of Cooking, a course that enrolled science majors who knew little about cooking and non-science majors who were excellent cooks. Shortly after the publication of the first edition of this book, Kelly transitioned in an administrative role as Provost and Dean of the College at Gustavus Adolphus College. She has appreciated remaining involved in the creation and updating of this 2nd edition, as it fulfils her vocation as a STEM teacher and advocate.

Ashley L. Corrigan Steffey, M.S., is an adjunct chemistry instructor who has taught the science of food and cooking at the University of San Diego for several years, blending academic expertise with personal passion. Growing up on a family dairy farm in central Minnesota, Ashley developed a deep connection to food science by helping grow and preserve fruits and vegetables, milking cows, and processing animals for meat. This hands-on experience, combined with a love for baking and recipe experimentation, fuels her approach to teaching. In addition to their work at the University of San Diego, Ashley is an adjunct instructor at Miramar College in San Diego, CA.

Mark A. Wallert, Ph.D. is a Professor Emeritus of Biology at Bemidji State University, in Bemidji, Minnesota. Mark retired in August 2023 following a 33-year career in the Minnesota State University System, which included 25 years at Minnesota State University Moorhead and eight years at Bemidji State University. Mark's career highlights include having over 240 undergraduates participate as members of his research team, having a 26-year research collaboration with Joseph J. Provost, and being named Carnegie Foundation Council for Advancement and Support of Education Minnesota College Professor of the Year in 2005.

About the Companion Website

Instructors can find additional materials, including guided-inquiry activities to accompany book chapters, and laboratory exercises, on the companion website: www.wiley.com/go/provost/food_science_2e.

1

Atoms, Elements, Compounds, and Molecules

Guided Inquiry Activities (Web): *1, Elements, Compounds, and Molecules; 2, Bonding; 3, Mixtures and States of Matter; 4, Water; 8, pH*

Learning Objectives

1. *Name and describe the particles that make up an atom.*
2. *Describe the relationship of atoms and elements to the periodic table.*
3. *Define and recognize ions: cations and anions.*
4. *Recognize and identify ionic bonds and covalent bonds.*
5. *Recognize and distinguish between ionic compounds and covalent molecules.*
6. *Describe what holds ionic compounds together and how that is different from what holds covalent molecules together.*
7. *Describe the types of attractions that occur between compounds and molecules: electrostatic, hydrogen bonding, and hydrophobic attractions, and be able to identify examples of these interactions.*
8. *Describe the intramolecular and intermolecular bonds that make up water, and be able to explain the intermolecular forces that cause freezing, melting, and vaporization of water.*
9. *Describe what makes something acidic, and what molecules are acids.*

1.1 Introduction

The process of cooking, baking, and preparing food is essentially an applied science. Anthropologists and historians venture that cooking originated when a pen holding pigs or other livestock caught fire or a piece of the day's catch of mammoth fell into the fire pit. The smell of roasted meat must have enticed early people to "try it"; the curious consumers found culinary and nutritional benefits to this discovery. The molecular changes that occurred during cooking made the meat more digestible and the protein and carbohydrates more readily available as nutrients. Contaminating microbes were eliminated during cooking, which made the consumers more healthy and able to survive. Moreover, the food was tastier due to the heat-induced chemical reactions between the oxygen in the air and the fat, proteins, and sugar in the meat. Harnessing the knowledge of what is happening to our food at the molecular level is something that good scientists and chefs use to create new appetizing food and cooking techniques.

The Science of Cooking: Understanding the Biology and Chemistry Behind Food and Cooking, Second Edition. Joseph J. Provost et al.
© 2025 John Wiley & Sons, Inc. Published 2025 by John Wiley & Sons, Inc.
Companion website: www.wiley.com/go/provost/food_science_2e

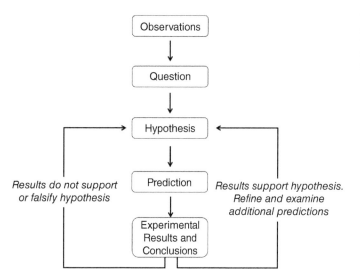

Figure 1.1 The scientific method. Scientists use a testable method originating from observations to generate a testable hypothesis to conduct their work. A cook or baker can also use this method to create more interesting food.

We are all born curious. Science and cooking are natural partners where curiosity and experimentation can lead to exhilarating and tasty new inventions. Scientific discovery is driven by hypothesis (see Figure 1.1 for a model of the scientific method). An observation of an event creates a question and/or a statement that explains the observation or phenomenon: the hypothesis. The hypothesis can then be tested by a series of experiments and controls that supports or falsifies the hypothesis, starting the cycle over again. For example, a scientist might observe that the growth rate of cancer cells in a petri dish slows when the cells are exposed to a sea sponge. The scientist may then hypothesize that a molecule found in the sponge binds to a protein in cancer cells. After adding the compound to a tumor, its growth slows and the cells die. Looking at how all of the individual molecules found in the sea sponge affect the growth of cancer cells can test this hypothesis. These experiments can lead to a more advanced hypothesis, testing, and eventually finding a new compound that can be used to fight cancer.

Cooking can also be a hypothesis-driven process that utilizes biology, chemistry, and physics. As you cook, you use biology, chemistry, and physics to create hypotheses in the kitchen, even if you aren't aware of being a scientist. Each time you try a recipe, you make observations. You may ask yourself questions about what you added to the concoction or how the food was baked or cooked. This creates a hypothesis or a statement/prediction that you can test through experimentation (your next attempt at the dish). A nonscientific idea is often approached as something to prove. That is different from hypothesis testing. A hypothesis is falsified rather than proven by testing. Cooking does just this; it will falsify your test rather than prove it. Tasting, smelling, and visualizing your results tell you if your hypothesis was supported or falsified. If wrong, you may create a new hypothesis that might be generated by the time you have washed the dishes from your first experiment. Learning more of the basic science behind food and cooking will help you appreciate the world around you and become a better scientist and a better cook, baker, and consumer.

1.2 Fundamentals of Food and Cooking

Bread baking provides a great example of the importance of having a scientific understanding of cooking and baking. Take a close look at bread. Notice that it is made of large and small caves surrounded by a solid wall (Figure 1.2).

The key to bread is making a way to trap expanding gases in the dough. Adding water to flour and sugar allows for the hydration and mixing of proteins and carbohydrates. Kneading the dough stretches a protein network called gluten, which allows for an interconnected network of protein ready to trap gas that is generated by the yeast. During the proofing step of making bread, the yeast converts sugar into energy-filled molecules, ethanol, carbon dioxide gas, and other flavorful by-products. The heat applied during baking allows the water to escape as steam, which expands the bread, links the gluten protein molecules further, and traps carbon dioxide gas. While this is happening, the heat catalyzes chemical reactions between proteins and sugars, creating a beautiful brown color, a dense texture, and over 500 new aromatic compounds that waft to your nose. Clearly, there is a lot of science that goes into making a loaf of bread.

Preparing food and drink is mostly a process of changing the chemical and physical nature of the food. Molecules react to form new compounds; heat changes the nature of how food molecules function and interact with each other, and physical change brings about new textures and flavors to what we eat. To gain a better appreciation for these chemical and physical processes, a fundamental understanding of the building blocks of food and cooking must first be understood. In the following chapter, we will study the biological macromolecules that make up the protein, carbohydrates, and fat in our diets.

Figure 1.2   Structure of bread. A close look at bread demonstrates the requirement of proteins and carbohydrates needed to trap expanding gases. gmeviphoto/Adobe Stock Photos

One of the most important building blocks of food is water; our bodies, food, and environment are dependent on the unique chemistry and biology of this molecule. Large biological molecules such as proteins, carbohydrates, and fats comprise the basic building blocks of food. Smaller molecules, including vitamins, salts, and organic molecules, add important components to cooking and the taste of food. Finally, the basics of plant and animal cells and cellular organization are key to understanding the nature of food and cooking processes. However, before we get into some of the science fundamentals, it is important to recognize and acknowledge the origins of the chefs who first embraced the science behind their profession.

1.2.1   Science, Food, and Cooking

Many chefs and bakers embrace the collaboration of science and food. Historically, one means whereby science has been utilized in the kitchen is in the area of food technology; the discipline in which biology, physical sciences, and engineering are used to study the nature of foods, the causes of their deterioration, and the principles underlying food processing. This area of food science is very important in ensuring the safety and quality of food preparation, processing of raw food into packaged materials, and formulation of stable and edible food. College undergraduates can major in "food science" or attend graduate studies in this area, working for a food production company where they might look at the formulation and packaging of cereals, rice, or canned vegetables. Recently, a new marriage of science and food, coined molecular gastronomy, has grown to influence popular culture that extends far beyond the historical definition of food science. A physicist at Oxford, Dr. Nicholas Kurti's interest in food led him to meld his passion for understanding the nature of matter and cooking. In 1984, Harold McGee, an astronomist with a doctorate in literature from Yale University, wrote the first edition of the influential and comprehensive book *On Food and Cooking: The Science and Lore of the Kitchen* [1]. This fascinating book is the basis for much of the molecular gastronomy movement and describes the scientific and historical details behind most common (and even uncommon) culinary techniques. Together with cooking instructor Elizabeth Cawdry Thomas, McGee and Kurti held a scientific workshop/meeting to bring together the physical sciences with cooking in 1992 in Erice, Italy. While there were more scientists than chefs attending, with a 5 : 1 ratio, the impact of the meeting was significant. It was at Erice that the beginnings of what was then called molecular and physical gastronomy became the catalyst for an unseen growth in science and cooking. Hervé This, a chemist who studies the atomic and subatomic nature of chemistry, attended the workshop and has been a key player in the growth of molecular gastronomy. Dr. This blames a failed cheese soufflé for sparking his interest in culinary precisions and has since transformed into a career in molecular gastronomy. Other participants of the meetings include chef Heston Blumenthal and physicist Peter Barham, who have collaborated and influenced many molecular-based recipes and projects. Finally, another scientist, biochemist Shirley O. Corriher, was present at these early meetings (Box 1.1). Shirley found her love for cooking as she helped her husband run a school in Nashville, nearby the Vanderbilt Medical School, where she worked as a biochemist. Her influence on science and cooking includes the friendship and advice of Julia Child and culminated in the many informative, scientifically-minded cookbooks she has written and published (Ms. Corriher, personal communication). We can also find evidence of modernist cooking in popular culture. For 13 years, Alton Brown brought a scientific approach to the culinary arts with the television series *Good Eats*. Through the work of these many talented scientist chefs, we can see the impact of scientific thought and practice in the kitchen, not only in the use of liquid nitrogen and the development of the *sous vide* cooking method, but also in the unique presentation and mixtures of flavors to challenge the daring foodie.

Box 1.1 Shirley Corriher

Shirley Corriher is one of the original scientists and cooks to bring scientific thought and reasoning into her cooking and baking. Using everyday language as a way to explain food science, Shirley authored the books *CookWise* [2] and *BakeWise* [3], which each paired classic recipes with the science behind the cooking or baking and turned regular people into kitchen success stories. Alongside her books, Shirley's friendship with Julia Child and her appearances on several of Alton Brown's *Good Eats* episodes helped to popularize the science of food, cooking, and baking. But Shirley wasn't always a brilliant cook, she began her career as a scientist. Shirley earned a degree in biochemistry from Vanderbilt University, where she worked in the medical school in a biomedical research laboratory while her husband ran a school for boys. She recalls her early attempt to cook for a large number of boys. Little did she know this experience would be the beginning of a new career. Shirley describes how she struggled with the eggs sticking to the pan and worrying that there would be no food for the students. Eventually, she learned to heat the pan before adding the eggs. The reason behind this solution was that the small micropores and crevices of the pan would fill with egg and solidify in the pan. This revelation sparked her desire to make connections between science and cooking. After a divorce, Shirley and her sons were forced into a financial struggle, where they had to use a paper route as a source of income. A friend, Elizabeth Cawdry Thomas, who ran a cooking school in Berkeley, California, asked Shirley to work for her cooking school. It was there that Shirley learned formal French cooking on the job. It wasn't long before Shirley found herself mixing with a group of scientists and chefs who appreciated the yet-to-be-studied blend of science and cooking. In 1992, the group, which included Thomas, Kurti, and Harold McGee, obtained funds to bring scientists and chefs together in Erice, Sicily, for workshops on non-nuclear proliferation. Shirley was a presenter at that first meeting, leading discussions on emulsifiers and sauces, and she continued as a participant in each of these early workshops (Ms. Corriher, personal communication). Corriher recalls that the term "molecular gastronomy" was voted on by the core group to reflect both the science and culinary aspects of the meeting. Shirley also talks about the mutual respect and friendship between her and leading food scientist Harold McGee. Shirley recalls reading his book and calling him to ask, "where had he been all this time?" She said, "You don't know me, but I and many other ladies in Atlanta are going to bed with you every night!" Shirley's books using science to explain how to become a better cook and baker are extremely popular. She teaches home cooks to not only trust the science of the kitchen, but also to trust in themselves, and her work has ultimately codified her essential role in American culinary society.

1.3   The Real Shape of Food: Molecular Basics

1.3.1   The Atom

What are the fundamental units of all food and cooking processes? Atoms and molecules. All living systems (animals, microbes, and smaller life forms) are made of atoms and molecules. How these atoms and molecules are organized, interact, and react provides the building blocks and chemistry of life. It makes sense that to best understand cooking and baking at the molecular level, you must first appreciate how atoms and molecules are put together and function. Let's start with the basics and ask: what is the difference between an atom and a molecule? The answer is simple: an atom is the smallest basic building block of all matter, while molecules are made when two or more atoms are connected.

An atom consists of three main components: protons, neutrons, and electrons. Simply put, protons and neutrons are found in the center or nucleus of the atom, while electrons orbit the nucleus of the atom (Figure 1.3). Protons are positively charged particles with an atomic mass of one atomic mass unit. Neutrons also have an atomic mass of 1 but do not have an electrical charge. Electrons have almost no mass and have an electrical charge of −1. Figure 1.3 shows an example of the periodic table entry for the element carbon. Carbon has 6 protons, which also dictates its number and position on the periodic table. Below the number 6 is the large capital letter C, which is the standard abbreviation for carbon used in all structural drawings. The number 12.011 beneath the C is the average atomic mass of carbon, or the weighted average mass of the atoms that naturally occur in a sample of the element.

What is an average atomic mass? All carbon atoms must have 6 protons, for the proton number is what defines an atom as a particular atom. Most carbon atoms have 6 neutrons, while a small number of carbon atoms have 7 neutrons, and an even smaller number have 8 neutrons. Protons and neutrons constitute the atomic mass of the atom because electrons are too small to add significant mass. Therefore, atoms with additional neutrons have additional atomic mass. A carbon atom with 6 protons and 6 neutrons has an atomic mass of 12, but a carbon with 6 protons and 7 neutrons has an atomic mass of 13. The average atomic mass is a weighted average of the mass carbon atoms found in nature. Nearly 99% of the carbon atoms found in nature have an atomic mass of 12, while just over 1% of the carbon in nature has an atomic mass of 13. Only the tiniest fraction of a percent of the carbon in nature has an atomic mass of 14. So the average atomic mass is the hypothetical mass of a carbon atom, that is, 99% with 12 atomic mass units, 1% with 13 atomic mass units and <1% with 14 atomic mass units. The result is an average atomic mass of 12.011. A similar process is followed to determine the average atomic mass of every element on the periodic table.

The elements of the periodic table are arranged and defined by the number of protons present within an atom of a given element. It is the number of protons that defines an atom, not the electrons or neutrons. A quick examination of a periodic table (Figure 1.4) shows that the types of elements are organized by the number of protons that an atom of that element has: from the smallest atom of the element hydrogen to the largest atom, which belongs to the element ununoctium. Any atom with 6 protons belongs to the element carbon; any atom with 7 protons is the element nitrogen. Thus, if a carbon atom gains a proton, it becomes a

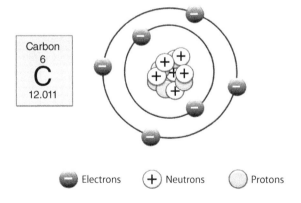

Figure 1.3   Atomic structure. Atoms are made of electrons in orbitals around the nucleus where protons and neutrons are found. The identity of an atom comes from the number of protons. On the left, the elemental symbol for carbon is shown. Carbon has 6 protons, which also dictates its number and position on the periodic table. The large capital letter C is the standard abbreviation for carbon used in all structural drawings. The number 12.011 beneath the C is the average atomic mass of carbon, which accounts for the naturally occurring isotopes.

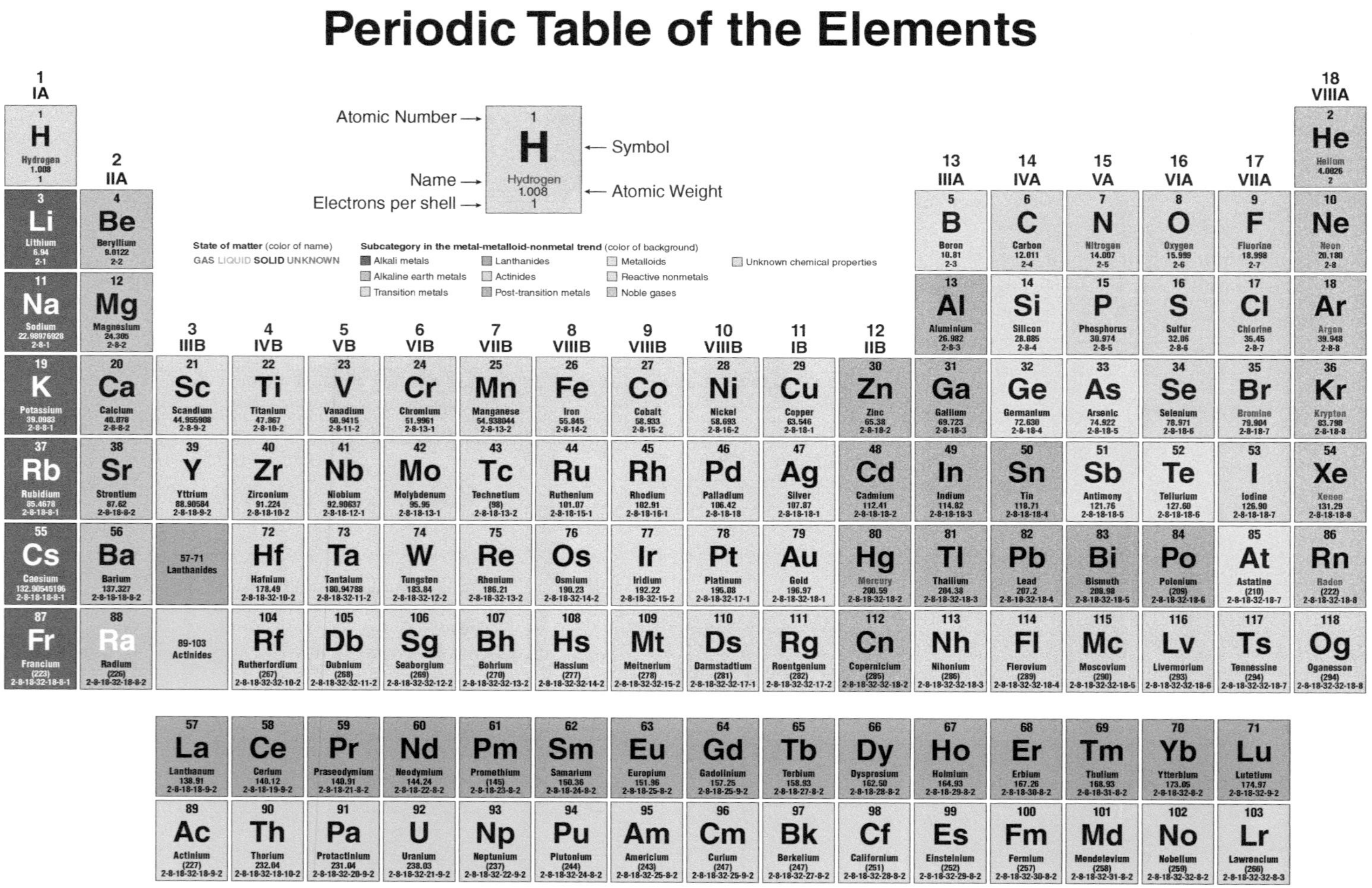

Periodic Table of the Elements

Figure 1.4 Periodic table. Each atom is arranged based on the number of protons (elemental number) increasing from left to right and top to bottom. Scientists use the periodic table to understand the physical characteristics of elements. Humdan/Adobe Stock Photos

nitrogen atom. However, if a carbon atom gains or loses an electron, it still is the element carbon, but now has a charge associated with it.

The total number of protons and electrons defines the charge of an atom. An atom of any element with an equal number of protons and electrons will have a net neutral charge; atoms that have gained an electron will have a negative charge and are called anions, and those that have lost an electron will have a positive charge and are called cations. Most of the atoms of the elements on the periodic table can gain or lose one or more electrons and become ions. Ions are very important in food; for example, table salt has the chemical name sodium chloride and is made of sodium cations and chloride anions. Both sodium and chlorine are found on the periodic table (Figure 1.4); sodium is at the far left end of row 3, while chlorine is at the far right end. When a sodium atom gives up an electron, it becomes a sodium cation. Chemists represent the sodium cation with the symbol Na^+, where the "Na" refers to the symbol for sodium, and the + sign indicates the +1 charge that the sodium cation carries since it is missing 1 electron. In the same way, chlorine accepts an extra electron and becomes the chloride anion, with the symbol Cl^-.

The numbers of neutrons within a given type of atom can also vary, and we call those isotopes. Isotopes are atoms that have the same number of protons but differ in the number of neutrons. Isotopes are interesting in chemistry but are not significant to the study of food, so we will not discuss about isotopes here.

Food is made of compounds and molecules, which are combinations of atoms and ions. So what about a compound or a molecule? How does a molecule differ from an atom or compound? A compound is a chemical substance made of different atoms or ions. Compounds can be made of ions held together by ionic bonds, atoms held together by covalent bonds, or combinations of both. Most of the compounds found in living things—and, therefore, in our food—contain the elements carbon, hydrogen, nitrogen, and oxygen. A group of other elements, including sulfur, magnesium, and iron, make up less than 1% of the elements in most living systems. Trace elements, such as copper, zinc, chromium, and even arsenic, although necessary for biological function, only make up a minute portion of an organism, less than 0.01% of all atoms and ions.

A molecule is a substance of two or more atoms connected by the sharing electrons through covalent bonds. Thus all compounds are molecules, but not all molecules are compounds. Molecules are often categorized further into organic, which are those molecules containing mostly carbon atoms, versus inorganic molecules, which are not primarily made of carbon. Since food is made of compounds and molecules, and compounds and molecules are made of atoms and ions that have chemically joined, let's talk a little bit more about the bonds that connect atoms together.

1.3.2 Ionic Compounds

We just learned that a compound is a chemical substance made of different atoms or ions. Compounds can be made of ions held together by ionic bonds, atoms held together by covalent bonds or combinations of both. As we begin our discussion of bonding, let's look first at the bonds that form between ions. Remember, ions are atoms that have either given up an electron to become a cation (e.g. Na^+) or gained an electron to become an anion (e.g. Cl^-). Ionic bonds are electrostatic; that means ionic bonds are electrical attractions of positive and negative charge that form between a cation and anion. The resulting molecule is called an ionic compound or a salt. This terminology is apropos because the salt that you sprinkle on your popcorn, NaCl, is an ionic compound consisting of a positively charged sodium cation (Na^+) and a negatively charged chloride anion (Cl^-). You should also notice that in the name of the compound—sodium chloride—the cation is listed first, and the anion is listed second. That is the naming convention for all ionic compounds. Potassium bromide is another ionic compound; it is a combination of a potassium cation (i.e. K^+) and a bromide anion (i.e. Br^-). Sodium and potassium are types of elements that chemists call metals. Metals are on the left-hand side of the periodic table, and atoms of metals readily give up electrons. That means that metals readily transform their atoms into positively charged cations. One of the ions in an ionic compound will have at least one metal element (Na, K, Ca, Al, etc.).

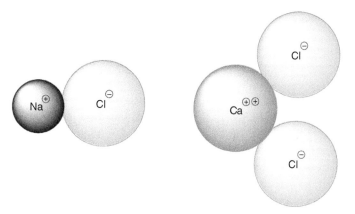

Figure 1.5 Ionic compounds: sodium chloride (i.e. NaCl) and calcium chloride (i.e. $CaCl_2$). A positively charged cation (Na^+) forms an ionic bond with a negatively charged anion (Cl^-) to form an ionic compound. Calcium gives up 2 electrons to form a cation of charge +2, and it is balanced by two chloride anions of charge −1.

The simplest ionic compounds are formed from monoatomic ions, where two ions of opposite charge act as the functional unit (Figure 1.5). A good example is table salt, or sodium chloride (NaCl). In sodium chloride, the positively charged Na^+ is made from one atom, and the chloride anion (Cl^-) is also made from a single atom. In addition to monatomic ions, a group of bonded atoms can also possess an overall charge; these are called polyatomic ions. Polyatomic ions are made of several atoms bonded as a group, and one or more of the atoms in that group carry a charge. Potassium nitrate, commonly called saltpeter and used in curing meat, is a complex polyatomic ion. Potassium nitrate has the chemical formula KNO_3, where the potassium ion (K^+) provides the positive charge and the nitrate ion provides the negative charge (NO_3^-). Nitrate compounds have been historically used to preserve meats and fish. The nitrate dries the meat by drawing the water out of the muscle tissue, leaving an inhospitable environment for bacteria to grow.

Ionic compounds or salts are essential to our experience of foods, cooking, and taste. Salts can often be the key to the demise or success of a given dish (Box 1.2). Thus, when we refer to salts throughout the rest of this text, we will specify whether we are using the scientific definition of salt (an ionic compound made up of a cation and anion) or the common definition of salt (meaning table salt, or NaCl).

Box 1.2 Figuring out food: Is pink Himalayan salt healthier than table salt?

Walking down the seasoning aisle in the grocery store, we are confronted with several types of salt: table salt, kosher salt, rock salt, sea salt, and the increasingly popular Himalayan pink salt. First, let's be clear: all these salts originate from the sea, whether via solar evaporation of seawater or mined from the remnants of ancient oceans that lie underground. All of these salts are made exclusively of sodium chloride (NaCl). The main difference between the salts lies solely in the size and shape of the crystal structure, which impacts how quickly the salt dissolves in a food dish or interacts with your taste buds. The crystal structure's size and shape also affect the salt's density or mass-to-volume ratio. Figure 1.6 shows that not all salts are created equal, so paying particular attention to the type of salt called for in a recipe is essential. Minor differences in the chemical makeup of salts arise from trace minerals (Ca, Mg, K, Cu, Zn, and Fe), soil sediments, and even organic compounds from algae.

Why do we use salt? Not only does salt have its unique taste, but it is also a flavorant: a flavor enhancer. Salt helps to balance a dish by reducing bitterness, it enhances the flavor profile of dishes rich in sugar or acid, and it even helps to release aroma compounds in food that our nasal receptors will detect. Not convinced that salt is a flavorant? Take two soft caramel candies and lightly salt one piece with flaky sea salt. Try the unsalted caramel candy first, followed by the salted caramel. You will be surprised that a small salt pinch can release a flavor burst.

So what is pink Himalayan salt? Pink Himalayan salt is a type of rock salt mined from the Salt Range mountains in Pakistan and is routinely portrayed as being healthier than table salt due to the impurities that give it its pink hue. However, a recent study concludes that while pink Himalayan salt contains small amounts of nutritive minerals, the quantity is insufficient to impact your daily nutrient intake (Table 1.1) [4]. Several brands of pink Himalayan salt were found to have trace amounts of nonnutritive minerals like lead, cadmium, mercury, and even arsenic [4]. If you want to add a delightful hue of pink salt to make a dish aesthetically pleasing, by all means, do! Just remember that pink Himalayan salt contains roughly the same amount of sodium chloride as table salt.

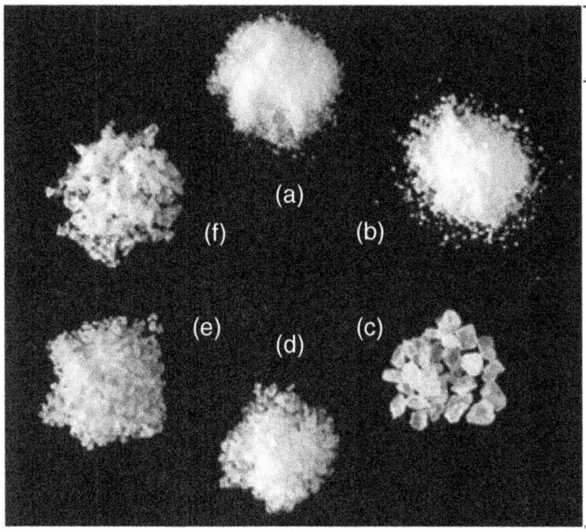

Salt Type	Mass per 1 Tablespoon
(a) Table Salt	15.3 grams
(b) Kosher Salt (Morton's)	12.9 grams
(c) Rock Salt (Morton's)	14.9 grams
(d) Coarse Sea Salt (Morton's)	14.7 grams
(e) Pink Himalayan Salt	13.5 grams
(f) Flake Sea Salt (Maldon)	5. 9 grams

Figure 1.6   Common salts. Six types of salts and corresponding mass per 1 tablespoon of each salt type.

Table 1.1   Mean % mass of nutrient and non-nutrient minerals of pink Himalayan sea salt and iodized table salt. Data converted from mg nutrient (or non-nutrient) per kg salt to % mass Adapted from [4].

% Mass (g nutrient/100 g salt)	Iodized table salt (%)	Pink Himalayan salt (%)
Sodium (Na)	42.8	39.5
Sulfur (S) (non-nutrient)	0.04	0.73
Calcium (Ca)	0.04	0.27
Potassium (K)	0.02	0.24
Magnesium (Mg)	0.008	0.27

1.3.3   Covalent Compounds and Molecules

We just learned that ionic compounds or salts are made from electrostatic attractions between ions of opposite charges. But there are also compounds made of uncharged or neutral atoms. Compounds made of chemically joined neutral atoms are called covalent or molecular compounds. In covalent compounds, atoms are held together by shared electrons, and the force that ties the atoms together is called a covalent bond. Covalent compounds are made up of nonmetal elements like carbon, oxygen, and nitrogen. As a group, molecular compounds are much more diverse in how their atoms are connected, and they are generally larger (i.e. contain more atoms) than ionic compounds. The main difference between ionic and covalent compounds is that covalent compounds are not held together by charges; instead, atoms are bonded together by mutually sharing electrons in a covalent bond.

The amino acid glycine is a component of protein and a great example of a covalent compound (Figure 1.7). In a molecule of glycine, each nitrogen, carbon, oxygen, and hydrogen atom shares electrons with neighboring atoms forming a covalent bond. The sharing of electrons that creates these covalent bonds has a particular order. Sharing two—or one pair of—electrons between atoms creates a single bond. Each atom contributes 1 electron to the shared pair. This single covalent bond is often shown by a single line drawn between the atoms. A double or triple bond is created when two or three pairs of electrons are shared between atoms (Figure 1.8).

Molecular compounds make up the bulk of our food and include water, sugars, fats, proteins, and most vitamins. Given their importance in food and cooking, let's look at two detailed examples of covalent compounds: fructose and acetic acid.

1.3.3.1 How Many Covalent Bonds?
The number of bonds that a nonmetal element will form is dictated by its position on the periodic table. Let's look at carbon as an example. Carbon is the most important element in our food because it is the element on which the molecules of living things are built. If we look at a periodic table, there are two large sections on the left and right

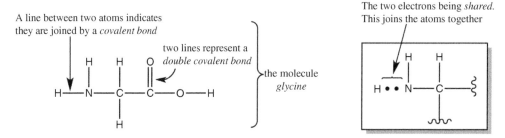

Figure 1.7 Covalent bonds have shared electrons. The sharing of 2 electrons forms a covalent bond. The straight line between atoms represents these electrons. Electrons are very tiny particles with a negative charge. Every atom of each unique element has a specific number of electrons. For example, every hydrogen atom has 1 electron.

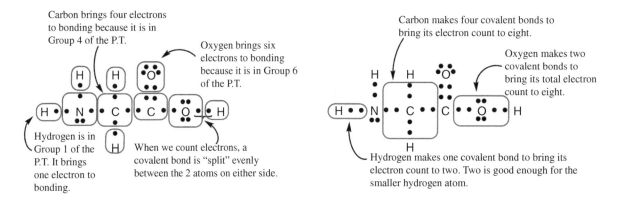

Figure 1.8 Counting electrons with covalent bonds. Shared electrons making a covalent bond are often drawn as pairs of dots. However, most molecular structures use single lines to represent the shared pairs of electrons.

separated by a smaller center section. The center section is not relevant to our discussion, so we can ignore it. Now, count the columns from left to right of the periodic table, skipping over the center section. If you start with the very first row (or period), you'll only count two columns. But starting with the second row (or period), you should get eight columns or groups.

Now, look again at glycine in Figure 1.9; how many lines or covalent bonds are extending from each of the carbon atoms? Each carbon has four lines. Now, look at the same molecule of glycine drawn a slightly different way in Figure 1.8. See the four black dots surrounding the central carbon? Those four black dots represent 4 electrons that carbon contributes to each of the four different covalent bonds. How do we know that carbon has 4 electrons to share in covalent bonds? Because carbon is in Group 4 or the fourth column of the periodic table. Carbon shares through four covalent bonds to give itself access to 8 total electrons. Eight is the number of completion in chemistry (notice how the periodic table ends at Group 8), and it also explains why oxygen is only making two bonds. Oxygen is in Group 6 of the periodic table, so it brings 6 electrons to the molecule. Oxygen needs only 2 more electrons to make 8 (the magic number!), and so oxygen forms two covalent bonds, sharing two of its six electrons so it can pick up 2 more electrons in the process of sharing. What about hydrogen? Hydrogen is so small that it reaches completion at two instead of eight. So hydrogen will share its 1 electron through one covalent bond in order to reach 2 total electrons, which is the last group in hydrogen's row of the periodic table.

Fructose is a sweet-tasting sugar found in fruit and honey (Figure 1.10). Looking at its molecular structure, the six carbon atoms are bonded (shown by the lines connecting atoms) to 7 of the 12 hydrogen atoms and 6 oxygen atoms. The remaining five hydrogen atoms are bonded to oxygens. Because of its atomic components, the molecular formula of fructose is $C_6H_{12}O_6$. This molecule is relatively large, has no overall charge, and all of the atoms are nonmetals.

Figure 1.9 An abbreviated periodic table that shows elements that are found in food, and numbers the columns or groups of the table.

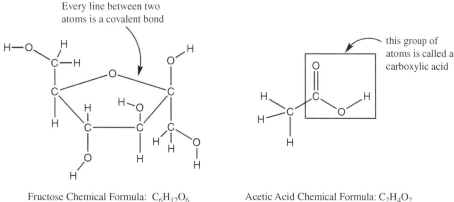

Fructose Chemical Formula: $C_6H_{12}O_6$

Acetic Acid Chemical Formula: $C_2H_4O_2$

Figure 1.10 Molecular structures of fructose ($C_6H_{12}O_6$) and acetic acid ($C_2H_4O_2$).

Clearly, at the molecular level, fructose is different from table salt. Fructose is organic or carbon-based, and is made of a particular arrangement of carbon, oxygen, and hydrogen. On the other hand, table salt is an ionic compound of sodium and chloride ions. Of course, we all know the difference between fructose and table salt just by taste! Most of us would probably prefer to eat a spoonful of honey over a spoonful of table salt! Also in Figure 1.8, acetic acid is responsible for the sour taste in vinegar. Interestingly, acetic acid is also made of carbon, hydrogen, and oxygen atoms. However, the arrangement and number of atoms between fructose and acetic acid are different. The carbon, oxygen, and hydrogen atoms are connected differently between fructose and acetic acid, and that is what gives the two covalent molecules very different chemical and biological properties. Acetic acid has two carbons. The first is bonded to three hydrogens, and the second is bonded to two oxygen atoms. One of the oxygen atoms is bonded to the carbon with a single bond, and the second has two bonds. This particular arrangement of atoms (i.e. —COOH) is called a carboxylic acid, and we will discuss carboxylic acids in more depth later. We have a great beginning to understanding the way food changes during cooking and where the flavor comes from, but we need a little more information on how the structures of molecular compounds are communicated (Section 1.3.3.2), how compounds behave and interact with other compounds and other basic concepts.

1.3.3.2 Drawing and Understanding Structures of Covalent Compounds

Scientists use a number of ways to represent chemical compounds. The simplest way to represent a molecule is the molecular formula. This is simply a count of each kind of atom in a molecule. The subscript describes the number of atoms in the molecule for the preceding element. While simple, it does not describe very much about the way the atoms are joined together (Figure 1.11). For example, both glucose and fructose can be described by the molecular formula $C_6H_{12}O_6$, as both a molecule of glucose and a molecule of fructose contain 6 carbons, 12 hydrogens, and 6 oxygen atoms. However, a molecular formula is often used for simple molecules to show how they react. For example, to produce caramel from table sugar ($C_{12}H_{22}O_{11}$), intense heat results in a loss of water and decomposition of sucrose to yield caramelen ($C_{36}H_{50}O_{25}$).

$$3\,C_{12}H_{22}O_{11} \rightarrow 8\,H_2O + C_{36}H_{50}O_{25}$$

A complete structural formula is used to depict the way atoms are bonded together and show every atom and every bond. Covalent bonds are illustrated as a line between atoms. For example, C—C shows that there is one (—) bond between two carbon atoms, where each bond is a pair of (i.e. two) shared electrons. Some com-

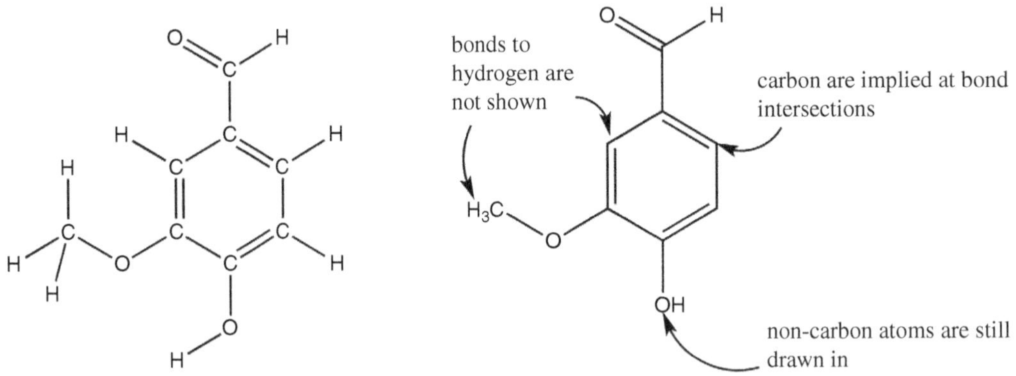

Vanillin - *complete structural formula* Vanillin - *Skeletal structure*

Figure 1.11 Structure of vanillin. On the left: structural formula of vanillin. Each atom is drawn and each bond is clearly marked. Notice the single and double bonds and carbon atoms bonded to H, O, and other C atoms. On the right: Skeletal formula of vanillin. Note the implied carbons at the intersection and end of each line. Groups of atoms are explicitly drawn. Double and single bonds are drawn the same as shown in a structural formula.

pounds have two bonding sets of electrons, a double bond shown as (=). Some molecules even have triple bonds, which involve 6 shared electrons (≡). Let's use vanillin, the molecule responsible for vanilla extract odor and flavor, as an example of different bonding arrangements. The molecular formula of vanillin is $C_8H_8O_3$. Vanillin contains both single and double bonds between atoms (Figure 1.11).

A third common way of depicting molecules with lots of carbon atoms is to draw a line structure formula, sometimes referred to as a skeletal formula. Skeletal formulas are useful in that they provide the information contained within a complete structural formula, but they are drawn with a few shortcuts or abbreviations. In a skeletal formula, a carbon–carbon bond is drawn without specifically showing the carbons and hydrogens, but all of the other atoms or groups of atoms are included. In these drawings, a carbon is implied at each bend and end of a line (the line represents the bond of a carbon atom); if any carbon atom doesn't have four covalent bonds, then there are hydrogens implied to ensure each carbon atom is involved in four covalent bonds.

1.4 Properties of Covalent Molecules

Now that you understand how atoms and ions create molecular and ionic compounds, the individual components of food molecules, we can start to complete the picture of what happens when egg protein curdles or clumps, when fat globules stick together during cheesemaking, or how flour can thicken broth into a gravy. In all of these processes and many others, it is the interactions between food molecules that cause the chemical and physical changes we observe in the cooking or baking process.

1.4.1 Interaction of Food Molecules

Forces that attract or repel two different molecules or compounds are called intermolecular forces. There are a number of different kinds of these forces, with different strengths and properties, but a key concept is that intermolecular forces are *not* bonds that hold atoms together; rather, intermolecular forces are weaker interactions that bring different molecules together or keep them apart. Once you know some details about intermolecular forces, you will have a better understanding of how to make a foam or emulsion, why adding lime juice slows down the browning of avocado, and why destroying the structure of a protein makes a solid in your cooked scrambled eggs.

1.4.1.1 Electrostatic Interactions

The first type of compound that we learned about was the ionic compound, remember (Section 1.3.1)? Ionic compounds are made of positively charged cations and negatively charged anions that are electrostatically attracted to one another. It's that electrostatic attraction that forms an ionic bond. Well, electrostatic attraction of positive and negatively charged ions can also occur between charged atoms that are part of separate molecules or on separate parts of very large molecules. Proteins are a type of very large, very important food macromolecule that we will learn about in Chapter 2. Proteins have parts of their structures that carry charges, including carboxylate groups ($-COO^-$) and amines ($-NH_3^+$). In Figure 1.12, there is an example of a carboxylate piece from one region of a protein macromolecule electrostatically interacting with the amino piece from another region of a protein macromolecule. One of the oxygen atoms of the $-COO^-$ carries the negative charge. On the other half of Figure 1.12, the nitrogen of $-NH_3^+$ is carrying the positive charge. The dotted line from positive to negative represents the electrostatic attraction.

Electrostatic attractions also stabilize the molecules of pectin used to thicken jams and jellies. Pectin is another very large, very important food molecule. The structure of pectin is complex, and we will learn about it in more detail in Chapter 7 on Fruits and Vegetables. For now, let's just focus on the region of pectin that is most relevant to our discussion of electrostatic attractions, and that is the $-COO^-$ group of atoms; the rest of the pectin molecule is simply abbreviated with a series of hexagons. See how in Figure 1.13 the $-COO^-$ group of atoms from two separate pectin molecules are joined via electrostatic attraction to a calcium cation (Ca^{++}). The calcium cation carries a charge of +2, so it is balanced by a −1 from each of two $-COO^-$ groups on separate pectin molecules.

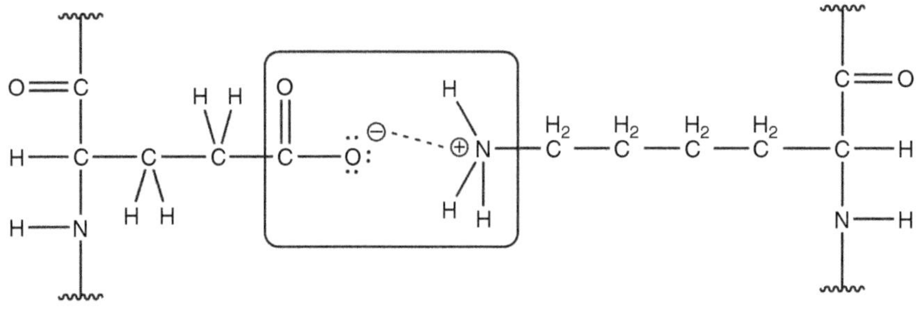

Figure 1.12 Electrostatic attractions between molecules or between parts of very large molecules are found frequently in food. The boxed region of the image depicts a —COO⁻ interacting with a —NH₃⁺. The negative charge of an oxygen anion attracts the positive charge of the nitrogen cation and is represented with a dotted line. This type of electrostatic attraction is important in creating higher order structures in protein macromolecules, as we will see in Chapter 2.

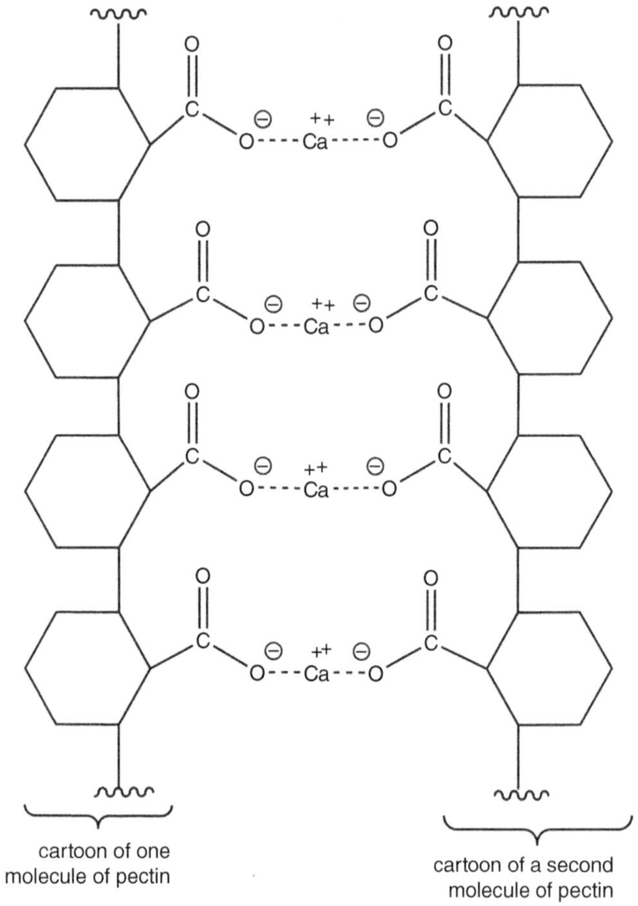

Figure 1.13 Electrostatic attractions between —COO⁻ and Ca⁺⁺ hold two molecules of pectin together. The dotted lines from negative to positive represent the electrostatic attraction.

cartoon of one
molecule of pectin

cartoon of a second
molecule of pectin

Lastly, electrostatic interactions also govern the behavior of a large protein macromolecule in milk, called casein. Macromolecules of casein have many —COO⁻ groups that coat their exterior with negative charges. Because of the many negative charges, the casein macromolecules will repel one another, and, as we will see in Chapter 4, it is this repulsion that keeps homogenized milk a stable mixture of water and fat.

1.4.1.2 Hydrogen Bonding

Some atoms, like oxygen and nitrogen in particular, have a high affinity for electrons, while other atoms, like hydrogen, have a low affinity for electrons. When atoms with differing affinities for electrons are bonded to one another, the high electron affinity atom (i.e. nitrogen or oxygen) pulls on the shared electrons more than the low electron affinity atom. Since electrons are negatively charged, the oxygen or nitrogen atoms become slightly negative, indicated by a partial charge ($\delta-$). At the same time, the hydrogen atom that has "lost" some of the shared electrons has a very weak positive charge ($\delta+$) (Figure 1.14). This unequal sharing of electrons creates a covalent bond with a partial negative ($\delta-$) end and a partial positive ($\delta+$) end. The opposite charges at either end of the bond are like the negative and positively charged poles of a magnet or a battery, and so chemists call these covalent bonds with partially positive and negative ends: *polar*.

The resulting partially positive and negative charged atoms can become attracted to and attract partially positive and negative atoms from nearby molecules or even within the same molecule. The resulting interaction between a partial positive (i.e. $\delta+$) component of one molecule and a partial negative (i.e. $\delta-$) component of another molecule is called a hydrogen bond. You can see some examples of hydrogen bonds in Figure 1.15. The hydrogen bond is also an electrostatic attraction between opposite charges, just like we saw in Section 1.4.1.1. However, a hydrogen bond is different from the electrostatic attractions we saw in Section 1.4.1.1 because in a hydrogen bond, the attraction is between *partial* charges. It is called a hydrogen bond because of the involvement of hydrogen as the low electron affinity atom.

While the periodic table contains many atoms with high affinities for electrons, for molecules involved in food and cooking, the high electron-affinity atom is a nitrogen or oxygen atom (Figure 1.15). Figure 1.15 contains typical oxygen and nitrogen-containing covalent bonds found in food and cooking that are polar due to the unequal sharing of electrons. For example, C=O is another oxygen-containing polar covalent bond. The oxygen has a higher affinity than carbon for the shared electrons, and the resulting C=O double bond is polar, which means the C=O double bond has a partial negative (i.e. $\delta-$) end at the oxygen and a partial positive (i.e. $\delta+$) end at the carbon. The same is true for the H–N bond; the nitrogen is partially negative (i.e. $\delta-$) and the hydrogen end of the bond is partially positive (i.e. $\delta+$), all because nitrogen has a high affinity for electrons and causes unequal sharing when bonded to low electron-affinity atoms like hydrogen and carbon.

Let's look at an example of hydrogen bonding in action. Given the hydrogen bonding potential of water (i.e. H_2O), and the presence of water in many foods and cooking processes, hydrogen bonding with water is

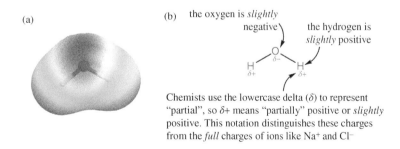

Figure 1.14 Oxygen has a high affinity for electrons. When sharing electrons in a covalent bond with a hydrogen atom, the oxygen pulls more of the shared electrons toward itself. This creates a small amount of excess negative charge on the oxygen. Consequently, the hydrogen has less of the shared electrons and becomes slightly positively charged. In part (a), the charge on the water molecule is shown as a continuum from blue, which is slightly positive, to red, which is slightly negative. Notice how the red or negative charge is centered on the side of the molecule with the oxygen atom, while the other side containing the hydrogen atoms is more blue or positive. In part (b), the water molecule is drawn with a typical line structure. The $\delta+$ and $\delta-$ symbols represent the partial negative and partial positive charges on the hydrogen atoms and oxygen atoms, respectively.

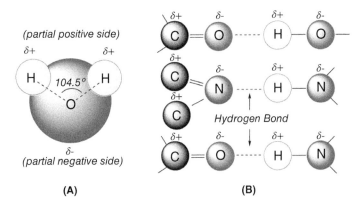

Figure 1.15 Hydrogen bonding. (A) The polar nature of the O–H bond means that the oxygen atom has more of the shared electrons, while the hydrogen atom has less of the shared electrons. (B) The electrical negative atoms N and O, when covalently bonded to the less electronegative H, result in a weak charge, which will form special weak bonds called hydrogen bonds between compounds.

a very important intermolecular interaction. Glucose is a common type of sugar also known as dextrose, and it is a component of table sugar. Glucose also contains many polar O–H bonds (Figure 1.16). If you mix a tablespoon of glucose into a glass of water, the white powdery solid will dissolve almost instantly. Why? The water has made many hydrogen bonds with the glucose, pulling the molecule into the water (Figure 1.16).

Now let's look at the example of starch. Starch is a white powder made of thousands of chemically joined glucose molecules, which means starch has tens of thousands of polar O–H bonds (Figure 1.17). When water is added to dried starch, the water molecules form intermolecular interactions with many O–H bonds of the starch. However, there are so many O–H bonds on the surface of starch granules that the water binds too tightly to the starch, causing the starch to form an almost solid gel. Additional structural changes cause the starch to expand and eventually contract, which happens at such a high rate with warm or hot water that an impenetrable blanket of water forms over the expanding granule of starch. So what is the take-home message? When making gravy, first mix your starch with cold water. The cold water slows down this process to allow controlled and complete hydration of the starch granules.

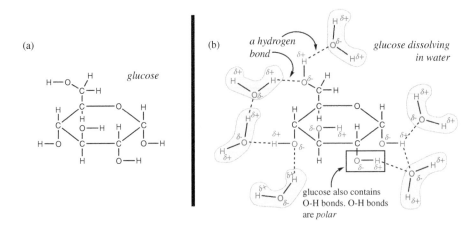

Figure 1.16 Glucose hydrogen bonding with water. (a) Glucose, (b) Glucose contains many polar O–H bonds that can interact via hydrogen bonding with the polar O–H bonds of water. The oxygen atoms of the O–H bonds are shown in red, and the hydrogen atoms in blue. Covalent bonds are represented with solid lines. Hydrogen bonds are represented with dashed lines between a partial negative (δ−) and partial positive (δ+) charge.

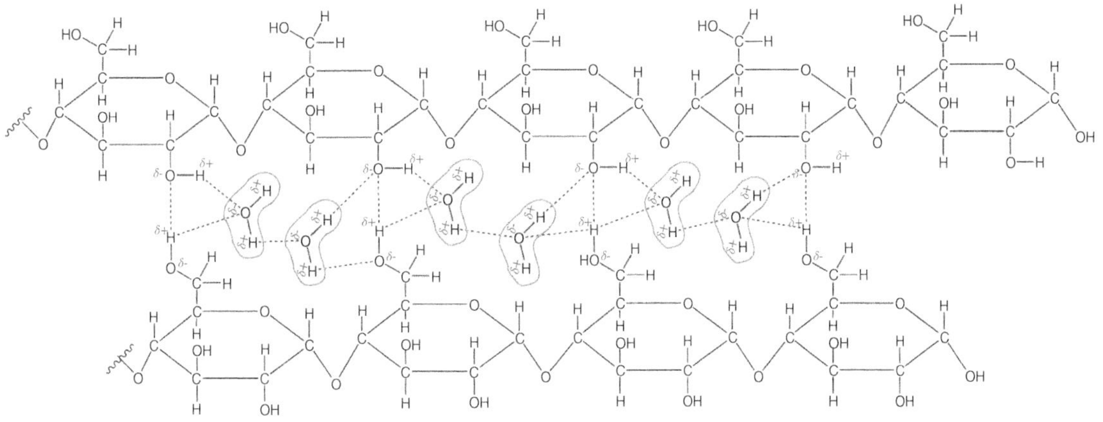

Figure 1.17 Hydrogen bonding in starch. Long chains of chemically joined glucose molecules make up starch. When mixed with water, the chains of starch form tangles of hydrogen-bonded strands—much like a tangle of yarn—which serve to thicken a gravy. Hydrogen bonds are represented with dashed lines between a partial negative ($\delta-$) and partial positive ($\delta+$) charge.

1.4.1.3 Hydrophobic Interactions

Hydrophobic interactions are forces of particular importance for food molecules existing in a watery environment. The word "hydrophobic" comes from two Greek roots: "hydro" meaning water and "phobos" meaning fear. Hydrophobic interactions are forces that define the relationship of "water-fearing" molecules to the watery environment that is found in foods. Let's consider an example that is often encountered in the kitchen: oil and water. Imagine what happens when any oil is poured into a glass of water. The oil separates from the water, creating a boundary at the interface. The separation of oil from water is due to the hydrophobic effect. Why is water such a defining element of food? Plant and animal tissues are rich in water. Animal muscle is made of nearly 70% water, while the water content of plants ranges from 75% to 90% of total mass. Thus, the proteins, sugars, fats, and other compounds of living things are constantly exposed to and surrounded by water molecules. Chemists call the watery environment "aqueous," which comes from the Latin word "aqua" meaning water. Compounds that have a full or partial charge can interact with the water molecules of the aqueous environment by using electrostatic interactions (Section 1.4.1.1) or hydrogen bonding (Section 1.4.1.2); compounds with charges easily dissolve and remain suspended in this water or aqueous environment. However, hydrophobic molecules, like fats and oils, are largely uncharged and cannot hydrogen-bond or be involved in electrostatic interactions. When placed in water, these hydrophobic molecules clump or aggregate together to separate from the aqueous environment; this is the hydrophobic effect in action. Hydrophobic molecules or regions of molecules that have no charge are dominated by C–C and C–H bonds. These C–C and C–H covalent bonds are characterized by equal sharing of electrons. These bonds do not contain high electron-affinity atoms, and they do not have partial charges; therefore, C–C and C–H bonds are *nonpolar*. Hydrophobic interactions bring nonpolar C–C and C–H rich molecules together so they can avoid interacting with water molecules.

Let's examine the science behind the hydrophobic effect more closely. There are two parts to the hydrophobic effect:

1. The water molecules want to avoid the hydrophobic molecule and instead interact with each other.
2. The hydrophobic molecules prefer to stick to one another and avoid the water.

So when oil is poured into water, the first thing that happens is that each hydrophobic molecule of oil is initially surrounded by a shell or cage of water molecules (Figure 1.18). Why does the water form a cage? To avoid the hydrophobic molecule! The absence of favorable interactions (such as hydrogen bonding or electrostatic

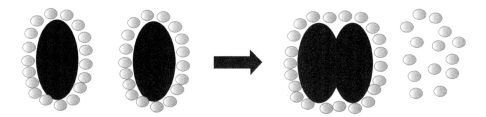

Figure 1.18 Hydrophobic effect is driven by entropy. Hydrophobic compounds shown in black are separate on the left. Because water (shown as small gray globes) cannot form hydrogen bonds to the water-fearing compound, they are forced to form a rigid shell of water where the water forms hydrogen bonds to itself. If the two hydrophobic compounds come together, there is a smaller surface area and less water needed to form the cage around the combined molecules. There is more order on the left-hand side of the image than on the right because the drive for disorder mediated the hydrophobic molecular interaction.

interactions) between the hydrophobic molecule and the water means that water molecules organize themselves in a cage to keep the number of water molecules that have to interact with the hydrophobe at a minimum. This allows the majority of the water molecules to favorably interact with each other. The "cage" of water molecules surrounding the hydrophobic molecule is a type of order; you can think of those water molecules as being organized. Those organized waters making the cage around the hydrophobe are in stark contrast to the remaining water molecules in the glass that can randomly interact with other waters as they please. Herein lies a scientific principle: ordered arrangements of molecules are always higher in energy than disordered molecules. The universe is always trying to reach a state of randomness or disorder because it is lower in energy; in science, we call this *entropy*. Think of a clean and orderly bedroom versus a messy and disorganized bedroom. Which requires more energy? The clean and orderly one, of course! Wherever there is order in nature, it takes energy to keep it that way. Therefore, disorder and randomness are easier and, therefore, preferable! So, how can a mixture of oil and water maximize the amount of disorder and randomness? By "clumping" all of the hydrophobic molecules together and separating them from the water, which will keep the caged or ordered water molecules to a minimum. Then, this is where part two of the hydrophobic effect comes in. When water and oil are mixed, the C–C and C–H bonds of the hydrophobic molecules begin to interact with one another via transient, weak, attractive forces called van der Waals attractions. Ultimately, the mixture of oil and water will separate because the hydrophobic oil molecules will stick to themselves via van der Waals attractions, while the drive for disorder will force the separation of water molecules from the oil. In summary, you can think of hydrophobic interactions as driven by the forces of nature that prefer disorder, with a minor contribution from the attractive van der Waals forces between hydrophobic molecules.

1.4.1.4 A Review of Important Vocabulary in the Interaction of Food Molecules

We have been using a lot of scientific terminology in our discussion to this point. It might be useful to clarify some of the terminology and summarize our discussion of intermolecular interactions here. The terms *hydrophobic* and *nonpolar* are often used interchangeably and describe molecules or regions of food molecules that possess only carbon and hydrogen atoms. Interactions of nonpolar molecules or nonpolar regions of large molecules are mediated by the hydrophobic effect (Section 1.4.1.3). Polar molecules or polar regions of large molecules contain partial negative charges on nitrogen or oxygen atoms, and partial positive charges on the hydrogen or carbon bonded to the nitrogen or oxygen; polar molecules are hydrophilic because they use their partial charges to interact with water via a special type of electrostatic interaction called hydrogen bonding (Section 1.4.1.2). Charged molecules or charged regions of large molecules have one or more atoms carrying a full positive or negative change, and because of their charged nature, they behave similarly to polar molecules. The charged atoms or ions interact with water and other polar molecules via electrostatic interactions (Section 1.4.1.1).

Hydrophilic, polar, water soluble		Hydrophobic, nonpolar, fat soluble	
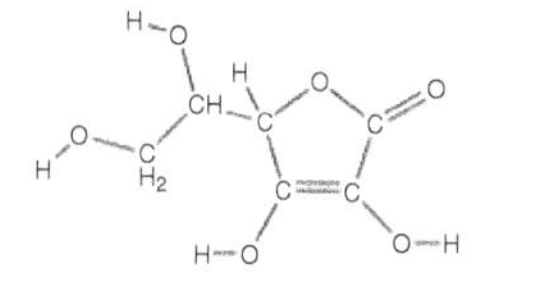	Vitamin C: found in fruits and vegetables including citrus fruits, peppers and strawberries; an antioxidant essential in collagen formation, wound healing and immune function	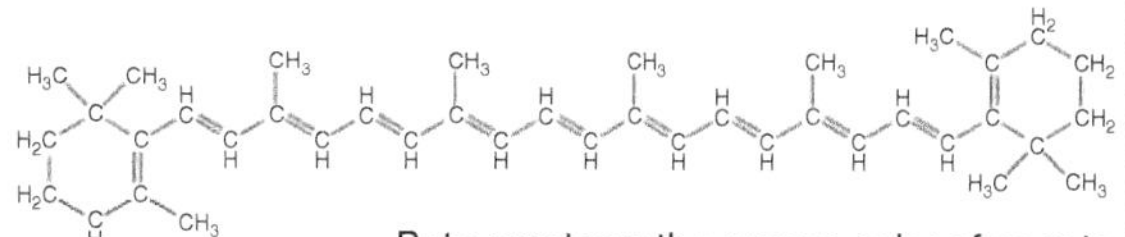 Beta carotene: the orange color of carrots	
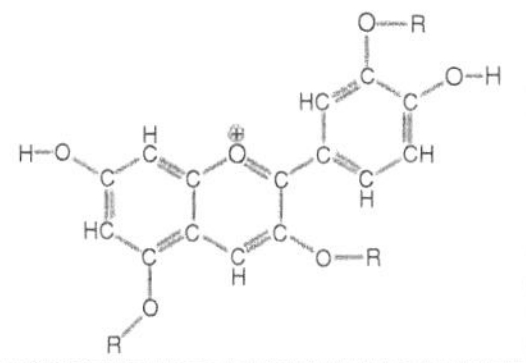	Cyanidin: an anthocyanin pigment that gives red-purple colors to blueberries, blackberries and cherries	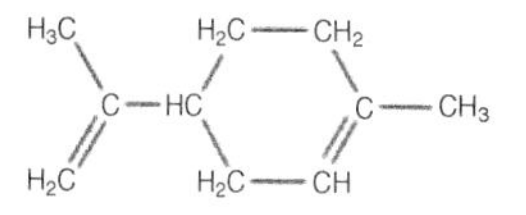	Limone: the primary component and fragrance of orange oil, also present in lemons and limes.
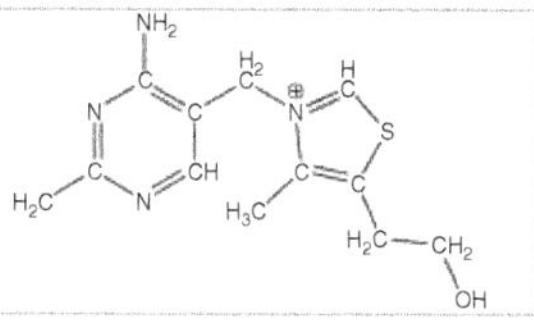	Vitamin B1: found in beans, nuts and in enriched bread products; essential for converting food into energy and for nervous system function.	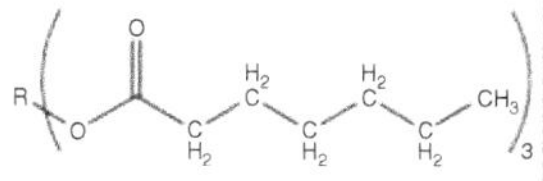	Caprylic acid as part of a triglyceride: the primary component of coconut oil

Figure 1.19 Hydrophobic, nonpolar, and fat soluble molecules versus hydrophilic, polar, and water-soluble molecules.

When trying to remember all of these types of molecules and their interactions, the phrase "like dissolves like" is useful (Figure 1.19). Nonpolar molecules interact with other nonpolar molecules via hydrophobic interactions. Nonpolar molecules (like olive oil) do not interact well with water (a polar molecule); thus, oil and water do not mix. On the other hand, water is polar, and polar molecules like glucose (Figure 1.16) will dissolve in water due to hydrogen bonding and electrostatic-like interactions.

1.4.2 Water

We just can't say enough about water! Water is arguably the most important covalent molecule in food and cooking: a cucumber is 95% water, an avocado is 73% water, and a chicken breast is 69% water. The properties of water must be considered when thinking about food and cooking. We want to understand how water dissolves polar and charged compounds, why water expands upon freezing, and why water-rich foods take longer and higher temperatures to heat up.

1.4.2.1 The Shape of Water

Water is composed of a single oxygen atom covalently bonded to two hydrogen atoms (H_2O). You already know about covalent bonds, so you know that each oxygen–hydrogen covalent bond consists of two shared electrons. However, the oxygen atom in water also contains two pairs of unbonded or "lone" electrons. These electrons are shown as a pair of "dots" or ".." in Figures 1.7 and 1.9. Chemists don't always draw in these "lone pairs" of electrons on an oxygen atom, but they are always there! Why does oxygen have two lone pairs? Because oxygen is in Group 6 of the periodic table and; therefore, brings 6 electrons to bonding (Section 1.3.3.1), it only needs to make two covalent bonds to reach 8 electrons, i.e. the number of completion. Since oxygen only shares two of the original 6 electrons it started with, the remaining 4 electrons group together as "lone pairs." Every neutral oxygen atom making two covalent bonds also has two lone pairs on the oxygen atom, and the oxygen atom in water is no exception. These 4 electrons making up the "lone pairs" on the oxygen atom in water may be invisible, but their presence makes the two bonds to hydrogen bend away. The result is a water molecule with a bent or "v" shape (Figure 1.20). How does this shape impact the properties

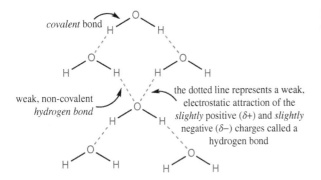

Figure 1.20 Water makes hydrogen bonds with other water molecules. In this image, the partially positive hydrogen atoms of water are colored blue, while the partially negative oxygen atoms are colored red. Covalent bonds are solid black lines. Weak, electrostatic hydrogen bonds are represented with gray dotted lines.

of water? You already know that the O–H bonds of water are polar; the hydrogen atoms have a partial positive charge (i.e. $\delta+$) and the oxygen atom has a partial negative charge (i.e. $\delta-$). The "v" shape actually adds to the polar nature of water. First, it creates a molecule with a partially negative side and a partially positive side. The molecule itself is polar. The bent shape allows other polar molecules (up to four) to more easily access and interact with water via hydrogen bonding.

How do you get four hydrogen bonds from one water molecule? The partial positive charge on each hydrogen atom attracts a partial negative charge oxygen atom in a different water molecule (this makes two). The partial negative charge oxygen in the same water molecule can attract partial positive charge hydrogens in two different water molecules. Thus, each water molecule can hydrogen-bond with four different water molecules at one time!

1.4.2.2. Water Dissolves

In Section 1.4.1.2, we discussed how water dissolves the sugar molecule glucose using hydrogen bonding. Glucose is a covalent compound with many polar O–H bonds that interact easily with the O–H bonds of water via complementary partial positive (i.e. $\delta+$) to partial negative (i.e. $\delta-$) electrostatic attractions that chemists call hydrogen bonds. But, the bent shape and polar nature of the water molecule also allow it to interact with the fully charged cations and anions of ionic compounds. Ionic compounds (Section 1.3.1) are solids at room temperature. In these solids, the ions are held tightly together in large networks called a lattice by the electrostatic attraction of opposite charges (Section 1.4.1.1). However, in water, the attractive force holding the ionic compound together is partly broken, and the ions separate from one another. You have seen this phenomenon with your very own eyes when a teaspoon of table salt (i.e. NaCl) dissolves in a glass of water. When dissolving table salt or any ionic compound in water, each charged ion is surrounded by a jacket or shell of water molecules. The partial negative end of the water molecules points toward the positive cation creating favorable electrostatic interactions (Section 1.4.1.1), while the partial positive ends of the water molecules point toward the negatively charged anion also creating favorable electrostatic interactions (Figure 1.21). The water separates and coats each cation and anion creating a "shell" of hydration, which is why a teaspoon of table salt seems to disappear into a glass of water. The coating or shell of water molecules acts to shield the electrostatic attraction between cation and anion, allowing them to separate from one another and dissolve in the water.

1.4.2.3 Water Activity

We have already seen how water can hydrogen bond to and dissolve polar covalent molecules like glucose (Section 1.4.1.2), and how water dissolves the ions of an ionic compound by surrounding them with favorable electrostatic interactions (Section 1.4.2.2). The capacity of water to engage in these types of interactions governs another important property of water in food: *water activity* (a_w). The concept of water activity states that, in food, some of the water is tied up and held rigidly in its hydrogen bonding or electrostatic interactions with various food molecules like sugars, proteins, and salts. The water that is held tightly in these hydrogen bonds

sodium chloride (i.e. table salt) is an ionic compound. It is made of 2 different types of atoms that are held together by a positive-to-negative attraction called an ionic *bond*

Na$^+$ Cl$^-$

+

H$_2$O

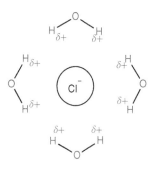

The sodium cation (Na+) is surrounded by a cloud of water molecules that are oriented to present their slightly negative oxygens toward the positively charged sodium

The chloride anion (Cl-)is surrounded by a cloud of water molecules that are oriented to present their slightly positive hydrogens toward the negatively charged chloride.

Figure 1.21 Salt dissolves in water. In water, the polar nature of water surrounds and reduces the attractive force between ionic compounds dissolving each ion into the water solution. The oxygen end of the water molecule (i.e. H$_2$O) has a partial negative charge, which is represented by the red color and $\delta-$ symbol. The hydrogen end of the water molecule has a partial positive charge, which is represented with the blue color and $\delta+$ symbol. The $\delta-$ oxygens orient themselves toward the Na+ cation, while the $\delta+$ hydrogens orient themselves toward the Cl- anion.

and electrostatic interactions is in contrast to the rest of the water molecules that are not close enough to engage in hydrogen bonding or electrostatic interactions with other food molecules. The water molecules that are free of hydrogen bonds or electrostatic interactions are called "bulk" or "free" water. If there are enough free water molecules, then the food item is said to have a a_w that can support the growth of bacteria, yeasts, and mold. Water activity is measured by determining the moisture content of food; this is essentially a measure of free water. Free water is available for microbes, enzymes, and other chemical reactions and can lead to spoilage. Dry goods like ground coffee beans, powdered milk, and even potato chips have a low a_w. Foods with low water activities have a long shelf life, as there is not enough water for the survival of microorganisms. To increase the shelf life of processed foods, manufacturers often add sucrose (table sugar) to packaged foods. This reduces the amount of free water available for bacteria to grow. Like glucose, sucrose forms hydrogen bonds with available water molecules, reducing the water activity of the food item. Many liquid or semisolid foods can also have low water activities, for example, jams and jellies range from 0.75 to 0.8, peanut butter is 0.35 or less, and honey ranges from 0.5 to 0.65. Peanut butter is typically shelf stable because the water activity is not high enough to support bacterial growth. In honey, which lasts a long time at room temperature, most of the water molecules are hydrogen bonded to a very high concentration of sugar molecules. This means there is not enough free or bulk water available for microbes to survive. The microorganisms that could grow in honey do not typically grow until a water activity of 0.60–0.62. Fresh meat and bread have a high water activity with a_w values greater than 0.9, and consequently, these foods have a short (or no) shelf life. Ice cream has a water activity >0.95 and has sections of both liquid (free) and solid (rigid) water molecules. The liquid water is filled with dissolved proteins, sugars, and salts, while the solid water is frozen in the form of ice crystals. Believe it or not, you have probably used the concept of water activity in your daily lives without knowing it. For example, have you ever dropped a cell phone in water and quickly submerged it in a rice container to dry out the charge port? If so, you have used the concept of water activity. This is because free water migrates from regions of high water activity (wet electronic $a_w > 0.5$) to low water activity (rice $a_w \cong 0.5$). We also use differences in water activities to soften dried-out cookies (aw ~0.6) and hard brown sugar ($a_w \cong 0.3$) by placing a piece of bread ($a_w \cong 0.9$) in the sealed container. The free water in the slice of bread migrates to the region of low water activity, thus softening the cookies or brown sugar in the process.

1.4.2.4 Physical Changes of Water: Vaporizing, Melting, and Freezing

The *state of matter* of a substance is governed by the intermolecular interactions of the individual molecules. What are states of matter? When thinking about food and cooking, there are three states of matter that are most relevant: solids, liquids, and gases. Since we are thinking about water in this part of our chapter, let's consider these three states of matter with respect to water. When we think of a glass of water, that water is in a liquid state. Solid water is ice, and the gaseous form of water is called steam or vapor. Ice, liquid water, and steam are each comprised of water molecules, and each has the chemical formula, H_2O. But ice, liquid water, and steam are the same molecules of water in different physical states of matter (Figure 1.22). Liquid water is not chemically different from ice and steam, but it is physically different. You can tell because ice *melts* into liquid water, and liquid water *freezes* into ice. In the same way, liquid water can *evaporate* into steam, and

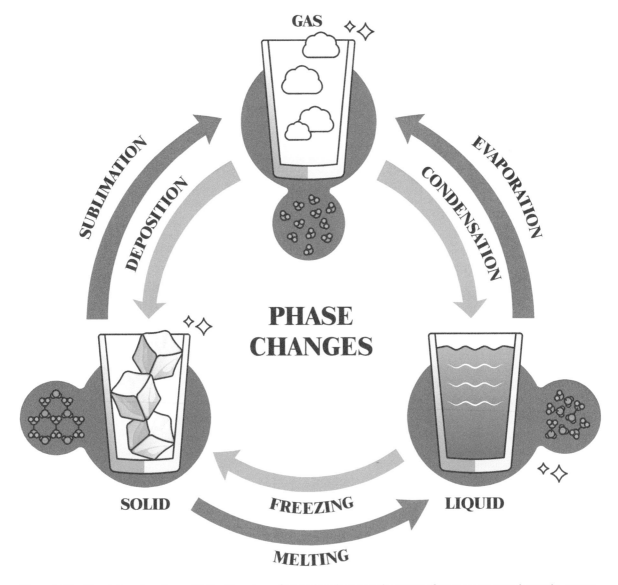

Figure 1.22   The states of matter: solid, liquid, and gas. Conversions between the states of matters are noted over the arrows. VectorMine/Adobestock

steam can *condense* into liquid water. One form of water can become another in what is called a change of state. So if ice is chemically the same thing as liquid water, what is changing about the water molecules when ice melts into liquid water? The changes are in the number of intermolecular attractions of the individual molecules.

Liquid water has enough heat from the surroundings to bend, rotate, and vibrate the bonds and molecules of water. So, the molecules themselves are moving around, and hydrogen bonds between molecules are continually breaking and reforming (Figure 1.23). As heat energy is removed from liquid water, each water molecule slows down enough to form stable hydrogen bonds, and the water molecules settle into a cage or lattice shape (Figure 1.23). Because of the lattice, ice has fewer water molecules per unit area than liquid water, which also makes solid water less dense than liquid water. That is why ice cubes float in your glass of water. If instead of removing energy, we add heat energy to liquid water, then the molecules start to move so rapidly that their intermolecular contacts are broken and the individual molecules escape as a gas (Figure 1.23). The molecules of water in the gaseous state are too far apart and move rapidly to form hydrogen bonds.

Let's think a little more about intermolecular attraction in the context of a state change. When solid water changes state (melts or sublimates) to a liquid or gas, heat energy must be added to the substance to break the intermolecular interactions and allow for more movement of the molecules. The same is true when liquid water evaporates to make water vapor or steam. The transition between melting and freezing or vaporization and condensation is a balance between the intermolecular forces holding the molecules together and the amount of energy required to free the molecules from these interactions. The greater the number and strength of the intermolecular forces, the more energy in the form of heat is required to cause a state change. Think of two blocks of wood covered in Velcro. The stickiness of the Velcro® represents intermolecular attractions, while the two blocks of wood represent two molecules. The more Velcro you have sticking the blocks of wood together, the more force is required to separate the blocks.

The transition of solid water directly to gaseous water might seem strange at first. But this very state change is behind "freezer burn"; the seemingly inevitable outcome of food that has spent too many months in the freezer. "Freezer burned" food isn't burned at all. When frozen food is exposed to the dry, cold air of the freezer, water molecules can directly escape from solid ice to the gas phase in a process called sublimation (Figure 1.22). The

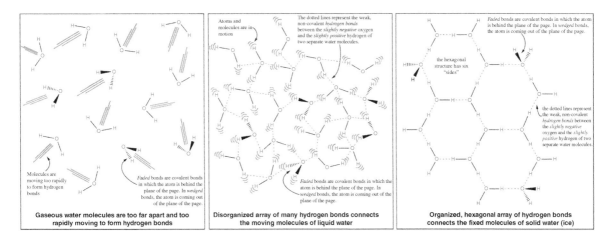

Figure 1.23 Water molecules in gaseous, liquid, and solid states. Water molecules are represented with a line drawing in their classic bent shape. The oxygen atoms are colored red to represent the partial negative charge (i.e. $\delta-$) on the oxygen, and the hydrogen atoms are colored blue to represent the partial positive charge (i.e. $\delta+$) on the hydrogen atoms. Covalent bonds are solid black lines, while non-covalent hydrogen bonds are represented with dashed gray lines and green lines represent molecular motion.

sublimation of water leaves the food dehydrated, the cells damaged, and the food more susceptible to reacting with oxygen in the air. The resulting dried patches of food have a different color and consistency. "Freezer burned" food might not look that appealing, but it is safe to eat.

When water freezes, it forms a stable lattice of hydrogen-bonded water molecules that is quite special. As shown in Figure 1.23, the lattice of molecules that make up ice takes up more space than molecules in liquid water. Think of it this way: it is possible to fit more molecules of liquid water in the same space than molecules of solid water. And if a sample of liquid water is frozen, those same molecules take up more space in the solid state than in the liquid state. A sample of water actually expands—or takes up more space—when it freezes into a solid. This expansion that occurs when water freezes to ice is problematic when freezing food. Plant and animal tissues have water inside each cell. So, if you freeze vegetables or meat, the liquid water inside those plant or animal tissues freezes into crystals of solid ice. The liquid water expands to fill more space as it freezes, and consequently, the solid ice ends up rupturing the cells of the plant and animal tissue. When the food thaws, the ice melts, leaving behind broken cells that lose their shape and leak water (Figure 1.24).

Because solid water freezes into an organized lattice of hydrogen-bonded molecules, anything that interferes with that lattice formation will make it harder for water to freeze. Pure water will freeze at 0°C or 32°F, but seawater freezes around 28°F or −2°C. Why? It's the salt in the water. Remember in Section 1.4.2.2, we examined how water forms a "cage" around the Na^+ and Cl^- ions of salt. When water molecules are clustered around ions, it is then harder for the water molecules to reorganize into a lattice and form the hydrogen bonds necessary to make ice. That's why salty water needs to be even colder to freeze. What else could interfere with water freezing? Well, anything that can dissolve in water in the first place! Remember, water can also readily dissolve sugar molecules (Section 1.4.1.2) because of all the hydrogen bonds the sugar molecule can form with the water molecules. Sugar dissolved in water will also interfere with the lattice arrangement necessary to form ice, which means sugar water will not freeze as easily as pure water. At the temperature of your kitchen freezer, a frozen berry is not as hard as an ice cube, and that is because of all the sugar dissolved in the water inside the fruit. While some ice certainly forms inside the plant cells, some of the sugary water doesn't freeze, giving the frozen berry a softer texture.

thawed fruit is soft and mushy

juice escapes

Berries at room temperature are juicy, but firm

Berries after freezing and thawing

Figure 1.24 Fresh versus frozen fruit. On the right, fresh raspberries are juicy and firm. On the left, fresh raspberries that were rapidly frozen and then thawed. Notice the watery liquid escaping from the thawed berries. Photo by Bill Keller.

1.4.3 Acids, Bases, and pH

Have you ever tasted pure lemon juice or a vinegar-based salad dressing and experienced the subsequent pucker of your lips and mouth? Your ability to taste sour foods like lemons, grapefruits, or vinegar is based on the ability of your mouth to detect a very special cation, H^+. This cation is so special, so important, that it gets its own name: chemists call the cation H^+ a *proton*. But that is a word we have already seen! Remember in Section 1.3 when we learned the parts of an atom? The proton was the atomic particle with a positive charge. Well, a hydrogen atom is so small that it is made of only 1 proton and 1 electron. When a hydrogen atom gives up its electron, it becomes a cation made of exactly 1 proton, and hence, the cation H^+ is called a *proton*.

What do foods like lemons, grapefruits, and vinegars all have in common that results in H^+? And if we can taste the H^+ cation, where does it come from? H^+ comes from molecules that chemists call acids; in fact, acids are molecules with the capacity to give up an H^+. Let's consider the molecule responsible for the sour taste (or H^+) of vinegar, acetic acid (Figure 1.24) (Box 1.3).

The term "acid" comes from the Latin "acidus," meaning "sour or tart." And the Latin word "acetus," means vinegar. Therefore, the molecule acetic acid, which is responsible for the H^+ and; therefore, the sour taste of vinegar, gets its name from the Latin words acetus and acidus. What is it in vinegar that creates acidity and conveys the taste of sour? In Figure 1.25, there is a group of atoms within the acetic acid molecule that are boxed: –COOH. This group of atoms (i.e. –COOH or –CO$_2$H) is called a *carboxylic acid*, so named because this group of atoms, bonded together exactly as shown in Figure 1.25, readily gives up H^+. The arrow going to the right, which is toward the products H^+ and the acetate anion, is larger than the arrow pointing to the left, which represents the recombining of H^+ and carboxylate to reform acetic acid. The larger arrow pointing to the right expresses what chemists observe about molecules like acetic acid: the carboxylic acid group of atoms readily breaks down into H^+ and the carboxylate anion. And as we've already said, acids are molecules or parts of molecules that readily give up H^+. Now, why does the carboxylic acid group of atoms within acetic acid readily give up H^+? That is a question best answered in a more advanced chemistry text, but suffice it to say that the carboxylate anion (i.e. –COO$^-$ or –CO$_2^-$) produced from the departure of H^+ is very stable. Acids are likely to give up an H^+ when the resulting anion is very stable.

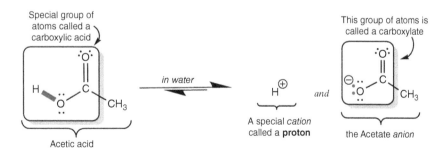

Figure 1.25 Acetic acid, the molecule responsible for the sour taste of vinegar, gives up a proton (i.e. the cation H+) and makes the acetate anion. The –COOH group of atoms is called a carboxylic acid. The O–H bond of the carboxylic acid (drawn in red) breaks, and the 2 electrons making the bond stay behind on the oxygen to generate a carboxylate anion. The H^+ cation generated from the breaking of the O–H bond of the carboxylic acid is the source of sour taste.

Box 1.3 Figuring out food: Apple cider vinegar

Apple cider vinegar has a passionate following by many who claim its usefulness in home remedies for maladies as diverse as teeth whitening, weight loss, and preventing diabetes. Distilled "pure" vinegar is 3–5% acetic acid diluted in water. Acetic acid is a simple carboxylic acid found in many foods (Figure 1.25). Traditional preparation of vinegar starts with the fermentation of plant sugars into alcohol (i.e. ethanol) by yeast and some strains of bacteria. Ethanol, in turn, is converted into acetic acid by gram-negative aerobic bacteria commonly found in nature. The production of acetic acid from ethanol is the cause behind the spoilage of open bottles of wine, as well as the source of the acetic acid in a bottle of red wine vinegar. When microbes ferment sugar from various plant sources, such as apples, grapes, or rice, into ethanol and then acetic acid we get different kinds of vinegars: balsamic, malt, rice, and apple cider vinegar. Yeast and bacteria can also ferment other sugars like maltose and lactose into the carboxylic acid-containing molecules, malic acid and lactic acid (Figure 1.26), and these acid products of bacterial fermentation contribute to the complex flavors of coffee and chocolate.

Can something as simple as "vinegar" have the numerous health benefits as many claim? A growing number of basic science and clinical medicine studies support some of these claims. For example, when given apple cider vinegar, obese rats on a high-fat diet exhibited a decrease in cholesterol, low-density lipoproteins (LDLs), and oxidative stress when compared to control mice [5]. Some studies on rats and humans even show the consumption of apple cider vinegar can reduce the blood sugar of diabetics after a meal by increasing the speed by which food moves through the stomach and gut, thereby limiting sugar absorption [6, 7]. The impact of these studies is interesting yet modest. Ultimately, any health benefits attributed to apple cider vinegar or any other vinegar are the result of molecules originating from the plant source. For example, fermentation of apple juice into apple cider vinegar produces more than just acetic acid; additional molecules are produced by the fermentation process, and even vitamins and minerals from the apple. While the full impact of apple cider vinegar on human health is not well understood, the drink does have some health benefits. But before chugging down a bottle, you should also be aware of the drink's problems. An increased intake of acidic vinegar will speed up the erosion of teeth, increase the risk of reflux, and may lead to kidney problems.

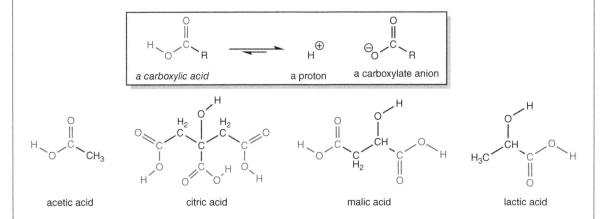

Figure 1.26   Common food acids contain carboxylic acid groups of atoms. The carboxylic acid group of atoms readily gives up an H^+ cation, which also creates the carboxylate anion. Some common acids are found in foods. Acidic food molecules can contain one or more carboxylic acid groups.

Figure 1.27 pH scale and foods. The pH scale from low pH (high H⁺ concentration) to high pH (low concentration of H⁺). The pH of a few common food and household items is noted.

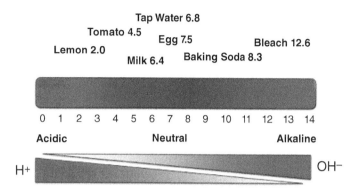

The carboxylic acid groups of atoms (i.e. –COOH or –CO₂H) are found as part of many other common acids that we encounter in foods, for example, malic acid, citric acid, and lactic acid all contain one or more carboxylic acid groups of atoms. Malic acid is the molecule responsible for the tartness of a green apple, citric acid is the sour taste of lemons, grapefruits, and other citrus fruits and lactic acid is responsible for the sour tang of yogurt and some cheeses (Figure 1.26). If a molecule has more than one carboxylic acid group, like citric acid and malic acid, the acid can give up more than one H⁺ or proton per molecule. The presence of H⁺ is an important feature in chemistry—and in food and cooking—that chemists have a method for measuring the amount of H⁺ in a watery solution. In fact, the method for measuring the amount or concentration of protons (i.e. H⁺) was developed by a biochemist engaged in brewing beer. The Danish biochemist, Søren Sørensen, was working at Carlsberg Laboratories in Copenhagen studying the proteins, enzymes, and yeast involved in making Carlsberg beer when he invented a scale to measure the level of acid, which was the proton concentration. The scale he invented is called the pH scale, where "p" stands for *puissance* in French or *potenz* in German, both words translating to power. Thus, pH stands for the power of hydrogen and is used to determine the acid content (i.e. the amount of H⁺) of a substance (Figure 1.27).

When using the pH scale, it is helpful to think about the amount of H⁺ relative to water. The pH scale spans 0–14, and pure water is neutral or the center of the pH scale. Even though pure water is neutral—it does not readily give up H⁺—there is a small amount of H⁺ formed just by a small number of H₂O molecules breaking down into H⁺ and ⁻OH (Figure 1.28). Therefore, the center of the pH scale is 7 and is equivalent to the amount of H⁺ as you would find in water. Some molecules are the opposite of acidic; these molecules don't release protons, instead, they take H⁺ from other molecules. Taking a proton from water creates an anion with a special name, hydroxide (i.e. ⁻OH). Substances with a pH higher than 7 have less H⁺ than water and instead produce hydroxide ions (i.e. ⁻OH); these solutions are considered basic or alkaline. Since alkaline or basic molecules produce very few H⁺ ions and consume H⁺ ions due to the presence of the hydroxide anion (i.e. ⁻OH), solutions of alkaline molecules have a pH greater than 7. Foods or drinks with a pH less than 7 are considered acidic and have a higher concentration of H⁺ compared to water. Most foods are neutral to acidic in pH, and very few are alkaline (Box 1.4). Acidic foods typically range in pH from 2 to 4, while the pH of more alkaline foods may approach 8 or even 9. Although pure water has a neutral pH of 7, tap water contains minerals and dissolved gasses that affect the pH, so in practice, the pH of tap water ranges between 6 and 8. The pH of most living cells is 7.2, and the pH of human blood ranges between 7.35 and 7.41.

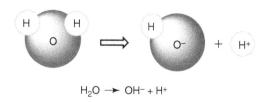

Figure 1.28 Dissociation of water. Water can dissociate into the ⁻OH (i.e. hydroxide anion) and H⁺ (i.e. proton) but only a small amount of the water molecules will do this.

Box 1.4 Figuring out food: Is alkaline water good for you?

Water is marketed with many interesting and unscientific health properties. "Memory water" was once sold to help promote health with the claim that diluted water molecules would "remember" the items once dissolved in the water and provide health benefits to the immune system and mental acuity. Alkaline water is water with a pH greater than 7, usually around pH 8–9, and some have claimed it helps with bone health, increases digestion of nutrients, supports hydration, and even fights cancer. With these kinds of benefits, who wouldn't want to drink alkaline water? By definition, alkaline water has a low concentration of hydrogen ions (H^+), less than you would find in regular, pure water. The very low H+ concentration means that alkaline water also has a significant amount of hydroxide ions (i.e. ^-OH). So far, so good. But for water to have an impact on the health of the rest of the body, water must first travel through the stomach. This is where we run into trouble. At rest, the average stomach has about 50 milliliters (roughly 3.5 tablespoons) of a concentrated, powerful acid called hydrochloric acid that is produced by specialized cells in the stomach. All that acid results in a very low pH of 1.5–3.0 in the stomach. The calculations show us the reality. There are 40 times H^+ in the stomach than ^-OH in a liter of alkaline water. Drinking even 10 large bottles of alkaline water will not significantly change the pH of the stomach fluids; those specialized acid-producing cells will continue to secrete hydrochloric acid into the stomach. So, drinking high-pH water cannot affect other parts of the body since the stomach will instantly neutralize the water (stomach acid H^+ and alkaline water $^-OH \rightarrow H_2O$).

Let's address one of the "facts" that advocates of alkaline water use to claim its anticancer properties. Tumor cells grow and divide rapidly. That rapid growth translates into a high demand for the metabolism of sugars, fats, and proteins to provide energy and building blocks for new cancerous cells. As a result of all of the high metabolic activity, cancer cells become acidic at a much higher rate than non-diseased cells, a phenomenon known to scientists as the Warburg effect. The cancer cells secrete this additional acid into the surrounding environment, which then supports the spread (invasion and metastasis) of tumor cells. The Warburg effect is real, and cancer cells are more acidic, but this is where science ends and supposition begins. Alkaline water claims to "fight" cancer by "neutralizing" the acid produced by tumor cells and even preventing cancer from starting. The problem with that theory is that alkaline water has already become acidic in the stomach. Even when alkaline water is injected into your veins bypassing the stomach, the body tightly controls the pH of extracellular fluid, blood, and cells through the use of buffers Adapted from [8].

Acids are very useful in cooking. For example, ceviche is a Spanish and South American dish that contains raw or partially cooked seafood treated with lemon and lime juice (Figure 1.29). Lemons and limes are citrus fruits that contain citric acid, a molecule that contains three carboxylic acid groups (Figure 1.26). The citric acid from the fruit juices provides flavor, partially breaks down the seafood protein, and tenderizes it, and slows the growth of some harmful microbes. Thus, acids in foods, in addition to their role in flavor, also act as preservatives (because microbes don't survive well in acidic environments), tenderizing, and hygroscopic agents. What are hygroscopic agents? Hygroscopic agents are molecules that absorb water. Acids help keep dry goods free-flowing because many carboxylic acids can form hydrogen bonds with water. Inclusion of the acid traps moisture (i.e. H_2O) in hydrogen bonds and, therefore, limits the clumping of dry goods.

Figure 1.29 Using carboxylic acids to make ceviche. Raw shrimp is made tender and tangy by carboxylic acids in citrus juices. fudio/Adobe Stock Photos

Science for the Chef: Peruvian Ceviche

Ceviche is a popular Latin American dish where raw fish or seafood is "cooked" by marinating it in acidic citrus juices, typically lime or lemon. The acid from the citrus denatures the proteins in the fish, giving it a firm texture and an opaque appearance, similar to that of cooked fish. There are several types of ceviche across different cultures, each with unique ingredients and preparation styles. Peruvian Ceviche is an excellent dish to explore the chemistry of acids and their role in food preparation. The process of acid-induced protein denaturation demonstrates fundamental biochemical principles in a culinary context, making it a delicious and educational experience. Here are some common types of ceviche:

Peruvian Ceviche: The most well-known type, typically made with fresh white fish (such as sea bass) marinated in lime juice, onions, cilantro, and aji peppers. It is often served with sweet potatoes and corn.

Ecuadorian Ceviche: Often includes shrimp marinated in lime juice and mixed with tomatoes, onions, cilantro, and sometimes ketchup. It is typically served with popcorn or plantain chips.

Mexican Ceviche: Usually made with fish or shrimp, marinated in lime juice, and mixed with tomatoes, onions, cilantro, and chili peppers. It is often served on tostadas with avocado.

Colombian Ceviche: Commonly includes shrimp or fish marinated in lime juice with tomato sauce or ketchup, onions, and cilantro. It is often sweeter due to the addition of ketchup or aji sauce.

Chilean Ceviche: Made with fish such as halibut or sea bass, marinated in lime juice, cilantro, onions, and sometimes ginger and garlic for a unique flavor profile.

Ingredients:
- 1 pound fresh sea bass or any firm white fish, cut into small cubes
- Juice of 6–8 limes (enough to cover the fish)
- 1 red onion, thinly sliced
- 1–2 aji amarillo peppers, seeded and finely chopped (or substitute with a mild chili pepper)
- 1/4 cup fresh cilantro, chopped
- Salt to taste
- Sweet potato, boiled and sliced (for serving)
- Corn on the cob, boiled and cut into rounds (for serving)

Instructions:
1. Marinate the fish:
   - Place the fish cubes in a glass or ceramic bowl (avoid reactive metal bowls as they can affect the flavor). Pour the fresh lime juice over the fish, ensuring it is fully submerged.
   - Add a pinch of salt, the sliced onions, and the chopped aji peppers. Gently mix everything together to ensure an even coating.
   - Cover the bowl and refrigerate for 15–30 minutes, or until the fish turns opaque and firms up.
2. Finish and serve:
   - Remove the ceviche from the refrigerator. Taste and adjust seasoning with more salt or lime juice if needed.
   - Stir in the chopped cilantro just before serving.
   - Serve the ceviche with boiled sweet potato slices and corn rounds on the side.

Science Behind the Recipe: The chemistry of acids in Ceviche

Protein denaturation: The primary chemical reaction in ceviche is the denaturation of proteins in the fish by the acidic lime juice. Proteins are complex molecules made up of amino acids linked in specific sequences. In their natural state, these proteins are folded into specific shapes that are stabilized by various interactions, including hydrogen bonds, ionic bonds, and hydrophobic interactions. When the fish is exposed to the acidic environment created by the lime juice, the pH around the protein molecules drops significantly. The high concentration of hydrogen ions ($H+$) from the citric acid in the lime juice interacts with the protein's bonds, particularly disrupting the ionic and hydrogen bonds that maintain the protein's structure. This causes the proteins to unfold, or denature, changing the texture of the fish from soft and translucent to firm and opaque, much like the changes that occur during cooking with heat.

Acid–base reaction: The lime juice (which is acidic, with a pH around 2–3) not only denatures proteins but also interacts with other components in the fish. For example, it can slightly alter the flavor profile by reacting with amines (basic compounds) present in the fish, neutralizing some of them and reducing any potential "fishy" odor.

Preservation: While ceviche is meant to be consumed fresh, the acidic environment also acts as a mild preservative by inhibiting the growth of some bacteria. However, it is important to note that the acidity in ceviche is not sufficient to kill all harmful pathogens, which is why the use of fresh, high-quality fish is crucial.

Flavor enhancement: The acid from the lime juice not only "cooks" the fish but also enhances its flavor. The sourness of the lime juice activates the sour taste receptors on the tongue, balancing the flavors and making the dish more refreshing. The addition of aji peppers adds a slight heat, while the onions and cilantro contribute aromatic compounds that enhance the overall flavor profile.

Key Concepts

1. All food is made of combinations of **elements** we can find on the periodic table. The **atoms** of each unique element have a unique number of **electrons** as defined by the periodic table. All unique chemical combinations of elements involve accepting, donating, or sharing electrons between atoms to create new substances that we call **compounds** or **molecules**.

2. The atoms of an element may accept or donate electrons to combine with other atoms into **ionic compounds**. Atoms that donate electrons become **cations**, and atoms that accept extra electrons are called **anions**. Compounds held together by electrostatic attraction of cations and anions are called ionic compounds or **salts**.

3. **Molecules** held together by atoms sharing electrons are called **covalent** compounds. The act of sharing electrons between two atoms is called a **covalent bond**.

4. Water is made of two hydrogen atoms joined to a single oxygen atom by covalent bonds. The covalent bonds of water are **polar** and allow water to engage in a type of intermolecular attraction called **hydrogen bonding**.

5. The presence of many **intermolecular attractions** between molecules of a substance will create a **solid**. When intermolecular attractions are weakened by the addition of energy, molecules separate from one another: **melting** into a liquid or **vaporizing** into a **gas**.

6. H^+ is a special cation called a **proton**, and the presence of H^+ in water is what makes an **acid**. Some molecules are especially able to give up H^+ cations, and those molecules are called acids.

References

1 McGee, H. (ed.) (2004). *On Food and Cooking*. New York: Simon and Schuster, Inc.
2 Corriher, S. (ed.) (1997). *Cookwise: The Secrets of Cooking Revealed*, New York: HarperCollins Publishers Inc.
3 Corriher, S. (ed.) (2008). *Bakewise: The Hows and Whys of Successful Baking with Over 200 Magnificent Recipes*. New York: Simon and Schuster Inc.
4 Fayet-Moore, F., Wibisono, C., Carr, P., Duve, E., Petocz, P., Lancaster, G., McMillan, J., Marshall, S., & Blumfield, M. (2020). An analysis of the mineral composition of pink salt available in Australia. *Foods*, 9(10), 1490.
5 Halima, B. H., Sonia, G., Sarra, K., Houda, B. J., Fethi, B. S., & Abdalla, A. (2018). Apple cider vinegar attenuates oxidative stress and reduces the risk of obesity in high-fat-fed male Wistar rats. *J Med Food*, 21(1), 70–80.
6 Hadi, A., Pourmasoumi, M., Najafgholizadeh, A., Clark, C. C. T., & Esmailzadeg, A. (2021). The effect of apple cider vinegar of lipid profiles and glycemic parameters: a systematic review and meta-analysis of randomized clinical trials. *BMC Complement Med Ther*, 21(1), 179.
7 Hiebowicz, J., Darwiche, G., Bjorgell, O., & Amer, L.O. (2007). Effect of apple cider vinegar on delayed gastric emptying in patients with type 1 diabetes mellitus: a pilot study. *BMC Gastroenterology*, 7, 46.
8 Fenton, T.R. & Huang, T. (2016). Systematic review of the association between dietary acid load, alkaline water and cancer. *BMJ Open*, 6(6), e010438.

Additional Readings

Jia, G., Chen, Y., Sun, A., & Orlien, V. (2022). Control of ice crystal nucleation and growth during the food freezing process. *Compr Rev Food Sci Food Saf*, 21(3):2433–245.

Kim, M. K., Lopetcharat, K., Gerard, P. D., & Drake, M.A. (2012), Consumer Awareness of Salt and Sodium Reduction and Sodium Labeling. *J of Food Sci*, 77(9):S307–S313.

Lorén, N., Niimi, J., Höglund, E., Albin, R., Rytter, E., Bjerre, K., & Nielsen, T. (2023). Sodium reduction in foods: Challenges and strategies for technical solutions. *J of Food Sci*, 88(3):885–900.

Syamaladevi, R. M., Tang, J., Villa-Rojas, R., Sablani, S., Carter, B. & Campbell, G. (2016), Influence of Water Activity on Thermal Resistance of Microorganisms in Low-Moisture Foods: A Review. *Compr Rev Food Sci Food Saf*, 15(3):353–370.

End of Chapter 1 Questions

1. Identify each as a solid, liquid, or a gas:
   a. A pure substance where the particles have no intermolecular forces with each other and maximum distance occurs between individual particles.
   b. A pure substance where the particles remain in fixed positions.
   c. A pure substance with a definite volume and the particles are able to move freely past one another.

2. Answer the following questions about subatomic particles (electrons, neutrons, and protons):
   a. Attracted to protons.
   b. Has the same mass as a neutron.
   c. These two particles make up the nucleus.
   d. Has no charge.
   e. Located outside of the nucleus.

3. Using the periodic table, report the symbol and elemental name for the following:
   a. Atomic number 8
   b. Atomic number 19
   c. Atomic number 26
   d. Atomic number 17

4. Which of the following are classified as cations?
   a. Cr^{3+}
   b. S^{2-}
   c. Br^-
   d. Cu^+
   e. C

5. Which of the following contain covalent bonds?
   a. $MgCl_2$
   b. CO_2
   c. $C_5H_{10}O_5$
   d. NaF
   e. K_2CO_3

6. Which of the following can be classified as ionic compounds?
   a. $MgCl_2$
   b. CO_2
   c. $C_5H_{10}O_5$
   d. NaF
   e. K_2CO_3

7. Determine the type of interaction (electrostatic, hydrogen bonding, or hydrophobic) occurring between the compounds.

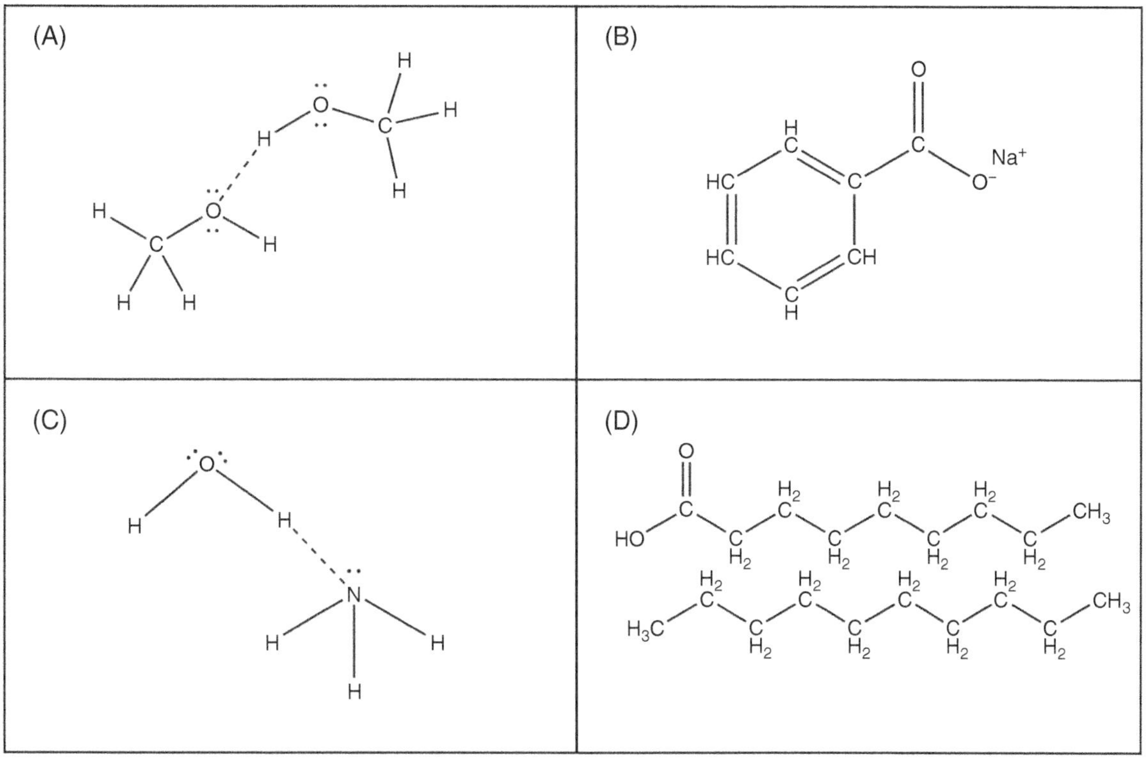

8. Ethanol is the "alcohol" in beer, and the ethanol in beer is dissolved in water. Using the structure of ethanol drawn below and what you know about *polar bonds*, draw a Figure that explains how water *dissolves* ethanol.

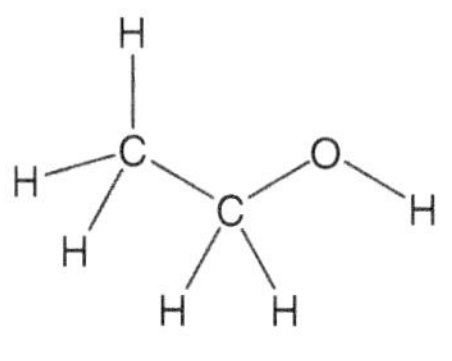

9. Adding table salt (i.e. sodium chloride) to liquid water makes it harder to freeze the water into a solid ice. Why does dissolving ions in the liquid water interfere with its ability to form the ordered, hexagonal array of hydrogen bonds found in ice?

10. Fruits and vegetables are comprised of cells, and each cell in the plant is full of liquid water. If you place a fruit or veggie in the freezer, the plant material will freeze, but upon thawing, the plant material is mushy and drips juice.

    a. If you were to look at the thawed plant material under a microscope, you would see that the cells of the plant were punctured from the inside out. Given the properties of liquid versus solid water. How did the freezing of the plant material end up rupturing the cells?

    b. It is possible to buy "flash frozen" fruits and veggies at the grocery store. According to the food manufacturers, these fruits and veggies were frozen *very quickly* to prevent the formation of large ice crystals. Consequently, the ice crystals formed in quickly frozen/flash frozen fruits and veggies are small. How might this improve the quality of the fruit or veggie upon thawing?

2

Macromolecules of Food and Cooking

Guided Inquiry Activities (Web): 5, Amino Acids and Proteins; 6, Protein Structure; 7, Carbohydrates; 9, Fat Structure and Properties; 10, Fat Intermolecular Forces; 11, Smoking Point and Rancidity of Fats; 12, Emulsions and Emulsifiers

Part I: Macromolecules

Learning Objectives

1. *Describe how a **macromolecule** is made from simple, monomer starting materials.*
2. *Recognize and identify the **monomer** units of proteins (amino acids), carbohydrates (sugars), and fats (fatty acids and glycerol).*
3. *Recognize and describe the **dehydration** reaction that joins monomer units into a macromolecule.*

2.1 Macromolecules (Proteins, Sugars, and Fats)

You have likely read about some of the different types of food molecules, like carbohydrates, proteins, and fats, on a nutritional food label. These large biological molecules, proteins, carbohydrates (sugars), and fats are the functional units of a cell and are key components of food and drink. Each is comprised of simple starting materials or building blocks, such as amino acids, simple sugars, and individual fatty acids, which are chemically combined to make a larger, more complex molecule that plays numerous functions within the cell and food.

How are these large molecules made? The chemistry of assembly and disassembly of the complex molecules is perhaps surprisingly similar, even if the details of the molecules are distinct. Polymerization is a process where smaller molecules, called monomers, are chemically combined to produce a larger chain known as a polymer (Figure 2.1). Starch, a carbohydrate polymer, consists of hundreds or thousands of individual sugar molecules that are connected to produce the final product. Chemically, two individual units are combined in a process called condensation or dehydration. In this example (Figure 2.1), a hydrogen atom is lost from one of the units, and an —OH group is removed from another unit, allowing the two units to combine or condense together in the formation of a new covalent bond, resulting in a larger linked growing

The Science of Cooking: Understanding the Biology and Chemistry Behind Food and Cooking, Second Edition. Joseph J. Provost et al.
© 2025 John Wiley & Sons, Inc. Published 2025 by John Wiley & Sons, Inc.
Companion website: www.wiley.com/go/provost/food_science_2e

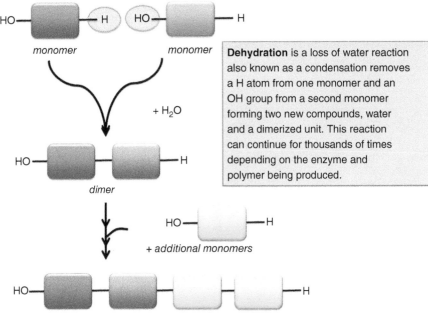

Figure 2.1 Growing a biological polymer. Adding individual building blocks called monomers into long strands of monomers creates polymers such as proteins, starches, or DNA molecules. Enzymes that dehydrate the monomers linking them together while generating a water molecule create most biological polymers.

polymer. The "lost" H and OH reform to generate a molecule of water as a side product. Depending on the molecule being created, this dehydration reaction can be repeated thousands of times, making a larger and more complex new molecule. When discussing polymers, we often give the polymer a name that indicates how many monomeric units have been condensed together. When two monomers are linked together, the growing polymer is called a dimer; the sugars lactose, maltose, and sucrose are dimeric polymers, also known as disaccharides. **Dehydration** is a loss of water molecules in the reaction, also known as condensation, which removes an H atom from one monomer and an OH group from a second monomer forming two new compounds: water and a dimerized unit. This reaction can take place thousands of times depending on the enzyme and polymer being produced.

As you can imagine, polymers can also be broken down, and individual monomeric units can be chemically removed. The process of removing a monomer from a polymer is called hydrolysis; this term means breaking (lysing) bonds through the addition of water. Hydrolysis reactions happen naturally in living cells, but in cooking, the presence of acid and heat often promote hydrolysis reactions and the breakdown or degradation of complex molecules like proteins, lipids, and sugars. You will see the theme of degradation throughout the book as we describe the production and breakdown of foodstuffs. In parts II through IV, we will look at the macromolecules that are important in our food: proteins, carbohydrates, and lipids or fats. These three types of macromolecules appear on every nutrition label. In Part V, we will look at mixtures of those macromolecules called emulsions. But what about DNA? DNA (deoxyribonucleic acid) is a biological macromolecule and a polymer of individual deoxyribonucleotides. While DNA and other types of nucleic acid are very important to the functioning of cells, tissues, and organisms, they are not a form of energy storage. We'll talk about nucleic acid and its roles in Part VII.

KEY CONCEPTS Part I: Macromolecules

1. **Macromolecules** are made of simple starting materials that are chemically joined to make larger, more complex molecules.
2. The **monomer** units of larger macromolecule **polymers** are joined by **dehydration** reactions in which a water molecule is removed.
3. Proteins are **macromolecules** made of chemically joined **amino acids**. Carbohydrates are macromolecules made of chemically joined sugars. Fats are macromolecules made of chemically joined fatty acids and glycerol.

Part II: Protein

The goals of part II pair well with Chapters 2, 3, 4, 6, 7, and 8

Learning Objectives

1. *Describe the essential building blocks of protein.*
2. *Recognize and identify the basic structure of an amino acid.*
3. *Explain how protein is a polymer.*
4. *Recognize and identify polar bonds within proteins that are capable of hydrogen bonding.*
5. *Recognize and identify nonpolar bonds within proteins that are capable of hydrophobic interactions.*
6. *Explain how chemical bonding and hydrophobic interactions are responsible for the folding of proteins into higher-order structures.*
7. *Explain how adding energy (heat) can disrupt the attractions holding proteins together and cause denaturation and coagulation.*
8. *Evaluate a nutrition label for protein content.*
9. *Describe how the calories for protein are determined for a nutrition label and explain how the percent daily value is calculated.*

2.2 Proteins

Proteins are found in every cell of every organism, plant, animal, or microbe. They do the work of the cell, provide structural support, allow cell movement when needed, carry oxygen, and are a source of energy and flavor. While some food is considered high or low in protein content relative to the fat or sugar content found in food, all food contains protein (Figure 2.2).

2.2.1   Protein Primary Structure

Proteins are made of individual building blocks (monomers) called amino acids. There are 20 common amino acids that make up every protein in the plant, animal, and microbe world. The difference between any two proteins is, in large part, the order

Figure 2.2   Protein-rich foods. Many foods are good sources of proteins including milk, cheese, meat, and fish. Pixelbliss/Adobe Stock Photos

and types of amino acids bonded together to make the protein. Every amino acid contains the following components: an alpha carbon (the center carbon in the structure), an amino functional group, a carboxylic acid functional group, and another group of atoms called the side chain (Figure 2.3). Each of the 20 different amino acids has a unique side chain group; thus, it is this group (often chemically notated as an "R group") that makes each of the 20 common amino acids different from the others. Chemically, the side groups can be organized by their chemical reactivity/properties. Some are hydrophobic or nonpolar, while others are polar or charged (the term hydrophilic might be used to describe these amino acids because of the side chain interaction with water). Some amino acids can also be described as acidic or basic, while several have other unique chemical qualities. In order to make a protein, individual amino acids are linked together via a covalent bond in a dehydration/condensation reaction that involves the carboxylic group of one amino acid and the amino group of another amino acid. The resulting covalent bond between the two amino acids is called a peptide bond. Biochemists call an amino acid that has undergone a chemical condensation with another amino acid to create a protein, a "residue," to distinguish from amino acids not in a peptide bond or free amino acids.

When a small number of amino acids are chemically joined together with peptide bonds, the new molecule may also be called a *peptide*. For example, five amino acids that have been chemically joined with peptide bonds are called a pentapeptide. You can see an example of a pentapeptide in Figure 2.4. But many proteins are made of thousands of chemically joined amino acid residues. A protein that large is impossible to draw with a traditional structural line drawing, as shown in Figure 2.4. To communicate chemical information about large proteins, chemists abbreviate the 20 possible amino acid residues with a single letter abbreviation.

The order of amino acid residues in a protein is determined by the genetic code that instructs a cell which amino acid to bond to the next amino acid. This linear sequence of amino acids in a protein is called the primary structure. Amino acids are counted from the amino acid on the "amino terminal end or the N-terminus" of a protein as the first residue in a polypeptide. The last amino acid of a protein is the "carboxyl-terminal end, or the C-terminus" (Figure 2.4). The order in which amino acids are connected determines how a protein folds into a three-dimensional structure and gives the protein its chemical properties.

Because many proteins are hundreds or thousands of chemically joined amino acid residues, the single letter abbreviations of amino acids are used in place of the chemical structure in order to communicate the structural information in an abbreviated way (Table 2.1). Therefore, a protein is often represented as a string of letters called the *primary sequence* or *primary structure* that uniquely identifies a protein. For example, insulin is made of 110 amino acids and is processed into a shorter 51 sequence of amino acids. The primary structure of insulin not only uniquely identifies insulin from other proteins in the body, but it also informs scientists about the function of insulin. Since the primary sequence is what determines the three-dimensional shape and the physical properties of a protein, it is the primary sequence that holds all the information a scientist needs to predict identity and function for a given protein. For instance, a particular sequence of 8 amino acid residues within a protein identifies that region as the part of a protein that binds

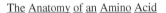

The Anatomy of an Amino Acid

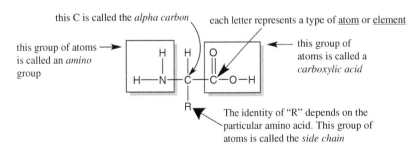

this C is called the *alpha carbon*

each letter represents a type of atom or element

this group of atoms is called an *amino group*

this group of atoms is called a *carboxylic acid*

The identity of "R" depends on the particular amino acid. This group of atoms is called the *side chain*

Figure 2.3 The anatomy of an amino acid. An amino acid has four main components; the R group is the portion unique for each of the 20 common amino acids.

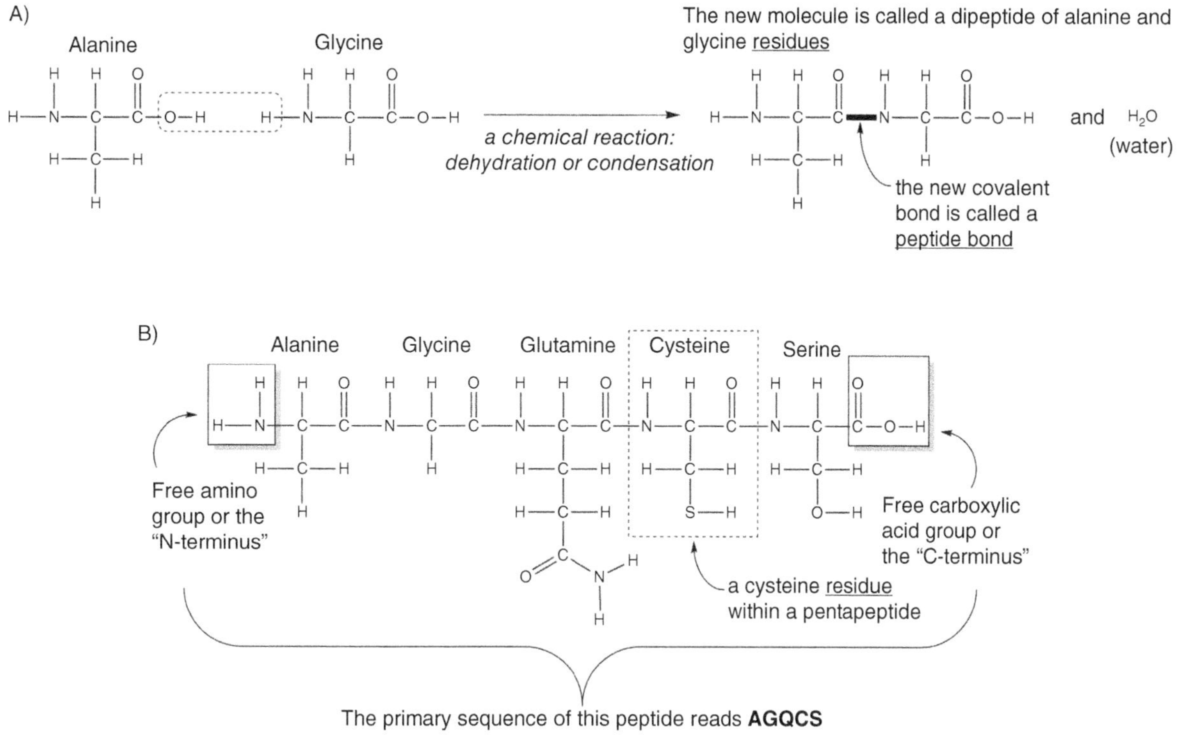

Figure 2.4 Amino acids are chemically joined in a condensation or dehydration reaction to make peptides. The order of amino acid residues in the peptide or protein determines the primary sequence or structure.

Table 2.1 The 20 amino acids that naturally occur in protein along with their three letter and one letter abbreviations.

Name	Three-letter abbreviation	One letter abbreviation	Name	Three-letter abbreviation	One letter abbreviation
Alanine	Ala	A	Leucine	Leu	L
Aspartic Acid	Asp	D	Lysine	Lys	K
Asparagine	Asn	N	Methionine	Met	M
Arginine	Arg	R	Phenylalanine	Phe	F
Cysteine	Cys	C	Proline	Pro	P
Glycine	Gly	G	Serine	Ser	S
Glutamic Acid	Glue	E	Threonine	Thr	T
Glutamine	Gln	Q	Tryptophan	Trp	W
Histidine	His	H	Tyrosine	Tyr	Y
Isoleucine	Ile	I	Valine	Val	V

the important activator, adenosine triphosphate, or ATP. Changes in the genetic code for a protein can lead to different amino acids being inserted into the primary structure, which then alters how a protein behaves. There are many examples of this in cancer, where one or two amino acid substitutions can lead to proteins causing cells to grow without control.

The proteins we encounter in food also have unique primary sequences. Glutenin and gliadin are two proteins responsible for the proteinaceous matrix gluten found in bread made from wheat. Glutenin is a large protein of >1000 amino acids. A short 30 amino acid portion of the glutenin primary sequence is shown here: ...PGQLQQPAQGQQGQQPGQAQQGQQPGQGQQ... Here is a short 30 amino acid portion of the gliadin primary sequence:...QLQPFPQPQLPYPQPQLPYPQPQLYPPQPQ....The single letters each correspond to an amino acid residue, for example—Q is glutamine, P is proline, G is glycine, L is leucine, and each amino acid residue differs in its side chain (Figure 2.3). Notice that the order and identity of the amino acid residues are different for glutenin versus gliadin, and that is what makes these two proteins different. The number, identity, and order of connected amino acid residues is what makes proteins different from one another and gives each their unique properties. As we will see in Chapter 9, the primary sequences of glutenin and gliadin help explain why gluten gives bread its stretch and strength.

Human genetic material codes for about 20,000–25,000 different proteins. Each of these proteins is made from combinations of the same 20 amino acids. A typical protein is between 500 and 900 amino acids in length. Humans can make some of our amino acids from scratch; however, we need to obtain nine amino acids from our diet since we cannot synthesize them and they are required to make the proteins that our cells need to support life. Proteins that are high in these nine "essential" amino acids are called high-quality proteins. Meat, fish, eggs, and dairy products are good sources of proteins that contain these essential amino acids. Incomplete protein sources are foods that contain proteins, but the proteins present are low in one or more of the essential amino acids (this means that the essential amino acid is present or isn't present in high amounts in the protein). Rice, dry beans, potatoes, and some other plant-based foods are foods that contain proteins with a limited amount of one or more of the essential amino acids and are thus considered incomplete. Therefore, individuals with a diet that is lacking in meat, fish, and eggs must combine plant foods from a variety of cereals and grains to achieve enough of the essential amino acids in their diet (Table 2.2).

2.2.2 Protein Secondary and Tertiary Structure

Once synthesized from amino acids, each protein molecule folds into a unique and special shape, a shape that is influenced by the order, number, and chemical nature of the amino acids present. How a protein folds and stays in this folded shape is based on the intermolecular forces imparted by the amino acids and the interactions of their side chains (Figure 2.5). Many of the hydrophobic amino acids are found clumped or aggregated together on the inside of the protein structure, where they can avoid the water that fills the cells. Positively and negatively charged amino acid side chains attract each other in electrostatic interactions or ionic bonds. Polar amino acid side chains tend to be involved in hydrogen bonding interactions with other polar amino acid side chains or water. The amino

Table 2.2 Essential amino acids and some food sources.

Food	Missing essential amino acid
Eggs, fish, meat	None
Quinoa, soy, buckwheat	None
Beans	Methionine, tryptophan
Corn	Lysine, tryptophan
Wheat and rice	Lysine
Peas	Methionine
Almonds and walnuts	Lysine, tryptophan

Figure 2.5 Forces maintaining the protein structure. The backbone of a protein (shown as a black line) is folded in its native state by the chemical interactions of the side chains.

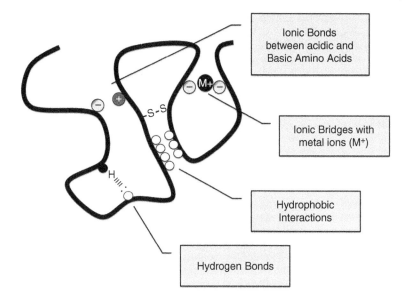

Ionic Bonds between acidic and Basic Amino Acids

Ionic Bridges with metal ions (M+)

Hydrophobic Interactions

Hydrogen Bonds

acid called cysteine contains sulfur in the side chain, and the side chains from two cysteines can come together to form a sulfur–sulfur covalent bond known as a disulfide bond. If the amino acid side chains on the surface of a protein are charged, depending on the overall charge (positive or negative charge), proteins can attract other proteins of the opposite charge or repel proteins of the same charge.

Protein secondary structure is the three-dimensional shape of local regions of a protein sequence. The most common types of secondary structure are an alpha helix and beta-pleated sheet (Figure 2.6). These shapes are held in place by weak forces (hydrogen bonding, ionic interactions, hydrophobic interactions) and form into structures/shapes seen throughout proteins. Alpha helices are found in many proteins, including those found in eggs and milk. The spiral shape of the helix is held together by hydrogen bonds between the N−H and C=O groups of the peptide backbone. The beta-pleated sheet is a folded structure with hydrogen bonds between different peptide strands. Ovotransferrin from chicken egg white contains beta strands organized into beta-pleated sheets alongside regions of structure with alpha helices (Figure 2.7). You can find more information on how chains of amino acids form secondary structure in Chapter and Section 8.5 when we learn about eggs.

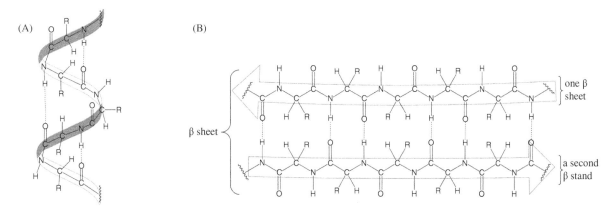

Figure 2.6 Alpha helices and beta strands are two common types of protein secondary structure. Hydrogen bonds hold elements of secondary structure in place.

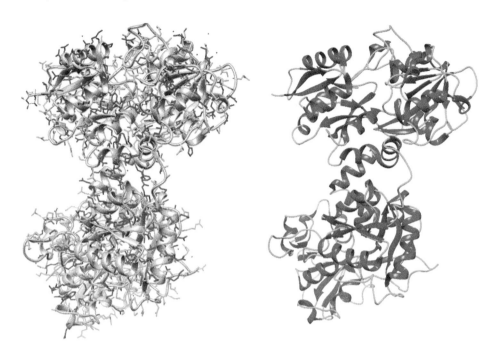

Figure 2.7 The three-dimensional structure of chicken ovotransferrin is shown in two representations. On the left, the amino acid residues of ovotransferrin are shown as line structures. Amino acid residues participating in alpha helices are colored red, those participating in beta-pleated sheets are colored blue, and all other amino acid residues are colored green. The secondary structure is drawn as a gray ribbon that is superimposed over the amino acid residues. The peptide backbone takes the shape of the alpha helix spiral and beta strands. On the right, only the secondary structure is shown with the same color scheme: alpha helices in red, beta strands in blue, and all other structures in green.

Secondary structures fold together bringing the protein into its final, mature three-dimensional shape. A single protein's mature overall structure is called the tertiary structure. Many weak forces dictate the tertiary structure of a protein. Once a protein is in this final folded shape, it is considered to be "native." If two or more folded proteins associate together, the combined set of proteins is described as a quaternary structure. When proteins exist in a complex with other proteins, often called subunits, they possess a quaternary structure. For example, in blood, four different proteins come together to form the protein hemoglobin. This protein is responsible for transporting oxygen throughout the bloodstream. Hemoglobin is made of four proteins and thus has a quaternary structure. Myoglobin, on the other hand, has only one protein. While it also binds oxygen (but in muscle cells), myoglobin is a single protein and does not have a quaternary structure. Both proteins have primary, secondary, and tertiary structures, but only hemoglobin has a quaternary function. Later in Chapter 7, you will learn how myoglobin is responsible for the color of red and even white meats.

2.2.3 Protein Denaturation

As described earlier, when a protein is folded into its functional shape, it is called a native folded protein. When proteins are subjected to heat, acids, or bases, the intermolecular forces holding the protein in its native tertiary and secondary structures are broken and the protein unravels; we call these proteins denatured (Figure 2.8). Ovalbumin, one of the proteins in egg whites, in its native structure is suspended/dissolved in the water present in the egg white, so light can pass through the liquid egg white and it looks transparent or clear. Heat denatures the proteins (ovalbumin and others) in the white, causing the individual protein molecules to aggregate or clump

Figure 2.8 Denaturing protein. Extreme conditions can lead to protein unfolding causing a mesh network of insoluble protein. Cheese curds (from acid) and egg whites (from heat) are two classic examples of denaturing proteins and food.

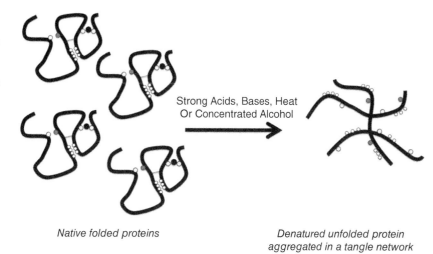

Strong Acids, Bases, Heat Or Concentrated Alcohol

Native folded proteins

Denatured unfolded protein aggregated in a tangle network

together. As a result, the egg white solidifies, and light is reflected off the egg white, creating a solid white appearance. In contrast, the proteins in yolks denature more slowly (or require more heat to be denatured) due to differences in the amino acids and intermolecular interactions that maintain the protein structure. As an example of this diversity, let's look at the three egg proteins: ovalbumin, conalbumin, and ovomucin. Ovalbumin, the major protein component of an egg at approximately 54%, denatures at 80°C. Conalbumin has fewer intermolecular forces that hold the protein in its native structure and denatures at 63°C. By contrast, ovomucin has a larger number of cysteine amino acids, which can form disulfide bonds that stabilize the protein's native structure and denature at a higher temperature. The thin part of an egg white has very little ovomucin; you can observe this by watching the thin egg white solidify first when frying an egg.

Acids and bases can change the charge state on the side chain groups of amino acids; a negatively charged carboxylic acid side chain might become neutral (uncharged) in the presence of an acid, while a positively charged amine might become neutral in the presence of a base (Figure 2.9). This change in charge state also leads to an unraveling or denaturation of protein molecules. As an example, the acid produced by lactobacillus bacteria causes negatively charged milk proteins to become neutral. In turn, the milk proteins aggregate and curdle into yogurt or soft cheese.

Proteins can also denature due to agitation or stress. In the kitchen, you can agitate proteins with a whisk (this is what you do when you make meringue). The mechanical agitation of the egg white proteins introduces air into the proteins, causing the proteins to distort, stretch, and denature. The long tangles of denatured proteins then coalesce, forming a cage around the air bubbles, and the whites expand, creating a foam.

2.2.4 Enzymes Are Proteins

Proteins play a diverse role in living cells. One special group of proteins, called enzymes, aids in chemical reactions. Enzymes reduce the energy needed for a chemical reaction, thus increasing the rate of the reaction. In other words, enzymes serve as biological catalysts. Some enzymes are involved in making new molecules needed for a growing cell, others are involved in breaking down molecules for cellular energy, and others are released from digestive organs to help break down food. Enzymes also play an important role in cooking and baking. Some enzymes chemically cut (or cleave) proteins into smaller pieces. Other enzymes chemically modify sugars or fats, converting them into new compounds with different flavors. For example, aged steaks are enhanced in flavor due, in part, to the enzymes released by dying cells that break down some of the connective tissue, which makes a tenderized meat, and generate flavorful amino acids. Aged cheese contains enzymes that chemically alter the

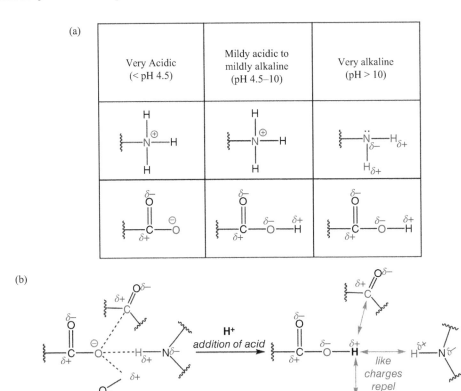

Figure 2.9 Acidic conditions frequently denature proteins. (a) Some amino acid side chains react with the protons (H$^+$) in acid. Amino acid residues like lysine, aspartate, or glutamate have side chains that gain or lose a proton (H$^+$) depending on the pH– these atoms are said to be protonated (gain of H$^+$) or deprotonated (loss of H$^+$). (b) The anionic oxygen of protein carboxylate is electrostatically interacting with other protein atoms to fold and stabilize protein structure. After the acid is added, the carboxylate oxygen is protonated and not negatively charged anymore, which disrupts the interactions and destabilizes the protein fold.

protein, sugars, and fats in the cheese, creating a more mature and stronger-tasting food. However, enzymes can also create havoc during cooking and baking. Lysozyme, an enzyme found in egg white, degrades large carbohydrate sugar polymers into much smaller sugar molecules. Thus, the presence of egg white in an egg yolk that is being used to make a custard or some pastries will result in disaster.

One complicating factor when working with enzymes in the kitchen is that enzymes are proteins. Yes, you knew that. But, for any protein to remain functional, that is, for a protein to "do its job," the protein must retain its folded, native structural form. Heat from cooking, dehydration, the addition of lemon juice, or whisking a mixture may compromise the protein's structure, leading to denaturation and an inactive enzyme/protein. Enzyme activity can also be slowed or reduced by storing food at cool temperatures in the refrigerator or freezer. Thus, either chilling or cooking foods can lessen enzyme activity. This is the basis for some recent controversy over the health benefits of a raw food diet. Raw food enthusiasts advocate that cooking food denatures the enzymes needed for digestion and better health. Promoters of a raw food diet believe that we can replace the digestive enzymes naturally produced in the human body with plant enzymes if the food is eaten uncooked. However, our stomach and intestinal digestive enzymes as well as the acid content of our stomach will denature most or almost all of the proteins long before the protein from the raw food will cross the intestine, travel through our circulatory system, and get to our organs.

2.2.5 Proteins and Water

Another important role of proteins in cooking is their capacity to hold water. As described earlier, the native structure of most proteins keeps the hydrophobic amino acids tucked away inside the structure where water is excluded, leaving the charged and polar amino acids on the surface of the protein free to form hydrogen bonds with water. When a protein denatures, the hydrophobic portions of the unraveled protein are exposed, driving the protein to interact with other protein molecules instead of water. Thus, native proteins help to retain water in foods, while denatured proteins allow water to be released. This phenomenon plays a critical role in food texture in meat, milk, plant products, and baked goods. Native proteins in meat allow the tissue to remain moist after cutting or grinding. However, once heated, the meat proteins denature and have fewer interactions with water. The juices leaked from a cooked steak are mostly water that is no longer retained by the myosin proteins of meat and is colored with other molecules released from the tissue. Resting a steak after cooking allows some of the water to find new interactions with proteins; thus, less juice leaks out upon cutting. Later, we will see that starch plays a similar role in holding water in baked goods, keeping them moist.

Box 2.1 Figuring out food: Calories

You may have noticed that a nutrition label assigns a certain number of Calories to protein, carbohydrates, and fats. What is a Calorie and where do those numbers on a nutrition label come from? The word "Calorie," with an uppercase C, is what chemists call a "kilocalorie," or 1,000 calories, with a lowercase "c". To distinguish between the two, sometimes, the nutritional calorie is written as a Calorie—as opposed to a calorie. One Calorie (i.e. kilocalorie) is the amount of energy it takes to raise 1 kg of water 1°C at sea level. While the energy needed to raise 1 kg of water at 1°C at sea level is the standard measurement all chemists compare to, you can think of a Calorie as a measure of energy. When a human body consumes food, that food is converted into energy, and that energy can be measured in the form of heat.

In the late 1800s, an agricultural chemist named Wilbur O. Atwater devised a method to measure the amount of energy that different food molecules release when they are broken down by the human body. In order to make these measurements of energy or Calories, Atware built a special Calorie-meter or "calorimeter." Atwater's calorimeter was a 4 foot by 8 foot box that a person could step inside of. Once a person was inside the calorimeter, Atwater could measure very precisely the amount of heat energy that was released by the human inside the box after they consumed different foods. Atwater could also measure the amount of oxygen consumed and carbon dioxide given off during the oxidation of food molecules, or their conversion into energy and waste products. All foods can be fully oxidized to yield energy for the organism, along with the waste products carbon dioxide and water. Some food molecules, like protein, also yield sulfur- and nitrogen-containing waste products as well.

Using this method, Atwater was able to determine that carbohydrates and proteins both release 4 Calories for every gram of food, while fats yielded 9 Calories for every gram. If alcohol (i.e. ethanol) is present, it contributes 7 Calories for every gram. However, not all molecules release energy in the human body; for example, dietary fiber is typically assigned no nutritional Calories in Atwater's system. Atwater's "general factor" 4-9-4 system of Calorie measurement for carbohydrates, fats, and proteins is still often the basis for how nutrition labels are calculated today. The Food Data Central Database (https://fdc.nal.usda.gov/) will sometimes report calorie calculations using Atwater "specific factors" that break down calories produced per gram for the *type of food* consumed within a certain category of carbohydrate, protein, or fat. For more information on Atwater "specific factor" calculations, see Annabel Merrill and Bernice Watt's 1973 report titled: Energy Values of Food ... basis and derivation, published by the United States Department of Agriculture.

2.2.6 Protein Nutrition

As we have learned in this section, protein consists of long chains of amino acids that have been chemically joined together by peptide bonds. Because of the nitrogen atoms in each amino acid and the sulfur atoms present in some amino acid side chains, it is more work for your body to break down protein. That means that out of the major macromolecule types (i.e. protein, carbohydrates, and fat), protein has the largest "diet-induced thermogenesis." In other words, it requires more work from your body to break down and extract energy from protein [1]. Some have argued that the large "diet-induced thermogenesis" for protein makes it the "healthiest" of the macromolecules to consume.

How do food scientists measure the amount of protein in your food? Well, they measure the nitrogen in the form of NH_3 or ammonia. Remember, we just learned that protein consists of chemically joined amino acids, and each amino acid residue contains a nitrogen atom that is part of an amide bond with a neighboring amino acid. Using the Kjeldahl method, the food is heated in strong acid to chemically break all the bonds and release the nitrogen. The nitrogen atoms are released and counted as ammonia or NH_3. Once the amount of nitrogen has been counted, it is equivalent to the amount of protein that was in the original sample. Finally, to make a nutrition label, the grams of protein are converted to Calories using Atwater's formula of 4 Calories per gram of protein (Box 2.1).

Protein is more work for your body to consume, and therefore, less likely to end up as excess energy that must be stored. So, how much protein should you consume? The United States Food and Drug Administration (FDA) has created the percent daily value (i.e. % DV) to help consumers assess how much of a particular macromolecule they should eat. The % DV is simply how many grams a given food item contributes toward the total amount of that macromolecule that is recommended. The recommendation is called a Recommended Dietary Allowance, or RDA, and comes from the Institute of Medicine, which issues guidelines on how much of a particular macromolecule (i.e. protein, carbohydrate, or fat) should be consumed [2]. For protein, the RDA for all adults is 0.80 g of "good quality protein" per kilogram of body weight (i.e. 0.36 g per pound of body weight) each day and is based on careful evaluation of scientific data by a panel of experts. This recommendation translates to ~55 g of protein for the average-sized sedentary adult. The protein needs will change based on one's physical activity and health. If a food item contains 10 g of protein, then those 10 g would be ~18% DV for a ~150 pound adult Now, despite this RDA for protein from the Institute of Medicine, there is typically no % DV listed on a nutrition label; that is, unless the food item makes a protein-related claim, such as "high in protein." Unless the food is meant for use by infants and children under 4 years old, no % DV is required on the label, and that is because current scientific evidence indicates that protein intake is not a public health concern for adults and children over 4 years of age.

KEY CONCEPTS Part II: Proteins

1. Proteins are polymers or long chains of **amino acids** joined together at **peptide bonds**. When two or more amino acids are joined together, there is a loss of water and the formation of a peptide bond. The amino acids that have been chemically joined with a peptide bond are called **amino acid residues**.
2. Each of the 20 amino acids found in proteins contains an **amino group**, a **carboxylic acid group**, an **alpha carbon**, and a **side chain**. The side chain is unique for each of the 20 amino acids and may contain polar and/or nonpolar bonds.
3. Long chains of chemically-joined amino acids typically **fold** into three-dimensional **globular** shapes so that **hydrophobic** parts are hidden from the watery environment and **polar** parts are on the outside and in contact with the water. These three-dimensional shapes are held together by **hydrogen bonding** and **electrostatic** interactions between amino acid residues and their side chains.
4. The addition of energy in the form of heat or agitation breaks the attraction between amino acid residues and their side chains; this causes unfolding or **denaturation** of the **globular** structure. Once the **hydrophobic**

interiors of the unfolded proteins are exposed, the **hydrophobic** regions stick together in a process called **coagulation**.

5. The protein on a nutrition label describes a mixture of individual protein macromolecules.
6. Consuming protein yields **4 Calories per gram**. The recommended daily value of protein is not required for most nutrition labels, but it is 0.8 g of protein for every 2 pounds of body weight, or ~50 g for an average adult.

Part III: Carbohydrates

The goals of Part III pair well with Chapters 2, 5, and 12.

Learning Objectives

1. *Describe the essential building blocks of carbohydrates.*
2. *Recognize and identify examples of simple sugars.*
3. *Recognize and identify monosaccharides from disaccharides and larger polysaccharides.*
4. *Recognize and identify the glycosidic bond of carbohydrate polymers.*
5. *Describe the differences between digestible carbohydrates and the indigestible carbohydrates that make up fiber.*
6. *Recognize and identify the polar bonds in sugar molecules that interact well with water and other polar substances.*
7. *Evaluate a nutrition label for carbohydrate content.*
8. *Describe how the calories for carbohydrates are determined on a nutrition label and explain how the % DV is calculated.*

2.3   Sugars Are Carbohydrates

Sugar, saccharides, polysaccharides, complex carbohydrates, simple sugars, starches, pectins, fiber, and gums all refer to the same family of biomolecules called carbohydrates. Carbohydrates, or "hydrated carbon," as the name suggests, contain carbon, hydrogen, and oxygen atoms, often arranged in a manner in which a chain of bonded carbon atoms is each bonded to —OH and —H groups. Because of the presence of multiple —OH groups in carbohydrates, they are soluble in water; —OH groups are also important in making the carbohydrates sweet (Figure 2.10).

2.3.1   Monosaccharides, Disaccharides, and Oligosaccharides

The simplest carbohydrates are the monosaccharides (Figures 2.10 and 2.11). The root word "saccharide" comes from the Latin "saccharum," which means "sugar." There are a multitude of monosaccharide sugars; however, only a few are involved in day-to-day cooking and baking. Glucose or dextrose is a monosaccharide also known as blood sugar; glucose is a key energy source for mammalian organisms. Glucose is found in grapes, berries, and some sports drinks, but the main source of dietary glucose comes from the breakdown of larger carbohydrates and starches. Fructose is a very sweet-tasting monosaccharide sugar found in sugar cane, sugar beets, honey, and corn. Galactose is a less sweet-tasting sugar that is important for the development of neural systems in youth. The primary source of dietary galactose is lactose, as this disaccharide breaks down into galactose and glucose. However, some foods, including papaya, tomato, persimmon, and watermelon, all contain significant amounts of galactose. A severe inherited disease called galactosemia is due to the inability of infants to use

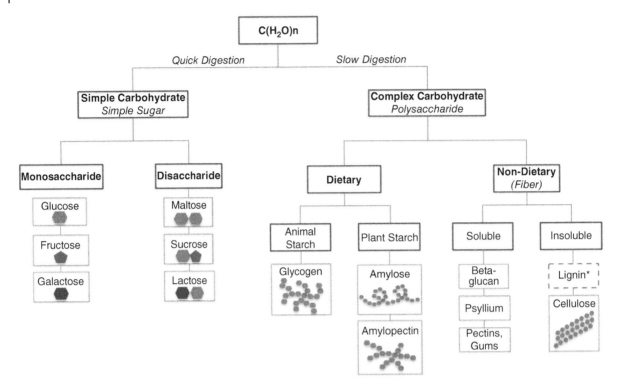

Figure 2.10   Organization of the types of carbohydrates in food.

galactose because of a genetic defect producing enzymes involved in metabolizing the sugar. Those with the disease must avoid foods with or that will produce galactose, or the patient will suffer from vomiting, diarrhea, enlarged liver, and impaired cognitive function. Ribose, another monosaccharide, which was first characterized from the sap of a gum plant, is important for the production of vitamins like riboflavin and is one of the key components of our genetic material. Unlike fructose, ribose is not very sweet, but as we will see in Chapter 3, it plays an important role in making the brown crust of baked goods and grilled meats.

When two monosaccharides are linked together, they form a disaccharide (Figures 2.11 and 2.12). Like other polymers, disaccharides are formed by a condensation/dehydration reaction between two simple monosaccharide

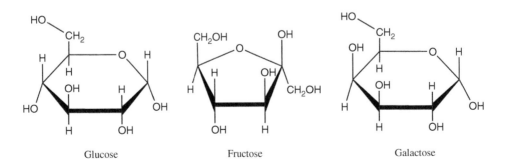

Figure 2.11   Common monosaccharides. Simple sugars (monosaccharides) are found in food and drink. Unless specified otherwise, there is a carbon atom implied at every intersection of two or more bonds.

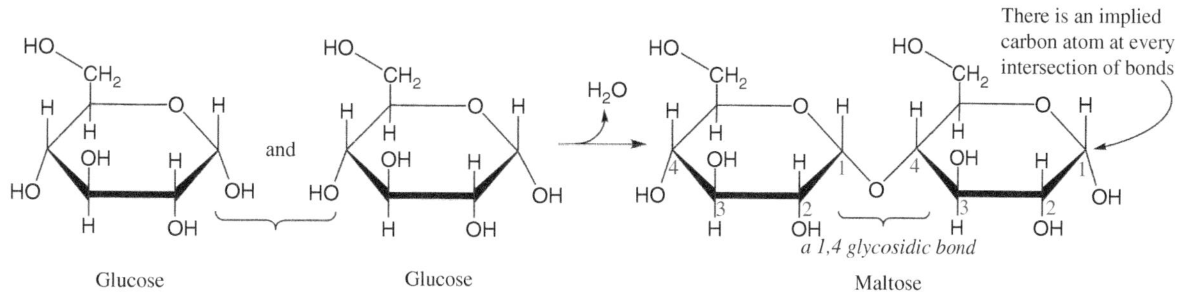

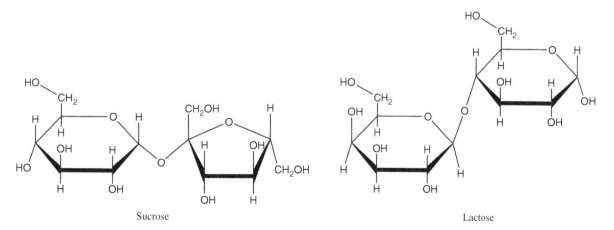

Figure 2.12 Disaccharides and glycosidic bonds. Maltose shown on the right is a disaccharide made of two glucose molecules chemically joined by a special bond called a glycosidic bond. Unless specified otherwise, there is a carbon atom implied at every intersection of two or more bonds.

units (Figure 2.13). The bond between the linked sugars is called a glycosidic bond. The bond can be formed in two configurations: an α-glycosidic bond and a β-glycosidic bond (Figure 2.12). α-Glycosidic bonds are formed when the oxygen atom between the monosaccharides falls below the carbon atoms. β-Glycosidic bonds happen when one of the linking carbon atoms is positioned above the oxygen atom in the bond. The α- or β-character of the glycosidic link is important to the function and structure of the saccharide polymer. Disaccharides cannot be used by the body in the disaccharide form for energy; thus, they must be metabolized into individual monomer sugars, or monosaccharide units, for biological use. However, disaccharides do play an important role as key ingredients in many cooking and baking recipes. Dietary examples of disaccharides include lactose and sucrose. Joining glucose and galactose together via a β-glycosidic bond makes lactose, often called milk sugar. You will learn more about lactose in the chapters on milk, cheese, and metabolism/ fermentation. Sugar beet and sugar cane plants produce sucrose, often called table sugar, as the plant cells trap energy from the sun by chemically combining carbon dioxide and water. A molecule of sucrose is formed when glucose and fructose monosaccharides are linked together in an α-glycosidic bond. If purified, there is no difference in sucrose between the two sources. Maple syrup and sorghum also contain sucrose. The conversion of sucrose into its single sugar components, glucose and fructose, is catalyzed by enzymes secreted into

Figure 2.13 Common disaccharides. Sucrose (i.e. table sugar) and lactose (i.e. milk sugar) are disaccharide sugars commonly found in food and drink. Unless specified otherwise, there is a carbon atom implied at every intersection of two or more bonds.

human saliva. The enzyme invertase splits the indigestible sucrose into the usable fructose and glucose in the mouth. Boiling sucrose for an extended time can also hydrolyze or break apart the two simple sugars from sucrose. This process is enhanced in the presence of acids like lemon juice or tartaric acid. Inverted sugar or "invert sugar syrup," is just sucrose that has been reduced into monomer sugars and should no longer be considered sucrose. This is a trick some winemakers use when fortifying their grapes with an additional boost of carbohydrates. Bakers sometimes use inverted sucrose (called invert syrup) because of the increased sweetness of fructose compared with sucrose.

Once you begin to link more than two sugars together, the terminology used to describe a sugar or carbohydrate may be unclear to the novice. What exactly do we mean at a molecular level when using the terms simple sugars, oligosaccharides, or complex carbohydrates? Now that you know a little more about the molecular structure of mono- and disaccharides, these terms are easily clarified by examining the number of monomer saccharides incorporated into the molecule. Both monosaccharides and disaccharides are considered "simple sugars"; these sugars are easily absorbed by the body and are readily available as an energy source or building block for other biomolecules. Carbohydrates that contain less than 100 monosaccharide units, called oligosaccharides, are found in dried beans, peas, and lentils. These molecules are poorly digested by the human body and typically pass through your digestive system unaltered. However, once in the gut, your intestinal bacteria metabolize the oligosaccharides and produce gases. Complex carbohydrates consist of even longer chains of monosaccharide units called polysaccharides. Polysaccharides are on the order of hundreds or thousands of monosaccharide building blocks. The individual monosaccharide units in complex carbohydrates often participate in multiple glycosidic bonds, which gives them a structure that is much more complex than the shorter-chain carbohydrates. Moreover, like proteins, these long polymers bind tightly to water through hydrogen bonds; thus, they readily absorb water. Complex carbohydrates can be placed into two major classes: starches, which are nutritionally available, and nutritionally unavailable complex carbohydrates, which include gums, fibers, and pectins. These complex carbs are certainly important in cooking, even though they lack nutritional value; for example, it is pectin that is often added to fruit puree to thicken it into a jam or jelly. Let's look more in depth at both types of complex carbohydrates.

2.3.2 Complex Carbohydrates: Starch and Dietary Fiber

Starches are long polymers of glucose and serve as a source of glucose storage that naturally occurs in plants and animals. There are three main forms of starch: glycogen, amylose, and amylopectin. Glycogen is the form of starch that is used in animals as our glucose reserve; it is made and stored in the liver and red muscle tissue. Glycogen is a good example of a branched polymer; this means that approximately every 10-glucose monomer has two glucose monosaccharide units linked by a glycosidic bond. The "second" link serves as a branch point to grow another polymeric chain. The energy storage molecule in plants is called starch; it also consists of long polymers of glucose molecules that serve as energy stores of glucose in seeds and the roots of rice, corn, wheat, potatoes, beans, and cereals. Plant starch consists of two types of molecules: amylose and amylopectin (Figure 2.14). Amylose, which makes up about 20% of plant starch, consists of unbranched chains of 200–4,000 glucose molecules that structurally form a coil. The remaining 80% of most plant starches are amylopectin. Amylopectin is branched, like glycogen, but the branch points occur less frequently, approximately every 25 glucose units. In both glycogen and plant starch, each glycosidic bond that connects one glucose to another is an α-glycosidic bond. Enzymes in our saliva can break down (hydrolyze) the α-glycosidic bonds in these large complex carbohydrates so that we can use the remaining smaller pieces as an energy source. By contrast, cellulose is also a complex carbohydrate made of glucose and a component of indigestible fiber. Cellulose is used by plants to provide rigid strength in cell walls in wood and fibrous plants and has a linear, extended (noncoiled) structure, which contributes to its role as a structural protein.

Furthermore, the linear arrangement of the carbohydrate allows for hydrogen bonding between different cellulose chains, thus creating a strong cross-linked fiber. However, the glucose units of cellulose are linked via a β-glycosidic bond. Our salivary and digestive enzymes cannot break down cellulose because of the β-glycosidic bonds, so cellulose is unavailable to humans as an energy source. However, cows, goats, and termites (to name a few) have symbiotic bacteria living in their gut that can break down the cellulose to yield glucose for use as an energy source; thus, cows and goats can survive by eating grasses and termites thrive on wood.

Cellulose is one component of another class of complex carbohydrates called dietary fiber. Dietary fibers, also called roughage or just fiber, are poorly digested plant polymers that contain a diverse mixture of monosaccharide components, many of which are chemically modified versions of glucose or fructose. Fiber comes in two forms: soluble and insoluble. Fiber that readily dissolves in water is soluble. Soluble dietary fiber tightly binds water through hydrogen bonds, swells, and turns into a gel. The thick water-soluble fiber gel slows down digestion; thus, foods that contain soluble fiber create a feeling of being "full." Moreover, some soluble fibers bind cholesterol and aid in carrying it through the intestinal system. Good sources of soluble fiber include oatmeal, lentils, apples, pears, celery, and carrots. Insoluble fiber does not dissolve in water; instead, it speeds digestion and transit of molecules through the digestive system. Sources of insoluble fiber include whole wheat, whole grains, bran, seeds and nuts, dark leafy vegetables, grapes, and tomatoes. Both soluble and insoluble fibers are important to food, nutrition, cooking, and flavor. Let's look at some examples.

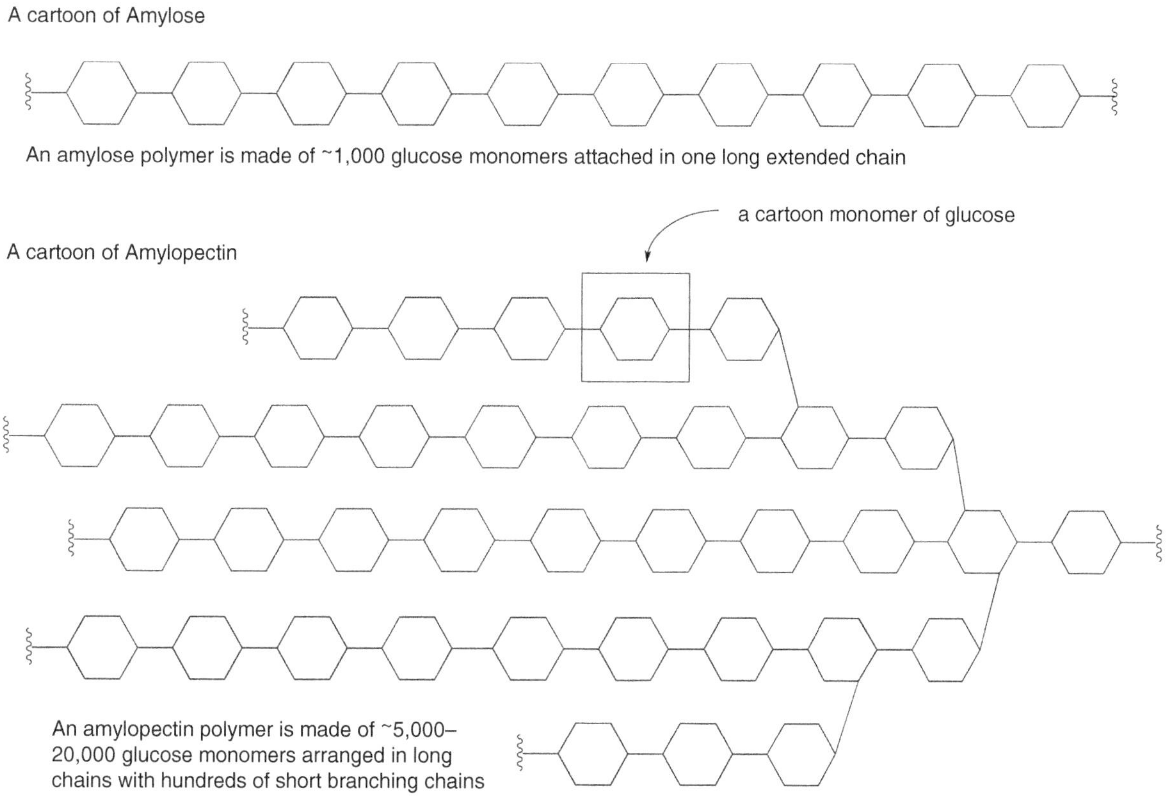

A cartoon of Amylose

An amylose polymer is made of ~1,000 glucose monomers attached in one long extended chain

A cartoon of Amylopectin

a cartoon monomer of glucose

An amylopectin polymer is made of ~5,000–20,000 glucose monomers arranged in long chains with hundreds of short branching chains

Figure 2.14 Plant starches. Two glucose polymers amylose and amylopectin are depicted in cartoon form. Amylopectin is an unbranched polymer, while amylopectin is branched with a tree-like structure.

Box 2.2 Figuring out food: Simple versus complex sugars: Why are they bad for you?

Sugar or "refined sugar" has gained a reputation as being "bad" for you. Refined sugar is simply purified sucrose, fructose, or glucose that comes from sugar cane, sugar beets, or other plants such as corn. There is nothing intrinsically bad about the sugar molecule when it is present in food or used in cooking. In fact, consuming 1 g of sucrose or fructose that is added to a food in cooking will have the same effect as eating 1 g of the same sugar from fruit or other plant sources. As an example, one banana contains about the same amount of simple sugar as a prepared food that contains four tablespoons of granulated sugar per serving! However, excess consumption of sugar, like any food, results in poor health consequences. Spikes in glucose (or blood sugar) occur after you consume a supersized candy bar that contains 40 g of carbohydrates (in the form of simple sugars). These sugars are easily absorbed into the body, causing a spike in blood glucose levels from 70–100 milligram per deciliter of blood (mg/dL) to 120–200 mg dL^{-1} depending on the individual. Having chronic high blood sugar leads to a number of negative health consequences, including glaucoma and nerve damage in the extremities. Why do you get a "crash" after eating a large quantity of sugar? Once glucose levels rise, the body releases a hormone called insulin from the pancreas. Insulin allows the glucose to move from the circulatory system into the muscle tissue. The amount of insulin released is directly proportional to the levels of glucose in the blood. Thus, abnormally high amounts of sugar, due to mega candy bar consumption, cause abnormally large amounts of insulin to be secreted by the pancreas, resulting in high amounts of glucose being transported into the muscle tissue. This phenomenon results in a serious reduction in blood sugar to a level that is much lower than a premeal level. Because the brain gets most of its energy from blood glucose, low levels of blood glucose cause you to feel tired, weak, confused, and even dizzy (Figure 2.15). Interestingly, eating the same mass of complex carbohydrates avoids this spike and crash in blood sugar levels because the starch/glycogen polymers have to be broken down by the digestive enzymes in the gut before the individual glucose molecules can be transported from the intestine into the bloodstream. This takes time and allows for the gradual release of glucose into the circulatory system. The result is a slow, gradual increase and decline in blood sugar levels due to the modest and controlled release of insulin. The rate at which a food spikes the blood glucose level is called the glycemic index. Foods that contain high amounts of simple sugars have a high glycemic index. In contrast, complex carbohydrates including starches and fibers have a much lower glycemic index.

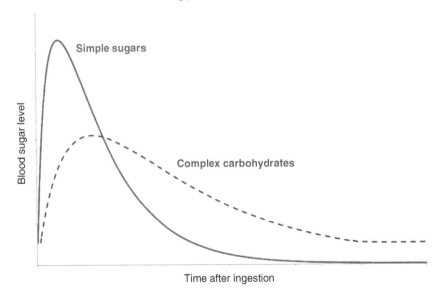

Figure 2.15 Simple versus complex carbohydrate. The impact on blood sugar (glucose) levels after eating simple mono- or disaccharides versus complex carbohydrates.

Because of the gel-like character that is imparted by soluble fiber in water, soluble fiber is often used as a thickening agent in cooking. Pectin, a soluble fiber found in plant cell walls, is made of the monosaccharide galacturonic acid (Figure 2.16). An immature apple contains chemically unmodified pectin, which gives the fruit a hard, rigid consistency. As the fruit ripens, enzymes modify pectin by adding methyl groups; this causes the fruit to soften (due to the gel formation and water retention). Moreover, in cooking, a more stable gel can be formed during the preparation of applesauce or apple pie. If you have ever tried to make an apple pie with unripened apples, the resulting watery mess was due to the lack of gel-forming pectin found in the fruit. Natural (from within the fruit) or added pectin is key to form an effective gel in jellies and jams. The pectin "gel" consists of long strands of carbohydrate polymer that are tangled together into a rigid mass that tightly retains water. A great example is the gel in jelly. In acidic conditions, when citric acids are included, they interact with mono- and disaccharides to form a stable gel that swells several times its dried volume. Mature fruits provide pectin from the cell walls to form an effective gel in making jellies and jams. The gel-like nature of pectin is also used to stabilize yogurts (Figure 2.16).

Oats contain a type of soluble fiber called beta-glucan, a polymer of glucose units that are also linked with β-glycosidic bonds. Beta-glucan reduces the amount of cholesterol the human body absorbs from food, and can therefore reduce the risk of cardiovascular disease [3]. Psyllium is another type of soluble dietary fiber that lowers blood cholesterol. Psyllium is a polymer of xylose monosaccharides found in some plant cell walls. Similarly to beta-glucans, when included regularly in the diet, psyllium fiber can bind and reduce the absorption

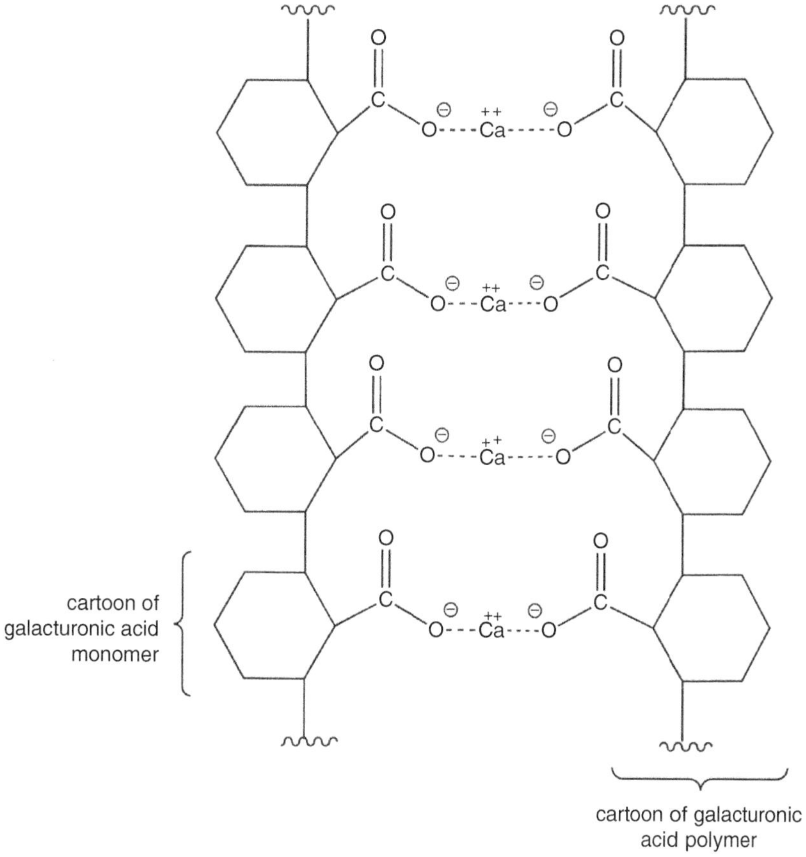

Figure 2.16 Pectin. A cartoon of pectin, a polymer of galacturonic acid. The positively charged calcium helps bind strands of pectin to form a gel.

of cholesterol [4]. Studies have also found that psyllium in the diet can reduce the amount of glucose absorbed from food consumed by Type 2 diabetics [5].

Onions, garlic, and lettuce contain inulin, a short polysaccharide primarily made of fructose. Inulin is a soluble dietary fiber whose use in cooking and food preparation has increased because of its ability to form viscous solutions at low concentrations. Because inulin is not digestible with human enzymes, there is a very low caloric value to the complex carbohydrate (1 kcal/g) and is often used as a low-calorie substitute for fat and simple sugars in cheese, frozen desserts, whipped cream dairy products, and processed meats.

Plant gums are another diverse family of complex carbohydrates that are utilized in foods and cooking. In plants, gums act as a mortar to keep plant cells glued together and are secreted into gaps in damaged plants and fruits. Like pectins, plant gums are used in the food industry for their thickening characteristics and come from a variety of sources, including the hardened sap from Acacia trees (gum arabic), seaweed extracts from brown seaweed (alginate) and red seaweed (carrageenan), and microorganisms (xanthan gum) (Figure 2.17).

What makes a gum different from a fiber? Technically, gums are a type of soluble dietary fiber; they are complex carbohydrates that cannot be degraded by human digestive enzymes and they dissolve in water to form a gel. In the food industry, gums are a special kind of fiber because gums can produce a large increase in the viscosity of a liquid; they can form a gel or solid depending on the amount of water present. How are they used in food preparation? Gums are used to improve mouthfeel, emulsify liquids, and trap or encapsulate flavor molecules. Mouthfeel is the way the food or drink feels thick or thin as it is eaten. When someone says a sauce is too thin, this is mouthfeel. The gum arabic present in a can of Mountain Dew acts as an emulsifier, helps to keep the oil in suspension, and improves mouthfeel. Some gums produce very smooth textures in processed foods and tightly hold onto water and reduce ice crystal formation during freezing/thawing. Because of these properties, gums are often added to process frozen, prepared microwave meals to prevent them from separating into water and solids. Agar agar is a gum used by molecular gastronomy-inspired cooks and chefs. Agar agar (also just called agar) is an extract of algae that has been used historically for many years by microbiologists as a solid gel infused with growth compounds to culture bacteria. This use, of course, was influenced by cooks who had used agar in Java, as well as in other Asian countries, for centuries. Agar agar is an efficient gelling agent that readily forms a semisolid consistency, can be easily infused with flavors (or can infuse flavors), can

provide a firm, solid shape and mouthfeel, and can suspend or encapsulate liquids and solid food particles. Unlike other gels, once formed, agar agar will retain its solid consistency even when heated at 90°C.

The complex carbohydrates that make up fiber cannot be broken down into their monosaccharide units by human digestive enzymes, so fiber does not contribute energy to our bodies in the same way that simple carbohydrates and starches do. But, the complex carbohydrates that make up fiber do travel to the large intestine where these long polymers meet our gut microbiota. The gut microbiota includes many species of bacteria, and these bacteria have a very large arsenal of degradative enzymes. Where human digestive enzymes fail, bacterial enzymes succeed in breaking down a variety of soluble fibers into *short-chain fatty acids* (SCFAs) such as acetate, a two-carbon fatty acid; propionate, a three-carbon fatty acid; and butyrate, a four-carbon fatty acid. These SCFAs travel from the large intestine

Figure 2.17 Red carrageenan. Red seaweed is dried and the cell wall carbohydrates are used as additives (gums) in foods. Hedvika/Adobe Stock Photos

to other parts of the human body and exert a variety of effects on the immune system and the central nervous system, including the brain [6, 7]. Consequently, soluble fibers do contribute some nutritional value; depending on the fiber, it could be 1–2 Calories for every gram.

2.2.3 Carbohydrate Nutrition

At 4 Calories/g, carbohydrates are essential to our daily nutrition and provide a vital energy source. However, as evident on a nutrition label, not all carbohydrates are equal. The **total carbohydrates** listed represent the sum of **starch** (digestible complex carbohydrates), **dietary fiber** (nondigestible complex carbohydrates), and **sugars** (digestible simple carbohydrates). Although the amount of starch is not directly listed on a nutrition label, it can be calculated. In the example shown, one serving contains 34 g of total carbohydrates, comprised of 7 g of dietary fiber and 5 g of total sugar. To calculate the amount of starch, subtract the dietary fiber and total sugar from the total carbohydrates:

$$34 \text{ g total carbohydrates} = \textbf{X g starch} + 7 \text{ g dietary fiber} + 5 \text{ g total sugar}$$

Therefore, one serving size contains 20 g of starch, a digestible complex carbohydrate.

The nutrition label distinguishes between **total sugars** and **added sugars** (Figure 2.18). The total sugar represents the sum of natural sugars found in the food and any added sugars included during food processing. Natural sugars include fructose or sucrose found naturally in fruit, lactose in milk, or maltose in corn. In 2016, the FDA changed the nutrition label requirements to ensure consumers were aware of the amount of sugar added to packaged food items. On an ingredient list, added sugars may be listed as sugar, sucrose, table sugar, syrups, high

Figure 2.18 Carbohydrates and nutritional labels. Look closely at the nutrition label to see, how to assess total carbohydrates, dietary fibers, and sugars? maradaisy/Adobe Stock Photos

Nutrition Facts	
6 servings per container	
Serving size	**1 cup (230g)**
Amount Per Serving	
Calories	**250**
	% Daily Value*
Total Fat 12g	**14%**
Saturated Fat 2g	**10%**
Trans Fat 0g	
Cholesterol 8mg	**3%**
Sodium 210mg	**9%**
Total Carbohydrate 34g	**12%**
Dietary Fiber 7g	**25%**
Total Sugars 5g	
Includes 4g Added Sugars	**8%**
Protein 11g	
Vitamin D 4mcg	20%
Calcium 210mg	16%
Iron 4mg	22%
Potassium 380mg	8%
*The % Daily Value (DV) tells you how much a nutrient in a serving of food contributes to a daily diet. 2,000 calories a day is used for general nutrition advice.	

fructose corn syrup, molasses, honey, concentrated fruit juice, maltose, and dextrose. To calculate the amount of naturally occurring sugar in a food item, subtract the added sugars from the total sugars on the nutrition label.

5 g total sugars = **X g natural sugars** + 4 g added sugars

In this example, one serving size contains 1 g of natural sugar and 4 g of added sugar.

The % daily value (% DV) for total carbohydrates helps you identify whether or not a serving of food is a low or high source of carbohydrates. For example, following a 2,000-calorie diet, the % DV for total carbohydrates is 300 g/day. To calculate the % DV, take the total carbohydrates per serving, divide by 300 g, and convert it to a percentage. In the example, one serving contains 34 g of total carbohydrates.

% DV total carbohydrates = (34 g/300 g) × 100 = 6.66% ≅ 7%

The FDA has set a % DV of 28 g/day for dietary fiber. Since fiber is not digestible and has many reported health benefits, including increasing bowel movement frequency, improving bowel health, lowering cholesterol, and even stabilizing blood glucose levels, the % DV is the minimum one should ingest. Therefore, one should aim for more than 28 g of dietary fiber daily. Foods like legumes, barley, lentils, berries, chia seeds, and avocados are considered high-fiber sources.

While the FDA has not established a % DV for total sugars, the FDA recommends that one eats less than 50 g of added sugars per day. Diets high in added sugars have been linked to low levels of vitamins and minerals, high triglycerides, and adverse health impacts, such as diabetes, heart disease, and obesity [8]. The American Heart Association recommends that women consume no more than 25 g and men no more than 38 g of added sugar per day [8]. Considering the average American consumes 17 teaspoons (68 g) of added sugar each day, there is cause for alarm [9]. While moderation is critical, the FDA recommends limiting foods that contain more than 20% DV of added sugars per serving.

KEY CONCEPTS Part III: Carbohydrates

1. **Carbohydrates** contain carbon, hydrogen, and oxygen atoms, typically arranged in a manner in which a chain of bonded carbon atoms is each bonded to —OH and —H groups.
2. Carbohydrates can be **simple** or **complex**. Simple carbohydrates or simple sugars include **monosaccharides** like glucose, fructose, and galactose. If two simple sugars are joined by a dehydration reaction, a **glycosidic** bond is formed that connects the two halves of the **disaccharide**. Simple **disaccharides** include lactose, sucrose, and maltose.
3. Complex carbohydrates are long chains of monosaccharide units that have been chemically joined by **glycosidic** bonds. Complex carbohydrates include plant and animal **starches** and **fiber**.
4. Starches are long chains of glucose monosaccharide units that are chemically joined via an **α-glycosidic bond**. When cleaved by human digestive enzymes, the **α-glycosidic bond** releases glucose **monosaccharides**, which makes starches a source of nutrition.
5. Complex carbohydrates also include soluble **fiber** such as **pectin, beta-glucans, psyllium,** and plant **gums,** and insoluble fiber such as **cellulose**. Complex carbohydrates that are classified as fiber are made of **glycosidic** linkages and **monosaccharide** units that cannot be broken down by human digestive enzymes, so fiber does not contribute calories to the same extent as simple carbohydrates and starches.
6. Total carbohydrates listed on a nutritional label represent the sum of **starch** (digestible **complex** carbohydrates), dietary **fiber** (nondigestible **complex** carbohydrates), and **sugars** (digestible **simple** carbohydrates).
7. Consuming digestible carbohydrates yields **4 Calories per gram**. The recommended daily value of digestible carbohydrates is 300 g/day, while the recommended daily value for dietary fiber is 28 g/day.

Part IV: Fats

The goals of Part IV pair well with Chapters 3, 4, 7, and 12.

Learning Objectives

1. *Describe the essential building blocks that make fat or triglyceride.*
2. *Explain how triglyceride is made of fatty acids and glycerol.*
3. *Recognize and identify triglycerides and the nonpolar bonds within triglycerides that are capable of hydrophobic interactions.*
4. *Recognize and distinguish between saturated and unsaturated fatty acid chains that are free or chemically joined within a triglyceride.*
5. *Recognize and distinguish between cis and trans double bonds found within unsaturated fatty acid chains that are free or chemically joined within a triglyceride.*
6. *Explain why a saturated fat is solid at room temperature while an unsaturated fat is a liquid at room temperature.*
7. *Explain the difference between hydrophobic and hydrophilic.*
8. *Describe the breakdown of triglycerides by heat to produce free fatty acids and explain the impact of free fatty acids on smoke point.*
9. *Evaluate a nutrition label for fat content.*
10. *Describe how the calories for fat are determined on a nutrition label and explain how the % DV is calculated.*

2.4 Lipids (Fats, Oils, Waxes, Phospholipids, and Fatty Acids)

Besides the well-publicized nutritional role fats and oils play in our food, this third biological macromolecule is important in cooking and the taste of the foods we eat. Lipids are a class of molecules composed mostly of carbon and hydrogen atoms and are poorly insoluble in water but soluble in solvents like chloroform or ether. Fats, fatty acids, oils, waxes, cholesterol, membrane lipids, and other molecules all belong to the lipid family. Lipids, unlike other biological macromolecules, are not long polymers (repeating units of monomers); however, many of the lipid molecules are composed of several smaller molecules bonded together to form a larger functional compound. We will look at the structure and chemical nature of several important lipids and investigate how lipids impact cooking and baking (Figure 2.19).

2.4.1 Fatty Acids

The simplest lipids are the fatty acids. Fatty acids are long chains of carbon atoms bonded to hydrogen atoms ending with a carboxylic acid group of atoms (Figure 2.20). The long chain of carbon atoms in fatty acids is bonded to one, two, or three hydrogen atoms, making fatty acids very nonpolar, and water-insoluble molecules. Free fatty acids do contain a carboxylic acid at one end that is a hydrophilic group of atoms, but the number of nonpolar C—C and C—H bonds in the carbon chain of the fatty acid far outnumbers the hydrophilic C=O and O—H bonds. The end result is that fatty acids are nonpolar and water-insoluble. In fact, by definition, lipids are insoluble in water, and as a type of lipid, fatty acids are also insoluble in water. Remember, to be soluble means that two molecules will dissolve in one another to form a homogeneous mixture; this is often called the principle of like dissolves like. When compounds are insoluble, the combination forms a

Figure 2.19 Coconut oil. Coconut oil is composed of nearly all saturated fats (12 carbon lauric acid with lesser amounts of 14 carbon myristic acid and 16 carbon palmitic acid), giving the oil a relatively high melting point of 25.5°C (78°F). Coconut oil will be a soft solid at room temperature and will easily melt to a liquid oil at warmer temperatures. 5ph/Adobe Stock Photos

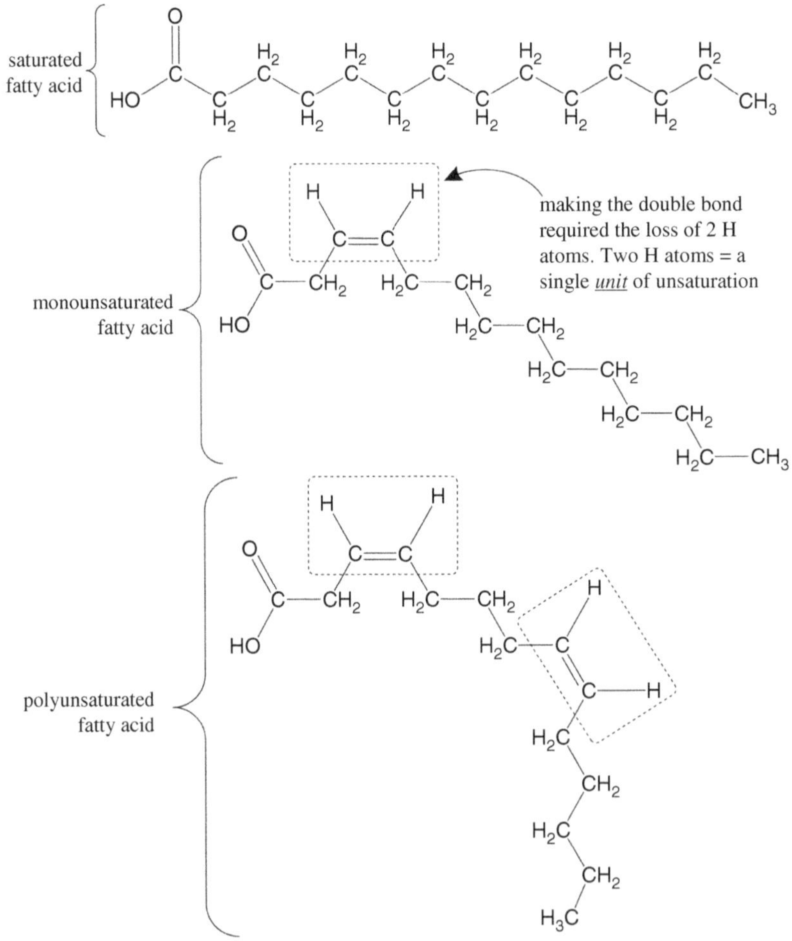

Figure 2.20 Saturated, unsaturated, and polyunsaturated free fatty acids. In fatty acids, the number of nonpolar C—C and C—H bonds outnumbers the polar C=O and O—H bonds, so the overall molecule is nonpolar and hydrophobic. The loss of two hydrogen atoms is called an unsaturation and results in a double bond. The unsaturated fats shown here contain cis double bonds, which is a particular orientation of atoms around the two doubly bonded carbons.

heterogeneous mixture. When a lipid (e.g. oil) is mixed with water, you will see boundaries form between the two phases—literally, the two cannot mix (Figure 2.21). Polar compounds can mix with or dissolve/are soluble in water (hydrophilic) to form homogeneous mixtures (i.e. sugar dissolving in water, lemon juice dissolving in water, vinegar dissolving in water). Nonpolar compounds can mix with or dissolve/are soluble in oils (hydrophobic) to form homogeneous mixtures (e.g. vanilla extract dissolving in oil, melted butter mixing with olive oil). Figure 2.22 highlights the differences between polar and nonpolar bonds.

Fatty acids are naturally found in plants and animal tissues and come in a range of carbon chain lengths. Fatty acids can be very short with four carbons (butyric fatty acid) or very long with 20 or more carbon atoms in their chain. Table 2.3 gives examples of different fatty acids. Short fatty acids are often involved with flavors of butter and cheese, and when volatilized by heat, they provide the flavor and smell of bread and cooked meat.

Fatty acids whose carbon chains have single bonds to other carbons or hydrogen atoms are considered saturated. That is, the carbon has four single bonds leaving the atom "saturated" with bonds to hydrogen and carbon atoms. Removal of two hydrogen atoms is called a "unit of unsaturation," and that is why fatty acids with

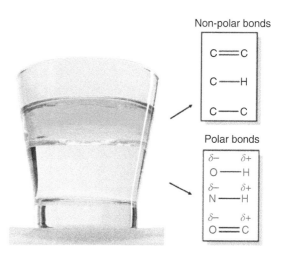

Figure 2.21 Oil in water. Polar and nonpolar bonds. Separated oil in water and the polarity of molecules that makeup each layer. The polarity of a molecule is determined by the separation of charge between its atoms. In polar molecules, most atoms are connected by polar bonds. In nonpolar molecules, nonpolar bonds connect most atoms. The oil layer is composed of molecules containing mostly nonpolar bonds, while the water or aqueous layer is made of molecules containing polar, partially charged bonds. imagedb.com/ Adobe Stock Photos

carbon–carbon double bonds (i.e. C=C) will have fewer hydrogen atoms. A fatty acid with one carbon–carbon double bond is a monounsaturated fatty acid, while fatty acids with two or more carbon double bonds are polyunsaturated. Unsaturation dramatically impacts the orientation of atoms around the C=C double bonds. The two possible configurations for unsaturated fatty acids are cis and trans. In a cis double bond orientation, when you look down the axis of the bond, the two carbon atoms or the two hydrogen atoms are on the same side of the double bond. This creates a bend in the fatty acid chain at the point of unsaturation. If you look down the axis of the double bond and the carbon (or hydrogen) atoms are on opposite sides, that is the trans configuration. Fatty acids with trans C=C double bonds have a small bend, but mostly a straight chain shape. The cis or trans shape of the fatty acid can have a significant impact on the melting point and chemical reactivity (Figure 2.22).

2.4.2 The Story of Trans Fats

In the early 1900s, less expensive liquid vegetable oil was used to create solid margarine and shortening. The process called hydrogenation involves heating fats with high heat in the presence of a nickel catalyst while bubbling hydrogen gas (i.e. H_2 [g]) through the mixture. The resulting chemical reaction adds two hydrogen atoms to every double bond, turning it into a single bond with an additional C−H bond per carbon (Figure 22). Changing the cis unsaturated fats to saturated fats by adding hydrogen created an inexpensive way to make liquid fat into a solid, and therefore an easy way to make foods like margarine. A by-product of hydrogenation, especially partial hydrogenation, was that instead of adding hydrogen atoms to every single

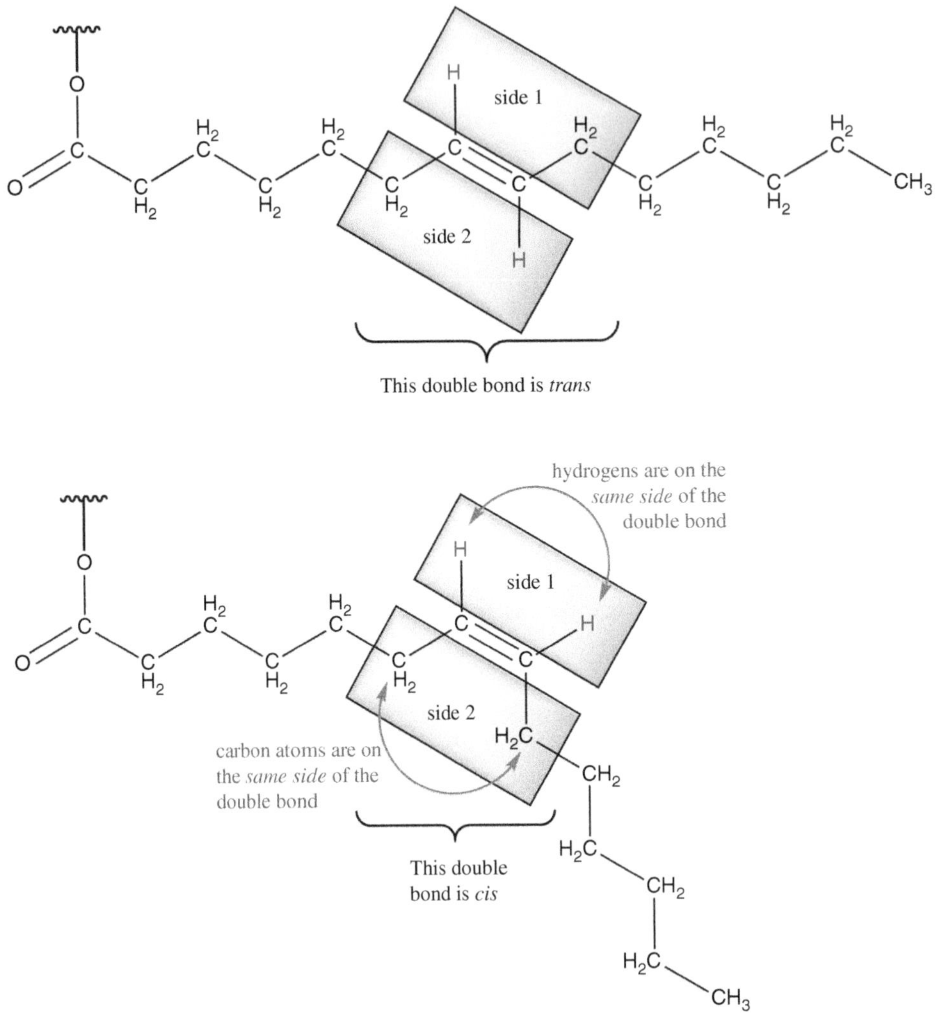

Figure 2.22   Cis and trans fatty acids. The loss of two hydrogens in a fatty acid chain results in the sharing of electrons, forming a double bond. The arrangement of the atoms around the double results in the hydrogens being on the same (cis) or opposite (trans) side of the double bond. Note the impact of cis and trans desaturation on the shape of the fatty acid chain.

cis C=C bond, some of the cis double bonds were converted to trans double bonds. The unintentional creation of trans fatty acid chains during the production of margarine and shortening was initially considered a bonus effect of hydrogenation. This was because trans fatty acids, due to their shape, are more resistant to reaction with oxygen and are poor food for bacteria. These trans fats have a much longer shelf life, are less likely to become rancid (a product of reaction with air or bacterial degradation), and, with the straight chain shape, are solid at room temperature. In the 1970s, margarine with these trans fats sold for nearly twice the price of butter consisting mostly of saturated fatty acid chains. Part of the reason for using trans fats in foods was to avoid the negative effects of saturated fats on cardiovascular disease. Unfortunately, in the 1990s, a relationship between trans fats and heart disease was discovered. A lawsuit in 2003 against Kraft caused the food company to eliminate the use of partially hydrogenated vegetable oils in foods like

Table 2.3   Common fatty acids. The length of the fatty acids is indicated by the number following the letter "C." C12 is a fatty acid made of 12 carbons. The number after the semicolon (:) indicates the number of unsaturated carbon double bond(s). For example, C10 : 2 $\Delta^{2,4}$ is a polyunsaturated fatty acid with 10 carbons and four double bonds. The delta (Δ) symbol indicates at which carbon from the carboxyl end of the double bond is located.

Common name	Abbreviation	Typical sources
Butyric	C4 : 0	Dairy fat
Capric	C10 : 0	Dairy fat, coconut, and palm kernel oils
Lauric	C12 : 0	Coconut oil, palm kernel oils
Palmitic	C16 : 0	Most fats and oils
Stearic	C18 : 0	Most fats and oils
Arachidic	C20 : 0	Peanut oil
Behenic	C22 : 0	Peanut oil
Lignoceric	C24 : 0	Peanut oil
Palmitoleic	C16 : 1 Δ^9 (omega-6)	Marine oils, macadamia oil, most animal and vegetable oils
Oleic	C18 : 1 Δ^9 (omega-9)	All fats and oils: especially olive, canola, sunflower, and safflower oils
Erucic	C22 : 1 Δ^{13} (omega-9)	Mustard seed and rapeseed oil
Linoleic	C18 : 2 $\Delta^{9, 12}$ (omega-3)	Most vegetable oils
Arachidonic	C20 : 4 $\Delta^{5, 8, 11, 14}$ (omega-6)	Animal fats, liver, egg lipids, and fish

Table 2.4   Fatty acid composition of common cooking fats and oils. Data from the US Department of Agriculture National Nutrient Database [10]. The fatty acid composition was normalized to 100%.

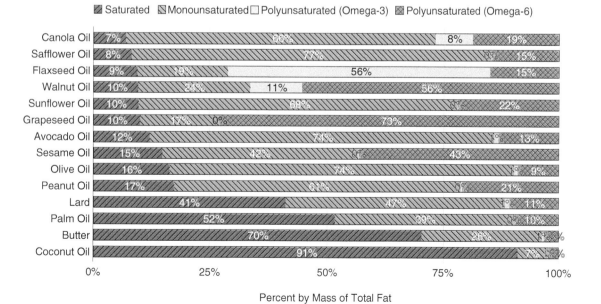

Oreo cookies. In 2006, Starbucks followed suit and eliminated trans fats in making its baked goods. While low levels of trans fatty acids are naturally found in some foods like pork, lamb, and milk, the problem with too many trans fatty acids is the higher incidence of coronary disease. Trans fats raise the levels of low-density lipoproteins (LDL), which can contribute to blood flow blocking fatty deposits in arteries and causing one type of white blood cells to change, collect cholesterol from LDL, and harden, further blocking blood flow to the heart.

2.4.3 Triglycerides

Fats and oils are essentially the same molecule: a triacylglycerol, also known as a triglyceride. In animals, triglycerides are stored in specialized fat cells called adipose cells and make up the major form of stored energy. To create a triacylglycerol, three fatty acids are bonded to a short organic molecule called glycerol through a dehydration reaction similar to what we have seen with proteins and carbohydrates (Figure 2.23). Fatty acids can be from 4 to 35 carbons long, but 14–20 carbon fatty acids are most common in food. When fatty acids and glycerol combine, bonds are broken and formed in a chemical reaction to produce a triglyceride and three molecules of water. In that process, a new group of atoms is formed called an *ester*. Overall, the new molecule is very hydrophobic and nonpolar. The complete triglyceride contains three fatty acids bonded via an ester to the glycerol backbone; the basic structure of a triglyceride/triacylglycerol is shown in Figure 2.23. The properties of a given triglyceride depend upon the chemical structure of the three fatty acids it contains, and the properties of a lipid depend upon the particular mixture of triglycerides it contains.

A fat is a triacylglycerol that is solid at room temperature (Figure 2.24). Solid fats are readily found in animal products such as meat and milk. Oils are liquid at room temperature. Liquid fats or oils are readily found in plant sources. Essentially, the only difference between fat and oil is the length and number of double bonds on the fatty acids. However, these are significant differences both chemically and for cooking.

Melting of a solid fat like butter is a change in the state of matter. Melting butter takes place when heat is applied, and the state of matter moves from solid to liquid. When in the solid form, fat molecules will stack

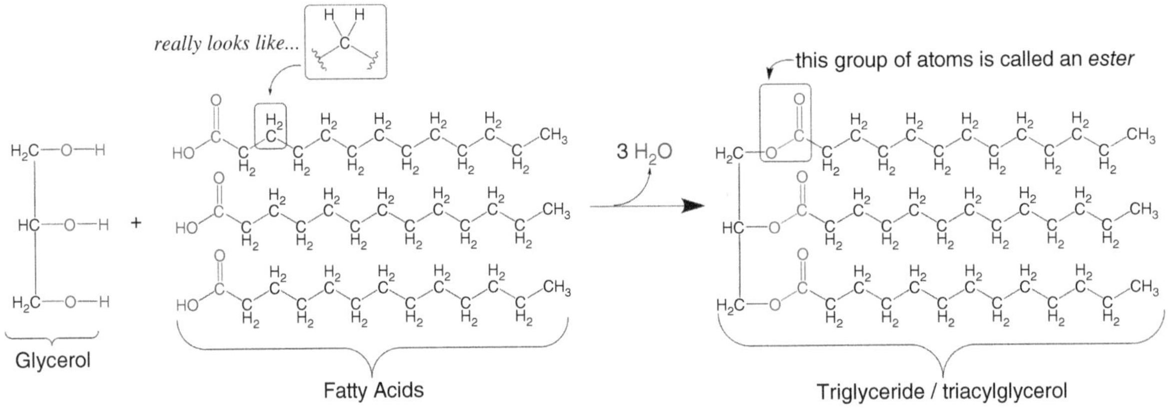

Figure 2.23 Glycerol and fatty acids combine to make a triglyceride. Like other polymer reactions, dehydration/condensation reactions result in the joining of two different molecules. The glycerol chain is the backbone of the fat. One, two, and three fatty acids bound to a glycerol molecule result in a mono-, di-, or triacylglycerol, also called a triglyceride, more commonly known as fat.

and interact with each other with weak but numerous non-covalent, hydrophobic interactions between fat molecules. The more physical contact between fat molecules, the more intermolecular (i.e. between molecules) hydrophobic interactions can form, and the more stable the fat will be in a solid state. When heat is applied, the added energy shifts, moves, and vibrates the molecules far enough apart from each other, so that they feel very little attraction. With fewer attractions between the heated fat molecules, the fats behave like liquids and are free to move around. The melting point is the temperature at which molecules of a substance shift from a solid configuration to a liquid state. Fats that melt at a higher melting temperature require more heat to break their intermolecular attractions than fats with a low melting point. The reason for this is the number of contacts between individual molecules of the fat. Fats or triglycerides made from saturated fatty acid chains pack very

Figure 2.24   Oil. Unsaturated and polyunsaturated fatty acids are liquid at room temperature. tolism/Adobe Stock Photos

efficiently and densely with many hydrophobic contacts between molecules. Longer fatty acid chains result in more hydrophobic interactions between fatty acid chains. Thus, it takes more energy to separate fats made from longer fatty acid chains than fats composed of shorter fatty acid chains. On the other hand, unsaturated fats or triglycerides contain carbon–carbon double bonds that impact the shape of the fatty acid chains, and therefore, the capacity of the triglyceride for intermolecular, hydrophobic attractions. Cis unsaturated fatty acids create the bent fatty acid shape discussed earlier. This bent shape of an unsaturated fatty acid chain keeps the triglyceride chains from packing efficiently and reduces the amount of heat or energy needed to separate and melt the solid fat. Trans unsaturated fatty acids have a small kink, but overall their shape is quite similar to saturated fatty acids, which means that trans fats pack like saturated fats. A simple approach is to remember that when comparing chains length, the longer fatty acid chain will have a higher melting point, and when comparing saturated versus unsaturated chains, the presence of cisunsaturated double bonds will lower the melting point.

Box 2.3 Figuring out food: Buttergate, the mystery of Canada's "hard butter"

On 29 December 2020, food researcher Dr. Sylvain Charlebois tweeted, "Is it me, or is #butter much harder now at room temperature?" Dr. Charlebois wasn't the only one who noticed that Canada's butter was no longer spreadable at room temperature. Cookbook author Julie Van Rosendaal posted a similar tweet in February 2021, where numerous Canadians agreed that something was very different about their current butter supply. The food controversy dubbed "Buttergate" spread on social media and reached media outlets across the globe.

   Butter is a water-in-oil emulsion composed of 80–90% fat, whose fatty acid composition is mainly saturated (Table 2.5). Butter's mixture of fatty acids, each with its unique melting point, makes butter spreadable (semisolid) over a wide range of temperatures. Butter's spreadability is most desirable to us between 59° and 65°F (15°–18°C) [11].

(Continued)

Box 2.3 (Continued)

Table 2.5 Average fatty acid composition of conventional butter [11] and corresponding melting points [12].

Fatty acid	Weight %	Structure	Melting point
Myristic (C14 : 0)	12	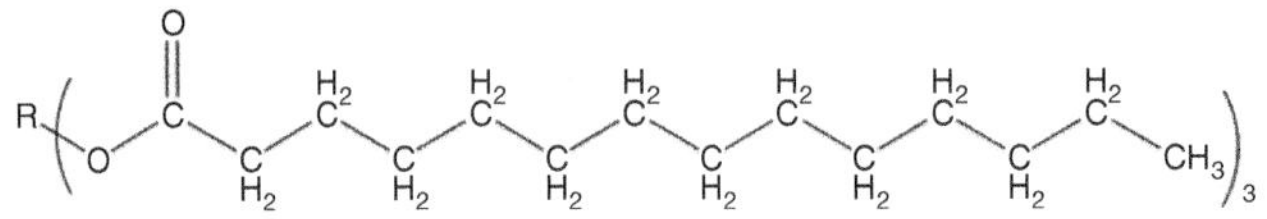	129°F/54°C
Palmitic (C16 : 0)	32	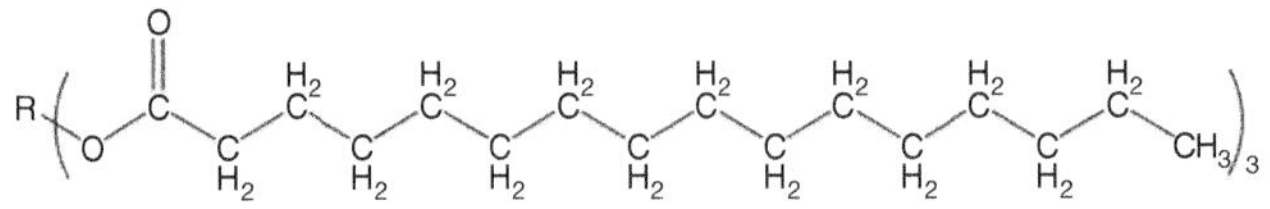	145°F/63°C
Stearic (C18 : 0)	10	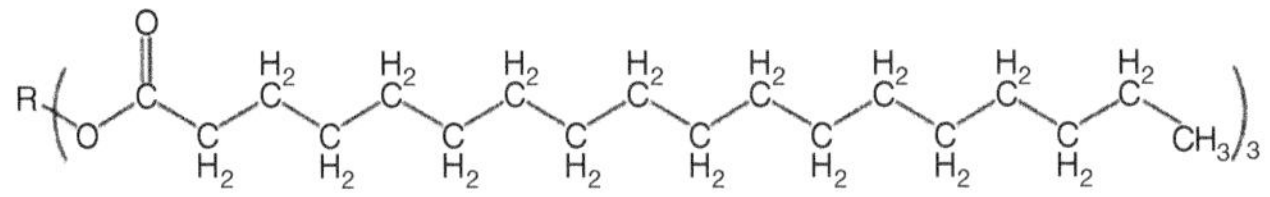	158°F/70°C
Oleic (C18 : 1)	18	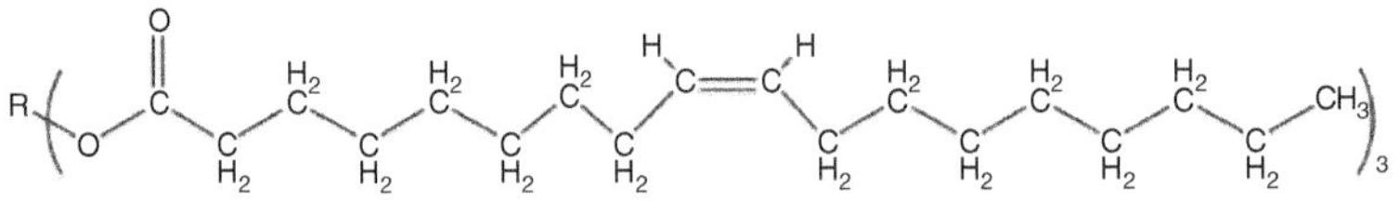	61°F/16°C
Other	28		

So why did Canadian butter not spread at room temperature? Food scientists believe using palm oil, rich in palmitic acid, as a dairy feed supplement is causing the melting point of butter to increase. Palmitic acid is the most prevalent fatty acid found naturally in butterfat, even for grass-fed cows [13]. However, the fatty acid composition of butterfat can be significantly influenced by a cow's diet, lactation cycle, breed, and even season. Thus, a diet supplemented with palm oil will change butterfat's fatty acid composition and melting point.

At 9 Calories/g, tropical oils like palm and coconut are used as feed supplements in the dairy industry to bolster Calories and thus milk production. As bypass fats, these feed supplements increase the butterfat percentage in dairy milk. Tropical oils are used when farmers need to meet butterfat quotas in dairy milk due to feed deficits, harsh winters, or low butterfat supply issues.

Fats provide special flavors and characteristics to foods. The lubricating properties of lipids create a slippery smooth mouthfeel to foods and make some foods seem moist. Fats are often reported to tenderize food. Think of the difference between a lean and marbled steak. The cut marbled with fat is easier to chew and more tender. In baking, fat acts as a shortening agent, that is, fat coats some of the proteins and starches in flour, limiting the network and keeping a crust together. Adding fat, for example, in the form of shortening, to a pie crust recipe creates a softer smaller crumb with thin layers of pie crust rather than a solid thick sheet of baked dough (Figure 2.25). SCFAs have a special flavor, creating a complex flavor for several foods. Microorganisms used to make some cheeses create sharp cheese flavors and some of the odors of cheeses. When cooked at high heat, the fats of meat will change chemically to gaseous molecules that impart the flavor and smell of meat. Raw meat has a little aroma and a very simple, blood-like taste. Phospholipids and fatty acids in meat react with oxygen to form smaller molecules called aldehydes, unsaturated alcohols, ketones, and lactones. Each provides a single note of flavor and smell of cooked meat. Certain fat-like molecules derived from plants called terpenoids provide strong flavors to cooked foods. Examples include cinnamon, cloves, and mint.

2.4.4 Cooking with Fats and Oils

Cooking with fats and oils provides a unique challenge. Heating fats and oils can create a smelly, smoky mess that leaves food tasting bitter. Fats will melt into oils when warmed, but unlike watery liquids, oils do not boil. Before the fat can reach the boiling point, it smokes and breaks down. The breakdown of fat at high temperatures is due to several factors. At high temperatures, oxygen in the air will oxidize the double bonds of unsaturated fatty acid chains, creating rancid and smelly products. Impurities in the oil, including sugars and proteins will burn in the oil, producing dark colors and off-tasting molecules. Water present in the fat will chemically react with the ester bonds and break the triglycerides apart. Some free fatty acids are naturally present in fats and oils in very

Figure 2.25 Role of shortening in pastry crust. The Greek pastry, baklava. Notice the sheets of flaky crust due to the addition of shortening. The fat in the shortening limits the access to water necessary to make gluten. Joshua Resnick/Adobe Stock Photos

small amounts, but the amount of free fatty acids increases as the fats/oils are heated and trace water reacts with the triglycerides, breaking apart the ester bonds and releasing free fatty acids. The smoke point of oil is the temperature at which heating causes the fat to give off smoke. At this point, the triglyceride chemically breaks down and releases free fatty acids from the glycerol backbone. The released glycerol further reacts with oxygen and heat to form acrolein and water. Acrolein is a toxic compound that irritates soft tissues. The quantity of free fatty acids within the oil is what influences its purity. The highest-quality fats and oils will have very low concentrations of free fatty acids and other impurities and, therefore, the lowest smoke points. Let's examine the effect of refinement or purification on the smoke point of canola oil from the rapeseed plant. As shown in Table 2.6, unrefined canola oil smokes at 107°C, while refined or purified canola oil smokes at 205°C, almost 100 degrees higher. In the same way, butter contains proteins and sugars from milk along with 10–15% water. The milk solids in butter readily burn and are the reason why overheating butter leads to a mess. Also in Table 2.6, butter has a smoke point of 175°C; but if butter is *clarified* to remove the milk solids and water, the smoke point increases to 230°C. Safflower oil (12% monounsaturated and 75% polyunsaturated) has one of the highest smoke points. To sauté at a higher temperature, use a vegetable oil like canola or safflower oil.

Olive oil poses a special challenge when cooking. There are several grades of olive oil, each depending on the way the oil is extracted and stored. One way of making olive oil is to simply squeeze or press the fruit in a process called "cold-pressing." The resulting oil is separated from large solids by filtering or by spinning the sample very quickly in a process called centrifugation. Sometimes hot water is then added, and the oil is extracted from the remaining paste to recover more oil. Another method involves dissolving the ground olive in a chemical solvent to extract the oil. To be called a "virgin" olive oil, no chemical purification can be used in its production. Extra virgin olive oil is cold-pressed, has less than 0.8% free fatty acids, and retains a good flavor. This olive oil is often used to add flavor to foods and not for cooking. Refined olive oil is extra virgin olive oil that does not qualify for the extra virgin status because it has been chemically treated to remove impurities. Some oils, simply labeled "olive oil," are a blend of refined and virgin olive oil and will have some of the flavors but will contain less of the impurities. Unrefined olive oil has a very low smoke point with a mix of impurities (320°F/160°C). Extra virgin olive oil has a slightly higher smoke point of 185°C/365°F, while processed purified, refined olive oil without the particles and acids from the olive paste will have a much higher smoke point of 450°F/232°C and serve as a good cooking medium.

Table 2.6 Smoke points of common cooking oils.

Fat	Oil grade/quality	Smoke point
Butter	Unsalted	350°F/175°C
Olive oil	Extra virgin	350–410°F/177–210°C
Sesame oil	Refined	350–410°F/175–210°C
Vegetable shortening	Hydrogenated	360°F/180°C
Canola oil	Refined	400°F/205°C
Clarified butter	Removed sugar, water, and protein	450°F/230°C
Corn oil	Refined	450°F/230°C
Peanut oil	Refined	450°F/230°C
Safflower oil	Refined	510°F/265°C
Avocado oil	Refined	520°F/271°C

2.4.5 Fat Nutrition

As we have learned in this section, fats and oils are made primarily of triglycerides, also known as triacylglycerols. Triglycerides contain three fatty acid chains that have been chemically joined to a glycerol molecule via an ester bond (Figure 2.23). The fatty acid chains of the triglyceride may or may not contain C=C double bonds. Remember, when a fatty acid chain contains one or more double bonds, it is called unsaturated. When a nutrition label lists Calories for fat, it is referring to triglycerides. For every gram of fat, the human body can extract 9 calories. That is more than twice the calories that the human body gains from carbohydrates or proteins, which makes fat the most "calorie dense" of the macromolecules. It makes sense for the human body to store energy primarily as fat because it holds more Calories per gram.

When we look at a regular nutrition label, the section on fats contains "total fat" as well as "saturated" and "trans fat" sections. The saturated fats on the nutrition label are triglycerides made of saturated fatty acid chains. Trans fats are a type of unsaturated triglyceride that we learned about earlier (Section 2.4.3). You'll notice that when adding up the grams of trans fat (i.e. 3 g) and the grams of saturated fat (i.e. 3 g) on the label in Figure 2.26, those numbers do not add up to the total fat (i.e. 12 g) listed; there are six grams of fat that appear unaccounted for on the label, and those six grams belong to cis unsaturated fats. Unsaturated fats can have a single cis C=C double bond, which is called cis monounsaturated, or multiple double bonds, which is called cis polyunsaturated. Since cis monounsaturated and cis polyunsaturated fats are the most naturally abundant unsaturated fats, often these fat types are referred to as just monounsaturated and polyunsaturated; the cis is implied.

Even though triglycerides made of unsaturated fatty acid chains are naturally abundant in plant oils and fish, neither monounsaturated nor polyunsaturated fats are typically listed on a nutrition label. Why? Nutrition labels are designed to give you information about food molecules that could be detrimental to your health if you consume too much. Saturated and trans fats are both associated with a higher risk of cardiovascular disease, which is why their consumption has a recommended daily value. The Dietary Guidelines for Americans [14] state that less than 10% of your daily calories should come from saturated fats, which is 20 g of saturated fat for a 2,000 calorie diet. Similarly, the Dietary Guidelines for Americans recommend that trans fats should be as low as possible. Partially hydrogenated oils used to be a common source of added trans fats in foods (Section 2.4.3), but since June of 2018 in the United States, partially hydrogenated oils are no longer considered safe because of the trans fats they contain, and therefore, partially hydrogenated oils may not be added to foods. However, trans fats do occur naturally in some animal-derived foods, so it is possible to encounter them in your diet. In

Nutrition Facts

Servings Per Container 2

Serving Size 1 cup (228g)

Amount Per Serving

Calories 250

	% Daily Value*
Total Fat 12g	**18%**
Saturated Fat 3g	**15%**
Trans Fat 3g	
Cholesterol 30mg	**10%**
Sodium 470mg	**20%**
Total Carbohydrate 31g	**10%**
Dietary Fiber 0g	**0%**
Sugars 5g	
Protein 5g	

Vitamin A	4%
Vitamin C	2%
Calcium	20%
Iron	4%

* Percent Daily Values are based on a 2,000 calorie diet. Your Daily Values may be higher or lower depending on your calorie needs.

	Calories	2,000	2,500
Total Fat	Less than	65g	80g
Sat Fat	Less than	20g	25g
Cholesterol	Less than	300mg	300mg
Sodium	Less than	2,400mg	2,400mg
Total Carbohydrate		300g	375g
Dietary Fiber		25g	30g

Figure 2.26 Nutrition labels contain entries for total fat, saturated fat, and *trans* fat. Cis monounsaturated and polyunsaturated fats are not typically listed on a nutrition label. magr80/Adobe Stock Photos

contrast, monounsaturated and polyunsaturated fats can reduce the risk of cardiovascular disease when eaten in place of saturated fats, and some polyunsaturated fats are absolutely essential to the human diet. Linoleic acid (LA) is a polyunsaturated fatty acid (i.e. PUFA) that contains two, cis C=C double bonds, and alpha-linolenic acid (ALA) is a PUFA with three cis, C=C double bonds (Figure 2.27). In LA, the second double bond is six carbons away from the end of the chain, and since omega is the end of the Greek alphabet, LA is called an omega-6 fat. ALA is called an omega-3 fat for the same reason. These PUFAs are essential to human growth and development and must be consumed in our food. Natural sources of omega-6 fats are easy to find in a Western diet; they include vegetable oils like corn, safflower, or peanut. Natural sources of omega-3 fats include fish oil, especially cold water fish like salmon, and flaxseed oil (Table 2.4).

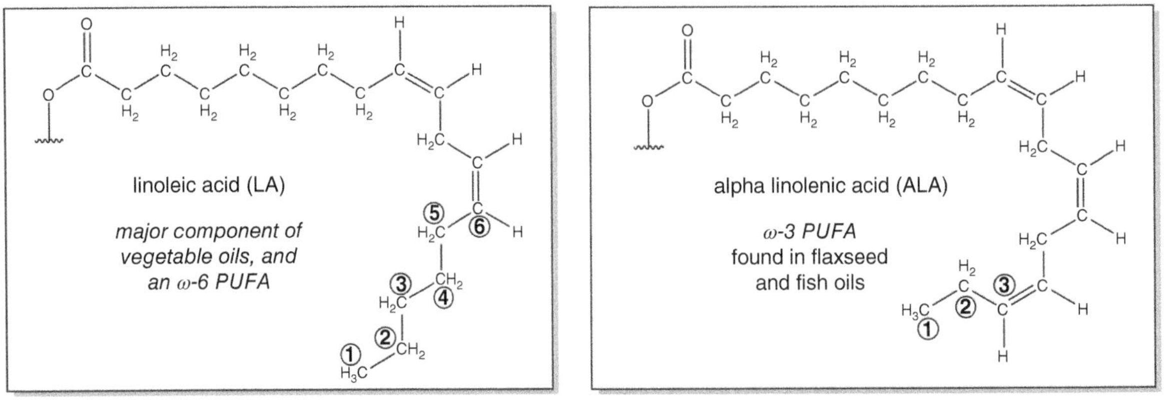

Figure 2.27 Essential polyunsaturated fats (PUFAs), linoleic acid (LA) and alpha linolenic acid (ALA). Linoleic acid is known as an omega-6 PUFA, while alpha linolenic acid is known as an omega-3 PUFA.

Cholesterol is another lipid, or fat-derived molecule. The human body makes its own cholesterol, and humans can consume cholesterol in their diet. Cholesterol moves through the human body in packages called "lipoproteins." Excess cholesterol travels through the body in lipoproteins called low-density lipoproteins (LDLs). The excess cholesterol in LDLs can be deposited on the walls of blood vessels and elsewhere in the body; this is why LDL is referred to as "bad" cholesterol. Because excess cholesterol is a significant risk factor for cardiovascular disease, the Dietary Guidelines for Americans recommend consuming as little cholesterol from food as possible. Many experts agree that cholesterol consumption should be no more than 300 mg for healthy individuals.

Box 2.4 Figuring out food: Inflammatory oils

Search Reddit or other social media about inflammation and oils, and it won't take long to find people posting ideas and conflicting information about how some oils aren't "good for you." For example, threads about how some processed oils from seeds are harmful to you, how vegetable oils can shorten life, and that some oils cause heart problems, promote cancer, or make arthritis worse are all common threads. Inflammation is the "first responder" when the body is damaged from injury or is invaded by foreign agents like viruses, bacteria, or toxic chemicals. Sadly, several diseases and conditions are associated with long-term (chronic inflammation), such as rheumatoid arthritis, type 2 diabetes, heart disease, lupus, and cancer. So it makes sense that people are concerned about overstimulation of the immune system. Some oils (more appropriately termed lipids) are key to how the body signals to start, sustain, and terminate inflammation. One class of lipids called eicosanoids is made from a 20-carbon polyunsaturated fat called arachidonic acid found in our diet and created by our cells. Some eicosanoids are part of the signal that brings about the redness, heat, swelling, and pain associated with inflammation due to tissue injury or infection. Other lipids, including some of the eicosanoids, help stop the fire of inflammation by signaling to cells to end the process.

Thus, lipids are part of the healing process. One of the concerns is that oxidized lipids support or can even cause some diseases. Recall that unsaturated fatty acids have carbon–carbon double bonds. These double bonds are easily oxidized to carbon-single bonds. The lipid oxidation happens slowly during storage with oxygen and UV light and by processes inside the cells. Tissues of patients with arthritis produce a high level of free radicals, as the immune system can cause the double bonds of some lipids to become oxidized as the lipids become single-bonded carbons. Some oxidized lipids are associated with increased cancer, worsening cardiovascular disease, and continued arthritic damage.

Confusing the issue is that some oxidized lipids are helpful in anticancer and other processes. One popular answer is to avoid oils rich in saturated and polyunsaturated fats. However, such a simple answer is problematic. For many years, we've known that a diet high in saturated fats increases the risk of heart disease. We've also known for some time that a diet high in polyunsaturated fats reduces the LDLs associated with blockage of heart vessels, can reduce the blood sugar over time, lower blood pressure, and reduces the risk of some cancers, heart disease, and diabetes. Simply following some advice to decrease polyunsaturated foods and oils in favor of saturated fats and oils could cause more harm than good. The simple answer is that the type of fat to avoid is critical. Linolenic acid is an omega-6 fatty acid associated with inflammation and is rich in Western diets. On the other hand, the unsaturated omega-3 fatty acids inhibit inflammation. Treating your dietary intake with balance and not focusing on a single issue like inflammation is the best approach when creating a healthy food plan.

KEY CONCEPTS Part IV: Fats

1. A fat or **triglyceride** is made from three **fatty acid** chains that have been chemically joined to a backbone **glycerol** molecule. When a **fatty acid** is joined to the **glycerol** backbone, there is a loss of water and the formation of an **ester** bond.
2. The fatty acid chains of a **triglyceride** are composed of **nonpolar** C—H and C—C/C=C bonds; therefore, triglycerides are **hydrophobic** by nature.
3. **Unsaturated** fatty acid chains contain carbon–carbon **double bonds** (C=C). These C=C double bonds can be **cis** or **trans**. Plant-derived, **unsaturated** fatty acid chains are **cis**.
4. **Saturated** fatty acid chains are shaped like long tubes or straight columns. **Cis** carbon–carbon double introduces a bend in the **unsaturated** fatty acid chain, while **trans** double bonds create a kink in the chain.
5. Triglycerides that are heated will break down into their component **free fatty acids** and the **glycerol** backbone. The presence of **free fatty acids** in a fat or oil will change the flavor and lower the **smoke point** of the fat/oil.
6. The fat on a nutrition label describes a mixture of triglycerides/triacylglycerols made of fatty acid chains that may be saturated, monounsaturated, or polyunsaturated.
7. Consuming fat yields **9 calories per gram**. The recommended daily value of fat is 78 grams.

Part V: Emulsions and Emulsifiers

Learning Objectives

1. *Recognize amphiphilic molecules and identify their hydrophobic and hydrophilic parts.*
2. *Explain how amphiphilic molecules like phospholipids stabilize emulsions.*

2.5 Emulsions and Emulsifiers

2.5.1 Defining an Emulsion

Proteins, lipids, and complex carbohydrates can each be used to prepare a critical technique for food and cooking called an emulsion. Emulsions are evenly mixed or homogeneous dispersions of two components that repel each other. Examples of food emulsions include sauces, dressings, or mayonnaises. In chemical terms, the two phases are immiscible. A common culinary emulsion is a mixture of oil and water (Figure 2.28). If there is more water

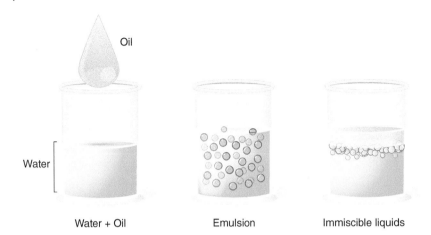

Water + Oil   Emulsion   Immiscible liquids

Figure 2.28 Water and oil emulsion. If water and oil are mixed, they will eventually separate into two distinct layers. But if the mixture of oil and water is stabilized, then it is called an emulsion. designua/ Adobe Stock Photos

than oil, the emulsion is called oil-in-water emulsion, and if there is more oil than water, the emulsion is a water-in-oil emulsion. Mayonnaise is an oil-in-water emulsion while butter is a water-in-oil emulsion. Either way, the basic concept of cooking with an emulsion is to create small enough droplets or fragments of the oil or water to be evenly dispersed through the mixture. Small droplets are created by a physical force called shearing, which is most commonly performed using blenders, whisking, or food processors. Industrial kitchens use fine-gauge strainers and force the fluids to create an emulsion. Consider shaking a mixture of oil and water. Shake or mix hard enough, and the tiny spheres of oil will seemingly mix within the water. However, the two phases quickly separate from each other. That is the challenge of making an emulsion keeping the solutions mixed (Figure 2.29).

2.5.2 Emulsifiers

Emulsifiers are food additive compounds that help create emulsions by separating them back into two distinct layers. For oil-in-water emulsions, an emulsifier must have both a water-loving (i.e. hydrophilic) and water-fearing (oil-loving) or hydrophobic region. The emulsifier lecithin is one of an entire class of biological emulsifiers

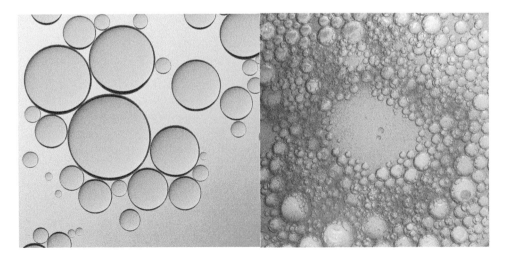

Figure 2.29 Different water and oil emulsions under the microscope. The circular droplets of fat are dispersed in the water phase. The image on the right has smaller fat droplets and is, therefore, a more stable emulsion. RDVector/Adobe Stock Photos, Pelevina Ksinia/Shutterstock

called phospholipids. Like the triglyceride (i.e. triacylglycerol), phospholipids have a glycerol backbone, but instead of three fatty acid chains, phospholipids contain two nonpolar, hydrophobic fatty acid chains that are bonded via esters to only two positions on the glycerol backbone (Figure 2.30). The third —OH of glycerol is bonded to groups of atoms that contain a phosphate group. Because of the charged atoms bonded to the glycerol at the third position, this region of the phospholipid is polar and hydrophilic. Therefore, the entire phospholipid as a whole is *amphipathic* or *amphiphilic*: the fatty acid esters create a hydrophobic region, and the phosphate-containing portion is charged and hydrophilic. Amphiphilic molecules like phospholipids can interact with both fat/oil and water to stabilize emulsions (Figure 2.31). In fact, layers of phospholipids form into spherical globules to enclose milk fat and keep the fat stably mixed with the water milk serum. The amphipathic nature of phospholipids drives the hydrophilic and hydrophobic regions to align together creating globules or complicated micelles and even into sheets of lipids called membranes. Phospholipids stabilize emulsions by surrounding oil droplets with the hydrophobic portion of the molecule and aligning the hydrophobic region of the molecule to face the water (Figure 2.31). A second additive called a stabilizer is also often used in making emulsion to help keep the small droplets prepared in the emulsion from reuniting into a larger mass and finally fully separating into two layers. Some molecules can act as both emulsifiers and stabilizers.

Common ingredients used for emulsifiers and stabilizers are egg yolks. The key compound in egg yolks used as an emulsifier is the phospholipid lecithin. The fatty acid component of the phospholipid buries itself into the oil droplets while the charged, polar group of atoms faces outward where it can interact with water. This coats the newly formed oil droplet with a negatively charged compound that allows the droplet to stabilize interactions with water and avoid other oil droplets.

Lecithin, formally called phosphatidylcholine, is a special phospholipid with a choline head group. Lecithin is found in high amounts in soybean and egg yolk and is used for cooking. The nonstick cooking spray PAM is primarily lecithin and water and is utilized for its oil-like properties. A quick search of your pantry and refrigerator

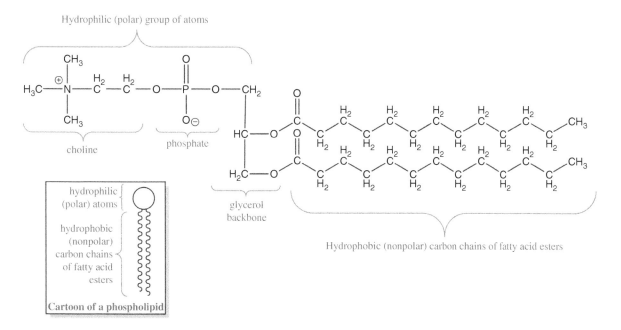

Figure 2.30 Phospholipid. A phospholipid is similar to a diglyceride or diacylglycerol, but with a polar group of atoms (i.e. a phosphate and choline) linked to the glycerol backbone in the third position. Because of their polar and nonpolar parts, phospholipids are amphiphilic; they have the ability to interact with both hydrophilic and hydrophobic molecules.

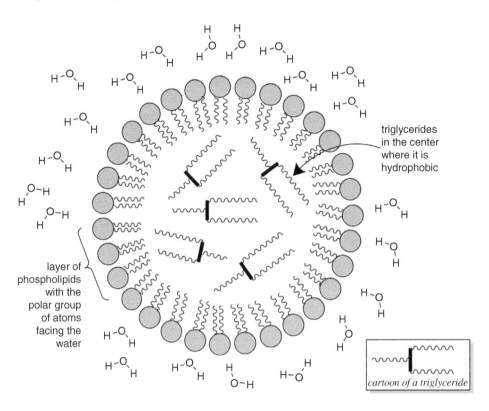

triglycerides
in the center
where it is
hydrophobic

layer of
phospholipids
with the
polar group
of atoms
facing the
water

cartoon of a triglyceride

Figure 2.31 Phospholipids interact with both nonpolar triglycerides and water and help to stabilize emulsions—mixtures of water and fat/oil.

will find lecithin in a number of items, including salad dressing, chocolate, and a number of interesting molecular gastronomy-inspired dishes and preparations. A curious dessert is liquid popcorn with caramel froth. Popcorn, sugar, and water are used with syrup and dry lecithin to create a two-phase drink where lecithin plays a role in keeping caramel cohesive.

Phospholipids are one type of emulsifier, but other classes of emulsifying compounds are also important in food. For example, sugar sorbitol is not only a sweetener. Sorbitol can also be converted to sorbitan by the loss of a water molecule; then, sorbitan is conjugated to long fatty acid chains via ester bonds (Figure 2.32). The result is a sugar–fat hybrid molecule that is a fantastic emulsifier. An example of a fatty acid-sorbitan emulsifier is polysorbate 80, a molecule used to emulsify ice cream. The carbon chain of the fatty acid interacts with the lipid, while the carbohydrate end of the sorbitan interacts with water. Polysorbate 80 not only keeps the fats separated, but it also keeps the ice cream smooth and slows down the "liquefaction" or melting of the ice cream. There are a variety of polysorbates used in foods and other consumer products. The number in the name (i.e. polysorbate 20, 40, 60, or 80) changes based on the length of the fatty acid chain that is chemically attached to the sorbitan.

Many emulsifiers are added to foods in order to increase access, extend shelf life, and decrease the overall costs of food. Without emulsifiers, bread, liquids, and other foods would break down and change flavor and texture, resulting in inedible consequences. Yet the benefits of emulsifiers come with a cost. New studies are finding that some emulsifiers, like polysorbate 80 and carboxymethylcellulose, change the ratios of helpful to harmful bacteria in the human gut, and these emulsifiers may contribute to inflammation, irritable bowel syndrome, and other health issues. The true impact of emulsifiers on gut bacteria is still under investigation; any negative outcomes must be considered alongside the positive effects of emulsifiers, such as increasing access to food.

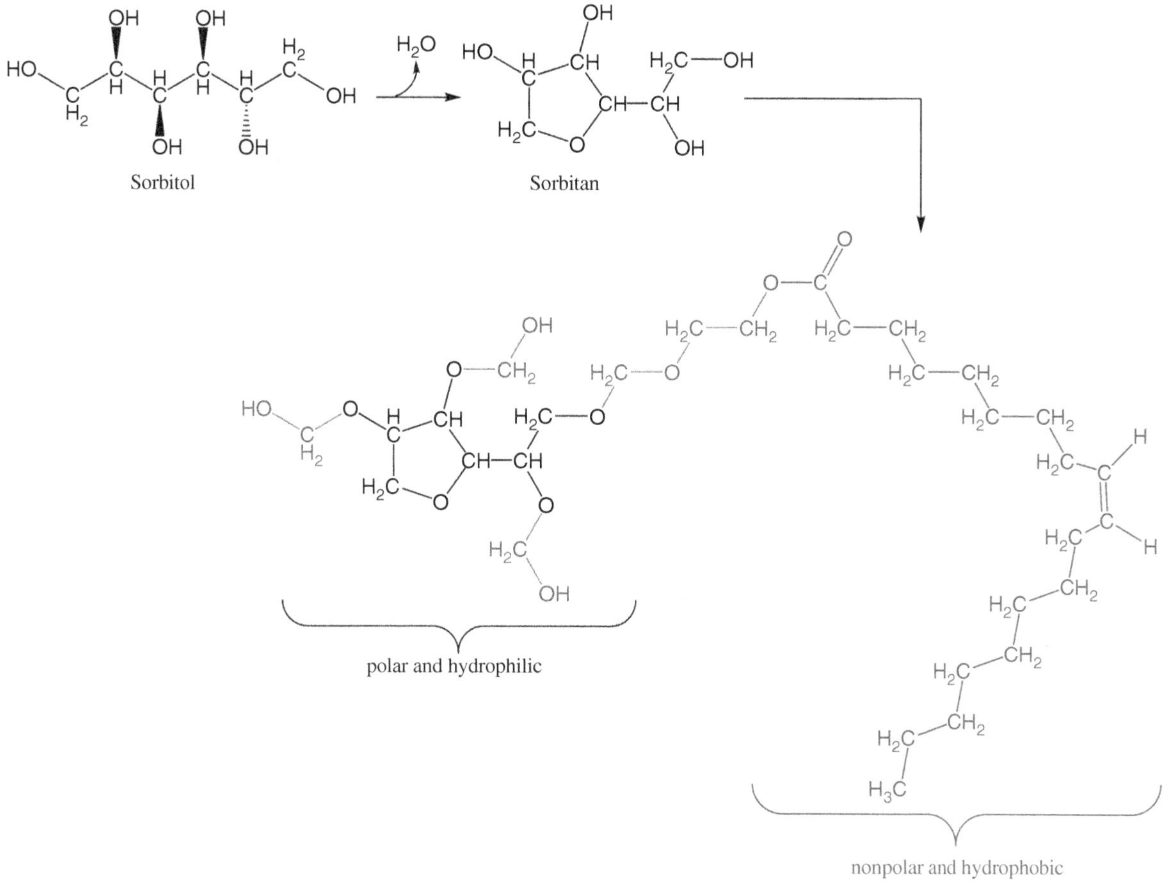

Figure 2.32 Sorbitol, sorbitan, and polysorbate 80. Sorbitol is a sugar alcohol that is converted to sorbitan by a loss of a water molecule. Sorbitan is then conjugated to ethoxyl groups (in gray) and to a fatty acid (in blue) via an ester. The resulting polysorbate is amphiphilic with both polar and nonpolar regions.

Some proteins also serve as effective emulsifiers. As we will see in Chapter 3, fresh milk separates into a translucent watery layer separated from a thick, opaque cream. But homogenized milk, like the kind you find in your average grocery store, is a smooth, opaque, white liquid; in other words, it is a stable emulsion. Casein, one of the main proteins found in milk, coats the globules of milk fat in a way that is similar to how lecithin coats oil droplets. The hydrophobic portions of casein interact nicely with the oil, and the charged component of the protein faces water-to-hydrogen bond, keeping the droplet intact. When those droplets of fat are stably and evenly distributed in the water, the light is scattered by the many droplets of fat, and the liquid appears white and opaque.

2.5.3 Emulsions in the Kitchen

Milk is probably not the only opaque liquid you have seen in the kitchen. Mayonnaise, sauces, and gravies are also smooth, opaque liquids. Think of the smooth, light yellow, opaque liquid of a hollandaise sauce (Figure 2.33). Hollandaise is a notoriously difficult sauce to make. The fat of clarified butter is stably mixed with the water of lemon juice and/or vinegar with egg yolks as the emulsifier. The stable emulsion scatters light and appears opaque to the eye. Similarly, a traditional homemade mayonnaise is a stable emulsion of oil, vinegar or lemon juice, and

Figure 2.33 Hollandaise sauce is an emulsion of clarified butter dispersed in vinegar or lemon juice with egg yolks as the emulsifier. losinstantes/Adobe Stock Photos

egg yolks. The cook starts by whisking the egg yolks with the vinegar or lemon juice and a bit of mustard (more on the mustard later). Then, the cook slowly streams in the oil while whisking vigorously. If the mayonnaise emulsion is successful, a thick, opaque, and off-white liquid will form. And just like milk and hollandaise sauce, the fat droplets suspended in the watery phase scatter light, and the mayonnaise appears opaque and creamy.

In making a suitable emulsion (sauce, mayonnaise, etc.), Harold McGee, the influential author of *On Food and Cooking* [15], writes that "the cook has made one of three mistakes: he has added the liquid to be dispersed too quickly to the continuous liquid, or added too much of the dispersed liquid, or allowed the sauce to get either too hot or too cold." The essential rules to making an effective emulsion start with the largest volume of liquid (oil or water) in the bowl or mixer. Add the lesser liquid to the greater liquid and add slowly to allow the droplets to form and become surrounded by the emulsifier. Keep the proportions correct; the larger phase should be three times that of the dispersed liquid. Ensure there is enough emulsifier for the dispersed liquid. If there isn't enough emulsifier, the droplets will not be covered. High temperatures can cause proteins to denature fully, and emulsifiers and stabilizers will stop working and begin curling. At colder temperatures, the oils will begin to solidify. Commercial mayonnaise is refrigerated and remains stable because more polyunsaturated oils are used that remain fluid at lower temperatures.

So if Hollandaise sauce is really an emulsion, are all sauces and gravies emulsions? Culinarily speaking, a sauce is one of a number of liquids of varying thicknesses or viscosities. A sauce will enhance a food such as meat, fish, or vegetable by deepening and broadening the dish; it is not the main point, but instead, the sauce should complement the dish. Gravy is a sauce made from watery juices and liquid fats collected from cooking meat. Gravies are often thickened with starch or some other carbohydrate polymer. A little food science can make both gravies and sauces easy to make; all it takes is understanding how to stabilize an emulsion. Remember, an emulsion is the combination of two liquids that don't want to mix. To make a successful emulsion, the oil or fat, and water must be dispersed evenly in one another to make a homogeneous colloidal mixture.

Once an emulsion is formed, even in the presence of emulsifiers, the mixture is doomed and will eventually separate into distinct layers, creating a watery, runny, and oily sauce. Hollandaise sauce will separate into water and fat phases over the course of a week, while an oil and vinegar salad dressing can separate within minutes. How is the cook to create an emulsion that lasts? Complex carbohydrates like starches and gums (Section 2.3) interact and bind to emulsifiers and are often used to thicken a liquid and stabilize an emulsion. The key job of a stabilizer is to keep the droplets of oil or water separated so they cannot collect to form larger droplets and eventually separate into two different phases. When added to a liquid, the thickening effect of starches and gums slows the movement of molecules, making it harder for the fat or water droplets to recombine with other fat or water droplets. This principle of using starches and gums as an emulsion stabilizer is illustrated in a variety of common sauces and condiments. For example, the starches and gums released from tomatoes during preparation act as stabilizers in pastes and ketchup. Dried ground mustard seed contains a considerable amount of gum carbohydrates, which serve well as emulsion stabilizers in hollandaise sauce and in homemade mayonnaise. Agar and carrageenans are used as stabilizers of emulsions in prepared foods like cheese products. A quick examination of Velveeta® processed cheese will reveal alginate and whey proteins on the list of ingredients—both are used to emulsify and stabilize the cheese. Thickening the emulsion limits the diffusion or movement of the oil or water droplets so they cannot come together.

Using starches or gums to thicken and stabilize an emulsion also explains some of the science behind roux sauce. A roux is a butter and flour paste that has been lightly cooked. The fat coats the proteins in the flour and allows the particles to disperse evenly as the watery liquid is added. The addition of watery stock, broth, milk, etc. with heat hydrates the starch into a gel that thickens the liquid. A roux is used in three of the five "mother sauces" of classical French cooking: the béchamel sauce, the velouté sauce, and the Espagnole sauce. What is "mother sauce"? Well, depending on the culinary school of thought, there are three, five, or even nine "mother sauces." French cooking schools describe five or sometimes seven sauces, Asian cooking schools often describe nine sauces, and African cooking describes nine or more base sauces. A mother sauce is the basic sauce from which other flavored sauces are created. For example, a béchamel sauce is made of milk, flour, and butter, where the flour and butter are used to make a white roux, and the final sauce is a stable, thick, creamy liquid. The base béchamel recipe can be made into a cream, mornay, or cheese sauce with the addition of other ingredients or flavors. Other cuisines also use the principle of the roux. For example, Japanese curry uses a roux of curry powder, flour, and oil that is added to stewed meat and vegetables. The outcome is a thickened, stable emulsion of watery juices and oil with spices for flavor.

Figure 2.34 A Béchamel sauce. FomaA/Adobe Stock Photos

KEY CONCEPTS Part V: Emulsions and Emulsifiers

1. **Amphipathic** molecules, such as phospholipids, are comprised of **hydrophilic** parts made of **polar** bonds that can interact with water and **hydrophobic** parts made of **nonpolar** bonds that interact with other **hydrophobic** regions.

2. An **emulsion** is a mixture of water with fat or oil. An **emulsifier** is typically an **amphipathic** molecule that can use its **hydrophilic** parts to interact with the water and its **hydrophobic** parts to interact with the fat/oil. The **emulsifier** bridges the interface between the water and fat/oil so the two can form a stable **emulsion**.

3. Emulsions can be stabilized by gelled starches or gums, which thicken the emulsion and make it more difficult for the droplets of water and fat to recombine and separate into distinct layers.

Part VI: Nucleic Acid

Learning Objectives

1. *Describe the essential building blocks that make up nucleic acids: DNA and RNA*
2. *Identify the nitrogenous base, the ribose sugar, and the phosphate within a nucleobase.*
3. *Describe the role of DNA and RNA in the cell.*
4. *Explain why nucleic acid does not appear on a nutrition label.*

2.6 Nucleic Acid: The Fourth Biological Macromolecule

Whether a plant or an animal, food was once a living thing. The protein, carbohydrates, and fats we find in food all played important roles in the functioning of the plant or animal before it became food. As we saw in Part I of this chapter, biochemists classify proteins, carbohydrates, and fats as biological macromolecules. Nucleic acid is the fourth of the biological macromolecules. It is present in every living thing, and it contains all the information necessary to make the organism function Just like architectural plans explain how to build a building, nucleic acid is a biological code that explains how to build an organism (Figure 2.35).

2.6.1 The Structure and Function of Nucleic Acids

Nucleic acid comes in two main types: ribonucleic acid (RNA) and DNA, and these nucleic acids are themselves a combination of a sugar—called ribose—and a nitrogenous base, which contains...you guessed it...no shortage of nitrogen atoms! Adenine, guanine, and cytosine are nitrogenous bases found in both DNA and RNA, while thymine is found only in DNA and uracil is found only in RNA. DNA and RNA monomers are assembled into a polymer, where the nitrogenous base component of each monomer is different and the order of monomer units is significant. The monomer units of DNA and RNA are often represented by just a single letter, signifying the identity of the nitrogenous base within the monomer unit. The sequence of letters in the polymer, for example—ATCCGA—represents a piece of a polymer chain, in which the identity of the nitrogenous base is adenine, then thymine, then cytosine, cytosine, guanine, adenine and so on. You might recognize this type of pattern as similar to the way proteins are constructed (section 2.2.1). In the same way that a protein is a polymer of amino acids in which the side chain of each monomer unit is one of twenty possible options, DNA and RNA are a polymer of deoxyribonucleotides or ribonucleotides—also called nucleobases—in which the nitrogenous base of each monomer unit is one of four possible options. While a specific *sequence* of amino acid residues makes a unique protein,

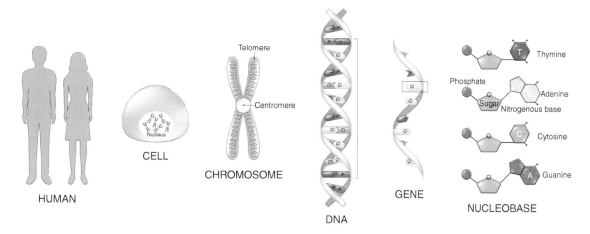

Figure 2.35 The relationship of DNA to cells and organisms. DNA is a polymer of deoxyribonucleic acid in which the nitrogenous base is one of four possibilities (T—thymine, A—adenine, C—cytosine, G—guanine). DNA monomers are assembled into a polymer where the nitrogenous base component of each monomer is different and the order of monomer units is significant. The monomer unit of DNA is often represented by just a single letter signifying the identity of the nitrogenous base within the monomer unit. A single polymer chain of DNA pairs with another complementary chain to make a double helix shape. The order of the pairing is also significant in preserving and propagating the code—or sequence—of DNA bases that communicate information. Adenine-containing monomer units pair with thymine, and guanine-containing monomer units pair with cytosine in the double helix. Dee-sign/Adobe Stock Photos

a specific sequence of nitrogenous bases within a DNA polymer will code for a specific piece of biological information, and that region of code is called a *gene*, and the information it encodes is called *genetic* information. Every organism contains thousands of genes; for example, a human contains ~20,000 genes, while a typical bacteria like *Escherichia coli* contains just over 4,000 genes, and rice contains over 40,000 genes.

A single polymer chain of DNA pairs with another complementary chain to make a double helix shape. The nature of the pairing is also significant in preserving and propagating the code—or sequence—of DNA bases that communicate information. Complementary pairing means that adenine-containing monomer units pair with thymine, and guanine-containing monomer units pair with cytosine in the double helix.

Figure 2.36 The parts of a nucleic acid monomer are also called a nucleobase. For DNA, the monomer is called a deoxyribonucleotide, while for RNA, the monomer is called a ribonucleotide. They differ in a single spot on the ribose ring. In RNA, there is an –OH in that spot, while in DNA, it is an –H.

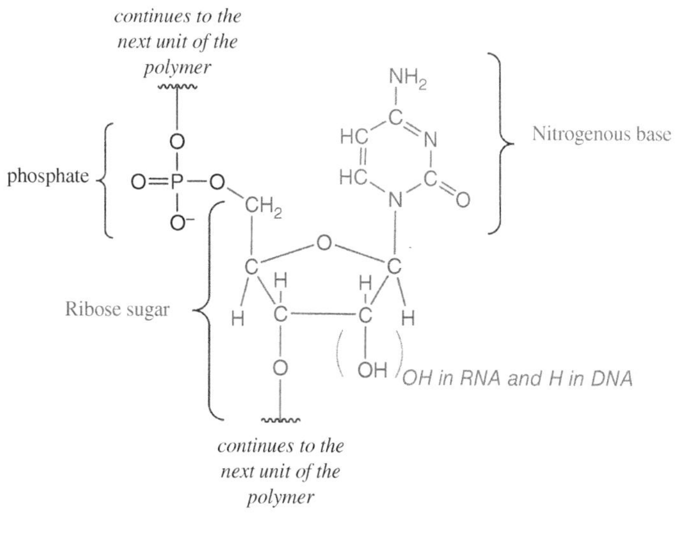

BIOMOLECULES

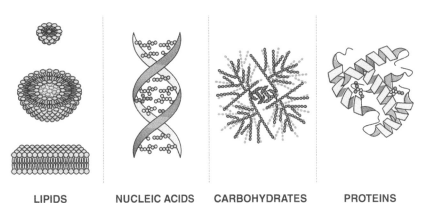

Figure 2.37 The four types of biomolecules that are found within the cell. VectorMine/Adobe Stock Photos

LIPIDS   NUCLEIC ACIDS   CARBOHYDRATES   PROTEINS

Genes are therefore specific sequences of DNA that are held within the smallest unit of every living thing—the cell. We will learn more about cells in Chapter 6 on Fruits and Vegetables, Chapter 7 on Meat and Fish, and then again in Chapter 14 on Taste and Smell, and Chapter 15 on Metabolism. Every living thing is made of one or more cells—tiny compartments that hold DNA and RNA and manufacture all the types of biological macromolecules: protein, fat, and carbohydrate that make up an organism. And while the order and pairing of DNA and the role of RNA in copying and translating genetic information within the cell is incredibly interesting, it is beyond the scope of this book. You can find the basics on the storage and communication of genetic information in an introductory biochemistry course or text.

2.6.2   Nucleic Acid and Nutrition

If nucleic acid is the fourth type of biomolecule, why doesn't it appear on a nutrition label? It is because, in living things, nucleic acid is not a molecule used to store energy, nor is it broken down *for* energy. In short, nucleic acid is not nutritional. In living things, nucleic acids are used to store and transmit information, not energy.

Now, that does not mean that we don't eat nucleic acid! Nucleic acid is present as DNA and RNA in the cells of every living thing: plants and animals. A strawberry is made of cells containing biological macromolecules, including DNA and RNA, and the same is true for a piece of salmon. We can eat nucleic acid, and polymers of nucleic acid can be broken down by our digestive systems; in most cases, those nucleic acid breakdown products are excreted by our bodies. In contrast, when we eat a piece of salmon, the fats, proteins, and carbohydrates are broken down in our digestive systems, and the component parts are either used for energy production or recycled, reassembled, and stored as fat, protein, or carbohydrate in our own bodies. In cooking, the nitrogen atoms common to nucleic acid are important contributors to the flavorful molecules produced in Maillard browning (see Chapter 3).

KEY CONCEPTS Part VI: Nucleic acid

1. Nucleic acid is a type of biological macromolecular polymer that comes in two forms: **DNA** and **RNA**.
2. DNA and RNA monomers are also called deoxyribonucleotides and ribonucleotides, or nucleobases.
3. Deoxyribonucleotides and ribonucleotides, or nucleobases, are assembled from a deoxyribose or **ribose sugar**, a **nitrogenous base,** and a phosphate.
4. DNA and RNA are polymers of monomer deoxyribonucleotides or ribonucleotides, where each monomer unit differs in the identity of the nitrogenous base.
5. The nitrogenous bases adenine, guanine, and cytosine are common to DNA and RNA. Thymine is present only in DNA monomers, and uracil is present only in RNA.

6. Genes are specific sequences of DNA—where the sequence is dictated by the order and identity of the nitrogenous bases within the polymer.

7. DNA and RNA store and communicate genetic information. They are not a significant source of energy, nor are they a form of energy storage; therefore, nucleic acid does not appear on nutrition labels.

Science for the Chef: Let's make mayonnaise to understand an emulsion

Mayonnaise is an emulsion of egg yolks and oil, creating a velvety texture with a magical ability to transform dishes. Its rich consistency envelops ingredients, elevating flavors and creating a harmonious symphony on the palate. A versatile condiment, mayonnaise binds salads, enriches dressings, and lends moist indulgence to sandwiches and burgers. With its subtle tanginess, mayonnaise balances contrasting tastes and unites sweet, salty, sour, and savory elements. In the realm of culinary artistry, mayonnaise stands as a testament to the fusion of ingredients, enhancing dishes, and turning ordinary bites into extraordinary culinary experiences. By immersing yourself in the intricate scientific ballet underpinning each step and meticulously adhering to these instructions, you'll triumphantly craft homemade mayonnaise via the mesmerizing process of emulsification.

Ingredients:
- 1 large egg, at room temperature
- 1 tablespoon Dijon mustard
- 1 tablespoon white wine vinegar or lemon juice
- Salt and freshly ground black pepper, to taste
- 1 cup neutral-flavored oil (such as vegetable, canola, or safflower), chilled
- Optional: a pinch of sugar for balance

Instructions:
Prepare the ingredients: Allowing the egg to reach room temperature is essential because this enables the egg proteins to be more fluid. As for the chilled oil, its cold temperature serves a critical role in the emulsification process. Cold oil is thicker and less likely to spread rapidly, making it easier for the hydrophilic (water-attracting) and hydrophobic (oil-attracting) ends of the emulsifiers in the egg yolk to surround the oil droplets and stabilize the emulsion.

Combine the egg, mustard, vinegar, and seasonings: Introduce the egg, Dijon mustard, white wine vinegar or lemon juice, salt, and pepper into the food processor or blender. The mustard is enriched with natural emulsifiers, specifically lecithin. These emulsifiers work by arranging themselves around oil droplets, facilitating their dispersion in the water-based mixture. The addition of vinegar or lemon juice, being acidic, weakens the protein bonds in the egg yolk, generating an environment conducive to emulsification. Should you opt for a pinch of sugar, it harmonizes the flavors for a well-rounded taste.

Blend the initial mixture: Pulse the mixture several times at the start. This pulsing initiates the breakdown of the egg proteins. As the blender's blades swiftly move through the mixture, they exert mechanical force on the proteins. This force disrupts the protein structures and begins to unfold them. This initial "unfolding" prepares the proteins for the crucial role they will play in binding the oil and water phases of the emulsion.

Slowly add the oil: With the food processor or blender running, meticulously drizzle the chilled oil into the mixture in a deliberate, unhurried stream. Commence this process with a drop-by-drop approach. The gradual incorporation of oil is fundamental because it grants ample time for the lecithin molecules, found in the egg yolk, to assemble themselves around the oil molecules. Lecithin's hydrophilic ends orient themselves toward the water-based components, while the hydrophobic ends snugly embrace the oil droplets. This dual-layer arrangement encases the oil, prohibiting droplets from congregating and ensuring their even distribution throughout the mixture.

Watch the emulsion form: As the oil amalgamates, the mixture's consistency transforms into a lush, creamy texture. This captivating alteration signifies the inception of an emulsion. Lecithin, the emulsifier, orchestrates this by crafting minute structures known as micelles. Micelles are spherical assemblies where the hydrophilic heads of lecithin face outward toward the watery surroundings, while their hydrophobic tails encase the oil droplets. This arrangement enshrouds and stabilizes the oil-water interface, thwarting any inclination to separate.

Continue adding oil: Gradually escalate the rate at which you add the oil, maintaining a measured, thin stream while the appliance is operational. This incremental increase in oil fortifies the stability of the micelles. As more oil is integrated, the hydrophobic tails of lecithin have an expanding array of oil droplets to embrace. This augmentation further strengthens the micellar structures and consolidates their grasp on the dispersed oil, ensuring an enduringly velvety and cohesive emulsion.

Achieve desired consistency: Continue blending until the mayonnaise attains your envisioned texture. The ceaseless blending process serves a dual purpose: it perpetuates the disintegration of oil droplets into smaller entities, leading to a smoother texture, and it encourages the intertwining of protein structures with the emulsified oil, contributing to the overall creaminess.

Store: Carefully transfer the homemade mayonnaise to a pristine, hermetic receptacle and nestle it within the refrigerator. The steadfast micellar formations crafted during the emulsification process persist, averting any inclination toward separation. This ensures that your homemade mayonnaise remains indulgently creamy and delightful for approximately 1–2 weeks.

References

1 Barr, S. B., & Wright, J. C. (2010). Postprandial energy expenditure in whole-food and processed-food meals: implications for daily energy expenditure. *Food & Nutrition Research*, *54*(10), 5144.

2 Institute of Medicine (2005). *Dietary Reference Intakes for Energy, Carbohydrate, Fiber, Fat, Fatty Acids, Cholesterol, Protein, and Amino Acids*. Washington, DC: The National Academies Press.

3 21 CFR 101.81 Health Claims: Soluble fiber from certain foods and risk of coronary heart disease (CHD) Retrieved 25 August 2022. https://www.ecfr.gov/current/title-21/chapter-I/subchapter-B/part-101

4 Williams, P. G. (2014). The benefits of breakfast cereal consumption: a systematic review of the evidence base. *Advances in Nutrition*, *5*(5), 636S–673S.

5 Matilde, S., García, J. J., Fernández, N., Diez, M. J., & Calle, A. P. (2002). Therapeutic effects of psyllium in type 2 diabetic patients. *European Journal of Clinical Nutrition*, *56*(9), 830–842.

6 Smith, P. M., Howitt, M. R., Panikov, N., Michaud, M., Gallini, C. A., Bohlooly-y, M., Glickman, J. N., & Garrett, W. S. (2013). The microbial metabolites, short-chain fatty acids, regulate colonic Treg cell homeostasis. *Science*, 341(6145), 569–573.

7 Fung, T. C., Olson, C. A., & Hsiao, H. Y. (2017). Interactions between the microbiota, immune and nervous systems in health and disease. *Nature Neuroscience*, *20*(2), 145–155.

8 Johnson, R. K., Appel, L. J., Brands, M., Howard, B. V., Lefevre, M., Lustig, R. H., Sacks, F., Steffen, L. M., Wylie-Rosett, J., & American Heart Association Nutrition Committee of the Council on Nutrition, Physical Activity, and Metabolism and the Council on Epidemiology and Prevention (2009). Dietary sugars intake and cardiovascular health: a scientific statement from the American Heart Association. *Circulation*, *120*(11), 1011–1020. https://doi.org/10.1161/CIRCULATIONAHA.109.192627

9 US Department of Agriculture (2020). *What We Eat In America, NHANES 2017-2018*. https://www.ars.usda.gov/ARSUserFiles/80400530/pdf/FPED/tables_1-4_FPED_1718.pdf (accessed 25 August 2022)

10 United States Department of Agriculture (n.d.). *US National Nutrient Database*. Fooddata Central. https://fdc.nal. usda.gov (accessed 24 July 2022)

11 Riel, R. R. (1960). Specifications for the spreadability of butter. *Journal of Dairy Science*, 43(9), 1224–1230.

12 Gunstone, F. D. (ed.) (1996). *Fatty Acid and Lipid Chemistry*. Weinheim: Blackie Academic & Professional.

13 Pustjens, A. M., Boerrigter-Eenling, R., Koot, A. H., Rozijn, M., & van Ruth, S. M. (2017). Characterization of retail conventional, organic, and grass full-fat butters by their fat contents, free fatty acid contents, and triglyceride and fatty acid profiling. *Foods*, 6(4), 26.

14 Department of Health and Human Services. Dietary Guidelines for Americans 2022-2025. https://www.dietarygu idelines.gov/sites/default/files/2021-03/Dietary_Guidelines_for_Americans-2020-2025.pdf (accessed 25 August 2022)

15 H McGee,. (ed.) (2004) *On Food and Cooking*. New York: Simon and Schuster, Inc.

Additional Readings

Jadhav, H.B., Annapure (2023) U.S. Triglycerides of medium-chain fatty acids: a concise review. *Journal of Food Science and Technology*, 60, 2143–2152.Munialo, C.D. (2023), A review of alternative plant protein sources, their extraction, functional characterisation, application, nutritional value and pinch points to being the solution to sustainable food production. *International Journal of Food Science & Technology* https://doi-org.sandiego.idm.oclc.or g/10.1111/ijfs.16467

Nagpal, T., Sahu, J. K., Khare, S. K., Bashir, K. and Jan, K. (2021), Trans fatty acids in food: A review on dietary intake, health impact, regulations and alternatives. *Journal of Food Science* 86, (12), 5159–5174.

Urbanus, B.L., Cox, G.O., Eklund, E.J., Ickes, C.M., Schmidt, S.J., and Lee, S.-Y. (2014), Sensory differences between beet and cane sugar sources. *Journal of Food Science*, 79(9), S1763–S1768.Wang, Ruican, & Hartel, Richard W (2021). Understanding stickiness in sugar-rich food systems: A review of mechanisms, analyses, and solutions of adhesion. *Comprehensive Reviews in Food Science and Food Safety*, 20(6), 5901– 5937.

End of Chapter 2 Questions

1. Using the drawing of the macromolecule:

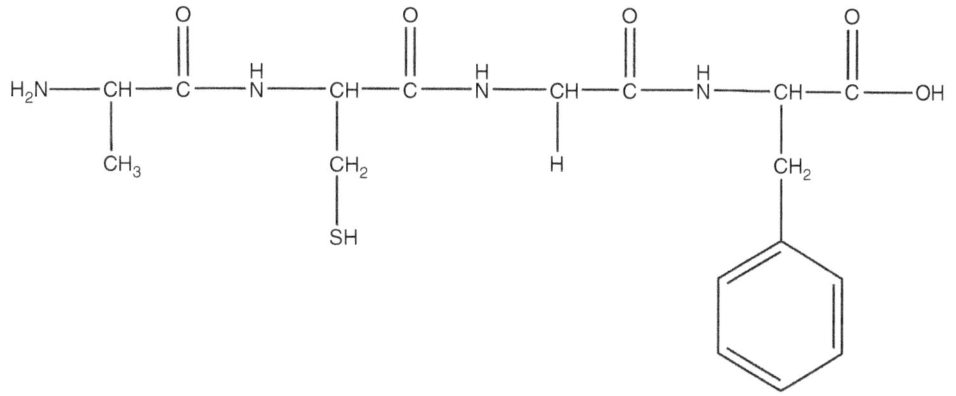

  a. Identify the type of monomer used to make the macromolecule. Fatty acid, monosaccharide, or amino acid?
  b. What type of bond (be specific) connects the monomers in this macromolecule?
  c. The macromolecule is composed of how many monomers?

2. Consider the structures of the two fatty acids. Which fatty acid do you expect to have a higher melting point? Explain.

(A)

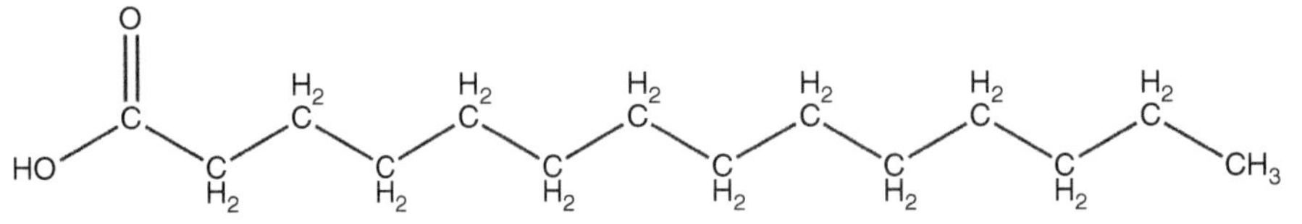

(B)

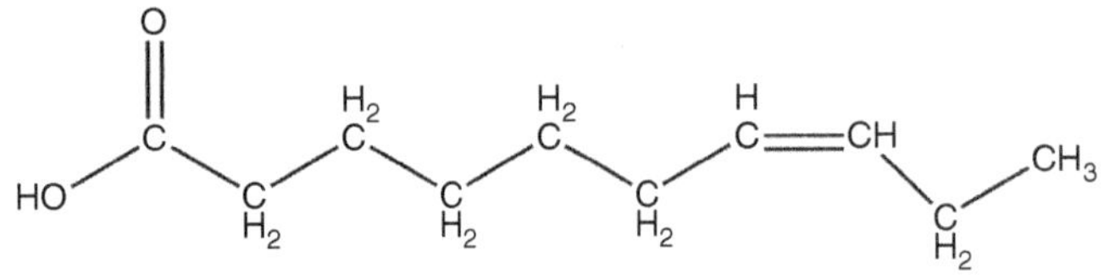

3. In stir-frying, vegetables are chopped small and cooked on a scorching metal surface (>400°F), with a bit of oil and constant stirring (to keep them from burning). What types of oils would be essential for safe and tasty stir-frying? Explain why.

4. Compare and contrast the structure of amylose and cellulose. How are amylose and cellulose structurally similar and different? Explain why humans are able to digest amylose, but cannot digest cellulose.

Amylose

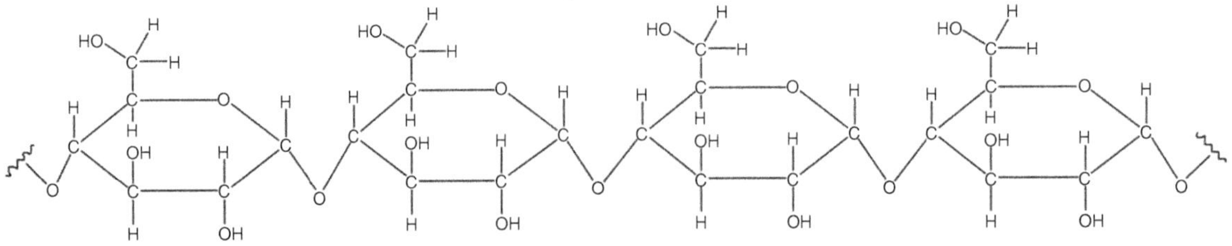

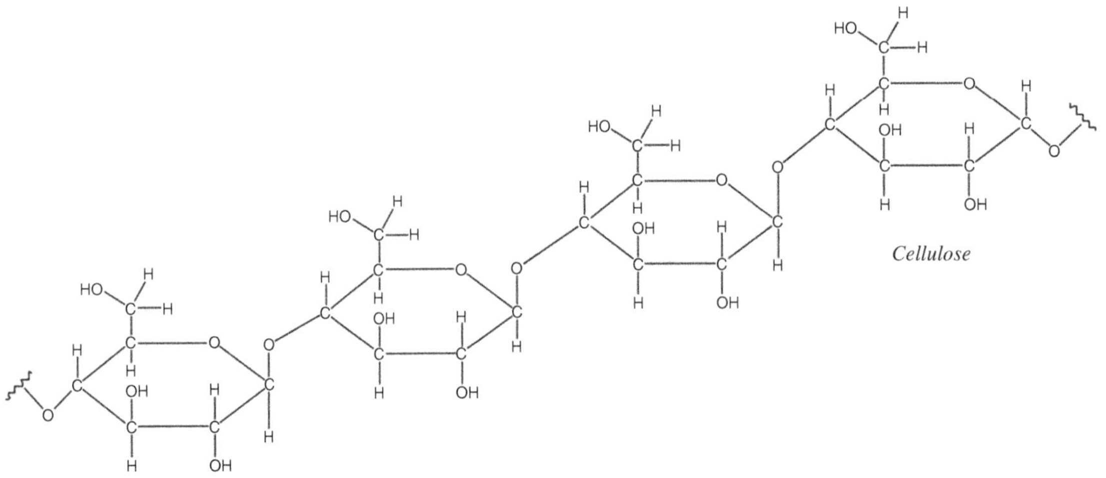

Cellulose

5. What levels of protein structure are damaged during denaturation? When cooking, what are two visual cues that tell us protein denaturation has occurred?

6. At neutral pH, the side chain interactions of aspartate and lysine are shown below:

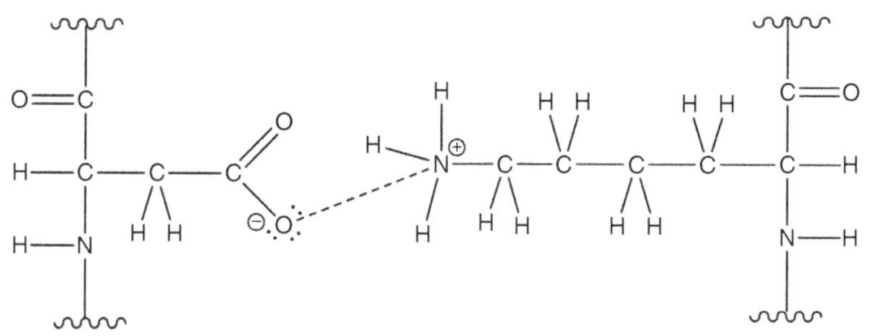

   a. What type of interaction is represented by the dotted line?
   b. If acid (H^+) is added to the protein, explain what happens to the noncovalent interaction.

7. Using the drawing of the phospholipid, show how the hydrophilic (polar) group is able to interact with polar water molecules via noncovalent hydrogen bonds. Include three water molecules in your drawing.

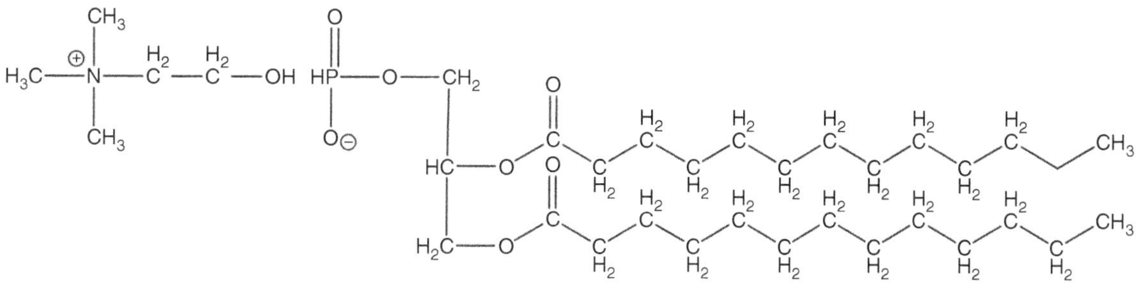

8. Mayonnaise is an emulsion made of three main ingredients: oil, vinegar, and egg yolks (lemon juice and mustard are sometimes added as well). Explain how the three main ingredients work together to make the stable emulsion.

   Use the nutrition label to answer the following questions:

   a. How many calories are from fat?
   b. How many grams of starch are in one serving?
   c. What types of complex carbohydrates likely make up the starch?
   d. Does the breakfast bar contain any natural sugars?
   e. Is this breakfast bar a good source of dietary fiber?

BREAKFAST BAR
Nutrition Facts
Serving Size = 1 Cookie (30g)

Calories	130
	%Daily Value
Total Fat 4.5g	6%
Saturated Fat 2.5g	12%
Trans Fat 0g	
Cholesterol <5mg	1%
Sodium 120mg	5%
Total Carbohydrate 22g	8%
Dietary Fiber 5g	18%
Total Sugers 9g	
Includes 9g Added Sugars 18%	
Protein <1g	

9. Use the following word bank and select the appropriate word(s) to categorize each molecule.

Amino acid	Disaccharide	Fatty acid (saturated or unsaturated)	Simple carbohydrate
Carbohydrate	Emulsifier	Monosaccharide	Sugar
Complex carbohydrate	Fat	Polysaccharide	Triglyceride

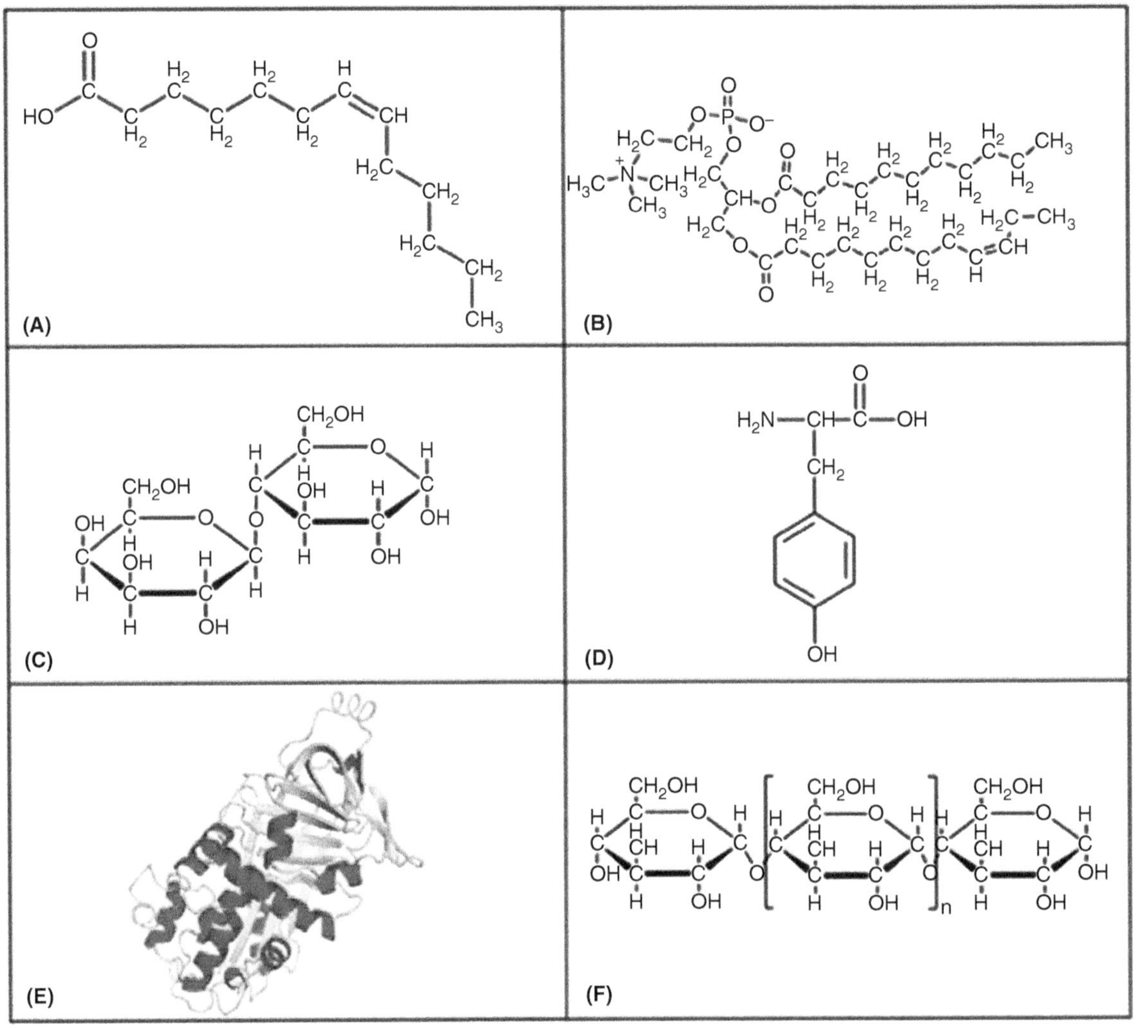

(A)

(B)

(C)

(D)

(E)

(F)

3

Flavor and Color in Food and Drink: Browning Reactions

Guided Inquiry Activities (Web): *5, Amino Acids and Proteins; 7, Carbohydrate; 17, Browning; 23, Meat Cooking*

Learning Objectives

1. *Understand the definition of a chemical reaction and know reactants versus products.*
2. *Distinguish between the energy of the starting reactants and the final products of a reaction.*
3. *Describe the impact and importance of a catalyst in a chemical reaction.*
4. *Know what an enzyme is and its role as a catalyst.*
5. *Describe the two reacting components of the Maillard reaction and the importance of the anomeric carbon in the reaction.*
6. *Identify the components of a reducing sugar.*
7. *Describe the basic steps involved in the Maillard reaction.*
8. *Relate the impact of water and pH on the rate of the Maillard reaction.*
9. *Describe some of the products of the Maillard reaction and where they are found.*
10. *Describe the process of caramelization.*
11. *Recall the foods that are browned by caramelization.*
12. *Interpret the role of water loss in the initial step of caramelization.*
13. *Relate the size and characteristics of the classes of caramel.*
14. *Explain the classification of caramel used in food and beverage.*
15. *Know how ascorbic acid browning is both similar and unique from the Maillard and caramelization reactions.*
16. *Know the reactants and their location in the cell of enzyme-catalyzed fruit and vegetable browning reactions.*
17. *Understand the role phenol oxidase plays in the browning process.*
18. *Explain how fruit and vegetable browning can be prevented.*
19. *Discriminate the difference between each of the browning processes.*

3.1   Introduction

We will describe chemistry/molecular changes in cheese browning, the chemistry of Maillard browning of proteins, the chemistry of caramelization browning of sugars, and enzymatic browning, including some introductory chemistry and biochemistry of enzymes.

Brown food can be both incredibly appetizing and unattractive. Grilling produces a delicious flavor that cannot be matched by poaching, while a browned apple slice is something we try to avoid eating (Figure 3.1). The chemical

The Science of Cooking: Understanding the Biology and Chemistry Behind Food and Cooking, Second Edition. Joseph J. Provost et al.
© 2025 John Wiley & Sons, Inc. Published 2025 by John Wiley & Sons, Inc.
Companion website: www.wiley.com/go/provost/food_science_2e

reactions of browning create pleasing flavors and odors in many foods and even drinks. For example, chocolate cocoa is bitter and astringent until reactions leading to brown color and flavor are created. Coffee and beer are better for the browning that occurs during their making. Plant enzymes catalyze a different set of browning reactions, in this case, to provide first aid for an injured plant and to detract animals from eating and further damaging the plant. There are thousands of flavors and aromas produced by these reactions; a good scientist and cook will appreciate and understand the process by which these compounds are formed. Furthermore, by understanding these browning reactions, one can learn to enhance the flavorful reactions while deterring unwanted ones in the kitchen and ultimately become a better cook.

This process that creates a fantastic diversity of new molecules is often called browning. However, this term is not entirely accurate, as many chemical products are not brown but yellow or even colorless. Moreover, browning reactions can be organized in several different ways. First are the two browning reactions created by heat and "without a catalyst." These reactions are the caramelization and Maillard reactions. Fruit browning is another browning reaction different from the Maillard reaction and caramelization as this reaction is catalyzed by an enzyme (Figure 3.2).

Figure 3.1 Browning is better. The impact of browning reactions on the taste and attractiveness of our food. Both are cooked to a safe temperature but one is much more tasty.

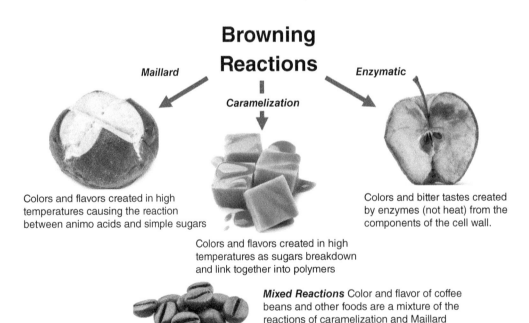

Browning Reactions

Maillard

Enzymatic

Caramelization

Colors and flavors created in high temperatures causing the reaction between amino acids and simple sugars

Colors and bitter tastes created by enzymes (not heat) from the components of the cell wall.

Colors and flavors created in high temperatures as sugars breakdown and link together into polymers

Mixed Reactions Color and flavor of coffee beans and other foods are a mixture of the reactions of caramelization and Maillard

Figure 3.2 Food browning reactions. The reactions that make food brown and change its flavor can be organized by catalyzed or non-catalyzed as well as the reactants and products of each type of browning reaction. rdnzl/Adobe Stock Photos, margo555/Adobe Stock Photos, Yay Images/Adobe Stock Photos, grey/Adobe Stock Photos

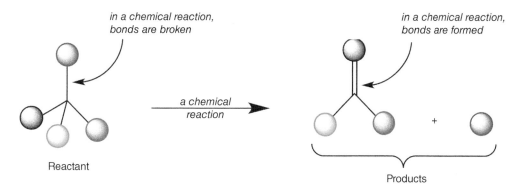

in a chemical reaction, bonds are broken

a chemical reaction

in a chemical reaction, bonds are formed

Reactant

Products

Figure 3.3 The anatomy of a chemical reaction. A chemical reaction (chemical change) takes place when a chemical bond is broken or made.

A closer look reveals that each type of browning is a chemical process—bonds are broken and formed to make new molecules. In some instances, the browning occurs with the aid of enzymes, and other browning reactions occur without a catalyst, also called nonenzymatic browning. To better understand the difference between an enzyme- and non-enzyme-catalyzed reactions, it will help to consider how a chemical reaction proceeds. A chemical reaction is a process by which one or more chemical substances will change as bonds are made and/or broken, resulting in one or more new molecules (Figure 3.3). The chemical equation for a reaction shows one or more starting substances called reactants forming new molecules called products. How fast the reaction happens is measured as the rate of the reaction.

3.2 Chemical Reaction Kinetics

When a chemical reaction involves the collision of two reactants, the reacting atoms or molecules must collide with enough force and correct geometry to change the bonds. When a chemical reaction involves breaking one molecule into two or more, the reactant must vibrate with enough thermal energy to break the bonds. How fast the reactants collide or vibrate to form products depends on several parameters. Heating increases the energy of the molecules involved in a reaction. Increasing the energy of a system makes molecules vibrate more and move around more rapidly. When a reaction involves the collision of two molecules, increasing the temperature increases the number of collisions over time. Compared with reactions at room temperature, every 10°C/50°F increase in temperature nearly doubles the reaction rate. Increasing the concentration of reactants also helps to increase the number of collisions of reacting molecules over time.

One class of browning reactions, the Maillard reaction, can occur at room temperature, albeit very slowly. Heating food increases the number of successful collisions between sugar and protein molecules that make up the Maillard reaction, and thus browning occurs more effectively at higher temperatures. To better understand this process, it is helpful to investigate a concept called activation energy (Figure 3.4). Activation energy is the amount of energy needed to start and continue a spontaneous reaction, that is, a reaction where there is less total energy in the products than the starting reactants.

At the start of a chemical reaction, the reacting molecules absorb energy from the surroundings and collide. Appropriately oriented collisions between reaction molecules may not result in a chemical reaction unless there is enough energy in the collision. The collision's minimum energy is called the reaction's activation energy. The energy diagram in Figure 3.4 often shows this. The inherent energy of a substance (for our purpose, a molecule from food) is shown on the vertical or Y-axis, while the horizontal or X-axis describes the reaction process, also called a reaction coordinate.

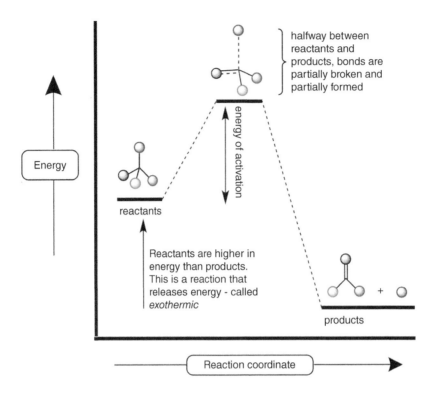

halfway between reactants and products, bonds are partially broken and partially formed

energy of activation

Energy

reactants

Reactants are higher in energy than products. This is a reaction that releases energy - called *exothermic*

products

Reaction coordinate

Figure 3.4 Energy diagram for a chemical reaction. The progress of a chemical reaction is shown on the horizontal axis (reaction coordinate), and the free energy of the substances involved in the reaction is shown on the vertical axis. The energy required for the reaction to reach the intermediate (also called a transition state) of the reaction is the energy of activation. A spontaneous reaction will end with less free energy than the starting compounds possessed.

Let's consider the reaction between sugar and oxygen, producing CO_2 and H_2O. The inherent energy of the starting reactants is higher than the energy of the products. Thus, energy is given off, and the reaction is considered exothermic or spontaneous. However, the bag of sugar in your pantry has yet to react with oxygen in the air! That is a good thing; there is a lot of potential chemical energy in a bag of sucrose if all of the molecules react simultaneously! So what stops the reaction from happening as the sugar bag sits in your kitchen? The answer is the activation energy. From the energy diagram, you can see there is an intermediate or activated complex of the reaction. To produce the activated complex requires additional energy—called the energy barrier, an amount of energy that is given off as the reaction proceeds. The activated complex (also called a transition state) is short-lived, and the overall energy of activation is released as the activated complex forms products. The energy barrier requirements prevent sugar and oxygen from reacting without adding heat. Thus, for sugar to combust, one has to add heat to overcome the energy barrier and produce a nice brown crust on a crème brûlée (Figure 3.5).

Heating food increases the number of molecules with the required energy to react. At room temperature, only a few molecules have

Figure 3.5 Heat increasing the rate of a reaction—crème brûlée. Heat from the flame provides the energy needed for oxygen to react with sucrose sprinkled on the top of a desert. Richard Villalon/Adobe Stock Photos

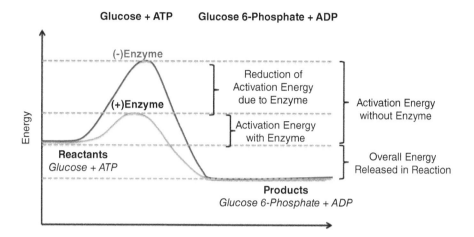

Figure 3.6 Effect of an enzyme catalyst on the energy of a reaction.

enough energy for sugars and proteins to collide and form brown compounds. Thus, browning happens at a low temperature but very slowly. Increasing the temperature increases the number of molecules with productive collisions and reactions. The products of browning reactions are one of the reasons why we heat our food to create more brown molecules.

Enzymes are protein catalysts. Catalysts reduce the activation energy for a reaction (Figure 3.6). Thus, in a catalyzed reaction, it takes less energy to overcome the activation complex/energy barrier, and more product molecules can be formed at a lower temperature. Note that a catalyst does not change the total energy of the reaction. Heat can also speed up an enzyme-catalyzed reaction. The additional energy added to a system when cooking, as in a non-catalyzed reaction, results in more molecules with enough energy to overcome the activation energy barrier. However, remember that enzymes are proteins. Heat can also impact the structure of the protein. Many proteins will denature at around 45–65°C /113–149°F. Therefore, heating an enzyme-catalyzed reaction will increase the rate until the protein is heated above its denaturation point, at which time the reaction will slow down with increased temperature.

So we catalog browning reactions in two categories: enzyme-catalyzed reactions (vegetable and fruit browning) and non-enzyme-catalyzed reactions. These reactions include caramelization, the Maillard reactions, lipid oxidation, and ascorbic acid browning.

3.3 The Maillard Reaction

The Maillard reaction is a complex series of reactions between some sugars and the amino groups of amino acids and proteins (Figure 3.7). This reaction creates the pleasant odor and flavor of baked bread, cooked meat, and buttered popcorn flavor, among hundreds of others. The reaction is also responsible for the brown color of beer, roasted meat, coffee, and chocolate. The reaction was first described and eventually named after the French physician/scientist Louis Camille Maillard. In modern days, he would be considered a biochemist. At the age of 16, Maillard entered the University of Nancy in Lorraine, France, where he earned a master's degree in science at the age of 19, became a physician, and began research on metabolism in 1903. There he studied the metabolism of biological molecules, particularly the synthesis and breakdown of proteins. At the age of 34, he published the first article describing reactions between sugars and amino acids, starting with glycine and glucose [1]. Maillard established the formation of a class of dark-colored compounds he named "melanoidins." Eventually, he studied and found the order of reactivity for the kinds of sugars that react with different amino acids, defining the reactivity

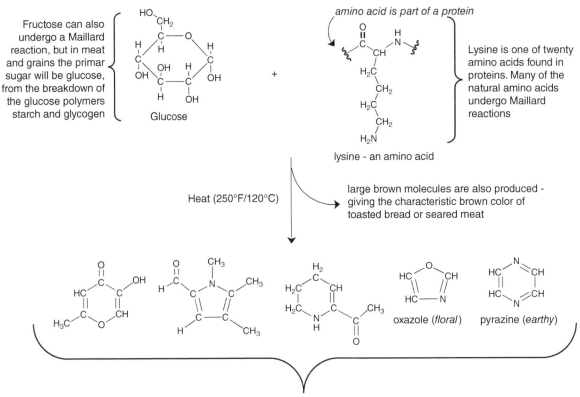

Figure 3.7 The Maillard browning reaction. Maillard browning is a reaction between select sugars (reducing) and nitrogen atoms of many amino acids.

for each. While he eventually published 14 papers on reactions between sugars and amino acids, specifics on how the reactions were taking place remained elusive until the early 1950s, when chemist John Hodge from the US Department of Agriculture (USDA) defined the chemical steps in the early stages of the browning process. Interestingly, Maillard referred to the reactions that create brown flavorful molecules as "my reaction." However, the convention of calling these complex reactions the "Maillard reaction" had appeared in the literature (published in a review by General Foods and several chemical publications) by the early 1950s (Box 3.1). Thus, the large set of nonenzyme-catalyzed reactions between reducing sugars and amino groups are now called the Maillard reaction.

The initial steps of the Maillard reaction are the most straightforward and, therefore, the most accessible place to begin. For those interested in cooking, this is where the magic happens. Heat, sugars, and proteins or amino acids result in the brown, tasty, and aroma-filled goodness of cooked food. A closer look at the chemistry shows that the reaction occurs at the C=O bond (a carbonyl group) of a simple sugar and the nitrogen of an amino group found on amino acids and proteins. Under higher temperatures (to surmount the activation energy barrier), around 140–160°C/284–320°F, we begin to observe the formation of significant amounts of browning products (Figure 3.7). However, pH and other conditions can significantly reduce the reaction's temperature at which appreciable products are formed.

Sugar molecules can take on two different structures or shapes: a straight or open chain and a ring or closed form of the molecule. In sugars such as glucose, the ring-opening process produces a C=O or carbonyl at the anomeric carbon. See Figure 3.8 for examples of several reducing sugars and the anomeric carbon. When dissolved in our cells' water and food, these sugars can be found in the ring-opened and closed conformations. The carbonyl group formed upon ring opening is what participates in the Maillard reaction, and sugars that can "ring open" to form a C=O group are called "reducing sugars." However, in other sugars—such as fructose or sucrose—ring opening is not possible, and these sugars do not participate in the Maillard reaction. Examples of monosaccharides that can open and close to form carbonyl groups are glucose, ribose, galactose, and ribulose (Figure 3.8). Disaccharides can be reducing sugars if at least one of the anomeric carbons can convert to the open-chain format. For example, the disaccharide found in milk, lactose, is composed of galactose bonded to glucose and has two anomeric carbons. The anomeric carbon of galactose is tied up in two C—O bonds and unable to convert to reducing sugar. However, the anomeric carbon of the glucose portion of lactose can open and close and thus react to reduce other compounds chemically. Maltose (formed from the breakdown of starch found in sweet potatoes and cereals and used to make candies, syrups, and ketchup) is another example of a disaccharide that is a reducing sugar—foods with milk and potatoes brown for this reason. However, don't try to use table sugar (sucrose) to brown your food. Sucrose is a disaccharide of glucose, and fructose has both of the anomeric carbons bonded to each other and is not a reducing sugar (Figure 3.8).

Not all sugars work well in the Maillard reaction with amino acids. The carbonyl carbon of the aldehyde group is required for the reaction. Sugars that contain an accessible aldehyde are called "reducing sugars." A reducing sugar can act as a "reducing agent" in several chemical reactions, including the Maillard reaction. They are given this name because of the reaction between the C=O and another compound. Under the right conditions, the C=O will react and cause the second compound to be "reduced" in its electron state. Thus, the sugar causes the other compound to be "reduced" and is called "reducing sugar." This carbon is also called anomeric carbon for reasons we will not worry about here. Simply put, the sugar has to have the ability to open and close the ring structure and is thus called a reducing sugar (Figure 3.8).

Proteins and amino acids found in food are the second reactants in the Maillard reaction. The atom involved is the nitrogen of amino groups within amino acids, small proteins (peptides), or the amino groups of proteins themselves. In the reaction, the amino nitrogen bonds to the anomeric carbon of a reducing sugar. Combining the reducing sugar (at the carbonyl carbon) and the nitrogen of amino groups is called a dehydration reaction and takes place with the loss of a water molecule. Very quickly, the dehydration product rearranges to form a new compound of both the sugar and an amino acid called an Amadori compound (Figure 3.9).

At this point, you can see the diversity of new molecules that the Maillard browning reaction can form. Many reducing sugars are found in meat, vegetables, and fruits. Each of the 20 amino acids has an amino group, and one amino acid, lysine, has a side chain that contains an additional amino group. Some food scientists report that arginine, which also includes a second amino group, can also form the Amadori compound of a Maillard reaction. Considering that 18 of the amino acids can react once and two of the 20 common amino acids can react with three different outcomes (the amino group, side chain, or both), the number of Amadori compounds possible with cooking is significant. If we limit the reducing sugars to eight of the more common reducing sugars, we can predict the possible number of initial products of the Maillard reaction. Looking at just the 18 amino acids that can only react with one of the eight sugars, there are $18 \times 8 = 144$ initial products. If two of the amino acids have three different combinations of sugar reactions, there are an additional $2 \times 3 \times 8 = 48$ possible products. Therefore, in our simple calculation that ignores many other sugars, peptides, and proteins that also react to form flavor molecules, 192 Amadori compounds can be made from these reducing sugars and amino acids. When you consider that there is a minimum of three different paths that each Amadori product can take to create the final compound, there are an additional $192 \times 3 = 576$ different compounds from eight different sugars and 20 amino acids. That is a lot of flavor, color, and aroma compounds! When you factor in the many additional reactions that follow the

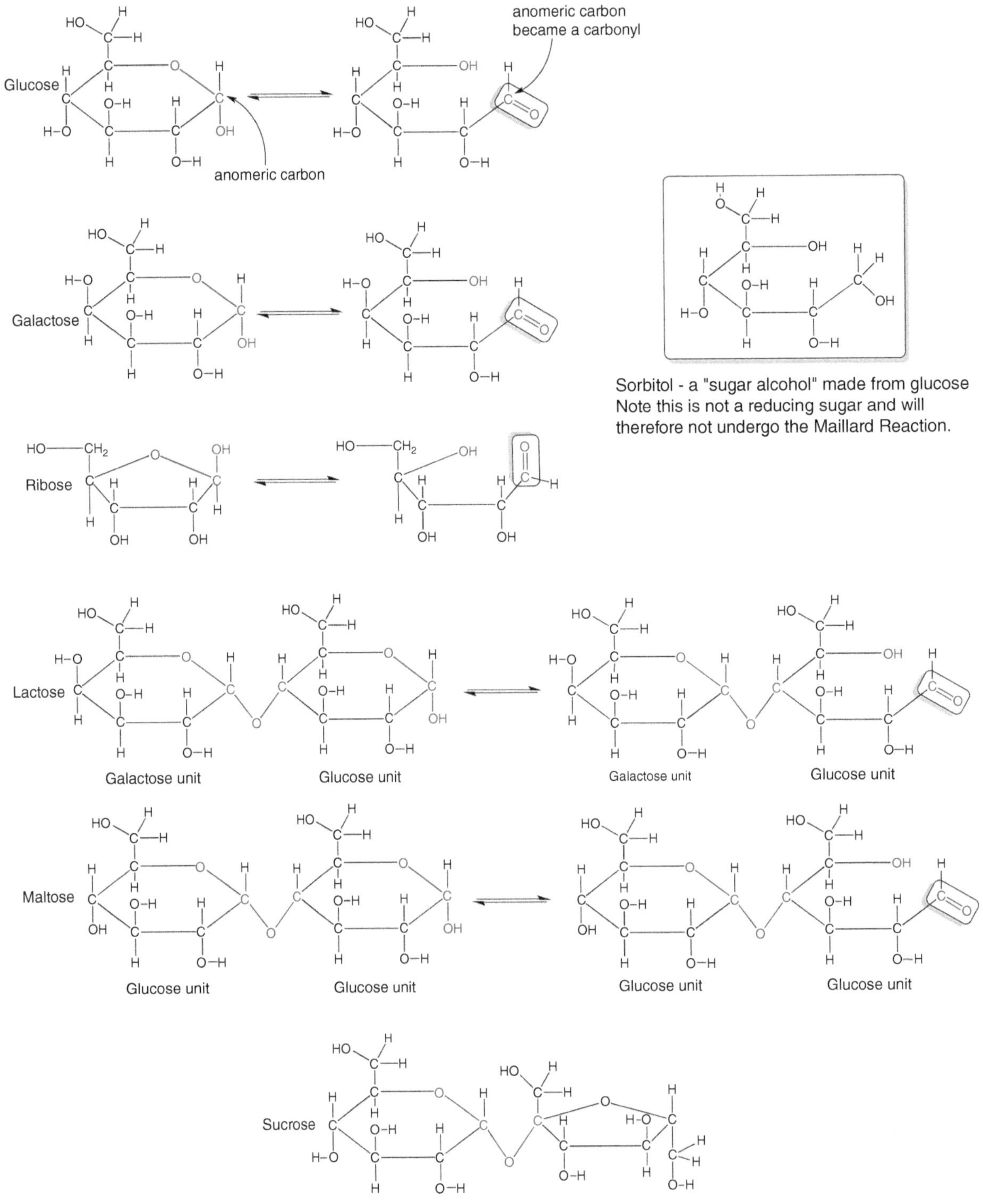

Figure 3.8 Anomeric carbons and reducing sugars. Some sugars have ring-opened forms in which the anomeric carbon becomes a carbonyl group (C=O), while in other sugars the anomeric carbon is trapped by two C—O bonds and cannot ring open.

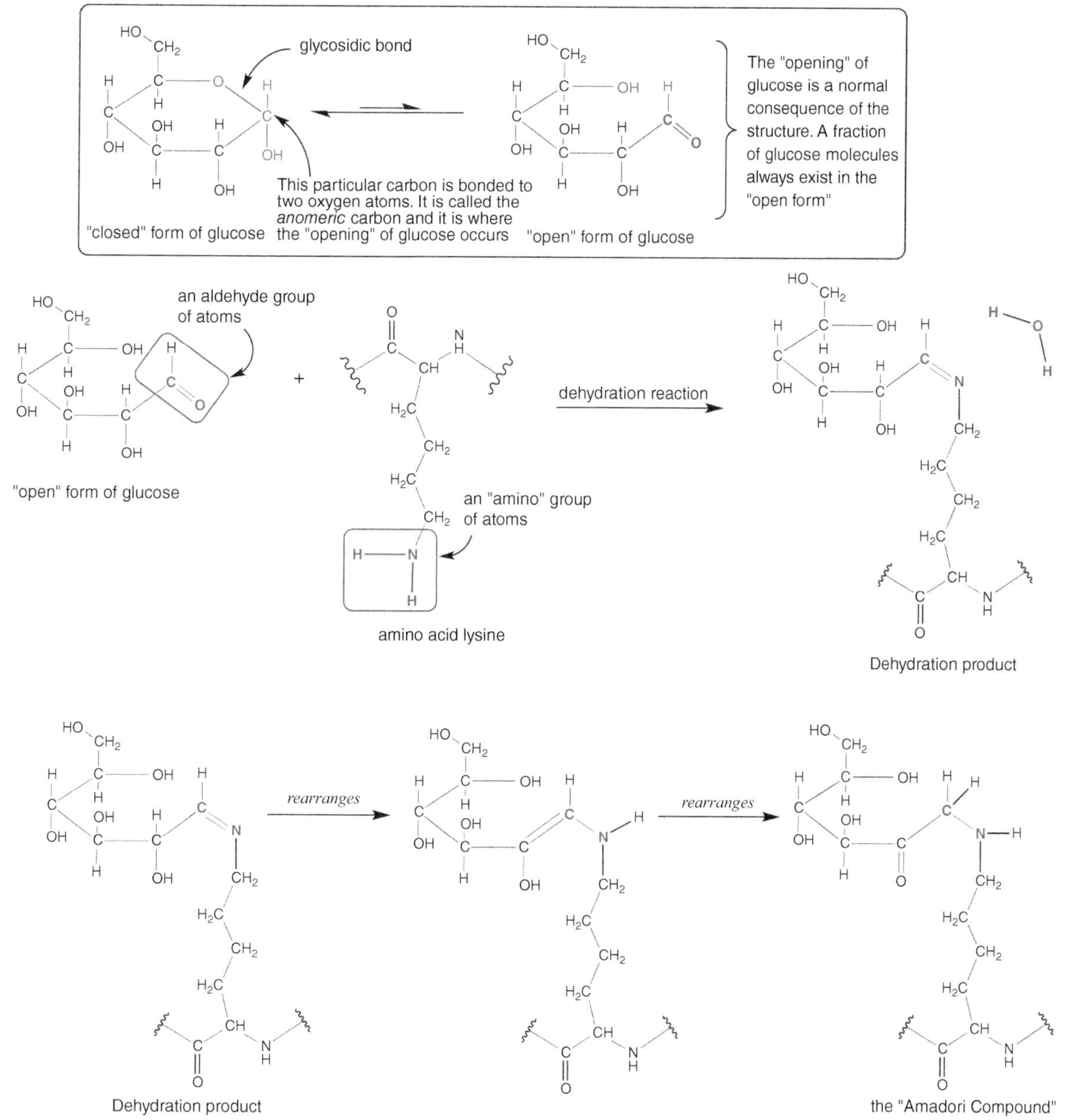

Figure 3.9 The Maillard reaction. The Maillard browning reaction requires a reducing "open" form of a sugar and the amino group of an amino acid. After the dehydration (also called a condensation) reaction, the compound will rearrange to form the intermediate "Amadori compound."

initial Amadori step, one can begin to comprehend the thousands of possible new products and the complexity of food color, flavor, and aroma.

There is an order of reactivity for both sugars and amino acids for the Maillard reaction. The most reactive reducing sugars are the smallest-sized sugars. Five-carbon sugars such as ribose react faster than six-carbon sugars such as glucose, galactose, and fructose. Sorbitol and other sugar alcohols used as low-calorie substitutes do

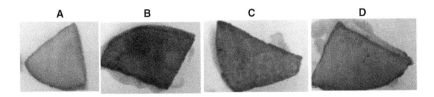

Figure 3.10 Browning potatoes. Potatoes dipped in a dilute solution of sugar (ribose) and an amino acid were lightly fried for 3 minutes and 45 seconds. (A) Water control—no sugar or amino acid, (B) ribose–lysine, (C) ribose–leucine, and (D) ribose–glycine.

not react, and if used in baking or cooking, will not brown. These sugars can "burn" or combust, but that is a completely different reaction with oxygen and not the same as the Maillard reaction.

Free amino acids react faster than peptides or amino acid residues within proteins. Lysine is the most reactive amino acid and browns very nicely with foods rich in ribose. Foods made with wheat, such as bread, are rich in lysine-containing proteins and produce a rich flavor and dark color when toasted. The whey protein in milk is also rich in lysine and is often used as a browning agent to increase browning in foods. Cheese and aged meat whose proteins have been degraded to peptides and amino acids and have a high concentration of sugars are particularly quick to brown. Ribose in meat is created by the breakdown of adenosine triphosphate (ATP) and nucleotides making up DNA and RNA molecules. In baking, egg washes contribute to baked goods' flavor and browned color by providing amino acids, including lysine. The amino acid cysteine reacts well with ribose, creating a robust meaty flavor and aroma. In fact, through the Maillard reactions, over 200 different volatile aroma compounds from the reaction between cysteine and reducing sugars have been found in cooked meat. See Figure 3.10 for an example of potatoes cooked in different sugar amino acid compounds (Table 3.1).

3.4 Factors that Impact Maillard Reaction Browning: pH, Temperature, and Time

Several exciting factors impact the rate and completeness of the Maillard browning reaction, the most impressive of which is pH. The acidity of food can dramatically alter the rate and ability of a food to brown. For example, old-fashioned pretzels' characteristic dark brown color is made by dipping the dough in lye and then cooking. Lye is a strongly basic/alkaline solution of sodium or potassium hydroxide (NaOH or KOH) and is used in many dishes, including lutefisk and some types of noodles. In the case of the pretzels, the lye created a basic/alkaline pH outside the pretzel dough, increasing the rate and quantity of Maillard browning reactions during heating. Rates of browning reactions can be increased severalfold when the pH is above 7. Conversely, lower, more acidic pHs inhibit the Maillard reaction (Figure 3.11). The explanation for this becomes evident if one considers the chemistry for the first stage of Maillard browning. The open/closed formation of reducing sugars is dependent on pH. In acidic solutions (low pH), reducing sugars are primarily in the ring conformation and unable to react. In more basic/alkaline conditions (high pH), more sugar molecules are in the ring-opened form, which allows the anomeric carbon to react with the amino group. In addition, when the pH is low, the amino group becomes a protonated ammonium ion, which is very unreactive toward the reducing sugar. When the pH is raised, the proton is removed from the amino group, and the reaction can occur. The pH and how it alters the nature of the two reactions of the Maillard reaction explains why most marinades, which include an acid to break down proteins in tougher cuts of meat, take longer times and higher temperatures to brown. Figure 3.12 shows the acid/base chemistry impacting the sugars and amino acids in the Maillard reaction.

Table 3.1   Intensity of Maillard browning of common amino acids and sugars.

Amino compound	Absorbance at 420 nm			
	+D-Glucose	+D-Fructose	+D-Ribose	+α-Lactose
Lys	0.947	1.04	1.22	1.23
Gly	0.942	1.07	1.34	1.49
Trp	0.826	0.853	0.972	1.32
Tyr	0.809	0.857	0.951	1.06
Pro	0.770	0.783	0.792	0.876
Leu	0.764	0.747	0.895	1.11
Ile	0.746	0.797	0.870	0.986
Ala	0.739	0.792	0.945	1.06
Phe	0.703	0.751	0.800	0.941
Met	0.668	0.669	0.828	0.888
Val	0.663	0.800	0.772	0.900
Gln	0.602	0.644	0.633	0.639
Ser	0.600	0.646	0.679	0.751
Asn	0.560	0.578	0.565	0.560
His	0.535	0.573	0.529	0.609
Thr	0.509	0.601	0.590	0.600
Asp	0.353	0.426	0.378	0.336
Arg	0.335	0.331	0.370	0.312
Glu	0.294	0.338	0.341	0.320
Cys	0.144	0.202	0.150	0.273

Amino acids and sugars were mixed at a 1:1 ratio with a final concentration of 5 mM dissolved in 40 mM carbonate buffer pH 9.0. Each solution was autoclaved at 121°C for 10 minutes. The absorbance after each sample was treated was taken at 420 nm [2]. A higher absorbance number indicates more Maillard reactions progressed during the 10-minute reaction time. The more Maillard product the darker (more absorbance) the sample. This data shows the most reactive pairing is lactose with glycine (Gly) or tryptophan (Trp), and the least reactive amino acid is cystine (Cys) with any of the sugars tested.

Temperature also impacts the rate of the reaction. As presented earlier in this chapter, increasing the temperature increases the number of collisions and the number of molecules with enough energy to overcome the energy barrier to reaction, allowing a spontaneous reaction to proceed. However, most foods will have some browning at lower temperatures, just not at an appreciable rate. Most Maillard reactions begin above 100°C/212°F, and the reaction will continue even faster if heated above 140°C/284°F. Aged champagne, which is not allowed to warm significantly, still develops a yellow color due to the proteins and sugars found in the grapes undergoing Maillard reactions. But because the temperature wine is stored at is cool, the rates of the reactions producing the yellow Maillard compounds are slowed down.

Another way through which heating food increases the rate of Maillard reaction is by evaporating water. Water content slows the reaction, partly because water absorbs heat during cooking as it steams and evaporates. Thus the energy from cooking is spent cooking off the water instead of overcoming the energy of activation of the Maillard reaction. As long as the food is wet, the temperature of the food will stay at the boiling point of water,

100°C/210°F, and the Maillard reaction is slow to proceed. However, as the water content at the surface of the food decreases as water evaporates, the temperature of the food can then rise to the temperatures needed for Maillard browning to occur. For example, fried or grilled meat can have a thin layer of browned tissue that touches the heat source, but the reaction is limited to this semi-dried area of the food. Similarly, toasted or baked bread is dried and brown on the surface but moist and unreacted in the middle. Heat transfer is one of the reasons why oil is used for frying food. Oil efficiently transfers heat from the pan or heating element to the surface of the food, creating the right environment for the Maillard reaction.

One of the significant challenges when browning food is how long and how high one heats the food to achieve the Maillard reaction but not dry out the food. No one wants to eat a dried piece of brown bread or meat. While boiling, poaching, or sous vide cooking can produce moist food, food cooked with these

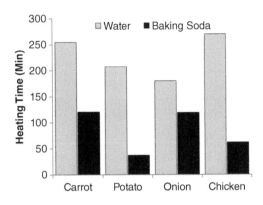

Figure 3.11 Basic/alkaline conditions speed browning time. 2.5 cm cubed sections of food were dipped in water or a dilute solution of baking soda (sodium bicarbonate) and lightly fried for the indicated time. Note that decreasing the pH with baking soda decreased the time needed to brown each food item.

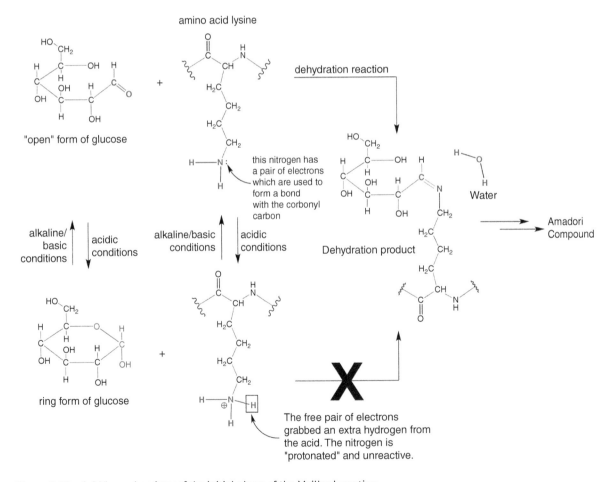

Figure 3.12 Acid/base chemistry of the initial phase of the Maillard reaction.

methods will not brown, nor will it have the flavor bouquet of the Maillard reaction. Simply dropping the food onto a hot pan may not be the correct fix, as the moisture at the surface of the food will inhibit the browning until the high temperatures sufficiently dry the tissue. While it might be enticing to dip food into lye, the flavor impact of the strong base can be problematic and unsavory, and the use of strongly basic solutions can be dangerous. One way to use our understanding of science and cooking is to use a dilute solution of a weak base such as baking soda (aka sodium bicarbonate) to raise the pH and add a wash of lysine-containing food and a bit of ribose (purchased from health stores) to the surface of a food item for quick browning. The mixture of amino acids and basic/alkali pH is why tempura batters include baking soda—for quick and tasty browning during cooking.

3.5 Maillard Is Complicated

The Maillard browning process is actually a very intricate series of reactions that start with the formation of the Amadori complex. In the early 1950s, USDA scientist John Hodge created a three-phase scheme describing the possible pathways by which the Maillard reactions could form many final products [2]. Three possible pathways combined with many different sugars and amino acids available to react in most foods give us an insight to the many possible products and flavors of browning (Figure 3.13). See Table 3.2 for examples of some of these products [3].

Shorter chain products are formed by the fission pathway. This pathway involves single electron compounds called free radicals. While complex in nature, this pathway can result in shorter two- and three-carbon compounds. These compounds can react with other sugars or amino acids, giving rise to interesting compounds including diacetyl (the smell of butter) and acetone (a fruity smell).

The dehydration pathway involves losing water and a nitrogen (amine) group. This pathway can produce ring structures, including the furan-based molecules, which give roasted coffee smells, meat flavors, and, if sulfur is present, a burnt bread odor and flavor.

The Strecker degradation is a third pathway providing many aromatic and brown-colored compounds, especially during roasting. This reaction can occur among the products of the Maillard reaction but can also occur apart from Maillard chemistry. In this reaction, dicarbonyl species (two C=O bonds) produced from the breakdown of the Amadori product or elsewhere react with free amino acids to create aminoketones. These aminoketones condense, and pyrazines are formed after oxidation, typically from oxygen in the air (Figure 3.13). Pyrazines have intense flavors that add to the complexity of the Maillard bouquet.

In addition to their fabulous aromas and flavors, many of the final products from any one of these pathways can further react in polymerization reactions to create brown, high molecular weight pigments called "melanoidins." These pigments provide the dark color for beer and coffee as well as the brown color associated with browned meats and breads.

3.6 Caramelization: Browning Beyond the Maillard

Caramelization involves another complex set of nonenzymatic browning reactions. While both Maillard reactions and caramelization create flavorful brown compounds, caramelization differs from the Maillard reaction in several fundamental ways. Firstly, in caramelization, the reaction occurs only between sugar molecules; no proteins or amino acids are involved in this reaction. Secondly, the required temperature needed to induce the reaction is higher for caramelization (starting at 160–180°C/320–356°F) than the Maillard reaction (typically 140–160°C/284–320°F). Finally, the reaction for caramelization is an oxidation reaction producing long polymers of sugars with some shorter volatile compounds. The flavors and color of most cooked dishes have the Maillard and caramelization reactions to blame. Braised beef, beer, and chocolate are just a few foods and drinks that owe much of their

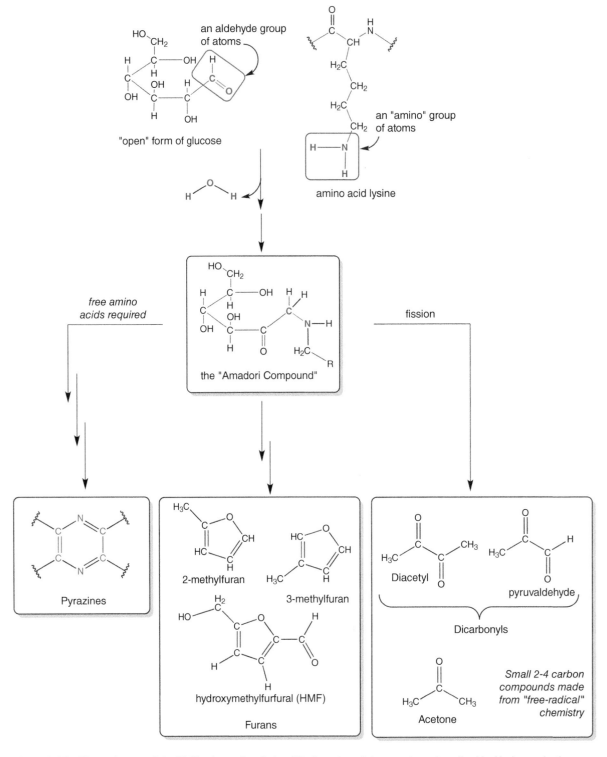

Figure 3.13 The pathways of the Maillard reaction. A simplified version of the reactions described by Hodge and others Adapted from [2].

Table 3.2   Select Maillard products.

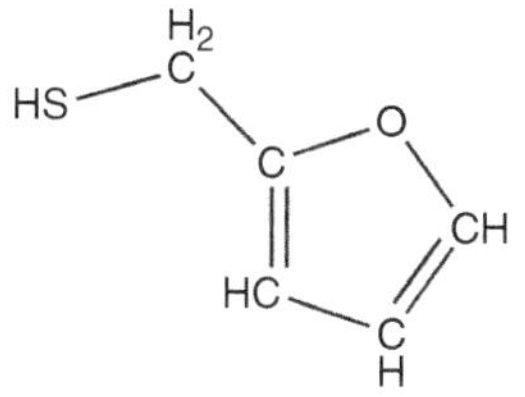 2-furanylmethanethiol	This is one of the browning compounds responsible for the aroma of freshly roasted coffee.
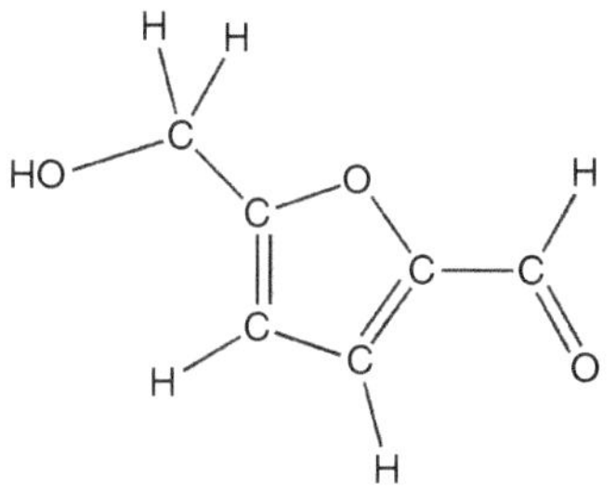 hydroxymethylfurfural (HMF)	Hydroxymethylfurfural (HMF) is an intermediate of the Maillard reaction and is found in corn syrup.
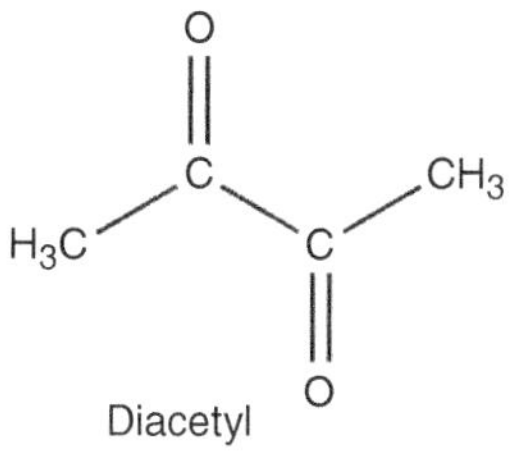 Diacetyl	Diacetyl is a compound produced from fats in butter and gives a strong buttery and butterscotch flavor. This compound is also a common product of browning through the fission Maillard reaction.
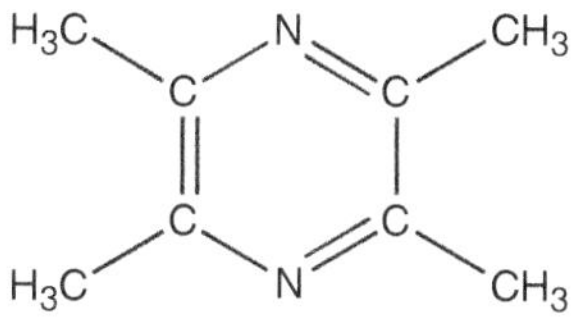 Tetramethylpyrazine	Methyl furans (some containing sulfurs, not shown here) are responsible for meaty flavors and are often the result of heating onion and garlic-like flavors. This is different from the caramelization flavor and aroma also found with onions.

Box 3.1 Figuring out food: Maillard and cancer

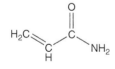

Figure 3.14   Structure of acrylamide.

Excuse me, but there is a strange chemical in my French fries! Acrylamide (Figure 3.14) is a compound that, when polymerized, can be used by chemists and biologists to analyze protein size and make plastics.

In 2002, much to the surprise of those who use acrylamide, scientists researching the Maillard reaction found small amounts of acrylamide in foods cooked at high temperatures. The plastic acrylamide was recently discovered in French fries and potato chips. Acrylamide is not found in raw potatoes. While not a "natural" compound, acrylamide is the product of the Maillard reaction. The reaction is between asparagine—one of the 20 amino acids found in nearly every food—and several common reducing sugars. In general, acrylamide is used

(Continued)

Box 3.1 (Continued)

in water treatment and the production of paper and clothing. In food, acrylamide is produced when foods are heated above 120°C/284°F by frying, broiling, or baking. As a monomer, at high concentrations, acrylamide is a known neurotoxin. However, after reacting as a polymer, it is nontoxic. Studies in rodents using concentrations of monomer acrylamide that were a thousandfold higher than those found in food demonstrated no increase in cancer risk; however, it is unclear how well these studies translate into humans. Classified as very toxic, the probable oral lethal human dose is between 50 mg/kg and 500 mg/kg or between 1 teaspoon and 1 ounce for a 150 lb person. One study of French fries from over 35 restaurants found acrylamide averaged about 300 µg of acrylamide per 1 kg of French fries [4]. This is well below the European Commission of 750 µg/kg (the FDA has not established limits at this time). It turns out, lots of foods and drinks contain acrylamide. Other common sources include fried potato chips, crackers, bread, several types of cookies, oats and mixed cereals, toasted dry cereals, canned black olives, prune juice, and coffee. Tobacco smoke contains a significant amount of acrylamide.

The impact of acrylamide on cancer risk is still under investigation, but there is some hint of an increased risk for some types of cancer. The full impact of acrylamide on the human diet is still unknown. While one could imagine avoiding acrylamide intake, the initial study found the compound in 40% of the American diet. Cooking methods now used by industry and the restaurant industry are working on ways to reduce the heat that generates the Maillard-driven production of acrylamide.

---

taste to a combination of Maillard reactions and caramelization. Many caramelization products react with proteins in a more complicated version of the Maillard reaction. Caramels made solely from sugar are used in puddings or desserts, including nougats, caramel brittles, and custards (Figure 3.15). In contrast, caramelizing sugars in protein-containing foods contribute to the flavors one associates with browned onions, carrots, and coffee/cocoa beans. Browning describes both flavoring processes, but a careful cook needs to know the difference. Once both processes are understood, there is a world of opportunity to create interesting flavors and foods.

Caramelization degrades mono- and disaccharide sugars to form large polymers and volatile flavor molecules. These reactions happen when

Figure 3.15 Making caramel. Manuel Capellari/Wirestock/Adobe Stock Photos

sugar is heated at or above its melting point. Like the Maillard reaction, caramelization requires heat to drive the complex, multistep reaction. The first part of the reaction is to melt the solid sugar crystals into liquid form. Then the sugars are further heated, where the sugars begin to lose OH groups and C atoms as lower molecular weight and often volatile compounds. Further heating will induce the collisions necessary to join many of the degraded sugars into large, viscous, dark-colored polymers. It is possible to overheat the sugar; extended heating will create a burned carbon (charcoal-like) residue. Complete oxidation (burning) of sugars will convert all the carbon, hydrogen, and oxygen atoms to CO_2 and H_2O, leaving nothing behind.

For solid, crystalline sugars—in contrast to liquid sugars like honey or syrups—there are two common techniques to make caramel candies: wet and dry. The dry method is simple and begins with slow heating of solid sugar, such as sucrose (i.e. table sugar), until the solid is turned into a liquid. As a solid is heated, the non-covalent intermolecular interactions holding the sugar molecules in the rigid crystal lattice are defeated. Sugar then melts and can move more freely as a liquid. This melting occurs as the sugar molecules acquire enough energy to overcome the intermolecular forces binding them into an orderly crystalline lattice. Melting does not occur instantaneously because molecules bound to each other in a lattice must absorb the energy and physically break the intermolecular forces. Typically, the outside of a crystal will melt faster than the inside of the sugar crystal. Heating the solid sugar takes time for the heat to penetrate the center of the sugar crystal.

For many years, the conventional thinking was that adding heat first melted sucrose into a liquid, and then the liquefied sugar broke down via chemical changes to a caramel product. However, Dr. Shelly Schmidt, a food chemist at the University of Illinois, published interesting research findings in several papers [5–8]. This work demonstrates that when heated slowly, some sugar molecules begin to chemically change into caramel products while still in a solid, crystal lattice form. This chemical change seems to occur before the molecules separate from solid into liquid form and not after melting, as is the case for most solids. Dr. Schmidt showed that at lower temperatures and with a slow rate of heating, the sugar molecules began to chemically change and liquefy at 145°C/290°F while still solid, instead of at the expected temperature (190°C/380°F) and with higher rates of heating used to make the liquid sugar. The caramelized sugars made at this lower temperature did not contain the undesirable compounds made by caramelization at higher temperatures. Some caramelization products are flavorful and have a pleasing odor, while others impart a bitter and burnt taste to food. Anyone who has caramelized sucrose too quickly at high temperatures understands the poor tasting, burned result. While still being investigated, this is a shift in the traditional understanding of how solids transition from a solid to a melted form. How this all happens (what scientists call "mechanism") has yet to be determined. Dr. Schmidt's work impacts how food is processed to control the kinds of caramelization products formed; applying this new understanding of how to make one set of compounds over the other by carefully applying heat and time yields more predictable results (Box 3.3).

The first step of caramelization chemistry requires reducing sugars. The reaction proceeds as reducing sugars lose a water molecule (dehydration) and rearrange into new compounds. This reaction requires the presence of the carbonyl (C=O) of a reducing sugar. Sucrose, as you may recall, is made of glucose and fructose and is not a reducing sugar. Therefore, sucrose must first be converted to its individual components: glucose and fructose. The breakdown of sucrose can happen in a few ways. In solution, sucrose will very slowly break down into glucose and fructose monosaccharides. This process is increased by heat (thus the melting of sucrose or dissolving of sucrose before heating) and accelerated by acids, including tartaric and lemon juice. Two enzymes—invertase from plant cells and sucrase from animal cells—can also break the glycosidic bond of sucrose into glucose and fructose. Converting sucrose to its fundamental units is called inversion and will produce what is commonly called "inverted sugar." Examining the temperatures required to start the caramelization reaction reflects the need to "invert" sucrose. Table 3.3 also explains why foods cooked and baked with fructose (including honey) brown faster and darker than those with sucrose or glucose.

Table 3.3 Temperature required for caramelization of sugars.

Sugar	Reaction temperature
Fructose	110°C / 230°F
Glucose	160°C / 320°F
Sucrose	160–180°C / 320–356°F
Maltose	180°C / 356°F

Box 3.2 Controversies in science

Dr. Schmidt's work challenged the accepted ideas regarding the melting and decomposition of solids, and there is disagreement within the food science community over this work, highlighting the nature of science and critical peer review. In a commentary published in a scientific journal, several experienced food science experts published a series of critical questions and comments about the published work of Dr. Schmidt [9]. The reasoning for the work of Dr. Schmidt and the arguments by scientists led by Dr. Roos is that the published melting points for sucrose vary from source to source. Dr. Schmidt begins her work by explaining that these variations occur because of the degradation/decomposition of the sucrose molecules before and during melting. The group opposing these conclusions believes the differences in melting point are due to impurities, differences in crystal sizes, and water mixed at the surface of the crystals. At the heart of the dispute is that many of the conclusions made by the Schmidt group are incorrect, and it is further argued that such incorrect conclusions about melting versus decomposition devalue food science and mislead the popular media. The latter has taken place, as several food-related blogs discuss how sugars don't melt as they heat, but instead, they caramelize while or instead of melting.

The commentary by Dr. Roos is filled with specific and very detailed scientific reasoning to question the decomposition phenomena. In the same journal, published alongside the initial questions, Dr. Schmidt rebutted each point to clarify her explanation of terminology and added new information to support her conclusions that "the loss of crystalline structure in sucrose is caused by the onset of thermal decomposition, not thermodynamic melting" [10]. So a nonscientist is left asking, "who is right"? The more appropriate question is, "what is actually happening?" Scientists are more focused on interpreting and discussing experimental results to understand the nature of the world around us. This is a good example of how different scientists or groups of researchers present and peer-review publish opposing viewpoints and conclusions to complicated experiments. More work may be needed to understand the implications of both sets of scientists.

Like the Maillard reaction, the diversity of products of the caramelization reaction is immense and depends on the sugars involved in the reaction and the chemical environment of the reaction. As we saw in Chapter 2, individual sugar molecules can react with the reducing C=O of another sugar to form a sugar anhydride after a water loss. The reaction is a condensation reaction, also called a dehydration reaction (Figure 3.16), where two molecules are joined in a covalent bond by the loss of a water molecule. The result is a larger molecule. Depending on the time, temperature, and process, this reaction will continue with some sections of the molecule breaking from the parent molecule (fragmentation), forming small aromatic molecules. For example, fructose will begin this reaction by forming difructose anhydride (Figure 3.17). As the sugar polymer continues to grow by dehydration condensation, carbons, oxygen, and hydrogen molecules can be transformed into different structures. Heating the sugars in acidic or basic conditions leads to different products and colors where the protons and hydroxyl ions serve as catalysts for the chemical rearrangement of the intermediate caramelization compounds.

Several organizations, including the US Food and Drug Administration, classify the colors and flavors of the process based on the chemical solutions used during caramelization. When a sugar solution is heated under different conditions, the resulting caramel colors are grouped into four classes as shown in Table 3.4 for different products and uses.

The caramelization reaction will continue to form both small volatile compounds and very large polymers of the sugar monomers. Initial heating will decompose and dehydrate the sugar monomers, and a population of the

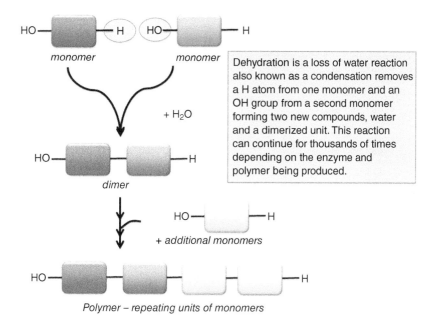

Figure 3.16 Polymerization reaction.

compounds will also give rise to smaller volatile molecules. Fructose, and to a lesser extent, glucose can decompose to the nutty, buttery-smelling 2-hydroxymethylfurfural (HMF). Diacetyl is another small volatile decomposition product responsible for the buttery flavor and smell of caramel. Heating fructose, glucose, or sucrose for 1–2 hours causes a loss of one water molecule per carbohydrate monomer. Continued heating of the sugar leads to additional loss of water molecules and condensation of the dehydrated sugar monomers into larger polymers. The building blocks of these polymers arise from the modification of fructose to difructose dianhydrides (DFA) (Figure 3.17).

Continued heating of sugars will create a complex mixture of small volatile molecules and larger nonvolatile polymers. Using a sensitive analytical technique called mass spectrometry, over 300 distinct major compounds and up to several thousand minor compounds have been identified. Often polymers of up to six fused glucose or fructose molecules are chemically altered to produce the large, nonvolatile, colored caramel products. While the complete chemistry of the reactions is not well understood, caramelization polymers are classified into three complex mixtures first described based on their size, color, and flavor characteristics (Figure 3.18):

1. *Caramelans* are a class of caramelization products with the formula $C_{12}H_{18}O_9$ formed by the loss of 12 water molecules. Caramelans have a bitter taste and a nutty-brown color.
2. Darker brown compounds, the caramelens ($C_{36}H_{50}O_{25}$), are produced after an additional hour of cooking where each sugar loses about eight water molecules and condenses with other reactants.
3. Longer heating of sugar will result in a darker and deeply flavored compound made of larger particles called *caramelin* ($C_{125}H_{188}O_{80}$). This molecule is very dark and does not dissolve well in water.

Caramel is one of the oldest and most widely used flavor compounds, with over 50 metric tons of caramel produced and consumed in foods yearly. As indicated in Table 3.4, the products of caramelization are used frequently in food and drink for many different properties, including taste. For example, colas owe their color

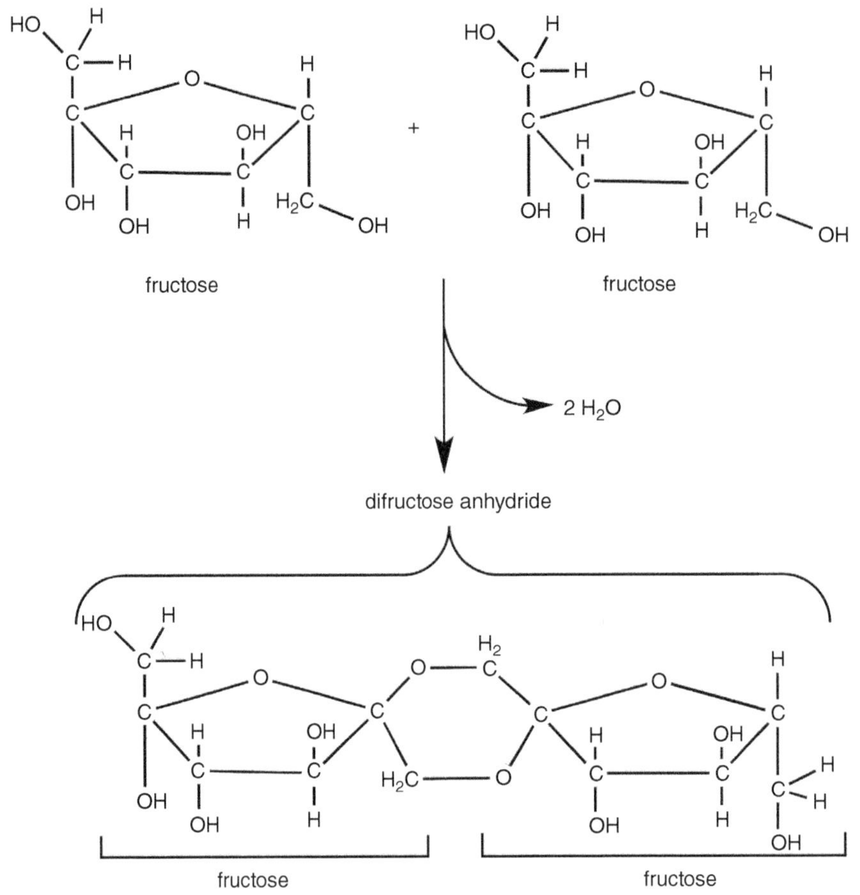

Figure 3.17 Dehydration reaction forming difructose anhydride.

and some of their flavors to caramelization products. Some caramelization products can act as emulsifiers and help keep molecules in solution. Several caramels found in coffee and honey have antibacterial properties. One compound produced by both caramelization and via the Maillard reaction, 4-methylimidazole, or 4-MEI,

Table 3.4 Classification of caramel used in food and beverage.

Class	Classification	Preparation	Uses
I	Plain or spirit caramel	No ammonium or sulfur compounds can be used	Distilled high-alcohol spirits such as whisky
II	Caustic sulfite caramel	High pH (NaOH) and sulfite (SO_3^{2-}) used	Beer, malt bread, sherry, and malt vinegars
III	Ammonia caramel	No sulfites but ammonium compounds can be used	Beer, sugar candies, and soy sauce
IV	Sulfite ammonia caramel	Both sulfite and ammonium can be used	Widely used for soft drinks and in acidic solutions

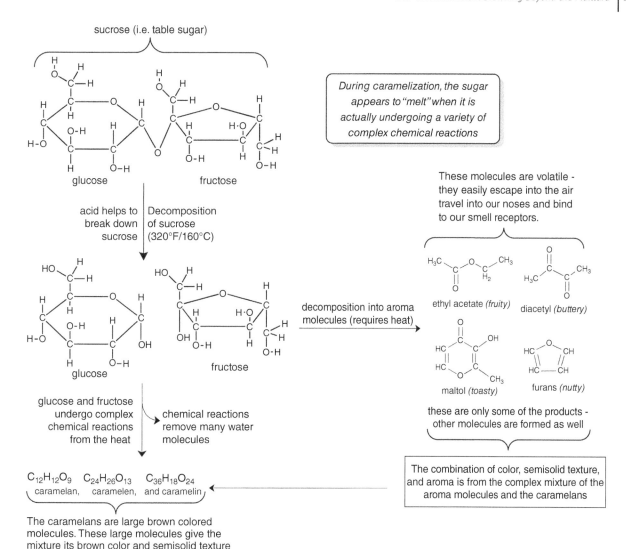

Figure 3.18 The many reactions of caramelization.

is found in dark beers, coffee, and roasted meats and is part of the caramel coloring and flavor in some cola soft drinks. Even though caramelization is a natural reaction, some people are concerned about the potential danger of ingesting dark-flavored molecules. When mice were exposed to very high concentrations of 4-MEI—levels that far exceed the average consumption of a human—a moderate but statistically significant increase in tumors was documented in the animals. The modest increase in tumor risk has led to concern about drinking colas, coffee, or even dark beer. While California now includes 4-MEI on its list of probable carcinogens, one would have to drink tens of thousands of cans of soft drinks a day to approach the concern for the compound.

Complicating our understanding of the process and products of caramelization is that these reactions occur alongside the Maillard reaction. Many of the intermediates from both browning reactions can react to form new aromas, flavors, and colors. However, the distinction between the two processes remains simple. Maillard reactions occur at lower temperatures than caramelization, and the reaction is between a reducing sugar and the amino group of an amino acid or protein. Caramelization proceeds at higher temperatures than the Maillard reaction and is the degradation and combination of two or more sugar monomers.

A review of grandmother's caramel recipe shows how much kitchen chemistry she was conducting when making the soft, chewy candies. Her basic recipe included sugar, milk or whipping cream, and butter. Her selection of sugar (corn syrup, honey, table sugar, or fructose) would provide different starting monosaccharides producing different possible flavors. Sometimes she would add vanilla extract for an additional flavor, but now we know she didn't need to, as both browning reactions create plenty of flavors. At times, she might mix all the ingredients before heating, producing a mixture of the two browning reactions. At other times her instructions would direct her to heat the sugar before adding the milk or cream, allowing the caramelization reaction to produce its compounds without competing with Maillard browning. Both methods will produce a fine candy, but each flavor profile is very different, as we now understand. We now know why the directions also included heating the mixture to either 120°C/250°F for lighter-colored candy or 180°C/350°F for a darker, more richly flavored candy. The addition of milk, evaporated milk, or cream provided fats for texture and proteins for Maillard browning. The combination of caramelization and the Maillard reaction produces flavors, including diacetyl, the odor, and flavoring of butter and butterscotch. Whew, what a scientist she was!

Finally, a discussion of caramelization would not be complete if one only considered table sugar! The cooking of many different foods produces caramelization reactions. For example, green coffee beans are roasted, providing caramel color and flavors from the sugars in the coffee bean or seed. Roasting coffee beans first requires drying the beans at 190–220°C/375–425°F. Steam from this heating swells the seed and produces the "first crack" as the bean splits its shell. Roasting for more extended and higher temperatures will produce a darker bean with solid flavors. Of course, the flavors are a combination of browning reactions. Seeds contain many complex and simple sugars, proteins, and amino acids to react and form colorful flavors and odors. Braising tough cuts of meat is a way to slow-cook beef, poultry, or pork for a more tender result. Braised meat is typically browned on the stovetop by Maillard reactions before placing it into a tight-fitting pot or kettle with a small amount of liquid. The addition of sugars (corn syrup or brown sugar) and vegetables high in sugar content, such as carrots and onions, are often used to add flavor as the meat is simmered at or near the temperature of caramelization.

Carrots, corn, potatoes, and onions have high sugar content and are great vegetables for caramelization. Adding a small amount of fructose to low-sugar vegetables like broccoli and cauliflower allows them to brown during roasting. Baking potatoes doesn't get hot or dry enough to induce caramelization at 150°C/300°F. However, slices of potatoes brown nicely if warmed enough to release the steam (unpacking the complex starches) and then heated sufficiently to get the caramelization rolling at a higher temperature. Carrots and onions are exceptionally high in the reducing sugars fructose and glucose, which promote browning reactions. Remember to get the reaction right; it takes time and high temperatures (up to 180°C/356°F) for the reactions to occur. Directions for onion and carrot caramelization stress the importance of heating the vegetable for nearly an hour. Of course, an impatient cook might use too much heat and burn the food instead of caramelizing it. Again, this is where you can use a little science of cooking to ease the process. Remember that caramelization occurs between reducing sugars that can form open or straight chains. The equilibrium between closed and open chains is shifted to the open form in basic or alkaline conditions. A sprinkle of baking soda helps to raise the pH of onions and carrots to the open chain, where the higher concentration of the open chain reducing sugar can dehydrate and polymerize into caramel goodness. Of course, the Maillard reaction is also taking place at a higher rate at alkaline pH, and more flavor is always better, isn't it?

Box 3.3 Figuring out food: Is grilling food bad for you?

Understanding the question about what might cause tumors to form should begin with a definition of cancer. Cancer is a disease where normal cells lose the regulation of their normal growth. When damage to a cell's DNA changes the coding of select proteins that control cell growth, a cancerous tumor cell is made. Tumor cells grow unchecked, escape their normal confines, and spread throughout the body, where new tumors take over. Compounds that produce changes in our DNA come from several sources. Some are due to natural processes and others via our environment and diet. The question is, do some of the compounds produced during grilling get into our cells and alter the DNA?

The accurate measure of if something is cancerous is how much a compound increases the risk of a cell "going bad." The main risk of grilling is from the Maillard reaction products further reacting with each other and oxygen in high heat found in the flame of a hot grill (Figure 3.19). Two key cancer-promoting compounds are heterocyclic aromatic amines (HCAs) and polycyclic aromatic hydrocarbons (PAHs) [11]. Both compounds are formed when the Maillard products from meat interact with the smoke and oxygen in the flames of a high-heat grill. Several forms of HCA and PAH compounds can enter the cell and alter the DNA leading to an increased risk of cancer. The amount of HCA and PAH found in grilled food depends on the level of doneness. Blackened or well-done meats will have more compounds than medium or rare grilled meat [12].

Cancer is found at higher levels in cells or rodents fed with extremely high amounts of PAH and HCA than in control cells or animals (Figure 3.20). The levels of these compounds used in these studies are hundreds to thousands of times greater than those found in any grilled food. An actual link between the compounds found in human cancer is unclear. Some studies show a mild risk, and others show no impact on grilled meats and cancer. The problem is that PAH and HCA compounds are found everywhere in our environment, including soil and water. These compounds come from smoke, industrial processes, forest fires, and volcanoes. The International Agency for Research on Cancer (IARC) determined consumption of red meat to be "probably carcinogenic to humans," based largely on data from epidemiologic studies and the evidence from published studies [13]. Because there is no proven information about the safest method of cooking meat, like most things, moderation might be the best approach.

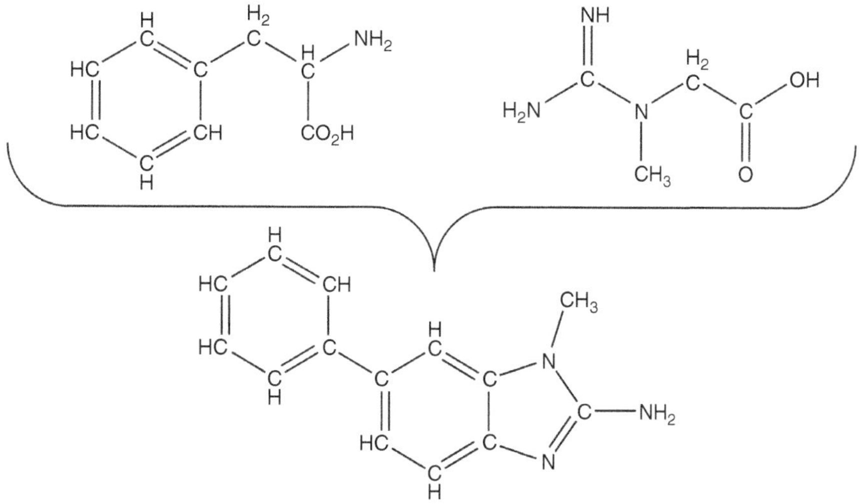

Figure 3.19 Generation of HCA. 2-amino-1-methyl-6-phenylimidazo[4,5-b]pyridine (PhIP) one of the most common HCAs found in grilled meat created from two compounds found in muscle (phenylalanine and creatine).

(Continued)

Box 3.3 (Continued)

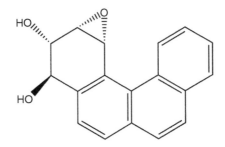

benzo[c]phenanthrene

a polyaromatic hydrocarbon

(−)-1R,2S-epoxy-3S,4R-dihydroxy-1,2,3,4-
tetrahydrobenzo[c]phenanthrenes (B[c]Ph DE)

the epoxide metabolite of benzo[c]phenanthrene

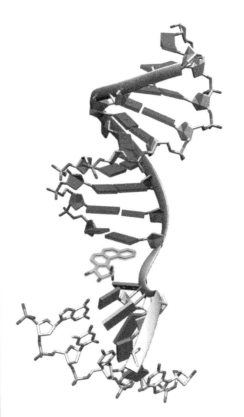

DNA

adduct of B[c]ph DE with guanosine

Figure 3.20 Polycyclic aromatic hydrocarbons. Benzo[c]phenanthrene is a common polycyclic aromatic hydrocarbon and its epoxide metabolite can react to form an adduct with a guanine DNA base. This combination of the adduct and DNA leads to genetic mutations that can lead to tumor formation. The intercalation of the adduct (green) in the center of a DNA helix (PDB: 2I9G). Notice the disruption to the helix shape Adapted from [14].

3.7 Ascorbic Acid Browning

Fresh-squeezed citrus fruit juices will brown under 3 days if left alone. Sugars and amino acids are involved in this browning reaction that also involves vitamin C (ascorbic acid). Ascorbic acid is a weak water-soluble acid, also known as vitamin C, naturally found in citrus fruits and many vegetables. Citric fruit juice is especially rich in ascorbic acid, and the breakdown of ascorbic acid generates brown pigments and off-flavors over time. This browning, like caramelization and the Maillard reaction, is a nonenzymatic browning process.

Now that we've studied the complex chemistry of amino acid and sugar browning reactions, it will surprise no one to learn there are also many possible products for the breakdown of ascorbic acid, and they depend upon the conditions of the juice during degradation. Up to 17 different ascorbic acid breakdown products have been identified, but two of these products are present. There are two essential degradation products of ascorbic acid, furfural, and dehydroascorbic acid. In the presence of oxygen (aerobic conditions), ascorbic acid is converted to dehydroascorbic acid and further degraded to brown compounds. Dehydroascorbic acid is produced in ascorbic acid degradation and then reacts in the Maillard pathway taking part in the Strecker reaction and producing brown pigments. In acidic and low (anaerobic) conditions (that found in concentrated, canned, and fresh citrus juice), furfural is one of the main components of ascorbic acid breakdown. This compound is then further degraded to form brown pigments after reacting with amino acids in the fruit juice (Figure 3.21).

Ascorbic acid degrades to dehydroascorbic acid and furfural, which, through very different pathways, react with amino acids to create bitter off-flavor brown pigments, which decrease the quality of the juice. Several researchers in the food industry have found that treating the juice to remove amino can stop the final few steps of each browning pathway. While the browning reaction can occur with or without oxygen, it proceeds more slowly in the absence of oxygen. Reducing headspace (the gap above the juice in the container) or exchanging the air for nitrogen gas can limit the juice's exposure to oxygen. One of history's oldest food additives, sulfites, are used to compete with ascorbic degrading reactions, sparing the vitamin from breakdown. For example, sodium metabisulfite ($Na_2S_2O_4$) is used in wine, beer, and other foods and reduces browning severalfold even when juices are heated to over 100°C/212°F. In addition, the sulfite inhibits the oxidation of ascorbic acid to dehydroascorbic acid, which prevents later reactions leading to the production of brown pigments.

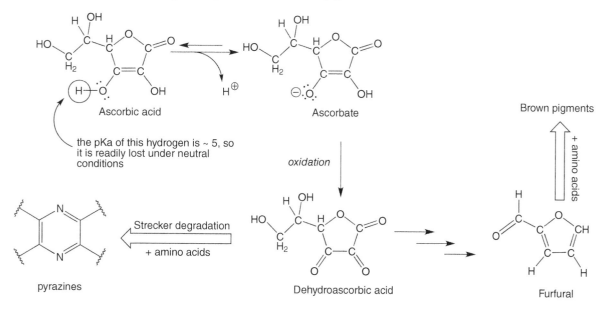

Figure 3.21 Ascorbic acid and browning reactions common to fruit juices.

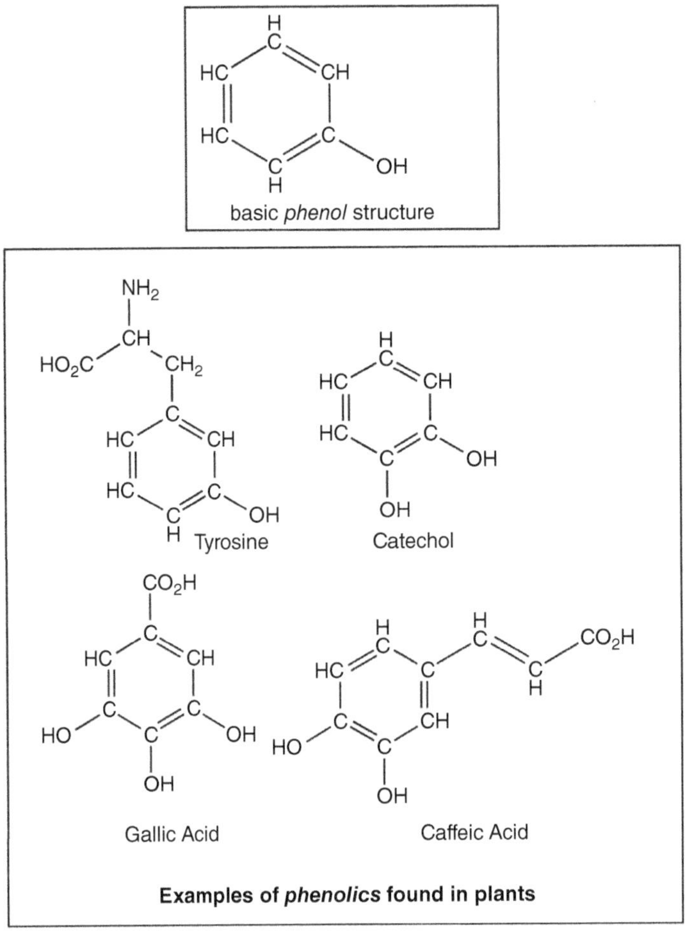

Figure 3.22 Phenols.

3.8 Enzyme-catalyzed Browning

Cut an apple or banana and leave it on the counter for an hour or so. The once enticing food is now brown and bitter tasting. Likewise, vegetables, fruits, and even some shellfish will all brown after being cut or damaged. Unlike caramelization or the Maillard reaction, this browning reaction is unpleasant and catalyzed by enzymes, not heat. Plant browning occurs when plant compounds called phenols react with oxygen to form large, polyphenolic brown pigments.

Phenols are rings of carbon atoms with alternating single and double bonds and with at least one hydroxyl group (—OH) coming off the ring. For example, the amino acid tyrosine, found in almost every protein in nearly every organism, is a phenol (Figure 3.22). Polyphenols are compounds with several phenol rings bonded together.

Phenols and polyphenols are found in nearly every living organism. They have interesting biological roles in various organisms, including mold, bacteria, algae, and higher animals. Plants have over 8,000 different phenolic compounds with just as many diverse roles, including the color and flavors of fruit, bark, and leaves. Tannins are a family of polyphenols found in plants; they provide color and function as pesticides for the plant. Some tannins are plant hormones regulating root development, fruit softening, and flowering; tannins also have an astringent taste. Tannins are responsible for the dry pucker that accompanies a sip of strong black tea or red wine.

The enzyme polyphenol oxidase (PPO) binds to the plant phenol and polyphenol compounds (Figure 3.22) found in the cell wall and oxygen to start the reactions that lead to bitter brown compounds responsible for fruit and vegetable browning (Figure 3.23). The primary function of enzymes is to decrease the activation energy needed for a reaction, which is why this browning reaction does not require heat for the reaction to occur. Another important characteristic of enzymes is that they are proteins that bind very specifically to their reactant (also called a substrate), producing a single product. Thus PPO will bind and react only with the plant phenols. Overall, the browning reaction in fruits and vegetables is complex, but like other browning reactions we've studied, the first step is key for producing the final compound (Figure 3.23).

A more accurate description of PPOs is that they are a family of enzymes found in all plants and animals. The oxidases catalyze the formation of quinones from mono- or diphenolic molecules and are responsible for the first reactions in the browning process. There are three types of PPO (tyrosinase, catechol oxidase, and laccase); each is a different protein made from different genes, but each protein binds and catalyzes the same primary reaction adding oxygen to a phenol. Tyrosinase catalyzes the addition of an —OH group onto monophenols and converts the product diphenol to quinones. Tyrosinase is also responsible for synthesizing the neurochemical dopamine and melanin, the pigment in the skin, hair, and eyes. Laccase also converts diphenols to quinones. Finally, catechol oxidase (aka diphenol oxidase) converts diphenols and, to a lesser extent, monophenols into a quinone compound. All three enzymes are called PPOs in the scientific literature and for the rest of this chapter. The bottom line is that these enzymes alter phenols to quinone compounds, which starts the browning reaction producing complex melanin products.

Plant cells are filled with various phenols, polyphenols, and two or more types of PPO enzymes (Figure 3.24). Yet unbruised or broken fruit and vegetables don't brown. This is because the PPO enzymes are inactive until vegetables or fruits are cut or damaged. Polyphenol oxidase is inactive because the plant cell has done a brilliant job of separating the phenols from the enzyme until it is beneficial to the plant to activate the enzyme. If the enzyme can't reach its substrate (phenols and polyphenols), no reaction (and no browning) can occur. Many of the phenols used in the browning reaction are sequestered in cellular compartments called vacuoles. Damage to

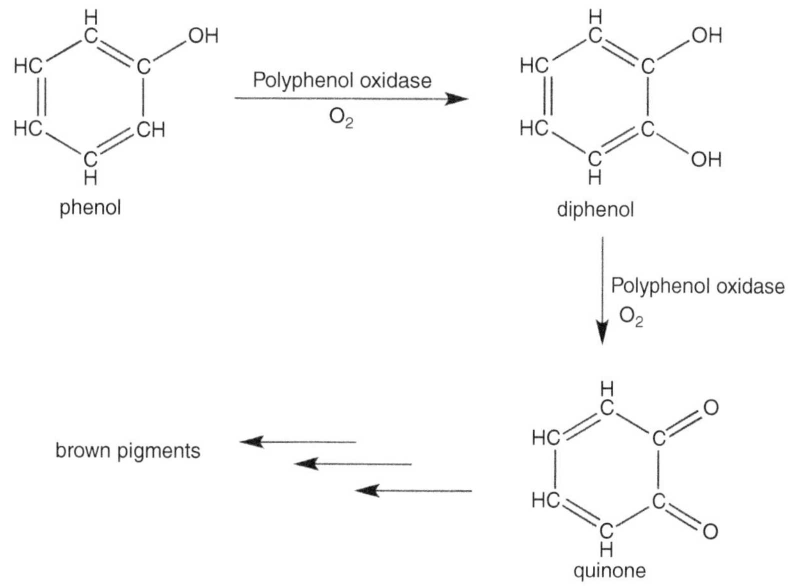

Figure 3.23 Enzyme-catalyzed browning. The actions of PPO on many of the phenols found in plants lead to the addition of oxygen atoms and a brown bitter-tasting pigment.

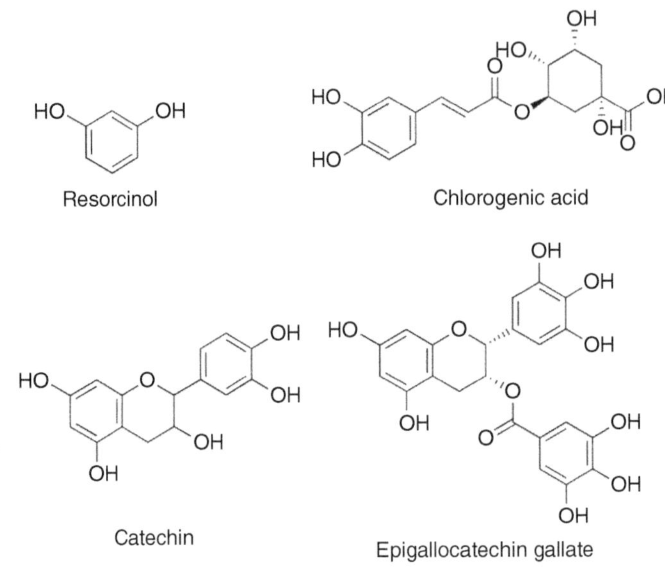

Resorcinol

Chlorogenic acid

Catechin

Epigallocatechin gallate

Figure 3.24 Examples of plant phenols and polyphenols.

the fruit or vegetable breaks the vacuoles releasing the phenols, where the enzyme and oxygen are now free to bind and react with the phenols starting the browning process (Figure 3.25).

The reaction of browning the phenols is quite complex. First, monophenols react with oxygen and PPOs to form diphenol compounds. The oxidase enzymes will then use a second molecule of oxygen to oxidize each alcohol of the diphenol to a carbonyl group (C=O) of the product quinone. Once formed, the quinones react with other

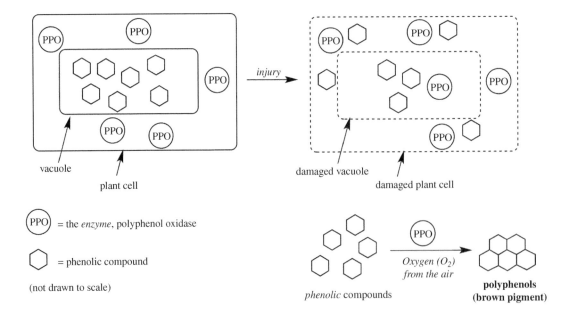

Figure 3.25 The browning of plants. Damage to plant cells (fruits or vegetables) causes the release of the enzyme PPO, which can then interact and react with its substrate, the phenolic compounds found on the plant cell wall.

compounds, including amino acids, to form large complex brown polymers. In animals, the brown melanin pigments found in skin and hair are formed by similar reactions that start with tyrosine.

The natural role of phenols and PPO enzymes depends on the plant. Plants use some polyphenols to generate plant hormones and others to strengthen the cell walls, while other polyphenol products create color and UV light protection. Tea leaves, figs, and cocoa plants all utilize PPO-catalyzed synthesis of polyphenols to create a darker color and new flavors. Several interesting scientific studies suggest that the brown compounds produced by PPO enzymes benefit the plant. When insects bite or other physical damage occurs to the plant, the resulting wound allows microbial pathogens to infect the plant tissue. In this case, polyphenols are a defense mechanism, a kind of plant immune response. In tomatoes, tobacco, and other plants, wounding activates the PPO enzymes. The resulting enzyme-catalyzed browning produces polyphenol compounds that inhibit the parasitic growth of mold, fungus, and other microbes while promoting wound closure. The bitter taste of the brown compounds is thought to discourage further attacks on the plant by insects and herbivores.

In addition to being less attractive to insects, brown food is bitter and unattractive to humans. As a result, nearly half of our vegetable and fruit crop is lost or wasted each year due to browning. Fool loss like this represents a significant waste of farming resources and agriculture. One interesting approach to avoiding such loss is the creation of genetically modified organisms (GMOs), in this case, a genetically modified plant that will not produce brown polyphenols. The Arctic apple, a GMO product, is a non-browning apple where eight genes of various PPO enzymes have been silenced. This approach used to reduce the production of PPO in the apple uses a technique called RNA interference. RNA interference or molecular therapies like CRISPR are being developed to treat cancer, tissue transplantation, liver disease, and other human diseases. In this case, silencing the gene for PPO means the damage that occurs between harvest to your household will be minimal, creating more food at a lower cost.

Unless you can use RNA interference to stop browning food in your kitchen, the average person will have to use an understanding of enzymatic browning to block the reaction at home. Recall that inhibition of PPO will block the formation of browning compounds. Polyphenols, oxygen, and enzyme: stop any of these three components from mixing, and you've just created a way to keep food from browning. Limiting access to oxygen is easy but only mildly effective. Immersing vegetables and fruits in water before and during cutting reduces the availability of oxygen to bind to the enzyme and will slow browning. Potatoes, which quickly brown, are quickly washed after cutting during the industrial production of French fries and potato chips. Some companies have experimented with packaging fruit in carbon dioxide gas to eliminate oxygen but with limited success. Several compounds bind and block the ability of PPO to react with polyphenols. As classic enzyme inhibitors, many of these compounds have nearly the same structure as the phenolic substrates of PPOs. These substrate-mimicking compounds compete with the natural enzyme substrate inhibiting the polyphenol reaction. Guaiacol, resorcinol, and phloroglucinol are all examples of such "competitive inhibitors."

Blanching vegetables and some fruits is an effective way to inhibit the enzyme and soften the cell wall. Blanching is the short submersion of a vegetable in boiling water for a short period. Some, but not all, PPO enzymes are denatured and rendered inactive after short periods of 100 °C/212°F heat of boiling water. However, longer exposures are needed to kill several forms of PPO and would result in cooking the vegetable (Box 3.5).

Some acids are also effective inhibitors of PPO. The pH optimum for PPO activity lies between 6.0 and 6.5. Almost no activity remains if the pH exceeds 1 unit above or below these points. Phosphate buffers are sometimes used as they can buffer the pH at either end of this range and are considered food-safe. Citric acid, malic acid, and ascorbic acids are effective inhibitors of the enzyme, less for their acidic effects but other properties. Both citric acid and malic acid can bind and strip copper away from PPO enzymes. Polyphenol oxidase requires copper to be enzymatically active. However, each acid will only partially inhibit the production of brown compounds in apples and other fruits or vegetables. While ascorbic acid is involved with browning in citric juices, the acid is an effective inhibitor of PPO function and does not impart taste to food like malic and citric (sour and tart) (Figure 3.26).

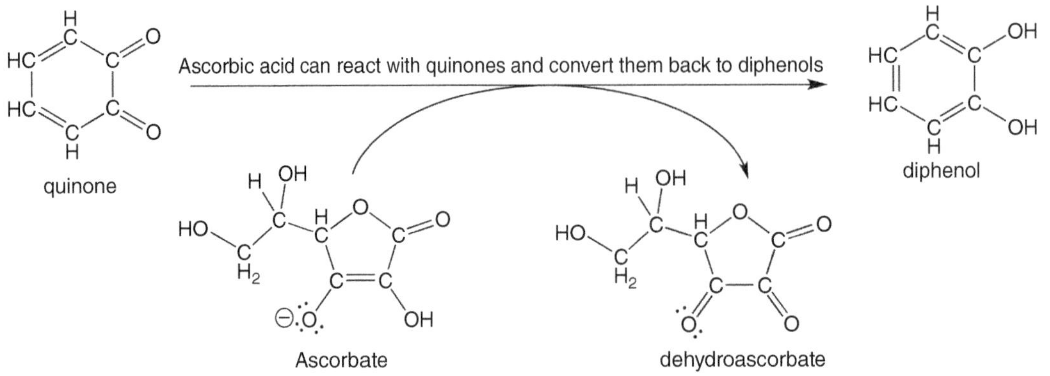

Figure 3.26 Ascorbic acid is an inhibitor of PPO. Ascorbic acid can reverse the production of the brown pigment forming quinones.

Ascorbic acid will react with quinone compounds, reversing the PPO reaction. As long as ascorbic acid is present in excess, the quinone products cannot continue the reaction to form brown pigments. However, browning can occur once the ascorbic acid is consumed in the reaction.

Box 3.4 Polyphenol oxidase inhibitors

Polyphenol oxidases include tyrosinase, catechol oxidase, lactase, and others. There are very specific inhibitors for each enzyme, many of which are used in food preparation. Tyrosinase and catechol oxidase are inhibited by tropolone, cinnamic acid, and salicylhydroxamic acid but not laccase. Laccase is inhibited by azide, cyanide, thiocyanate, fluoride, and sulfhydryl compounds. New compounds with promise include kojic acid (found in fermented Japanese foods) and hexylresorcinol (also used in throat lozenges).

Calcium chloride is another modest inhibitor of PPO activity. When used alone, any one of these PPO inhibitors is moderately effective, but combinations of inhibitors are even better at preventing browning. Several combinations of the acids mentioned previously and calcium ions have a strong effect on the PPO activity. NatureSeal is a company producing a range of products to inhibit browning in packaged salads, fruit, and other produce. NatureSeal is a mixture of calcium and ascorbic acid. Used as a dip, the combination can inhibit browning in apples for up to 14 days. Using this technology, one fast-food restaurant sold nearly US$0.5 billion in just sliced apples in 2010 (Box 3.5).

Box 3.5 Tyrosinase, animal noses, and turning colors

Have you ever wondered why some dogs' noses turn from a dark brown/black color to light tan or pink? The same basic phenomena take place with Arctic rabbits and Siamese cats. In each case, the PPO enzyme tyrosinase is involved. The enzyme variant in these animals is active in warmer temperatures. The mutation (two glycines are replaced by arginine or tryptophan amino acids) renders the enzyme inactive at lower temperatures. This "temperature-sensitive" PPO found in these animals serves as a camouflage of sorts. Rabbits and Burmese and Siamese cats both lose the pigment production of tyrosinase in cold temperatures and become white, blending in with the snowy background. In dogs, this effect is called "winter nose" and is more common in larger dogs such as Labrador Retrievers and Huskies. The biological significance of this phenomenon in dogs is unclear but troublesome for those living in northern climes. A pink or "Dudley Nose" in a dog is an unwanted characteristic that can disqualify a dog from a show (Figure 3.27).

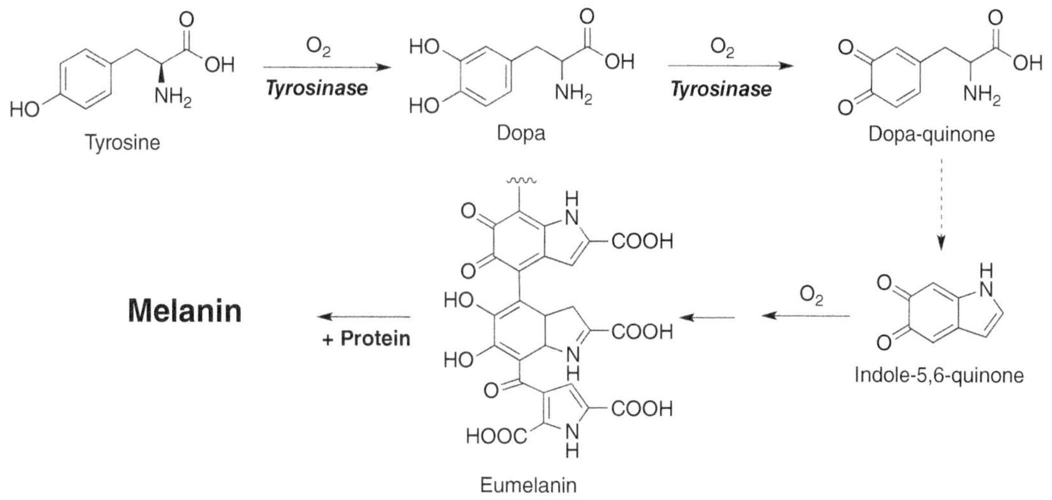

Figure 3.27 Tyrosinase starts the conversion of tyrosine to the pigment melanin. One of the metabolic fates of the amino acid tyrosine is to be converted to the dark pigment melanin. The activity of tyrosinase can be influenced by temperature.

3.9 Overview

Overall, browning reactions and food and cooking are complex, and a good scientist-cook who can understand the concepts and requirements for each reaction can use this skill to be a much better creator in the kitchen (Figure 3.28). Keep in mind that two reactions, the Maillard and caramelization, are not catalyzed by an enzyme and need heat for the reaction to proceed. Maillard is the reaction between the oxygen on a simple mono- or disaccharide and the nitrogen of amino acids and proteins. Caramelization is a polymerization reaction that results in small fragments and long, complex brown compounds. Some of the products of caramelization are light brown and bitter, while others are dark brown/black and have deep rich flavors. Ascorbic acid browning is a combination of the Maillard reaction with vitamin C and is found mostly in fruit juices rich with vitamin C. Enzyme-catalyzed reactions primarily involve the enzyme PPO. In plants, enzymes catalyze the reactions between oxygen and phenols found on the cell walls. The result of this reaction is a bitter brown-tasting compound that serves as a way for plants to fight off fungal and microbial infection. By applying this newfound knowledge of browning reactions to your kitchen creations, you can carefully promote the flavorful browning reactions (Maillard and caramelization) while discouraging the unwanted, bitter products of ascorbic acid and fruit browning reactions.

Browning Reactions in Food and Drink

Maillard Reaction

Reaction: between O of reducing sugars and N of amino acids & proteins

Reaction Conditions: Minimal heat ~ 140–160°C/284–320°F

Reaction Results: 100s of small brown and flavorful aromatic compounds – some long polymers

Caramelization

Reaction: between mono- and di-saccharide sugars

Reaction Conditions: Minimal heat ~ 160–180°C/320–356°F

Reaction Results: 100s of compounds-carmelans (small and bitter dark brown), caramelens (mid sized – mild sweet), and caramelin (large polymer dark, deep flavor and poorly water soluble)

Polyphenol Oxidase

Reaction: between oxygen and phenols/polyphenols from the cell wall

Reaction Conditions: Catalyzed by enzyme: polyphenol oxidase. Reaction is slow below 4°C/40°F

Reaction Results: bitter brown pigments often reacting with other compounds and can act as antimicrobial agents

Figure 3.28   Simple overview of browning reactions.

Science for the Chef: Understanding the science behind bulgogi

One interesting recipe that highlights the Maillard reaction in a non-Western cuisine is Korean bulgogi. Bulgogi is a traditional Korean dish made with thinly sliced beef marinated in a mixture of soy sauce, sugar, garlic, and other seasonings. When the marinated beef is cooked on a hot grill or pan, the Maillard reaction occurs, leading to the caramelization and browning of the meat, resulting in a delicious umami flavor and enticing aroma. The Maillard reaction plays a crucial role in enhancing the taste and overall appeal of this popular Korean dish.

Ingredients:
- 1 pound thinly sliced beef (such as ribeye or sirloin)
- 1/4 cup soy sauce
- 2 tablespoons brown sugar
- 2 tablespoons sesame oil
- 3 cloves garlic, minced
- 1 tablespoon grated ginger
- 2 tablespoons toasted sesame seeds
- 2 green onions, thinly sliced
- Optional: 1 tablespoon pear or apple puree (acts as a natural meat tenderizer)

Instructions:

1. In a bowl, combine soy sauce, brown sugar, sesame oil, minced garlic, grated ginger, and optional pear or apple puree. Mix well to dissolve the sugar.
2. Add the thinly sliced beef to the marinade and toss to coat the meat thoroughly. Allow the beef to marinate for at least 30 minutes, or refrigerate it for up to 24 hours for a more intense flavor.
3. Preheat a grill or a pan over medium-high heat.
4. Cook the marinated beef in batches on the hot grill or pan for 2–3 minutes on each side until it is well browned and caramelized. The high heat triggers the Maillard reaction, resulting in the development of rich flavors and a nicely browned crust on the meat.
5. Once cooked, remove the beef from the heat and sprinkle toasted sesame seeds and sliced green onions on top for added flavor and garnish.
6. Serve the bulgogi with steamed rice and side dishes like kimchi, pickled vegetables, and lettuce wraps for a traditional Korean dining experience.

Science Behind the Recipe

The Maillard reaction is a chemical reaction that occurs between amino acids (from proteins) and reducing sugars when exposed to heat. In the case of bulgogi, the soy sauce, which contains amino acids, reacts with the natural sugars present in the marinade (such as brown sugar or fruit puree). When the marinated beef is cooked at high temperatures, the Maillard reaction takes place, resulting in the browning and caramelization of the meat's surface.

During the Maillard reaction, a complex series of chemical reactions occur, generating new flavor compounds, aromas, and a desirable brown color. This reaction is responsible for the characteristic umami flavor and appealing aroma of the cooked bulgogi. The Maillard reaction not only enhances the taste of the beef but also contributes to the overall visual appeal of the dish.

Key Concepts

1. A **chemical reaction** is a process of rearranging (making or breaking) bonds to make a substance. As two molecules or **reactants** collide with enough force, the transformation into **products** takes place. The energy required to make the chemical change is the **activation energy**. This is why heat is required to make a reaction happen. **Enzymes** are **catalysts** that decrease the energy input are needed for a reaction to happen.

2. Enzymes are proteins that act to decrease the energy of activation for a chemical reaction. Enzymes bind specifically to their reactants (also called **substrates**), but a protein must be native to be active. Heat can unravel or **denature** enzymes, resulting in an inactive catalyst. Thus enzymes often work at much cooler temperatures than one observes in other **non-catalyzed** browning reactions.

3. The **Maillard reaction** is an example of a non-catalyzed reaction that takes place at moderately high temperatures. The reaction occurs between the **anomeric carbon** of a simple sugar (mono- or disaccharide) whose anomeric carbon can "open" to the reducing form and an amino group from amino acids or proteins.

4. Because many different sugars and amino groups in food are available for the Maillard reaction, hundreds of unique flavors and aromas are produced. The reaction begins with a **dehydration reaction**, followed by a **rearrangement** into an **amadori product**. This intermediate product can follow many chemical reaction **pathways** giving rise to the various products of the Maillard reaction.

5. Because water absorbs the energy of cooking as it evaporates, wet foods will brown poorly. The anomeric carbon of reactive sugars is in **equilibrium** with **open and closed forms**. A more alkaline (higher pH) condition shifts the equilibrium to the reactive open form of the sugar making the reaction with amino acids occur faster and at lower temperatures.

6. **Caramelization** is a reaction between sugars. The reaction starts with **dehydration** as two sugars form the beginning of a **polymer**. As the reaction continues, more and more sugars will react, making very large caramel molecules. Through the process of heating, some of the polymers will degrade into several minor compounds that are aromatic and add to the flavor and aroma of this browning product. Like the Maillard reaction, caramelization requires heat to input enough energy to overcome the activation barrier.

7. Caramelans are classified by size, from small to large: **caramelans**, **caramelens**, and **caramelins**. In addition, a different **classification** in their use in food and beverage is organized by how the caramel is prepared and used.

8. Enzyme-catalyzed reactions primarily refer to the reaction between phenols found in the plant cell wall and oxygen. The **enzyme polyphenol oxidase (PPO)** catalyzes the reaction. There are several forms of the enzyme found in both plants and animals. Plant phenols and polyphenols are stored away from the enzyme in compartments called **vacuoles**. Once the cell wall integrity is lost by damage or cutting, the enzyme, phenols, and oxygen mix, and the reaction begins.

The browning reaction catalyzed by PPO does not require heat because the enzyme reduces the activation energy needed to allow the reaction to take place. Different processes reduce the enzyme activity, reducing the product's unpleasant brown color and bitter taste. The simplest thing is washing away the enzyme after cutting the vegetable or fruit. Another is to store the food at low temperatures where the reaction is much slower, even with the enzyme present. Ascorbic acid inhibits the reaction and, along with other chemical inhibitors of PPO, is used to extend the shelf life of cut vegetables and fruit.

References

1 Hodge, J. (1953). Chemistry of browning reactions in model system. *J Agric.Food Chem*, *1*, 928–943.
2 Ashoor, S. H. & Zent, J. B. (1984). Maillard browning of common amino acids and sugars. *J Food Sci*, *49*, 1206–1207.
3 Somoza, V. & Fogliano, V. (2013). 100 years of the Maillard reaction: why our food turns brown. *J Agric Food Chem*, *61*, 10197–10197.
4 American Society of Agronomy (ASA), Crop Science Society of America (CSSA) (2015, December 2). Fries with a side of acrylamide. ScienceDaily. Retrieved July 19, 2024, from https://www.sciencedaily.com/releases/2015/12/151202160332.htm
5 Schmidt, S. J. (2012). Exploring the sucrose–water state diagram. *Manuf Confect*, *92*(1), 79–89.
6 Lee, J. W., Thomas, L. C., & Schmidt, S. J. (2011). Investigation of the heating rate dependency associated with the loss of crystalline structure in sucrose, glucose, and fructose using a thermal analysis approach (Part I). *J Agric Food Chem*, *59*, 684–701.
7 Lee, J. W., Thomas, L. C., Schmidt, S. J., Feng, H., Cadwallader, K. R., & Schmidt, S. J. (2011). Investigation of thermal decomposition as the kinetic process that causes the loss of crystalline structure in sucrose using a chemical analysis approach (Part II). *J Agric Food Chem*, *59*, 702–712.

8 Lee, J. W., Thomas, L. C., & Schmidt, S. J. (2011). Can the thermodynamic melting temperature of sucrose, glucose, and fructose be measured using rapid-scanning differential scanning calorimetry (DSC)? *J Agric Food Chem*, 59(7), 3306–3310.

9 Roos, Y. H., Franks, F., Karel, M., Labuza, T. P., Levine, H., Mathlouthi, M., Reid, D, Shalaev, E., & Slade, L. (2012). Comment on the melting and decomposition of sugars. *J Agric Food Chem*, 60(41), pp. 10359–10362.

10 Lee, J. W., Thomas, L. C., & Schmidt, S. J. (2012). Response to comment on the melting and decomposition of sugars. *J Agric Food Chem*, 60(41), 10363–10371.

11 Cross, A. J., & Sinha, R. (2004). Meat-related mutagens/carcinogens in the etiology of colorectal cancer. *Environ Mol Mutagen*, *44*(1), 44–55.

12 Jägerstad, M., & Skog, K. (2005). Genotoxicity of heat-processed foods. *Mutat Res*, *574*(1–2), 156–172.

13 International Agency for Research on Cancer [IARC] (2015). [Accessed August 22, 2022]; *Q&A on the carcinogenicity of the consumption of red meat and processed meat.*

14 Batra, V. K., Shock, D. D., Prasad, R., Beard, W. A., Hou, E. W., Pedersen, L. C., Sayer, J. M., Yagi, H., Kumar, S., Jerina, D. M., & Wilson, S. H. (2006). Structure of DNA polymerase β with a benzo [c] phenanthrene diol epoxide-adducted template exhibits mutagenic features. *Proc Natl Acad Sci U S A*, *103*(46), 17231–17236.

Additional Readings

Ajandouz, E. H., Chickpea, L. S., Dalle ORE, F., Benajiba A., & Puigserver A. (2002) Effects of pH on Caramelization and Maillard Reaction Kinetics in Fructose-Lysine Model Systems. *J Food Sci*, 66, 926–931.

Alandes, L., Quiles, A., Pérez-Munuera, I., & Hernando, I. (2009), Improving the Quality of Fresh-Cut Apples, Pears, and Melons Using Natural Additives. Journal of Food Science, 74(2), S90–S96.Beans Moon, J.K., & Shibamoto T. (2009) Role of Roasting Conditions in the Profile of Volatile Flavor Chemicals Formed from Coffee *J.Agric Food Chem*, 57, 5823–5831.Botram D. S., Wedzicha, B. L., & Dodson, A. T. (2002) Acrylamide is formed in the Maillard reaction. *Nature*. 419, 448–449.

Ertugrul, U., Tas, O., Namli, S., & Oztop, M. H. (2021), A preliminary investigation of caramelisation and isomerisation of allulose at medium temperatures and alkaline pHs: A comparison study with other monosaccharides. *Int J Food Sci Technol*, 56(10), 5334–5339.Qiao, L, Han, X, Wang, H, Gao, M, Tian, J, Lu, L, & Liu, X. (2021) Novel alternative for controlling enzymatic browning: Catalase and its application in fresh-cut potatoes. *J Food Sci.*, 86(8), 3529– 3539.

End of Chapter 3 Questions

1. What heat induced reactions involve amino acids and reducing sugars?
    a. Caramelization
    b. Fruit browning
    c. Maillard
    d. Ascorbic acid browning

2. The first product to form when sugars undergo dehydration in a caramelization reaction is?
    a. Quinone
    b. Amadori compound
    c. Diacetyl
    d. Difructose anhydride

3. When compared to dry meat, wet meat will brown _____ when placed in a hot frying pan.
   a. Faster
   b. Slower
   c. Never
   d. Always

4. Consider the steps of the Maillard reaction. Identify the correct order by placing a #1 by the first step, #2 by the second step, etc.
   a. ___ Rearrange to form the amadori compound
   b. ___ Dehydration reaction forming a C=N dehydration product
   c. ___ Rearrange to form a C−N dehydration product
   d. ___ Dehydration of the amadori compound to form furans and other aroma compounds

5. When you make caramel sauce, what is chemically responsible for the color and stickiness of caramel?

6. You come across a chocolate chip cookie recipe that uses "brown butter." The instructions tell you to melt butter in a saucepan until the butter turns golden brown. Once the color is achieved, then add it to your wet ingredients. What reaction occurs, and how will this batch of cookies taste differently from adding the "brown butter?"

7. An egg is often washed onto the surface of bread dough before it is placed in the oven. Explain why bakers do this.

8. Explain why a wash in baking soda would make a food (meat, fish, or vegetables) brown quicker than an untreated food.

9. Scientists describe fruit browning reactions as a means plants can engage in "chemical warfare" or "chemical defense." What does this mean?

10. You make a fresh batch of guacamole. Identify three things you can do to slow down or prevent your leftover guacamole from turning brown.

4

Milk and Ice Cream

Guided Inquiry Activities (Web): *7, Carbohydrates; 8, pH; 9, Higher Order Protein Structure; 10, Fat Intermolecular Forces, Solids, and Oils; 12, Emulsion and Emulsifiers; 16, Milk*

Learning Objectives

1. *Understand the role of carbohydrates, particularly lactose, in milk as a source of energy and its impact on the flavor of milk, as well as the enzymatic process of lactose digestion and nutritional benefits.*
2. *Comprehend the role of lactobacilli and lactococci bacteria in the fermentation of lactose to lactic acid and their impact on milk.*
3. *Recognize the nutritional significance of milk as a source of essential fats, proteins, and carbohydrates, and common human allergies to milk.*
4. *Understand the composition of milk proteins and differentiate between the two main classes of proteins found in milk: casein and whey.*
5. *Describe the role of amino acids in protein structure and recognize the essential amino acids that must be obtained through diet.*
6. *Explain the process of curd formation in milk and understand the difference between acid soluble and acid insoluble milk proteins.*
7. *Explain the scientific principles and physical changes that occur in the processes of homogenization and pasteurization.*
8. *Explain the formation of foam in milk and the role of proteins and fat. Understand the process of butter production, the role of different fatty acids, and the factors affecting its spreadability.*
9. *Explore the flavor and aroma compounds present in butter, their formation during production, and their impact on the taste of baked goods.*
10. *Explain the scientific principles behind the formation and composition of ice cream as a colloidal mixture, including the role of emulsifiers and air pockets in its structure.*
11. *Describe the impact of freezing point depression on the freezing process of ice cream and its role in controlling the formation of ice crystals, emphasizing the importance of proper mixing techniques and temperature control.*

The Science of Cooking: Understanding the Biology and Chemistry Behind Food and Cooking, Second Edition. Joseph J. Provost et al.
© 2025 John Wiley & Sons, Inc. Published 2025 by John Wiley & Sons, Inc.
Companion website: www.wiley.com/go/provost/food_science_2e

4.1 Introduction

In this chapter, we will use our understanding of science to investigate the molecular composition of milk, including lactose, fats, and proteins, and the impact of these macromolecules on taste, cooking, pasteurization, and homogenization. The ability of different people to drink milk will be investigated based on genetic variation and diversity for lactose intolerance. We will include the nature of protein and lipid cages of foams and whipping cream, acid-producing microbes, ice cream, and freezing point depression. Finally, the information in this chapter includes milk's physical properties and chemical changes that occur during butter and cream preparation.

4.1.1 The Fundamentals of Milk

Milk is a basic nutritional component for all mammals, providing nutrients and immune protection from the mother to the newborn for over 4,000 species. Animals such as goats, buffalo, sheep, and cows have been used to provide milk and dairy products for humans. The earliest evidence directs us to time periods as long as 8000–10,000 years ago, when sheep and goats were farmed for their milk in ancient Iran and Iraq. Unique cheeses, creams, and other products are still made from goat and buffalo milk, while cows produce drinking milk for many.

So just what exactly is milk? The official US Code of Federal Regulations (CFR) defines milk as "the lacteal secretion obtained from one or more healthy milk-producing animals, e.g. cows, goats, sheep, and water buffalo, including, but not limited to, the following: low-fat milk, skim milk, cream, half and half, dry milk, nonfat dry milk, dry cream, condensed or concentrated milk products, cultured or acidified milk or milk products" [1]. While technical, it is evident from this description that milk is a complex mixture of components. Simply put, milk is a liquid secreted by the mammary glands after the birth of her young. The milk produced for the first few days is called colostrum, a pale yellow to clear fluid with a high concentration of antibodies from the mother to protect the newborn against disease. After a few days, the mature white milk that we are more familiar with is produced and harvested by humans for drinking and cooking.

The essential components of milk are water, fat, milk proteins (primarily casein and whey), and sugar (lactose). The relative concentration of each component of milk can change depending on the animal (Table 4.1). The components of milk can be divided further into two phases: the slightly acidic, aqueous (i.e. water) phase and the oil or fat phase (Figure 4.1).

Table 4.1 Milk components across species.

Species	Water	Fat	Casein	Whey	Lactose
Human	87.1	4.6	0.4	0.7	6.8
Cow	87.3	4.4	2.8	0.6	4.6
Buffalo	82.2	7.8	3.2	0.6	4.9
Goat	86.7	4.5	2.6	0.6	4.4
Sheep	82.0	7.6	3.9	0.7	4.8
Horse	88.8	1.6	1.3	1.2	6.2
Rat	79.0	10.3	6.4	2.0	2.6
Ass	88.3	1.5	1.0	1.0	7.4
Reindeer	66.7	18.0	8.6	1.5	2.8
Camel	86.5	4.0	2.7	0.9	5.4

Adapted from http://evolution.berkeley.edu/evolibrary/news/070401_lactose; http://www.sciencedaily.com/releases/2005/06/050602012109.htm;

As you may recall, water and oil/fat do not mix—they are immiscible and nonhomogeneous. The water or aqueous phase is often called the serum and contains dissolved proteins and sugars. In fact, the bulk of the milk proteins are found in the aqueous phase of milk and are classified as casein or whey proteins. The fat phase is dispersed in many tiny fat droplets, also called globules, which are coated with proteins and two phospholipid membranes with a cytoplasmic layer between them (Figure 4.2). If left to stand, the fat phase of raw milk will eventually separate from the watery serum.

A glass of milk provides much-needed nutrients for humans; thus, milk use throughout the world remains high. Russia, Finland, and Sweden consume the most milk (130–180 liters per person per year), while the United States consumes 83 liters of milk per person per year. In milk, lactose (a disaccharide sugar) serves as an important source of energy for the human body. Individuals in the United States typically under-consume calcium, vitamin D, and potassium; milk provides an easy way to make up for deficiencies in these important nutrients. Milk also provides two of the "essential" fats that are required for healthy living but are not made by humans. Moreover, the whey proteins in milk are rich in branched-chain amino acids that are often sold as a dietary supplement for their potential role in supporting muscle recovery and preventing mental fatigue. Not insignificantly, protein from milk provides amino acids used as building blocks for

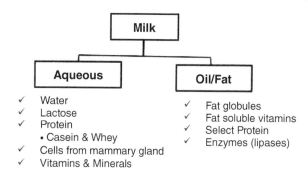

Figure 4.1 Composition of milk. Milk is made of two phases: aqueous (mostly water) and oil.

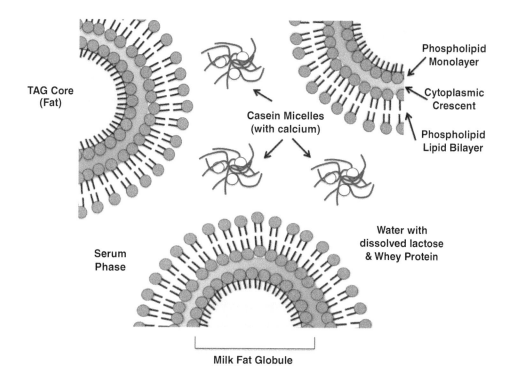

Figure 4.2 The structure of milk. Fat globules with three layers of lipid membranes, phospholipid monolayer, and bilayer, respectively, encase the triacylglycerol fat core. The serum (aqueous) phase with dissolved lactose and whey proteins contains the casein micelles coordinated or bound with calcium ions.

a diverse range of biological molecules, including fats, sugars, new proteins, and nucleic acids, the building blocks of DNA. With all of these benefits, one would think that milk should be the primary liquid in one's daily diet. However, there are drawbacks to the heavy consumption of milk. Whole milk is high in fat, and whole milk and dairy products have high levels of cholesterol. In addition, some of the proteins found in cow, goat, and sheep milk are suspected allergy-causing compounds for 2.5% of infants and some adults. Later in this chapter, you will learn about lactose intolerance and the difficulty that many people throughout the world have in digesting milk sugar.

Box 4.1 Figuring out food: Can you milk almonds or oats?

Almond milk, oat milk, soy milk, and rice milk. This is a short list of "alternative" milk sources whose demand is growing as a milk replacement [2]. What are these plant-based alternative choices for milk? Let's again review the composition and definition of milk. Milk is a mixture of fat and protein globules suspended in a watery protein solution. So essentially, milk is fat globules floating in the water. Almond, oats, rice, and soy are the watery extracts of the plant with added fats and oils called lecithins, which provide milk-like fat texture and stabilize and emulsify the plant-extract oils and proteins. Plant-based milk includes several additives like thickeners (e.g. agar, guar gum, or xanthan gum), minerals (e.g. calcium carbonate and phosphate), and vitamins (e.g. vitamin D and E) to make them more "milk-like" in terms of nutrition, taste, and mouthfeel. Plant-based milk alternatives are produced by either a wet or a dry method. The wet method involves grinding the seed into hot water and soaking the raw material to extract the proteins and lipids. The dry method uses a mill to grind the seed into a fine powder, which is later hydrated with water. Starches and cellulose from the seeds are degraded by enzymes that are made from the seeds. The enzymes are made and activated by a similar mechanism used to sprout barley in beer production called malting.

   For those looking for an animal-alternative food source, these milk choices can help supplement the diet and, sometimes, can substitute while cooking and baking. Another reason some choose plant-based milk is a question of sustainability. As noted in the production of plant-based milk choices, plenty of water is used. The amount of water used to make alternative milk and grow these crops significantly affects the environment. According to the University of California San Francisco Office of Sustainability, growing the almond crop is a high water demand affair [3]. Most of the almonds produced in the United States are grown in California, where droughts and perennial water shortages make the burden of water demand to grow almonds (approximately 15 gallons of water to produce 16 almonds worth of milk) challenging to justify and sustain. Soy milk has advantages and disadvantages as well. Some populations of women with a large soy intake have a decreased risk of breast cancer. At the same time, some soy sources contain estrogen, which will have significant health risks. Oat milk is lower in allergens found in cow's milk and milk made from some seeds and nuts. But oat milk, while high in fiber, is deficient in most other nutrients found in other types of milk.

   The difficulty of sustainable dairy farming has been challenging to achieve. It is important to note that a larger population is fed by farmers using less land than ever. Sustainability is essential to everyone, especially farmers. According to the United States Department of Agriculture, 91% of all dairy farms are considered small family farms, most of which work to reduce their footprint to survive economically. In recent years, dairy farms have made significant strides in becoming less resource-intensive. Through the adoption of advanced technologies and sustainable practices, such as precision feeding and improved waste management systems, farmers have been able to optimize resource utilization, reduce energy consumption, and minimize water usage. Additionally, the implementation of innovative farming methods like vertical farming for animal feed production and the integration of renewable energy sources has further contributed to the reduction of resource intensity in the dairy industry. Furthermore, decreasing energy resources in dairy milk farms remains a crucial aspect of mitigating climate change. Traditional dairy farming heavily relies on energy-intensive processes like cooling systems, milking machines, and transportation. By investing in renewable energy sources such as solar panels or wind turbines and adopting energy-efficient technologies, dairy farms can substantially reduce their carbon footprint, contributing to both climate change mitigation and long-term cost savings for farmers. Today's dairy farms produce more milk with less resource demand than ever. UC Davis reports that over the last 70 years, dairy farms have reduced their land use by 90%, feed usage by 77%, and water use by 65% per glass of milk produced [4].

So are almond milk and other plant-based milk alternatives truly milk? No. Do they serve a purpose for those avoiding animal products? Absolutely. Are plant-based kinds of milk healthier and more sustainable options? This answer isn't so obvious. Skeptical and critical thinkers will use this information to learn more when making their choice of milk or milk alternatives.

4.2 Biology and Chemistry of Milk: Sugar, Protein, and Fats

4.2.1 Milk Sugar

Carbohydrates are organic molecules made up of carbon, oxygen, and hydrogen (Figure 4.3). Simple sugars are carbon-based ring structures more formally known as monosaccharides, while disaccharides are made of two monosaccharides. One of the benefits of drinking milk is the energy available from carbohydrates. Nearly 5% of cow's milk is carbohydrate, with most of the sugar in the form of lactose. Lactose provides almost half of the calories of milk and gives milk its sweet flavor. Lactose is a disaccharide made of two simple sugars, glucose and galactose, linked together by a special covalent bond called a glycosidic bond (Figure 4.4). Glucose is an important monosaccharide that makes

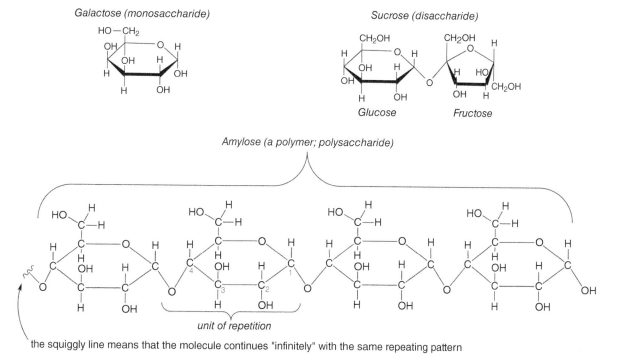

Figure 4.3 The forms of carbohydrates. Carbohydrates, primarily of hydroxylated carbons (hydrocarbons aka carbohydrates), are found in mono-, di-, and polysaccharide forms.

up the primary source of potential energy for tissues and cells. The brain uses about 120 g of glucose (about 1/2 cup of solid sugar) each day! Fortunately, milk is a good source of glucose but is not available until lactose is digested into the two monosaccharides. Galactose is also an important nutritional component used by the brain, not for energy but as one of the building blocks of nerves and brain tissue. Galactose is an important component of the specialized cells that insulates, the nerves and is a component of connective tissues. However, before lactose can be used for either glucose or galactose, the glycosidic bond holding the two sugar molecules together must be broken. An enzyme produced in the gut called lactase binds lactose and a water molecule to break the disac-

A disaccharide molecule of lactose (i.e. milk sugar)

1-4 α glycosidic bond

This half of lactose is made of Galactose

This half of lactose is made of Glucose

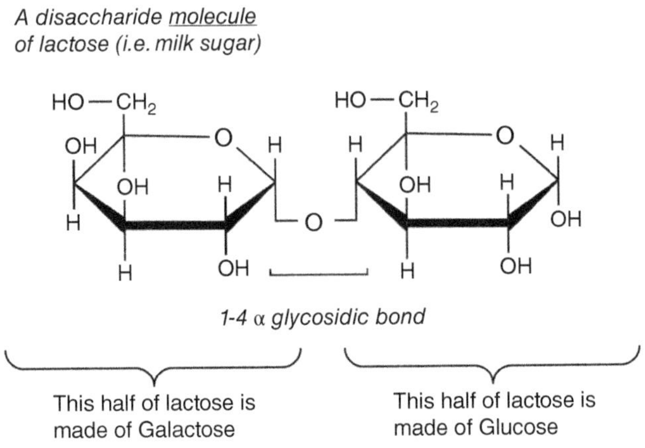

Figure 4.4 Lactose. The disaccharide lactose is composed of galactose and glucose joined by a special glycosidic bond.

charide, producing glucose and galactose for the body to absorb and use. In adults, galactose also supports immune function, as this sugar contributes to how antibodies function. Some commercial bakeries use galactose to sweeten foods and reduce the tartness of some acids.

Even though lactose is only one-fifth as sweet as other sugars (e.g. table sugar), there is a good reason for milk to use lactose as its flavor agent. That's because of milk's evolutionary or primary role to provide nutrients for newborns. Because infants are particularly susceptible to illness and disease, it is important for the milk to remain safe to drink. Many environmental bacteria can use simple sugars in food, including glucose, to grow; these bacteria then contaminate the food with toxins. However, many microbes must produce the enzymes needed to digest lactose. This takes time. Once exposed to lactose, microbes can take several hours to a day to produce the needed enzymes to digest the sugar, while most infants already produce plenty of lactase. Thus, the lactose in milk is not a readily available food source for most microbes, and the milk remains safe for newborns to drink. Other sugars found in milk also impact health and bacterial growth. Recent analysis finds that some of the components in milk support the growth of important bacteria that help an infant grow and thrive. Nature at work for your safety!

One exception to this rule is the two strains of bacteria found in the environment primed and ready to metabolize lactose, called *lactobacilli* and lactococci. Once these bacteria have been introduced to milk, they readily convert lactose to lactic acid, causing the milk to sour or curdle (the beginning of yogurt and cheese). Unlike many other bacteria, these bacteria do not produce significant toxins, and ingestion of these microbes is safe. As *lactobacilli* and lactococci make the milk more acidic by the production of lactic acid, the milk is less livable for many of the dangerous microbes.

4.2.2 Lactose Intolerance

What happens if a person doesn't produce lactase and continues to drink milk and eat ice cream? If lactose is not broken down in the gut, water rushes into the intestines via osmosis, creating a bloated feeling and watery stool. If this wasn't bad enough, the natural bacteria found in the human intestine could digest the lactose, producing carbon dioxide, methane, and hydrogen gases. Of course, this all results in flatulence, cramps, bloating, and diarrhea. This syndrome, initially called lactose intolerance, occurs in most adults in the world.

Lactose is the primary sugar in milk and a vital energy source for infants. Lactase (more formally known as lactase-phlorizin hydrolase [LPH]), is coded by the lactase (*LCT*) gene and is expressed (produced) in the small intestine. Lactase will bind and cleave (hydrolyze) lactose into glucose and galactose (Figure 4.5). Glucose is an essential monosaccharide for energy, and galactose is used as a building block of the brain and neural tissue of a growing infant. Most children produce the enzyme; however, lactase production decreases with age. After infancy, the levels of enzyme lactase are less than 10% at their peak for most people (Figure 4.6).

The reasoning for this is pretty straightforward. Milk is primarily food for infants. As humans age, their diet provides glucose and other carbohydrates from starches and fruits, reducing the need to use lactose. Over time, genes coding for lactase are naturally switched off (no longer activated), resulting in fewer and fewer of the enzymes being produced. Limiting protein production to only proteins needed for the body makes sense because making protein takes energy. Avoiding protein production that is not needed is a way the body saves energy in creating unnecessary protein. After weaning from milk, most humans stop making most or even all of their lactase, known as lactase nonpersistence (LNP). The inability to digest milk is also known as lactose intolerance.

LNP is the ancient or "original" case. Approximately two-thirds of the world's population are LNP (Figure 4.7). The rest of the population has gained five or more genetic

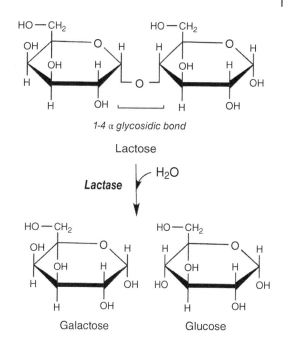

Figure 4.5   The activity of lactase.

LACTASE HOTSPOTS
Only one-third of people produce the lactase enzyme during adulthood, which enables them to drink milk.

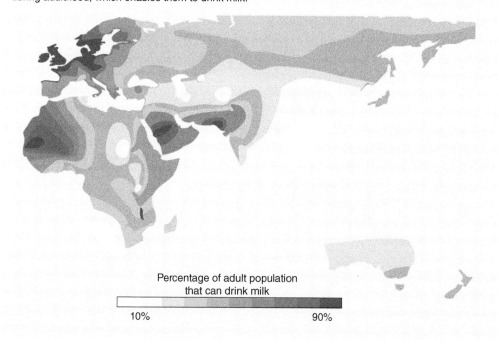

Figure 4.6   Distribution of people with lactose tolerance. Reproduced with permission from Ref. [5] / Springer Nature

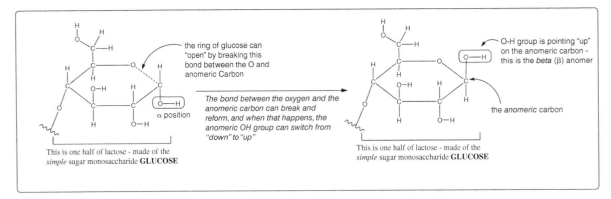

Figure 4.7 Lactose-flipping anomers in ice cream.

mutations that allow the LCT gene to remain active and produce lactase. This genetic trait is known as lactase persistence (LP). The level of lactose tolerance or LP, is high in Northern European populations, less in Southern Europe and the Middle East. There are pockets of areas where LP is high. These areas correlate to pastoralized populations (where people have raised domestic cattle).

Domesticated animals produce milk and provide a good source of nutrition and energy for adults who can use the food. Yet, as late as 7,000 years ago, a genetic mutation in ancient Europeans arose that allowed for the continued production of lactase. Because of the more temperate climate, dairy cattle could be maintained year-round, and milk and dairy products could be safely stored and consumed long after milking. The genetic difference allowed those people with the mutation to take advantage of the energy and nutrients stored in milk long after breastfeeding and provided a natural advantage to those carrying this mutation [2, 3]. Strangely, the trend of colder climates and persistent lactase production does not totally align with the observation of lactose tolerance in nomadic tribes from Africa and the Middle East. However, we see that migration and the mixing of genes and cattle raising in these warm climates support the hypothesis of how the LP trait was maintained. This also makes sense why over long periods of time, the LP mutations did not spread to ancient peoples without cattle as these mutations would not confer any metabolic advantage.

So what to do if your stomach doesn't agree with ice cream, milk, or other lactose-rich foods? Avoid those foods or take Lactaid! The enzyme lactase is purified from human-friendly yeast such as *Aspergillus niger* and can be purchased in pill form or premixed in milk or other dairy products. Lactaid includes a small dose of the enzyme in each pill or is already added to milk to predigest the lactose. The milk is then safe to drink and will not cause the lactose effect on those with lactose intolerance syndrome.

4.2.3   Lactose and Ice Cream

Lactose affects the flavor of milk and can cause metabolic problems, but lactose can also have an important effect on the texture of milk-based food. Ever wonder why ice cream that has been partially melted and refrozen has a gritty feel? The sandy texture is due to lactose that has crystallized. Lactose can flip its —OH on the glucose portion of lactose. When dissolved, lactose converts between these two shapes, α-lactose and β-lactose. The versions of lactose are called anomers, and while they are chemically the same, there are a few differences. When the —OH is below the ring of glucose, lactose is considered to be in the α-anomer. When the —OH is above the ring, lactose is in the β-anomer. At room temperature, there is more β-lactose than α-lactose. At high cooking temperatures, the β-lactose is less soluble and falls out of the solution, leaving more α-lactose in the food. At low temperatures, the α-lactose anomer crystallizes into very hard solid grains, giving refrozen ice cream that gritty feel. Therefore, milk that has been cooked, then cooled, and later frozen will have the gritty texture of crystallized α-lactose (Figure 4.7).

4.2.4   Milk Proteins

About 3.3% of milk content is protein, and like lactose, protein is found in the aqueous or liquid phase of milk. While there are thousands of milk proteins, most often milk proteins are easily thought of as divided into two basic classes of proteins, casein protein and serum or whey protein. If you were to add vinegar or another acid to milk, you would create white clumps of fat (i.e. the curds) surrounded by a yellowish-white liquid (i.e. the whey)—in other words, you would make cottage cheese! Curds are the fats and proteins that become insoluble when milk is acidified. Whey is the mixture of water and proteins still in solution after the milk has been acidified and curdled. Milk proteins are divided into acid-soluble (those proteins that are stable at pH 4.6 or greater and remain in solution) and acid-insoluble (those proteins, mostly caseins, which denature and precipitate as the milk turns acidic). You will learn more about curds in Chapter 5.

4.2.5   Proteins

Proteins are long polymers of hundreds of individual building blocks called amino acids. When bonded together, proteins fold into long tangles, forming specific shapes, giving the protein many of its valued properties. There are 20 common amino acids that make up protein chains. All but nine of these amino acids are produced by the human body, and the remaining nine amino acids (called the essential amino acids) must be included in the diet. Milk provides all 20 amino acids with whey and other proteins. Milk proteins are divided into two categories: proteins that remain in solution when the milk is acidified (acid soluble) and proteins that are not soluble, which then aggregate and precipitate when the milk is acidified (acid insoluble) (Figure 4.8).

When proteins are folded correctly, they are considered to be in their native state. When heat, acid, and chemical reactions alter the ability of the protein to hold its shape, the protein is unraveled or denatured. Depending on the protein and the conditions, denatured proteins can coagulate and form highly tangled webs of proteins that are insoluble precipitates (Figure 4.9). For milk, this last precipitation is how curds (coagulated insoluble denatured proteins) are formed. Whey proteins remain in their folded, native state when milk is acidified, while those proteins that unravel and denature (i.e. the caseins) form clots of proteins we know as curds. However, both casein and whey proteins stay native and folded at high temperatures, giving milk the ability to withstand cooking and boiling without clumping. This high-temperature stability is why milk is used for creams and sauces and in hot drinks like coffee, tea, and hot chocolate.

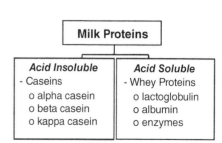

Figure 4.8   Organization of milk proteins.

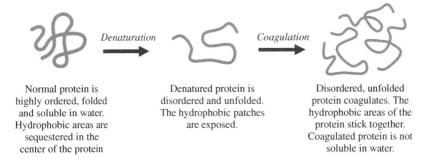

Normal protein is highly ordered, folded and soluble in water. Hydrophobic areas are sequestered in the center of the protein

Denatured protein is disordered and unfolded. The hydrophobic patches are exposed.

Disordered, unfolded protein coagulates. The hydrophobic areas of the protein stick together. Coagulated protein is not soluble in water.

Figure 4.9 Native and denatured proteins. Heat, acidic, alkali, or other conditions will lead to the loss of protein structure called denaturation. Some proteins, when denatured (unraveled), can tangle together forming insoluble aggregates. Reproduced with permission from Ref. [6] / John Wiley & Sons

4.2.6 Casein Milk Protein

Eighty-two percent of all of the milk protein is casein, while the remaining are whey proteins. Casein proteins are a group or family of closely related proteins, which include three types: α-casein (α-s1 casein and α-s2 casein), β-casein, and κ-casein (kappa casein) proteins. Separate genes found on bovine chromosome six code for each type of casein, and it is the specific sequences and ratios of these casein proteins that determine milk production and quality. Each casein protein is a long polymer of about 209 amino acids, and the different forms are about 80% similar in sequence. That means, when lined up head to tail, the proteins have the same amino acids at the same positions 80% of the time. The extent of this amino acid similarity indicates that these proteins are highly *homologous*—that is, each of the individual casein proteins likely came from a common ancestral gene coding for a protein found in ancient milking cows. Cow's milk contains all four caseins, while humans do not produce the α-casein.

Casein proteins differ from whey proteins in a number of ways. First, none of the casein family proteins include the amino acid cysteine, while whey proteins contain several cysteine amino acids. Cysteine has a side group that includes an important sulfur atom involved in maintaining protein structure (Figure 4.10). Some of the flavors and smells of milk come as a result of the reaction of sulfur with other molecules while cooking and are thus not due to casein. Another difference between whey and casein protein is that casein has a high phosphate content. Unlike whey proteins, each casein protein is *phosphorylated*; that is, several of the casein amino acids are covalently bonded to phosphate groups (PO_4^{3-}), which gives the protein an overall negative charge and allows casein to bind to the positively charged calcium ion. The casein–phosphate–calcium complex is critical for calcium delivery in humans. Noncomplexed calcium is poorly absorbed in the gut of humans and is eliminated from the body. For calcium to be absorbed (the ability of a molecule to be transported through the intestine and into the body is called bioavailability) through the gut and into the bloodstream, the calcium ion must be complexed with another molecule. Most frequently, this is through interactions with casein phosphate or by either vitamin D or citrate (a molecule found in fruit).

Casein proteins are found in milk's aqueous (water) phase, forming large clusters called micelles. A micelle is a globular complex of amphipathic molecules found in a water environment (Figure 4.11).

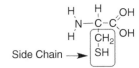

Side Chain →

Figure 4.10 The amino acid cysteine.

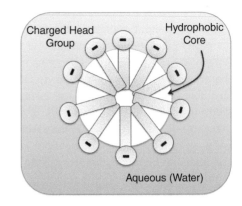

Charged Head Group

Hydrophobic Core

Aqueous (Water)

Figure 4.11 Model of a micelle.

The molecule's water-fearing or hydrophobic part faces the micelle's middle or interior. The charged, water-loving (hydrophilic) portion of the molecule found on the outer portion of the micelle interacts with the water and other molecules suspended in the milk. Casein proteins form micelles or are currently thought of as a mixture of nanoclusters of calcium phosphate/casein in the inner core of the micelle. The calcium acts as a bridge between the negative phosphate charges of adjacent casein proteins. It acts as a glue to hold the cluster together. On the extreme outer surface of these micelles are the κ-casein proteins (Figure 4.12). These proteins form a hairy layer and play a critical role in keeping micelles from clumping together and forming curds. How caseins work to keep micelles from clumping together is an important feature of milk and will be further discussed in Chapter 5.

4.2.7 Whey Milk Proteins

An acid-stable set of proteins in milk are collectively called whey proteins. Whey proteins have a high cysteine amino acid content and therefore a relatively high concentration of sulfur atoms and little or no phosphorus atoms and are much more difficult to acid-denature than casein proteins but are more susceptible to heat denaturation than casein. β-Lactoglobulin makes up about half of the whey proteins and is a major human allergen, but its function in milk is not clear. Albumins, antibodies, and enzymes make up the remaining portion of whey proteins. Each has an important role in nutrition and supports immune function. When denatured by heat, whey proteins can form very tight protein curds, such as those found in ricotta cheese. The strong rotten egg and ammonia smell of cooked milk is the sulfur and nitrogen atoms found in whey proteins reacting under heat with oxygen and water to form hydrogen sulfide gas and ammonia.

4.2.8 Milk Fat

The rest of the milk (the nonwater portion) is 3.5% total fat and made of various-sized droplets of fat and some proteins. This "milk fat" is formed into globules filled with different kinds of fats and covered with a membrane skin studded with protein. The membrane is itself made of a kind of fat called phospholipids, and it is the chemical nature of these phospholipids and of the proteins surrounding the milk fat globules that keep milk fat in solution. Imagine what happens when an oil and water mixture (think of an oil and water Italian salad dressing) is allowed to sit for a short time. The oil coalesces (comes together into one phase) and separates from the water.

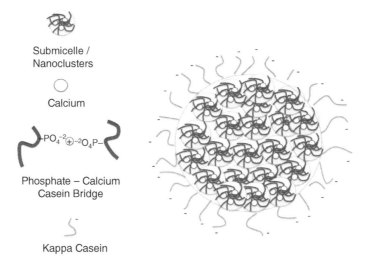

Submicelle /
Nanoclusters

Calcium

PO_4^{-2} ⊕ $^{-2}O_4P$–

Phosphate – Calcium
Casein Bridge

Kappa Casein

Figure 4.12   Casein micelle and its components.

While the fat and water will separate if raw milk is left alone, the fat globules are maintained, and the cream is an accumulation of the less dense fat globules separated from the more dense water layer of milk. The reason the globules don't meld together to form one big oil layer is that the phospholipids and proteins that make up the membrane form a strong cover for the fat droplet and repel each other. It is this very tough membrane that maintains the integrity of the globule in very harsh environments. The phospholipid and protein-coated fat droplets allow milk to be dehydrated or condensed by heating and with remarkable stability. To better understand these forces, one needs to examine the fats in the droplet and the phospholipids and proteins that make the membrane.

Fats, lipids, and oils all describe molecules that do not mix well with water, are hydrophobic, and are primarily made of carbon structures with hydrogen atoms (Figure 4.13). Lipids and fats are essentially a generic name for single carbon chains (fatty acids) or groups of fatty acids and other atoms, all bonded to a three-carbon molecule called glycerol. Three fatty acids bonded to a glycerol backbone are called triglycerides. While the fat that makes up most of the milk fat is composed of fatty acids and triglycerides, hundreds of different kinds of fats have been identified in milk. Fatty acids are carbon chains with an acidic carboxyl group on one of the carbon chains. The length of the carbon chain and the nature of bonding can vary greatly. Short-chain fatty acids are those with four and six carbons that are produced by the cow later in lactation and give milk its buttery character. Longer chains (8–16 carbons long) are either produced by the lactating animal or come from the diet of the cow. The saturation refers to the bonded hydrogen atoms to each carbon on a chain. Thus, mono- and polyunsaturated fatty acids refer to the number of double bonds found on the carbon chain of a fatty acid. Milk contains about 65% saturated, 30% monounsaturated, and 5% polyunsaturated fatty acids (either free fatty acids or fatty acids bound to glycerol).

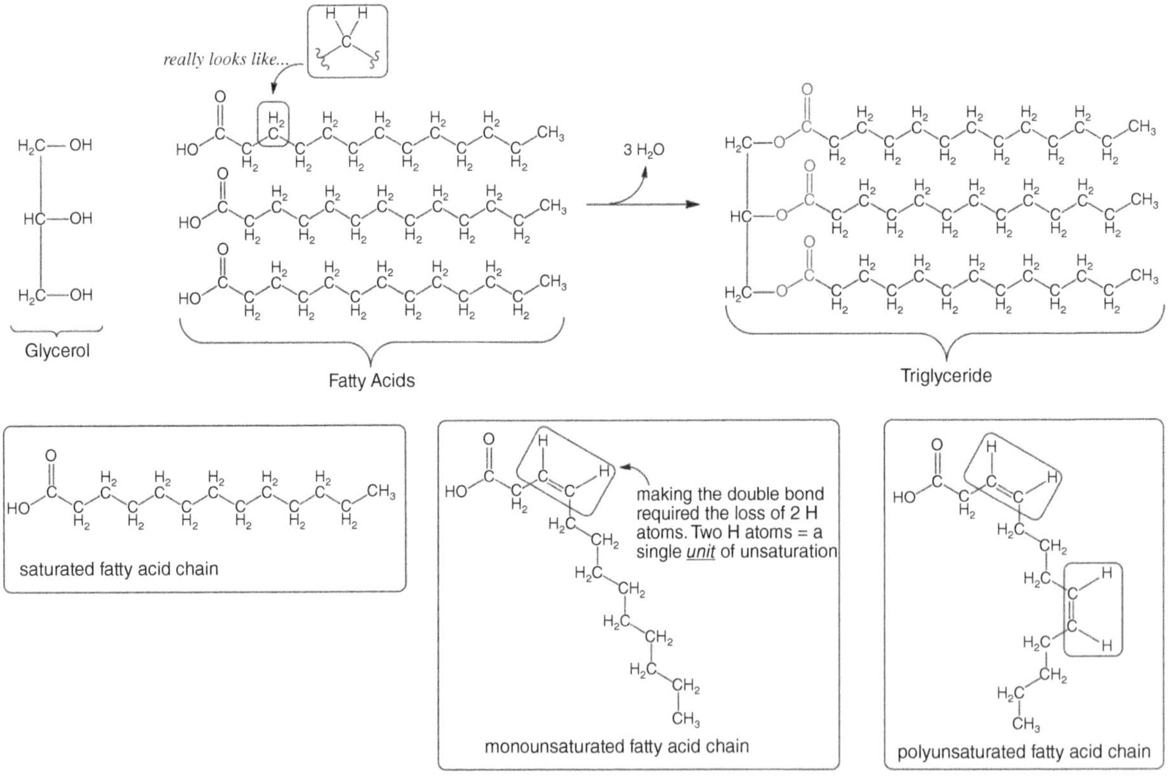

Figure 4.13 Structure of fatty acids and triglycerides.

Triglycerides and phospholipids are similar molecules with two or three fatty acids linked to a three-carbon glycerol molecule. Triglycerides make up most of the milk fats and have three fatty acids bonded to the glycerol. Phospholipids differ in their structure as these lipids have two fatty acids bonded to the central glycerol molecule with the third —OH of glycerol bonded to a phosphate-containing group. While the triglycerides are hydrophobic and found in the center of the milk globule, the milk fat globule membrane is made primarily of two kinds of phospholipids: phosphatidylcholine and sphingomyelin (Figure 4.14). These amphipathic molecules form the membrane by orienting their charged and polar groups toward the watery phase and their hydrophobic tail to the interior of the globule, where the nonpolar triglycerides reside. The membrane provides a durable coating that slows down the coalescing of the fat into a single phase. This concept is important when cooking with milk or making creams and butter.

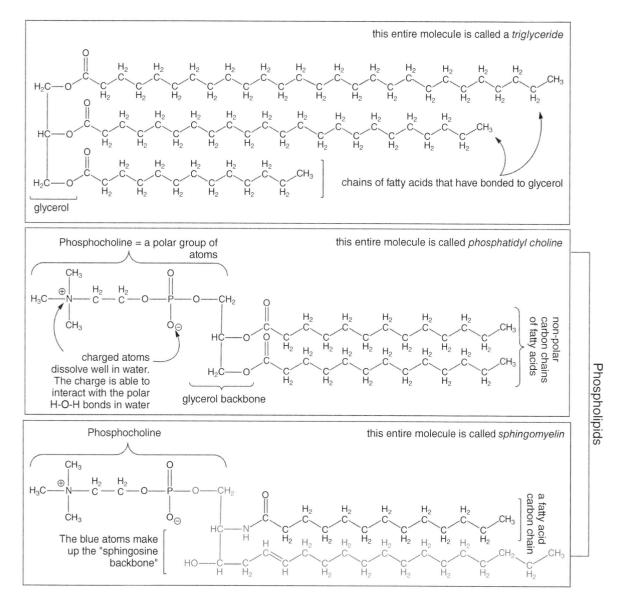

Figure 4.14   The phospholipids involved in milk membranes.

When raw milk is left alone, separation does occur. As the saying goes, "The cream rises to the top." While the fat globules remain intact, the density of fat is less than the density of water. This process of aggregating fat globules at the top of the milk is called creaming, and the resulting fat globule layer is often separated to be used to make butter, whipped cream, or soft cheeses. If the rate at which milk fat globules rise in raw milk is measured, you might be surprised to learn that it happens faster than should occur due to density alone. Some of the proteins coating the fat globules attract each other, causing the fat globules to aggregate and rise more quickly. This process of accelerated creaming is called cold agglutination and happens as the immune proteins in milk bind to each other and proteins on the fat globules bring about clustering and quicker separation than density differences of oil and water alone can account for.

4.2.9   Homogenization and Pasteurization

To avoid creaming and prevent contaminating microbe growth that will acidify the milk and potentially cause disease, raw milk is quickly shipped to local processing centers where raw milk is homogenized and pasteurized. These are two different processes that allow milk to be stored, shipped, and safely consumed long after milking the cow. Homogenization is a process that breaks the milk fat globules into much smaller and more uniformly sized fat globules. Fresh from the cow, milk fat globules range from 1 to 10 microns (μm); after homogenization, the globules range from 0.2 to 2.0 μm in diameter. Milk is homogenized as raw milk is forced at pressure through very small diameter nozzles. Large milk globules are forced through a narrow opening, which causes the membranes to sheer, essentially breaking the tough globules' coverings while mixing together and creating many smaller fat globules. The result of homogenization is the many smaller globules whose fat droplets are only partially covered by membranes. The exposed lipid droplets quickly become covered with casein proteins from the liquid phase of the milk. All of this results in smaller, casein-coated milk fat globules that are less likely to separate from the rest of the milk. The negative charge from the casein proteins ensures that the milk fat globules do not combine and form a solid layer of fat (Figure 4.15).

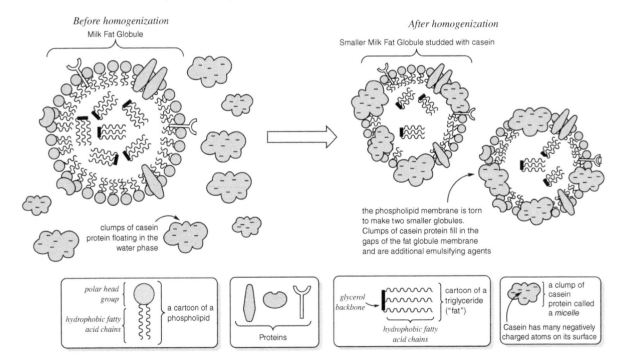

Figure 4.15   Impact of homogenization on milk fat globules. Forcing milk through small pores creates a sheering force that tears the micelle into smaller globules and encourages the integration of formerly independent casein micelles.

Pasteurization is the process of quickly heating and then cooling the milk. Milk is heated to temperatures high enough to kill any contaminating bacteria or other microbes present in the raw milk but not enough heat to destroy its nutritive properties. The heat-stable proteins like casein and tough membranes of fat globules make milk very heat stable, and pasteurization is effective in sterilizing milk for long-term storage. The process of pasteurization was first used to process sake by Buddhist monks and formally used or invented for beer and wine by the French chemist Louis Pasteur in the 1860s. The concept of pasteurization is to heat the food or drink enough to kill microbes but not cook the food. As a form of sterilizing foodstuffs, pasteurization is about how high a temperature the food is exposed to and for how long. High temperatures are capable of killing most microbes after only a short period of time. Modern pasteurization heats milk to temperatures below the boiling point where nutrients degrade and proteins denature. Most milk is consumed regionally and doesn't need a long shelf life; therefore, it is pasteurized using high temperatures for a short time. Raw milk is heated to 71.6°C for 15 s by pumping the milk through heated pipes and then chilled to the appropriate temperature. Treating milk with higher temperatures for slightly longer times (161°C for 20 s) followed by placing the milk into sterile containers creates sterile milk that has a very long shelf life and can be stored for up to 6 months without refrigeration. Bacteria, spores, and other methods of disease transmission are eliminated. This method called ultrahigh-temperature processing (UHT) results in a partial loss of some of the vitamins (riboflavin, vitamin C, folic acid, and a few others), and the higher temperatures denature and cook some of the proteins (Maillard reaction—see Chapter 6) and alter some of the fats producing what some consider a slightly off-flavor. UHT milk is popular in regions without local dairy farms as it holds its usefulness longer. Many European countries use UHT milk for more than half of their milk needs due to space and reduced energy requirements, as the milk does not have to be refrigerated (Box 4.2).

Box 4.2 Common kinds of milk found on the market

1. Vitamin D or whole milk: Homogenized and pasteurized milk packaged with additional vitamin D added. None of the fat has been removed prior to packaging (3.5% fat).
2. Low-fat or skim milk: Milk in which some or nearly all of the milk fat has been removed. These milks range from 2% fat to less than 0.5% fat (nonfat or skim milk). Because the body of the milk is more watery without the fat, this kind of milk is often supplemented with whey protein.
3. Condensed milk: Sweetened or unsweetened whole milk with much of the water boiled away. This milk was created to serve as a concentrated form of milk and to fight food poisoning during the US Civil War in 1865. Now, it is commonly used for a range of different cooking and baking purposes. Originally, sweetened condensed milk had added table sugar to limit bacterial growth.
4. Whipping and heavy creams: Cream is the fat globule layer from milk that has been creamed. Differences between heavy (30%) and whipping cream (36–40% fat) are primarily in the concentration of fat. Both creams can be used to make whipped cream, although the more fat the better the resulting foam. Half and half is a mixture of milk with cream for a lower percentage of fat (10–18%).

4.2.10 Whipped Creams and Foams

Milk can be whipped into a foam, which is a type of colloid. To form a foam colloid, one substance (in this case, air) is trapped in another. If low-fat skim milk is foamed (<2% fat), as in the foaming of milk for a cappuccino, it is the denaturing proteins that trap the air bubbles much like egg white foam. But if heavy cream (>35% fat) is foamed, as in the preparation of whipped cream, it is the fat that traps the air bubbles. The resulting creams and foams are unique for milk and different from other foams, such as egg white foams.

Figure 4.16 Whipped cream. Paylessimages/Adobe Stock Photos

Both are types of **colloids**—where one substance (in this case, air) is trapped in another (cages of fat—whipped cream or bubbles of protein and some phospholipids) (Figure 4.16).

4.2.10.1 Whipped Creams

The directions for whipped cream are pretty simple. In a large chilled bowl, add one cup of cold heavy cream and whip until peaks are stiff. Add sugar or vanilla, just as the peaks are almost stiff. Have you ever whipped cream that was warm or overwhipped the cream? You likely found cream whipped in this nature was oily or tasted kind of like butter. To better understand what is going on while making whipped cream, one has to understand what is happening to the fat globules of heavy cream during the whipping (Figure 4.17). Mechanical agitation by the mixer beaters causes the membranes covering the globules to break and shear, forming smaller globules with sections of exposed fat (oil) droplets. Whipping also introduces small air bubbles to the cream, and the interaction of air with the globule also helps to disrupt or break apart the membrane. While whipping, the air bubbles become

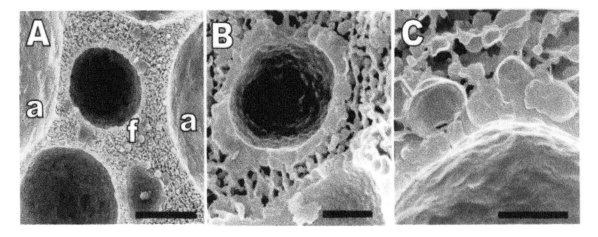

Figure 4.17 Electron microscopy of whipped cream. (A) Overview of (a) air and (f) fat globules. (B) Internal structure of air bubble highlights the partially coalesced fat. (C) Interaction of fat globules within the fat layer. Reproduced with permission from Douglas Goff, University of Guelph.

smaller, and the sections of exposed fat within the globules clump or coalesce to form cages surrounding the small air bubbles. The trapping of air increases the volume of the now "whipped cream." As the fat globule membrane is disrupted, some of the casein protein will coat some of the exposed fat surface, which supports the fat–air interface. Stop whipping too early, and the cream remains unconnected milk fat globules from which air will easily disperse. Whip too long, and the globules will break down into many smaller fat droplets with too little membrane to cover them. This will result in the beginning of butter, as the fat droplets are mostly naked oil and can readily coalesce into a solid mass of fat. The increase in volume of the whipped cream foam due to the trapped air is called **overrun**, a term often used when making ice cream. A high-fat cream will create a stiff foam capable of trapping a large volume of air bubbles; therefore, the higher the content of fat in the cream, the thicker the whipped cream and the greater the overrun or increase in volume. Thus, heavy cream or country creams with more than 36% total fat will allow for a stiffer, higher-volume cream, and a low-fat cream such as light cream will have a thinner, liquid-like whipped cream consistency. The UHT pasteurized creams are difficult to foam because of the chemical changes that occur in the milk fat globule membrane and proteins during pasteurization. Foaming a UHT pasteurized cream requires the addition of surface-active agents and stabilizers such as whey or gelatin proteins or complex carbohydrates such as gums to provide the needed additional support. So why do most whipped cream recipes direct you to use chilled cream and a cold bowl? Once the milk fat globule membrane is torn and the fat droplet is exposed, the exposed fat needs to remain solid to maintain the cage around the air bubbles. Like any other fat, once the material warms enough, the fat turns to liquid oil, the cage collapses, and the oil coalesces into a solid mass, giving the overwhipped or warmed foam a greasy, buttery feel to the tongue.

4.2.10.2 Foams

Imagine a latte or cappuccino and the milk foam sitting on top. The foam can make or break your drinking experience. In this context, the **foam** is made of denatured proteins that form a thin protein film holding small air bubbles in place. In the case of milk foams, the milk is "steamed" into a foam by forcefully bubbling hot steam into the milk. The whey proteins are denatured by the heat of steam and the physical agitation of mixing with the heated water vapor and air (Figure 4.18). Under these conditions, the whey proteins lose their native shape and unravel such that the water-fearing, hydrophobic interior of the protein ends up facing the air, and the water-loving, hydrophilic portions of the proteins align themselves to face the water portion of the milk. The result is a thin film of protein surrounding small bubbles of air—a foam. The steam doesn't introduce the air as much as it

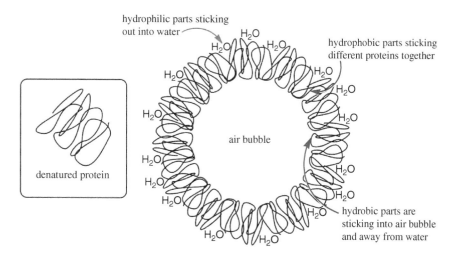

Figure 4.18 Denatured protein and foam. The layer of denatured protein forms a tenuous cage around the air pocket of milk foam.

mixes air with the milk and provides the thermal energy needed to denature the whey proteins. The delicate foams of lattes and cappuccinos not only require denatured protein but also an absence of fat. Fat itself is hydrophobic, and when mixed with denatured protein, the exposed hydrophobic regions of the denatured proteins will preferentially cluster around the fat, which will in turn inhibit the formation of the protein film around the air bubbles. Because of the higher content of protein and low fat content, low-fat and skim milk make some of the best milk foams.

4.2.11 Butter

Butter is the result of overwhipping a heavy cream. The milk fat globules, now mostly stripped of the protective membrane, aggregate and form into a solid mass of fat. Water and proteins left over from the cream are squeezed out of the solid fat (the resulting solid is now called butterfat) with small pockets of water distributed throughout the solid butter. The milk fat from which butter is produced is made of a range of different triglycerides composed of different fatty acid chain lengths ranging from 18 carbons with one double bond (oleic acid—32% of butter fat) to the short saturated chain butyric acid representing 3% of the butterfat (see Table 4.2). This complex mixture of fats gives butter its important characteristic of "spreadability." The physical change of melting requires that the interactions between fatty acids be disrupted by the addition of energy in the form of heat. When the butter temperature is low, the carbon chains of the fatty acids stack tightly against one another, held in place by a type of weak attractive van der Waals interactions called London forces. Heat provides enough energy for the forces holding the fats together to be disrupted as the fat begins to vibrate, rotate, and move more freely with the increased added thermal energy, and as more of the intermolecular forces are broken, the fat turns from a solid into a liquid oil. The longer and straighter the fatty acid chain (saturated fatty acids are straight, while unsaturated are kinked or bent), the more contacts each chain makes with the other fatty acid chains and the more heat required to defeat the forces holding the fats in place. A fat that is made from one type of triglyceride (a triglyceride is comprised of three fatty acids bonded to a glycerol backbone) will have a sharp melting curve—there will be a definable temperature at which the intermolecular forces holding the solid together give way and the solid melts. In contrast,

Table 4.2 Fat Composition in Butter.[a]

Fatty Acid	Structure	% Total Fatty Acid in Butter
Oleic acid	$CH_3(CH_2)_7CH{=}CH(CH_2)_7COOH$	31.9
Myristic acid	$CH_3(CH_2)_{12}COOH$	19.8
Palmitic acid	$CH_3(CH_2)_{14}COOH$	15.2
Stearic acid	$CH_3(CH_2)_{16}COOH$	14.9
Lauric acid	$CH_3(CH_2)_{10}COOH$	5.8
Butyric acid	$CH_3CH_2CH_2COOH$	2.9
Caproic acid	$CH_3(CH_2)_4COOH$	1.9
Capric acid	$CH_3(CH_2)_8COOH$	1.6
Caprylic acid	$CH_3(CH_2)_6COOH$	0.8
Linoleic acid	$CH_3(CH_2)_4CH{=}CHCH_2CH{=}CH(CH_2)_7COOH$	0.2
Linolenic acid	$CH_3CH_2CH{=}CHCH_2CH{=}CHCH_2CH{=}CH(CH_2)_7COOH$	0.1

From http://antoine.frostburg.edu/chem/senese/101/consumer/faq/butter-composition.shtml Good link on evolution and lactose intolerance http://evolution.berkeley.edu/evolibrary/news/070401_lactose and http://www. sciencedaily.com/releases/2005/06/050602012109.htm
[a] Reproduced with permission of 1997–2010 by Fred Senese.

butter, which is made from triglycerides with a range of different lengths of fatty acids, will have a broad melting curve. The triglycerides made of shorter-chain fatty acids will melt at the low-temperature end of the curve, while the triglycerides made of longer-chain fatty acids are still solid. At higher temperatures, the longer-chain triglycerides will have begun to melt, but due to their large size, they will slow down the faster-moving smaller fatty acids, creating a gentler slope to that part of the curve. A fat consisting of one kind of triglyceride will quickly make a transition from solid to liquid at the melting point, while butter with its mixture of triglycerides and component fatty acid chains slowly changes from solid to liquid over a much wider temperature range. Within this temperature range, the butter will be both a liquid and a solid and therefore the most spreadable. The bottom line is that butter is spreadable because it doesn't have one type of fat giving a sharp melting point from solid to liquid. See Table 4.2 for the different types of fatty acids in milk fat.

4.2.11.1 Making Butter

Butter can be made from sweet cream or cultured cream. Sweet cream is a term used to distinguish untreated cream from the cream separated from milk after bacteria have been introduced and grown to acidify and partially metabolize milk, providing a sour or cultured taste. The cream is often concentrated by heating. To make cultured butter, the cream is inoculated with acid-producing bacteria, such as *Streptococcus cremoris, Lactococcus lactis,* and *Leuconostoc.* These microorganisms use the fats and sugars found in food to produce lactic acid and another flavorful compound, diacetyl, which gives milk its slightly sour (lactic acid) and rich, buttery taste (diacetyl). Historically, the cream would be allowed to stand and sour from environmental bacteria. Now it is easy to introduce the correct living bacteria by adding cultured yogurt to fresh cream, as cultured yogurt contains the proper living bacteria. Fermentation by these same bacteria also produces sauerkraut from cabbage and the sour in sourdough bread.

The cream is then cooled and aged to allow some of the fat to crystallize. Once churning begins, crystals seed the growth of additional crystals and help to tear the milk fat globule membranes. As the cream is churned, the milk fat globules are broken, and the fats coalesce first into granules and then into larger solid masses. Finally, the residual buttermilk liquid is separated, and the fat is kneaded to remove pockets of remaining buttermilk and to evenly distribute the fat. At this time (or sometimes during the churning process), salt is added as a 10% solution to end up with a 1–2% final concentration. In addition to enhancing the flavor, many harmful bacteria and other microorganisms that can use butterfat as a food source are not able to grow in salty conditions; in this way, the added salt is a preservative, increasing the shelf life of butter as it inhibits bacterial growth.

The buttermilk is the low-fat (0.5%), high-protein liquid remaining after the milk fat has solidified. Modern buttermilk is made by adding *Streptococcus lactis* bacteria to low-fat milk. As before, the bacteria will acidify the milk by producing lactic acid, which causes curdling of some of the proteins and produces the tangy flavor associated with buttermilk. Sometimes citric acid is added to aid in the process, and a second bacterium is added to create diacetyl to give the buttermilk a more buttery taste.

Butter adds a rich flavor to baked goods. Much of the flavor of butter comes from over 120 different compounds, including the familiar fatty acids in triglycerides, lactones, methyl ketones, and diacetyl (Figure 4.19). Short-chain fatty acids with four to six carbon chains contribute to flavor, while longer-chain fatty acids have little taste or aroma. While some of these fats are from the diet of the cow from which the milk came, many are produced when enzymes called lipases in the milk itself and from the bacteria used to culture the cream use water to cleave the free fatty acid from the glycerol backbone. One fatty acid, butyric acid, makes up only a small fraction (2.9%) of the total fat but is responsible for the rancid flavor and smell of rotten milk and butter. Depending on the time of year and what the cow was eating when milked, short-chain fatty acids can be more abundant in the triglycerides of the butter. Over time, the enzyme lipase can remove butyric acid from the glycerol backbone, causing the dairy product to smell and taste rancid.

While the triglycerides found in milk and therefore cream and butter are primarily comprised of fatty acids bonded to a glycerol backbone, a small percentage of triglycerides contain hydroxy acids (~0.3%) and keto acids

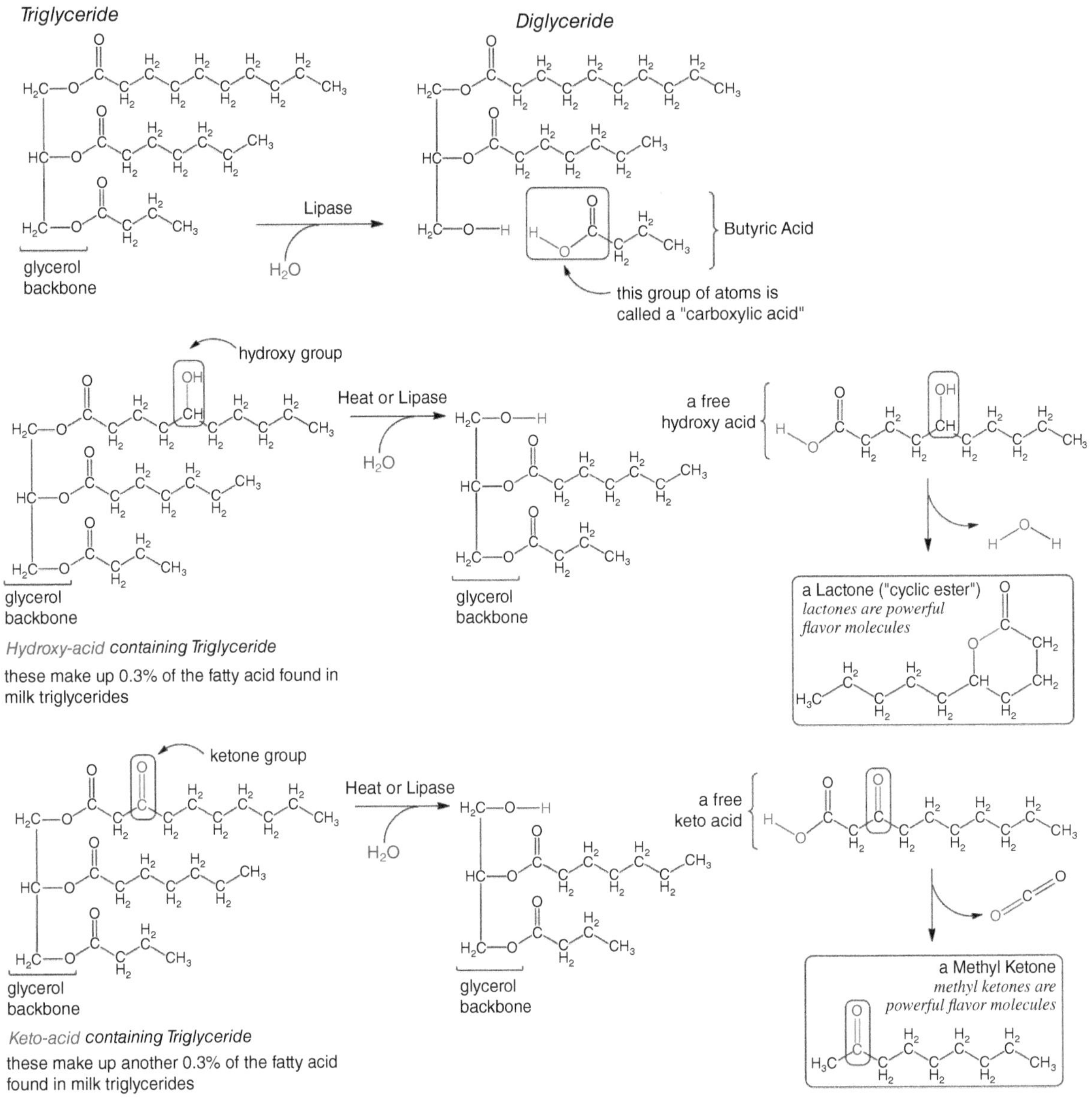

Figure 4.19 Generation of some of the aroma and flavor compounds of butter.

(~0.85–1.3%) bonded to the glycerol. While the fraction of hydroxy acid and keto acid containing triglycerides is very small, these acids can produce powerful flavor molecules. When heat or a lipase cleaves a hydroxy acid from the triglyceride, the product hydroxy acid can lose a water molecule and cyclize into a lactone. When a keto acid is enzymatically or thermally freed from the glycerol backbone, the product keto acid loses carbon dioxide and forms a methyl ketone. Only a very small amount of these molecules is required to be noticed by the human senses. For example, the flavor threshold of small molecules containing methyl ketones is 0.02 ppm when in oil. When milk or butter is heated in making baked goods or ghee (clarified butter), lactones provide the nutty and fruity aromas and flavors of butter sauces. Lactones and methyl ketones are produced readily when butter is heated in cooking. The enzymes in bacteria used to culture cream can also catalyze these reactions and impart flavor to the butter. In the same way, blue cheeses owe much of their flavor and aroma to methyl ketones produced as fats that are consumed by the mold *Penicillium roqueforti*.

In the production of buttermilk by bacterial fermentation and in the fermentation of "cultured" butters, bacteria metabolize citric acid to diacetyl. Diacetyl can be smelled at very low concentrations (1–2 ppm) and has a characteristic nutty and buttery aroma. The presence of diacetyl is the major difference between cultured and sweet cream butters. Diacetyl is also a primary component of artificial butter flavors.

Butter also has other important uses in cooking and baking. Many of the organic flavor and aroma molecules of herbs and spices are hydrophobic and, as such, will not be easily released into a watery environment. The fat in butter is also hydrophobic, and in liquid form as an oil, butterfat, is an effective solvent for hydrophobic flavor molecules; butter acts as a flavor carrier. The crumbling of baked cookies, bread, and other foods is due to dried and crystallized starch. Amylose, one of two complex carbohydrates found in starch granules, can form a gel during baking and eventually crystallize after cooling. Over time, these crystals dry and the result is a crusty, crumbly food. Butter helps trap the moisture surrounding the sugar crystals maintaining a softer crumb. Cakes, breads, and biscuits often use butter as a key ingredient. Butter coats gluten, a wheat protein responsible for forming the proteinaceous matrix that gives baked goods their structure. If mixed with the flour before adding the liquids, butter helps to coat the starch granules of the flour, limiting the penetration of water and the formation of the elastic gluten. It is this property that makes butter critical to the flakiness of pastry dough. In addition to limiting gluten formation, the incorporated butter forms thin layers separating the dough, and when baked, it melts and keeps the pastry flaky. Butter can also reduce the length of the protein strands that form the gluten matrix, resulting in a more tender baked good. However, a note of caution should be considered. Salt has the opposite effect on gluten and tightens the protein strands to make a more dense baked good. Using unsweetened cultured or sweet butter is a way to avoid the impact of salt on gluten.

In addition to fat, butter will still have remnants of water, whey protein, and lactose sugars. The protein and sugar biomolecules are considered milk solids, which, when heated, fall out of solution and are easily burned. The sugar also provides a source of energy for microbes that shorten the shelf life of butter. One way around this problem is to remove the milk solids and water to produce clarified butter. The smoke point for butter is around 65°C (150°F), but removal of the sugars and proteins raises that smoke point to 450°F/230°C for clarified butter. Clarified butter is made by warming the butter gently to boil away the water, and the resulting protein floats to the top of the butter where it can be skimmed off, while the sugar and some of the proteins precipitate to the bottom of the melted butter, where they can be separated by pouring. The resulting yellow liquid is pure butterfat and can be used for dipping lobster, coating steak, and forming the base of a sauce (roux) and can take the high temperatures of sautéing. Without the sugars and proteins of butter, clarified butter is a more efficient storage of butterfat without refrigeration. Ghee, a clarified butter traditionally made from cow or buffalo curdled milk (yogurt), is widely used in Indian foods where it has been utilized for its storability.

I can't believe it's not... When margarine first came to market as a butter substitute, it was made with buttermilk and animal fat or tallow; modern margarine as a butter substitute is primarily produced from vegetable oil. Margarine is a water-in-fat emulsion of solidified vegetable fat, water, and in some cases skimmed milk for protein and cookability. The solidified vegetable fat begins as vegetable oil, produced from the seeds of corn, sunflower, flax, canola, and other plants. These oil-containing seeds are pressed for the oil and further processed by extraction with a volatile solvent such as hexane. The solvent is removed and the oil is refined by distillation (see Chapter 11 for more information on distillation). The resulting liquid is filled with mono- and polyunsaturated fats (Figure 4.20).

These kinked fatty acids do not stack well and are easily melted at room temperature. In order to convert the liquid vegetable oil into a solid fat (margarine), the double bonds of the fatty acids have to be converted to saturated single-bonded carbon. The process of "partial hydrogenation" is the addition of hydrogen atoms to the double-bonded carbons. The conversion of liquid oil to solid fat is accomplished by bubbling hydrogen gas at high temperatures in the presence of a nickel catalyst. The result is an incomplete conversion of triglycerides containing unsaturated fatty

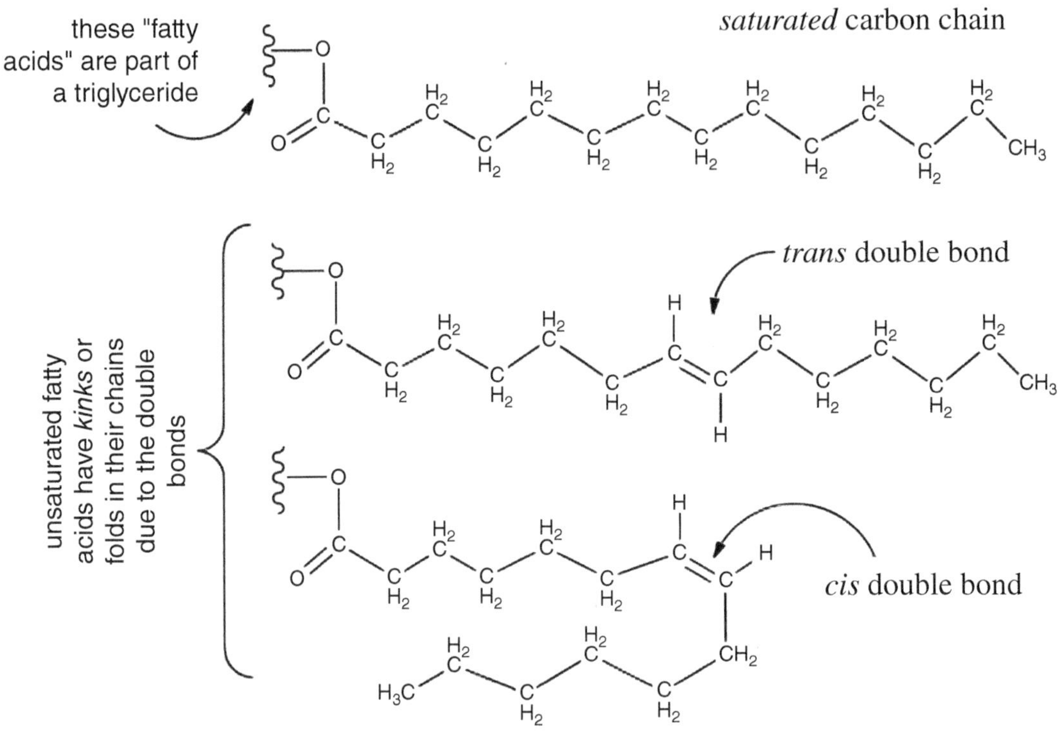

Figure 4.20 Unsaturated and saturated fatty acids.

acids that are liquid at room temperature to the solid forming saturated fatty acids (Figure 4.21). This partial hydrogenation is an inexpensive way to produce a butter substitute. The partially hydrogenated oil is then mixed with water, salts, and emulsifiers for taste and cooking, in addition to gums that serve as thickening agents.

Some of the controversy in the consumption of margarine is due to a side product of the reaction. The double bond found in most plants and animals is arranged in what is called a *cis* configuration. This causes the carbon

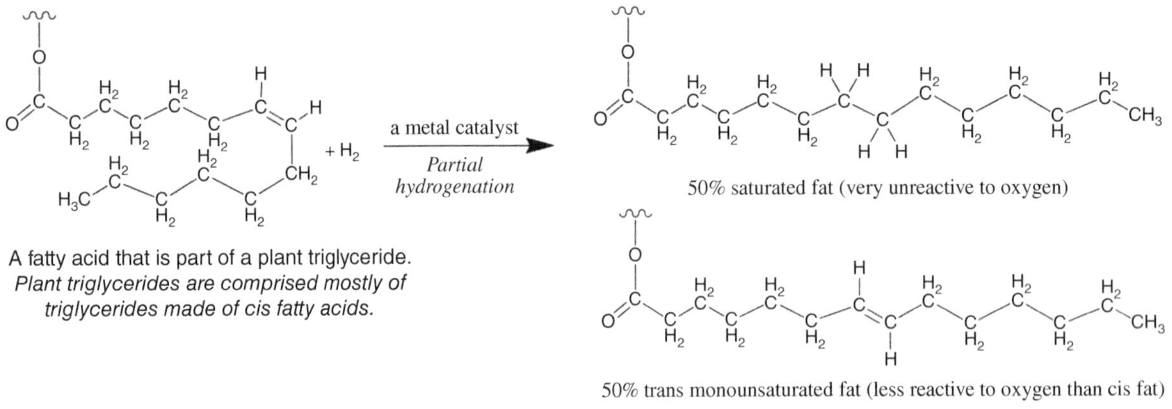

Figure 4.21 Partial hydrogenation. Heat and a metal catalyst help to add hydrogen atoms across the double bond of a plant saturated fatty acid. The result is a partial mixture of solid at room temperature of saturated fatty acid and a monounsaturated (trans) product.

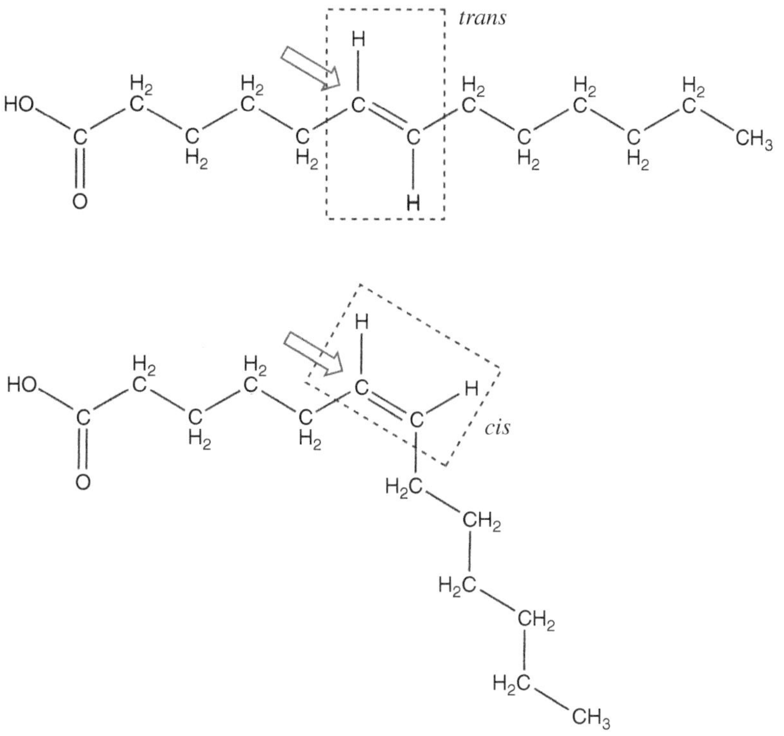

Figure 4.22 Cis and trans fatty acids. Notice the location of the hydrogen atoms on either side of the carbon double bond.

atoms on either side of the double bond to be on the same side (Figures 4.21 and 4.22). The prefix *cis* is based on the Latin "cisalpine," meaning "on the near side of." This bond creates a bend or kink in the carbon chain and alters the physical character and melting point of the fat. The high temperatures used in hydrogenation create a handful of fats that are in the *trans* configuration. Here the carbons are still in a double bond, but instead of the bend, the *trans* double bond containing carbon chains is straighter—more like a saturated fatty acid. *Trans* fats are potent antimicrobials, and, unlike cis fatty acids, they resist reaction with oxygen, which results in rancid fats and spoiled food. However, an increase in *trans* fatty acid consumption has been associated with a number of health risks. In particular, a diet high in *trans* fat increases the LDL, or "bad cholesterol," responsible for forming plaques and heart disease. Modern hydrogenation processes use different pressures, times of reaction, and temperatures to give rise to fats free of *trans* fats.

Artificial or low-fat butter is another butter substitute that works well for spreading on food but not for cooking. These spreads are a mixture of vegetable oil emulsified with whey protein, buttermilk, and water. Depending on the product, many are very high in starches, gums, and milk proteins, all of which burn easily, making cooking with these butter substitutes difficult.

4.3 Ice Cream

While the true origins of ice cream are somewhat unclear, there are several stories describing how ice cream was first created in ancient China via Marco Polo and other tales of the Roman Emperor Nero sending enslaved people to the mountains to collect snow for an ice cream-like treat. Ice cream's rise in popularity and availability certainly correlate to key scientific advances, particularly the concept of freezing point depression. The principle of freezing point depression states that adding salt (solute) to an ice water mixture (solvent) reduces the temperature at which the mixture freezes (Figure 4.23).

Ice cream is far from being a solid frozen bowl of cream. In fact, ice cream is a mixture of solids (ice and partially frozen milk fat), liquid (unfrozen cream and sugar water), and pockets of air trapped in the freezing mixture by mixing. These three phases are scattered among each other, forming a colloid. A colloid is a mixture with properties of homogeneous and heterogeneous mixtures. It is formally defined as a microscopically dispersed mixture in which dispersed particles do not settle out. Ice cream can be described as an emulsion and a foam, both examples of colloids. The formation of an emulsion (the solid phase of frozen fat globules and ice water distributed through the liquid phase of sugar water and cream) is typically unstable. Still, proteins and lipids coat the fats and stabilize the mixture from collapsing into two separate fat and water phases (Box 4.3). These mediators between fat and water phases are called emulsifiers. The foam nature of ice cream is due to the trapped pockets of air created as the freezing cream is mixed. Overrun is the increase in the volume of the ice cream before and after mixing due to this trapped air. Some ice cream can have an overrun of nearly twice the volume of the ingredients before freezing and mixing.

Figure 4.23   Ice cream. A careful control of chemistry (freezing point depression) and biology fatty globule composition can help make a better ice cream. pilipphoto/Adobe Stock Photos

The ratio of each phase of the colloid is critical for the mouthfeel and creaminess of the ice cream. Too much fat and the ice cream will have the consistency of butter, too much sugar or milk solids create a weak ice cream, while the emulsifiers limit the amount of crystals keeping the ice cream from becoming crunchy.

Federal standards (21 CFR § 135.110) require that ice cream contain a minimum of 10% milk fat and 20% milk solids—the solids refer to proteins and sugars like lactose or sucrose. Most ice creams include stabilizing emulsifiers to minimize the formation of ice and fat crystals that decrease the taste of ice cream. Fat is important for taste, providing both a creamy feel to the tongue and sweetness. The proteins and sugar add body or chewiness to the ice cream. A number of different stabilizers or emulsifiers can be found in ice cream. Added whey protein or gelatin protein from muscle tissue is used to coat the fat and provide the body. Custards use egg yolks, which have the phospholipid lecithin as an emulsifier. Another commonly used emulsifier is Polysorbate 80. This is a complex carbohydrate with a long unsaturated fatty acid bonded to it. As an emulsifier in ice cream, Polysorbate 80 can be found in fairly high concentrations where it keeps the ice cream scoopable. The carbohydrate portion of the molecule interacts with water and protein, while the fatty acid tail of Polysorbate 80 hydrophobically interacts with the fat globules. This coating keeps the fat and water phases together. Stabilizers include complex carbohydrates (starches and gums) and are commonly found in the ingredient list of commercial ice cream. A common additive used to reduce the formation of ice crystals is alginate. Also a complex carbohydrate, alginate is isolated from the cell walls of algae. Alginate contains many —OH functional groups and readily binds water through hydrogen bonding. The extensive hydrogen bonding of alginate limits the flow of water and forms a gel that acts as a thickener. The organization of the water–carbohydrate complex also defeats the formation of ice crystals. The cell wall carbohydrate from red algae (seaweed), carrageenan, is used in place of alginate in many foods (Figure 4.24).

Box 4.3 Figuring out food: Ice cream that doesn't melt

Who wouldn't love to take time to savor a tasty ice cream treat on a hot day? What stops us from enjoying such a treat is that heat energy starts to break the bonds between water molecules (solid water or ice) that hold water or ice together. As the heat energy ultimately disrupts enough of these "lattice" bonds, the water, along with the fat, protein, and sugars suspended in the water, becomes liquid. If one could keep the fat and water solution from separating into a fluid—voila, non-melting ice cream (well, sort of). Such a discovery was reported in the kitchen of Japan's Biotherapy Development Research Center in Kanazawa City. While mixing polyphenols from strawberries with ice cream to make the ice cream healthier by adding in the antioxidant, the food scientists found the unexpected; the ice cream didn't melt! Polyphenols are one of the natural compounds found in the cell walls of plants and have antioxidant activities. When mixed with milk, sugar, and flavor, the phenols act as an emulsifying agent and are likely to interact with the milk proteins to form a near-solid-like gel. Think of a thick, whipped topping with the flavor of ice cream. Hmmm, living better through chemistry?

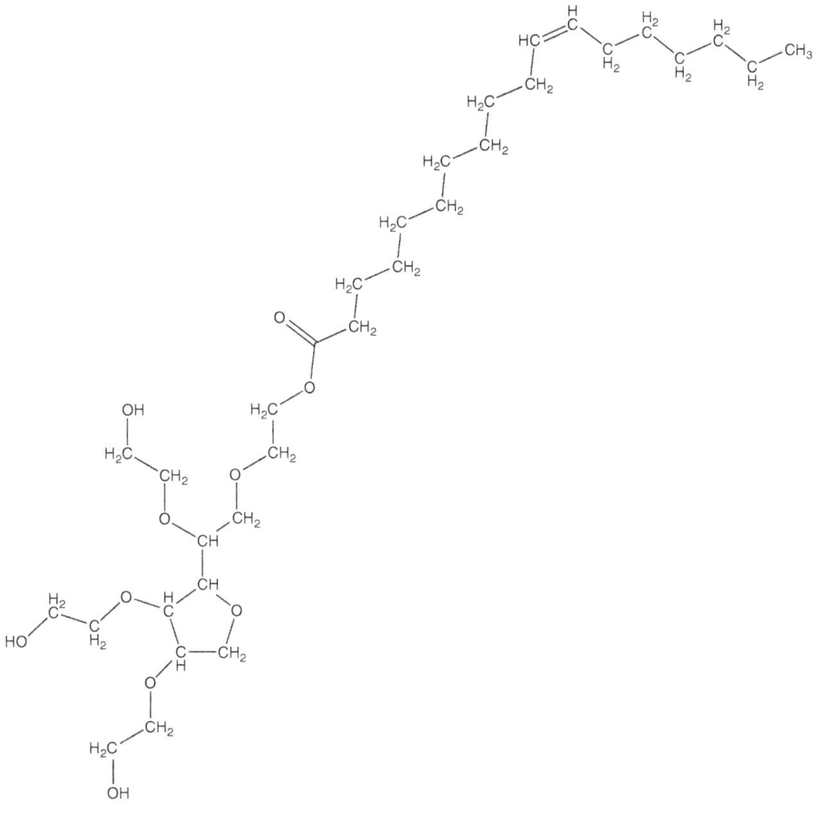

Figure 4.24 Polysorbate 80. An emulsifier often used in ice cream and other food and cosmetic compounds is made from fatty acids and other organic compounds.

Ice cream can come in many confusing grades and styles. Superpremium and premium ice cream have low overrun and high fat content with the best quality ingredients. Standard ice cream has more overrun (air) than superpremium or premium ice cream and meets the minimum requirements of 21 CFR § 135.110. Fat-free ice cream has less fat than the CFR standard and must have less than 0.5% fat per serving. In contrast, light ice cream is a description of the amount of calories coming from fat. Light ice creams must have less than half of their total calories per serving from fat. Low and reduced fat ice creams fall somewhere between light and fat-free in their fat composition. Standard vanilla ice creams, also called Philadelphia-style ice cream, differ from French vanilla in that French-style ice creams, like custards, use egg yolks as an emulsifier, while standard or Philadelphia-style ice creams (also called New York) use no egg or just the egg whites. Gelato is a frozen ice cream-like dessert that has higher fat and almost no overrun. Sherbet stretches the ice cream-like properties with fruit juice and some milk fat, whereas sorbet is not an ice cream at all! Sorbet contains no milk or cream and is instead a frozen puree of fruit with added alcohol or wine to reduce freezing temperature. Soft serve ice cream is low fat (3–6%) with up to 60% air overrun.

Making ice cream is pretty straightforward, and while an ice cream maker helps, it can be done without a machine. A simple base recipe is a combination of milk, heavy cream, sugar, and salt. From this base, flavorings, including vanilla and chocolate, can be added and are as diverse as there are ice cream creations. Richer custard or French-style ice creams include adding egg yolks as emulsifiers, followed by heating and cooling the mixture. Cream and milk are added to a mixture of egg yolk and sugar, which is then cooled before freezing. With your ice cream mixture complete, you are ready to freeze it, but now comes the work! Air must be introduced, crystallization must be limited, and the fat and liquid phases must be kept together while freezing. This is all accomplished by mixing. Mixing can be accomplished by hand by placing the liquid ice cream into a larger container of ice, water, and salt. The salted ice bath will have a lower temperature than ice water alone (see Box 4.5), allowing the sugar and fat water "ice cream" to freeze. Ice cream makers maintain a constant mixing as the liquid ice cream mixture begins to freeze. Once frozen, the ice cream can be eaten or left in the freezer to "harden." At freezer temperature (−4°F/−20°C) about only 75% of the water is frozen, and the rest is a liquid sugar–water mixture. Rapid and deep freezing causes most of the liquid water to freeze without forming unwanted crystals. Partial thaw and refreeze cycles will increase the amount of the liquid phase, and larger crystals will form, giving the ice cream an off-taste and crunchy tooth feel (texture).

Box 4.4 Baggie ice cream

One fun way to make ice cream at home is to use two strong sealable baggies. The inner bag is mostly filled with liquid ice cream, sealed with strong tape, and placed inside a larger one-gallon-sized plastic bag filled with crushed ice and a cup of salt. Roll or shake the bags for 10 or 15 minutes, and then remove the inner bag and enjoy.

Box 4.5 Colligative properties: Freezing point depression or why the ice and salt bath?

A very practical requirement for making ice cream is to freeze the solution of fat, water, and sugar. Ice cream mixtures have been immersed in ice and salt baths since the first ice cream makers in 1843. Under normal conditions, pure water freezes at 32°F/0°C, and a simple ice bath would not be cold enough to freeze the liquid ice cream. Milk freezes at about 31.1°F/0.5°C, and with the added sugar and other components, a typical ice cream solution won't begin to freeze until five or so degrees colder. Therefore, a colder-than-ice temperature is needed to make ice cream, and a mixture of salt and ice will do the trick. Temperatures of −4°F/20°C can be reached with enough table salt (sodium chloride) and −40°C with calcium chloride hexahydrate salt.

Mixtures of ice and salt do not "melt" the salt, but instead, the combination of the two lowers or depresses the melting point of the ice water. This is called a colligative property. These are characteristics or properties of a solution that depend on the number of particles of solute (in this case salt) in a solvent (water or ice). These properties are not impacted by the chemistry of the compounds dissolved in the water or the size, just the number of particles. For freezing, the more particles dissolved in the solvent, the greater the impact on freezing point depression. Salts are ionic molecules composed of a cation and an anion attracted to one another by opposite charges. The ions will separate in water as each is solvated by water molecules. Sodium chloride will dissolve into the sodium cation (Na^+) and chloride anion (Cl^-). Thus, for each molecule of NaCl dissolved in water, both sodium and chloride particles are available to impact the freezing point depression. This can be measured using the following equation:

$$\Delta T_f = i K_f c_m g$$

where ΔTf is the change in freezing point, i is the number of ions present, K_f is a constant for the solvent, and c_m is the concentration of particles dissolved in the water.

From this equation, you can see that more particles mean a higher total concentration, which creates a greater change in freezing point. Sucrose, lactose, and other sugars do not dissociate into ions when dissolved in water, so one molecule of NaCl will have twice the impact on freezing point than sucrose. This is how an ice–salt bath can achieve a temperature low enough to absorb heat from the ice cream.

Science for the Chef: Matcha milk

Creamy matcha milk tea, also known as matcha latte, is a popular beverage made by combining matcha green tea powder with milk (dairy or plant-based). Matcha is a finely ground powder made from shade-grown green tea leaves, which gives it a vibrant green color and a unique flavor profile. Matcha milk tea offers a balance of earthy and slightly bitter notes from the matcha, combined with the creaminess and richness of the milk. It is enjoyed for its distinct flavor as well as the energizing and calming effects associated with matcha's unique blend of caffeine and amino acids.

Ingredients:
- 2 teaspoons matcha green tea powder
- 2 tablespoons hot water
- 2 cups milk (dairy or plant-based)
- 2 tablespoons honey or sweetener of choice (optional)
- Ice cubes
- Whipped cream (optional)
- Matcha powder for garnish (optional)

Instructions:

1. In a small bowl, whisk the matcha green tea powder with hot water until it forms a smooth paste.
2. In a saucepan, heat the milk over medium-low heat until it is steaming but not boiling. Stir occasionally to prevent scorching.
3. Pour the steamed milk into a blender or use a frother to create foam. This step is optional but will give the drink a creamy and frothy texture.
4. In a serving glass, combine the matcha paste with the milk, stirring well to ensure it is evenly mixed.
5. Sweeten the milk tea with honey or your preferred sweetener, if desired. Adjust the amount based on your taste preferences.
6. Add a handful of ice cubes to the glass to chill the drink.
7. If desired, top the matcha milk tea with whipped cream for an extra creamy indulgence.
8. Optional: Dust the whipped cream with a sprinkle of matcha powder for an attractive garnish.
9. Stir the drink gently to incorporate all the ingredients.
10. Serve the creamy matcha milk tea immediately and enjoy the refreshing and rich flavors.

Science Behind the Recipe

The creaminess of matcha milk tea is achieved through the use of milk or a plant-based alternative. Milk contains a significant amount of fat and proteins, which contribute to its creamy texture and mouthfeel. When heated, the proteins denature and the fat globules melt, resulting in a smoother consistency. Matcha green tea powder is rich in antioxidants and has a unique flavor profile. Mixing the matcha powder with hot water creates a concentrated paste that can be easily incorporated into the milk. The combination of the earthy matcha flavor and the creamy milk provides a harmonious taste experience. By using a blender or frother to create foam, you introduce air into the milk, giving the matcha milk tea a velvety and frothy texture. This process, known as aeration, helps to trap air bubbles within the milk and creates a light and airy mouthfeel.

Key Concepts

1. **Carbohydrates**, particularly **lactose**, play a vital role in milk as a source of energy while also influencing its flavor.

2. The **fermentation** of lactose in milk involves the participation of ***lactobacilli*** and ***lactococci*** bacteria. These bacteria convert lactose into lactic acid, which can lead to the souring or curdling of milk. This fermentation process creates an acidic environment that inhibits the growth of harmful microbes in milk.

3. Milk holds nutritional significance as it provides essential fats, proteins, and carbohydrates. However, certain proteins found in cow, goat, and sheep milk can trigger allergic reactions in susceptible individuals.

4. Milk proteins consist of two main classes: **casein** and **whey**. Understanding their composition helps differentiate between these proteins and their respective roles in various milk products.

5. **Amino acids**, the building blocks of proteins, play a crucial role in protein structure. It is essential to recognize the **essential amino acids** that the body cannot synthesize and must be obtained through diet.

6. When milk is acidified, the process of **curd formation** occurs. **Acid-soluble** and **acid-insoluble** milk proteins differentiate in their behavior during this process.

7. **Homogenization** and **pasteurization** are two important processes in milk treatment. Homogenization involves physical changes to milk fat globules to achieve a uniform distribution, while pasteurization employs heat to destroy harmful microorganisms.

8. Milk proteins, such as caseins and whey proteins, play a crucial role in foam formation by unfolding and rearranging at the air–water interface, creating a protective layer around the air bubbles. The fat content in milk, composed of **triglycerides**, acts as an **emulsifier**, enhancing foam stability by providing a continuous phase that surrounds the air bubbles and prevents their coalescence.

9. Butter production is a complex process that involves the transformation of cream into butterfat, which consists of various **fatty acids**. The composition of these fatty acids, including their **chain length** and **saturation** level, greatly influences the properties of butter, such as its spreadability. Temperature also plays a significant role in butter production, affecting the crystallization and solidification of the butterfat, which ultimately impacts the texture and consistency of the final product.

10. Butter contains small volatile, nonpolar flavor and aroma compounds that are formed during its production. These compounds significantly impact the taste of baked goods.

11. Ice cream is a **colloidal** mixture with a complex composition. Emulsifiers and air pockets play crucial roles in determining the structure and texture of ice cream.

12. **Freezing point depression** has a significant impact on the freezing process of ice cream. Proper mixing techniques and temperature control are essential to control the formation of ice crystals, thereby ensuring a desirable texture and consistency in the final product.

References

1 Title 21, Vol. 8, Ch. 1, Pt 1240, subpart A, Section 1240.3(j), Release 13.
2 Dhankhar, J. & Kundu, P. (2021). Stability aspects of non-dairy milk alternatives. In: *Milk Substitutes - Selected Aspects* (ed. M. Ziarno), London: IntechOpen, March 4th, 2021. https://doi.org/10.5772/intechopen.
3 Fleischer, D. (2018). UCSF sustainability stories. In Title of the Book, pp. 1–713. https://sustainability.ucsf.edu/1.713 (accessed June 2023).
4 UC Davis CLEAR Center (n.d.). How Dairy Milk Has Improved Its Environmental and Climate Impact. https://clear.ucdavis.edu/explainers/how-dairy-milk-has-improved-its-environmental-and-climate-impact (accessed June 2023)
5 Andrew, C. (2013). Milk revolution. *Nature*, *500*, 20–22.
6 Boyer, R. F. (2005). *Concepts in Biochemistry*, 3e, p. 110. Hoboken: Wiley.

Additional Readings

Effect of milk homogenisation and foaming temperature on properties and microstructure of foams from pasteurised whole milk. K. Borcherding, W. Hoffmann, P.Chr. Lorenzen, K. Schrader. *Food Science and Technology* 41 (2008) 2036e2043

The Effect of Overrun, Fat Destabilization, and Ice Cream Mix Viscosity on Entire Meltdown Behavior Biqing Wu, Dieyckson O Freire, and Richard W. Hartel. *Journal of Food Science* Vol. 84, Iss. 9, 2019

Hebling e Tavares, J. P., da Silva Medeiros, M. L., & Barbin, D. F. (2022). Near-infrared techniques for fraud detection in dairy products: A review. *Journal of Food Science*, 87(5), 1943–1960.Labeling of Vanilla Type Affects Consumer Perception of Vanilla Ice Cream Parker A.R., Penfield M.P. *J. Food Sci*. Vol. 70, Nr.8, 2005 S553.

Li, H., Zhao, Y., Jiang, Y., Li, J., Yang, J., Xue, Y., Kang, Z., Zulewska, J., Li, H., & Yu, J. (2023). Extensive hydrolysis of milk protein and its peptidomic antigenicity analysis by LC–MS/MS. *Journal of Food Science*, 88(6), 2655– 2664.

Ramsing, R., Santo, R., Kim, B.F. *et al.* Dairy and plant-based milks: implications for nutrition and planetary health. *Curr Envir Health Rpt* (2023). https://doi-org.sandiego.idm.oclc.org/10.1007/s40572-023-00400-z.

Warren, M.M. and Hartel, R.W. (2014), Structural, Compositional, and Sensorial Properties of United States Commercial Ice Cream Products. *Journal of Food Science*, 79(10), E2005-E2013.

End of Chapter 4 Questions

1. Which of the following is the primary carbohydrate found in milk?
   a. Glucose
   b. Sucrose
   c. Fructose
   d. Lactose

2. What is the function of lactase in the digestive process?
   a. Break down fats in milk
   b. Convert lactose into glucose and galactose
   c. Convert lactose into sucrose
   d. Enhance the flavor of milk

3. Which of the following bacteria are responsible for the fermentation of lactose in milk?
   a. *Escherichia coli*
   b. *Lactobacilli*
   c. *Staphylococcus aureus*
   d. *Salmonella enterica*

4. Milk is a significant source of essential nutrients. Which of the following components is primarily responsible for providing energy in milk?
   a. Essential fats
   b. Proteins
   c. Carbohydrates
   d. Vitamins

5. Identify the two basic classes of proteins found in milk.
   a. Casein and lactose
   b. Casein and serum
   c. Casein and whey
   d. Whey and lactose

6. Which type of protein remains soluble in milk when it is acidified?
   a. Casein proteins
   b. Whey proteins
   c. Lactose proteins
   d. Serum proteins

7. What is the primary function of casein proteins in milk?
   a. To provide essential amino acids
   b. To give milk its sweet flavor
   c. To support the stability of micelles
   d. To support immune function

8. What is the main purpose of homogenization in milk processing?
   a. To kill bacteria and other microbes present in raw milk
   b. To break down lactose into smaller sugar molecules
   c. To separate milk into cream and whey
   d. To break the milk fat globules into smaller and more uniformly sized particles

9. Which statement accurately describes pasteurization?
   a. It involves boiling raw milk to eliminate all microorganisms
   b. It heats milk at temperatures sufficient to kill harmful microorganisms while preserving its nutritive properties
   c. It preserves all the nutrients and proteins in milk
   d. It is a process invented by Louis Pasteur to produce beer and wine

10. What is the primary factor responsible for trapping air bubbles in foamed milk?
    a. Whey proteins
    b. Fat
    c. Denatured proteins
    d. Heat of steam

11. Why does heavy cream produce a different type of foam compared to low-fat skim milk?
    a. Heavy cream has a higher fat content than skim milk.
    b. Skim milk has more whey proteins than heavy cream.
    c. Heavy cream contains more denatured proteins than skim milk.
    d. Skim milk has a higher fat content than heavy cream.

12. Which of the following is a characteristic of unsaturated fats?
    a. Solid at room temperature
    b. High melting point
    c. Double bonds in their fatty acid chains
    d. Long shelf life

13. How does the process of hydrogenation affect the properties of fats?
    a. It increases the level of saturated fats.
    b. It decreases the stability of fats.
    c. It reduces the shelf life of fats.
    d. It converts liquid fats into solid fats.

14. What is the role of emulsifiers in food products?
    a. They enhance the flavor of the product.
    b. They provide a smooth texture to the product.
    c. They increase the nutritional value of the product.
    d. They stabilize the mixture of immiscible ingredients.

15. What compound is primarily responsible for the characteristic aroma of butter?
    a. Lactose
    b. Casein
    c. Lactic acid
    d. Diacetyl

16. What is the purpose of emulsifiers in ice cream?
    a. To increase the volume of the ice cream
    b. To stabilize the mixture and prevent the separation of fat and water phases
    c. To enhance the freezing process and reduce ice crystal formation
    d. To improve the creaminess and mouthfeel of the ice cream

17. How does freezing point depression contribute to the making of ice cream?
    a. It increases the sweetness of the ice cream
    b. It enhances the flavor and aroma of the ice cream
    c. It helps in achieving a smooth and creamy texture
    d. It reduces the freezing time required for making ice cream

18. Which of the following statements best describes the role of air in ice cream?
    a. Air adds body and chewiness to the ice cream
    b. Air increases the fat content of the ice cream
    c. Air stabilizes the mixture and prevents crystal formation
    d. Air enhances the freezing point depression effect

5

Cheese

Guided Inquiry Activities (Web): 6, Higherorder Protein Structure and Denaturation; 7, Carbohydrates; 8, pH; 9, Fats, Structure, and Properties; 10, Fats Intermolecular Forces; 14, Cells and Metabolism; 16, Milk

Learning Objectives

1. *Understand the composition of milk and its two phases or components.*
2. *Explain the structure and characteristics of casein and its role in the formation of cheese curd.*
3. *Describe the amphipathic nature of casein and its ability to form micelles in milk, and how this allows for the retention of fat and moisture in cheese.*
4. *Analyze the molecular interactions within the casein micelle, including the role of calcium and the impact of binding of κ-casein on the micelle's surface.*
5. *Identify the distinct structural and solubility characteristics of whey proteins.*
6. *Describe and differentiate between the three methods of curd formation in cheese production. Explain the molecular basis for acid-mediated curd formation.*
7. *Explain the process of lactic acid fermentation in cheese production, including the role of lactobacteria and the impact of milk pasteurization.*
8. *Describe the role of bacteria and their corresponding enzymes in the ripening process of cheese.*
9. *Compare and contrast the ripening processes of different cheese types, such as hard and semihard cheeses (e.g. Cheddar), soft cheese (e.g. Camembert), and blue cheeses.*
10. *Analyze the chemical changes that occur during cheese ripening, including the formation of flavorful and aromatic molecules, and explain how they contribute to the complexity and uniqueness of cheese flavors.*

5.1 Introduction

The components of dairy critical to cheese production will be emphasized in this chapter: curds and whey, lactose as a bacterial energy source, and fats. The science of making cheese, the involvement of different types of bacteria (starters and finishing), effects of acid or temperature, and the role of proteins and fats in flavoring and aromas will be discussed in detail. Processed cheeses, both their components and the science of melting will be covered in the context of using different cheeses for cooking purposes, such as melting, gratins, and soups.

The Science of Cooking: Understanding the Biology and Chemistry Behind Food and Cooking, Second Edition. Joseph J. Provost et al.
© 2025 John Wiley & Sons, Inc. Published 2025 by John Wiley & Sons, Inc.
Companion website: www.wiley.com/go/provost/food_science_2e

5.1.1 An Overview of Cheese

Its mention may conjure up thoughts of melted, stringy mozzarella on pizza, sharp, hard cheddar on a cracker, or pungent, crumbly gorgonzola in a lettuce salad. Few other single foods have such diversity in texture, flavor, aroma, and appearance. The diversity, derived from variation in the components, the details of the cheesemaking process, and the creativity of the cheesemaker, results in hundreds of different varieties of cheese that are produced for both economic and culinary values.

However, early ventures in cheesemaking were about something other than economics or taste. It is hypothesized that approximately 5,000 years ago, the peoples of the Middle East and Central Asia discovered that milk soured and curdled upon exposure to the hot sun and warm temperatures. This state of milk is not particularly palatable to most (think about pouring "chunky" milk on your breakfast cereal); however, when this sour milk was drained and salted, a tastier and, more importantly in the ancient Middle East, less perishable food that resembled yogurt resulted (Figure 5.1). It was also discovered that the milk curd became more firm and cohesive when the curdling took place in the presence of a piece of animal stomach. It may sound disgusting, but 5,000 years ago, you had to use what was available for your cereal bowl.

During the Roman Empire, when longer-distance travel was more common, cheesemaking and the cheese trade industry began to flourish. As such, types of cheese expanded significantly due to differences in local climate, animals, and feed, even within the same country. The result was a remarkable diversity of traditional European cheeses, which number from 20 to 50 in most countries and several hundred in France alone. Although the European Union still leads the world in cheese production, export, and consumption [1], cheese is produced and enjoyed around the world (Figure 5.2).

In this chapter, we will study the science behind the process of cheesemaking: the molecular basis for milk **curding**, what is **drained** from the milk curd when it is **pressed**, the flavorful aromatic molecules that are formed during salting and **ripening,** and the time in which a cheese might sit for days, weeks, months, or years before it is eaten. In addition, we will study the differences in the cheesemaking process and components that result in various cheeses. By the end of this chapter, you will understand why Velveeta™ melts beautifully in your macaroni and cheese, while stringy mozzarella is the best cheese for the top of your pizza (Figure 5.3).

Figure 5.1 Cheese. A durable form of milk and nutrition and a tasty way to eat with crackers. volff/Adobe Stock Photos

Figure 5.2 Cheese as an ancient food. The fourteenth century pictorial evidence of the cheesemaking process from an Arab medical health book. *Preparing and serving cheese* by an unknown master from the *Tacuinum Sanitatis*, the fourteenth century. Public Domain

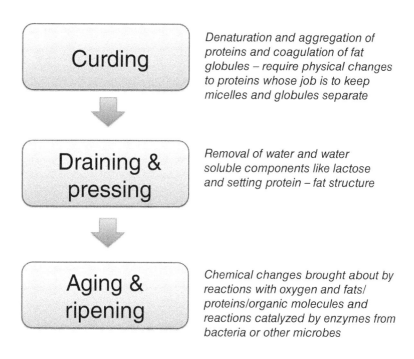

Curding

Denaturation and aggregation of proteins and coagulation of fat globules – require physical changes to proteins whose job is to keep micelles and globules separate

Draining & pressing

Removal of water and water soluble components like lactose and setting protein – fat structure

Aging & ripening

Chemical changes brought about by reactions with oxygen and fats/ proteins/organic molecules and reactions catalyzed by enzymes from bacteria or other microbes

Figure 5.3 The process of cheesemaking.

5.2 Milk Curdling and Coagulation

Have you heard a nursery rhyme about Little Miss Muffet (Figure 5.4)? Have you ever thought about what Little Miss Muffet was eating? What exactly are curds and whey? Miss Muffet's curds and whey could have looked like a bowl of long-outdated, sour, chunky milk, which sounds disgusting. However, you might enjoy the products of controlled coagulation every day, like cheese, yogurt, and sour cream. What exactly comprises a milk curd, that soft, coagulated (chunky) part of soured milk, and why does it form? To start, it is important to remember that milk is made of two phases or components. The milk globules, made up of fats and proteins, are one part of milk. These milk globules are suspended in an aqueous or watery phase that is filled with different proteins. Milk stays in this form until something happens to the proteins associated with the globules that keep the globules from coalescing into curds. Simply put, curds form and separate from the rest of the liquid milk because of the properties, characteristics, and interactions of the milk protein that coats the fat globules: casein.

5.2.1 The Milk Proteins: Casein and Whey

As you recall from Chapter 2, proteins are polymers of amino acids that are connected to one another via dipeptide bonds. Proteins, in their active native and functional form, have a distinct structure that typically maximizes intermolecular interactions with molecules with similar properties (e.g. polar interacts with polar) and minimizes interactions of unlike components (e.g. polar and nonpolar). This distinct structure is called the tertiary or quaternary structure (Figure 5.5).

Milk contains two types of protein: casein (mixed in and on the fat globules) and whey (the protein in the aqueous phase). These protein types are very distinct in

Figure 5.4 Cheese and Mother Goose. Little Miss Muffet sat on her tuffet eating her curds and whey. Unknown origin.

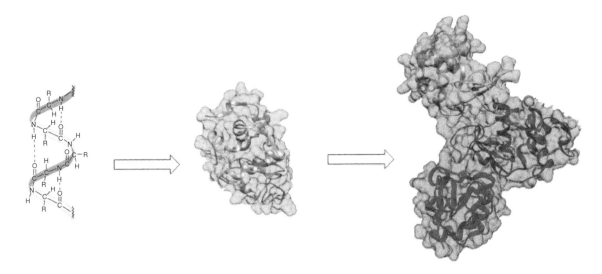

Figure 5.5 Level of protein structure. Primary structure, secondary structure, tertiary structure, and quaternary structure.

structure and their impact on cheesemaking and the final cheese product. Generally speaking, casein provides the basis of a cheese's solid structure and texture and governs the retention of fat and moisture in the cheese. In contrast, the whey proteins remain soluble in the milk liquid and are largely removed from the final cheese product during draining. You might ask, What is special about casein? What makes it so different from whey?

5.3 Casein

In Chapter 4, you learned that casein, the most abundant protein in milk, is actually a family of related proteins organized into four different subtypes: α(s1)-casein, α(s2)-casein, β-casein, and κ-casein. The casein proteins play a key role in human development as they help to transport calcium (in the form of calcium phosphate) to the bones. Caseins' mechanism for transporting calcium is related to cheese production, as you will learn below.

Casein proteins have an amphipathic character. This means that some parts of the structure of casein are hydrophilic, while other portions of the molecule are hydrophobic. Amphipathic molecules, like fatty acids, often aggregate or clump into a spherical shape and form a large molecular complex called a micelle (Figure 5.6). The micelle structure maximizes interactions between "like" chemical groups and minimizes interactions between, unlike chemical groups. In contrast, the hydrophobic nonpolar groups are contained inside the micelle sphere, where there is little to no interaction with water (or other polar groups). Figure 5.6 is an illustration of a micelle comprised of fatty acid molecules in which the molecules' hydrophilic, polar, or charged head groups are directed toward and interact favorably with the polar water; Figure 5.7 illustrates a milk-fat globule, which is a type of micelle. In the milk-fat globule, amphiphilic phospholipids form the barrier of the globule, where the polar head of the phospholipid interacts with the watery environment, and the fatty triglycerides are on the hydrophobic

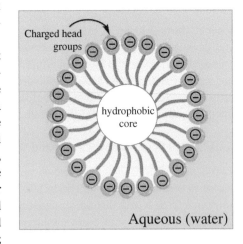

Figure 5.6 Amphipathic nature of a micelle. Charged, polar head groups (shown with a (−)) face the water solvent, while the hydrophobic portion of the compound aggregate due to hydrophobic forces.

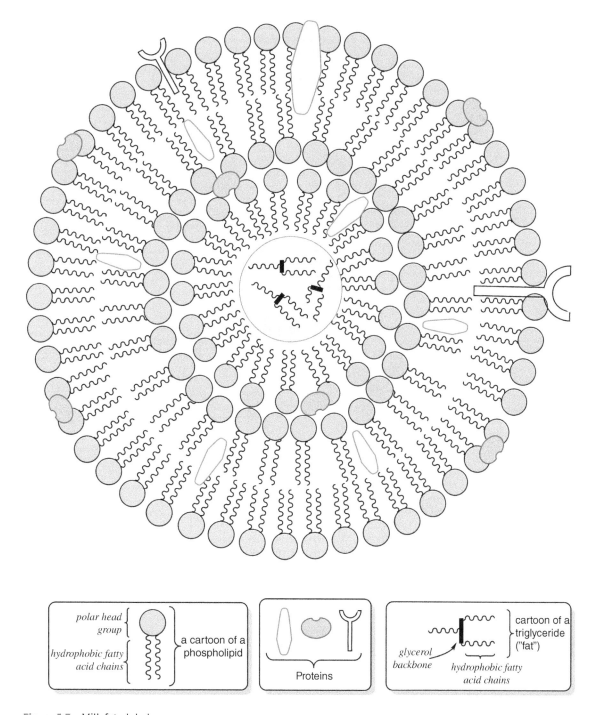

Figure 5.7 Milk fat globule.

interior of the globule. See Chapter 2, Section 2.5.2, and Figures 2.30 and 2.31 for more chemical detail on phospholipids and how they can stabilize mixtures of hydrophilic and hydrophobic components.

Casein protein molecules form protein micelles in milk; however, in order to form this structure, the micelles need some help. Protein micelles are a bit more complicated than fatty acid micelles, as the individual proteins are larger and more structurally complex than phospholipids, and the polar and nonpolar components of a protein are interspersed throughout the protein sequence and structure. However, protein micelles can still form with a largely nonpolar interior and a polar/charged exterior. Calcium is a cation that supports the formation of the casein micelle structure. Thus, calcium in milk does more than just help to make your bones strong. It actually holds the four types of casein proteins tightly in the micelle complex. Furthermore, this protein–calcium complex helps to keep milk fat globules from coalescing into curds (Figure 5.8). How does this work at

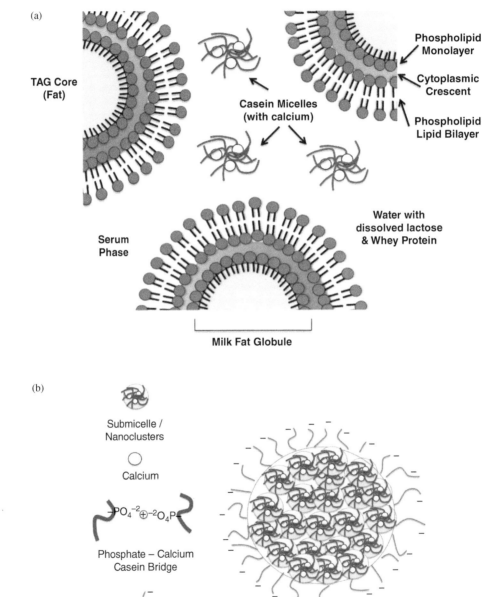

Figure 5.8 The composition of milk with casein micelles. (a) The composition of milk is shown as the three-phospholipid membrane of the milk-fat globules and the water-soluble casein micelles. (b) The calcium and phosphates help maintain the inner structure of the casein micelle, while the carboxyl group on κ-casein repels micelles keeping them in solution.

the molecular level? One of the casein proteins, κ-casein, is a negatively charged protein (i.e. it has many deprotonated carboxyl groups, COO⁻) that sits on the outside of the micelle. The κ-casein causes the micelle exterior to be negatively charged, which limits the size of the protein complex as the negative charges repel each other, preventing too many negatively charged proteins from joining the complex (Figure 5.8). Because the micelles are relatively small, there are many of them in the milk solution, so they interfere with/prevent the milk fat globules from getting too close to one another. These conditions are true for liquid milk at its normal pH. However, any disturbance in pH or other chemical changes that remove the negative charges from the κ-casein proteins will result in the aggregation of milk fat and protein, leading to curdling. To produce cheese, a cheesemaker disturbs the casein micelles to encourage the aggregation of the casein proteins into a solid-like mass that forms the basis of the cheese curd.

Box 5.1 Casein proteins: A detailed look

There are four sub-types of casein proteins. It is thought that the α- and β-casein proteins aggregate together to form submicelles, with the hydrophobic parts of α- and β-casein directed to the inside of the micelle and the hydrophilic and negatively charged components are directed to the outside. The negatively charged groups, which are serine amino acids that have been phosphorylated (i.e. phosphoserines) (Figure 5.9), attract and bind to the positively charged calcium of calcium phosphate particles. However, a negative surface remains due to the phosphate component. To combat the negative surface, another layer of α-casein binds to the negative surface via its positively charged amino acids, leaving the phosphoserines exposed to another layer of α-casein (Figure 5.8). The sub micelles are thought to grow in size in this stepwise manner and interact with other sub micelles via calcium phosphate bridges to form the larger micelle of about 150–300 nm in size until the kappa casein protein, κ-casein, binds to the surface of the micelle. κ-casein has one hydrophobic yet positively charged section (amino acid residues 1–95) which interacts with the other casein proteins that make up the "core" of the micelle and the bound calcium phosphate. The other section of κ-casein (amino acid residues 113–169) carries a negative charge and contains mainly polar residues; this portion of κ-casein decorates the micelle surface.

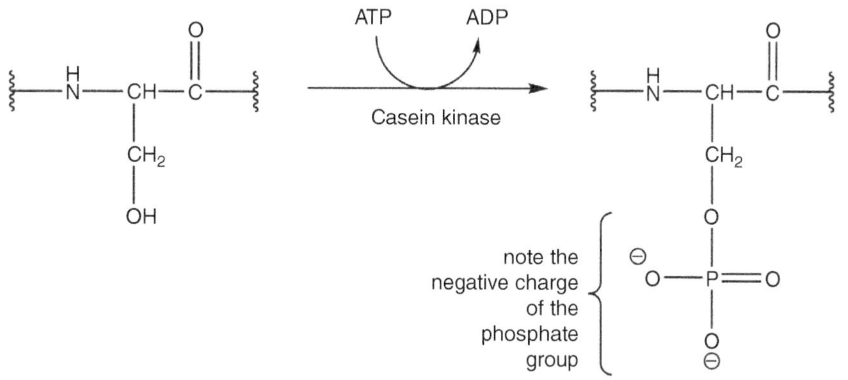

Figure 5.9 Serine and phosphoserine. The enzyme, casein kinase, transfers a phosphate group from adenosine triphosphate (ATP) to the side chain of serine in casein to create a negatively charged phosphoserine.

(Continued)

Box 5.1 (Continued)

Interestingly, once κ-casein binds, the micelle does not get any larger. How does κ-casein prevent the casein micelle from growing? Evidence suggests that it has something to do with κ-casein's inability to bind calcium phosphate and perhaps a structural change that occurs when it binds to the growing micelle to contain more β-sheet character. In addition, the negatively charged amino acids of κ-casein interact favorably with the water, allowing the micelle to be soluble (negative charges and polar molecules interact well). However, since negative charges repel one another, κ-casein effectively prevents two micelles from interacting and keeps the micelles small enough to remain soluble. Limiting the size of the micelle is important because when a single molecule or aggregate of molecules becomes too large, it may no longer be soluble in a solution and precipitates or "falls out" of the solution. When you make cheese, this is exactly what occurs!

5.4 Whey

Whey comprises all of the non-casein protein components of the milk, about 20% of the total milk proteins. Approximately 20–40 different proteins make up the family of "whey proteins," which have numerous different roles within an organism ranging from defensive proteins and transport proteins to enzymes. The most abundant whey proteins are:

- β-lactoglobulin (unknown function).
- α-lactalbumin (functions in lactose synthesis).
- Serum albumin (a transport protein).
- Immunoglobulins (defensive proteins).

However, what primarily distinguishes the whey proteins from the caseins is that they do not interact with other proteins to form large complexes or micelles in milk but fold into compact, globular structures that are soluble in milk water under most conditions (Figure 5.10). In other words, they do not aggregate together during cheese production and are not part of the cheese curd. As such, whey is typically considered a by-product of cheese production, and approximately 630,000 tons of whey protein are produced annually from the worldwide manufacture of cheese. That being said, in a few cheeses and dairy products like ricotta and yogurt, whey proteins play a critical role in production, texture, nutrition, and the overall final product.

5.5 More Milk Curdling

Now you know that the main components of cheese curd are the casein proteins and that casein proteins are soluble in milk but become insoluble during cheese production. The next thing to understand is how curd formation is induced. Does it vary in different cheese types? Are there different ways to induce curd formation?

Curd formation is commonly induced in one of the below three ways:

1. Increasing the acidity of the milk
2. Increasing the temperature of the milk
3. Through the use of an enzyme, commonly called rennet

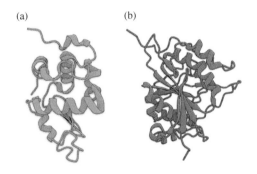

(a)　　　(b)

Figure 5.10 Lactalbumin. Two types of whey proteins are (a) α-lactalbumin and (b) β-lactoglobulin.

Depending on the type of cheese being made, any one, two, or all three of these methods may assist in the production of a particular cheese type.

5.5.1   Acid-mediated Curd Formation

In acid-mediated curd formation, a cheesemaker makes the milk more acidic through the direct addition of an acid, like lemon juice, or bacteria fermentation production of lactic acid. This is the only method used for curd formation in many fresh cheeses, such as Chévre, cream cheese, goat cheese, and ricotta, while some cheeses use acid-mediated curd formation in combination with another method. Let's learn about how acid induces the formation of cheese curd at a molecular level. Then we will go into some detail about the different acid-mediated curd formation processes and the resulting cheeses produced from each type.

5.5.2   The Molecular Basis for Acid-mediated Curd Formation: Why does Acid Cause Milk to Curdle?

Milk curds form when casein micelles are disrupted. One way to cause molecular disruption is by changing the pH of a solution, in this case, the milk. As discussed in Chapter 2, a protein's structure and characteristics are usually susceptible to pH; when the pH changes, the protonation state of acidic and basic amino acids within the protein changes. Generally speaking, a decrease in pH (a more acidic solution) will cause some amino acids to become more protonated. In contrast, an increase in pH (a more basic solution) will cause some amino acids to become more deprotonated. As the pH continues to rise, the interaction between acidic and basic amino acids is lost as the amino groups lose their proton, and there is no longer a positively charged group (NH^+) to interact with the negatively charged carboxyl group (COO^-) at pH greater than 9.

How does this work? The native structure of a protein is often stabilized by intermolecular interactions between its component amino acid side chains. Suppose an interaction depends upon an amino acid having a particular charge (positive or negative). As soon as the pH of the solution is altered, the relative amount of the amino acid that has that charge will change, potentially destabilizing the protein's structure. As an example, an interaction might occur between a negatively charged amino acid (i.e. glutamate) and a positively charged amino acid (i.e. lysine) within a protein structure at a particular pH (Figure 5.11). If there is a decrease in the pH of the solution (more acidic), more of your glutamate residues will be protonated, and the interaction with lysine might be weaker or less likely to occur. Thus, some protein molecules might denature/unfold in the solution.

If we relate this concept to the formation of cheese curds, milk has a pH of approximately 6.5. At pH 6.5, the casein and whey proteins are soluble in the milk. The caseins are largely folded and incorporated into the micelle structure described earlier, with the negatively charged domain of κ-casein at the surface of the micelle. The negatively charged domain of κ-casein is due to a high number of deprotonated glutamic acid (8) and aspartic acid (2) amino acids within the surface domain at pH 6.5. At the same time, the whey proteins have their compact, globular individual protein structures.

As milk pH decreases to a value of 5.5, the surface κ-casein protein and other casein proteins begin to lose some of their negative charges due to the protonation of these (previously) negatively charged aspartate and glutamate amino acids, which makes them neutral. This "neutralization" impacts the casein micelle in two ways. With the loss of negative charges, the interaction between calcium and casein is reduced; thus, some calcium phosphate is lost from the casein micelle structure. With less calcium, the sub-micelles that comprise the larger micelle complex are not as attracted to one another and dissociate from one another. As a result, the micelle begins to fall apart. Furthermore, as the milk pH continues to drop, the fragmented micelles become even less negatively charged, and the repulsion between one micelle fragment and another is reduced even further. The fragmented casein micelles, other free non-micelle proteins in the milk, and milk fats begin to aggregate to prevent exposure of the hydrophobic protein and fat components to the polar milk water. The resulting large networks of protein and fat precipitate out of the milk liquid as a soft, fragile gel, also known as a curd.

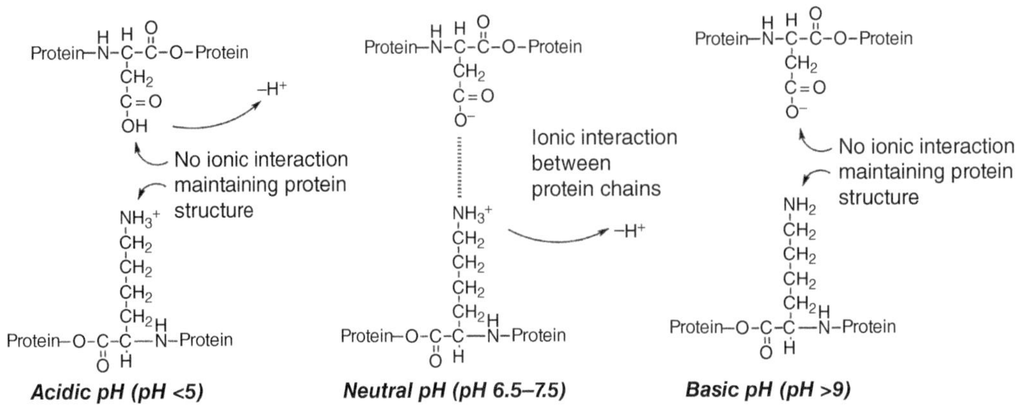

Figure 5.11 The impact of pH on ionic interactions and protein structure.

Box 5.2 Mascarpone: Acid-curdled cheese

Mascarpone is an Italian fresh cheese originating from the Lombardy region, made by curdling milk cream with citric acid or acetic acid. The whey is removed from the casein curds without any type of pressing; thus, the cheese has a high moisture content and is thick and soft. Because it is made from cream, it has a very high fat content ranging from 60 to 75%. The texture of mascarpone ranges from smooth and creamy to buttery, depending on how it is processed during cheesemaking. Making the cheese is so simple that many people easily make their own mascarpone at home.

Recipe
- Five hundred milliliters of whipping (36%) pasteurized cream (not ultrapasteurized)
- One tablespoon of fresh lemon juice

Bring 1 inch of water to a boil in a wide skillet. Reduce the heat to medium-low so the water is barely simmering. Pour the cream into a medium heat-resistant bowl, and then place the bowl into the skillet. Heat the cream, stirring often, to 190°F/89°C. It will take about 15 minutes of delicate heating. Add the lemon juice and continue heating the mixture, stirring gently, until the cream curdles. All that the whipping cream will do is become thicker, like sour cream. The back of your wooden spoon will have a thick layer of cream, and you will see just a few clear whey streaks when you stir. Remove the bowl from the water and let it cool for about 20 minutes. Meanwhile, line a sieve with four layers of dampened cheesecloth and set it over a bowl. Transfer the mixture into the lined sieve. Do not squeeze the cheese in the cheesecloth or press on its surface (be patient; it will firm up after refrigeration time). Once cooled completely, cover with plastic wrap and refrigerate (in the sieve) overnight or for up to 24 hours. Then, enjoy your homemade mascarpone by itself or in another recipe.

5.5.3 How Is Milk Made More Acidic?

There are two mechanisms whereby milk can be made more acidic. The easiest way is simply by adding an acid! There are many acids that are used in cooking and are effective enough to instigate the curdling process, such as lemon juice, citric acid, or vinegar. If you have ever eaten ricotta (often used in lasagnas), mascarpone (used in the dessert Tiramisu), or pannier (common in South Asian cuisine), you have eaten cheeses that are made by the direct addition of an acid to milk. What is really neat about these cheeses is that you can prepare them in your kitchen and they don't really melt! They simply get drier and stiffer when you heat them. The chemical explanation behind the non-melting characteristic is the aggregation of the casein proteins in the presence of acid. When the acid curd is heated, the casein aggregates have strong interactions, while the milk

Figure 5.12 Glycolysis and NAD$^+$/NADH. NAD$^+$ is quickly used during respiration/fermentation and is replaced in bacteria and some animal cells through the production of lactate.

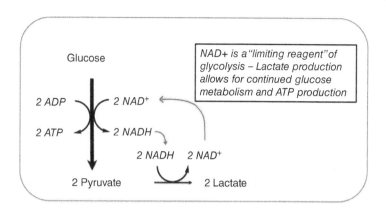

water is disrupted. The water evaporates and the casein proteins become more concentrated and dried out, rather than stringy and more liquid-like (which requires the presence of water). Because of this trait, acid-curdled cheeses retain shape upon exposure to heat and can be fried.

The second mechanism used to make milk more acidic is more indirect and utilizes the process of bacterial fermentation that you first learned about in Chapter 3. Here are some key points as it relates to milk coagulation in the cheesemaking process. In fermentation, bacteria utilize the enzyme lactate dehydrogenase to convert the glycolysis products of pyruvate and NADH to lactic acid and NAD$^+$ (Figure 5.12). While fermentation is critical for the organism's survival in that it regenerates the NAD$^+$ necessary to produce adenosine triphosphate (ATP) through glycolysis, the lactic acid produced also makes the environment more acidic! Thus, if you allow bacteria to grow in milk used to produce cheese, you have an internal source of acid, the lactic acid produced by the growing bacteria. Moreover, the lactic acid produced doesn't just affect the formation of the milk curd; the presence and amount of lactic acid generated also impact the taste and aroma of cheese. Depending on the cheese being made, two strains of bacteria may be used in this process. A starter bacteria (such as lactobacteria) begins acid production through fermentation, while a second strain of bacteria, commonly called a nonstarter or ripening bacteria (Figure 5.13) may be used in the latter stages of cheesemaking to "finish" or ripen the cheese. The finishing bacteria vary depending on the type of cheese being made, some produce gas to make Swiss cheese, while others produce other flavor molecules as the bacteria metabolize protein, fat, and sugars, producing new flavorants. The common characteristic of finishing bacteria is that they can thrive in the more acidic pH produced

Figure 5.13 Bacterial culture and cheese.

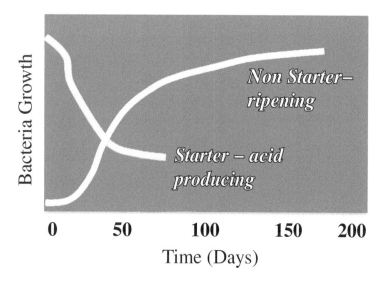

by the starting cultures. Many of these bacterial strains can also handle the higher heat that may complete cheese production.

5.6 Lactobacteria and Fermentation

In order for lactic acid fermentation to occur, you need a source of sugar and an organism that can utilize the sugar for energy. As you learned in Chapters 1 and 4, the primary sugar found in milk is lactose. Lactose is composed of a glucose molecule and a galactose molecule, linked together via an *o*-glycosidic bond (Figure 5.14). Given that both glucose and galactose, are sugar sources for glycolysis, you might think that lactose is an ideal sugar source for every organism—two sugar molecules for the price of one disaccharide. However, lactose cannot enter glycolysis in its disaccharide form. Lactose must be cleaved at its glycosidic bond in order to produce the more "useful" glucose and galactose and there are only a limited number of bacteria that contain the necessary enzymes to carry out this cleaving or hydrolysis process.

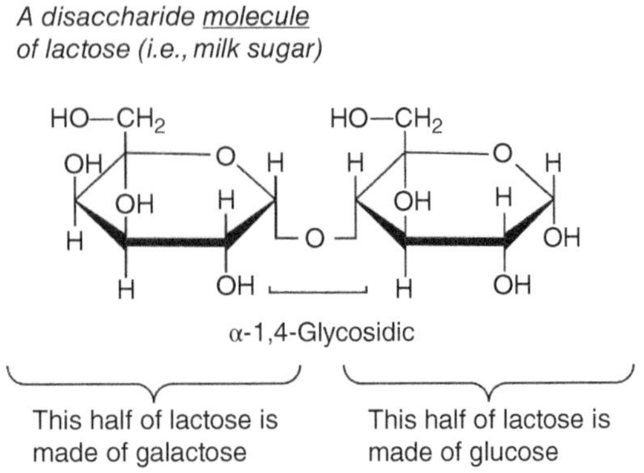

A disaccharide *molecule* of lactose (i.e., milk sugar)

α-1,4-Glycosidic

This half of lactose is made of galactose

This half of lactose is made of glucose

Figure 5.14   Lactose.

The type of bacteria that are utilized in the curdling phase of cheese production is called (broadly) the lactobacteria. Lactobacteria are naturally present and thrive in raw milk, as they have a nearly endless supply of sugar (~5 g lactose/100 g of milk). As the bacteria naturally present in the milk ferment (i.e. use the lactose as an energy source, making lactic acid), the milk becomes increasingly acidic and begins to curdle. You might be thinking, but isn't all milk pasteurized? Doesn't pasteurization kill all of the bacteria, lactobacteria, and other potentially harmful types? Pasteurization does kill harmful, disease-causing bacteria like *Salmonella, Listeria*, and *Escherichia coli*, which are associated with many foodborne illnesses, along with most all of the lactobacteria that are native to the milk. Thus, for cheeses that are produced from the lactic acid generated through bacterial fermentation, one or more "starter bacteria" are added to the pasteurized milk during cheesemaking to ferment lactose, thereby making the milk more acidic to induce curdling.

Broadly speaking, two groups of lactobacteria may be used to "start" or induce curdling: moderate-temperature (i.e. mesophilic) *lactococci* and the heat-loving (i.e. thermophilic) *lactobacilli* (Figure 5.15). These lactobacteria are both homofermenters, as they only produce lactic acid as a product of fermentation and not ethanol, which is

Figure 5.15   Dry and fresh yeast. daffodilred/ Adobe Stock Photos

beneficial from the perspective of cheese production. Moreover, some lactobacilli benefit human health, as you have heard advertised for dairy products with "probiotic" bacteria. The *Lactococcus* family of lactobacteria is relatively small, with only seven identified species, while the *Lactobacillus* species has approximately 50 members.

What makes a cheesemaker choose a particular bacterial strain or type? The type of bacteria used depends upon the specific steps involved in making a particular cheese type (like temperature), as well as the metabolic characteristics of the bacteria that allow for the production of other distinct molecules and by-products. Suppose milk is subjected to temperatures of 70°C during curd formation. In that case, *lactococci* will effectively ferment, produce lactic acid, and reduce the pH of the milk adequately to induce the formation of the curd. Most cheeses are acidified by the *lactococci*, as these mesophilic bacteria can survive and thrive under most typical cheesemaking conditions. However, for the cheeses that undergo a high-temperature cooking step, including mozzarella and hard Italian cheeses, thermophilic *lactobacilli* bacteria must be utilized since these bacteria thrive and survive under high-temperature conditions. Although starter bacteria are adequate to induce the formation of the curd, and to contribute to the cheese ripening and aging process, in the preparation of almost all cheeses, another agent is used to make the curd more robust, elastic, and strong—this agent is an enzyme called rennet.

5.6.1 Enzyme-mediated Curd Formation

A second method used to induce curd formation and coagulation of casein proteins involves an enzyme called rennet. Rennet is a general name for an enzyme or enzyme mixture that contains protease enzymes. Proteases cleave or degrade proteins into smaller pieces by breaking dipeptide bonds (Figure 5.16). Proteases are very important in biological system efficiency and play a key role in degrading proteins that are no longer needed, allowing for the recycling of amino acids. However, protease activity is also important and relevant in many aspects of food chemistry. In fact, protease activity is the basis for the warning label on your gelatin box about adding fresh pineapple, kiwi, or papaya.

Traditional rennet is made from the fourth stomach of a milk-fed calf or other ruminant animals. The effects of rennet on cheese production and curd formation are thought to have been discovered when ancient peoples stored their milk in pouches made from an animal stomach and discovered that the milk soured and coagulated more rapidly than when stored in a different type of natural container. Traditional rennet from a calf stomach is often required for traditional, artisanal European cheese production. However, through genetic engineering, a pure version

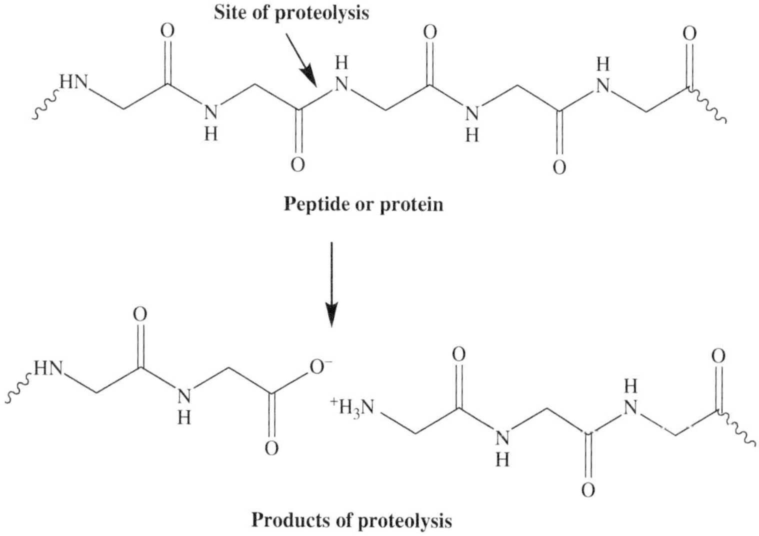

Figure 5.16 Hydrolysis of the protein backbone. Enzymes (proteases) or strong acids can break the protein backbone in a process generally called proteolysis.

Site of proteolysis

Peptide or protein

Products of proteolysis

of the calf enzyme critical to curd formation, called chymosin or rennin, is now produced in a bacterium, mold, or yeast and is purified from these model organisms. Most cheese in the United States is made with these engineered rennets, and this is the rennet that you can purchase at your local grocery store. To avoid the animal origin of rennet, a closely related alternative enzyme, vegetable rennet, isolated from thistle, may be used.

Some proteases are not very specific. They will break dipeptide bonds between any two amino acids or at amino acids that have a certain property (e.g. positively charged, aromatic). These non-specific enzymes are very useful for breaking a protein into its individual amino acid components during the process of biological protein degradation. Other proteases fulfill a specific role and are very specific for the types of amino acids that they will recognize and proteins that they will cleave. A protease called chymosin, one of the components of rennin, is critical in curd formation. Chymosin is a protease that catalyzes the cleavage of a single dipeptide bond between Phe105-Met106, phenylalanine and methionine amino acids at positions 105 and 106 in κ-casein (Figure 5.17).

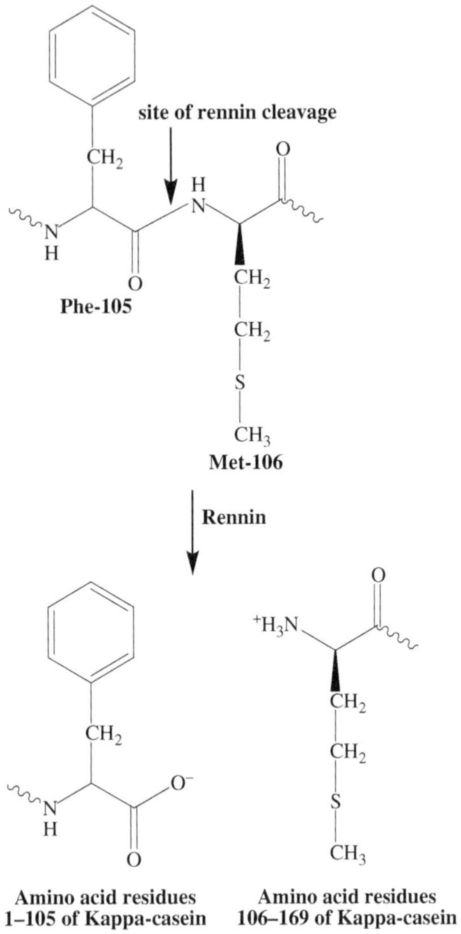

As you recall from the discussion about κ-casein earlier in this chapter, the negatively charged tail of κ-casein (i.e. amino acid residues 116–169) assists in keeping the casein micelle soluble and small. The proteolysis cleavage event that is carried out by chymosin effectively removes the negatively charged tail of κ-casein (called the κ-casein glycopeptide) from the casein micelle, leading to a similar molecular outcome that was discussed in acid-mediated curd formation. Without the negative charge, the casein micelles fragment, calcium is lost from the micelle, and the individual casein proteins aggregate, resulting in the formation of a rubbery, strong cheese curd. The cheese curd formed by chymosin is much stronger than that formed in acid-induced coagulation, as the κ-casein glycopeptide that effectively prevented the micelles from aggregating is soluble and is lost in the liquid whey. Without the negatively charged tail, the micelles easily aggregate, even more easily than in acid-induced coagulation, where aggregation must overcome the structural presence of the negatively charged κ-casein glycopeptide.

Figure 5.17 Actions of rennin on protein. The proteolytic action by rennin on casein results in its digestion into peptides.

Box 5.3 The effectiveness of rennet coagulation

In rennet coagulation, the curd forms in less than an hour and is firm enough to cut into pieces that are as small as a grain of wheat. Many cheeses use a combination of acid and rennet to yield a curd that is appropriate for the type of cheese being made. In the production of hard or semihard cheeses, such as Cheddar, Gouda, and Parmesan, rennet is the primary agent for coagulation. Cheeses of more moderate moisture content (which makes them softer) are curdled with a smaller amount of rennet in combination with acid. Soft cheeses, like ricotta, mascarpone, and paneer, are coagulated by acid without the presence of rennet, which leads to a very soft curd (that will take a long time to form at room temperature). Fortunately, heat is usually a component of many recipes. Read more in Section 5.6.2.

As stated earlier, the rennet used in cheese production typically contains a mixture of proteolysis enzymes, including chymosin, pepsin (an enzyme in the digestive system that is most effective at cleaving peptide bonds that involve a hydrophobic, aromatic amino acid), and lipase (an enzyme that catalyzes the hydrolysis of fats). The non-chymosin proteases cause the degradation of proteins and other molecules during curd formation to a lesser and less specific degree than chymosin. In addition to its role in curd formation, this protein degradation is also important to the texture and flavor of the finished cheese product.

5.6.2   Temperature-mediated Curd Formation

The final mechanism used to assist with curd formation in cheese production is temperature. By cooking the coagulating milk at a temperature in which the starter bacteria thrive, the microbes metabolize lactose and undergo fermentation more readily. Thus, at these more optimal bacterial-growth temperatures, more lactic acid is produced and the pH of the milk is decreased more rapidly. Proteins also more readily denature at higher temperatures, causing them to unfold and form nonnative structures. These denatured casein proteins disperse from a micelle and begin to aggregate, causing them to precipitate and curdle out of solution. Thus, in addition to supplementation with starter bacteria and rennet, increasing the temperature of pasteurized milk to between 78 and 104°F (26 and 40°C) promotes the coagulation of the milk and fermentation of the starter bacteria. For most cheeses, the optimal temperature lies between 86 and 95°F/30−35°C, while some goat cheeses are made at 68−77°F/20–25°C, while Pecorino cheeses are made with milk held at 95–104°F (35–40°C).

During the process of acid-mediated, enzyme-mediated, or milk sitting on your kitchen counter on a hot summer day-mediated curdling, the casein proteins clump together, forming a solid curd mass, while the whey and other solution proteins and peptides remain in solution. However, if the process stopped here, then most cheeses would resemble yogurt or cottage cheese: a curdled mass of protein sitting in a sea of whey, water, and other nonaggregated molecules. In order to prepare a more solid and drier cheese mass, the solid curd must be separated from the liquid through the process of cutting, draining, and pressing.

5.7   Removing Moisture From the Cheese

5.7.1   Cutting, Draining, and Pressing

The next cheese preparation phase involves removing moisture from the cheese curd, which significantly impacts the final product's texture (Figure 5.18). You may have noticed this difference in texture when you have eaten or worked with different types of cheeses in the kitchen. For example, Parmesan is a hard, dry cheese easily shaved

Figure 5.18   Cheese curds. Worker with cheese curds mixed in the watery whey. Belish/Adobe Stock Photos

Figure 5.19 Camembert cheese. Lukas Gojda/Adobe Stock Photos

into strips due to its texture. Mozzarella is a high-moisture cheese that is soft to the touch and is impossible to shave into thin strips. The difference in texture between these two cheeses and hundreds of others is based on the treatment of the formed curd (also known as the coagulum).

Gravity is the simplest method for draining liquid whey from the cheese curd. In draining by gravity, the whole curd is ladled into a cheese mold and is allowed to drain. This method takes several hours, days, or weeks depending upon the cheese type and is used for soft cheeses that retain a considerable amount of moisture, including Camembert (Figure 5.19). If you have ever eaten this kind of cheese, you can attest that it is very soft and is, in fact, sold in small round boxes for protection.

Some firmer cheeses also use gravity as a draining method; however, the curd is first "cut" into smaller pieces to allow for a greater surface area from which the whey can drain. As you would expect from the name, a cheese curd is cut into smaller pieces in a lengthwise and crosswise direction by a knife or a Swiss harp (Figure 5.20). This fragmentation of the curd mass provides a larger surface area from which the whey can drain. The moisture content of the final cheese product can be controlled by the size of the cheese curd pieces. For soft cheeses, the curd is cut into pieces that are the size of a walnut; for firm cheese, a medium-sized curd (the size of a corn kernel) is cut; for hard cheese, the curd is cut to the size of a grain of rice.

Although cutting the curd does lead to a firmer cheese, gravity alone does not adequately remove enough whey-containing liquid to prepare a hard or even semihard cheese. Thus, once cut, the curds are often stirred and

Figure 5.20 Cutting curds with a cheese harp. A worker uses a Swiss Harp to cut the curd into smaller squares. Olaf Speier/Adobe Stock Photos

Figure 5.21 Separating the curds from the whey. The use of a cheese cloth to drain the curd from the whey for Parmesan cheese. matteosan/Adobe Stock Photos

warmed to expel additional whey. What does cooking do to the curd? Through heating the cheese curd, additional liquid-containing whey is released from the curd, and the curd shrinks in size, thereby becoming more solid or firm. The increase in temperature also helps to further denature curd proteins. In the native state, the surface amino acids of most proteins will bind water molecules via hydrogen bonds. Once a protein is denatured, the protein unfolds, disrupting the water-loving amino acids on the surface and exposing the hydrophobic amino acids normally located in the interior of proteins. A denatured protein will have less water-binding capacity, and water will be released as the denatured protein molecules aggregate with one another.

This temperature increase also activates the finishing bacteria and increases the activity of enzymes released from bacteria into the curds. Activating the bacteria and enzymes in the cheese has two effects: the acidity of the curd is increased and the bacteria are activated, thereby aromatic and flavorful molecules that contribute to the flavor of the final cheese product (we will talk more about the chemistry of these molecules later). As you might expect, different cheeses are warmed to different temperatures; a higher warming temperature leads to greater shrinkage of the curd and a more firm cheese. For example, a Cheddar curd is warmed to a temperature of 100°F /38°C, the more firm Gruyere is warmed to 120°F /48°C, and the hard Parmesan and Romano cheeses are warmed to 130°F /54°C during cheese cutting.

After cooking the curd, the curd needs to be separated from the whey. The separation might be carried out by lifting the curds out of the cooking kettle with a piece of cheesecloth, allowing the curd/whey mixture to flow into molds (where the two separate), or placing the mixture into a larger rectangular vat to drain (Figure 5.21). However, for most cheeses, the cheese is still not adequately firm enough. After this initial separation step, the curd may be placed into a mold (if it is not already molded) and is pressed to expel more whey and draw the curd into a cohesive mass.

Simple or gentle pressing, which involves stacking one layer of curds on top of another layer of curds, is the only pressing method used for semihard cheeses that retain a high moisture content (e.g. manchego). However, when more whey needs to be removed, as is the case with hard cheeses, or to encourage rind formation, the cheese is pressed mechanically (Figure 5.22). As the cheese undergoes molding and pressing, the bacteria continue to acidify the cheese, which contributes both to whey expulsion and generation of flavorful molecules.

5.7.2 Cheddaring

Cheddar cheese undergoes a different method, called cheddaring, during the draining and pressing stage of cheese preparation. In the preparation of Cheddar cheese, the curd is left to sit (it is not separated from the whey) and is spread out laterally. During this time, it draws together into a mass called a curd cake. The curd cake is cut into pieces, which are stacked on top of one another for a period of time, and then they are relayered until the curds have a stringy texture, like chicken breast. This method expels whey, as the curd continues to be acidified by the active bacteria. The curd mass is then milled into small pieces, salted, transferred to molds, and pressed.

Figure 5.22 Cheese press. Mechanically pressing cheese before aging. Comugnero Silvana/Adobe Stock Photos

5.7.3 Salting

Perhaps you have looked at the nutritional values for a variety of cheeses and noticed the sodium content in a 1 oz serving (e.g. blue, 396 mg: Cheddar, 176 mg: part-skim mozzarella, 150 mg: and Parmesan, 433 mg). Where does this sodium come from, and what does it do? Cheese contains sodium because the cheesemaking process involves the addition of salt either during or at the end of production. Although salting contributes to the taste of many cheeses, salt does more than add taste and raise your blood pressure. Salting draws moisture out of the curds, inhibits the growth of bacteria (both the helpful lactobacteria and harmful microbes), and alters the activity of enzymes that are important to the ripening or finishing stages of cheese production.

First, how does salt impact moisture removal? Salt pulls moisture from the cheese through osmosis. Osmosis is the net movement of solutes (dissolved particles) from a region of high concentration to a lower concentration across a semipermeable membrane. Thus, if you have two objects that contact one another and one object is saltier than the other, the saltier object will pull moisture from the less salty object, and salt will move from the saltier object to the less salty one (Figure 5.23). You may have taken advantage of the power of osmosis if you have ever placed a piece of bread into a canister that contains hardened brown sugar. Initially, the bread is soft (i.e. high moisture content) and the brown sugar is hard. After some period of time, the moisture from the bread is pulled into the sugar, and when you open your canister, the bread is hard and the brown sugar is soft. But let's now think about osmosis in the context of salting the cheese.

Salt is placed on the surface of freshly formed cheeses or cheese curds. The cheese is less salty and has a higher moisture content than the salt. As such, the salt on the surface migrates into the cheese (very slowly), and some of the moisture in the cheese moves toward the salt (is pulled out of the cheese). After a period of time, the salt grains on the cheese's surface "melt" as they intermingle with the whey-containing liquid, and the cheese becomes less moist and saltier.

Cheddar and Colby are good examples of cheeses that are salted in the curd stage of cheese preparation. Because the salt is mixed into the curd (which can be mixed), a homogeneous balance of moisture and "saltiness" is attained within the mixture. In other cheeses, such as Pecorino and German blue, the molded

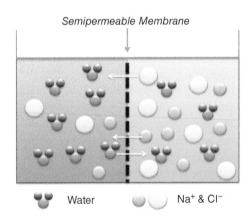

Figure 5.23 Osmosis. Water, sodium, and chloride ions (salt) will spontaneously move from a high to a low concentration. For cheese, a high salt solution will result in a lower water concentration, and the water molecules in cheese will flow from the cheese to the brine.

cheese is rubbed or sprinkled with salt, often multiple times. In these cases, the salt is more concentrated on the outer layer of the cheese and only gradually penetrates to the interior. Time is required to achieve a salt and moisture balance between the interior and the exterior of the cheese, leading to the formation of the rind. The rind not only holds the cheese together but also keeps the cheese from drying out. For Camembert, the salting process is complete after 4–6 days, and the formed rind holds the soft interior of the cheese in place.

Many types of cheese are salted by using a brine, a liquid solution with a salt concentration of 15–20%. It is an effective method, as it enables the cheese to be evenly salted (the cheese literally sits in this liquid, salty brine). The cheese might remain in the brine bath for anywhere from 30 minutes to several days. For example, a 10 kg piece of Cheddar will sit in brine for 40 hours.

Salt also reduces bacterial growth and enzyme activity. You may be thinking bacteria and enzymes need to flourish in order to make cheese, and the lactobacteria are our friends in the cheesemaking process. This is true; however, other bacteria, including disease-causing ones, also flourish in a moist, warm environment. Salt reduces the growth of unwanted and potentially harmful bacteria and molds and slows the growth and enzyme activity associated with the beneficial starter and ripening bacteria. This reduction in the function of the effective bacteria is an important step, as it controls the degree of acidity of the cheese and the molecules produced by the bacteria.

5.8   Ripening or Affinage

At this point, most cheeses look like a salty, crumbly, rubbery curd, which is not the most appetizing visually, aromatically, or in terms of taste. In the final step of cheesemaking, called the finishing phase, the cheese is allowed time to ripen. During ripening, the starter bacteria (still present) and other added bacteria (called the finishing or ripening bacteria), specifically their enzymes, transform this unappetizing mass into a delicious cheese over a period of 2 weeks (mozzarella) to two or more years (Cheddars and Parmesans). Temperature and humidity play a significant role in governing how and which bacteria grow and thrive, impacting the flavor, texture, and aroma of the cheese. The French term for ripening is *affinage*, which means "end" or "ultimate point." As such, this stage of cheesemaking may be carried out by an *affineur*, a cheese tenderer, or finisher. The affineur takes care of the cheeses in the cheese-ripening cellars until the cheese has ripened adequately for packing and sale. As has been a theme within this entire chapter, the ripening process varies considerably for different cheese types. Thus, first, we will discuss some of the basic techniques and chemistry involved in ripening and then go into detail about how ripening is carried out in the production of certain types of cheeses. Notably, not all cheeses undergo the ripening or aging process. In fresh cheeses that are prepared by acid or acid/heat coagulation methods, the cheese is ready to eat as soon as the curds are processed.

5.8.1   Work in the Ripening Cellar

Many cheeses are ripened within a special storage room that is specific for ripening cheese called a ripening cellar. A ripening cellar imitates the conditions of a cave, which was the historical location for this purpose (have you ever heard of the caves of Roquefort?). The cellar is controlled for humidity and temperature. The majority of cheeses ripen between 46 and 60°F (8–10°C), with a relative humidity of 85–95%. The climate of the cellar is controlled by ambient temperature and humidity, as well as the movement of air through the space.

In addition to temperature and humidity, each cheese receives a different type of care in the ripening cellar, carried out by the affineur. The affineur regularly brushes, rubs, or washes the surfaces of the ripening cheeses with salt brine and ripening bacteria or coats cheese surfaces with salt. All cheeses must be turned from time to time to ensure even development of the bacteria and to prevent deformities in the cheese shape. The affineur also determines when the cheese has acquired its proper appearance, aroma, taste, and texture, so that it can be packaged for sale.

5.8.2 The Chemistry of Ripening: How Does It Happen?

The chemistry of ripening is dependent upon the presence of bacteria and the corresponding activity of their various protease and lipase enzymes that break down the proteins, carbohydrates, and fats found in the milk to smaller, aromatic, and flavorful molecules. The bacteria, yeast, or molds that carry out this work have either been present throughout the entire cheese production process or are added during ripening. If ripening occurs via starter lactobacteria, then the cheese ripens in a homogeneous fashion, as the bacteria are present in the entire body of the cheese. This type of ripening is common in hard and semihard cheeses like Parmesan, Cheddar, and Gouda. However, the soft cheese Camembert is ripened from the outside inward due to a coating of the *Penicillium camemberti* mold (carried out by the affineur). In blue cheeses, the blue-green *Penicillium roqueforti* mold is added to the cheese by piercing the cheese mass and adding the mold to the middle. The cheese is then pierced further to allow oxygen access to the bacteria. The piercing speeds up ripening and causes the cheese to be ripened from the interior to the exterior.

5.8.3 The Chemistry of Ripening

We have been talking a lot about ripening, but what exactly happens during the ripening stage? What flavorful and aromatic molecules are produced during ripening, and where do they come from? The aromatic, flavorful, and distinct molecules that we smell in a particular type of cheese are due to the degradation of three types of molecules in milk: protein (casein), lipids/fats, and lactose (milk sugar). The specific molecules that are formed depend upon the microbes present during ripening, as well as the type of milk used.

Casein proteins can be degraded to their amino acid components by proteases in the ripening bacteria. Individual amino acids themselves have flavorful properties that range from sweet to savory. For example, alanine has a sweet taste, tryptophan has a bitter taste, and the sulfur-containing amino acids cysteine and methionine impart a "meaty" or "eggy" taste to dishes. The amino acid glutamate (also known as MSG) is often added to savory dishes to enhance flavor. When casein is degraded during ripening, all of these amino acids are produced and affect the flavor of the cheese (Figure 5.24). Moreover, while one of these amino acids will remain in that molecular state, many will be broken down further into smaller amines, including trimethylamine (which has a fishy taste), putrescine (smells of spoiling meat), or ammonia. While these flavors sound disgusting in this context, small amounts present within a particular cheese bring about a complex, rich, and unique flavor (Figure 5.24).

Fats also have an important role in the flavor and texture of cheese (Figure 5.24). In fact, the same ripening enzymes that degrade proteins will also metabolize and chemically change fats to produce a symphony of aromatic and flavorful compounds. Triacylglycerol fats from milk fat are broken down by lipases (i.e. enzymes that degrade lipids) into free fatty acids in the curd. The fatty acids then undergo additional oxidation/reduction chemistry. In beta-oxidation, the fatty acid is oxidized at the carbon that is "beta" to the carboxylic acid. This allows for a decarboxylation reaction, which reduces the fatty acid chain by two-carbon units, yielding a ketone or alcohol. In delta-oxidation, an oxygen atom is added to the carbon four units away from the carboxylic acid to yield a δ-hydroxyacid. These molecules can cyclize to a lactone. The fatty acid can also be converted to an ester. The relevance of this chemistry is that some short fatty acids have a peppery, sheep-like, or goat-like taste or aroma (Figure 5.24).

Finally, the short carboxylic acids, lactic acid and citric acid, can also be broken down to contribute to the flavor and aroma of cheese. These acids, both of which are products of bacteria fermentation, produce important molecules like diacetyl (a "buttery" taste), ethanal (also known as acetaldehyde), and ethanol (Figure 5.24).

In order to bring this ripening chemistry to life, let's discuss some of the molecular details of the ripening of several cheeses that have very different properties: blue, Swiss, Camembert, Cheddar, and Muster, and Limburger.

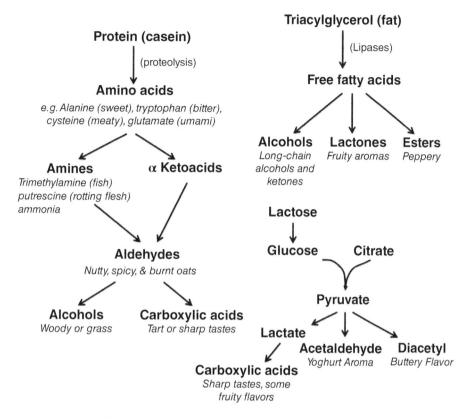

Figure 5.24 Aging cheese—proteins, fats, and lactose. Microbial cultures or the remnants of bacteria cells continue to metabolize biological molecules into new products, many of which have characteristic aromas and flavors.

5.9 Blue Cheeses, Molds, and Chemistry

Molds are microbes that require oxygen to grow, can tolerate drier conditions than bacteria, and produce a significant number of protease and lipase enzymes that impart a distinct texture and flavor to cheese. The standard ripening molds used in cheese production come from the genus *Penicillium*, from which the antibiotic penicillin is derived. The blue *Penicillium* are unique molds in their ability to grow in the low oxygen conditions found in small cavities and veins within cheese. *P. roqueforti* gives Roquefort cheese its veins of blue, its name, and its flavor (Figure 5.25). *P. roqueforti* and its close relative *Penicillium glaucum* also color the interior and flavor of Stilton and Gorgonzola. But where does the flavor of a blue cheese come from?

The typical blue cheese flavor comes from the *Penicillium* mold's metabolism of milk fat. *P. roqueforti* degrades 10–25% of the triacylglycerol fat in its cheeses,

Figure 5.25 Blue cheese. Roquefort cheese finished with *P. roqueforti* mold. rdnzl/Adobe Stock Photos

first liberating short-chain fatty acids from the triglyceride (Figure 5.24). These fatty acids give a peppery taste but are transformed further by the *P. roqueforti* to δ-hydroxyacids and β-ketoacids that give a peppery taste and then further transform the fatty acids to methyl ketones and alcohols that give the characteristic aroma of a blue cheese. Heptan-2-one and nonan-2-one are most often linked with a blue cheese smell; pentan-2-one is usually described as fruity. Other compounds, including alcohols and esters, round out the flavors; thus in gorgonzola, heptan-2-one and nonan-2-one are key impact molecules, but 1-octen-3-ol, 2-heptanol, ethyl hexanoate, and methylanisole are also important odorants (Figure 5.26).

5.9.1 Camembert Cheese, More Molds, and Chemistry

Penicillium camemberti is the mold microorganism that is smeared on the surface of Camembert and Brie cheese to form the characteristic white crust, buttery taste, ammonia-like aroma, and soft texture. Let's think about how a microorganism on the exterior of a cheese can do all of this chemistry!

The smeared *P. camemberti* breaks down lactic acid (the product of the metabolism of lactose) to carbon dioxide and water. Because of the loss of the acid, the pH of the cheese's surface increases to a value of approximately 7.5. At this relatively high pH (for a cheese), the calcium phosphate present within the casein curds becomes insoluble. Thus, the calcium phosphate actually precipitates on the surface of the cheese! This precipitation creates a calcium concentration gradient between the interior and exterior portions of the cheese, where the interior is at a much higher concentration than the exterior. Just like in osmosis, the calcium phosphate moves away from the interior, essentially breaking apart the casein micelles on the interior of the cheese (remember that the calcium phosphate helps to bind the micelles together). In doing so, as a Camembert cheese ripens, its interior softens (Figure 5.19). Moreover, as the calcium phosphate migrates to the exterior, it continues to precipitate, making larger the "white rind."

What about the taste and aroma of Camembert? The aromatic and flavorful molecules of Camembert primarily come from extensive lipolysis that is carried out by the *P. camemberti* molds, fatty acid catabolism, and proteolysis. The distinctive odor of Camembert is due to the production of several alcohols, aldehydes, ketones, and sulfur compounds that are the products of lipolysis, fatty acid metabolism, and proteolysis. As an example, 1-Octen-3-ol and 1-octen-3-one are produced by the activity of the *P. camemberti* on the fatty acid, linoleic acid; these molecules give a mushroom taste to the cheese. Butane-2,3-dione produced upon the metabolism of citrate gives the buttery taste to Camembert (Figure 5.26).

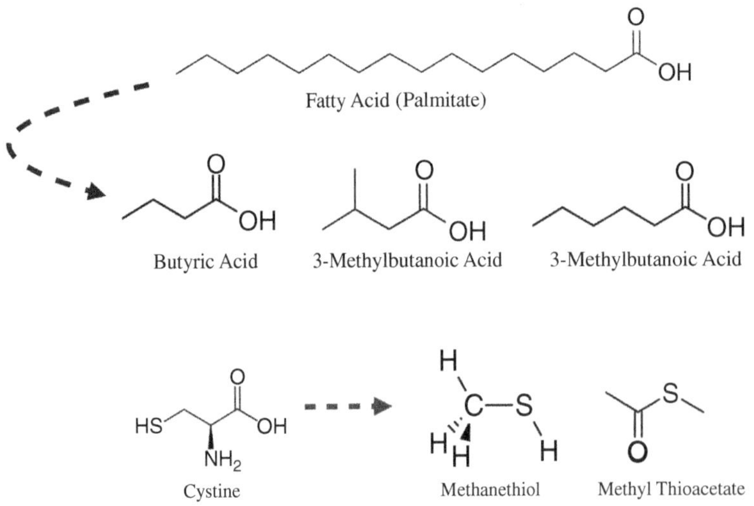

Fatty Acid (Palmitate)

Butyric Acid   3-Methylbutanoic Acid   3-Methylbutanoic Acid

Cystine   Methanethiol   Methyl Thioacetate

Figure 5.26 The "smelly" compounds of limburger cheese. Fatty acids are metabolized to small carboxylic acids whose chemical properties make them volatile. Similarly, the amino acid cystine is metabolized to other volatile sulfur-containing compounds that make a rotten egg or sweaty feet smell to the cheese.

Sulfur compounds are important in relation to the taste of many cheeses. Thanks to the ripening bacteria, *Geotrichum candidum* and *P. camemberti*, proteolysis of casein and further chemistry of the amino acid products lead to several molecules that are strong contributors to the taste and aroma of Camembert. The amino acid methionine is broken down to compounds that impart a flavor of garlic and cabbage. Methanethiol is an important Camembert odorant, and although by itself its boiled potato smell is not particularly pleasant, in combination with other volatiles, it contributes to the cheese aroma of Camembert. Although it is rare that one molecule carries the smell of a particular cheese, but *S*-methylthiopropionate, $CH_3CH_2C\!\!=\!\!O(SCH_3)$, does indeed smell like Camembert.

5.9.2 Swiss

An important bacterium in Swiss cheese cultures is *Propionibacterium shermanii*, the hole-making bacteria. The propionibacteria consume the cheese's lactic acid during ripening, converting it to a combination of propionic and acetic acids and carbon dioxide gas. The carbon dioxide gas creates the bubbles and the characteristic "eyes" or holes in the cheese due to the collection of the gas at the weak points within the curd. In addition, the copper cauldrons used to prepare Swiss cheese damage some milk fats, and the liberated fatty acids are modified to esters and lactones, which have aromas of pineapple and coconut (Figure 5.26).

5.9.3 Cheddar

Lactic acid bacteria that initially acidify the milk and initiate curdling persist in the drained curd and generate many of the flavor molecule products of ripening that persist in the semi-hard and hard cheeses like Cheddar, Gouda, and Parmesan. Although the number of bacteria often drops dramatically during cheese-making, the enzymes survive and continue to degrade proteins into savory amino acids and aromatic by-products. Thus, proteolysis and amino acid catabolism are responsible for the aromas associated with Cheddar cheeses. The lactic acid bacteria, along with added nonstarter bacteria, contribute to the formation of the aromatics that we associate with Cheddar. Amino acids are converted to keto and hydroxyl acids by *Lactobacillus*, while *Lactococcus* strains further convert the keto and hydroxyl acids to carboxylic acids. In the case of *Lactococcus*, the first step in the degradation of amino acids is transamination, leading to the formation of α-ketoacids. For example, the amino acid phenylalanine is converted to phenyl pyruvic acid and is further degraded to the flavor compounds phenyl lactate and phenyl acetate. While methanethiol, as a product of the degradation of methionine, is associated with Cheddar-type sulfur notes in good-quality Cheddar, it does not produce Cheddar flavor. The recognizable Cheddar aroma is produced by a mixture of 2,3-butanedione, methional, and butyric acid. Methional is considered to have a boiled potato-like aroma, while methanethiol adds garlic tones to the flavor of Cheddar. Butyric acid derived from lipolysis has a cheesy and sweaty odor, which is considered an important component of Cheddar flavor.

5.10 The Smelly Cheeses: Muster and Limburger

Smelly Limburger cheese originated in northern Europe and was a favorite in Germany and blue-collar eighteenth-century America for centuries. The bacteria that give Muster, Limburger, and other cheeses a pronounced stench that can clear a room (and contribute more subtly to the flavor of many other kinds of cheese) are *Brevibacterium linens*. *Brevibacteria* is another smear bacteria that are likely natives of salty environments and grow only in warmer conditions (optimum growth temperature of 20–28°C) at salt concentrations of up to 15%; these salty conditions inhibit the growth of most other microbes. However, unlike typical finishing bacteria, *brevibacteria* do not tolerate acidic conditions well and require oxygen to grow. Thus, these bacteria are added during the ripening process by wiping a cheese with a salt brine that causes the characteristic sticky, orange smear of *brevibacteria* to flourish. Due to the presence of the bacteria, extensive lipolysis and proteolysis occur at the surface; hence, it is considered to be a surface-ripened cheese. The chemistry imparted by the *brevibacteria* causes the significant formation of several carboxylic acids, including the volatile butyric, 3-methylbutanoic, and hexanoic (caprioc) acids, which have aromas that have been likened to "sweaty feet." Other aroma molecules that contribute to the smell of Limburger include methionine-derived methanethiol and methyl thioacetate made from fatty acids like palmitate and the amino acid cystine (Figure 5.26).

These cheeses have a sweaty feet smell because another habitat of these bacteria is the human skin. In fact, the human skin is quite likely the original source of the culture. Yes, someone, somewhere, likely transferred bacteria from their skin, grew the cheese in a salty, warm, and oxygenated condition, and ate it. Although this sounds odd now, this is how a lot of the food that we eat originated.

5.11 Cooking with Cheese

When used as an ingredient in cooking, cheese can add flavor, texture, and appearance to a food in a great diversity of ways. In the case of a great grilled cheese sandwich, you want a cheese to be stretchy and stringy upon melting. In fondues, the cheese should melt into a smooth, cohesive, and saucy texture. In some cases, you don't want the cheese to change texture at all, such as a pasta dish made with goat cheese. There is no recipe in which you desire a cheese blob that is an oily, lumpy mess spread over the top of your steamed broccoli. What causes different cheeses to be cooked and used in different ways in the kitchen? As we have seen before, cooking properties of cheeses are based on intermolecular interactions and acid–base chemistry.

5.11.1 Melting Cheese

Melted cheese is a staple ingredient in a number of popular culinary dishes that cross ethnic boundaries. Think about the importance of a good cheese melt to your lasagna, fondue, or quesadilla. However, even among these three dishes, the type of melt desired ranges from stringy to smooth and flowing. First, let's talk about what happens to the molecules during the melting process, and then we will go into why some cheeses melt in a stringy way, while other cheeses provide a better melt for sauces or soups (Figure 5.27).

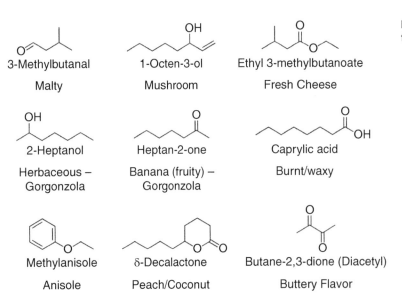

Figure 5.27 Some of the aroma and flavor compounds of cheese.

3-Methylbutanal — Malty

1-Octen-3-ol — Mushroom

Ethyl 3-methylbutanoate — Fresh Cheese

2-Heptanol — Herbaceous – Gorgonzola

Heptan-2-one — Banana (fruity) – Gorgonzola

Caprylic acid — Burnt/waxy

Methylanisole — Anisole

δ-Decalactone — Peach/Coconut

Butane-2,3-dione (Diacetyl) — Buttery Flavor

During the melting process, cheese makes a transition from a solid to a more liquid-like state in the presence of heat. Simply put, the cheese is able to "flow." With any state of matter change (i.e. solid to liquid, liquid to gas), interactions between molecules change. In the case of cheese, the interactions and molecules that are relevant to melting are the milk fats, proteins, water, and calcium. As you might expect, the interactions change depending on the melting temperature. Fats are the first molecules within the cheese to melt, occurring at 90–100°F (32–38°C). During the melting of fat, the intermolecular interactions between individual fatty acids become less robust or tight, which softens the cheese and beads of melted fat come to the surface. As the temperature continues to increase to 130°F/54°C (for soft cheeses), 150°F/68°C for Cheddar/Swiss hard cheeses, and 180°F/82°C for Parmesan, the intermolecular interactions between the individual casein proteins break (remember: this is what holds the casein curd together as a solid mass), and the protein mass begins to flow as a thick, viscous liquid.

You may be curious about why a higher temperature is required to melt Parmesan vs. mozzarella and the difference in "melted properties" in different cheeses. Some of these differences are due to the amount of water contained in the cheese matrix. In a low-moisture cheese like Parmesan, the interactions between individual protein molecules are powerful because there is very little water to interfere with the interactions. Thus, a higher temperature is required for melting; moreover, when melted, there is not a lot of "flow" due to the lack of water. Thus, individual Parmesan shavings don't combine; they remain separate within a baked pasta dish. In contrast, the shreds of a high-moisture mozzarella melt together (at a lower temperature) and readily flow (due to the presence of water).

Have you ever tried to melt Cheddar cheese in the microwave and ended up with an oily cheese blob? As mentioned, most cheeses expel some oil/fat during the melting process as the intermolecular interactions between fat molecules are reduced during the early stages of cheese heating. As the protein matrix begins to break down with the loss of protein–protein interactions at the higher melting temperatures, additional fat is leaked to the cheese's surface. The higher the fat content of the cheese, the more likely this event is to occur. To help avoid the oily mess phenomenon, use a lower-fat cheese, melt at lower temperatures, and stir often.

If you have ever tried to make a fondue with mozzarella cheese, you know that stringy cheeses don't work well, unless you want to string your fondue-dipped bread across an entire room. Melted cheese is stringy when enough calcium is present to link one intact casein molecule to another one, resulting in a long, stretchy fiber. This phenomenon requires a few conditions:

1. The casein molecules must be largely intact and not extensively degraded by protease enzymes. Thus, cheeses that are not aged for an extended time are stringier than those that are well-aged.
2. The cheese must be of low to moderate acidity, as with greater acidity comes a greater loss of calcium from the protein matrix.
3. The cheese must be of moderate moisture and salt content.

Cheeses are the stringiest at their melting point, or the point at which the hot dish has cooled enough to eat. They get stringier as the dish is stirred and stretched. The most common stringy cheeses are mozzarella, Emmental, and Cheddar. These cheeses are good in pizza. In contrast, cheese sauces and soups melt nicely into a smooth, even, and often creamy texture without any note of stringiness. How is this achieved? First, avoid using cheese that is prone to being stringy! In short, use a moist cheese like Colby or Jack or a well-aged grating cheese like Parmesan or Pecorino. In preparing the sauce or soup, the cook should grate the cheese finely prior to addition, heating the dish as little as possible after adding the cheese, and minimizing stirring.

Another way to cook with cheese (or create a cheese sauce) is to stop the melting fat from congealing together by melting the cheese in a blocker. Cooking cheese with a roux is a common method of using stringy cheese in your dishes and having a delicious and manageable final product. A roux is a classic French culinary technique that uses flour and butter to create a base (i.e. a roux) for cheese and other sauces or gravies. Flour and fat (often butter) are cooked to allow the starches to coat with fat and water. The proteins and expanded starch–fat complex interferes with the melting fat (from the cheese), preventing it from forming large interacting strings. The result is a smooth-flowing cheese sauce. You may recognize the consistency of a cheese roux in your favorite mac and cheese dish.

5.12 Processed Cheeses

Anyone who has used Velveeta to make nachos or macaroni cheese knows that it has great melting properties (Figure 5.28). Velveeta and many other cheeses that you purchase, receive in holiday gift boxes, and consume are known as processed cheeses. Why did the concept of processed cheese come about? Processed cheese was a practical solution to the problem of cheese deterioration that occurred during the shipment of cheeses to distant countries. As a solution to this problem, in 1911, Emmental was mixed with citric acid salts to produce a cheese that was much less perishable. A similar solution was developed in the United States in 1916 when Cheddar was treated with phosphoric acid salts. The results of these processes were cheeses that deteriorated less readily and melted without becoming rubbery or fat separated. The development of processed cheeses spread rapidly to other countries in the 1920s (Figure 5.29).

Today, processed cheese is a cheese that uses surplus, scrap, and unripened cheese products as its basis. Manufacturers combine a mixture of sodium citrate, sodium phosphates, and sodium polyphosphates with a

Figure 5.28 Melting cheese. The meltability of cheese depends on the type of cheese, protein, fat, and mineral content. hicheese/Adobe Stock Photos

Figure 5.29 American processed cheese. Bill/ Adobe Stock Photos

mixture of new, partly ripened, and fully ripened cheeses. The polyphosphates carry water into the cheese, remove calcium from the casein matrix and bind to the casein. This loosens the protein matrix, as protein–protein interactions between individual casein molecules and the calcium bridges that cross-link caseins are both reduced. The reductions in protein–protein interactions, along with the high moisture content, result in a cheese that melts into an even, smooth, and creamy blend.

Science for the Chef: Science of cheese and cooking

Cheesy acid-infused Mexican elote. Mexican elote is a delicious street food that beautifully combines the flavors of creamy dairy, tangy cheese, smoky corn, and the bright acidity of lime. The science behind the addition of acid, in this case, lime juice, helps transform the Cotija cheese, creating a more complex and enjoyable eating experience.

Ingredients:
- 4 ears of corn
- 1/4 cup mayonnaise
- 1/4 cup sour cream
- 1/4 cup Cotija cheese, crumbled
- 1/4 teaspoon chili powder
- 1/4 teaspoon paprika
- 1/4 teaspoon salt
- 1 lime, cut into wedges
- Fresh cilantro, chopped (for garnish)

Instructions:
1. Start by grilling or roasting the ears of corn until they are cooked and slightly charred. This adds a smoky flavor to the elote.
2. In a small bowl, mix together the mayonnaise and sour cream. These dairy products provide a creamy base for the elote and help balance the flavors.
3. Once the corn is cooked, spread the mayonnaise and sour cream mixture over each ear of corn, ensuring they are evenly coated. The creamy mixture adds richness and acts as a binding agent.
4. Sprinkle the crumbled Cotija cheese over the coated corn. Cotija cheese is a traditional Mexican cheese known for its crumbly texture and salty flavor. It adds a tangy and savory element to the dish.
5. Dust the corn with chili powder, paprika, and salt. These spices contribute to the overall flavor profile, providing a hint of heat and complexity.
6. Squeeze lime wedges over the corn to add acidity. The acid from the lime juice not only heightens the flavors but also plays a crucial role in the science of cheese. The acid helps break down the proteins in the Cotija cheese, enhancing its texture and flavor.
7. Finally, garnish the elote with fresh chopped cilantro for a touch of freshness and herbaceousness.

Science Behind the Recipe

Mexican elote involves several fascinating chemical processes. First, when corn is grilled or roasted, it undergoes the Maillard reaction, a complex series of chemical reactions between amino acids and reducing sugars. This reaction creates a range of flavorful compounds, resulting in the grilled corn's characteristic caramelization, browning, and smoky taste.

The addition of lime juice to the elote serves multiple purposes. Lime juice is acidic; this acidity plays a crucial role in flavor enhancement. The acid interacts with the acid receptors on our tongues, amplifying the perception of other flavors present in the dish. Furthermore, the acid in lime juice breaks down the proteins in Cotija cheese. This enzymatic breakdown softens the cheese's texture, making it easier to crumble and melt over the corn while intensifying its savory taste. Cotija cheese is an acidic traditional Mexican fresh cheese with a high salt content, thus, it is safe to store and has a unique, tangy flavor.

Mayonnaise and sour cream are key ingredients in elote, contributing to its creamy and rich texture. Mayonnaise contains oil, which is an emulsion of fat and water. The fat molecules in the mayonnaise coat the corn kernels, adding a smooth and velvety mouthfeel. Sour cream contains proteins that contribute to the overall creaminess and thickness of the sauce, enhancing the overall taste experience. The spices used in elote, such as chili powder and paprika, not only add a kick of heat but also interact with other ingredients on a molecular level. Capsaicin, the compound responsible for the heat in chili peppers, interacts with pain receptors, triggering a sensation of spiciness. The combination of heat and flavors from the spices complements the corn's natural sweetness and the cheese's savory notes.

Key Concepts

1. Milk is composed of two phases or components: milk **globules** containing fats and proteins and an aqueous phase filled with different proteins.

2. **Casein** is the most abundant protein in milk and is the primary protein component of the cheese curd.

3. Casein exhibits an **amphipathic** nature, meaning it has both polar and nonpolar regions. This property enables casein to form **micelles** in milk by interacting favorably with each other; this character allows for retention of fat and moisture in the curd.

4. Calcium plays a crucial role in stabilizing the micelle by cross-linking the casein proteins together in the micelle structure. κ-casein binds to the surface of casein micelles, preventing them from growing too large and keeping them soluble.

5. **Whey** proteins in milk have compact, globular structures and are soluble in the aqueous phase.

6. **Curd** formation in cheese production can occur through three methods: increasing acidity, increasing temperature, and the use of rennet. **Acid-mediated curd** formation is based on changes in pH that affect the protein structure, leading to coagulation and curd formation.

7. **Lactic acid fermentation** is a key process in cheese production, facilitated by **lactobacteria**, which utilize lactose as an energy source to produce lactic acid.

8. **Pasteurization** kills harmful bacteria in milk but also destroys the lactobacteria necessary for curd formation, requiring the addition of starter bacteria in cheesemaking.

9. Bacteria and their enzymes play a crucial role in the **ripening process** of cheese, breaking down proteins, carbohydrates, and fats (through proteases, lipases, and other enzymes) into smaller molecules that influence the flavor, texture, and aroma of the cheese.

10. The ripening progression and characteristics of different cheese types are influenced by specific microbes, including molds, which are added or naturally present during the ripening process.

11. **Chemical changes** occur during cheese ripening, involving the degradation of casein proteins and lipids/fats, leading to the formation of flavorful and aromatic molecules such as amino acids, amines, fatty acids, and lactones, contributing to the complexity and uniqueness of cheese flavors.

References

1 Dairy: World Markets and Trade Circular, July 2012, USDA—Foreign Agricultural Service, Office of Global Analysis, Washington, DC.

Additional Readings

A new technological approach for ripening acceleration in cooked cheeses: T Homogenization, cooking and washing of the curd. Véleza, M.A., Hynesa, E.R., Rodriguezb, G., Garittab, L., Wolfa, I. and Perottia, M.C. LTW Food Science and Technology Volume 112, September 2019, 108241

Giha, V., Ordoñez, M. J., & Villamil, R. A. How does milk fat replacement influence cheese analogue microstructure, rheology, and texture profile? *J Food Sci*, 2021;86(7):2802– 2815.Mefleh, Mefleh, M., Pasqualone, A., Caponio, F., and Faccia, M. (2022), Legumes as basic ingredients in the production of dairy-free cheese alternatives: a review. *J Sci Food Agric*, 102(1), 8- 18. Mohd Shukri, A., Alias, A.K., Murad, M., Yen, K.-S., & Cheng, L.-H. (2022). A review of natural cheese and imitation cheese. *J. of Food Proc. and Pres.*, 46(1), e16112. Pollard, A., Sherkat, F., Seuret, M.G., and Halmos, A.L. (2003), Textural Changes of Natural Cheddar Cheese During the Maturation Process. *J Food Sci*, 68(6), 2011-2016.Proteolysis in Irish farmhouse Camembert cheese during ripening Mane, A., and McSweeney, P. H. *J Food Biochem*. 2020;44: e13101

Sharma, P, Segat, A, Kelly, AL, Sheehan, JJ. (2020) Colorants in cheese manufacture: Production, chemistry, interactions and regulation. *Compr Rev Food Sci Food Saf.* 19(4), 1220– 1242.

Singh, T.K., Drake, M.A. and Cadwallader, K.R. (2003), Flavor of Cheddar Cheese: A Chemical and Sensory Perspective. *Compr Rev Food Sci Food Saf*, 2(4), 166-189.

End of Chapter 5 Questions

1. Which two components are required for lactic acid fermentation to occur?
   a. Lactose and calcium
   b. Lactobacteria and lactose
   c. Casein and lactose
   d. Rennet and lactobacteria

2. What is the primary sugar found in milk?
   a. Sucrose
   b. Fructose
   c. Glucose
   d. Lactose

3. What is the composition of lactose?
   a. Glucose and fructose
   b. Glucose and galactose
   c. Galactose and maltose
   d. Fructose and maltose

4. Which type of bacteria is utilized in the curdling phase of cheese production?
   a. Streptobacteria
   b. Lactobacteria
   c. Escherichia coli
   d. Salmonella

5. How does the milk change as lactobacteria ferment the lactose?
   a. It becomes sweeter.
   b. It becomes more acidic.
   c. It becomes thicker.
   d. It becomes less nutritious.

6. What is the purpose of pasteurization in milk processing?
   a. To enhance flavor and aroma
   b. To kill harmful bacteria and extend shelf life
   c. To reduce lactose content
   d. To increase the protein content

7. Which chemical process occurs during pasteurization of milk?
   a. Oxidation of lactose
   b. Hydrolysis of casein
   c. Denaturation of whey proteins
   d. Fermentation of lactobacteria

8. Which chemical process occurs during homogenization of milk?
   a. Hydrolysis of lactose
   b. Denaturation of casein
   c. Oxidation of whey proteins
   d. Emulsification of milk fat

9. What is the purpose of adding starter bacteria to pasteurized milk during the cheesemaking process?

10. Which of the following accurately describes the role of whey in cheesemaking?
    a. Whey is a coagulating agent that helps in curd formation.
    b. Whey is the liquid byproduct that is separated from the curd during cheesemaking.
    c. Whey is a type of bacteria that ferments lactose and produces lactic acid.
    d. Whey is a protein that provides structure and texture to the cheese.

11. Which of the following best describes how rennet contributes to the cheesemaking process as a protease?
    a. Rennet hydrolyzes lactose into glucose and galactose, facilitating bacterial fermentation.
    b. Rennet cleaves casein proteins, specifically κ-casein, to induce milk coagulation and curd formation.
    c. Rennet breaks down lipids into fatty acids and glycerol, enhancing the flavor and texture of cheese.
    d. Rennet promotes the growth of bacteria, aiding in the development of unique cheese characteristics.

12. Explain the enzymatic action of rennet during the cheesemaking process, including the specific protein involved and the resulting changes in the milk.

13. During cheese ripening, proteases in the ripening bacteria degrade casein proteins into their amino acid components. Which chemical process is responsible for this degradation?
    a. Hydrolysis
    b. Oxidation
    c. Reduction
    d. Condensation

14. Explain the chemical processes involved in the breakdown of lipids/fats during cheese ripening, including the role of lipases and the formation of aromatic and flavorful compounds.

6

Fruits and Vegetables

Guided Inquiry Activities (Web): *14, Cells and Metabolism; 18, Starch; 19, Plants; 20, Fiber and Cell Walls; 21, Plants and Color*

Learning Objectives

1. *Recognize and identify the components of a plant cell and their basic functions, including the cell wall, chloroplasts, vacuoles, and amyloplasts.*
2. *Understand how the connections of glucose monomers in the polysaccharide starches amylose and amylopectin differ from the connections of glucose monomers in cellulose and why starches are digestible and cellulose fiber is not.*
3. *Recognize and identify the glycosidic bond of a glucose polymer. For 1→4 glycosidic connections, recognize and identify the difference between alpha and beta (i.e. α(1→4) and β(1→4)) connections.*
4. *Identify the different polymeric components of the plant cell wall, including cellulose, pectins, hemicelluloses, and lignin, and understand the role of each polymer in the cell wall.*
5. *Explain the differences between climacteric and non-climacteric ripening of fruit and identify examples of each type.*
6. *Explain the difference between photosynthesis and respiration and which is more relevant to a harvested plant.*
7. *Explain the effects of heat on the breakdown of plant cell wall polymers hemicellulose and pectin and the texture of plant material.*
8. *Define and explain the gelation and retrogradation of starch in terms of hydrogen-bonded chains of amylose and/or amylopectin with water.*
9. *Recognize and identify C=C double bonds that are conjugated versus isolated or cumulated. Identify molecules that absorb visible light by the presence of many conjugated double bonds.*
10. *Recognize and identify the main classes of plant pigment molecules: chlorophylls, carotenoids, and anthocyanins, and rationalize the water vs. fat solubility of each class based on the structure.*
11. *Understand the role of phenols and polyphenol oxidase in the browning of damaged plant tissue.*
12. *Understand the relationship between plant color and the presence of antioxidants.*
13. *Understand the role of antioxidants in protecting cells from oxidative damage due to reactive oxygen species.*
14. *Recognize and identify terpenes and phenolics as classes of aroma molecules, and explain the effects of volatile aroma molecules on flavor.*

The Science of Cooking: Understanding the Biology and Chemistry Behind Food and Cooking, Second Edition. Joseph J. Provost et al.
© 2025 John Wiley & Sons, Inc. Published 2025 by John Wiley & Sons, Inc.
Companion website: www.wiley.com/go/provost/food_science_2e

6.1 Introduction

The world of edible plants not only includes the fruits that we know and love—apples, bananas, grapes, lemons, tomatoes, and so on—but also roots, leaves, flowers, and seeds. The soft, juicy peach, the dense, opaque flesh of a potato, and the firm crunch of a celery stalk are evidence that plants can make a variety of molecules for many, many uses. Small sugar molecules and volatile aroma molecules make the peach appealing to eat, while the dense starch-filled cells of a potato store energy to last many months. The tough fibrous celery stalk is made of long polymers that hold the plant upright as it grows. Human beings have been plant-eaters from the very beginning, and in fact animals depend on plants to construct complex molecules from the most basic raw materials: sunlight, minerals, air, and water. The energy of simple molecules and sunlight is stored in the more complex molecules that make up that plant material, and other organisms obtain that energy by consuming the plant. Furthermore, since plants cannot run away from predators or travel in search of the means to reproduce, they became remarkable chemists, making molecules that interact with our senses of taste, smell, and sight in delightful or unpleasant ways.

Think of when you've bitten into a fruit that has not yet ripened: a banana that is dense and not sweet, or perhaps an apple that is tart or even bitter. Fruit undergoes significant chemical changes as it ripens. We must understand the different molecules that make up the structure of the plant and the chemical changes they undergo to understand how certain fruits, like bananas or avocados, will ripen after being picked, while others, such as pineapples or grapes, do not ripen after picking. In the same way, understanding the science of plant colors and the chemistry of plant cell wall structure will allow you to prepare tender yet vibrant green beans and avoid the limp and dull appearance of an overcooked green bean.

This chapter will begin with the chemistry and biology of plants and their cells. We will first study the molecules responsible for the structure of a plant and the molecules used by the plant to store energy. We will also examine the effects of cooking on these structural, energy-storing, and colorful molecules during the preparation of delicious and appealing foods. Finally, we will consider the molecules responsible for plants' unique and appealing colors, tastes, and aromas.

6.2 Plant Parts and Their Molecules

Plants are living things that have solved the "problem of life" in a manner fundamentally different from animals (you and me!). Plants nourish themselves by building their own glucose out of water, air, and minerals and with the energy of sunlight. Animals cannot use sunlight for energy, and we cannot build our own glucose out of such simple materials. We need help making the essential molecules of life, and thus we eat them—by eating plants or other animals.

6.2.1 Plant Cells

Since plants are living things, they are also built from cells. As you learned in Chapter 2, the cell of a plant shares some similarities to the cell of an animal: both contain a nucleus that holds the genetic material, and both have a barrier or membrane that controls the flow of molecules in and out of the cell. In a plant cell, however, the cell membrane is accompanied by a thick and rigid cell wall that provides the plant with the structural support to stand upright. See Figure 6.1 for an annotated illustration of the cell. The primary cell wall surrounds a plant cell as it grows and is responsible for structural integrity and mechanical strength; it is made of a flexible layer of proteins and complex carbohydrates, including hemicelluloses, cellulose, and pectins. Sometimes, a stronger secondary cell wall is added to provide additional strength after the plant cell has stopped growing—for example, secondary cell walls are the primary component of wood. Finally, a gluey layer of pectin—also called the middle lamellae—forms the exterior layer of the plant cell wall and helps

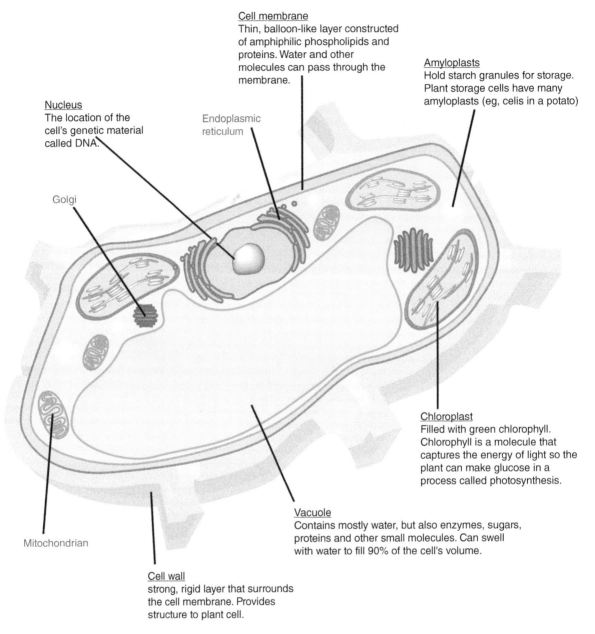

Cell membrane
Thin, balloon-like layer constructed of amphiphilic phospholipids and proteins. Water and other molecules can pass through the membrane.

Amyloplasts
Hold starch granules for storage. Plant storage cells have many amyloplasts (eg, cells in a potato)

Nucleus
The location of the cell's genetic material called DNA.

Endoplasmic reticulum

Golgi

Chloroplast
Filled with green chlorophyll. Chlorophyll is a molecule that captures the energy of light so the plant can make glucose in a process called photosynthesis.

Mitochondrian

Vacuole
Contains mostly water, but also enzymes, sugars, proteins and other small molecules. Can swell with water to fill 90% of the cell's volume.

Cell wall
strong, rigid layer that surrounds the cell membrane. Provides structure to plant cell.

Figure 6.1   Anatomy of a plant cell.

adjacent cells adhere to one another. Celluloses, hemicelluloses, and pectins are all carbohydrate-based polymers that will be discussed in more detail in subsequent sections.

6.2.1.1   Chloroplasts Contain Chlorophyll

Plants are autotrophs. They can take energy from sunlight and store that energy in carbohydrate molecules. The green molecule chlorophyll not only makes plants green, but it is the molecule that captures the light energy and allows plant cells to photosynthesize. Animal cells contain mitochondria for conducting the chemistry that produces the molecules of metabolism, while in plants, that role is played by the chloroplasts (Figure 6.2).

Figure 6.2 Photosynthesis and respiration in a plant. designer_things/Adobe Stock Photos

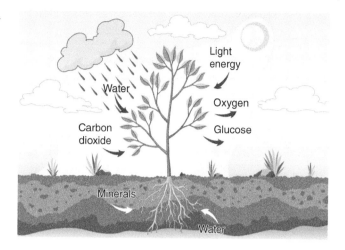

Figure 6.3 Plant cell, vacuole, and water. (a) The plant cell when water is low; water leaving the cell. Cell contents shrink away from the cell wall. Cell becomes limp and flaccid. Plant wilts. (b) Water in equilibrium. The water is entering and leaving the plant cell at the same rate. (c) Plant cell when water is abundant. Water entering the cell enlarges the vacuole. The cell becomes firm or turgid. The plant material is firm and crisp.

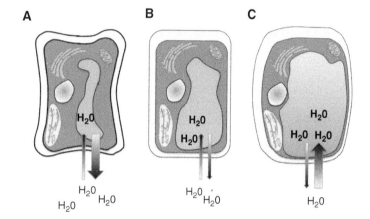

6.2.1.2 Vacuoles

The plant cell wall is a firm but flexible container that holds mostly water. When water is abundant (Figure 6.3), water molecules travel into the cell's vacuole (through the wall and membrane), and the vacuole enlarges. Plants do not require water as frequently as animals, mainly due to the water-filled vacuole that takes up most of the space inside the cell. Some vegetables contain so much water that they could almost be considered water delivery devices. Lettuce contains 98% water by weight, and carrots, even though stiff and crunchy, are 88% water by weight. All this water is contained within the balloon-like vacuole. When water is abundant and the vacuole enlarges, all the other contents of the cell are squished against the cell wall, which bulges under pressure. These water-filled, swollen cells make for a crisp, firm, turgid vegetable. If the cells are low on water, the vacuole shrinks, pressure disappears, and cells become limp and flaccid. The water-rich vegetable is crisp and juicy; the limp one is chewy and less juicy. If a vegetable has lost water, you can "reinflate" the water-depleted cells by soaking the vegetable in water for a few hours.

When plants are in the sunlight, they "breathe in" carbon dioxide (CO_2) and "breathe out" oxygen (O_2) while using light energy and enzymes to make glucose ($C_6H_{12}O_6$) from carbons and water. The glucose molecules are then used as the plant's food—the glucose is essentially how the plant stores the energy of light (Table 6.1).

Table 6.1 Metabolism of plants.

Photosynthesis	Respiration
A building-up process that stores energy	*A breaking-down process that releases energy*
$\underset{\text{(carbon dioxide)}}{6CO_2} + \underset{\text{(water)}}{6H_2O} + light \rightarrow \underset{\text{(glucose)}}{C_6H_{12}O_6} + \underset{\text{(oxygen)}}{6O_2}$	$\underset{\text{(glucose)}}{C_6H_{12}O_6} + \underset{\text{(oxygen)}}{6O_2} + \underset{\text{(carbon dioxide)}}{6CO_2} \rightarrow \underset{\text{(water)}}{6H_2O} + energy$

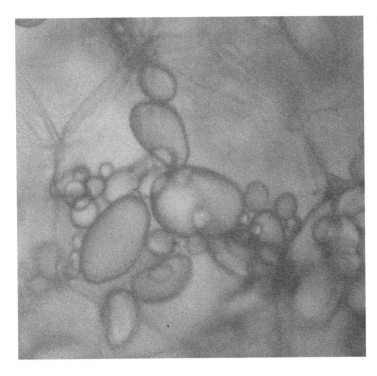

Figure 6.4 Amyloplasts inside potato cells. Markus Nolf.

At the same time, the glucose molecules the plant makes by photosynthesis are respired in plant metabolism in the same way most organisms (including humans) consume glucose to run metabolism (see Chapter 15 for a discussion of human metabolism). Plants take in O_2 and use it to oxidize the glucose via many enzymes—the result is the release of CO_2, water, and energy for metabolism.

6.2.1.3 Amyloplasts Store Starch

Starches are long polymers of glucose. These long, complex carbohydrates are how plants store glucose for later use, just like humans store energy as fat molecules. Plants deposit starch molecules as a series of concentric layers in microscopic, solid granules that are collected into amyloplasts (Figure 6.4). In plants, starch is found most often in one of two forms: amylose or amylopectin. The composition of plant starch is typically 20–25% amylose and 75–80% amylopectin, and the starch is packaged into granules of various shapes and sizes depending on the species of plant (Figure 6.5). The starch content and granule size also impact the cooking qualities of the starch.

Figure 6.5 Microscopic view of the starch granule. (a) A powerful microscope image of a starch granule. (b) A starch granule is exposed to the starch-degrading enzyme, alpha amylase. Kristian Peters – Fabelfroh / Wikimedia Commons / CC BY SA 3.0

Table 6.2 Plant organs by function, tissue, and cell type.

Organ		Function	Plant material and cell type	Example
Vegetables	Roots	Anchor plant into the ground; absorb nutrients	Tough fibrous material—cells have thick, cellulose-rich *cell walls*	Inedible
			Some roots swell up with storage cells full of *amyloplasts*	Carrots, parsnips, radishes, sweet potatoes
	Stems, stalks	Conduct nutrients to roots and leaves; give structural support	Fibrous material (stems, stalks)—cells have thick, cellulose-rich *cell walls*	Asparagus stems, celery stalks
	Tubers, rhizomes		Some stems swell up with storage tissue (tubers, rhizomes)—cells are full of *amyloplasts*	Potato, turnip, ginger
	Leaves	Produce sugar molecules by photosynthesis	Plant material is thin so gases can penetrate/escape. Almost no structural support—*cell walls* are thin and flexible. Cells have many *chloroplasts* and large air pockets between them for gases	Lettuce leaves, spinach leaves
	Flowers	Plant's reproductive organs	Brilliantly colored and scented. Cells are full of *chloroplasts and chromoplasts* (contain other colored molecules besides green), and the *vacuoles* contain molecules that give the flower a smell	Broccoli, cauliflower, artichoke
Fruits		Surround seed(s); sometimes the seed is a "pit"	Almost entirely storage tissue. Cells contain thinner *cell walls*. There are *amyloplasts* and storage *vacuoles* full of sugars and good smelling/tasting molecules	Apples, peaches, snow peas, and so on
Seeds		Contain plant embryo and its food supply	Tough outer layer—cells have thick cell walls with cellulose and lignin. Inner storage cells contain many *amyloplasts*, as well as proteins and oils	Wheat kernel, peas, corn, lima bean, nuts

Putting the organelles of a plant together with the plant organ or type of vegetable material, Table 6.2 describes the type of plant, its organs with function, and the type of material responsible for the actions of the plant.

6.3 Plants Are Comprised of Different Types of Complex Carbohydrates

Carbohydrates are molecules made up of one or more sugars. The most common sugar we find in food is glucose (i.e. dextrose)—$C_6H_{12}O_6$ (Figure 6.6). Starch is a carbohydrate and one of the polymers of glucose that you will find in plants (Figure 6.7).

6.3.1 Carbohydrates for Energy Storage

Living things need glucose for energy, and plants are no exception! Fruits and vegetables are parts of plants, and plants store their glucose in a chain form until it is needed for energy. Root and tuber crops are excellent examples of starch-containing plant organs. Beets, carrots, and potatoes store starch underground to survive the winter months.

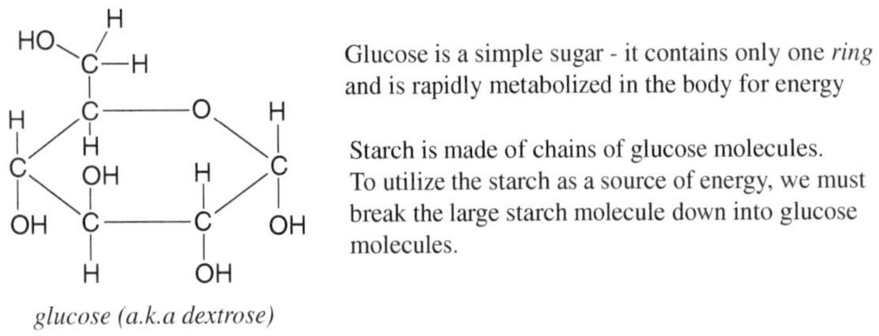

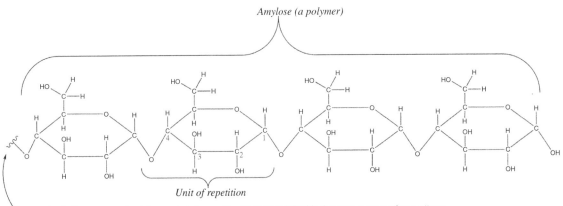

Figure 6.6   Glucose, a simple sugar.

Glucose is a simple sugar - it contains only one *ring* and is rapidly metabolized in the body for energy

Starch is made of chains of glucose molecules. To utilize the starch as a source of energy, we must break the large starch molecule down into glucose molecules.

glucose (a.k.a dextrose)

If you examine the structure of amylose or amylopectin in Figures 6.7 and 6.8, you will notice that each is a combination of repeating units. The same molecular piece is repeated over and over again. We first encountered this idea in Chapter 2. Amylose and amylopectin are *polymers*—where *poly* = "many." The unit of repetition is called a *monomer*. In amylose and amylopectin, the monomer unit is a glucose unit. Glucose polymers are also called *polysaccharides*—where the root *saccharide* comes from the Latin *saccharum*, meaning "sugar." Amylose and amylopectin are both polymers of the sugar glucose, but they differ in how the glucose monomer units are

Amylose (a polymer)

Unit of repetition

the squiggly line means that the molecule continues "infinitely" with the same repeating pattern

Amylopectin

Figure 6.7   Amylose and amylopectin. Two types of starch. The numbers indicate how chemists count the atoms in each glucose monomer and distinguish how amylose and amylopectin are connected.

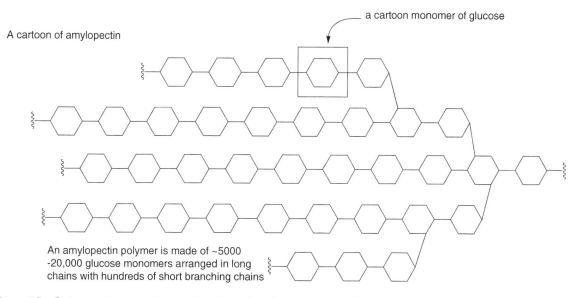

A cartoon of amylose

An amylose polymer is made of ~1000 glucose monomers attached in one long extended chain

a cartoon monomer of glucose

A cartoon of amylopectin

An amylopectin polymer is made of ~5000 -20,000 glucose monomers arranged in long chains with hundreds of short branching chains

Figure 6.8 Cartoons of large amylose and amylopectin polymers comprised of glucose monomers.

connected. In amylose, the glucose monomers are connected at the 1 and 4 positions or in a "head to tail" arrangement. In amylopectin, there are $1\rightarrow4$ or "head to tail" connections, but also $1\rightarrow6$ branches. As shown in Figure 6.8, a typical amylose polymer will be made of approximately 1000 glucose monomers, each connected $1\rightarrow4$ in a single, long chain. Similarly, a typical amylopectin polymer consists of 5000–20 000 glucose monomers connected in many $1\rightarrow4$ chains linked through $1\rightarrow6$ branches. The consequence is that an amylopectin polymer is very bushy. The enzyme alpha-amylase can break starch molecules down into individual glucose units. In addition, the human body can break down amylose and amylopectin into individual glucose molecules using *enzymes* called *glycosidases*. The resulting glucose monomers are further oxidized to produce energy. That is why carbohydrate yields 4 calories of energy for every gram consumed (see Section 2.3.2 in Chapter 2). As we will discuss in the following sections, the different shapes of the amylose and amylopectin polymers also impact the texture of the food when it is cooked.

Box 6.1 Hot potatoes

Could potatoes really be responsible for changing the course of world history? Potatoes are full of starch, a fuel source for animals, as well as some protein and vitamins. Historians claim that potatoes supported the caloric needs of ancient civilizations like the Incas of South America. The potato kept without spoiling for long periods, and the Incas dried the potato flesh to be eaten during food shortages due to crop failures. Some have argued that the nutrition provided by the potato was responsible for the rise of more advanced civilizations in the ancient Americas. The native South American potato made its way to Europe by the sixteenth century, and as

(Continued)

Box 6.1 (Continued)

historian William H. McNeill has claimed, "By feeding rapidly growing populations, [it] permitted a handful of European nations to assert dominion over most of the world between 1750 and 1950" [2]. Consumption of the potato coincided with the end of famine in Europe, and as Charles C. Mann observes, "Hunger's end helped create the political stability that allowed European nations to take advantage of American silver. The potato fueled the rise of the West."

Today, there are over 5000 varieties of potatoes, and potatoes are a major worldwide crop (fourth or fifth most important). Russet, red, Yukon Gold, or white potatoes come in various shapes, colors, and sizes. All potatoes are a type of plant organ called a tuber—or a swollen area of the stem that specializes in storing starch. The cells of a potato contain many *amyloplasts* full of starch granules. Picking the right potato for your cooking needs depends on the type and quantity of starch within the tuber. Potatoes can be divided into two basic types: *boiling* or *waxy* potatoes that hold together well when heated versus *baking* or *floury* potatoes that fall apart when cooked. The quantity and type of starch make the difference between waxy and baking potatoes.

Russet potatoes are the classic example of a baking potato—they are ideal for making fries and mashing. These potatoes are high in starch (up to 22%), and the starch of the baking potato contains 26% of the straight amylose and 74% of the branched amylopectin. Not only do the cells of baking potatoes have weaker cell walls, but the cells are also densely packed with starch granules that swell and gelatinize easily when they come in contact with water (starch gelation was introduced in Chapter 2 and is explained in Section 6.5.2 of this chapter). As baking potatoes are cooked, the cell walls and vacuole walls break open, leaking starch and water, and the cells separate from one another. The higher calcium content of a baking potato also helps the starch to readily form a soft solid or gel [3]. The result is the fluffy or mealy texture of a baked potato. In contrast, waxy potatoes maintain shape during and after cooking and are great additions to soups and salads. Since baking potatoes have a mixture of starch types and higher levels of total starch, they gel at lower temperatures (58°C/136°F) than boiling potatoes, which do not gel until 70°C (158°F). The low gelation temperature of the baking potato starch means the starch absorbs water and swells sooner in the cooking process, which puts pressure on the cell walls. The starch of a boiling potato has a different composition and a higher gelation temperature. A boiling potato also has more pectin and cellulose holding its cells together, so the cells do not burst open or separate as readily. These factors help the boiling potato maintain its shape during cooking. Some potatoes, like the yellow or Yukon Gold, have starch composition and cell wall strength that are in between classic baking or boiling potatoes, making these a versatile option for cooking. Now, when selecting a potato for baking, mashing, salads, or boiling, you can be sure to pick the right type!

6.3.2 Carbohydrates for Cell Walls and Cell Glue

The cell wall is responsible for the strength of the individual cells and, ultimately, the plant itself. The cell wall is made of complex carbohydrates (polysaccharides) that form an armored shell around the cell's contents. When coupled with the cell glue—also made of complex carbohydrates—the cell walls of many cells connect; the strength of these many connected cell walls allows the plant to grow vertically *against* the forces of gravity!

All three layers of the plant cell wall are made of some or all of the following four components: cellulose, pectins, and hemicelluloses—all carbohydrate polymers called polysaccharides—and a tough, woody polymer called lignin. The primary cell wall is typically a matrix of three-polysaccharide polymers (i.e. cellulose, hemicelluloses, and pectins), which provide structure and mechanical strength to the cell. The construction materials for making a reinforced concrete wall are often used to describe the plant cell wall and matrix. When building a reinforced concrete wall, steel bars are embedded within a matrix of pourable concrete. In the same way, the plant cell walls have "bars" and "cement" [4]. The bars are the tough fibers of *cellulose* that act as a structural framework. In some plant cells with a secondary cell wall, the very rigid *lignin* is also present, but most of the time, we don't eat lignified plants (i.e. you don't chew tree bark). The tough lignin polymer is added to create the secondary cell wall only after the cell has stopped growing. Lignified cell walls are thick and rigid and can persist even after the cell has died. The "cement" is a semisolid, flexible mixture that fills the space between the cellulose fibers and is made mostly of the soluble fiber components hemicellulose and pectin. When neighboring cell walls are close together (as in the tissue of a plant), a gluey cement material holds the cells together. The layer of cell glue, or the middle lamellae is mostly pectin.

We encounter the cell walls of plants as the fiber in our diets. The fibers in celery stalks, the stone surrounding a peach seed, and the seedpods of beans and peas are mostly cell wall material. Overall, fiber is defined as material in our plant foods that our digestive enzymes can't break down into absorbable nutrients. Nutritionally, we divide dietary fiber into soluble and insoluble types. Soluble fiber is dissolved in water, while insoluble fiber is not. Both forms do not break down in our stomach or intestines but pass through undigested. It may seem surprising that cell wall material is indigestible when in fact, it is comprised of carbohydrates, just like starch! As we learned earlier, cellulose forms the "bars" of our reinforced concrete wall and therefore helps to create the structure that holds the plant up. Cellulose may be one of the most abundant glucose polymers on earth, but it is also an insoluble *fiber*—an indigestible carbohydrate. So why can't we digest cellulose?

Cellulose is another polymer of glucose, a *polysaccharide*—just like the starch polymers amylose and amylopectin. If you examine the structures of amylose and cellulose in Figure 6.9, you can see that—once again—the type of connection between the glucose monomers differs between amylose and cellulose. Both are linear, 1→4 connections of glucose, but the direction of the 1→4 connection differs for the two polymers. The direction of the connection can be assessed from the *glycosidic carbon*—the carbon bonded directly to two oxygen atoms. In Figure 6.9, the glycosidic carbons are in boldface type. In amylose, the oxygen atoms make the 1→4 connection point "down" from each glycosidic carbon—this is called an alpha (1→4) or α(1→4) linkage. In contrast, the oxygen atoms making each 1→4 connection in cellulose point "up"—this is called a beta (1→4) or β(1→4) linkage. This seemingly insignificant difference between α(1→4) and β(1→4) linkages is the reason that the carbohydrate polymer amylose is a digestible source of nutrition for humans, while cellulose is an indigestible component of fiber. The β(1→4) linkage of cellulose polymers is a bond that human enzymes cannot break; therefore, humans cannot digest cellulose. Most animals cannot digest cellulose, but microorganisms like bacteria and fungi can, which is how plant material rots in nature.

Lignin is another component of insoluble fiber, but unlike the other fiber components, lignin is not a polysaccharide. Lignin is a polymer of one of three types of *monolignols*. Monolignols are made from the amino acid phenylalanine, so their structure fundamentally differs from the monomer glucose. The polymer lignin is very strong and tough; it gives mechanical strength to the cell wall, but lignified plants are not typically eaten due to their tough consistency.

Hemicellulose and pectins are soluble fiber—this means they dissolve in water to make a clear, jelly-like semisolid. Hemicelluloses and pectins are also polysaccharides, but the monomer units of these polymers are not glucose. Hemicelluloses are also made of β(1→4) linkages of sugars such as xylose. Again, the β(1→4) linkages make hemicelluloses indigestible. Pectins are mostly polymers of a modified galacturonic acid—another type of carbohydrate or sugar. There is so much pectin in the cell walls of fruits like apples and oranges that it can be purified as a powder and used to thicken jams and jellies!

Both the insoluble fiber (cellulose and lignin) and the soluble fiber (hemicelluloses and pectins) contribute to our health in different ways. Insoluble fiber increases the bulk of consumed food and helps it move through the large intestine more efficiently and quickly. The insoluble fiber also binds toxins in our foods—preventing them from being absorbed by our intestines. The soluble fiber makes intestinal contents thicker—slowing the mixing and movement of nutrients and toxins and increasing the rate at which they move through the intestines. According to the Food and Drug Administration, "Eating a diet high in dietary *fiber* promotes healthy bowel

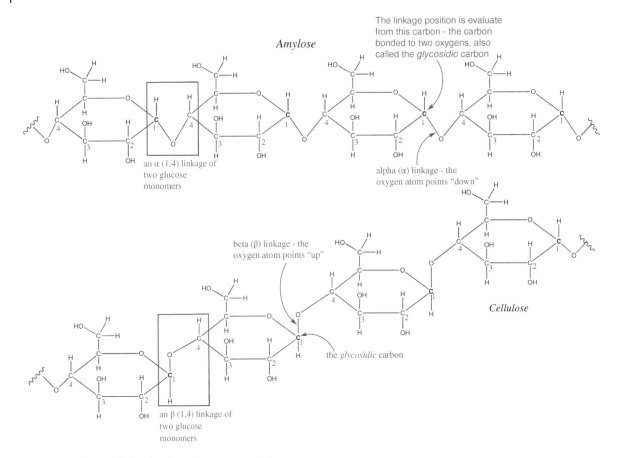

Figure 6.9 Glycosidic bonds of amylose versus cellulose.

function. Additionally, a diet rich in fruits, vegetables, and grain products that contain dietary fiber, particularly soluble fiber, and low in saturated fat and cholesterol may reduce the risk of heart disease" (http://www.fda.gov). It is recommended that we consume 20–35 g of fiber per day! (Table 6.3).

Insoluble fiber literally does not dissolve in the watery environment of our stomachs and intestines. The solid fibers of cellulose and lignin add bulk or volume to foods. Soluble fiber does dissolve in the liquid digestive environment. Since the dissolved fiber molecules are large polymers (i.e. very large molecules), they make the digested food thicker, which makes it move more slowly through the intestines (Table 6.3). While digestible carbohydrates like amylose and amylopectin yield 4 cal per gram, the indigestible carbohydrate in fiber is assumed not to yield any calories. While we cannot break fiber down for energy, the bacteria in our guts can. So, in reality, fiber can contribute 1–2 calories per gram (see Section 2.3.2 in Chapter 2). But when considering a nutrition label, the dietary fiber should not count toward the total calories.

Breaking down the cell wall and defeating the cell glue is an important process in cooking and eating plants. As we have seen, the cell wall and cell glue give the plant its fibrous structure and can make eating the plant difficult. Some vegetables, such as lettuce and tomatoes, have softer tissue and are easily eaten, while others must be cooked to defeat the cell wall and glue and make some of the content more accessible and tasty. Fruits designed to attract animals to eat the seed-containing fruit have weakened the cell wall's structure during ripening.

Table 6.3   The components of dietary fiber: indigestible plant material.

Component	Soluble/insoluble	Polymer of carbohydrate monomers	Common food sources
Cellulose	Insoluble	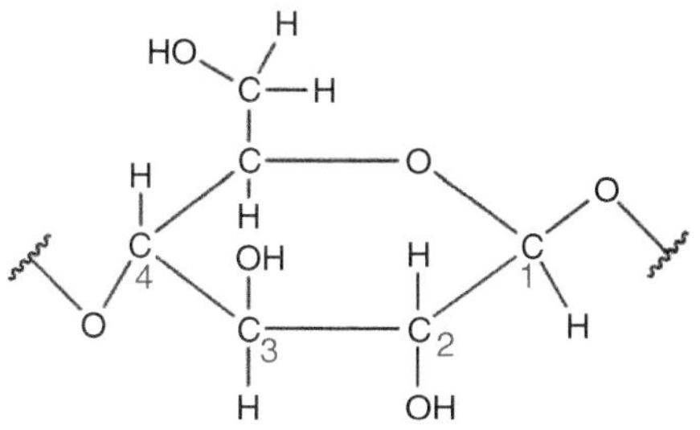	The stringy part of celery, wheat bran
Pectins	Soluble	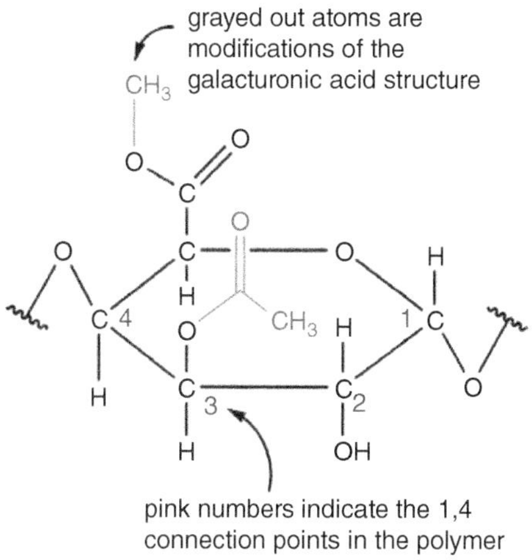	Fruit (e.g. apples and citrus fruits)
Hemicellulose	Soluble	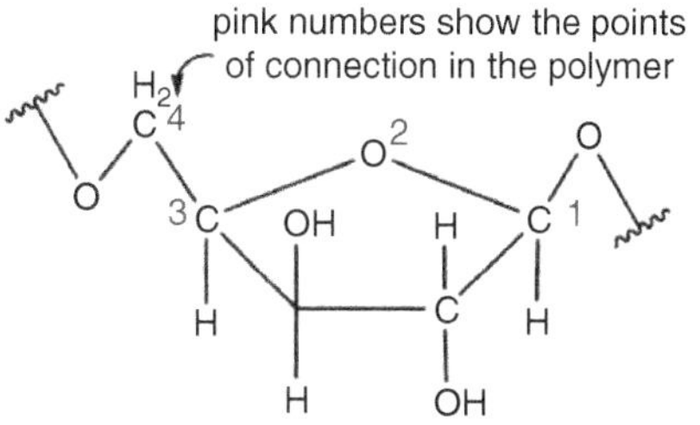	Many fruits, veggies, and grains
Lignin	Insoluble (major component of wood)	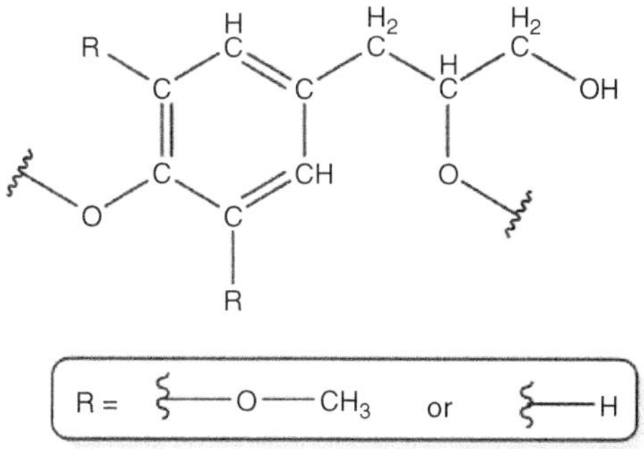	Woody asparagus and broccoli stems, the gritty parts of pear and guava flesh

6.4   Harvesting, Cooking, and Eating Plants

6.4.1   Respiration and Spoilage

When a plant is picked, harvested, or otherwise separated from the soil, water, or sunlight, *photosynthesis* stops. However, when a plant is picked or harvested, *respiration* continues. The plant continues to breathe for some time (Table 6.4). Plant cells are hardier than animal cells and can survive longer (sometimes for weeks or months!), but once they cannot *photosynthesize* nutrients, the plant will eventually respire away all its stored energy and accumulate waste products, and the cells will eventually die. We perceive this as the loss of flavor and changes in texture that occur with the deterioration and spoiling of produce. Bacteria, yeasts, and molds can also accelerate spoilage by attacking the weakened plant cells and consuming their cell contents. How quickly a plant succumbs to spoilage depends on how fast it respires after harvesting. The more perishable a fruit or vegetable, the faster its respiration rate will be. Plants with high rates of respiration will consume all their available stored energy (the glucose and glucose polymers, starch) quicker and die sooner.

The key to preserving the lifetime of the produce is to slow the rate of respiration, that is, how fast the plant cells take up oxygen and release carbon dioxide. The most effective strategies for slowing respiration are to keep the plant cold and limit its oxygen supply. Plastic bags can limit oxygen but also trap moisture, which can encourage mold growth. Food suppliers understand the impact of respiration on ripening (see Section 6.4.2) and will store and transport food in low oxygen conditions and cold temperatures to slow down the respiration of fruits and vegetables. Unfortunately, many tropical and subtropical fruits and vegetables cannot tolerate cold temperatures, resulting in bruised, brown fruit that may not ripen. However, many plants, including bananas, berries, cabbage, avocados, pears, and onions, survive at very low oxygen levels (the gas is often replaced with carbon dioxide or an inert gas such as nitrogen), allowing for commercial worldwide shipping.

6.4.2   Ripening

Eating immature fruit is not a pleasurable experience. The fruit is dense, hard, and not sweet tasting. Ripening is an intentional, dramatic, and final phase of life for a fruit; it is the process that makes a fruit edible. We see this process occur as the tart, hard fruit softens and sweetens. As a fruit ripens, enzymes convert some of the starch (a tasteless glucose polymer) into free glucose molecules. Some of the glucose is further metabolized into fructose and other monosaccharides by other metabolic enzymes. The conversion of starch into these smaller sugars (i.e. monosaccharides) gives the mature fruit its characteristic sweet taste. Other enzymes break down

Table 6.4   Postharvest respiration rates in CO_2 mL/kg/h at 59°F/15°C.

Potato (cured)	17	Pear (Bartlett)	16 (after 1 day in storage)
Grape	16	Pear (d'Anjou)	9 (after 1 day in storage)
Lemon/orange	18–19	Lettuce	39 (head), 63 (leaf)
Apple	15 (fall), 25 (summer)	Brussels sprouts	200
Cabbage	28	Banana (ripe)	140
Carrot (topped with greens)	40	Asparagus	235
Peach	87 (at 20°C)	Strawberry	114

Data from agriculture handbook number 66 (HB-66) by the USDA and the agricultural research station.

the pectin in the primary cell wall, softening the rigid structure. When you bite into a ripe fruit—for example, a ripe peach—the flesh of the fruit is easily crushed, and "juice" squirts and drips everywhere. When a fruit is ripe, the weakened cell walls of the fruit give way to the pressure of your teeth, and the water- and sugar-filled vacuoles burst (i.e. the "juice"). The color changes of ripening include the degradation of green chlorophyll that reveals the orange and red carotenoid pigments that give fruits like bananas and tomatoes their characteristic color and the formation of anthocyanin pigments that impart pink, red, purple, and blue colors to fruits such as strawberries and blueberries. The pleasant aromas associated with ripening are the result of molecules made by the plant during ripening and include aldehydes, ketones, alcohols, esters, and terpenes. These volatile aroma molecules are made from a variety of proteins, carbohydrates, lipids, and vitamins present in the plant, and each plant species has a characteristic blend of aroma molecules that gives it a unique flavor (see Section 6.6). For a more comprehensive discussion of flavor, please see Chapter 14.

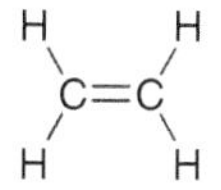

The energy needed for these chemical reactions comes from the energy produced in respiration (Table 6.1). These changes are in marked contrast to vegetables that do not ripen in the same way that plants do. After harvest, vegetable respiration is slow and eventually fades to very low levels. Since vegetables do not soften and release flavor molecules on their own, we use cooking to soften the plant cell walls and break open the cells, releasing flavor molecules.

The ripening of fruits occurs by one of two mechanisms. Those fruits for which respiration dramatically increases after detaching from the parent plant are considered *climacteric*. Bananas, apples, mangos, pears, peaches, avocados, and tomatoes are all examples of climacteric ripeners. Remember, a green banana will ripen to a yellow color while in your kitchen … far from the banana tree from which the fruit was harvested. Other fruits maintain the same rate of metabolic respiration after harvest, and these are called *non-climacteric*. Non-climacteric fruits include blueberry, cucumber, watermelon, peas, lemon, pineapple, strawberry, orange, grape, and grapefruit; these fruits will not ripen once harvested.

The rise in the production of CO_2 observed for climacteric fruits is due to dramatically increased respiration after harvest. This change in respiration is associated with the production of a simple gas called ethylene (Figure 6.10). Not only was ethylene the first plant hormone ever discovered, but it was also the first gas found to act as a hormone—essentially, ethylene is produced at one location in the plant, and then it travels through the plant to act on the fruit. Ethylene is made within the plant from the breakdown of the amino acid methionine, and it induces dramatic ripening. Once stimulated by ethylene, the fruit will increase its respiration rate severalfold, and the color, texture, and taste will change rapidly. The coordinated ripening of large amounts of fruit at one time is thought to attract and encourage animals to ingest and spread the seed found within the fruit. Farmers and grocers have found that climacteric fruits can be harvested while still unripe (i.e. "green") and will ripen on their own by producing ethylene gas. If the ethylene is trapped around the fruit or the plant is intentionally exposed to ethylene, it will ripen much faster. For example, a common recommendation to accelerate fruit ripening is to place the fruit in a sealed, brown paper bag. This trick only works for climacteric fruits and depends on trapping the ethylene gas around the fruit to accelerate its ripening. In the same way, bananas are grown in Central and South America and shipped to North American grocery stores while still green. Exposure of these green bananas to ethylene gas produces rapid ripening before the bananas are placed on your grocery store shelves. Those mystery green bags that keep fruit fresh? They are reported to bind ethylene produced by the plants, limiting the overripening. Unfortunately, these strategies will not work on non-climacteric fruits; they do not ripen after harvest.

In addition to ripening, ethylene controls other aspects of plant growth, development, wounding responses, and cell aging. The hormone signals to activate the production of enzymes and other plant components that alter the metabolism of fats, proteins, and carbohydrates.

Ethylene
(C_2H_4)

Figure 6.10
Ethylene, the ripening plant hormone.

6.5 Cooking Plants

6.5.1 Changes to the Cell Wall during Cooking

During ripening, when a fruit or vegetable is cooked, the changes in texture that we observe are caused by changes in the *cement* molecules—the hemicelluloses and pectin. If the cement is weak, the cells will easily separate from one another when chewed. In the wild, animals have developed sharp teeth for tearing, flat teeth to grind the fibrous plant matter, and strong jaw muscles to pulverize vegetable material. In the same way, humans have evolved modern strategies to break down plant material and make it easier to chew—heat and acid are two effective means of breaking down plant cell walls and turning flora into food.

When cooking a plant with heat, the plant cells break down in several ways:

- Above 140°F/60°C, the cell membrane is damaged and disintegrates, the vacuole ruptures, and the cell loses fluid to the surroundings. The cells deflate and become flaccid, but the cell walls are still intact and will give the material a firm texture while chewing. This is why after sautéing onion, carrot, and celery for a few minutes in a little olive oil, water appears in the skillet. This is called "cooking out" the vegetables or cooking vegetables until they "release their juices." If the vegetable is boiled, the sugars and flavor molecules found in the vacuole are lost to the boiling water. Adding table sugar to potato or corn while boiling can help retain some of the sweet flavors that are lost in this way.
- As the plant material is heated hotter and longer, the carbohydrate polymers of the cell wall begin to break down, and the cells begin to separate from one another.
- When the plant material contains *amyloplasts*, heating causes the starch granules within the amyloplasts to swell as the starch absorbs the water released by the disintegrating vacuole—such as the starch granules of potato, for example. This is why cooking potatoes results in a pasty or mealy texture—not juicy. Extended cooking can cause the starch to leak out of the damaged cells and/or gel into a soft solid.

When plant tissue is heated above >140°F/60°C, the cell walls begin to break down. The cellulose is resistant to heat and chemicals and remains largely unchanged. However, the polymers hemicellulose and pectin gradually break down into shorter molecules. These shorter pieces of hemicellulose and pectin dissolve in the cooking liquid, and the cell wall cement literally disintegrates. With no cement, the cells easily separate from one another, and the plant tissue becomes tender. If you keep boiling the plant material, all the cell wall cement will eventually be destroyed, leaving you with a puree.

The cell wall cement-dissolving part of fruit and vegetable tenderizing depends on the cooking environment: even though hemicellulose is a component of soluble fiber, it is not very soluble in acidic liquids. Therefore, hemicellulose resists breaking down into shorter molecules and dissolving in acidic conditions. For example, when vegetables/fruits are steamed, very little additional fluid is added. The plant material is exposed mostly to the mildly acidic cell fluid itself and to the steam, which has a slightly acidic pH of 6. Consequently, steamed veggies retain their firmness. The reverse is true for alkaline conditions, which accelerate hemicellulose breakdown. Fruits/vegetables cooked in alkaline conditions quickly become mushy (Table 6.5).

Another element of the cooking environment that can impact the breakdown of the cell wall is the presence of calcium ions. The calcium ion can cross-link pectin molecules together, strengthening their interactions and making the cell wall cement stronger. Pectin chains are made of galacturonic acid monomers joined together in a polymer chain. As shown in Table 6.5 and Figure 6.11, the galacturonic acid monomer of pectin is sometimes modified by esterification, but within a pectin chain, some of the monomers are modified and some are not. As we saw in Chapter 2, Section 2.4.2, an ester is formed from the reaction between a carboxylic acid and an alcohol. When a galacturonic acid monomer is not modified, a "free" *carboxylate* (Figures 6.11b and c) is left with a negatively charged oxygen called an oxygen anion, which can form ionic bonds with a positively charged cation. Calcium is a readily available positively charged cation with a + 2 charge, which is perfectly suited to interact with two negative charges—one oxygen anion

Table 6.5 The effect of cooking liquids on vegetable/fruit firmness.

Condition	Cooking liquids	Effect on vegetable/fruit
Acidic conditions (pH < 7)	Tomato sauce, fruit juices/purees	Maintains firmness for longer
Alkaline conditions (pH > 7)	Baking soda (NaHCO$_3$) added to water	Softens more readily
Neutral conditions (pH ~ 7)	Plain water	Softens readily
High calcium	"Hard" water (i.e. water that contains CaCO$_3$)	Maintains firmness for longer
High sodium	Table salt (i.e. NaCl) added to water	Softens readily

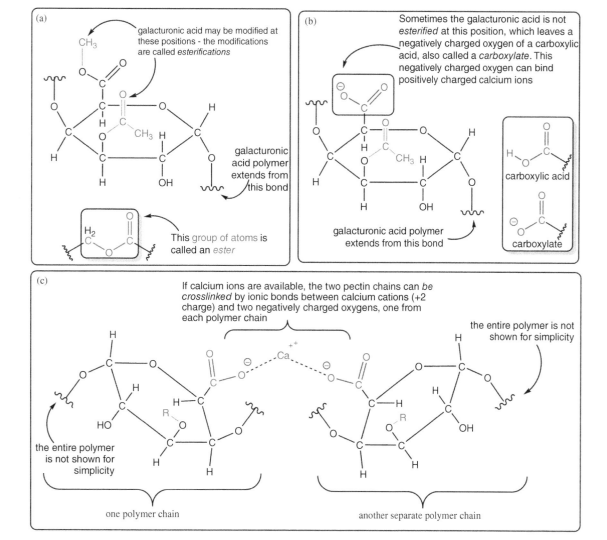

Figure 6.11 Modification of pectin. (a) Pectin is comprised of polymers of galacturonic acid, which are sometimes modified by esterification (gray atoms). (b) and (c) If the galacturonic acid is *not* modified by esterification, the free oxygen anions from carboxylic acid groups of two separate polymer chains can interact with positively charged calcium ions (Ca^{2+})—linking the two chains together.

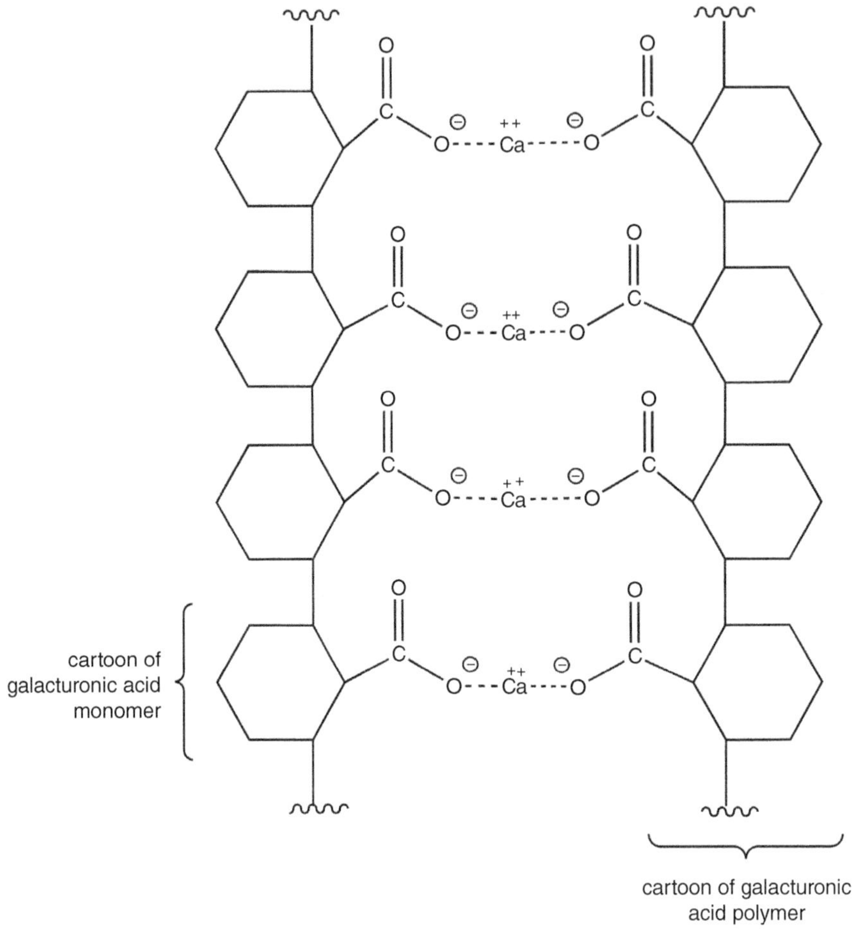

cartoon of galacturonic acid monomer

cartoon of galacturonic acid polymer

Figure 6.12 Calcium holds pectin to form a gel. The calcium cation (Ca^{2+}) links together two oxygen anions of a galacturonic acid monomer from separate chains of pectin.

from each pectin chain. Vegetables cooked in "hard water"—that is, water containing *calcium carbonate* ($CaCO_3$)—resist softening due to the excess *calcium ions* in the water that cross-link the pectin cell wall polymers together (Figure 6.12). Sodium ions (Na^+) displace the calcium ions found naturally in the plant cell wall, and since the sodium ions have only a + 1 charge, they do not promote pectin cross-linking.

A common demonstration of the impact of pH and the breakdown of cell walls and glue is described by the authors of *Modernist Cuisine: The Art of Science and Cooking*. In their book, the impact of pH on cooking beans is described by boiling raw beans under three different conditions: (i) water containing vinegar (acidic conditions), (ii) distilled water (slightly acidic to neutral), and (iii) water with baking soda (alkaline conditions). Under alkaline conditions, the cell wall cement and cell glue molecules—pectins and hemicelluloses—dissolve readily, and the beans quickly become soft and mushy. The beans cooked in an acidic solution only slightly soften as the cell wall cement and glue remain intact. Finally, the beans cooked in the distilled water (pH ~7) soften at the best rate as the cell wall cement slowly disintegrates; a brief cooking time is all that is necessary for a tender vegetable. To minimize the impact of the slightly acidic cell contents, the beans should be cooked in a large pot of water. This dilutes the plant acids as the cells break down and release their contents and thus reduces the time needed to soften the cell wall. As we will learn later in this chapter, acidic conditions will also degrade the green color of green beans, and cooking in a large pot of water can take care of both issues!

Box 6.2 Spherification

Turning a flavorful liquid into a sphere that pops in your mouth is an invention of modern kitchen *alchemy* that uses the principles of calcium cross-linking of polysaccharides, such as the calcium cross-linking of pectins discussed in this section. In the process of *direct spherification*, a flavorful liquid is blended with sodium alginate powder. Sodium alginate is an anionic polysaccharide similar to pectin that is derived from seaweed. In direct spherification, the liquid mixture of sodium alginate is dripped into a calcium chloride bath (Figure 6.13a). The calcium ions displace the sodium and cross-link the chains of alginate to make a thin, flexible solid around the sphere of liquid. The result is a liquid encapsulated in a calcium cross-linked film of alginate. In *reverse spherification*, a calcium-containing liquid is dripped into a sodium alginate bath (Figure 6.13b). Either method produces the liquid-filled spheres. In many ways, the spheres resemble the roe or fish eggs used to make caviar. The spheres are an attractive way to decorate dishes and add bursts of flavor in unexpected ways.

Figure 6.13 Modern chemistry in the kitchen. (a) Direct spherification. (b) Reverse or indirect spherification of liquid flavors. (c) Students preparing their own creation of a direct spherification. M.studio/Adobe Stock Photos

6.5.2 Changes to Starch During Cooking

6.5.2.1 Starch Gelation

When we eat fruits and vegetables—including starchy vegetables like potatoes or corn—we consume starch that our bodies later break down into glucose for energy. When we cook starchy vegetables like potatoes or corn and grains like rice, our cooking methods make the starch polymers more pleasant to eat.

When starch granules are mixed with water at room temperature, not much happens to the starch. But if the starch in the water mixture is heated, eventually, the temperature gets high enough that the granule begins to swell up as water penetrates the center (Figure 6.14). Eventually, the granule absorbs enough water and swells to such a degree that it breaks apart into a network of starch molecules mixed with water. These networks of long starch molecules with water in between are called *gels*. The temperature at which the starch granule falls apart into this matrix of water and starch molecules is called the gelation temperature (Table 6.6).

The gelation temperature of starch is dependent on the kind of starch it is. The amount of amylose, the size of the amylose chains, and the size of the starch granule are all factors in the thickening power. Amylose forms stronger gels than amylopectin—this is because amylose has a linear (thread-like) structure, while amylopectin

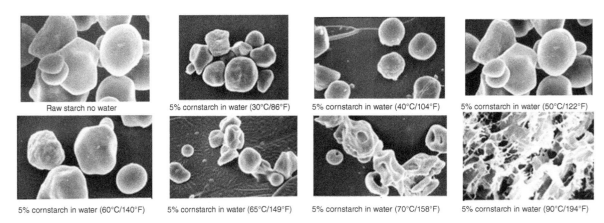

Figure 6.14 Starch in water at increasing temperatures. Images by Dr. ZoeAnn Holmes, Professor Emeritus. Reproduced with permission of Oregon State University.

Table 6.6 Types of common thickening starch and their properties.

Type	Gelation temperature	Thickening power	% Amylose	Avg. granule size	Additional factors that affect thickening power
Wheat	126–185°F, 52–85°C	+	26	17 μm (50%), 7 μm (20%), other (30%)	High-protein content (~10–12%); variable granule size inhibits good amylose gels
Corn	144–180°F, 62–80°C	++	28	14 μm	Opaque gels
Potato	136–150°F, 58–65°C	++++	20	36 μm	Amylose polymers are unusually long; improve gels
Tapioca (cassava root)	126–150°F, 52–65°C	++	17	14 μm	Forms clear gels but stringy
Arrowroot	140–187°F, 60–86°C	++	21	23 μm	Forms clear gels but stringy

20% amylose means a 20:80 ratio of amylose to amylopectin.

has a bushy, branched structure. The threads of amylose more readily form gels: three-dimensional networks of polymers that have trapped water molecules between the polymer chains. Weak, noncovalent attractions between the slightly negative oxygen atoms and the slightly positive hydrogen atoms of the O–H groups link the amylose chains together. These weak, noncovalent attractions are called **hydrogen bonds**, which we first encountered in Section 1.4.1.2. Amylopectin molecules will form gels, but the gels are softer and less stable because the bushy amylopectin molecules cannot stack closely and form stable hydrogen bonds with each other as easily. When the starch granules are large, they are easily trapped in the three-dimensional network of starch chains and water, increasing the liquid's viscosity. The tightly ordered clusters of amylose molecules require higher temperatures, more water, and longer cooking times to gel or separate into a network filled with water molecules. Clusters of amylopectin molecules gel at lower temperatures.

The gelation property of starch, in particular, is essential to the chemistry of roux-thickened sauce. A **roux** is a butter and flour paste that has been lightly cooked for only a few minutes. The cook then adds a watery liquid to the roux while stirring or whisking. The initial cooking of the flour and fat ensures that the fat coats the proteins in the flour and allows the particles to disperse evenly as the watery liquid is added. Adding watery stock, broth, milk, etc. with heat hydrates the starch within the roux into a gel that thickens the liquid. See *Science for the Chef* at the end of this chapter for a recipe on making the white roux sauce, béchamel.

6.5.2.2 Starch Retrogradation

When a cook combines starch with a liquid and heats it to form a gel, the starch molecules leak out of the starch granules, and the sauce thickens. When the sauce is thick enough, the cook will turn off the heat, and the mixture will cool. As the temperature falls, molecules have less energy to move and groove; consequently, they move less and form more stable hydrogen bonding networks with each other, and with the water molecules interspersed among the polymer chains. If the temperature gets low enough, the starch molecules will congeal or solidify into a solid gel. Amylose chains reform their tightly packed hydrogen-bonded structures quickly because their linear shape makes tight packing easy and hydrogen bonding strong (Figure 6.15). With little water, the amylose chains will pack into a hard crystalline solid instead of a moist gel. Amylopectin molecules take longer to reassociate into hydrogen-bonded networks upon cooling because their bushy shape makes tight packing difficult, and they form weaker networks (i.e. softer gels) than amylose.

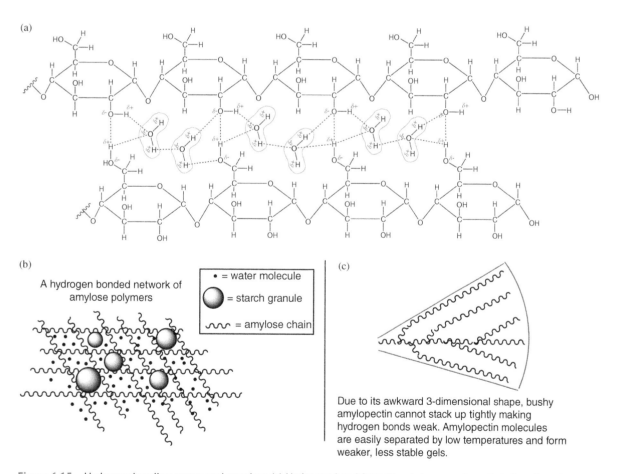

Figure 6.15   Hydrogen bonding, water, and starches. (a) Hydrogen bond formation between adjacent chains of amylose and water in the formation of a gel. (b) A cartoon depicting the gelation of amylose. (c) A cartoon depicting the bushy branched shape of amylopectin.

Table 6.7 Rice varieties and % amylose.

Rice	% Amylose
Long grain	22
Medium grain	18
Short grain	12
Waxy ("sticky") rice	0–1

All rice is cooked by boiling in water, which heats the starch granules up, causing them to swell and leak starch. In the table, 22% amylose means a 22:78 ratio of amylose to amylopectin. So a 0% amylose starch has 100% amylopectin.

This process of heating starch to gelation temperatures and cooling the gelled network into a solid is called retrogradation. The retrograded starch is more compact (i.e. harder) than the native starch. This happens when pie fillings, puddings, and other gel-like solids are made. It also explains why cooked rice turns hard in the refrigerator overnight.

Rice comes in three basic varieties shown later that vary in their amylose/amylopectin ratios (Table 6.7). Any starch is characterized by its "% amylose" content; therefore, the remaining starch is made of amylopectin. Long-grain rice is high in amylose starch (22%), and boiled long-grain rice will have a firm, springy texture when cooked and become inedibly hard when refrigerated overnight. Boiled "sticky rice" has a softer, stickier texture and hardens much less during overnight refrigeration; this results from the small percentage of amylose and the larger amount of amylopectin. The "bushy" amylopectin forms weaker, softer gels—so low % amylose (i.e. high % amylopectin) rice is "softer" rice.

In the same way, when potatoes are boiled, professional cooks recommend cooking the potato slices at approximately 320°F/160°C for about 20–30 minutes, then *cooling them down* while standing (i.e. don't mash or otherwise disturb the potatoes) for 30 minutes. You can then finish the cooking process by briefly reheating the potatoes by steaming or simmering, followed by gentle mashing. This process of heating → cooling → and briefly reheating avoids the gluey texture that mashed potatoes can sometimes get by first gelling and retrograding the starch into a tightly packed network.

6.5.2.3 Gelation with Other Plant Carbohydrate Polymers

Other carbohydrate polymers are capable of forming gels. The cell walls of some fruits (e.g. citrus fruits and apples) are especially rich in pectin. As we saw earlier, pectin is a polymer of carbohydrate monomers (the monomer is the sugar galacturonic acid; see Table 6.3 and Figure 6.11), and it forms the cell wall cement of the primary cell wall and the cell glue of the middle lamellae. When fruit is cut up and heated near to boiling point, the pectin chains break into smaller chains and dissolve into the surrounding fluid. Upon cooling, the pectin chains can gel in much the same way that the chains of amylose starch gel—a moist solid is created by the association of pectin chains into a three-dimensional network that traps water molecules. Boiling off water (to concentrate the diluted pectin chains) and adding sugar encourage better gelation of the pectins. The polar sugar molecules help to form the gels by using hydrogen bonds to connect neighboring pectin chains. Adding extra sugar is so effective at encouraging pectin gelation so that fruit cooked in a concentrated sugar solution will retain its firmness—the sugar helps the pectin gel within the structure of the fruit cell walls. Pectin can be purified from citrus fruits and apples, and the purified pectin powder can be added to crushed fruit (cooked or not) to give a firm gel. Food chemists have developed modified pectin that can gel with the aid of calcium ions that help cross-link the pectin chains together. The cook adds the calcium after the fruit and modified pectin have been cooked together. This innovation allows food manufacturers to gel pectin without sugar and produce "low-sugar" preserves and fruit jellies (Figures 6.11c and 6.12).

Table 6.8   Plant gums.

Plant gums (carbohydrate polymers)	Uses
Agarose and alginates: cell wall polymers from seaweed Acacia gum (gum arabic, from acacia trees) Guar gum: from seeds of *Cyamopsis tetragonobola* Xanthan gum: carbohydrate polymers produced by bacteria in industrial fermentation	These carbohydrate polymers can thicken and stabilize like starch, gel-like starch, and pectin. They are added to jams, jellies, custards, and sauces (salad dressings, ketchup, etc.), and used to create a smooth texture in ice cream.

There are other plant carbohydrates that can gel and/or thicken like pectin and starch. These carbohydrates are usually called gums, and they come from a variety of plant sources (Table 6.8).

6.6   Colorful and Flavorful Fruits and Vegetables

6.6.1   Where Does Plant Color Come From?

What makes color? Why do we perceive most plants and plant parts to be green in color while some are red, orange, yellow, purple, and so on? Molecules are the source of the colors we see. For example, when a ripening fruit loses its green color (think of green bananas...), it is the loss of the chlorophyll pigment that we see. Other plant pigments are also due to molecules.

All carbon-based molecules with color share one structural feature in common—multiple double bonds (Figures 6.16 and 6.17, and Table 6.9). The number and arrangement of those double bonds determine if the molecule can absorb visible light. Molecules that can absorb visible light have color.

Colored molecules contain many conjugated double bonds (Figures 6.17 and 6.18). It's not possible to easily correlate the structure of a molecule to its *exact* color. However, the presence of more than five conjugated double bonds is a good indicator that the molecule will be colored. Most plants contain the three main pigment molecules (chlorophylls, anthocyanins, and carotenoids) but in different concentrations.

While all colored molecules absorb visible light, not all colored molecules perform the same purpose within the plant. The chlorophyll in chloroplasts is essential for the light-harvesting reactions of photosynthesis, while colored fruits are the plant's strategy for appealing to hungry animals that might eat the fruit and disperse the seeds, and still, other colored molecules help defend the plant against intense light, harmful oxidation, or predators.

Anthocyanins are responsible for the red, pink, blue, and purple colors of flowers, leaves, fruits, and vegetables. Purple carrots, red cabbage, and red strawberries all owe their beautiful colors to anthocyanins. In fruits, the anthocyanins accumulate as the fruit ripens; this is obvious in the ripening of strawberries, blackberries, and blueberries. This is consistent with the primary role of the anthocyanin pigments within the plant: to attract birds, insects, and animals. Birds and insects are typical pollinators—they help the plant spread pollen around, a necessary part of plant reproduction. Animals help the plant by eating the fruit, digesting it, and then dispersing the seeds.

Anthocyanins all share the same basic structure shown in Table 6.3, but they can be broken into three main types: pelargonidins, cyanins, and delphinidins (Figure 6.19). The three types vary in the number of —O—H or —O—R groups attached to the main ring system and have slightly different colors. Most fruits and vegetables have combinations of these pigments in various concentrations, yielding the variety of colors we see in nature. When a fruit ripens and changes in color from green to red or yellow (think of ripening tomatoes and bananas), the fruit stops making chlorophyll and begins making carotenoid pigments instead. The absence of green chlorophyll also

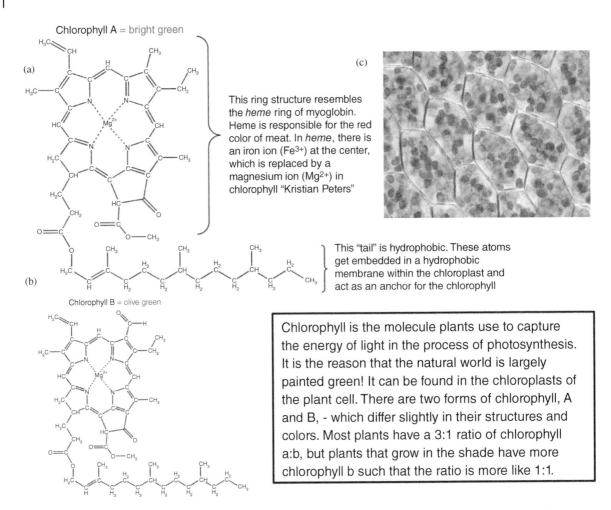

Chlorophyll A = bright green

(a)

This ring structure resembles the *heme* ring of myoglobin. Heme is responsible for the red color of meat. In *heme*, there is an iron ion (Fe^{3+}) at the center, which is replaced by a magnesium ion (Mg^{2+}) in chlorophyll "Kristian Peters"

(c)

This "tail" is hydrophobic. These atoms get embedded in a hydrophobic membrane within the chloroplast and act as an anchor for the chlorophyll

(b)

Chlorophyll B = olive green

Chlorophyll is the molecule plants use to capture the energy of light in the process of photosynthesis. It is the reason that the natural world is largely painted green! It can be found in the chloroplasts of the plant cell. There are two forms of chlorophyll, A and B, - which differ slightly in their structures and colors. Most plants have a 3:1 ratio of chlorophyll a:b, but plants that grow in the shade have more chlorophyll b such that the ratio is more like 1:1.

Figure 6.16 Why plants are green. The structures of (a) chlorophyll A and (b) B, are the green pigments in the chloroplasts of plant tissue. (c) Chloroplasts are plant organelles containing chlorophyll. (c) Kristian Peters.

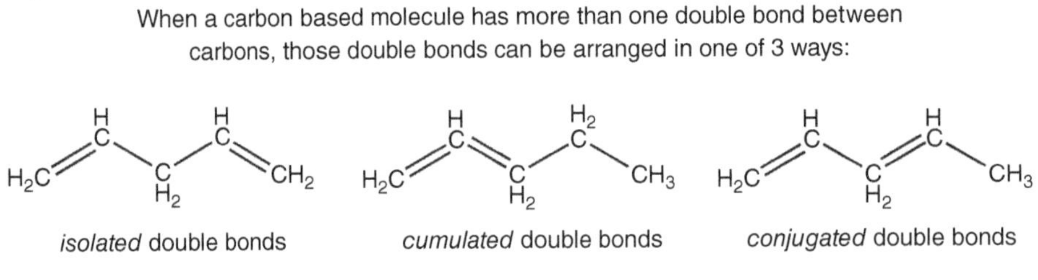

When a carbon based molecule has more than one double bond between carbons, those double bonds can be arranged in one of 3 ways:

isolated double bonds   *cumulated* double bonds   *conjugated* double bonds

Figure 6.17 Double bonds. The arrangement of carbon double bonds has a significant impact on the light-absorbing ability of the compound.

reveals the red and yellow carotenoids present in the fruit. In the plant, carotenoids also capture light energy for photosynthesis (not as much as chlorophyll, though), and they help protect the plant by absorbing excess light energy and quenching harmful *reactive oxygen species*. As we will see in a later section, carotenoids can do the same for us when we consume them in our diet!

Table 6.9 Plant pigment molecules, their classes, and common sources.

Type	Molecule	Common Sources
Anthocyanins	pelargonidin, 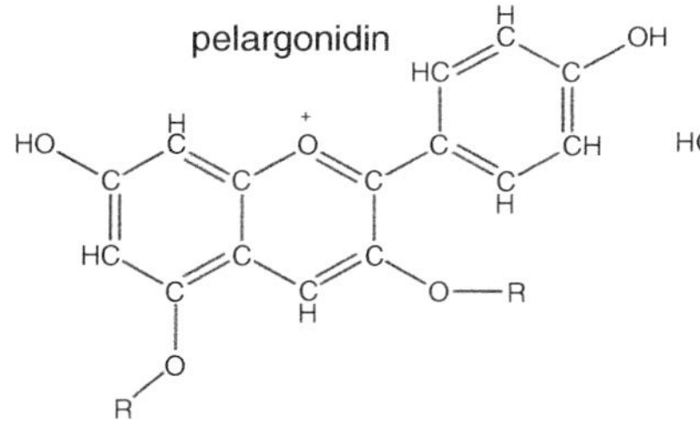 cyanidin, 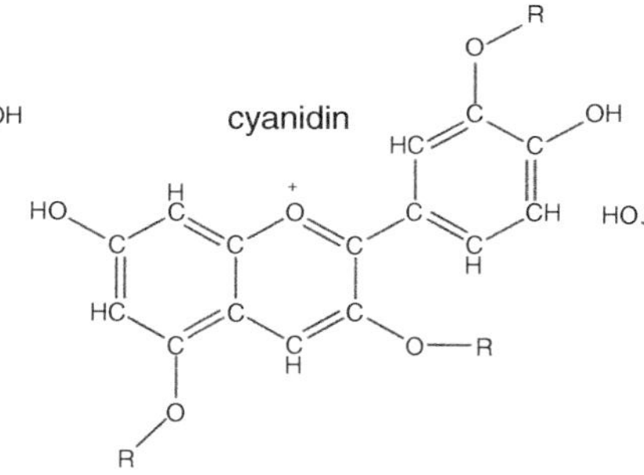 delphinidin 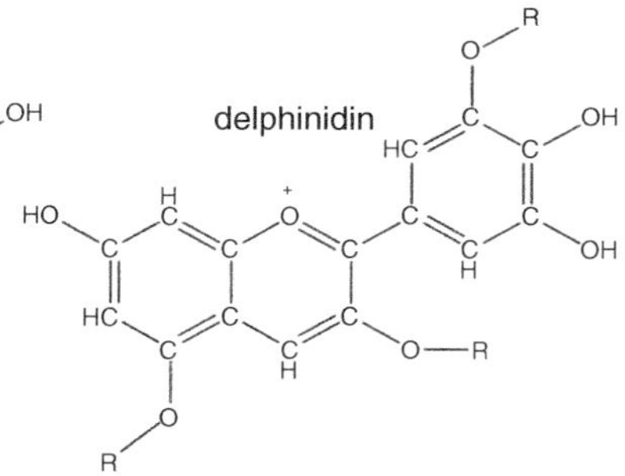	(Reds) Ripe raspberries and strawberries, blueberries, plums, cranberries, pomegranates, and red kidney beans (Purples) Grapes, blackberry, blueberry, cherry, cranberry, elderberry, acai berry, raspberry, red cabbage, and red onion
Carotenoids	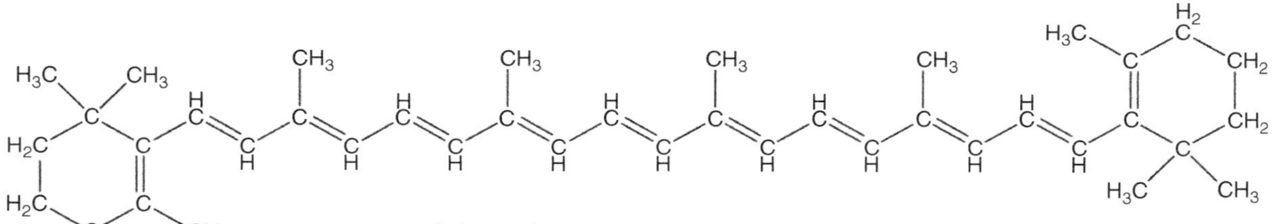 beta carotene = orange	Carrots, sweet potato, butternut squash, and pumpkin
	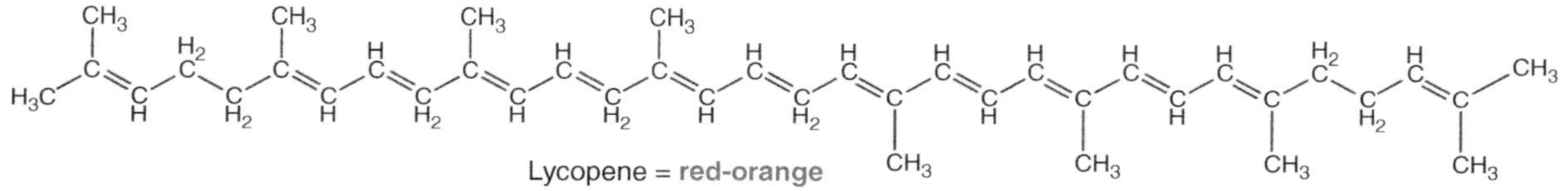 Lycopene = red-orange	Tomatoes, red carrots, watermelons, and papayas

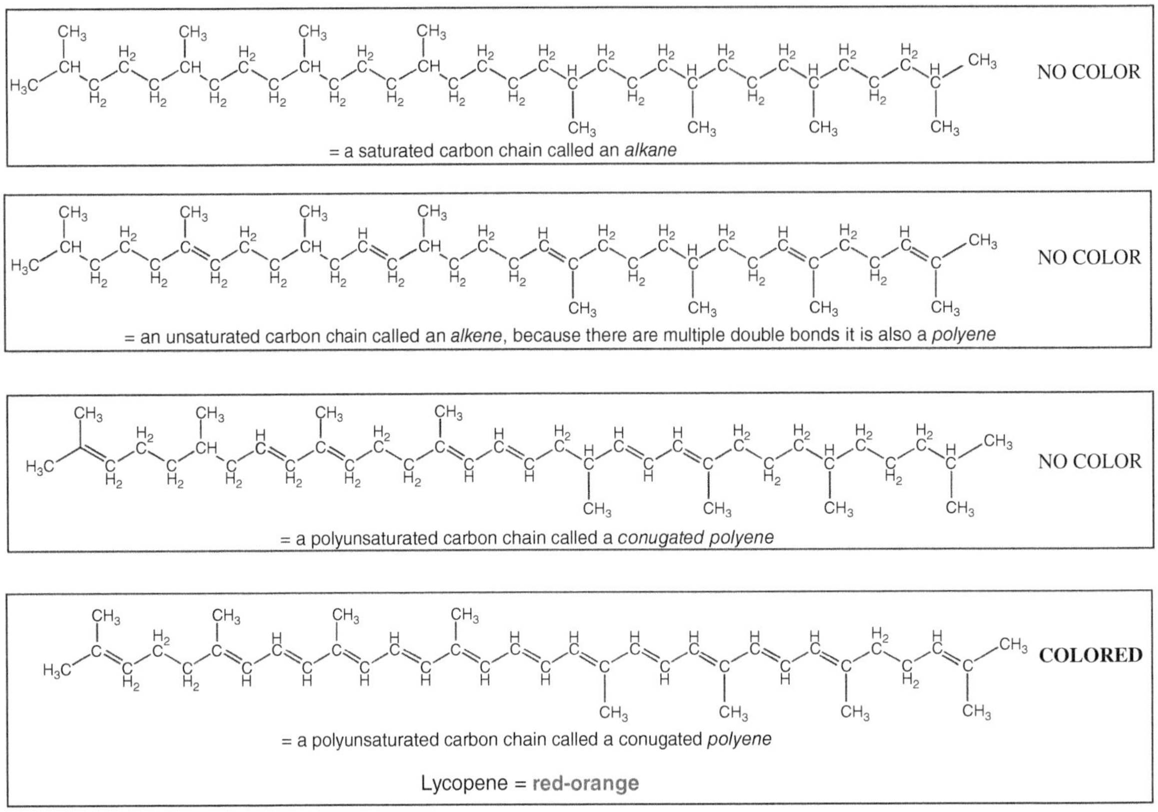

Figure 6.18 Impact of carbon double bonds on light absorption. Four different unsaturated carbon compounds are shown and only the conjugated compound lycopene has light-absorbing characteristics.

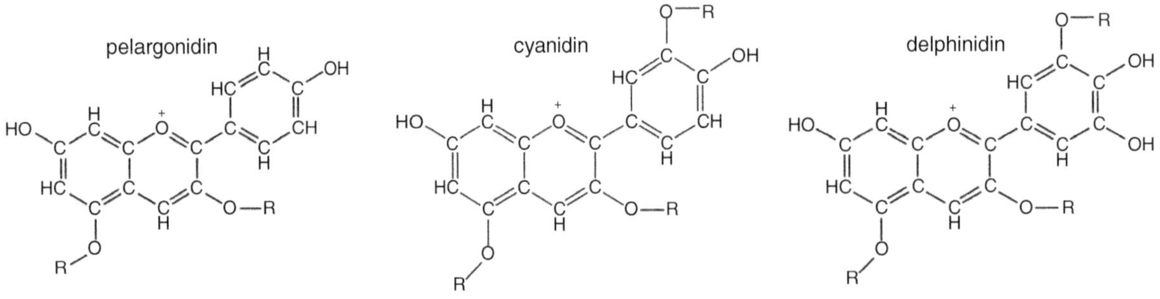

Figure 6.19 Anthocyanins: pelargonidin, cyanin, and delphinidin.

6.6.2 Cooking Colors

When green plant material is cooked, two main chemical changes occur to the chlorophyll. The heat breaks down the plant cell's membranes, including the chloroplasts. The chlorophyll pigment is chemically cleaved from the long hydrophobic tail by three different means: (i) acid conditions, (ii) alkaline conditions, or (iii) by the actions of the enzyme, chlorophyllase (Figure 6.20). Chlorophyllase is most active at 150–170°F/66–77°C, and so is only denatured and inactivated near the boiling point of water. The hydrophobic tail of the chlorophyll

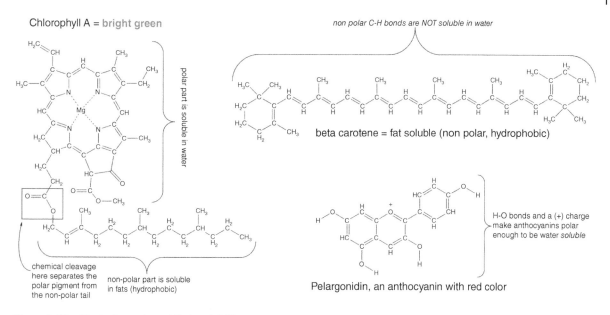

Figure 6.20   Plant pigments and their solubility.

anchors the chlorophyll molecule in the hydrophobic membranes of chloroplasts (membranes are mostly made of fat-like, hydrophobic molecules). With the chloroplasts damaged and the hydrophobic chlorophyll tail severed, the pigment can escape into the cooking liquid (if you are boiling) or the slightly acidic plant cell fluid (if you are steaming or sautéing). One of the reasons we blanch (briefly plunging the vegetable or fruit into boiling water and then quickly submerging it in cold water) is to denature and inactivate chlorophyllase and thus maintain the green color of the food. Acidic conditions are the main reason the pigmented part of the chlorophyll can lose its magnesium ion, resulting in the drab green color of cooked vegetables. The plant fluid itself is slightly acidic, as are some cooking liquids (Figure 6.21).

Cook the green veggie long enough (boiling/blanching, steaming, sautéing, or microwaving), and the chlorophyll will turn drab green. You can limit this transformation by avoiding acidic conditions and cooking

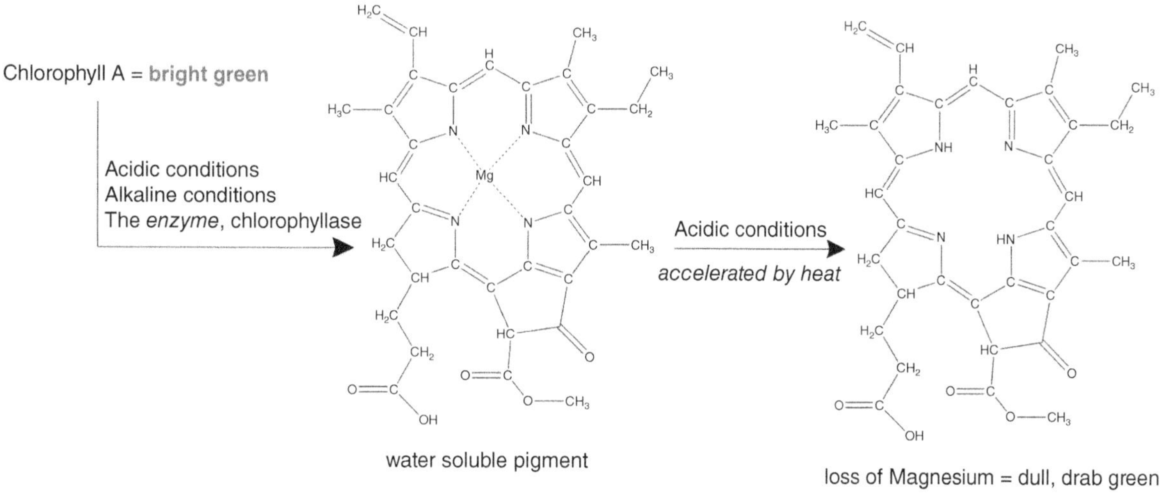

Chlorophyll A = **bright green**

Acidic conditions
Alkaline conditions
The *enzyme*, chlorophyllase

water soluble pigment

Acidic conditions
accelerated by heat

loss of Magnesium = dull, drab green

Figure 6.21   The loss of bright green chlorophyll in cooking.

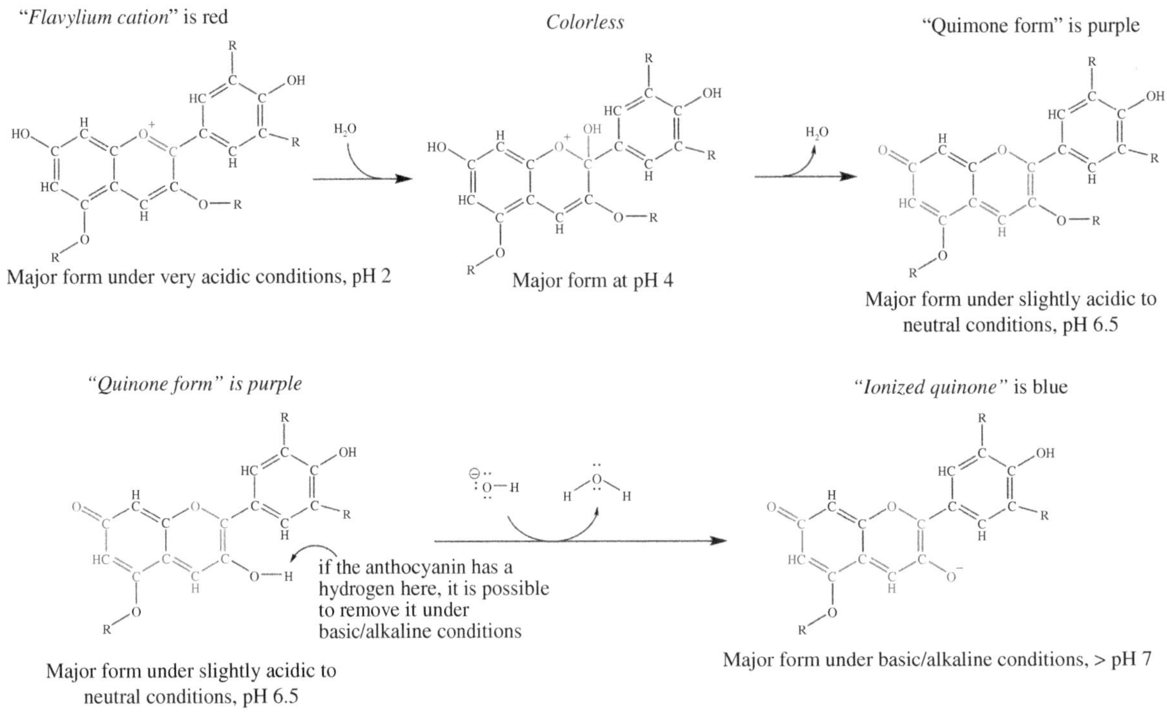

"*Flavylium cation*" is red

Colorless

"Quimone form" is purple

Major form under very acidic conditions, pH 2

Major form at pH 4

Major form under slightly acidic to neutral conditions, pH 6.5

"*Quinone form*" is purple

"*Ionized quinone*" is blue

if the anthocyanin has a hydrogen here, it is possible to remove it under basic/alkaline conditions

Major form under slightly acidic to neutral conditions, pH 6.5

Major form under basic/alkaline conditions, > pH 7

Figure 6.22 Changes to anthocyanins with pH. The flavylium cation form is present at low pH (very acidic conditions). As the pH increases, water reacts with the flavylium cation to produce the colorless carbinol. As the pH increases further, a purple quinone is formed, and water is released again. Finally, increasing the pH above 7 (alkaline/basic pH) will produce a blue ionized quinone. Structural changes are highlighted in red.

for as short a time as possible. Cooking in a large amount of boiling water also helps dilute the natural acids from the plant material.

Anthocyanins are very water-soluble and easily leach from the plant material when cooked in water. Anthocyanins are also sensitive to changes in pH. The color of anthocyanin molecules changes depending on the acidic or alkaline nature of the environment. Under acidic conditions, anthocyanins have the red-violet hue of the flavylium cation, while under slightly acidic to neutral conditions, the color changes to purple quinone (Figure 6.22). If the pH is basic/alkaline, a proton is removed, forming the blue ionized quinone (Figure 6.22). Regardless of the starting color (from red-violet to violet-blue), all anthocyanins change color with changes in pH.

For example, this is why blueberries can turn a greenish hue when added to pancake or muffin batter. If excess baking soda (i.e. the alkaline sodium bicarbonate) is used or incompletely mixed into the batter, the alkaline condition changes the color of the anthocyanin molecules in the blueberries. Similarly, mash raspberries with a bit of water. The acids (malic, oxalic, and citric) of the fruit make the pH acidic and the color of the anthocyanin red. Mix in a little baking soda, which raises the pH to be more alkaline, and observe the anthocyanins of the raspberry to a dark purple color.

The carotenoids are much more stable for cooking, and many natural carotenoid extracts are used as natural food colorings. Upon prolonged heating, carotenoids can undergo isomerization of trans double bonds to cis double bonds (Figure 6.23). The cis versions of the molecule are not as intensely colored. This is why overcooked yellow squash has a faded appearance; the yellow carotenoids have isomerized. Long-term storage of carotenoids can also cause fading of the color due to oxidation of the double bonds. In this case, the effect is the same; the loss of the double bond by reaction with oxygen causes a loss of color.

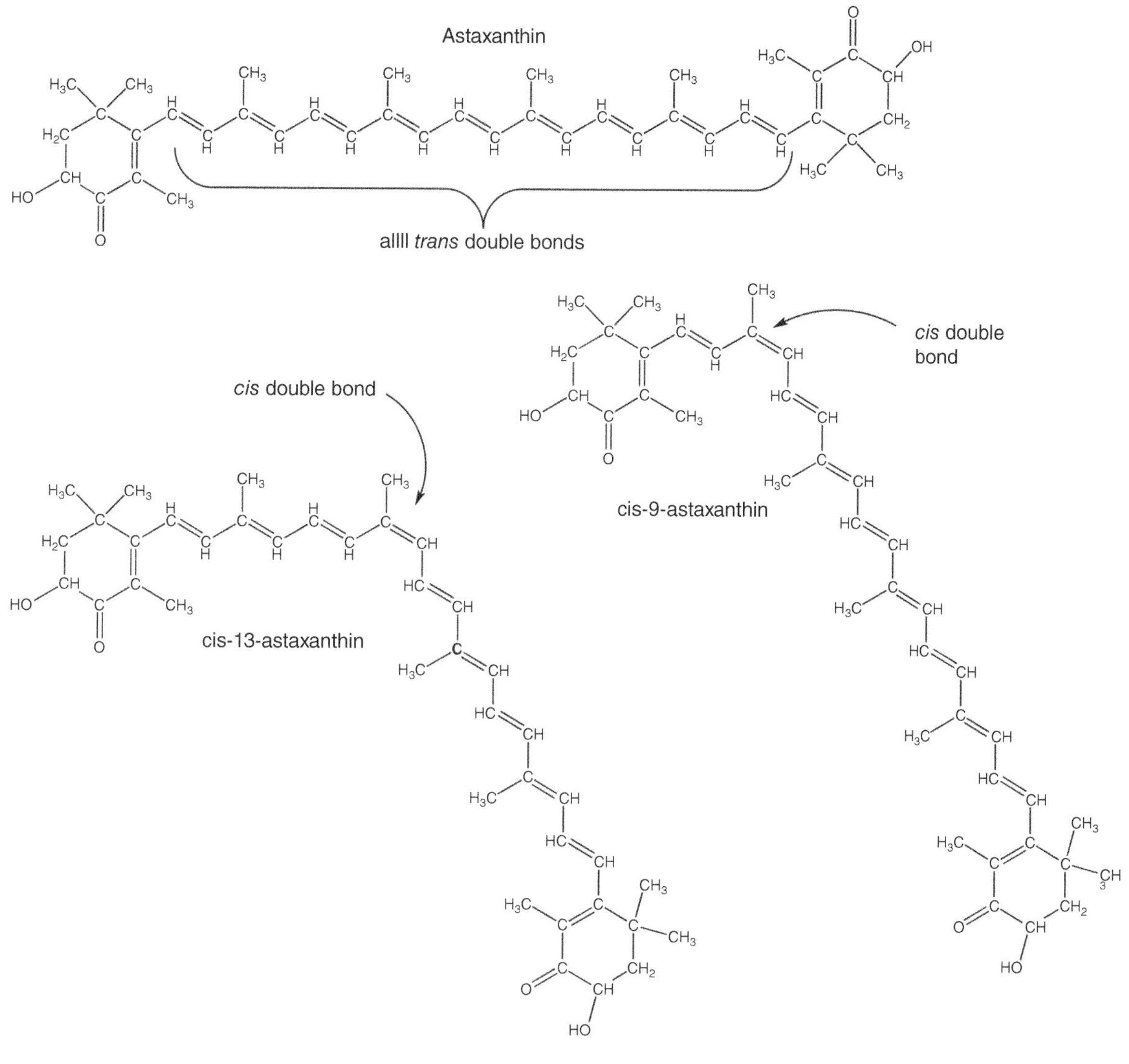

Figure 6.23 Prolonged heating causes isomerization of carotenoid. Astaxanthin is a carotenoid with a reddish-pink color. Upon heating, astaxanthin can isomerize at one of two trans double bonds to form either *cis*-9-astaxanthin or *cis*-13-astaxanthin. The cis isomers are less colored. This same type of isomerization can happen to any of the carotenoids.

6.6.3 Browning in Fruits and Vegetables

As discussed in the chapter on browning, another color change that occurs in food is the browning of fruits, particularly apples, avocados, pears, bananas, potatoes, and lettuce. However, this browning only occurs when the fruit/vegetable has been cut or otherwise damaged/bruised (Figure 6.24).

When plant cells are damaged and the plant tissue browns, the color change is due to the oxidation of phenolic compounds into a brown pigment. These phenolic compounds are part of the plant's chemical defenses. When the plant tissue is damaged, the brown pigments form barriers and have antimicrobial properties that prevent the spread of infection or bruising in plant tissues. Polyphenol oxidase (PPO) is the enzyme responsible for the browning reaction. PPO and phenolics mix with oxygen when the plant cell is damaged, and the vacuole breaks open, leaking the phenolics into the plant cell cytoplasm. The phenolics react with oxygen in an oxidation reaction that forms large, brown polyphenol molecules.

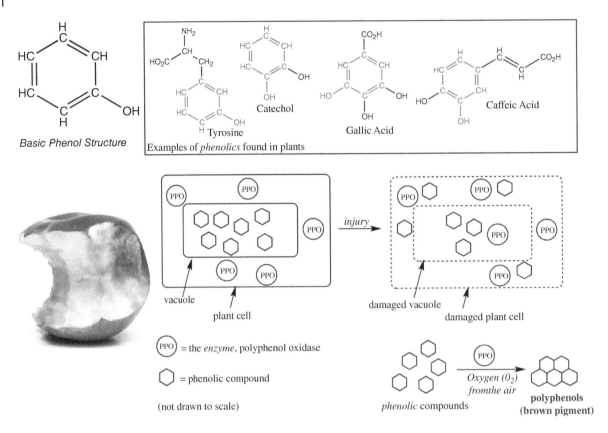

Figure 6.24 Phenolic compounds are oxidized by polyphenol oxidase (PPO), which browns the plant tissue in a chemical defense mechanism. antonsov85/Adobe Stock Photos

When cutting fruit to serve, the resulting brown pigments are unappealing. Several strategies can minimize browning reactions:

- *Limit oxygen*—immerse the cut pieces in cold water.
- *Kill the enzyme*—inactivate PPO by immersing the veggie in boiling water for 3 minutes at 115°C and then chill and bag. This works well for lettuce since it has more fibrous material and withstands the heat.
- *Slow the enzyme down*—refrigeration at less than 40°F and acidic conditions make the PPO enzyme work slower (enzymes hate acid). Douse the fruit or veggie in lemon juice, and put it in the fridge.
- *Chemically fight the oxidation*—add an *antioxidant* like vitamin C that blocks the oxidation reactions to produce the brown pigments. Vitamin C is an antioxidant found in citrus fruits. A vitamin C solution will slow the appearance of browning in cut fruit. But vitamin C also rapidly reacts with oxygen in the air and is destroyed by heat.

6.6.4 Fruits and Vegetables Are Great Sources of Antioxidants

Oxidation is a chemical process that is in part caused by the presence of oxygen in the air we breathe. Oxygen is essential for much of life on Earth. Animals need it, plants need it, and many microorganisms need it. Oxygen is necessary for respiration, but breathing oxygen has its consequences. To do the chemistry of respiration, cells produce reactive oxygen species as by-products. Reactive oxygen species are also sometimes called free radicals or oxygen radicals. Reactive oxygen species are so reactive that they destroy cell membranes, proteins, and genetic material (i.e. DNA). This oxidative damage that accumulates due to reactive oxygen species damages

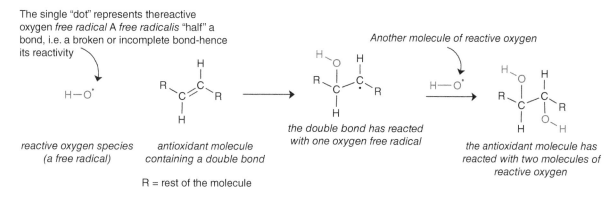

The single "dot" represents thereactive oxygen *free radical* A *free radicalis* "half" a bond, i.e. a broken or incomplete bond-hence its reactivity

Another molecule of reactive oxygen

the double bond has reacted with one oxygen free radical

the antioxidant molecule has reacted with two molecules of reactive oxygen

reactive oxygen species (a free radical)

antioxidant molecule containing a double bond

R = rest of the molecule

Figure 6.25 Double bonds are great for reacting with (i.e. neutralizing) reactive oxygen free radicals.

membranes, proteins, and genetic material and eventually kills cells and tissues. It makes you wonder why we breathe oxygen ... seems pretty dangerous!

One of nature's strategies to fight oxidative damage is to use antioxidant molecules that react harmlessly with reactive oxygen species and prevent them from damaging proteins and other cellular components. Our human bodies make some of their own antioxidant molecules—and these molecules are very important for health. Plants also make antioxidant molecules as part of their defense against reactive oxygen species made in photosynthesis, and we can consume plants! The more antioxidant molecules we have, our arsenal for combating reactive oxygen species is better equipped.

So, how can we know if a plant contains these beneficial antioxidant molecules? Antioxidant molecules have many double bonds that are great for reacting with reactive oxygen species (Figure 6.25). As we also know that the molecules that give fruits and vegetables their vibrant colors also contain many double bonds, we can understand which are best for use in our diet as antioxidants! (Table 6.10).

Therefore, colorful plants are antioxidant-rich plants. As we also saw earlier, a molecule can have double bonds and not be colored—some of these colorless antioxidants are important too. Each *antioxidant* molecule protects against different kinds of reactive oxygen species and the damage they create—no one antioxidant can protect against all types of oxidative damage. Therefore, it is best to eat a variety of colorful fruits and vegetables and consume various antioxidant molecules.

Grapes are very rich in antioxidant molecules, and the health benefits of moderate red wine consumption have been widely studied. Anthocyanins are responsible for the color of red grapes and wines. They have "antioxidant,

Table 6.10 Antioxidant molecules, their colors, and some sources.

Antioxidant molecule or class of molecules		Color	Examples of fruits and vegetables containing an abundance of the antioxidant molecule(s)
Beta-carotene		Orange	Carrots, cantaloupe, apricots
Lutein		Yellow	Corn, kiwi fruit, zucchini, pumpkin, spinach, kale, red grapes
Zeaxanthin	Carotenoids	Yellow	Orange pepper, corn
Lycopene		Red orange	Tomatoes (fruit and juice), red delicious apples
Chlorophyll		Green	Spinach, green leafy lettuce (i.e. romaine)
Phenolics (tannins)		Colorless (brown)	Grapes, tea, cinnamon, clove, vanilla, basil
Vitamin C		Colorless	Citrus fruits, bell peppers, kiwi fruit, grapefruit
Vitamin E		Colorless	Sunflower seeds, almonds, spinach, salad greens

Table 6.11 Other health benefits of plant pigment molecules: not an exhaustive list!

Molecules	Health Benefit	Sources
Anthocyanins (Figure 6.20)	Slow development of heart disease	Grapes, berries, plums
Beta-carotene (Figure 6.20)	Precursor to vitamin A, essential for normal vision, immune system function	Sweet potato, carrot, pumpkin
Lutein	Slows development of cataracts and macular degeneration	spinach, turnip greens, peas
Phenolics (Figure 6.24), terpenes	Inhibit the growth of cancer cells	Grapes, rye, many fruits and veggies

antimicrobial, and anticarcinogenic activity" and "exert protective effects on the human cardiovascular system." Anthocyanins have different biological functions in plant tissues, such as protection against damaging sun exposure, pathogen attacks, oxidative damage, and attack by reactive oxygen species. Polyphenols in the grape are generally responsible for antioxidant activity, but resveratrol—a particular phenolic present in grapes—is characterized by anticancer, antioxidant, anti-inflammatory, and cardioprotective activity. In the grapevine, resveratrol also defends the plant against attack by predators and pathogens (Table 6.11).

6.6.5 Flavor

Plants—especially fruits—can smell and taste wonderful. We experience flavor as taste—using our tongue—and aromas—using our nose (see Chapter 14 for a deeper discussion of this topic). Both components contribute to our perception of flavor. Taste is experienced by our tongues in five major categories: sweet, salty, sour, savory, and bitter. As we saw earlier in Section 6.4.2, the sweet taste of ripe fruit is due to the glucose, fructose, and other sugar molecules that are liberated from the stores of starch by enzymes during the ripening process. The bitter taste of cruciferous and leafy greens like chard, chicory, and endive is actually a sign these vegetables are good for you! The molecules that make plants so healthy to consume are often called phytonutrients, including phenols, flavonoids, isoflavones, terpenes, and glucosinolates. These molecules benefit human health, but are almost always bitter, acrid, or astringent. Glutamic acid is an amino acid whose concentration is increased in mature tomatoes and mushrooms, giving these foods an umami or savory flavor. The sour oxalic, ascorbic, and quinic acids are altered in maturing pears and other fruits. Metabolism of glucose into the Krebs cycle and different pathways will produce succinic, malic, citric, and ketoglutaric acids. Immature fruit and some vegetables have high concentrations of these acids making the plant taste sour and tart. Loss of these organic acids accompanies the decrease in starch content during ripening.

Aroma molecules (or odorants) are experienced in the nose, not the mouth, so the molecules must evaporate and travel among the gas molecules in the air. Molecules that readily travel in the air are called volatile. We can smell volatile molecules because they enter our noses with the air we breathe and travel up into our nasal passages when we crush plant material in our mouths. Using a technique called gas chromatography, chemists can actually measure the aroma molecules present in the air. This method demonstrated that bananas' floral and fruity aroma is due to over 40 different molecules! While isoamyl acetate (or "banana oil") is a significant contributor to the banana bouquet, it is not solely responsible for the complex aroma of the banana. Each fruit and vegetable has its own palate of molecules responsible for its characteristic smell and taste. There are over 400 volatile molecules accountable for the unique aroma of tomatoes! During ripening, enzymes produce a massive number of small molecules that volatilize to the smell receptors in our nose and make the fruit appealing to eat.

Aroma molecules are typically volatile liquids that evaporate a little at a time—that is why we can smell them. Since aroma molecules evaporate into the air over time, fresh plant material will smell differently (and taste

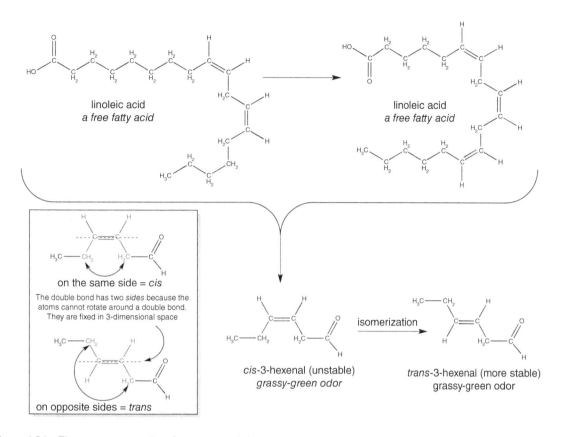

Figure 6.26 The grassy green odor of cut grass and ripe tomatoes is caused by *cis* and *trans*-3-hexenal. *cis*-3-hexenal is made from free fatty acids (linoleic and linolenic acids) by enzymes in the plant, but it is relatively unstable. The *cis* form isomerizes to the more stable *trans* form.

differently) than plant material cooked or dried. This fact explains why fresh basil smells and tastes so differently from dried basil, and it also explains why the flavor of bottled herbs and spices can change over time, limiting their useful shelf life. A bottled herb or spice is actually dried and pulverized plant material—leaves, seeds, flowers, and so on. The longer the plant material sits, the more volatile aroma molecules evaporate away.

Many aroma molecules are produced by the metabolism of amino acids and fats. For example, amino acids like isoleucine and valine are converted into volatile esters that contribute to the characteristic aroma of melon. In contrast, free fatty acids can produce fresh or grassy odors. *cis*-3-Hexenal is a product of the free fatty acids linoleic and linolenic acid, and it is the molecule responsible for the intensely grassy green scent of freshly cut grass. *cis*-3-hexenal also contributes to the aroma of ripe tomatoes. *cis*-3-hexenal is somewhat unstable and readily isomerizes to *trans*-3-hexenal, which has a similar odor but is much more stable (Figure 6.26).

Most aroma molecules fall into two structural categories—terpenes and phenolics. It's nearly impossible to predict exactly how a molecule will smell based on its structure, but molecules of similar structure generally give related aromas (Table 6.12). One plant will typically have many aroma molecules in different concentrations:

- Terpenes provide citrusy, fresh, and floral aromas (Figure 6.27). Terpenes are very volatile and, therefore, evaporate quickly when the plant material is heated. It's best to add fresh herbs, fruit juice, etc. to the dish after it's been cooked.
- Phenolics have more characteristic, distinct flavors that are generally warm or sweet (Figure 6.27). They are less volatile and mostly water soluble—which means they persist longer in cooking—and they will dissolve in the water of the foods we cook and eat and in our saliva as we chew.

Table 6.12 Examples of aromatic herbs, spices, and fruits and some of the molecules that contribute to each aroma.

Herb/spice/fruit	Terpenes	Phenolics	Herb/spice/fruit	Terpenes	Phenolics
Basil	Cineole, linalool	Eugenol	Cloves		Eugenol
Peppermint	Pinene, menthol		Ginger	Cineole, citral, linalool	
Sage	Cineole, pinene		Lemon	Limonene, linalool, pinene	
Thyme	Pinene, linalool	Thymol	Cinnamon	Cineole, linalool	Cinnamaldehyde, eugenol
Vanilla	Linalool	Eugenol, vanillin			

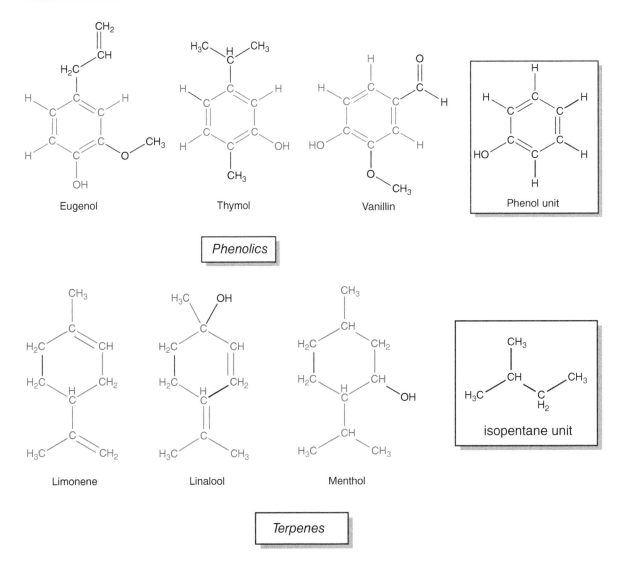

Figure 6.27 Examples of terpene and phenolic aroma molecules found in plants. Phenolics all share the basic structure of phenols, while terpenes are some combination of isopentane units.

The particular combination of aroma molecules gives the food its unique smell. Combine aroma molecules with other flavor molecules (taste, astringency, and pungency), and you get the unique flavor of a particular fruit or vegetable.

Science for the Chef: Let's describe how to make a white (béchamel) cheese sauce

Using starches or gums to thicken and stabilize gravy or sauce explains some of the science behind a roux sauce. A roux is a butter and flour paste that has been lightly cooked. The fat coats the proteins in the flour and allows the particles to disperse evenly as the watery liquid is added. Adding watery stock, broth, milk, etc. with heat hydrates the starch into a gel that thickens the liquid (Section 6.5.2.1).

A roux is used in three of the five "mother sauces" of classical French cooking: the béchamel sauce, the velouté sauce, and the espagnole sauce. What is a "mother sauce"? Depending on the culinary school of thought, there are three, five, or even nine "mother sauces." French cooking schools describe five or sometimes seven sauces, Asian cooking schools often describe nine sauces, and African cooking describes nine or more base sauces. A mother sauce is a basic sauce from which other flavored sauces are created. For example, a béchamel sauce is made of milk, flour, and butter, where the flour and butter are used to make a white roux, and the final sauce is a stable, thick, creamy liquid. The base béchamel recipe can be made into a cream, mornay, or cheese sauce with other ingredients or flavors. Other cuisines also use the principle of the roux. For example, Japanese curry uses a roux of curry powder, flour, and oil added to stewed meat and vegetables. The outcome is a thickened, stable emulsion of watery juices and oil with spices for flavor.

Figure 6.28   A Béchamel sauce. FomaA/Adobe Stock Photos

Ingredients:
- 1 tablespoon butter (14.2 grams)
- 2 tablespoons flour (15 grams)
- 1 1/4 cup warm milk
- Salt, black pepper, and cayenne pepper
- 1/2 cup grated cheddar cheese
- nutmeg to taste

Instructions:
Step 1: Create the roux

- **Measure out equal parts of oil/fat and flour:** Measuring volumes (cups/tablespoons, etc.) is less effective than measuring masses. Most professional kitchens will weigh out the dry or paste ingredients with a scale for accuracy and consistency rather than use a volumetric approach.

- **Melt the butter**: Do this slowly at a low temperature to avoid burning the casein proteins in the butter. The purpose of fat in a roux is to encase the starch granules of flour. As the carbohydrate starches hydrate in a sauce, clumping occurs when water tightly forms a shell around the hydrophilic starch particles. The fats from the butter will slow down this process. Because there is trapped water in butterfat, wait until the butter is melted and the water bubbles escape the liquid.
- **Stir in and cook the flour**: Cooking will start the browning process between the proteins in the flour and the simple sugars in the butter and the endosperm of the wheat flour (see Chapter 3). The longer and higher the temperature at which a roux is cooked, the darker and more complex flavor the roux becomes. A barely cooked roux is a white roux, which produces a neutral, nearly flavorless sauce. When a roux is cooked longer and at higher temperatures, Maillard reactions (Chapter 3) will give a rich, dark, nutty flavor (peanut butter or chocolate roux). A béchamel sauce uses a white roux and only needs to be cooked for a few minutes until the raw flour taste is gone and the roux is still a pale yellow.

Step 2: Thicken the liquid

- **Add milk warmed to the touch slowly to the roux**. Cooled milk will cause the fat surrounding the flour granules in the roux to solidify and expose the carbohydrates in the flour. Water will rush in and form a tight, nearly impenetrable shell, and clumping will happen. Too hot, and the fat will separate as liquid oil and again expose the complex sugars of the flour. More clumping... The ideal case is to slowly expose the starch to the water while mixing to allow gelation without aggregation. Not too hot, not too cold... just right. Continue to heat while stirring (you want to avoid that clumping, right?) until enough water has evaporated and the starch has escaped the granule and has formed a viscous tangled mass of complex carbohydrates!

Step 3: Add the cheese and nutmeg

- **Once the sauce is nearly thickened to the final consistency, add the grated cheese.** Freshly grated cheese will give a smoother, less gritty result. Shredded cheese from the grocery store is coated with flour or starch to prevent the cheese from sticking together. The added uncooked flour or starch will give the sauce a doughy flavor and aggregate into tiny lumps.
- **Grate some nutmeg into the sauce**. Nutmeg is a spice made from the dried seed of the nutmeg tree. Nutmeg contains the phenolic myristicin and the terpene 4-terpineol that contribute to its warm and woodsy flavor.

Key Concepts

1. Plant cells have a nucleus and cell membrane like animal cells, but they also contain a **cell wall** for structure and strength, **chloroplasts** for converting light energy and carbon dioxide into food, **vacuoles** to store water, and **amyloplasts** to store starch.

2. **Amylose** and **amylopectin** are two types of **starch**. Both amylose and amylopectin can be broken down by the human body into **glucose** monomers, which are a source of energy.

3. Amylose is a glucose polymer or polysaccharide with **1→4 glycosidic** connections that form a linear structure. Amylopectin is a **polysaccharide** of glucose monomers connected 1→4 and with **1→6 branches**, which gives amylopectin a bushy structure.

4. Plant cell walls are made of tough structural polymers of insoluble **fiber**, including the polysaccharide **cellulose** and **lignin**. A semisolid, flexible mixture of soluble fiber components, **hemicellulose** and **pectin**, fills the space between the cellulose fibers and reinforces the cell wall.

5. Plants can use chloroplasts and the energy of light to photosynthesize carbohydrates from carbon dioxide and water. When a plant is harvested, **photosynthesis** stops, but **respiration**—or the consumption of carbohydrates for energy—continues until energy stores are depleted.

6. Fruit **ripening** is either **ethylene** dependent (i.e. **climacteric**) or ethylene independent (i.e. **non-climacteric**) and is characterized by softening of the plant cell walls due to enzymatic breakdown of the cell wall **polymers** and sweetening of the fruit due to enzymatic breakdown of stored starch into sweet-tasting simple carbohydrates.

7. Cooking plant material results in the breakdown of hemicellulose and pectin from the cell wall and the **gelation** of starch from the amyloplasts. Pectin and other polysaccharides from plant **gums** can also gel and thicken a liquid. **Retrogradation** is a reforming of polymer-polymer hydrogen bonds that exclude water and stiffen the gel.

8. Colored molecules contain many **conjugated double bonds**, which allow the molecule to absorb visible light.

9. The primary plant pigment molecules are **chlorophylls**—responsible for green color; **carotenoids**—responsible for yellow to orange colors; and **anthocyanins**—responsible for red to purple colors.

10. Chlorophyll can be released into watery cooking liquids after the cleavage of its **hydrophobic** tail. Carotenoids are fat-soluble. Anthocyanins are water soluble, but their colors change with changes in pH.

11. The brown color of damaged plant tissue is due to the **oxidation** of **phenolic compounds** into a brown pigment by the enzyme PPO.

12. **Antioxidants** are molecules with double bonds that can react with and neutralize **reactive oxygen species**. Colorful plants are rich in antioxidant molecules.

13. Aroma molecules are typically **volatile**, which means they evaporate into the air, and that is why we can smell them.

14. **Terpenes** and **phenolics** are two classes of aroma molecules found in plants. Terpenes are more volatile and fat-soluble, while phenolics are less volatile and more water-soluble.

References

1 Meireles, Elaine, A., Cíntia, N. B., Carneiro, Renato, A., DaMatta, Richard, I., Samuels, and Carlos, P., Silva. (2009). Digestion of starch granules from maize, potato and wheat by larvae of the yellow mealworm, Tenebrio molitor and the Mexican bean weevil, Zabrotes subfasciatus. *Journal of Insect Science*, *9*(1), 43.

2 Mann, C. C. 1493. Uncovering the new world Columbus created, *Vintage Books, a division of Random House* New York, 2012.

3 Koch, K. and Jane, J.L. (2000) Morphological changes of granules of different starches by surface gelatinization with calcium chloride. *Cereal Chem*, *77* (22), 115–120.

4 McGee, H. (2001). *On Food and Cooking*.

5 Gonda, I., Bar, E., Portnoy, V., Lev, S., Burger, J., Schaffer, A. A., & Lewinsohn, E. (2010). Branched-chain and aromatic amino acid catabolism into aroma volatiles in cucumis melo L. *fruit. Journal of experimental botany*, *61* (4), 1111–1123.

6 Maoz, Itay, Efraim Lewinsohn, and Itay Gonda. (2022). Amino acids metabolism as a source for aroma volatiles biosynthesis. *Current Opinion in Plant Biology*, *67*, 102221.

7 Haytowitz, David, B., Ahuja, Jaspreet K. C., Wu, Xianli, Somanchi, Meena, Nickle, Melissa, Nguyen, Quyen, A., Roseland, Janet, M., Williams, Juhi, R., Patterson, Kristine, Y., Li, Ying, Pehrsson, Pamela, R., (2019). USDA national nutrient database for standard reference, legacy release. *Nutrient Data Laboratory, Beltsville Human Nutrition Research Center, ARS, USDA*.

Additional Readings

Cortez, R., Luna-Vital, D. A., Margulis, D. and Gonzalez de Mejia, E. (2017), Natural pigments: Stabilization methods of anthocyanins for food applications. *Compr Rev Food Sci Food Saf*, 16(1), 180–198.

Garcia, F. and Davidov-Pardo, G. (2021). Recent advances in the use of edible coatings for preservation of avocados: A review. *J. Food Sci.*, *86*(1), 6–15.

Jing, H., Nie, M., Dai, Z., Xiao, Y., Song, J., Zhang, Z., Zhou, C., & Li, D. (2023). Identification of carotenoids from fruits and vegetables with or without saponification and evaluation of their antioxidant activities. *J. Food Sci.*, 88(6), 2693–2703.

McClements, D. J., & Grossmann, L. (2021), The science of plant-based foods: Constructing next-generation meat, fish, milk, and egg analogs. *Compr Rev Food Sci Food Saf*, *20*(4), 4049–4100.

Oliveira, S. M., Brandão, T. R. S. & Silva, C. L. M. (2016). Influence of drying processes and pretreatments on nutritional and bioactive characteristics of dried vegetables: A review. *Food Eng Rev* 8, 134–163.

Sagar, N. A., Pareek, S., Sharma, S., Yahia, E. M. and Lobo, M. G. (2018). Fruit and vegetable waste: Bioactive compounds, their extraction, and possible utilization. *Compr Rev Food Sci Food Saf*, *17*(3), 512–531.

Sun, T., Xu, Z., Wu, C.T., Janes, M., Prinyawiwatkul, W. and No, H.K. (2007). Antioxidant activities of different colored sweet bell peppers (Capsicum annuum L.). *J. Food Sci.*, *72*(2), S98–S102.

Wang, S., Li, C., Copeland, L., Niu, Q. and Wang, S. (2015), Starch retrogradation: A comprehensive review. *Compr Rev Food Sci Food Saf*, *14*(5), 568–585.

End of Chapter 6 Questions

1. The main component of the wheat seed that contains starch is the _____
   a. Bran
   b. Germ
   c. Endosperm
   d. Carbohydrate center
   e. Embryo

2. Photosynthesis and respiration are chemical processes that occur in plants.
   a. When a plant (or piece of a plant) is harvested, picked, or otherwise separated from the ground (or the rest of the plant), what happens to photosynthesis?
   b. After harvest, grapes have a respiration rate of 16, while peaches have a respiration rate of 50. Which of these fruits will spoil faster, and why is the respiration rate a good measure of this (provide a chemical explanation)?
   c. What can you do to slow the respiration of a harvested plant?

3. The flesh of a cut, bruised, or bitten apple or avocado will eventually turn brown.
   a. If a peeled and cut apple is quickly heated to steaming, why do the pieces not turn brown (provide a chemical explanation)?
   b. If peeled and cut apple slices are coated in water containing vitamin C, they do not turn brown. Why? (provide a chemical explanation)
   c. Every guacamole recipe mashes avocado flesh with lemon or lime juice. What does this have to do with the reaction shown above?

4. Paper bags are a favorite tool of the produce manager at your local grocery store.
   a. You can take an unripe (hard) avocado and place it in a brown paper bag for 24–48 hours to ripen it. What is the chemical reasoning behind why this recommendation works?
   b. You can ripen a green tomato by placing it in a paper bag with some ripe bananas. What is the chemical reasoning behind why this recommendation works?
   c. The "paper bag" trick doesn't work for ripening strawberries, however. Why?

5. You are trying to boil carrots into a puree. You have the choices of acidic, alkaline, or neutral conditions. Rank these conditions according to how long it would take the carrots to disintegrate into a puree. Explain the chemical reasoning behind your ranking.

6. You can take cauliflower florets, boil them, and then puree into a fluffy white side dish. Potatoes can be peeled, boiled, and gently mashed to yield a similar-looking dish. A dish of mashed cauliflower has far fewer calories than a dish of mashed potatoes. To put it another way, you can eat three times the amount of cauliflower compared to potatoes for nearly the same number of calories.

Nutrition Facts		
Potato		
Serving size		100g
Amount per serving		
Calories		**77**
		% Daily Value*
Total Fat 0.1g		0%
Saturated Fat 0g		
Trans Fat 0g		
Cholesterol 0mg		0%
Sodium 6mg		0.3%
Total Carbohydrate 19.6g		7%
Dietary Fiber 2.1g		7.5%
Total Sugars 0.8g		
Includes 0g Added Sugars		
Protein 2.0g		
Vitamin D 0mcg		0%
Calcium 12mg		1%
Iron 8mg		0.6%
Potassium 425mg		9.0%

* The % Daily Value (DV) tells you how much a nutrient in a serving of food contributes to a daily diet. 2,000 calories a day is used for general nutrition advice.

Nutrition Facts		
Cauliflower		
Serving size		100g
Amount per serving		
Calories		**25**
		% Daily Value*
Total Fat 0.3g		0.4%
Saturated Fat 0g		
Trans Fat 0g		
Cholesterol 0mg		0%
Sodium 30mg		1%
Total Carbohydrate 5g		2%
Dietary Fiber 2g		7%
Total Sugars 2g		
Includes 0g Added Sugars		
Protein 2g		
Vitamin D 0mcg		0%
Calcium 22mg		1.7%
Iron 0.4mg		2%
Potassium 299mg		6%

* The % Daily Value (DV) tells you how much a nutrient in a serving of food contributes to a daily diet. 2,000 calories a day is used for general nutrition advice.

Macronutrient	Calories
1 gram Protein	4
1 gram Fat	9
1 gram Carbohydrate	4

Potato
(Food Data Central ID: 170026)

Cauliflower
(Food Data Central ID: 169986)

a. Using the Table of Calories above (review Calorie counting from Chapter 2) show how a 100 g serving of potato equals approximately 77 calories. Show all work.

b. Both cauliflower and potato are primarily carbohydrates. Why does a serving of cauliflower have fewer calories compared to an equivalent serving of potato? Provide a chemical explanation.

c. On the nutrition label for the potato, the total carbohydrate is 18 grams. Name all the types of carbohydrate that make up those 18 grams.

d. When making mashed cauliflower, you have to boil the cauliflower for at least 30 min (compared to 10 min for potato) and then puree it with a blender (while the potatoes can be gently mashed). Why does cooking the cauliflower take longer and require more aggressive mashing methods?

7. Suppose a recipe calls for 2 large red potatoes, boiled. Potatoes are a high-starch food. What happens to the potato when it is boiled? Explain what is happening chemically to the molecules when the potatoes go from inedibly hard to soft and chewable.

8. Suppose a lemon vinaigrette calls for chopped fresh thyme as an ingredient.
   a. What type of molecule is primarily responsible for the smell of fresh, chopped thyme?
   b. What would be the effects of substituting dried thyme in this recipe? (Include the chemical reasoning behind your explanation).
   c. Thyme and lemon are complementary. Thyme is often called a "lemony" herb. What does this tell you about the flavor profiles of thyme and lemon?

9. What is the natural role of fruit browning?
   a. To produce compounds that are toxic to bugs who may injure the fruit or vegetable
   b. To inhibit microbe growth in the rest of a plant when injured
   c. A form of plant chemical warfare
   d. All of the above
   e. None of the above

10. A branched starch found in potatoes, rice and grains and used as a source of energy in the human diet is _____?
    a. Amylose
    b. Amylopectin
    c. Cellulose
    d. Glucose
    e. Glycerole

7

Meat and Fish

Guided Inquiry Activities (Web): 5, Amino Acids and Proteins; 14, Cells and Metabolism; 22, Meat Structure and Properties; 23, Meat Cooking

Learning Objectives

1. *Understand the structure and function of muscle fibers.*
2. *Analyze the impact of muscle fiber density on meat tenderness and select appropriate cooking techniques based on the density of muscle fibers.*
3. *Identify the role of connective tissues, collagen, and elastin in meat tenderness.*
4. *Describe the organization of muscle fibers in different types of animals, such as fish and shellfish, and understand their implications for muscle structure and cooking.*
5. *Explain the structure of collagen and its role in providing support, strength, and structure to muscles.*
6. *Understand the different types of muscle fibers, including red (type I) and white (type II) fibers, and their characteristics in color, myoglobin content, glycogen content, fat content, and flavor/water content.*
7. *Describe the factors that influence the color of meat.*
8. *Understand the chemical and physical changes that occur in muscle tissue after an animal's death.*
9. *Analyze the effects of stress at the time of harvest on glucose and ATP levels and its influence on the onset of rigor mortis and meat quality.*
10. *Identify the chemical and physical changes that occur during aging muscle, including the enzymatic breakdown of proteins, the production of flavor molecules like IMP and ribose, and the denaturation of proteins and its impact on water holding capacity.*
11. *Understand the key components and techniques used in the development of plant-based meats.*
12. *Explain the main ingredients in producing plant-based meats, such as plant proteins, fats, flavor enhancers, and texturizers.*
13. *Evaluate the environmental and ethical implications of plant-based meats: Learners should be able to assess the environmental and ethical advantages of plant-based meats compared to conventional meat production and articulate the potential limitations or challenges associated with plant-based meats.*
14. *Explain the significance of proteases in tenderizing meat, the role of protease activity in releasing amino acids and creating new flavors, and the effects of wet aging and dry aging on meat flavor and tenderness.*

The Science of Cooking: Understanding the Biology and Chemistry Behind Food and Cooking, Second Edition. Joseph J. Provost et al.
© 2025 John Wiley & Sons, Inc. Published 2025 by John Wiley & Sons, Inc.
Companion website: www.wiley.com/go/provost/food_science_2e

7.1   Introduction

Texture, flavor, and the juiciness of meat are the key characteristics used to assess the quality of animal and fish meat. The chemical composition and changes in meat components as the tissue ages and the reactions of these compounds during cooking are responsible for its flavor. The transformation of uncooked meat following the process of aging and cooking exemplifies the characterization of meat as "muscle tissue that has undergone both chemical and physical alterations." This chapter will cover the intricate interplay of anatomy, physiology, and biochemistry of muscle to unravel the mysteries behind the captivating transformations that occur during cooking. By exploring the intricate details of muscle tissue, we will shed light on the fascinating scientific mechanisms underlying the changes in flavor, texture, and tenderness that turn raw meat into food. A basic understanding of the physiology and anatomy of muscle and knowing the key differences between cuts of meat and the unique characteristics of animal, fish, and shellfish muscle are important for getting the most out of your kitchen.

Around two million years ago, early man developed a larger brain and smaller digestive tract, and according to anthropologist Leslie Aiello, this is when our ancestors started eating meat. The switch in diet allowed for less energy devoted to digestion and more metabolic energy spent maintaining the larger, energy-demanding brains. Our ancestors were attracted to meat because of the high nutritional and caloric value owed to the protein and fat in the meat. However, some of these same components and their sources are also causes of controversy. Epidemiological studies show a causative relationship between red meat consumption and a significant increase in the risk of cardiovascular heart disease and colon cancer. Fat, as well as the products of cooking meat over high heat and smoke, is the likely culprit of health risks. Conversely, several studies suggest that lean red and other meats positively impact human health by providing nutrients, vitamins, and beneficial polyunsaturated fats when included in our diet in moderate amounts. One kind of healthy fat, conjugated linoleic acid, has several positive health benefits, including increasing immune functions and providing potential anticancer properties in animal studies. However, for some, the overshadowing philosophical question of the humaneness of raising living beings to be eaten is a continuing controversy. Despite these issues, over the past decade, yearly meat consumption in the United States has ranged from 110 to 116 lbs per person, slightly up from the 1970s when, on average, a US citizen consumed 104–106 lbs per year. Consumption of meat is not limited to the United States. China and India are also increasing meat consumption in their diets. In 1997, humans were responsible for consuming 235 million tons of meat, and the United Nations predicts that worldwide meat consumption will double by 2050.

The invention of imitation meat, that is, products that imitate fish, beef, and chicken, is a new and exciting application of science to meet health and animal use challenges. The energy and environmental resource consumption of raising fish, beef, and birds to meet the predicted 9.6 billion global population by 2050 is a worldwide issue. Three-dimensional printing of meat and extruded chicken with fibers similar to authentic meat is within sight. In his blog "Good Eats," Alton Brown enthusiastically wrote about a new mock chicken product with authentic consistency and fiber-like quality [1]. Now there are options for plant-based meat products. Whether creating a new "Earth-friendly" food or sitting down to cook a great steak, understanding the biology, chemistry, and biochemistry of meat and the changes that occur as the food is cooked is an important tool.

7.2   Muscle Motors: How Muscle Works

There are several types of muscle, smooth, cardiac, and skeletal, each of which can be found as food in some cultures. In most organisms, skeletal muscle performs the same fundamental task: to shorten or contract, creating movement. The flick of a tail, the closing of a shell, and the flexing of a leg are all movements caused by muscle shortening due to the sliding of protein filaments past one another. Muscles are highly organized groups of cells, extracellular protein glue, and connective tissues holding the muscle fibers together. Muscle cells—also called muscle fibers—are the basic unit of muscle tissue and collectively makeup what we think of as muscles

(Figure 7.1). Muscle fibers are nearly as thin as a thread of hair and often are as long as the muscle itself. Inside the muscle fibers are the proteins: actin, myosin, troponin, and tropomyosin. These proteins are responsible for the contraction and extension of muscle groups and are common among skeletal muscles in fish and mammals. As a muscle develops, these specialized cells stay the same. The increase in the size of a developed muscle group is instead due to an increased number of contractile protein filaments in the cells, making a well-exercised muscle grow in size. Unfortunately, the meat becomes tougher with more contractile filaments and larger fibers. Fibers are bundled and held together by connective tissue and filled with long polymerized proteins responsible for shortening the length of a muscle group. In fact, rigor mortis is the stiffening of muscle and loss of extension or relaxation as these key muscle fiber proteins lock in contraction against each other. Let's start understanding science and meat cooking by focusing on the contractile proteins at the center of muscle fibers.

Muscle cells, often referred to as fibers, have a unique structure that helps them work together to create movement. Inside each muscle cell, there are intricate bundles of proteins that look like braided ropes. These protein bundles are grouped into thin and thick threads, like two types of threads in a fabric. These threads are organized into sections called sarcomeres. Think of sarcomeres as the building blocks of muscle movement. The thin threads, known as "thin filaments," are made mostly of a protein called actin. This actin protein forms long chains that are like the core of the thread. Wrapped around this actin core are other proteins called tropomyosin and troponin, creating a complex arrangement that helps control muscle contraction (Figure 7.2).

In contrast, the "thick filaments" are made up of a different group of proteins. The most important protein here is myosin, which acts like a motor. Myosin can grab onto molecules called ATP and ADP, as well as actin, and use them to generate movement. Imagine myosin as a kind of ratcheting motor that connects the thin and thick filaments, allowing them to slide past each other and create muscle contraction. There's another protein called titin, which is incredibly large and acts like a spring and anchor. It interacts with actin and tropomyosin in the thin filaments, helping to control the movement between the thin and thick filaments. Think of titin as a vital part of the muscle's structure, ensuring everything stays connected and works smoothly to produce movement.

Let's delve into the intriguing process of muscle contraction, where nerve signals kickstart a highly coordinated sequence of events. Imagine calcium taking center stage as it is released within the muscle fiber, setting the scene

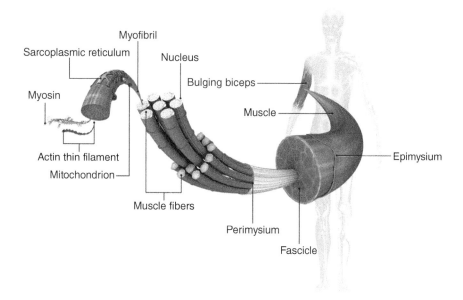

Figure 7.1 The structure of muscle and muscle fiber. The organization of tissue from actin to muscle. 7activestudio/Adobe Stock Photos

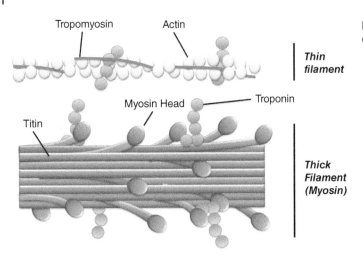

Tropomyosin   Actin

Thin filament

Figure 7.2   Muscle complex. Detailed illustration of actin, myosin, troponin, and titin proteins.

Myosin Head ————— Troponin

Titin

Thick Filament (Myosin)

for action (Figure 7.3). As the process unfolds, myosin, a crucial player, is poised with ADP in hand, yet to engage with actin. Here's the twist: before calcium's arrival, two proteins, troponin and tropomyosin, work together to conceal the interaction site between myosin and actin. However, when calcium enters the picture, troponin and tropomyosin shift their roles, revealing the binding site. This triggers a temporary connection, or cross-bridge, between myosin and actin. Picture myosin performing a controlled curling motion, much like a forearm movement, in what is termed the power stroke (Figure 7.3, Step 5). This action pulls the thin filament closer to the thick myosin filament, effectively reducing the space between the two ends of the filaments (note the role of titin in holding actin in place). During the power stroke, ADP is released from Myosin's binding site (Figure 7.3, Step 6).

Following the power stroke, myosin and actin remain linked until ATP, the energy currency, joins the scene (Figure 7.3, Step 1). ATP prompts myosin to detach from actin, disrupting their interaction (Figure 7.3, Step 2). This detachment is followed by a transformation of ATP, bound to myosin, into ADP and inorganic phosphate (Pi) (Figure 7.3, Step 4). Calcium makes its graceful exit as the curtain falls on this act, returning to its designated location within the muscle cell.

Hold on, the plot thickens! Visualize these intricacies not as isolated incidents but as a complex choreography, occurring thousands of times for each set of filaments. Keep in mind that each muscle fiber harbors an array of these filaments. The true protagonist here is a minuscule motor that orchestrates the delicate sliding motion, ultimately shortening the fiber and contributing to muscle contraction. This orchestrated performance unfolds repeatedly—hundreds to thousands of times within each muscle unit (sarcomere), fortifying the muscles and bestowing them with strength (Figure 7.4).

7.3 Muscle Organization

Muscle fibers are composed of smaller units called sarcomeres, the building blocks of muscle contraction. Sarcomeres are arranged end to end within muscle fibers, forming long chains known as myofibrils. A sarcomere is a tiny section of a muscle fiber, similar to a brick in a wall. Inside each sarcomere are two main protein filaments: thin filaments made of actin and thick filaments made of myosin. These filaments have a specific arrangement, with the thin filaments positioned between the thick filaments.

Myofibrils are collections of sarcomeres aligned end to end. They span the entire muscle fiber length, giving the muscle its striped appearance under a microscope. Myofibrils contain the necessary proteins and structures for the sliding action between actin and myosin filaments, enabling the muscle to contract and relax.

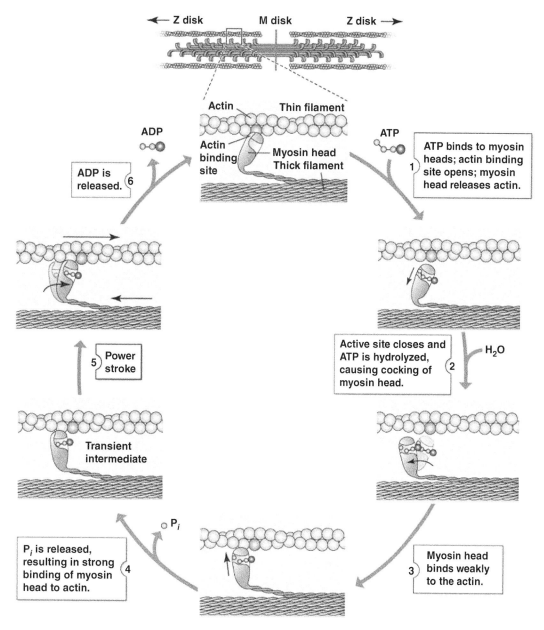

Figure 7.3 Mechanism of muscle movement. The myosin walks along the actin using the hydrolysis of ATP. Only one myosin head is shown. Voet Voet and Pratt, Figure 7.32. Reproduced with permission from Voet Voet and Pratt 4th Ed / John Wiley & Sons

The organization and density of muscle fibers in the meat are influenced by factors like age, species, and muscle use, which ultimately affect the tenderness of the meat. As animals age, the size and density of muscle fibers tend to decrease, making the meat less tender. Conversely, constant use of specific muscles, like those found in the shoulders and legs of standing animals, can increase the density of muscle fibers.

The density of muscle fibers plays a significant role in the culinary outcome of meat. Muscles with higher fiber density, such as those from heavily used or less tender parts like the shoulder or leg, tend to be tougher and require

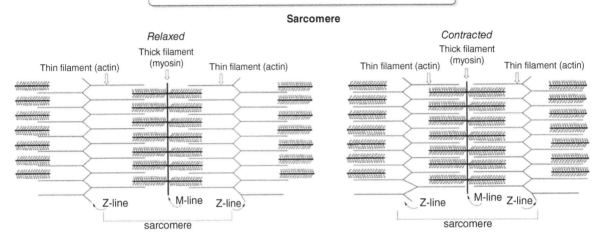

Figure 7.4 Sarcomere contraction. The steps involved in muscle fiber contraction and an illustration of the relaxed and contracted sarcomere.

longer cooking times and specific techniques to break down the connective tissues and collagen. Slow cooking methods like braising or stewing are ideal for these cuts as they help soften the connective tissues and achieve the desired tenderness. On the other hand, muscles with lower fiber density, like those found in more tender cuts such as tenderloin or ribeye, naturally require less cooking time and are already tender. These cuts are well-suited for quicker cooking methods like grilling, broiling, or pan-searing.

Understanding the density of muscle fibers can guide the selection of appropriate cooking techniques and determine the cooking times necessary to achieve the desired tenderness and texture of the meat. Slow cooking methods are effective for breaking down and tenderizing tougher cuts with higher fiber density, while shorter cooking times are suitable for more tender cuts with lower fiber density. By matching the cooking method to the specific characteristics of the meat, you can ensure a delicious and enjoyable culinary outcome.

Inspect a piece of beef or pork meat cut across the grain to see how muscle fibers are organized. The elongated arrangement of muscle fibers is the grain of the meat. Cut against the grain, and you can see how muscle fibers are grouped together. Long muscle fibers are bundled together into a fascicle. Fascicles are further bundled together to form muscles. Special connective proteins, collectively called the perimysium, glue the fascicles together (refer to Figure 7.1). The key protein of the perimysium is collagen; this is important for gelatin dishes, emulsions, and other cooking uses. Surrounding the muscle (bundles of fascicle) is a protective fibrous protein sheet called the epimysium. At the end of each muscle group are connective tissues called tendons, also made of a different form of collagen. The amount of connective tissue and collagen and the thickness of the fascicles influence the tenderness of a cut of meat. Feel the difference by rubbing the uncooked meat in your hand. Tender meat with less connective tissue, collagen, and smaller myofibrils will be noticeably smoother, whereas a tough cut of

Figure 7.5   Muscle structure of fish.

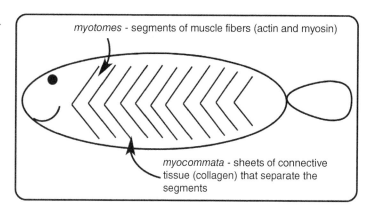

myotomes - segments of muscle fibers (actin and myosin)

myocommata - sheets of connective tissue (collagen) that separate the segments

meat will be much more firm, uneven, and grainy. We cut against the meat's grain to create shorter, less connected muscle fibers, making it easier to chew.

The flaky quality of fish skeletal muscle gives us insight into the different muscle needs of land and water animals. A terrestrial animal must propel itself and support its weight as it walks, climbs, jumps, and fights gravity. Fish (teleost), on the other hand, are suspended in a buoyant environment and therefore do not require strong muscles or well-connected muscle tissue. Like land mammals, fish muscle cells are filled with thin and thick actin and myosin fibers contracting and relaxing with the availability of calcium and ATP. However, fish do not possess the elaborate organization of beef and meat of other land animals. Individual muscles of a fish are organized into individual blocks of collagen sheaths in the shape of a "W." Each of these muscles overlaps the other, allowing for the undulating nature of a swimming fish. The coordinated contraction of each nestled into the next (via the thin and thick fiber contraction) allows for the swift flick of the tail and contortion of the fish's body in water. These individual sheaths, called myotomes, are evident as individual flakes of cooked fish (Figure 7.5).

Shellfish (animals with shells that live in water) are not technically fish, but they also have less defined muscle organization and rely on the contractility of muscle fibers in a similar way to finned fish and land animals. Except for crabs, most crustaceans have a head and a body region, and it is the abdomen or body portion that contains most of the muscle fiber used for food and cooking. Expanded to include octopus, cuttlefish, and squid, shellfish muscles are composed of thick groupings of muscle fibers connected to exoskeletons for support and mobility.

7.4   Tender Connections

The thickness of the muscle fiber is due to the number of thick and thin filaments. This is one of the factors that makes muscle tissues tough or tender. A tender piece of meat is easily chewed and cut. Meat with a high quantity of long interconnected protein fibers will be more rigid for teeth to cut than a cut of meat with fewer of these proteins. To test for meat tenderness, an instrument called the Warner-Bratzler tenderometer is used to measure the shear forces of muscle and meat. Essentially, this instrument is a guillotine connected to a computer to measure the pressure required to cut through the sample. Fibers and connective tissues combine to make a cut of meat tough or tender. Connective tissue holds muscle fibers and groups of fibers in place. It covers the muscle, acts as glue within the muscle cell, and fixes muscles to skeletal elements. Most of the connective tissues are made of a few kinds of proteins. Two of the major connective proteins are collagen and elastin. The space within and around muscle cells is filled with proteins that act as a kind of mortar holding bundles of cells in place. Epimysium is a protein membrane also known as silver skin. This is the thin, tough white, and silvery-colored connective tissue made of different kinds of collagen. Tendons, informally called sinews, are dense connective tissues attaching muscle to the bone. Similar to other connective tissue and cartilage, tendons are primarily made of collagen (80%

Figure 7.6 Structure of collagen fibril. The smallest unit is the triple alpha helix shown above. Three collagen strands in a helix combined to form a collagen molecule, which are then overlapped and cross-linked to form a tough collagen fibril. Dark grey spheres represent carbon atoms, blue spheres are nitrogens atoms and red spheres are oxygen atoms.

of total protein) and elastin (2–4% by weight). While not particularly tasty itself, understanding the composition of connective tissue can help you cook a more tender cut of meat, prepare a tasty soup, or even make an excellent dish of pho.

Collagen is a diverse family of proteins, each with a slightly different amino acid composition. Two main forms of collagen are found in muscle, types I and III. Collagen is made of three long, insoluble helical protein chains, each wrapped around three other collagen strands to form a triple helix, much like braided hair (Figure 7.6).

While collagen and muscles are described as "fibers," each plays a distinct role in the structure and function of muscles. Collagen fibers are part of the connective tissue framework surrounding muscle fibers, providing support, strength, and structure to the muscle. They contribute to the toughness and texture of meat and are rich in collagen protein. Unlike collagen, muscle fibers are the fundamental units of muscle tissue responsible for muscle contraction. These elongated and cylindrical cells contain specialized proteins such as actin and myosin, which enable the muscle to generate force and produce movement. Muscle fibers work in coordinated groups to create strength and power for various bodily functions. While collagen fibers contribute to the overall integrity and toughness of muscles, muscle fibers are primarily responsible for the contractile properties and movement of the muscle. Another way to view muscle fibers in the context of meat is shown in Figure 7.7.

Figure 7.7 Roast cut to highlight myofibril bundles and a cutaway showing muscle fibers and sarcomere. vitals/Adobe Stock Photos

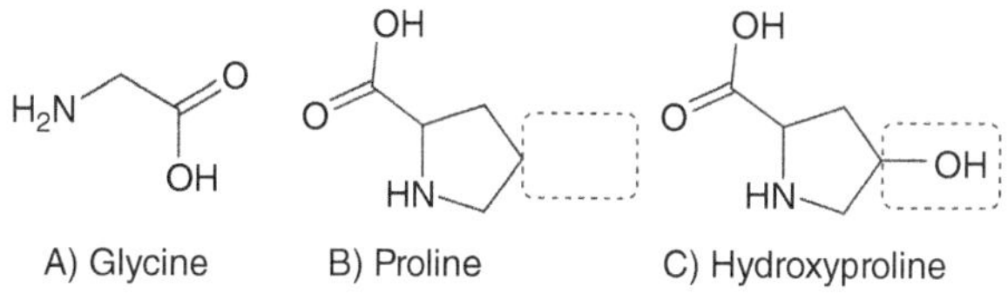

There are a few special characteristics of collagen that impact meat and cooking. First, each collagen protein is shaped into a long, narrow helix (Figure 7.8). The overall shape is reminiscent of a slinky toy pulled nearly straight. The tight collagen helix is a complicated shape for proteins to assume because of the severe turns between atoms in the protein backbone, causing atoms to clash spatially. However, collagen has a special repeating amino acid sequence of glycine/proline followed by any amino acid (often written as Gly–Pro–X; Figure 7.8). This amino acid

Figure 7.8 Key amino acids of collagen. A) The smaller amino acid glycine allows high flexibility while B) Proline limits the rotation. C) The prolyl-4-hydroxylase enzyme converts the dashed hydrogen in proline to an alcohol functional group (O-H), making hydroxyproline.

sequence allows for the unique shape and stability of each collagen strand. Because the side chain of glycine is tiny (Figure 7.8), the peptide backbone can easily twist upon itself and permit the protein to form into a long, extended, coiled chain. The high abundance of proline in collagen is very important as it stabilizes the collagen triple helix. The enzyme prolyl-4-hydroxylase is responsible for adding the stabilizing O—H group to proline. It is this O—H group that strengthens the collagen triple helix (Figure 7.8). The reaction catalyzed by oxidase prolyl-4-hydroxylase requires vitamin C. Scurvy is a collagen-related disease that occurs when people are deprived of citrus fruit, a source of vitamin C. Sailors and pirates on long tours at sea would lose teeth, have bleeding gums, and find spots or bruises (from leaking blood vessels) on their skin, all due to weak collagen.

Collagen is bundled into long fibers that allow for stretching and great tensile strength. Packed side by side, each triple helix is organized into parallel bundles of other collagen triple helices. Strengthening this arrangement are cross-links or covalent bonds between separate triple collagen helices. In addition to glycine and proline, the collagen of mammals and fish contains the amino acid lysine. An oxygen atom requiring enzyme found in and around connective tissue called lysyl oxidase links two adjacent lysine amino acids from separate helical strands of collagen, forming a strong bridge that holds the collagen fibrils together (Figure 7.9). Collagen fibers can be found at every level of muscle in land animals and fish. Think of collagen as interconnected, braided strands of steel woven throughout and around the muscle. The stronger and more connected the braids, the more difficult it will be to cut the meat with a knife or teeth.

Now that we understand the molecular structure of muscle cells and connective tissue, we can explain why some cuts of meat are more tender than others. Both actin and myosin fibers and collagen are major contributors to the tenderness of meat. Highly exercised muscles will have an increase in thick and thin filaments compared with lesser-used muscles. The force required to cut through the tissue increases as the concentration of collagen and the number of collagen cross-links increase. As an animal ages, the amount of total collagen and cross-links between collagen strands increases. The saying "tough old bird" now has a special meaning. A 20-day-old chicken has 12% less total collagen than a 1.5-year-old chicken, while the older chicken has almost 12 times more cross-linked collagen than the younger animal. This knowledge will also help you select the most tender cuts of beef or pork. Two factors impact the amount of collagen and level of cross-linked collagen: (i) the most heavily used muscles (shoulder and rear legs and hip muscles) and (ii) an abundance of thick and thin fibers, total collagen, and cross-linked collagen in older animals. The tail and legs (packed with collagen) are tough cuts of meat, whereas less used muscles, located further away from the legs, will be more tender. The beef tenderloin cut is from the psoas major muscle and is found along the central spine. This muscle does very little work, which makes this one of the most tender cuts of meat.

Fish muscle tissue, like beef and pork, has collagen serving important connective tissue roles in and around the muscle fibers. However, fish possess considerably less total collagen and fewer cross-links than beef or pork, resulting in softer muscle tissue. This should be intuitive as fish are buoyant and require less muscle support than land animals. As we will learn later, as collagen is heated, the protein softens and will solubilize (dissolve). This process is due to proteins losing their structure with increased heat. This loss of structure, or denaturing of

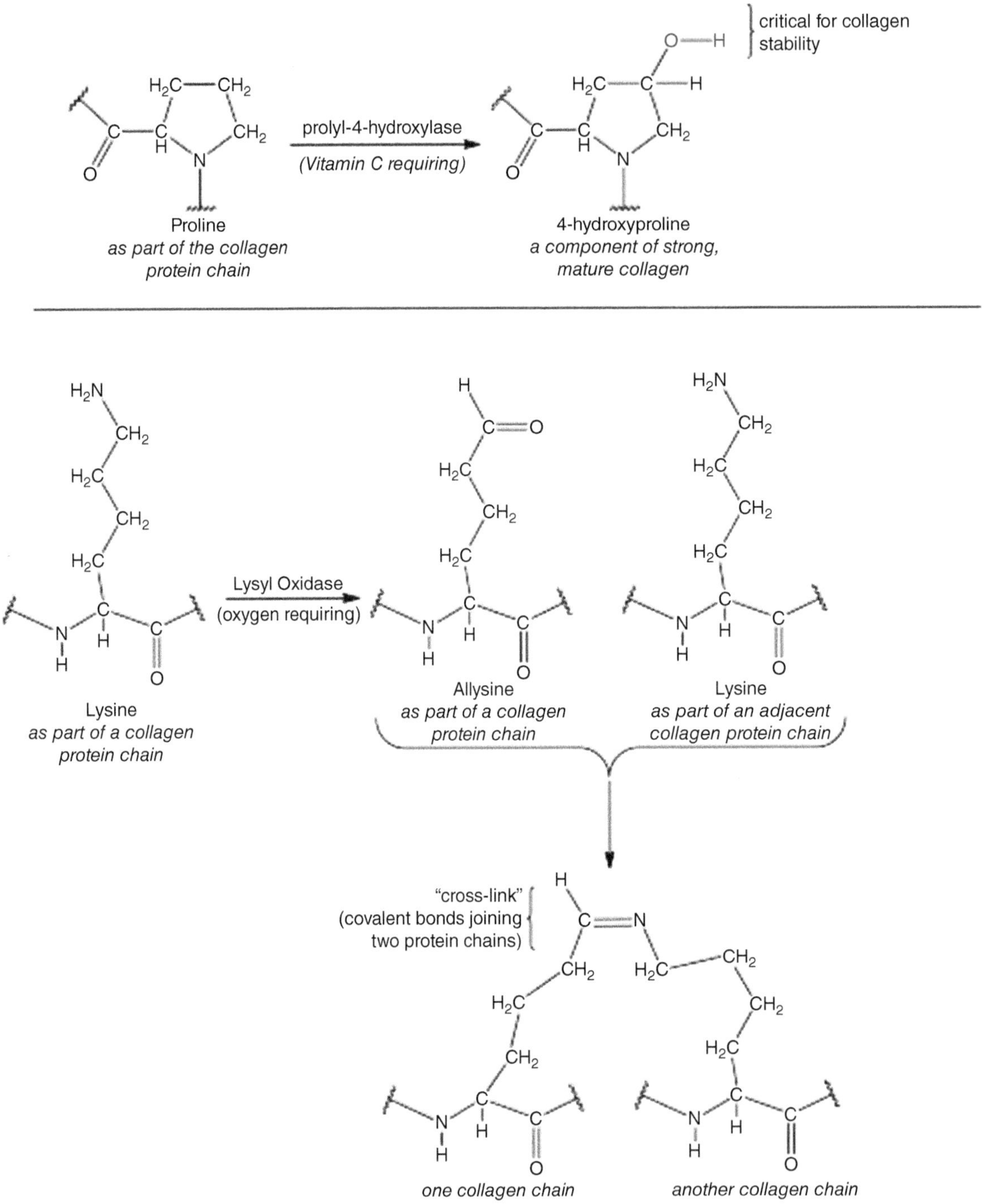

Figure 7.9 Important reactions modifying collagen for cross-linking strength. Proline and lysine modification and cross-linking reactions are shown.

the protein, decreases the ability of the collagen to act as connective tissue and muscle glue. This explains another difference between fish, beef, and other mammals: the amino acid makeup of collagen. Fish collagen is formed from the same triple helix using the glycine–proline–X sequence; however, the makeup of other amino acids differs in concentration between fish and beef, which impacts collagen's stability. The point at which human collagen melts and denatures (with a different total amino acid composition than beef or fish) is 118°F (48°C). Beef and pork collagen denatures at 98.6°F (32°C), and fish collagen will begin to melt and denature at an even lower temperature of 68°F (20°C). Thus, it takes less heat and less time to cook and soften the fish tissue than beef or pork.

Box 7.1 Modernist cooking: Sous vide

It takes a lot of heat to turn liquid water into gaseous water (steam) because of the strength of hydrogen bonding. Heat adds energy to the water molecules, giving them more kinetic activity until they are free from all or almost all hydrogen bonds, allowing them to escape into the gas phase (steam water). The energy required to vaporize liquid water is called the latent heat of vaporization; this will enable foods to cook at or near 100°C (212°F) as long as there is considerable water content in the tissue. Simmering large pieces of meat in water or stock allows the cook to keep a constant temperature; the water stays near boiling point because any excess energy (heat) is being spent as water molecules escape the water phase into the vapor phase. Modernist cooks, also called molecular gastronomists, use a method called sous vide, French, for cooking under vacuum. Imagine a fish with a large midsection and a long, tapering thin end. The thinner parts of the fish would reach a higher temperature earlier while cooking than the thicker middle portion of the fish. Sous vide-style cooking allows for a precise, even temperature throughout

Figure 7.10 Sous vide cooking. Using a controller as shown here to maintain a narrow temperature range while circulating water, food in a bag can be cooked to a precise heat throughout the food. *Source*: Photo by Jeff Rogers.

the food. This means the food is cooked to the same temperature in the core without excess temperature at the surface of the food. Essentially, sous vide cooking involves placing a food item and seasoning into an airtight, sealable plastic bag, removing the air, immersing the bag in a controlled water bath, and precisely heating the food to a controlled temperature (Figure 7.10). This cooking style allows the entire food item, regardless of the shape or different size, to be cooked to a specific temperature. Sous vide is more about controlling the temperature in a water bath than removing (vacuuming) the air in the bag. The bottom line is to "cook" the food, edge to edge, at a temperature perfect for maintaining moisture and denaturing connective proteins.

7.5 Red or White Meat

The color of meat is much more complex than picking a chicken leg or a breast. Muscle fibers are organized into different categories depending on the amount of myosin, cellular content, and potential to metabolize food and produce ATP. The myoglobin and mitochondrial content of the cells is responsible for the different colors and

Table 7.1 Properties of muscle fibers.

	Type I muscle fibers	Type II muscle fibers
Alternative names	Slow-twitch, slow oxidative, red fibers	Fast-twitch, fast glycolytic, white fibers
Physical role	Endurance and aerobic muscle work; small generated force and smaller in size	Fight or flight and anaerobic muscle work; large generated force and fast movement
Color	Red	White or pale
Myoglobin and mitochondrial content	High	Low
Glycogen content	Low	High
Fat content	High	Low
Flavor/water content	High/low	Moderate/high

tastes of each tissue. We will focus on two kinds of muscle fibers: red fibers called type I (i.e. slow oxidative or slow-twitch) and white muscle fibers called type II (i.e. fast glycolytic or fast-twitch). There is a third intermediate muscle fiber type, a hybrid of type I and type II fibers. Type I red muscle fibers use various food macromolecules to produce energy in the form of ATP, including carbohydrates, amino acids, and fat. Type I cells require oxygen and additional enzymes to meet the metabolic demand and produce a sustained supply of ATP. These components both add a red color to the tissue and, when cooked, provide several interesting, savory flavors to the meat. Type II or fast-twitch muscle fibers primarily produce ATP from the aerobic metabolism of carbohydrates. These fibers, also called white or glycolytic muscle fibers, are rich in glycogen (i.e. "animal starch") and the enzymes of the glycolytic pathway (see Chapter 2). As a result of the high concentration of glycogen in type II muscle fibers, these tissues hold more water and are considered juicier. Thus, endurance muscle groups will need a longer sustained ATP production for muscle contraction and will have a predominantly red color. Slower-moving or less-used muscles require less ATP except for short bursts of activity and will have more type II fibers and appear white (Table 7.1).

The red/brown color of meat comes from the predominance of type I muscle fiber-rich with metal-binding proteins called hemoproteins. Lighter-colored muscles primarily comprise type II fibers that lack appreciable levels of these proteins and appear pale or white. Hemoproteins (a type of metalloprotein) bind an iron cation within a carbon cage known as a porphyrin ring. The iron bound within the porphyrin is called heme, and this reddish-brown molecule is surrounded and held in place by the amino acids of a protein. Together, the metal, heme, and protein are known as hemoproteins and have a wide range of interesting chemistry.

Type I muscle fibers have a high mitochondrial content. Mitochondria contain many colorful hemoproteins involved in the oxidative metabolism of carbohydrates, amino acids, and fats. Cytochromes are a family of hemoproteins in the mitochondria that give some dark brown/red tints. These proteins are ultimately responsible for the transfer of electrons to oxygen in the generation of ATP.

Two additional major hemoproteins found in animals are hemoglobin and myoglobin. Hemoglobin is an iron-containing protein located in the red blood cells and is responsible for transporting oxygen throughout the circulatory system. Myoglobin also binds oxygen; however, it is limited to the muscle fiber, where it stores O_2 needed for muscle cell metabolism. Myoglobin is a monomeric (single-chain) protein made mostly of alpha helices (Figure 7.10). As a storage compartment of oxygen in muscle, it is critical for the exercise capacity of red muscle. The more oxygen a muscle fiber can contain, the more metabolism and ATP the cell can produce during long bouts of muscle use. The level of myoglobin can increase with exercise. Deep-diving animals have such a high myoglobin content, which gives their tissue a dark red color. Both hemoglobin and myoglobin are deep red when the porphyrin ring iron is bound to oxygen (Figure 7.11). However, hemoglobin has almost no impact on the

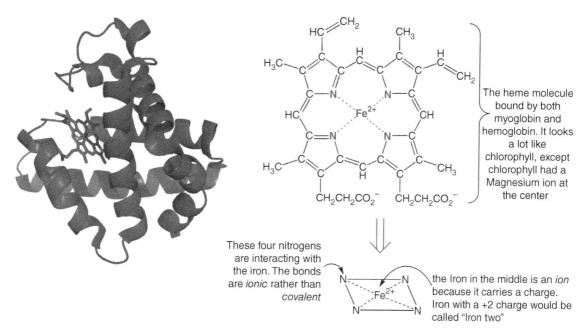

The heme molecule bound by both myoglobin and hemoglobin. It looks a lot like chlorophyll, except chlorophyll had a Magnesium ion at the center

These four nitrogens are interacting with the iron. The bonds are *ionic* rather than *covalent*

the Iron in the middle is an *ion* because it carries a charge. Iron with a +2 charge would be called "Iron two"

A simplified cartoon drawing of the heme molecule inside *myoglobin*

Figure 7.11 Myoglobin. (Left) A ribbon diagram of myoglobin represented with twirly ribbons (i.e. alpha helices) with the heme molecule (red) bound to the protein. There is one molecule of heme for every one molecule of myoglobin. (Right) The porphyrin ring or hemoprotein bound to an iron (Fe) in the center of the heme molecule.

meat color, as most of the hemoglobin is bled from the tissue before packing. While there is some hemoglobin in fish postharvest, myoglobin is the dominant pigment in muscle enriched with type I fiber.

The myoglobins and cytochromes are also excellent sources of bioavailable iron for humans. Iron is an essential mineral for the human diet—since humans also use the heme- (and therefore iron)-containing proteins hemoglobin and myoglobin to transport and store oxygen. It is not well understood, but it *is* well known that animal sources of iron are more digestible in humans. Plants contain iron, but we cannot obtain much of it from eating plants. The thought is that the heme molecule protects the iron and prevents it from being bound up by indigestible plant fiber.

The color of myoglobin and, therefore, red meat can change depending on what is bound to the myoglobin. Myoglobin can bind several small molecules, including oxygen, carbon monoxide, and water, through the heme-iron interaction. Each state of myoglobin (oxygen bound, carbon monoxide bound, and whether the iron's charge is +2 or +3) has a different color absorbance that will significantly impact the appearance of the meat. There are three common forms of myoglobin: (i) oxygen-bound myoglobin, which has a distinct cherry red color; (ii) deoxymyoglobin, where iron is in the +2 oxidation state and the iron is bound to water instead of oxygen, giving myoglobin a purple color; and (iii) metmyoglobin, which takes place when the iron is in the +3 oxidation state and oxygen has been converted to water. Metmyoglobin is a result of the oxidation of iron—that is, the loss of an electron, which takes the iron from a +2 to a +3 cation, and the result is a brown-colored myoglobin. This reaction occurs slowly over time. The formation of metmyoglobin is increased under conditions of high temperature and with the increase in acidity is sometimes seen if an animal is stressed at the time of harvest.

To preserve the red color of fresh meat, some meat producers use carbon monoxide (CO), a gas that, when present at the time of packing, will replace oxygen and water in the +2 iron/myoglobin to produce a deep red myoglobin and meat color that lasts twice as long as the red color of untreated meat. The use of CO to preserve the

A cartoon of the *heme* molecule inside *myoglobin*, the oxygen carrier protein

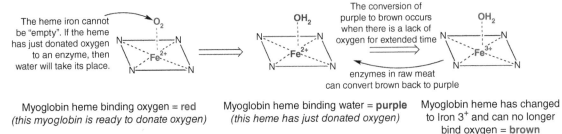

The heme iron cannot be "empty". If the heme has just donated oxygen to an enzyme, then water will take its place.

The conversion of purple to brown occurs when there is a lack of oxygen for extended time

enzymes in raw meat can convert brown back to purple

Myoglobin heme binding oxygen = **red**
(this myoglobin is ready to donate oxygen)

Myoglobin heme binding water = **purple**
(this heme has just donated oxygen)

Myoglobin heme has changed to Iron 3^+ and can no longer bind oxygen = **brown**

Figure 7.12 The oxidation state of Fe-heme in the muscle. The different colors of meat are due to the ligand bound to iron and the oxidation state of iron in the heme of myoglobin.

color of red meat has been used in several countries, including Canada and Norway. The Food and Drug Administration will allow a maximum concentration of 4.5% CO during processing and packing in the United States. Because actual spoilage of meat is not about the color but rather contamination by microbes or the oxidation of fats, color is a poor indicator of meat spoilage. The amount of CO in the meat is harmless and tasteless, but this process, called modified atmosphere packaging (MAP), is still controversial. Consumers are concerned about the masking of spoiled meat that looks attractive and red. However, others argue that CO-enriched packaging is preferable to other MAP processes where meat is packaged in a high-oxygen environment. High O_2 MAP will provide a more "natural" red-colored meat, but high concentrations of oxygen gas will also support harmful oxygen-requiring (aerobic) bacteria. Adding CO inhibits such dangerous oxygen-requiring bacteria providing an additional safety factor in meat production (Figure 7.12).

Muscles are considered red or white due to the fiber type and amount of myoglobin found in the tissue. However, type I and II muscle fiber distribution varies within and between organisms. Chickens possess 10% red fiber in the white breast muscle, while the migratory duck, goose, and quail have 75–85% red muscle fiber in their breast muscle. There is very little muscle where all fiber is type I or II, and most muscle is a mixture of the two. While considered the "other white meat," pork is actually made of 15% or more red muscle fibers than white chicken muscle. Domestic pork is "white" because of the mostly sedentary lifestyle that limits the development of the red fibers. Fine differences are observed with the darker pork leg. A muscle supporting a bone requiring more stamina is darker, whereas the outer muscle is made of more glycolytic type II muscle fiber. Beef, while a red meat, mostly made of intermediate red fast-twitch fiber type IIA (a hybrid of type I and II muscle cells), still has about 10% type I and 15% type II fiber.

Fish and shellfish are very different in red and white muscle makeup. Figure 7.5 shows a cartoon showing a cross section of the fish body and the muscle fiber composition from tilapia fillets. The myotomes are evident in the V-pattern along the fillet, and the "red muscle" fibers are the dark stripe down the center of the fillet. Fish, with smaller muscle segments or myotomes, have mostly white aerobic muscle tissue arraigned for fast bursts of speed (Figure 7.5). Red muscle tissue is located just under the skin, particularly along the middle of the fish that powers the slower steady swim movements (Figure 7.13). Bottom-dwelling fish (demersal) drift along with the current; most are not active swimmers and do not have as much red fiber. Active swimming top-current fish (pelagic) have more red muscle fiber, adding a rich taste to the meat. Tuna, salmon, and shark, fish that swim long distances, often at high speed, are rich in mitochondria, myoglobin, and type I fibers and have darker or red-colored flesh. Because of the endurance exercise, the meat of these fish is highly marbled with fat, providing energy directly to produce ATP needed for actin and myosin contraction. Shellfish, however, are mostly made of white meat as they require short bursts of energy to close a shell or scuttle to a hiding location.

As we have seen, myoglobin content differs between red and white muscle fibers, but myoglobin content in the muscle tissue also varies by species and the animal's age. These factors also contribute to meat color. Younger animals have less myoglobin in their muscles than older animals. For example, veal is very pale brownish pink

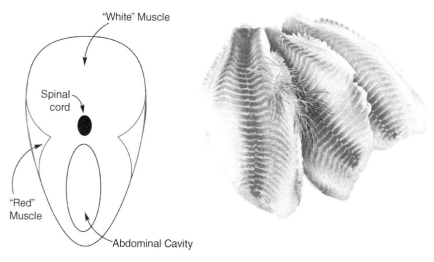

Figure 7.13 A cartoon showing a cross section of the fish body and the muscle fiber composition. The myotomes are evident in the V-pattern along the fillet, and the "red muscle" fibers are the dark stripe down the center of the fillet. The structure of the muscles can be seen in the tilapia fillets on the right. kostrez/Adobe Stock Photos

(veal has 2 mg of myoglobin per gram of meat), while young beef is cherry red (8 mg of myoglobin per gram of meat). Across species, beef has the highest myoglobin content, followed by lamb and then pork. Fish and fowl (chicken and turkey) have even less myoglobin. This trend also follows the level of physical activity we might expect from these animals. For example, farm-raised pork is from a sedentary pig, while beef is from a slightly more active cow. Game animals—wild turkey, venison, and so on—have more myoglobin in their muscles than their domesticated counterparts due to their more physically active lifestyles [2].

7.6 Death and Becoming Meat

There is a point after harvesting the tissue when the muscle is considered meat. Fresh muscle is difficult to chew, and while fresh meat possesses some interesting flavors, most of the savory taste associated with meat occurs as the chemical and physical changes of the tissue occur hours and days after the animal's death. These changes take place in three phases:

1. The continued metabolism by cells in the muscle tissue.
2. Followed next by rigor mortis, and finally.
3. The chemical and physical changes in the meat after rigor.

When combined with the chemical reactions between the molecules catalyzed by heat in muscle tissue during postmortem and cooking, muscle has changed and is now called meat. Like cheese and other foods, the enzymes responsible for metabolism and other processes in a cell remain active after the organism's death. The actions of these enzymes set the stage for important chemical changes in fish and land animal muscle.

One crucial postmortem change in carbohydrate metabolism is catalyzed by enzymes that work for a while after the animal's death. Glycolysis converts glucose into pyruvate or lactate, providing ATP for muscle cells. Glycogen, a polymer of glucose, serves as a reserve fuel for muscle fibers. Stress hormones accelerate the breakdown of glycogen into glucose. Glucose is further metabolized to produce ATP for muscle movement. Fish generally have lower glycogen levels than land animals, resulting in smaller fuel reserves after death and why fish will become stiff (rigor) sooner than terrestrial animals.

During rigor mortis, myosin filaments cross-link with actin filaments, forming strong connections that prevent the relaxation of the muscles. The lack of ATP prevents the detachment of myosin from actin, effectively freezing the muscles in a contracted state and resulting in the characteristic rigidity observed during rigor mortis.

Rested and well-fed animals have higher glycogen levels, allowing for extended ATP and pyruvate/lactate production. Stress at the time of harvest affects glucose and ATP levels, leading to a quick onset of rigor mortis and resulting in dark-colored meat known as "dark cutters." The duration of postmortem glycolysis (and thus the production of ATP and other breakdown products) depends on several factors, including the level of glycogen before harvest. If fish struggle during the catch, they will quickly use their small glycogen stores, leaving only a short time before carbohydrate metabolism is finished and ATP production ceases. Cattle are often rested and fed before slaughter to increase the glycogen level in the tissue before harvest. Chicken breast glycogen can decrease by nearly half if the animal struggles during harvest. Stress at the time of harvest influences the level of glucose and ATP in the muscle. Adrenaline, the fight-or-flight hormone, is produced if an animal is stressed and activates the glycogen breakdown enzymes. When combined with increased muscle activity in stressed muscle, the tissue has a low ATP level and will enter rigor mortis quickly. This results in "dark cutter" meat because of its dark appearance (Figure 7.14).

The dark color in "dark cutters" is due to changes in pH. Normally, glycolysis produces lactic acid, lowering the pH and creating a cherry red color in myoglobin when exposed to oxygen. In glycogen-depleted tissue, less lactic acid is produced, resulting in a higher pH and a darker color when myoglobin binds with oxygen. Despite the dark appearance, this meat is safe and palatable.

High stress before harvest can lead to two common meat quality issues: pale, soft, exudative (PSE) and dark, firm, dry (DFD) meat. PSE meat is caused by rapid protein denaturation due to acidic conditions and warm temperatures in exercised muscle. DFD meat occurs when the pH doesn't drop sufficiently, resulting in a higher pH, rapid rigor, and reduced protein denaturation. Interestingly, certain strains of pigs are more genetically susceptible to PSE. These pigs have a gene encoding for a protein that releases more calcium into the muscle fiber cells during stress. When combined with the lower pH, increased muscle calcium increases the protein degradation rate and mushy tissue production.

Both conditions (DFD and PSE) can affect meat quality and increase the risk of bacterial contamination. The pH in tissues with DFD is 0.5–1.5 pH units higher than the expected 5.6. In this case, prolonged stress before harvest reduces the glycogen, but the animal recovers from excessive lactic acid production. After slaughter, the tissue does not have enough glycogen to produce more lactic acid and has a lower pH and expected levels of ATP production. This results in meat that enters rigor quickly, a high pH that produces darker red myoglobin, and protein that does not denature and tightly holds on to water. DFD meat that is dark in color, tough without

Figure 7.14 Bright red steak. Alexstar/Adobe Stock Photos

protein denaturation, and with less acidic conditions is more likely to spoil with bacterial contamination.

A famous agricultural scientist, Dr. Temple Grandin, whose autism gave her a unique insight into how cattle were stressed during handling before slaughter, has been vital in advancing how cattle are handled. Dr. Grandin has created a system of how to humanly slaughter beef and other animals to reduce stress and create a higher-quality product.

The duration of rigor varies depending on the animal type. Fish with redder muscle fibers have longer rigor contractions than those with more white fibers. Rigor sets in within a few hours after death for beef and pork, while fish can experience longer rigor contractions. It is best to handle fish, beef, and pork after rigor has passed to avoid muscle fiber tears that can make the

Figure 7.15   Cartoon of protease action. A protease, like a pacman, will "chew" proteins into smaller pieces by hydrolysis.

meat soft and mushy. The length of rigor differs with the animal type. Fish with redder muscle fiber will have a more prolonged rigor contraction than fish with more white fiber. Maximum rigor also varies from 6 to 8 h for fish with red muscle and 4.5–6 h for fish with white muscle. Well-fed, farmed salmon will go into rigor at 10 h postharvest and last for nearly 60 h. Rigor will be set in beef and pork in 1.0–2.5 h postmortem. The handling of fish, beef, and pork is best left until rigor has passed. Freezing or cutting the rigor-contracted muscle leads to large-scale tears in the muscle fiber, leaving space for water to leave and causing the tissue to become soft and mushy (Figure 7.15).

Eventually, rigor is lost, and muscle groups extend and loosen. Two factors are responsible for the loss of rigor: the pH and the release of enzymes from the muscle fiber. Over time, the lower pH from glycolysis will create enough lactic acid to denature the contractile proteins, releasing myosin from actin. As muscle cells age, a special set of enzymes called proteases are released from the cells. These proteases bind and cleave the peptide backbone of other proteins. The action of many proteases is enhanced by the conditions found in postmortem tissue. These proteases will attack and cleave many proteins, including actin and myosin. This enzymatic protein destruction loosens the muscle fibers and ends rigor (Figure 7.16).

The changes prior to and during rigor mortis dynamically impact each other and create transformations after rigor that generate new flavorful molecules. After reactions with proteins like myosin, ATP is rapidly converted to ADP and AMP (adenosine monophosphate). Enzymes in the cell will remove an ammonia group from the adenosine base, creating inosine monophosphate (IMP). IMP is further degraded by a series of muscle enzymes to inosine and hypoxanthine. Both inosine and IMP are key components giving meat its "meaty" flavor and smell. IMP, like monosodium glutamate, has an umami flavor and is often added to foods like soup for additional flavoring. Hypoxanthine, however, has been reported to have a neutral to bitter taste (Figure 7.17).

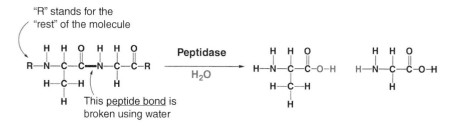

Figure 7.16   Proteolytic cleavage. Protease hydrolysis of a protein substrate.

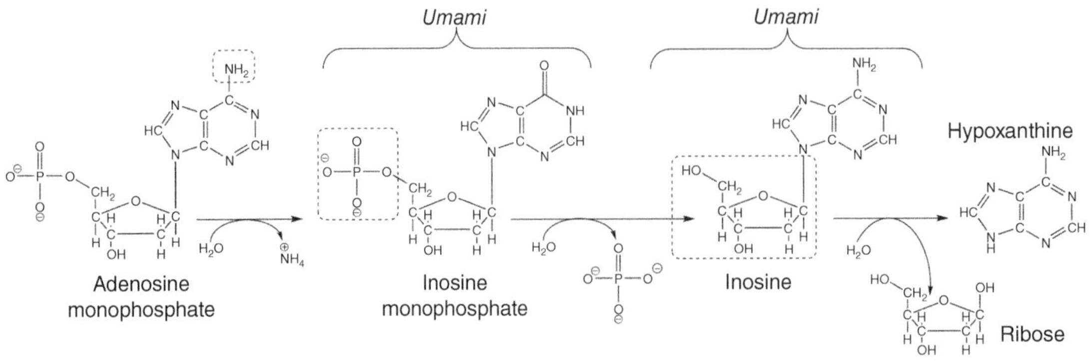

Umami *Umami*

Hypoxanthine

Adenosine
monophosphate

Inosine
monophosphate

Inosine

Ribose

Figure 7.17 Production of umami compounds from AMP. Production of umami compounds from AMP. The enzymatic degradation of AMP produces inosine, ribose, and other umami-flavored compounds.

The ribose portion of ATP is also an important flavoring component of meat. Ribose and ribose-5 phosphate (both breakdown components of ATP) increase in aged meat and contribute to the roasted aroma of cooked chicken. Other sugars resulting in glycogen and glucose metabolism are produced during meat aging by a host of different enzymes. These simple sugars, including ribose, glucose 6-phosphate, and fructose, undergo diverse reactions with proteins during heating meat, creating hundreds of different flavors and aroma molecules.

One last chemical and physical change that occurs with aging muscle takes place with the proteins in and around the muscle fibers. The fall of pH induces proteins to unravel and denature (Figure 7.18). This results in the loss of interconnected proteins within muscle fibers and begins to soften or tenderize the tissue. Protein denaturation also impacts the ability of muscles to "hold" water. Natively folded proteins are organized to keep the water-fearing or hydrophobic amino acids within the globular shape of the protein away from water. The charged and hydrophilic amino acids are folded to face the exterior of the protein, where these functional groups can form hydrogen bonds with water (Figure 7.19). Thus, native proteins can form noncovalent interactions with water molecules, holding them in place. Denaturing proteins by pH opens up the folded structure, exposing the hydrophobic portions of the protein to the watery environment of the cell. This loss of folded, native structure also results in fewer noncovalently bound water molecules. Therefore, protein denaturation by acidic conditions results in a loss of the water holding capacity (WHC) of meat (Figure 7.20).

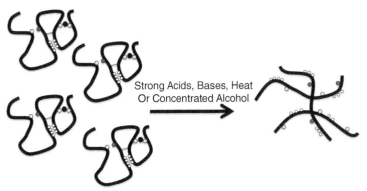

Strong Acids, Bases, Heat
Or Concentrated Alcohol

Native folded proteins

*Denatured unfolded protein
aggregated in a tangle network*

Figure 7.18 Protein denaturation. Heat applied to meat during cooking will denature the proteins. This will alter the water holding capacity of the tissue.

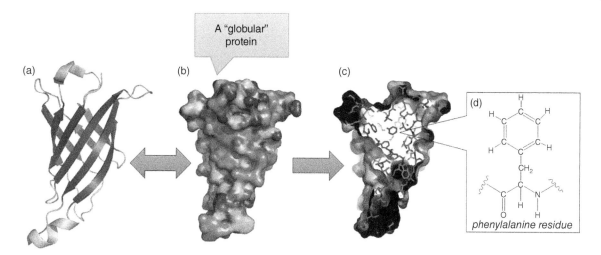

Figure 7.19 Representations of a globular protein. (a) A globular protein with ribbons representing alpha helices and beta sheets, (b) the same protein in (s) represented in a "space-filling" model of the actual space occupied by all the atoms. (c) is a slice right through the middle of protein (b)—so we can see what the protein looks like on the inside. Hydrophobic residues like (d) phenylalanine are found in the interior of the folded protein structure.

In addition to acid-induced denaturation of some of the proteins, the proteases released by cells also alter the structure of meat proteins. Proteases play a major role in tenderizing aged meat. Calpains digest the contracting fiber proteins, while cathepsins are proteases that bind and cleave the protein backbone of several muscle proteins, including collagen and other connective tissues. These proteins degrade many connective proteins, such as tropomyosin, actin, myosin, and collagen. The proteases reduce the connective fibers and myofibrils making the tissue easier to cut with teeth, and protease activity also imparts important new flavors. Longer digestion by proteases cleaves whole proteins into small peptides and individual amino acids. Some of these amino acids, glutamate and aspartic acid, produced by proteases, add to the savory flavor, while others have a sweet or sour taste. Many of these peptides and amino acids will react with carbohydrates like ribose and glucose during browning, again producing a complex mixture of the flavor and aroma of cooked meat. Meat tenderizers such as Adolph's Meat

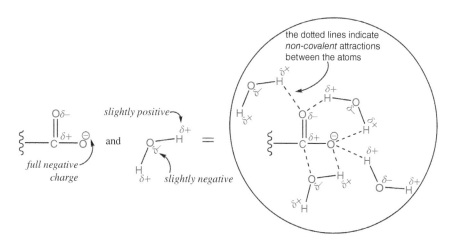

Figure 7.20 Water holding capacity of proteins.

Tenderizer are powdered preparations of protein proteases. Meat is prepared by soaking in proteases, cleaving the connective proteins, and creating umami flavors. Most of these proteases are from tropical plants, often papaya or pineapple.

Wet aged meat is left in vacuum-sealed plastic so that the aging process can occur in the meat department of your favorite grocery store. Aging meat on a foam tray wrapped in plastic leads to a more intense blood flavor than dry-aged beef. Dry aging (air-exposed in a cool, dry location, not in the foam-wrapped tray) results in water loss and can take up to 45 days. During that time, the meat will lose a significant fraction of its weight in water from dehydration. The result is a concentration of flavor molecules. Aging will also cause the tissue to have a significant protease activity releasing amino acids and tenderizing the tissue. Exposure to oxygen also increases the reaction of oxygen with fat molecules. In addition, the longer process of dry-aged meat means that more of the connective tissue has been digested by proteases, yielding a more tender piece of beef.

7.7 Flavor in Fish and Shellfish

Several interesting chemical and biological reasons exist for the molecules that create flavor in meat. We have already seen that the metabolism of carbohydrates, ATP, and proteins makes flavor molecules. Several other exciting molecules are deceptively simple yet provide intense flavors to meat. These include simple amino acids found in tissues and molecules derived from the food of the animals we eat. Additional flavors come from reactions that take place during cooking. In Chapter 3, you can learn about the Maillard reaction that imparts flavor during the cooking of many foods, including meat and fish. Fish and shellfish also have distinct flavors that come from their watery habitats.

Shellfish live in a unique environment. The brackish water that many shellfish make as their home requires them to have a special ability to tolerate a wide range of salinities (saltwater concentration). Osmosis is the net movement of solvent molecules, such as water, through a semipermeable membrane from a region of low solute concentration to an area of higher solute concentration. The tissues of most organisms are semipermeable to water (and impermeable to salt ions and other compounds) due to the plasma membrane surrounding each cell. Particles such as salt ions, proteins, and carbohydrates contribute to the solute concentration inside the cell. However, the solute concentration inside the cell is still lower than that of the saltwater environment, a so-called hypertonic solution. For many organisms, this would mean that the cellular water would, through osmotic pressure, move from the cell into the higher solute-containing salt water. This would cause the cells to shrink and die. To survive the saline living conditions, shrimp, prawns, and other shellfish have high concentrations of free amino acids. Specifically, glycine and alanine are found in high concentrations in the muscle tissue of shrimp. Scallops adjust to the high concentrations of glycine and the complex carbohydrate glycogen. Besides being a way to respond to high salt conditions, the meat of these shellfish is particularly flavorful and sweet because of the glycine and some of the simple sugars found in the tissue after glycogen breakdown. Some of the sweetness of shrimp is lost during heating due to chemical changes. Many of the amino acids become entangled by heat-denatured proteins or react with other compounds, changing the structure of glycine.

The flavor and aromas of freshwater and saltwater fish come from naturally occurring compounds, microbial spoilage, reactions with oxygen, and processing reactions. Lipase enzymes bind and combine unsaturated fats with oxygen and water molecules to create fresh fish smells. Lipoxygenases break fish fats into shorter six-, eight-, and nine-carbon chain compounds now made of aldehydes, ketones, and alcohols. Each has a unique and short-lived fresh fish aroma and flavor. In freshwater, but not saltwater, fish have enzymes that produce *trans*-2-hexenal and *cis*-3-hexenal to give a green, plantlike smell. Other fatty acid metabolites provide plantlike flavor and odor to fish and are found in some vegetables. So the freshest fish is often compared to the smell of fresh leaves.

Figure 7.21 Aquafarming. Farming fish even in the ocean will not produce a "fishy" taste without additional bromophenols in the diet. markobe/Adobe Stock Photos

Saltwater fish, as well as crustaceans and mollusks, have a unique flavor and odor associated with them caused by a family of related bromophenol compounds. This odor is missing in ocean farming (Figure 7.21). Several studies on the organoleptic quality of aquacultural raised fish have reported distinct differences and lower flavor quality in the fish. 2,6-Dibromophenol seems responsible for shrimp and crab flavors, while the singly brominated phenols 2-, 3-, and 4-bromophenol enhance the seafood quality of finned seafood (Figure 7.22). Interestingly, none of these compounds has been detected in freshwater fish, and the bromophenol content in wild saltwater fish is 1,000 times higher than that found in farm-raised ocean fish. These differences indicate that fish do not produce bromophenols on their own. Instead, marine worms and algae synthesize the compounds that are then transferred and stored in the fatty tissue of predator fish. Including either the algae or low concentrations of bromophenols in the fish food has increased the attractiveness of the fish flavor in farm-raised ocean fish.

Microbial contamination of fish and shellfish contributes to the offensive fishy smell and flavor of older fish. This happens as sulfur-containing compounds are released, the bacteria modify phenols, and fatty acids with putrid aromas and flavors are generated. The robust flavor and odor caused by microbial growth are due to the conversion of trimethylamine oxide (TMAO) to trimethylamine (TMA) (see Figure 7.23). Saltwater, but not freshwater, fish (except Nile perch and tilapia) possess relatively high levels of TMAO, which acts as an additional regulator for osmosis and stabilizes protein structure in finned fish. Deep-sea fish have particularly high levels of TMAO, giving strong, recognizable smells and tastes. The related sharks and rays use the amino acid breakdown product urea instead of TMAO. Bacteria and some fish produce an enzyme that decomposes both urea and TMAO. TMAO decomposes to TMA, providing the strong

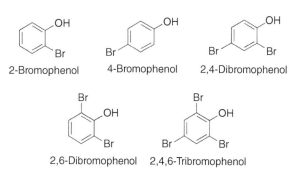

Figure 7.22 Simple bromophenols in shellfish. Mono-, di-, and tribromophenol structures involved with ocean fish aroma.

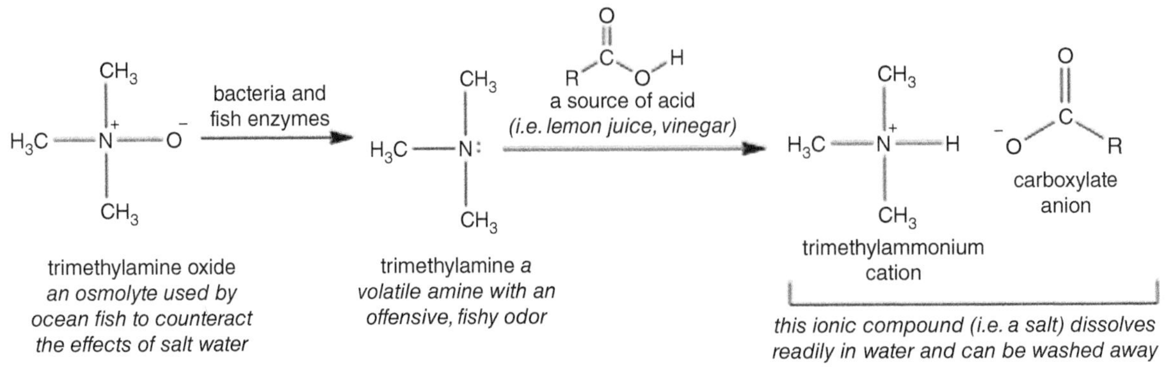

Figure 7.23 Amine compounds and acidic solutions.

fishy smell of old fish, while urea is converted into ammonia by the bacteria, giving older shark flesh the pungent odor of cleanser. Cooking fish with an acid such as lemon juice, tomato, or vinegar can reduce the distinctive fish smell. The acid combines with TMA to form an organic salt that is less able to volatilize and be detected by nasal receptors. Rubbing your hands with lemon is an excellent way to remove the fish smell from your hands and utensils in your kitchen!

Sugars like ribose and amino acids produced artificially or naturally occurring in meat are prime targets to react under heat to create an amazing array of new flavor and odor molecules. The special reaction primarily between amino acids, sugars, and lipids under dry heat is the Maillard or browning reaction. The Maillard reaction was first described over 100 years ago and is responsible for the smells and flavors of cooked meat. The aging of meat and the metabolic changes of proteins, carbohydrates, and ATP all provide the necessary starting materials for the Maillard reaction. Each time you brown meat-like fish, beef, or pork, the reaction between sugars and amino acids creates hundreds of new, flavorful molecules. For example, ribose from ATP will react with the amino acid cysteine from protein degradation, making over a dozen different products that can be smelled even when there are only 10 or so molecules per trillion air molecules.

However, meat must be relatively dry to reach the high temperatures necessary for Maillard reactions (around 300°F/149°C), and thus boiled or braised meats will not have some of the nutty and meaty flavor of browned meat. Boiled or poached fish and beef will have a very different flavor than when cooked and browned. The surface of the meat must be heated high enough to evaporate or boil off the water and allow the temperature of the tissue to get above 212°F/100°C.

Maillard reactions are also a challenge for the modern sous vide cooking method. For this modern molecular approach to cooking, food is placed in bags, often with a vacuum to remove air, and placed in a controlled circulating water bath. The idea is to cook the entire meat to a uniform temperature. However, the food must be removed from the bag and quickly browned at the surface to generate the Maillard products. The perfect combination of sous vide and Maillard browning can be found in this hamburger recipe. First, chopped lean meat is formed into a thick burger and cooked in the sous vide water bath in a sealed bag to 133°F (56°C). This allows the large beef patty to be evenly cooked from the edge through to the middle without drying or overcooking the edges of the meat. This is followed by a quick dip in liquid nitrogen. A fast exposure of the cooked burger to the low temperature of liquid nitrogen (−321°F/196°C) serves as a coolness factor and freezes the meat's surface without cooling the interior. The patty can then be deep-fried or, if you are brave, subjected to a flame from a gas torch for just a minute, creating an evenly cooked and browned modern burger.

7.8 Plant-based Meat Alternative

The science of plant-based meats represents a fascinating inter-section of food technology, biology, and consumer demand for sustainable and ethical food options. Plant-based meats are inno-vative food products designed to mimic traditional animal-based meats' taste, texture, and nutritional composition, all while entirely derived from plant sources. By carefully selecting and combining plant proteins, fats, and other ingredients, scientists and food technologists have made remarkable advancements in creating plant-based meats that offer a compelling alternative to animal products. This field of study encompasses a wide range of scientific disciplines, including protein biochemistry, food pro-cessing, flavor chemistry, and sensory science, all working together to develop plant-based meats that are delicious and nutritious and address the environmental and ethical challenges associated with conventional meat production. As consumer interest in plant-based eating continues to soar, the science of plant-based meats holds tremendous potential to revolutionize how we think about and consume meat, paving the way for a more sustainable and compassionate future.

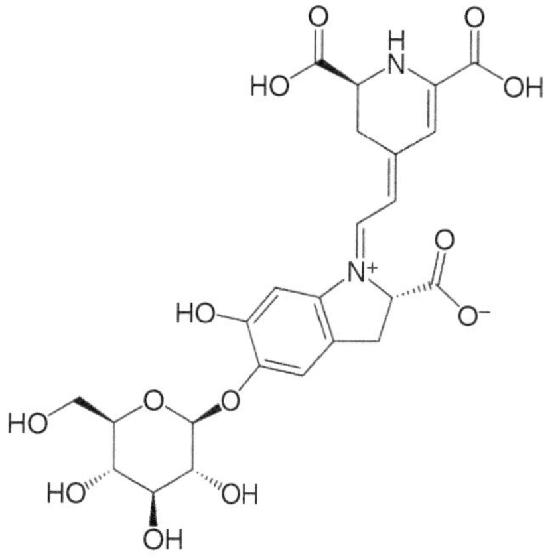

Figure 7.24 Betanin. One of the compounds from beet juice that gives Beyond Burger its red color.

The Beyond Burger and Impossible Burger are two popular examples of plant-based meat alternatives that have gained widespread attention for their ability to closely resemble the taste and texture of traditional beef burgers (Table 7.2). These innovative products rely on a combination of plant-based ingredients carefully selected to mimic the flavor, appearance, and mouthfeel of real meat.

The Beyond Burger is primarily composed of a blend of pea protein isolate, mung bean protein, and rice protein. These plant proteins provide the essential building blocks for creating a meat-like texture and protein content. Additionally, the Beyond Burger contains various fats to enhance juiciness and mouthfeel, such as coconut oil and canola oil. Beet juice extract is used to give the burger a red "meat-like" appearance. The key compound in beet juice that gives the color is betanin (Figure 7.24). Beyond Burger also uses potato starch to help bind the material together and give an appropriate texture. To impart a rich, savory flavor, the Beyond Burger incorporates natural flavors, including yeast extract and various spices.

Table 7.2 Plant-based meat main ingredients.

Beyond burger	Impossible burger
Water, pea protein, expeller-pressed canola oil, refined coconut oil, rice protein, natural flavors, cocoa butter, mung bean protein, methylcellulose, potato starch, apple extract, pomegranate extract, salt, potassium chloride, vinegar, lemon juice concentrate, sunflower lecithin, beet juice extract (for color).	Water, soy protein concentrate, coconut oil, sunflower oil, natural flavors, potato protein, methylcellulose, yeast extract, cultured dextrose, food starch modified, soy leghemoglobin, salt, tocopherols, soy protein isolate, vitamins and minerals (zinc gluconate, thiamine hydrochloride (B1), niacin, pyridoxine hydrochloride (B6), riboflavin (B2 & B12).

The Impossible Burger, conversely, centers around a key ingredient called soy leghemoglobin, which is derived from the roots of soy plants. Soy leghemoglobin is responsible for the burger's "bloody" appearance and contributes to its meaty flavor. The protein base of the Impossible Burger is primarily made up of soy protein concentrate, along with additional plant-based ingredients like coconut oil and sunflower oil to provide the desired fat content and juiciness. Methylcellulose, a plant-derived compound, helps with binding and texture. Like the Beyond Burger, the Impossible Burger also incorporates various flavors and seasonings, such as natural flavors, yeast extract, and spices, to enhance its taste.

Protein Selection: Both burgers use plant-based proteins as their main ingredient. These proteins, such as pea protein isolate, mung bean protein, and soy protein concentrate, are chosen for their ability to provide a meat-like texture and mouthfeel. The specific combination and processing of these proteins help create a structure resembling real meat's fibrous texture.

Fat Content: Fats are crucial in delivering meat's rich and juicy taste. Coconut oil, canola oil, and sunflower oil are commonly used in burgers to provide the desired fat content and satisfy the mouthfeel. These fats also help enhance the flavor and create a juiciness similar to traditional meat.

Flavor Enhancers: Various natural flavors, spices, and seasonings are added to both burgers to recreate the savory and umami taste characteristic of meat. Ingredients like yeast extract, which contains compounds like glutamic acid, contribute to the savory flavor known as umami. Spices and herbs are carefully selected to mimic the flavor profile of beef, providing a familiar taste experience.

Heme: The Impossible Burger utilizes soy leghemoglobin, a molecule derived from soy plants, to mimic the meat flavor. Soy leghemoglobin contains heme, a compound found in animal muscle, which contributes to the characteristic "bloody" flavor associated with beef. By incorporating heme from plant sources, Impossible Burger aims to recreate the distinct taste of meat.

Texturizers and Binders: Both burgers use texturizers and binders to ensure the plant-based patties hold their shape and have a satisfying texture. Ingredients like methylcellulose and potato starch help provide the desired firmness and binding properties, allowing the burgers to cook and hold together like traditional meat patties.

Both the Beyond Burger and Impossible Burger rely on advanced food science techniques to create a meat-like experience using plant-based ingredients to deliver a satisfying culinary experience while reducing the environmental impact of conventional meat production. A few specific approaches include the amino acid cysteine and a heme.

Cysteine, an amino acid commonly used in plant-based meat products, is crucial in generating flavor compounds through various chemical reactions. Cysteine contains a thiol (-SH) group, which is involved in several reactions during cooking. This thiol group can react with reducing sugars, such as glucose or fructose, through a process known as the Maillard reaction. The reaction begins with the condensation of the thiol group with a sugar carbonyl group, forming a Schiff base. This intermediate compound further undergoes rearrangements, dehydration, and polymerization reactions, forming various flavor compounds, such as pyrazines, thiazoles, and furans. These flavor compounds contribute to the rich, savory, and meat-like flavors found in plant-based meat products. By incorporating cysteine into their formulations, manufacturers can enhance the depth and complexity of flavors, making the plant-based meats more appealing to consumers seeking a meat-like experience.

Iron-hemes, on the other hand, are responsible for the distinct "bloody" flavor and reddish color of meat. Plant-based burgers like the Impossible Burger utilize a plant-derived heme molecule called soy leghemoglobin, which contains iron-hemes. These iron-hemes contribute to the authentic meaty flavor and help recreate the sensory experience of eating meat (Figure 7.25).

Plant-based meat producers initially attempted to extract heme, a key component responsible for the meaty flavor and aroma, from the roots of plants by utilizing bacteria. Specifically, they focused on a class of bacteria called nitrogen-fixing bacteria, which form a symbiotic relationship with leguminous plants like soybeans. These

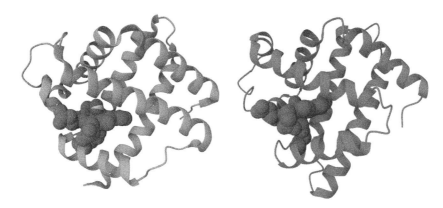

Figure 7.25   Hemes for color and flavor. Both images represent the structure of a globin molecule bound to heme with oxygen present. The combination of the globin protein and heme gives the complex a red color. On the left in green, is soybean leghemoglobin bound to heme (in red). On the right in blue, is the structure of myoglobin with an oxygen bound heme in red. The structure of both protein-heme complexes is very similar in shape, but different in amino acid sequence.

bacteria reside in nodules on the plant roots and possess an enzyme called nitrogenase, which allows them to convert atmospheric nitrogen into a usable form for the plant. Researchers identified that heme could be found in the nitrogenase protein complex of these bacteria. By isolating and purifying this protein complex, they were able to extract heme, which could be used as a crucial ingredient in plant-based meat products. This heme extraction process from the roots of plants allowed producers to incorporate a key flavor element into their products, enhancing their meat-like qualities.

However, it's worth noting that plant-based meat producers have since developed alternative methods for heme production, such as using genetically modified yeast to produce heme through fermentation processes. These advancements have provided more efficient and scalable means of obtaining heme for use in plant-based meat production, reducing the reliance on the extraction process from bacterial sources.

GMO stands for "genetically modified organism." It refers to an organism whose genetic material has been altered using genetic engineering techniques. In plant-based meats like the Beyond Burger and Impossible Burger, genetic modification produces specific components that contribute to the meat-like flavor and texture.

Specifically, GMO technology is used to introduce genes into microorganisms or yeast, such as Saccharomyces cerevisiae, to enable them to produce heme, which is a crucial flavor compound found in meat. Traditional heme extraction methods, such as obtaining it from the roots of plants through bacteria, were less efficient and scalable. To address this, scientists have genetically modified yeast cells to produce heme through fermentation processes. By introducing specific genes into the yeast, they can instruct it to produce heme, which can then be harvested and used as an ingredient in plant-based meat products.

Therefore, the Beyond Burger and Impossible Burger, which incorporate heme produced through genetically modified yeast, can be classified as GMO products. The genetic modification involved in making heme does not directly alter the plant ingredients used in these products (e.g. soy protein, pea protein), but it does involve genetically modifying microorganisms to produce a specific compound (heme). It's important to note that the safety and regulatory aspects of GMOs are subject to ongoing scientific and governmental scrutiny to ensure that consumer health and environmental considerations are addressed.

The juiciness of plant-based meat must also be created. Combining coconut and sunflower oils in plant-based meats, manufacturers can achieve a balance between a solid texture and moistness, closely resembling the experience of consuming traditional meat. Both oils play a crucial role in enhancing the water-holding capacity of plant-based meats. WHC refers to the ability of a food product to retain moisture during cooking and processing, which is important for maintaining a juicy and tender texture. Coconut oil, being a highly saturated fat, solidifies at room temperature. When incorporated into plant-based meats, it can contribute to a firmer texture and improve the overall mouthfeel of the product, similar to the saturated fats found in red meat. Additionally, coconut oil can help trap and retain moisture on the surface and within the meat substitute, preventing it from becoming dry and rubbery during cooking. Sunflower oil, on the other hand, is a liquid oil and contains unsaturated fats. It can add a lighter and more fluid consistency to the plant-based meat mixture. Sunflower oil helps to lubricate the ingredients, allowing them to bind together more effectively and retain moisture. This helps create a desirable juicy texture in the final product.

Soy is commonly used in plant-based meats like the Beyond Burger and Impossible Burger due to its versatility and high protein content. Soy protein provides a meat-like texture and helps bind the ingredients together to create a cohesive product. In terms of flavor, soy also plays a role. It has a relatively neutral taste that can be easily manipulated and enhanced with seasonings, spices, and other flavorings to mimic the savory and umami flavors associated with meat. The goal is to create a sensory experience similar to consuming traditional animal-based meat.

When it comes to the potential risks of soy to human health, it is a topic that has been widely debated. Soy contains compounds called phytoestrogens, which are plant-derived compounds that can mimic the hormone estrogen in the body. Some concerns have been raised regarding the potential effects of phytoestrogens on hormonal balance and reproductive health, particularly in individuals with certain medical conditions or hormone-related disorders. The scientific evidence on the health effects of soy is complex and varied. While some studies suggest potential benefits of soy consumption, such as cardiovascular health and reducing the risk of certain cancers, other studies have raised concerns about some possible negative effects on some cancer types and bone diseases [3, 4]. It is worth noting that moderate consumption of soy products as part of a balanced diet is generally considered safe for most individuals. Ultimately, the decision to consume soy-based products, including plant-based meats, is a personal one that should consider individual health considerations, preferences, and dietary needs.

Mouthfeel refers to the texture and sensations experienced in the mouth while eating. Meat has a fibrous texture, offering resistance and chewiness. It also has juiciness, which varies based on the type of meat and cooking method. The presence of fat contributes to tenderness and richness. Replicating these mouthfeel characteristics is important in creating satisfying plant-based meats. Overall, methylcellulose gel plays a crucial role in improving the mouthfeel of plant-based meats, making them more enjoyable to consume and enhancing their resemblance to traditional meat products. Methylcellulose gel is a type of hydrocolloid used as a thickening and gelling agent in various food applications, including plant-based meats. It is derived from cellulose, a naturally occurring polysaccharide found in the cell walls of plants. Methylcellulose is produced by chemically modifying cellulose through a reaction with methyl chloride.

To make methylcellulose gel, methylcellulose powder is typically dispersed in cold water while being agitated. As the powder hydrates, it forms a gel-like substance with high viscosity and the ability to hold water. This gelatinous texture gives methylcellulose its functionality as a thickener and binder. The gel acts as a binder, helping to keep the ingredients together and create a structure that mimics the fibrous texture of meat. It can contribute to a juicier and more succulent mouthfeel by retaining moisture within the meat substitute. Additionally, methylcellulose gel can give a smoother and more consistent texture to the product, preventing it from being overly crumbly or dry.

The science behind plant-based meats has paved the way for innovative and sustainable alternatives to traditional meat products. Through careful formulation and utilization of various plant-based ingredients, these products aim to mimic the taste, texture, and nutritional profile of meat while reducing environmental impact. A direct comparison of the environmental impacts attributable to a 1/4 pound Beyond Burger with a 1/4 pound beef patty on production, packaging, and distribution finds that Beyond Burger generates 90% less greenhouse gas emissions and requires 46% less energy, 99.5% less water (in competition), and 93% less land use (as characterized by ecosystem damage potential) than production, packaging, and distribution of US beef [5].

While ongoing research and development continue to refine these products, plant-based meats offer a promising option for individuals seeking a more sustainable and compassionate approach to their dietary choices.

7.9 Cooking Meat

7.9.1 Searing To Seal In the Flavor—Not!

Many recipes call for food to be browned on a hot skillet to *seal in the juices* or *lock the flavor into the meat.* This old saying could not be further from the truth. Subjecting the meat to high enough heat to cause the Maillard reaction will also damage the tissue of the meat. As we will learn in Section 7.10, heat causes the collagen to shrink, disrupting the integrity of the muscle fiber. The inevitable loss of water is detectable by the sounds of water boiling, sizzling during browning, and water leaking from the meat after browning. However, browning the meat before roasting does create, not lock in, the flavor molecules.

7.9.2 Stages of Cooking Meat

The collagen type and quantity within meat (remember that fish have much less than beef or pork) are reflected in the final temperature of the cooked meat. Fish requires a lower temperature as the muscle's connective tissue is more easily degraded than pork or beef collagen. See Figure 7.26 for an interesting comparison between fish and meat cooking. There are predictable phases of cooking meat. At 58–60°C/135–140°F, myosin has already begun denaturing and coagulating. The denaturation of myosin and other muscle fiber proteins decreases their ability to

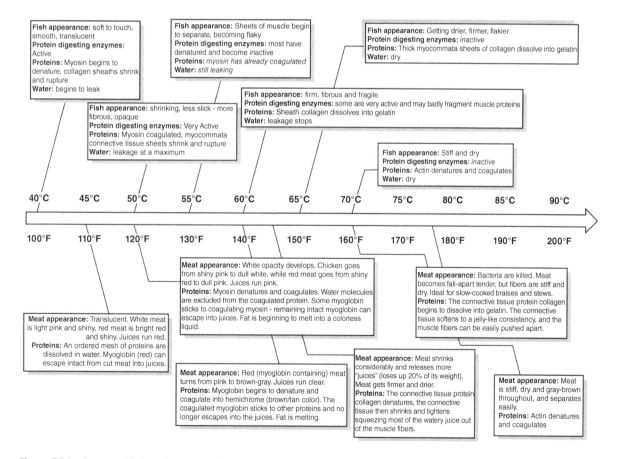

Figure 7.26 Stages of fish and meat cooking.

hydrogen bond to water, and the newly exposed hydrophobic portions of myosin will begin to aggregate, firming up the meat. Collagen also begins to denature. However, due to the shape of the long intertwined strands of collagen protein, denaturation causes collagen to shrink.

When the proteins in meat are subjected to heat (temps around 50–90°C), they start to denature. When that happens, the protein loses its shape and no longer holds onto water. At higher temperatures, collagen and other connective tissues start to break down and create more tender meat. Meat heated to this level (medium rare) will be on the cusp of moving from juicy to dry. The meat's slow, continual heating will continue to denature the collagen-containing connective tissue. At temperatures near 140°F/60°C, collagen continues to denature but slowly decomposes and dissolves into gelatin. Thus, slow cooking at a low temperature can reduce tough, collagen-rich meat into softer, more tender meat as the collagen is dissolved. Continued cooking will drive off the water, raise the internal temperature of the rest of the meat, and dry out the outside portion of the meat. Meat above 165–170°F/74–77°C will have lost most of the water, all of the protein will be denatured, and the myoglobin will be converted to brown hemichrome. This is the USDA definition of "well-done" meat.

Collagen is unique among most proteins because, in its native state, collagen is an insoluble fibrous protein. When denatured, the protein loses some of its three-dimensional structure and becomes water soluble. At about 160°F/71°C, beef collagen will denature and convert into liquid gelatin. Lower temperatures suffice for collagen from pork or fish. If carefully heated, gelatin will form a gel upon cooling; however, if overheated, the long collagen strands will completely denature and aggregate into an insoluble mess. JELL-O™ is the product of collagen conversion to gelatin. Gelatin has several culinary uses: it provides the soft, chewable texture of gummy bears, marshmallows, and candy corn and is often used as a stabilizer acting to maintain emulsions much like agar or pectins. Soup bones with connective tissues are often simmered for long periods to convert collagen in the tissue and bones to gelatin and create a thick base for soup. Tougher, more collagen-connected meats can become tender when the collagen is heated enough to convert the protein to gelatin. The conversion of collagen to gelatin can be increased under mildly acidic conditions. Adding vinegar or citrus juices helps this process along and is a key component of many barbecue recipes. Winning barbecue cooks learn how to convert collagen with low, slow heat while retaining as much juice as possible.

7.9.3 Let It Rest

Once the proteins denature and collagen shrinks, water is squeezed from the muscle fiber out of the tissue. Most cookbooks will encourage the chef to rest the meat after cooking. Resting allows some water to redistribute into the fibers as the tissue cools. Hydrogen bonds can re-form, retaining some of the liquid. One experiment showed that for a roast cooked to an internal temperature of 140°F/60°C, cutting the roast resulted in over 10 tablespoons of cell water lost. Much of the fluid is lost as the tissue is opened at the ends of the muscle fibers. Yet resting the meat for 10–20 minutes resulted in only a couple of tablespoons of lost water (Figure 7.27)

7.9.4 Marinating, Brining, Smoking, and Curing

There are several ways to treat meat to make it more tender, juicy, or flavorful. Marinades help decrease the toughness of meat by immersing the tissue in an acidic solution. Brining brings flavor and juiciness to meat by osmosis and diffusion, while smoking and curing add flavor and an element of safety with chemical changes to the meat by inhibiting pathogenic microbial growth.

Figure 7.27 Resting grilled beef steak. Halfpoint/Adobe Stock Photos

A quick review of the many posts for "the best marinade ever" reveals the true nature of the treatment. The common component is something that can change the pH of the meat. Vinegar (acetic acid) or lemon and lime juice (citric acid) are used in most acid marinades to reduce the pH to less than 4.5. Soaking the meat in an active yogurt culture will eventually acidify the meat and is popular with Indian food (vindaloo curry, tikka masala). Marinades with sodium bicarbonate or alkaline phosphates such as sodium triphosphate (also called tripolyphosphate [TPP]) raise the pH of the marinade from 8 to 10. The Scandinavian tradition of marinating white fish in lye (very alkaline sodium hydroxide, NaOH) and butter produces a smelly, soft, translucent favorite of Midwest Lutheran Church winter dinners called lutefisk. The alkaline lye soak deconstructs most collagen and connective tissues, forming a gelatinous, off-white fish product. Whatever your pleasure, ensure your recipe is more than a dash of acid or base to effectively reduce the pH. A squeeze of lemon or lime juice or a pinch of bicarbonate will not significantly bring the pH to the point of action on the meat protein.

The primary function of a marinade is to weaken the surface connective tissue by denaturing proteins with acidic or alkaline/basic solutions. The additional ingredients of marinades add to the flavor and salt concentration of the meat. Marinating fish, chicken, beef, or pork has the potential to soften the meat and add flavor, but strangely, marinating can also result in a dry piece of cooked meat. Why would this be? Acids and bases each effectively have the same impact on protein. Adding moderate amounts of acid or base will change the native structure of the protein and reduce the amount of water held by the charged groups of the protein. At higher concentrations of either acid or base, the protein will begin to degrade and hydrolyze (cleaving the protein backbone into smaller peptides), thus reducing the toughness of the meat.

To explain the action of a marinade, we need to review some acid–base chemistry. As was first introduced in Chapter 1, in cooking, pH can vary from acidic (i.e. <4.5) to mildly acidic (i.e. 4.5–7) to neutral (i.e. ~7) to mildly alkaline (i.e. 8–9) to very alkaline (>10). Across this spectrum of pH, side chains of amino acid residues can adopt different charge states depending on the availability of H+ (i.e. protons). At lower pH, carboxylate groups will shift from a negatively charged ion to a neutral polar compound ($R-COO^- + H^+ \rightarrow R-COOH$). Side chains containing amino groups are relatively unaffected by the lower pH, but at a pH above 8.0, they will shift from a positively charged compound to a neutrally charged group ($R-NH_3^+ \rightarrow R-NH_2 + H^+$). The impact of changing the pH below 5.0 or above 8.0 is a loss of WHC of the protein, changes in interactions between amino acid residues, and ultimately, the denaturing of the protein as described in Chapter 2. While the charged side chain of an amino acid residue can bind salt and water through ionic interactions and hydrogen bonds, neutral side chains are less able to do so. See Figure 7.28 for a detailed look at these interactions.

One of the problems with marinades is the loss of water or juiciness from the meat. The WHC of meat is a measure of how much water is held in place in meat tissue through noncovalent bonds between water and protein. Most of the water in a muscle fiber cell is tied up with the charged carboxylate and ammonium side groups of amino acids within meat proteins. These two chemical groups, when charged, can interact with other ions and hydrogen bond with water. This is an effective way to trap water in the meat tissue. However, when an acidic marinade lowers the pH, carboxylate ions can become neutral carboxylic acids, while raising the pH with an alkaline marinade turns the ammonium ion into a neutral amine. These neutralized amino acid side chains now have less hydrogen bonding potential. No longer held in place by noncovalent interactions, water is free to flow out of a ruptured muscle cell. In addition, as the pH moves to more acidic or alkaline/basic conditions, the protein will begin to denature, exposing the hydrophobic interior of the protein. The unraveled proteins aggregate in this condition, and hydrophobic interaction potential increases.

Another impact of denaturing protein is the loss of structure within thick and thin myofibril filaments. As the proteins denature and aggregate due to hydrophobic interactions, the myofibrils (thick and thin filaments) tend to shrink, and because these long protein fibers are connected to the cell wall, this shrinking squeezes water out of the cell. A filamentous protein called calpain binds the actin and myosin fibers to the cell membrane. When the fibers denature and shrink, they are still connected to the cell wall. Thus, the muscle cell fiber shrinks. An interesting marinade would include a two-step process: first, marinade in an enzyme-containing solution (proteases,

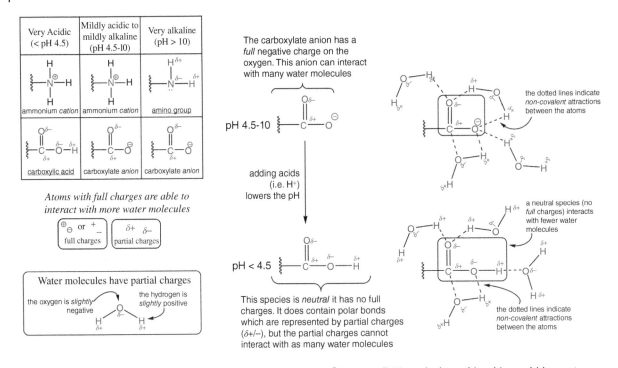

Figure 7.28 Charged Amino Acid Residues that can Interact with ("Hold onto") Water. Amino acid residues within meat proteins like lysine, aspartate, or glutamate have amino groups or carboxylic acid side chains that gain or lose hydrogen depending on the pH. Another name for H^+ is a proton. So these atoms are said to be protonated (gain of H^+) or deprotonated (loss of H^+).

or tenderizers) to partially denature the connective tissue, followed by an acidic or alkaline/basic marinade to further denature the proteins. The initial protein digestion step would allow you to avoid the squeezing of calpain and collagen. When experimenting in the kitchen, try including an emulsified oil with your marinade. Shirley O. Corriher reports that this leads to a deeper marinade in less time than without the oil.

Brining is the process of soaking meat in a seasoned saltwater solution and is often used to impart additional flavors and increase the amount of water in the tissue. Brining combines osmosis (movement of water as discussed earlier) and diffusion (of salts, spices, and flavors). There is quite a bit of confusion and some controversy about the science of brining.

As we learned with marinades, some water in a cell is associated or "bound" to proteins, while some cellular water is considered "free water." Free water can move around and be the solvent of the cell. Osmosis is the net movement of water through a semipermeable membrane (e.g. a muscle fiber membrane) from a low to high salt solution. Still, the complex nature of cellular water complicated the understanding of the brining process.

Most brines will have a 3–10% NaCl solution by mass. The Na^+ concentration in muscle cells is about 12 mM (millimolar), and the intracellular Cl^- concentration is 4.2 mM. This translates to a 0.026% Na^+ and a 0.015% Cl^- concentration. Thus the salt concentration of typical brine is over 650x higher outside of the cell than inside. Diffusion drives ions from a high concentration (the brine) to a low concentration (the muscle). However, the muscle membrane is pretty impermeable to Na^+ and Cl^- ions. Under these conditions, the net water movement by osmosis would be from the cell to the brine. This should lead to a dry piece of meat. Yet, anyone eating a brined turkey on Thanksgiving will tell you how juicy the meat is. A well-brined piece of meat can increase its weight by 10% or more... something has to be going on here.

While the muscle fiber plasma membrane is impermeable mainly to ionic compounds, the concentration of sodium and chloride ions outside the cell (i.e. in the brine) is so great that some sodium and chloride ions are driven across the membrane by diffusion. The salt ions then bind to and disturb the three-dimensional structure of muscle fiber proteins. This results in partial denaturation of the protein. Most of this work has been attributed to the chloride anions that bind to positive charges of the myofibril proteins. The altered filament protein structure breaks its contact with other proteins, allowing the filaments to swell. A well-brined piece of red meat will have a final concentration of 4.5% to nearly 6% sodium chloride. Osmosis (or water movement into the muscle cell) occurs when the new muscle fiber salt concentration exceeds the brine. The increase in chloride anions also helps the muscle fiber protein bind more water.

Muscles from different organisms, or, for that matter, different muscle groups from the same animal, will have very different starting salt concentrations and osmolarity. Fish brines are very different from brine for pork or poultry. A rule of thumb is to start with a lower salt concentration of around 1%. Remember that a 1% salt solution will still be hundreds of times more concentrated outside than inside a typical muscle cell. Protein changes and swelling seem to occur at about 5.5% NaCl in chicken breast and leg muscles. Using different concentrations of salt up to 10% will provide a solid range to test small pieces of meat for their ability to swell and retain water after cooking.

Salting or curing is a very old method of meat preservation before the time of refrigeration. The meat was salted (with NaCl) to remove moisture and therefore inhibit the growth of bacteria. With the advent of refrigeration, salt curing is unnecessary to preserve meat for months at a time, but salted cured meats are still made for their flavor alone. Often the meat is injected or immersed in brine solutions for a short time—the quick curing time means these modern "salt-cured" meats must be refrigerated and cooked to ensure meat safety since the curing was not strong enough or long enough to kill bacteria.

Curing evolved as a way to store the harvest of a hunt for longer than could be done with fresh or even cooked meat. Various meat cures are available, but all are trying to prevent bacterial growth and spoilage of the meat tissue. Many cure recipes use salts or sugar to draw water out of the meat. This leaves the tissue inhospitable to harmful microbes that thrive on uncooked meat. Some curing recipes allow for beneficial bacteria, such as *Lactobacillus*, also used for cheese production, to grow and produce lactic acid. These cures include sugar to serve as food for the bacteria to reduce the surface pH to about 4.5. The combination of an acidic pH with a low concentration of water and a high salt concentration leaves meat dry and able to be stored for long periods of time.

As in brining, salt draws water from the meat. However, unlike wet brining, salt curing requires much higher salt concentrations. Wet brines range from 15% to 30% salt. Dry brining requires the meat to be packed in a salt mixture. The term corned beef traditionally refers to a brisket (a tough cut of beef from the lower chest of cattle) placed in a barrel of coarse salt granules. Corned comes from the number of kernels or seeds of granular salt used to pack the beef. Over time, more and more water will move from the muscle fiber to the salt cure. The proteins will fully denature and entangle with each other. When combined with the dry texture, this gives dried meat its characteristic firmness and chewiness.

In addition to the use of rock or table salt (NaCl) in curing meat, sodium nitrite ($NaNO_2$), sodium nitrate ($NaNO_3$), and potassium nitrate (commonly called saltpeter or potash) have been used for centuries to inhibit harmful bacterial growth in cured meat. In the sixteenth to seventeenth centuries, cooks realized that using saltpeter (which turned out to be potassium nitrate, KNO_3) in addition to salt improved the color, flavor, safety, and storage life of meat. In the 1900s, chemists determined that the active ingredient in saltpeter (KNO_3) was a small amount of nitrite or NO_2. The nitrite reacts with the meat to form NO (nitrous oxide), which binds the iron (Fe) ion in myoglobin—giving it a permanent pink color—and the nitrite also inhibits the growth of bacteria. Meat cured with nitrates and nitrites retards the development of rancid off-flavors due to oxidation of fat during storage and also has the advantage of preventing the growth of *Clostridium botulinum*, the bacteria responsible for botulism.

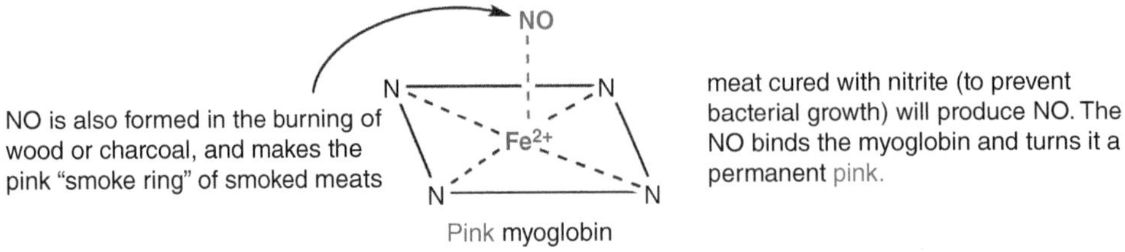

NO

NO is also formed in the burning of wood or charcoal, and makes the pink "smoke ring" of smoked meats

meat cured with nitrite (to prevent bacterial growth) will produce NO. The NO binds the myoglobin and turns it a permanent pink.

Pink myoglobin

Figure 7.29 Nitrate and meat. The replacement of oxygen by nitrogen compounds and the subsequent binding to iron give smoked meat its pink color.

Early curing used primarily sodium and potassium nitrates at high levels. However, salt-tolerant bacteria were found to be responsible for converting nitrates to more reactive nitrites. The nitrite ion (NO_2^-) reacts to form nitric oxide (NO), a short-lived but highly reactive gas. NO slows the oxidation of fat that normally takes place with the iron bound by myoglobin. NO binds very strongly to the heme iron in myoglobin, eventually forming a pink NO–Fe myoglobin complex that remains pink even after cooking (Figure 7.29). For short-term storage, nitrites are added to meats and sausages for safety concerns. Only for longer-term storage will nitrates be used, and then a culture of the transforming bacteria is included during processing.

There are concerns about including nitrites or nitrates in food. Nitrates and nitrites can react with the nitrogen-containing amino groups of amino acid side chains to form nitrosamines. Nitrosamines are a family of hundreds of closely related compounds, most of which are carcinogenic in animal studies. Nitrosamines bind to DNA, causing mutations, some of which can induce tumor formation. The early studies on nitrosamines and cancer were a little misleading as they used high concentrations of the compound for many days in a row to induce tumor formation. The formation of nitrosamine primarily happens under high heat or in the highly acidic conditions of the human stomach (pH ~2; Figure 7.30). Most meats have low concentrations of nitrates/nitrites, and the addition of the antioxidant vitamin C (i.e. ascorbic acid) inhibits nitrosamine formation. Modern regulations limit nitrate concentrations in meat to very low amounts (i.e. 200 ppm), and the inclusion of ascorbic acid is required for all commercially prepared meats. The National Academy of Sciences estimates that, on average, US citizens are exposed to less than a microgram of nitrosamines per day. Most of the nitrosamines come from bacon and beer in our diet. The latter is produced during malting. Thus, cooking with lower heat and ascorbic acid decreases the formation of nitrosamines, particularly when eating bacon. Limiting your intake of beer is a personal choice.

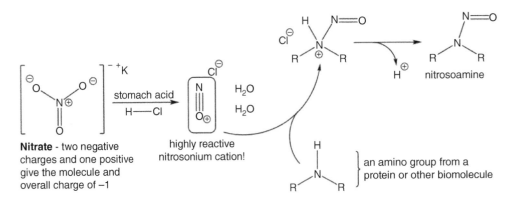

Figure 7.30 The conversion of potassium nitrate to nitrosamine in the acidic stomach.

Science for the Chef: The ultimate juicy tender steak: Using sous vide cooking

Sous vide cooking involves vacuum-sealing the steak in a plastic bag and immersing it in a temperature-controlled water bath for precise cooking (Figure 7.31). Before sealing the steak, you can enhance the flavor by seasoning it with salt, pepper, and desired herbs or spices. Salt is crucial for flavor and its scientific effect on meat. Salt helps to denature the proteins in the steak, allowing them to retain moisture and resulting in a juicier and more tender final product.

Once seasoned, place the steak in a vacuum-sealed bag, ensuring it is adequately sealed to prevent any water from entering. Preheat the sous vide water bath to the desired temperature for the doneness level you prefer. For medium-rare top sirloin steak, a temperature of around 137°F (58.3°C) is recommended.

Gently lower the sealed bag into the water bath, ensuring it is fully submerged. The controlled temperature of the water bath allows for even cooking throughout the steak. The sous vide technique is based on precise temperature control, ensuring that the steak reaches the desired doneness without overcooking or undercooking.

Meat is typically incubated in the sous vide circulator for 1–3 hours. This extended cooking time allows for the enzymes in the meat to break down and denature the muscle fiber proteins, resulting in a more tender and flavorful steak. The longer cooking time also gives the fat within the meat an opportunity to render, enhancing the overall taste and juiciness.

Once the steak has cooked sous vide, it is important to sear it to achieve a delicious crust and enhance the Maillard reaction. Preheat a cast-iron skillet or grill to high heat. Remove the steak from the bag, pat it dry with paper towels to ensure a good sear, and sear it on each side for a short time, about 1–2 minutes, until a golden brown crust forms. The Maillard reaction, a chemical reaction between amino acids and reducing sugars, occurs during searing, creating complex flavors and aromas.

Science Behind the Recipe

Sous vide cooking offers a scientific approach to achieving optimal results of tenderness, juiciness, and precise doneness by harnessing precise temperature control, protein denaturation, enzymatic tenderization, and the Maillard reaction. The science behind sous vide cooking can be explained as follows:

Precise temperature control: Sous vide cooking involves cooking the meat in a water bath at a controlled temperature. This temperature control is crucial because different proteins in the meat denature at specific temperature ranges. By setting the water bath to the desired temperature for medium-rare, you can achieve precise doneness without the risk of overcooking. Consistent temperature ensures that the meat is evenly cooked from edge to center, eliminating the possibility of overcooked or undercooked spots.

Protein denaturation: Proteins in the meat undergo denaturation when exposed to heat. Denaturation causes the proteins to unravel and reconfigure, resulting in changes in texture and moisture retention. Sous vide cooking allows for a longer cooking time at a

Figure 7.31   Sous vide cooking. mahony/Adobe Stock Photos

lower temperature, allowing the proteins to denature gradually and evenly throughout the meat. This results in a tender, juicy, and evenly cooked steak.

Enzymatic tenderization: During the extended cooking time in the sous vide water bath, enzymes within the meat break down connective tissues and proteins. This process, known as enzymatic tenderization, enhances the tenderness of the meat. Enzymes naturally present in the meat become activated at higher temperatures, allowing them to break down collagen and other tough components. Sous vide cooking at lower temperatures for extended periods allows optimal enzymatic tenderization, resulting in a more tender and succulent final product.

Maillard reaction: While sous vide cooking ensures perfectly cooked meat, it may lack the desired browning and flavor development on the surface. To achieve the desired crust and flavor, the meat is often seared or grilled after the sous vide process. Searing the meat on high heat triggers the Maillard reaction. The Maillard reaction is a chemical reaction between amino acids and reducing sugars in the meat, resulting in the formation of flavorful compounds and the characteristic brown crust. Sous vide cooking preserves the internal moisture of the meat, allowing for a longer and more controlled searing process without overcooking the interior.

Key Concepts

1. **Muscle fibers**, composed primarily of proteins like **myosin** and **actin**, exhibit a complex structure and function influencing meat properties at the molecular level.

2. **Sarcomeres**, the basic functional units of muscle fibers, consist of overlapping thin and **thick filaments** made of actin and myosin, respectively. The sliding mechanism between these filaments during muscle contraction affects meat tenderness.

3. Muscle fiber density, determined by the arrangement and packing of muscle cells, impacts **meat tenderness**. Dense muscle fibers require specific cooking techniques to break down the tough connective tissues and tenderize the meat.

4. **Connective tissues**, including **collagen** and **elastin**, are responsible for the support and structure of muscle fibers. Collagen, a fibrous protein, contributes to meat tenderness through its conversion to **gelatin** during cooking. Elastin provides elasticity to muscle tissues.

5. Muscle fiber organization varies among different animals, such as fish and shellfish, affecting muscle structure and cooking characteristics. Fish muscle fibers, for example, have a distinct organization with shorter sarcomeres, influencing texture and cooking times.

6. Collagen, a triple-helical protein, forms a structural framework in muscle tissues, providing support, strength, and structure. Its amino acid composition, cross-linking, and **denaturation** impact meat quality and tenderness.

7. Muscle fibers can be categorized into different types, such as **red (type I) and white (type II) fibers**, based on their physiological and biochemical properties. These fibers differ in color, myoglobin content, glycogen content, fat content, and flavor/water content, influencing meat characteristics.

8. Meat color is primarily determined by the oxidation state and binding of **myoglobin**, a pigmented protein in muscle tissues. The chemical reactions of myoglobin with oxygen, iron, and pH changes influence the color spectrum from red to pink to brown.

9. After an animal's death, chemical and physical changes occur in muscle tissue. The depletion of energy-rich compounds like glucose and ATP, along with changes in pH and enzymatic activities, initiates **rigor mortis** and affects meat quality.

10. Aging muscle involves the enzymatic breakdown of proteins by **proteases**, producing flavor molecules such as IMP and ribose. Additionally, protein denaturation and changes in **water holding capacity (WHC)** contribute to developing desirable meat flavors and tenderness.

11. Plant-based meats utilize key components and techniques to mimic the texture, flavor, and appearance of traditional meat products.

12. Plant proteins, such as soy, wheat, and pea proteins, are the primary protein sources in plant-based meats. These proteins undergo various processing techniques to create a meat-like texture.

13. Fats and oils are used in plant-based meats to provide juiciness and enhance mouthfeel. Emulsification techniques are employed to replicate the fat distribution in traditional meat.

14. **Flavor enhancers**, including natural and artificial ingredients, are added to plant-based meats to recreate the savory taste associated with meat. These may include yeast extracts, spices, and umami-rich compounds.

15. **Marinating** meat involves soaking or coating the meat with a marinade containing various ingredients, such as acids, enzymes, and spices, to enhance flavor, tenderness, and juiciness through chemical reactions and enzymatic activities.

References

1 *Alton Brown on the end of meat as we know it. Alton Brown blog post. Science Wired.* Available at http://www.wired.com/2013/09/fakemeat/ (accessed July 5, 2015).

2 Hedrick, G. B., Aberle, E., Gorrest, M. J. and Merkel, R. ed. (1989) *Principles of Meat Science*, 3e. Kendall/Hunt Pub. Co., Dubuque, IA, pp. 126–131.

3 Allred, C. D., Allred, K. F., Ju, Y. H., Virant, S. M., Helferich, W. G. (2001). Soy diets containing varying amounts of genistein stimulate growth of estrogen-dependent (MCF-7) tumors in a dose-dependent manner. *Cancer Res*, *61*, 5045–50

4 Du, M., Yang, X., Hartman, J. A., Cooke, P. S., Doerge, D. R., Ju, Y. H., et al. Low-dose dietary genistein negates the therapeutic effect of tamoxifen in athymic nude mice. *Carcinogenesis*. (2012) 33:895–901. 10.1093/carcin/bgs017

5 Keoleian, Gregory, A., and Martin C. Heller. (2018). Beyond meat's beyond burger life cycle assessment: a detailed comparison between a plant-based and an animal-based protein source." https://deepblue.lib.umich.edu/handle/2027.42/192044.

Additional Readings

Dry Aging Beef. Perry, N. Int J. Gastronomy Food Scie. 1, 1, 78–8-, 2012

Effect of meat marinating in kefir, yoghurt and buttermilk on the texture and color of pork steaks cooked *sous-vide* Latoch, A. Ann Agricul Scie 65, 2, 129–136, 2020

Enzymes in the Meat Industry. Sing, P.K., Shrivastava N., and Ojha B.K. Enzymes in Food Biotechnology. pp 111–128 Chpt 8 2019

He, J., Evans, N.M., Liu, H., and Shao, S. (2020) A review of research on plant-based meat alternatives: Driving forces, history, manufacturing, and consumer attitudes. *Compr Rev Food Sci Food Saf*, 19(5), 2639 – 2656.

Hughes, J.M., Clarke, F., Purslow, P., Warner, and R.D. (2020) Meat color is determined not only by chromatic heme pigments but also by the physical structure and achromatic light scattering properties of the muscle. *Compr Rev Food Sci Food Saf*, 19(1), 44 – 63.

O'Quinn, T.G., Brooks, J.C., and Miller, M.F. (2015), Consumer Assessment of Beef Tenderloin Steaks from Various USDA Quality Grades at 3 Degrees of Doneness. *J. Food Sci.*, 80(2), S444–S449.

Recent Advances in the Processing and Manufacturing of Plant-Based Meat. Wang, Y. Li, L., Gao, Yane., and Lai K.H. K/ Agroc Fppd Cje, 2023. 71. 3, 127–1290

Van Loo, E., Caputo, V., Nayga, Jr., R.M., Meullenet, J.-F., Crandall, P.G., and Ricke, S.C. (2010), Effect of Organic Poultry Purchase Frequency on Consumer Attitudes Toward Organic Poultry Meat. *J. Food Sci.*, 75(7), S384–S397.

Wang, Y., Tuccillo, F., Lampi, A. M., Knaapila, A., Pulkkinen, M., Kariluoto, S., Coda, R., Edelmann, M., Jouppila, K., Sandell, M., Piironen, V., and Katina, K. (2022). Flavor challenges in extruded plant-based meat alternatives: A review. *Compr Rev Food Sci Food Saf*, 21(3), 2898– 2929.

Yu, T.-Y., Morton, J.D., Clerens, S. and Dyer, J.M. (2017), Cooking-Induced Protein Modifications in Meat. *Compr Rev Food Sci Food Saf*, 16(1), 141–159.

End of Chapter 7 Questions

1. Which of the following best describes the role of collagen in meat tenderness?
   a. Collagen provides elasticity to muscle tissues.
   b. Collagen converts to gelatin during cooking, contributing to tenderness.
   c. Collagen determines the color of meat.
   d. Collagen breaks down into flavor molecules like inosine monophosphate (IMP).

2. What is the primary determinant of meat color?
   a. Glycogen content in muscle fibers
   b. Myosin protein in sarcomeres
   c. pH changes during rigor mortis
   d. Oxidation state and binding of myoglobin

3. Plant-based meats aim to replicate traditional meat by:
   a. Utilizing plant proteins, fats, and flavor enhancers.
   b. Extracting collagen from plants for structure and strength.
   c. Utilizing only soy protein as the primary protein source.
   d. Emulating the exact molecular structure of animal muscle fibers.

4. Which of the following molecular changes occurs to an animal's muscle tissue as it ages?
   a. Increased collagen cross-linking and stiffness.
   b. Decreased enzymatic breakdown of proteins.
   c. Reduction in the production of flavor molecules like IMP.
   d. Improved water holding capacity due to protein denaturation.

5. Meat will stiffen a few hours after eating. This is called rigor mortis. What is the result of freezing meat in rigor?
   a. The ATP remaining will be locked in the meat giving a more savory flavor
   b. The actin and myosin will break down more efficiently, giving a softer meat
   c. The tissue will tear, giving a mushy feel to the tooth
   d. The bundles of tissue will decrease in their overlapping

6. Ocean fish have a different savory or umami flavor due to the salt in the water. What is the actual cause of this flavor?
   a. Low water content in the meat?
   b. Increased amino acids to counter the salt in the water
   c. Higher salt concentrations of the tissue
   d. Trimethylamine oxides found in ocean fish
   e. Decreased levels of myoglobin

7. Fish and shellfish have additional compounds that impact their flavor. These are often due to _____?
   a. decreased collagen content
   b. a lack of requirement for terrestrial work
   c. the need to balance the salt content of their environment, limiting osmotic water loss
   d. higher metabolism of fat and protein molecules
   e. none of the above

8. Fish muscles are arranged differently than land animals. They form into what arraignment?
   a. Collagen-free tissue
   b. Myotomes
   c. Fatty connected tissue
   d. Slow twitch muscles

9. What is the primary factor influencing muscle fiber tenderness?
   a. Fat content and marbling
   b. Color and myoglobin content
   c. Density of muscle fibers
   d. Glycogen levels and flavor compounds

10. What is the main purpose of marinating meat?
    a. Adding moisture and preventing dryness
    b. Enhancing color and appearance
    c. Increasing cooking time and temperature tolerance
    d. Enhancing flavor and tenderizing the meat

11. What is the primary role of proteases in tenderizing meat?
    a. Breaking down proteins into amino acids and peptides
    b. Promoting browning reactions and flavor development
    c. Enhancing water retention and juiciness
    d. Converting collagen to gelatin during cooking

12. How does the depletion of ATP during rigor mortis contribute to changes in the muscle structure and overall meat quality after an animal's death?

13. How do plant-based meat manufacturers mimic the texture and mouthfeel of animal-based meat using plant proteins?

14. Discuss the use of the term GRAS (Generally Recognized as Safe) by Impossible Foods to argue against the need for a GMO label for their plant-based meat products. Present arguments in favor of and against this decision, taking into consideration the potential implications for consumer awareness and transparency.

15. Why does adding lemon juice or some other acidic liquid to fish reduce the fishy odor of saltwater fish while cooking?

8

Eggs, Custards, and Foams

Guided Inquiry Activities (Web): *5, Amino Acids and Proteins; 6, Higher Order Protein Structure; 8, pH; 9, Fats Structure and Properties; 12, Emulsions and Emulsifiers; 25, Eggs; 26, Custards and Egg Foams*

Learning Objectives

1. *Identify the role of a chicken egg in reproduction.*
2. *Identify and define the components of a chicken egg.*
3. *Identify the macromolecular components of egg yolk, egg white, egg membranes, and shell.*
4. *Describe the impacts of egg aging on pH and water content and, consequently, on egg white, yolk, and membranes.*
5. *Recognize the main types of secondary structure as present in egg white proteins: alpha helices, beta strands and sheets, and loops, and understand the relationship between amino acid sequence (primary structure), hydrogen bonding, and secondary structure.*
6. *Understand why proteins "fold" and how added energy in the form of heat or agitation can destabilize protein fold, causing denaturation as in the cooking and setting of an egg or the whipping of a meringue.*
7. *Identify hydrophobic and hydrophilic components of egg yolk and understand the need for emulsifying lipoproteins.*
8. *Explain the chemical processes of denaturation and coagulation in the context of cooking an egg.*
9. *Explain the impact of dilution on denaturation and coagulation of egg protein as in custards and how starch can stabilize a dilute protein network.*
10. *Explain the effects of metal ions and acid in the denaturation and coagulation of egg white proteins and the formation of an egg white foam.*
11. *Explain how characteristic odors and colors form in cooked eggs.*

8.1 Introduction

From fluffy omelets to the French dish of egg white battered Chicken Française and delicate meringues, eggs can be found in recipes from breakfast to dinner and dessert. Their perfect combination of water, protein, and fats makes eggs an excellent source of nutrition, a means of thickening, and a source of moisture in baked goods.

The egg houses all the ingredients needed to make a living, breathing creature. It is the essence of life itself. The yolk is a densely packed orb of fuel and raw materials. As Harold McGee unpacks in his tome *On Food and Cooking* [1], the word "yolk" comes from the old English for "yellow," which itself is derived from the Indo-European root

The Science of Cooking: Understanding the Biology and Chemistry Behind Food and Cooking, Second Edition. Joseph J. Provost et al.
© 2025 John Wiley & Sons, Inc. Published 2025 by John Wiley & Sons, Inc.
Companion website: www.wiley.com/go/provost/food_science_2e

Figure 8.1 A Partridge Cochin cockerel. *Source:* Sammydavisdog / Flickr / CC BY 2.0

meaning "to gleam, to glimmer." This same root gives us our words *glow* and *gold*. Even the ancients recognized the similarity between the egg's yolk and our sun; these yellow orbs are responsible for life.

Historians suppose that eggs have been part of the human diet from the very beginning of our existence. Eggs were cooked into dishes consumed by Romans in the first century, while omelets can be found in French cookbooks from the fourteenth and fifteenth centuries. In fact, the French monarch Louis XIV was fond of boiled eggs for breakfast, and Parisians were said to marvel as Louis knocked off the small end of a boiled egg with one swift stroke of his fork [2].

Despite the egg-loving Louis XIV, it was not until the nineteenth century that the chicken and its egg became the fascination of Europeans and Americans alike. In 1834, chickens of the Chinese breed (known then as "cochin") arrived in England as a present for Queen Victoria. These chickens were superior in egg production (and meatier) than the European and American varieties of the day. The now-familiar bright red comb on the chicken's head was so shocking in 1834 that it inspired chicken mania. When cochins arrived in the United States at the Boston poultry show of 1849, they attracted crowds that numbered in the thousands. This excitement catalyzed what has become known as *The Century of the Chicken* [3], also the century of the egg. While farmers were breeding chickens for egg and meat production, eggs were incorporated into cooking like never before. By the 1940s, egg production had become industrialized to support the mainstreaming of eggs in the modern diet. Many animals lay eggs, but the chicken egg is the most commonly eaten around the world, so this chapter will focus on chicken eggs (Figure 8.1).

8.2 What Is An Egg?

Any egg is a type of *cell* that is specialized for sexual reproduction, the process by which two parents contribute genetic material (genes) to make a new individual (Figure 8.2). Germ cells (e.g. egg and sperm) contain *half* the genetic material necessary to make an individual, the cells are **haploid**. Only eukaryotic organisms undergo sexual reproduction, and not even *all* eukaryotic organisms, only the most complex.

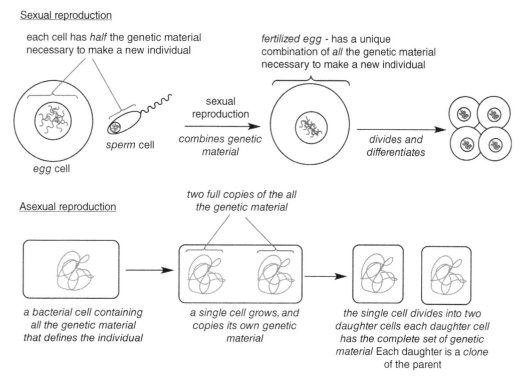

Figure 8.2 A comparison between sexual and asexual reproduction.

The egg is the larger, less mobile one of the two reproductive germ cells that combine their genetic material (genes) to make a new individual. The egg cell receives the sperm cell (the carrier of the other half of the genetic material), accommodates the joining of the two haploid sets of genetic material, and subsequently divides and differentiates into the new embryonic organism.

Asexual reproduction is how bacteria, yeast, and other simple organisms multiply. By definition, the product cells of asexual reproduction are genetically identical to the single, original parent, that is, the daughter cells are *clones* of the parent. On the other hand, sexual reproduction combines two sets of genetic material into a new, unique individual. Chicken eggs are created by female hens. The hen will produce the eggs whether fertilized or not; mass-produced, grocery store eggs are not fertilized. There is no nutritional difference or noticeable difference in physical appearance between fertilized and unfertilized chicken eggs. Folklore indicates that a "blood spot" in an egg indicates fertilization, but this is incorrect. The blood comes from the rupture of a blood vessel during the formation of the egg and has nothing to do with fertilization. Blood spots are not harmful, nor do they affect taste. They can simply be removed with a spoon. A chicken will lay eggs (fertilized or not) until she has accumulated a certain number of eggs in her nest. If the eggs are removed—perhaps by a predator or a human—the hen will lay another to replace it and may do so indefinitely.

The familiar chicken egg has a *yolk* surrounded by egg white, contained within a hard shell (Figures 8.3 and 8.4):

- The yolk accounts for approximately one-third of the weight of an intact chicken egg. Comprised of mostly fats and proteins, it carries 75% of the calories and most of the iron, thiamin, and vitamin A. Its purpose is to provide food for a developing chick.

- The white accounts for approximately two-thirds of the weight of an intact chicken egg. It is 90% water, the rest being protein. There are only traces of minerals, fatty material, glucose (carbohydrate), and vitamins. The white provides essential proteins and water and protects a developing chick.
- The shell is made of calcium carbonate ($CaCO_3$) and protein and is riddled with pores (tiny holes) that allow gases to pass in and out of the egg (Table 8.1).

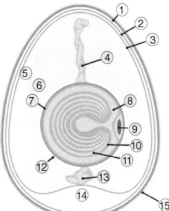

1. Eggshell – hard calcium carbonate and protein
2. Outer membrane - antimicrobial protein layer
3. Inner membrane – antimicrobial protein layer
4. Chalaza – protein cord that anchors yolk
5. Exterior albumen (outer thin albumen, less protein)
9. Middle albumen (inner thick albumen, more protein)
7. Yolk membrane – surrounds and protects yolk
8. Primordial white yolk: the first yolk to surround the germ cell
9. Germ cell (i.e. the egg cell) – *not actually red in real life*
10. Yellow yolk – fats and protein for germ cell
11. White yolk – less dense, high in iron
12. Internal albumen – the coating from which the chalazae extend
13. Chalaza - protein cord that anchors yolk
14. Air cell – air for chick to breathe
15. Cuticle – protein coating that gives egg color

Figure 8.3 Anatomy of an egg. *Source:* Benutzer: Horst Frank, https://commons.wikimedia.org/wiki/File:Anatomy_of_an_egg_c-m.svg. Used under CC BY SA 3.0 https://creative commons.org/licenses/by-sa/3.0/deed.en

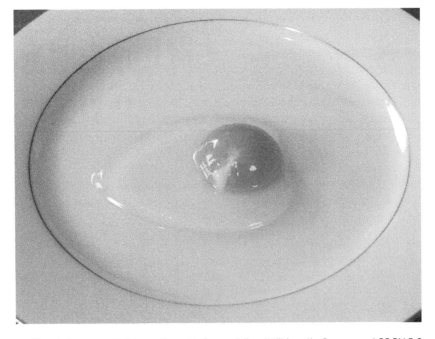

Figure 8.4 Raw egg. The chalazae are visible on the yolk. *Source:* Miya / Wikimedia Commons / CC BY 3.0

Table 8.1 The composition of a US large egg.[a]

	Whole egg	Egg white	Egg yolk
Weight	55 g	38 g	17 g
Protein	9.9 g	3.9 g	2.7 g
Carbohydrate	0.9 g	0.3 g	0.3 g
Fat	9 g	0	9 g
Monounsaturated	2.5 g	0	2.5 g
Polyunsaturated	0.7 g	0	0.7 g
Saturated	2 g	0	2 g
Cholesterol	213 mg	0	213 mg
Sodium	71 mg	92 mg	9 mg
Calories (cal)	84	20	94

a.   Adapted from [1].

8.3   Inside An Egg

An egg contains everything you need to make a chick. Eggs are unmatched as a balanced source of amino acids, and they include a plentiful supply of linolenic acid—an essential polyunsaturated omega-3 fatty acid—as well as several vitamins and minerals. Eggs also contain **cholesterol**, a hydrophobic molecule. Cholesterol is also considered a **lipid** because fatty acids can be converted into cholesterol using many enzymes. In humans, high blood cholesterol does increase the risk of heart disease, a fact that long-made medical professionals recommend limiting egg yolk consumption to two to three per week. However, recent studies show foods—including egg consumption—have little effect on blood cholesterol, rather, saturated fats have a far more powerful effect on raising blood cholesterol. In addition, the phospholipids in the egg's yolk interfere with our ability to absorb cholesterol, so we don't have to count our eggs after all.

8.3.1   The Yolk

The yolk (Figure 8.3) surrounds the germ cell (Figure 8.3). The germ cell contains half the genetic material needed to make a chick. This germ cell is the haploid egg cell surrounded by the yolk. The yolk is much larger than the germ cell. The yolk membrane (Figures 8.3 and 8.7) surrounds and protects the yolk and indicates egg freshness.

The yolk is made of about 50% solids (proteins, lipids, and some carbohydrates; see Chapter 2 for explanations of these macromolecules). There is about twice the amount of lipids as proteins in egg yolk. To the naked eye, an egg yolk looks like a homogeneous yellow solution. However, microscopic analysis shows us something very different. The yolk is made of two parts: white (or light) yolk called the plasma (Figure 8.3) and yellow (or dark) yolk called the granule (Figure 8.3). Under a microscope, the yolk appears as islands of semisolid dark yellow granules surrounded by a lighter-colored solution. The white yolk is mostly made of lipids like LDL (low-density lipoproteins that we will learn about later) and some proteins and is less dense and especially rich in iron, while the yellow yolk is denser and rich in fats and proteins. The dark yellow granules are made of tightly aggregated proteins enriched in HDL (or high-density lipoprotein) and are more protein-rich than the plasma, but still contain plenty of lipids, including lecithin (phosphatidylcholine). The color of the yellow yolk depends on what the hen eats. Yellow pigments in hen food include xanthophylls, which the hen obtains mostly from corn-based feed.

The high concentration of iron in the white yolk prevents it from fully setting when the egg is hard-boiled, while the protein complexes in the granule resist heat treatment, which is why it takes higher heat or longer cooking times to turn the yolk solid. However, adding salt to eggs before cooking causes yolk proteins to fall out of the solution and solidify (denature) much more easily.

Two interesting proteins are found in the yolk. Phosvitin is a protein where almost half of the amino acids are serine. Many of these serines are phosphorylated ($-PO_4^{2-}$). This is highly unusual, and the higher content of phosphoserines gives it a high resistance to heat denaturation and proteolytic cleavage. The charges of the phosphorylated serine amino acids give phosvitin a natural ability to bind positively charged metals. The major water-soluble protein is livetin accounting for 30% of the plasma proteins. These proteins are very similar to human antibodies and are a source of allergenic hypersensitivity for some people. Together, the three forms of livetin provide passive immunity (the transfer of antibodies from the yolk to the developing embryo).

After water, most of the yolk is made of lipids; 92% of the yolk is comprised of different types of lipids. Triglycerides (95%), phospholipids (31%), and cholesterol (4%) make up the main types of lipids in egg yolk, with some free fatty acids bound to proteins and other components. Lecithin is the old, traditional name for the amphiphilic molecule known to chemists as phospholipid phosphatidylcholine. This amphiphilic phospholipid is a minor but important component of egg yolk, making up about 0.9% of the lipids. As described in Chapter 2 (Section 2.5.2), the amphiphilic lecithin (i.e. phospholipid phosphatidylcholine) is both water-loving and water-fearing and used in cooking to make hollandaise and mayonnaise among many other foods. Together, the proteins and lipids in the yolk are used for their essential emulsifying properties.

8.3.2 The White

The egg white is mostly water (~90%) and a mixture of proteins collectively called albumen (Figures 8.3, 8.8, and 8.9). The albumen not only nourishes the chick, but it is also a biochemical shield against infection and predators. The *chalazae* (Figures 8.3 and 8.4) are dense elastic cords made of albumen that anchor the yolk to the ends of the shell and allow it to rotate while suspended in the middle of the egg. Chalazae are visible in a raw egg when cracked open (Figure 8.4). We will investigate the egg white proteins more closely in later sections.

8.3.3 Membranes and Shell

The membranes (Figures 8.3 and 8.7) line the inside of the shell and are made of antimicrobial proteins. The *shell* (Figure 8.3) is made of calcium carbonate and protein, and since the developing chick needs to breathe, the shell has thousands of tiny *pores* or holes. These pores are invisible to the eye. The cuticle (Figure 8.3) is a thin protein coating on the shell. This coating initially blocks the pores to slow the loss of water and prevent the entry of bacteria. When the chicken deposits the proteinaceous cuticle, pigment molecules are deposited into the shell, giving the egg its color. The pigments deposited are dependent on the type of chicken. White, brown, and even blue eggs and yellow-spotted eggs result from the genetic makeup (the breed) of the chicken. For example, Rhode Island Reds lay brown eggs. There is no nutritional difference between white, brown, or even blue eggs. The air cell (Figure 8.3) provides the developing chick with its first breaths of air and is also an indicator of egg freshness.

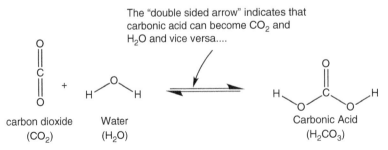

The "double sided arrow" indicates that carbonic acid can become CO_2 and H_2O and vice versa....

carbon dioxide (CO_2) Water (H_2O) Carbonic Acid (H_2CO_3)

Figure 8.5 Reaction generating carbonic acid.

8.4 Egg Freshness

Although an egg can remain edible for weeks if kept intact and cool, egg quality does deteriorate over time. When an egg is freshly laid, it contains carbon dioxide dissolved in the white and yolk. When carbon dioxide is dissolved in water (and egg white is 90% water), it forms carbonic acid (Figure 8.5).

As we learned in Chapter 1 (Section 1.4.3), acids dissolve in water, release their acidic proton, and make the water *acidic*. This can be measured by the pH scale (Figure 8.6). On a pH scale, 7 is neutral pH, below 7 is acidic, and above 7 is alkaline. The more acid in the water, the lower the pH, while adding acid to an alkaline solution will lower the pH.

As eggs age, the carbon dioxide dissolved in the white and yolk gradually escapes through the pores in the shell. Since carbonic acid is essentially carbon dioxide and water (Figure 8.5), as carbon dioxide leaves eggs, the white and yolk lose carbonic acid and become more alkaline. This change in pH changes interactions between egg white albumen proteins (the proteins interact less), and the egg white albumen (Figures 8.3, 8.5, and 8.7) is consequently runnier.

In addition to carbon dioxide escaping through the pores of the shell, water molecules also escape, making the overall contents of the egg shrink. This shrinking allows air to travel in through the pores in the shell and enlarge the air cell. In a typical refrigerator, an egg will lose 4 mg of water a day.

A final indicator of egg freshness is the yolk membrane (Figures 8.3 and 8.7). Because the yolk contains less water than the white, water gradually crosses the membrane from the white into the yolk. In the refrigerator, the yolk gains about 5 mg of water per day. This increase in water makes the yolk swell (enlarge) and the yolk membrane weaken. The grade an egg is given is due to the quality as determined by the way the yolk and white membranes function (Figure 8.7).

Poaching is the process of cracking a raw egg into a pot of boiling water. The white coagulates or sets in the boiling water before the yolk. Chefs recommend that only the freshest grade AA eggs be used for poaching; otherwise, you may find filmy strings of coagulated white floating in the water instead of an intact, perfectly poached egg. Since fresh, grade AA eggs are not too alkaline yet (from the escaping CO_2), the egg white protein—the albumen—is not too runny, and the yolk membrane is tight. These facts, along with a gently boiling pot of water, help in making the best poached eggs.

Figure 8.6 pH scale, a measure of acidity and alkalinity.

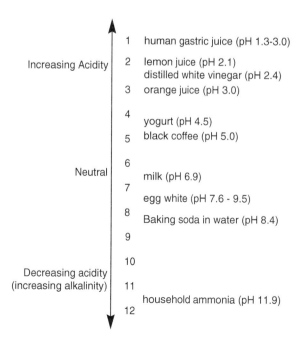

Increasing Acidity	1	human gastric juice (pH 1.3-3.0)
	2	lemon juice (pH 2.1) distilled white vinegar (pH 2.4)
	3	orange juice (pH 3.0)
	4	
		yogurt (pH 4.5)
	5	black coffee (pH 5.0)
	6	
Neutral		milk (pH 6.9)
	7	
		egg white (pH 7.6 - 9.5)
	8	Baking soda in water (pH 8.4)
	9	
	10	
Decreasing acidity (increasing alkalinity)	11	
		household ammonia (pH 11.9)
	12	

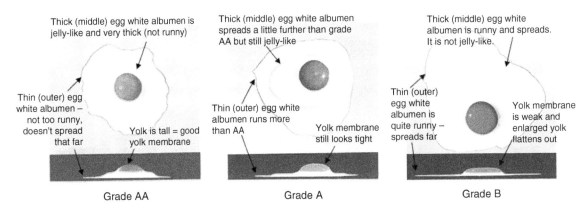

Figure 8.7 Egg grade is a function of egg white and yolk membrane.

8.5 Egg Protein

8.5.1 Proteins in Egg White

Egg white albumen is made of mostly water and protein. Ovalbumin is the most abundant protein; as we will see later, it is largely responsible for how egg whites cook. Another interesting protein, lysozyme, is the same protein found in human tear ducts. In both cases, the protein serves to act in an antibacterial function. The protein will cleave protein–carbohydrate bonds on the surface of bacteria and fungi cell walls. This leaves the bacteria damaged and unable to survive. Lysozyme has been used as a preservative, stopping microbial growth in meat, fruits, and vegetables. In the egg white, it prevents microorganism contamination that may penetrate the eggshell (Table 8.2).

8.5.2 How and why Egg Proteins Fold

As we learned in Chapter 2 (Section 2.5), proteins are amphiphilic molecules; they have both polar (hydrophilic) and nonpolar (hydrophobic) parts. In nature, we find proteins in water-based environments. Because not all parts of the protein love the water (some parts are hydrophobic), the protein folds in a three-dimensional way that buries the hydrophobic regions on the inside of the structure and exposes the hydrophilic parts to the outside, where

Table 8.2 The proteins in egg white albumen.[a]

Protein	Percentage of total (%)	Natural function
Ovalbumin	54	Nourishment for chick, may block digestive enzymes. Contains six cysteine residues. Two are engaged in a disulfide (—S—S—) bond
Ovotransferrin	12	Binds iron
Ovomucoid	11	Blocks digestive enzymes
Globulins	8	Plug defects in membranes and shell
Lysozyme	3.5	Enzyme that digests bacterial cell walls
Ovomucin	1.5	Thickens albumen, inhibits viruses
Avidin	0.09	Binds the vitamin biotin
Others	10	Bind vitamins, block digestive enzymes

a. Adapted from [1].

they can interact with the watery environment. It is easy to obtain egg white protein, so it was among the first proteins studied for its molecular structure. Modern biochemistry has allowed scientists to "see" how proteins fold at a molecular level. What they found was quite beautiful (see Figure 8.8).

But how can Figure 8.8 depict a protein? There are no visible amino acids or peptide bonds, just twirly ribbons and curvy arrows. What are these images showing anyway? As we learned in Chapter 2 (Section 2.2), the cartoons of Figure 8.8 and Figure 8.9a depict each α-helix ("alpha" helix) as a red twirly ribbon, while each yellow curvy arrow is a β-strand ("beta" strand). Many curvy arrows lined up lengthwise next to each other make a β-sheet. The squiggly green parts are loops. Suppose we represent a structure of the molecule ovalbumin using the same color scheme as you see in Figs. 8.8 and 8.9a, but instead, we replace the squiggles, twirly ribbons, and curvy arrows with atoms. In that case, the structure looks like Figure 8.9b. In Figure 8.9b, those same helices (ribbons), sheets (arrows), and loops are depicted using amino acid residues (the amino acids joined by peptide bonds; see Chapter and Section 2.2.1). The last version of ovalbumin (Figure 8.9c) is called a space-filling model. Here the surface of each atom (the edge of where the electrons are found most of the time) is shown in this presentation to allow researchers to get a feel for what happens at the surface of a protein.

Let's look more closely at the amino acid atoms making a helix in Figure 8.10. The twirly ribbons in Figure 8.10c and e are α-helices, and the loops that join them are really cartoon representations of how the peptide backbone of the protein looks in three-dimensional space. The bonds holding the amino acids together (called the peptide backbone) can be seen in Figures 8.10a, b, and d. Notice where the amino acid side groups are located in a helix—outside like the bristles on a round brush, the side groups pointing outward from the center of the helix. The forces holding the helix in place are the hydrogen-bonded atoms within this backbone, shown in Figures 8.10d and e. The peptide backbone is the C=O, alpha carbon, and nitrogen of each amino acid residue along the protein chain (Figure 8.11).

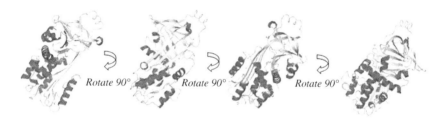

Figure 8.8   The protein ovalbumin (from chicken egg white) as viewed rotating around a vertical axis.

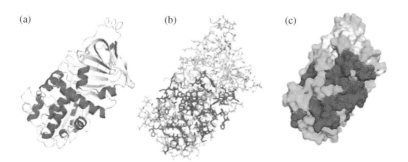

Figure 8.9   The same orientation of the ovalbumin protein molecule represented as a cartoon (a) and then as using *amino acid residues* (b) the amino acids joined by peptide bonds, and (c) *space-filling* representation of ovalbumin.

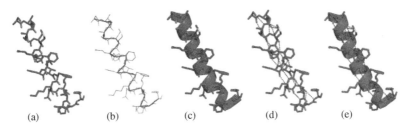

(a)   (b)   (c)   (d)   (e)

Figure 8.10 A single α-*helix* from ovalbumin drawn five ways: first as *amino acid residues*, amino acids joined by peptide bonds (a), and then in (b) a thin line is tracing the three-dimensional pattern of the peptide bonds between each amino acid (called the *peptide backbone*), which makes the helical ribbon shape. In (c), a cartoon ribbon has replaced the thinly traced line. In (d) and (e), we see that hydrogen bonds (dashed lines) hold the amino acid residues in this special helical shape.

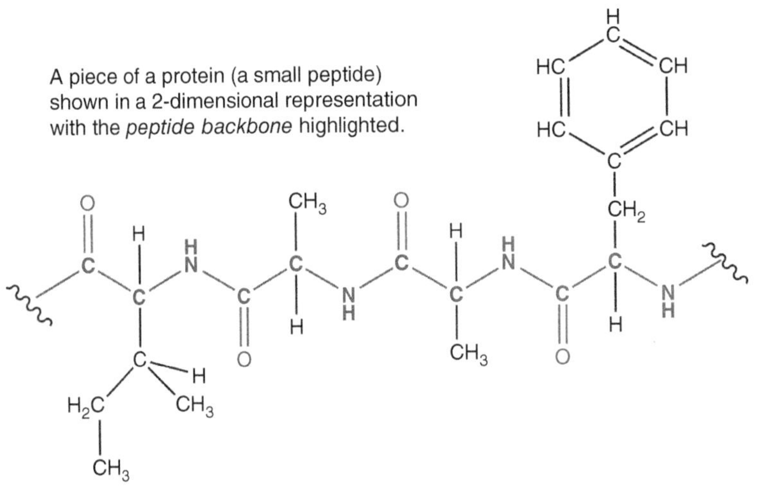

A piece of a protein (a small peptide) shown in a 2-dimensional representation with the *peptide backbone* highlighted.

Figure 8.11 Peptide backbone. The peptide backbone is shown with the bonds connecting the amino acids (side chains alternating above and below) the backbone.

In ovalbumin (as with almost any protein), some groups of amino acids fold into α-helices, some into β-strands, and some into *loops*. The primary sequence (see Chapter and Section 2.2.1) of the amino acid residues determines which structure they form. Figure 8.12 depicts how an α-helix, for example, is created by an arrangement of amino acid residues stabilized by hydrogen bonds—remember, hydrogen bonds are weak, non-covalent bonds that form between partially positive and partially negative atoms (Chapter and Section 1.4.1.2). β-sheet and loop structures are also held together by hydrogen bonds.

Figures 8.7 and 8.8 show that the protein ovalbumin has many α-helices and β-sheets all folded on top of and around each other to make a globular structure. In fact, when you fill in all the atoms using a type of space-filling representation, we can see the lumpy, three-dimensional, globular-looking protein (Figure 8.8c). The α-helices, β-sheets, and loops also interact with each other by hydrogen bonds. β-sheets are stabilized by the same atoms of the peptide backbone; however, the arrangement is very different (Figure 8.12). Unlike in a helix, the hydrogen bonds are between the same atoms of the peptide backbone, but between two strands rather than within the same strand (Figure 8.13). Each amino acid will form hydrogen bonds with neighboring chains.

Figure 8.12 Backbone of an α-helix. The helix is in a right-handed turn where hydrogen bonds form between every four amino acids in the same chain.

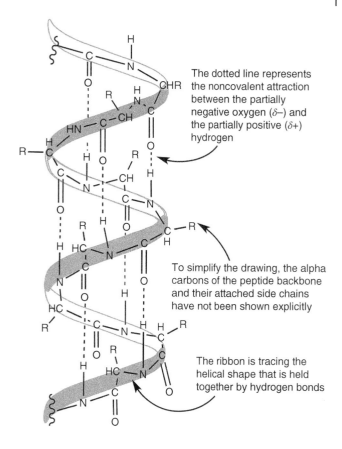

The dotted line represents the noncovalent attraction between the partially negative oxygen ($\delta-$) and the partially positive ($\delta+$) hydrogen

To simplify the drawing, the alpha carbons of the peptide backbone and their attached side chains have not been shown explicitly

The ribbon is tracing the helical shape that is held together by hydrogen bonds

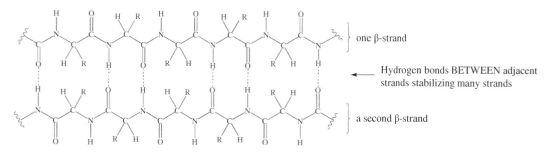

one β-strand

Hydrogen bonds BETWEEN adjacent strands stabilizing many strands

a second β-strand

Figure 8.13 β-sheets. In β-sheets, the peptide backbone is extended rather than coiled in a helix. In this form, hydrogen bonds form between the amide and carbonyl groups of neighboring strands.

Box 8.1 Egg allergies

Ovalbumin is the most abundant protein in eggs and also the protein most likely to cause egg allergy in humans. Since developing an allergy requires exposure to the allergen (e.g. ovalbumin), medical professionals recommend that babies do not eat egg whites until after they are 1 year old. However, since ovalbumin is only present in egg white, it is considered safe for babies to eat egg yolk.

8.6 Egg Fats

The egg yolk contains some protein, all of the fat and cholesterol, and greater than 75% of the calories for the whole egg. In fresh liquid yolk, over 50% of the yolk is water. The remaining yolk material is mostly fats (or lipids), protein, and smaller amounts of minerals and carbohydrates. Since water and fat do not mix, the yolk fat is suspended in the water as lipoproteins. To understand how these compounds behave, it is important to recognize whether they are polar or nonpolar (or if significant components of larger biomolecules are polar or nonpolar). A polar bond is due to an unequal sharing of electrons between atoms covalently bonded together (Figure 8.14). A lipoprotein is a spherical particle where the center is made of lipids—such as fats/triglycerides and cholesterol—and the outside is coated in phospholipids and proteins. Phospholipids and proteins are amphiphilic; they have both hydrophobic and hydrophilic parts. Hydrophobic groups of atoms are comprised of nonpolar bonds and *do not* interact with (literally, "fear") water. Hydrophilic groups of atoms are comprised of polar bonds and/or contain charged atoms; these groups of atoms interact with (literally, "love") the polar bonds of water via their charged or partially charged atoms (Figure 8.15). See Chapter 1 (Section 1.4.1.4) for a review of hydrophobic and hydrophilic (Figure 8.16).

The proteins and phospholipids of lipoproteins give egg yolk emulsifying properties. As discussed in Chapter and Section 2.5, emulsifiers can mix with water and fat and bring the two phases together in a fine, cream-like, stable mixture. When fats/oils and water form a stable mixture (i.e. one not separated into two phases), it is called an emulsion.

In its simplest form, mayonnaise is an emulsion of oil and vinegar, where eggs are the emulsifier. Mayonnaise does not separate into the oil phase (hydrophobic) and the water phase (the vinegar) (hydrophilic) because the amphiphilic molecules in the egg stabilize the emulsion. The yellow color of egg yolk is due to small amounts of carotenoid pigments ingested by the hen in her feed. Carotenoids are light-capturing pigments made by plants. Learn more about carotenoids in Chapter 6.

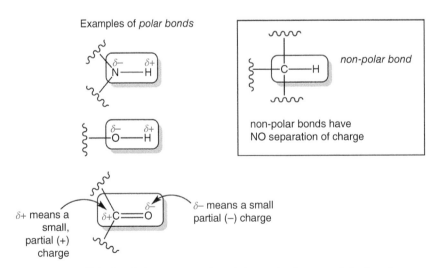

Figure 8.14   Polar bonds.

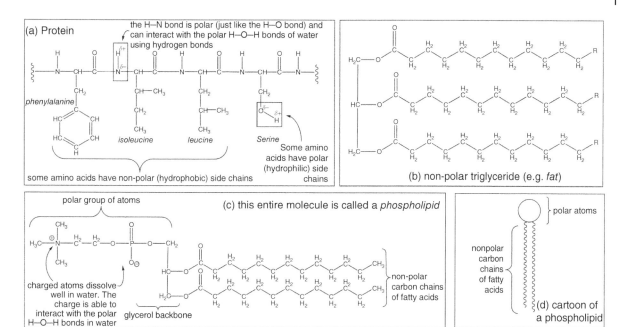

Figure 8.15 The major components of lipoproteins. (a) Protein contains hydrophobic (nonpolar) and hydrophilic (polar) groups of atoms. Proteins are amphiphilic and can mix with water by folding in a way that exposes the hydrophilic parts to water. (b) Triglycerides are nonpolar (hydrophobic) and do not mix with water. (c) Phospholipids are a modification of a triglyceride in which one of the fatty acid chains is replaced with a very polar group of atoms; consequently, phospholipids are amphiphilic. (d) A simple cartoon representation of a phospholipid.

8.7 Cooking Egg Protein

When a protein denatures, energy (e.g. in the form of heat) added to the mixture of proteins makes the molecules move around. Add enough energy, and the proteins move and groove enough to break all the hydrogen bonds holding the α-helices, β-sheets, and loops together, and the three-dimensional globular protein structure unravels (Figure 8.17). When the protein unfolds, all the hydrophobic parts buried inside the protein are exposed to the watery environment. The hydrophobic parts *hate* the water, and instead, they'd like to find another hydrophobic place to be. The exposed hydrophobic parts of a protein join together with exposed hydrophobic parts of other proteins, clumping together in a process called coagulation. The unfolded, coagulated protein is now a large, aggregated complex that can't stay dissolved in the water, so the coagulated protein solidifies, trapping water molecules between the unfolded proteins.

The proteins of egg white each have a unique denaturation temperature dictated by their specific amino acid sequence and three-dimensional fold (Figure 8.18). Ovotransferrin is the least stable of the egg white proteins and, in a pure form, will denature at 57°C/135°F. Globulins, ovalbumin (the most abundant egg white protein), and lysozyme denature in their pure forms at 72°C/192°F, 71.5°C/191°F, and 81.5°C/179°F, respectively. The mixture of proteins in egg white results in a composite denaturation temperature of approximately 93°C/145°F with coagulation above 99°C/151°F. The network of denatured and coagulated egg white protein is stabilized by the interaction of exposed hydrophobic areas and disulfide bond formation between protein chains.

The abundance of ovalbumin makes its denaturation very important to cooking egg white. Ovalbumin unfolds in two steps. The half-unfolded ovalbumin is called S-ovalbumin; it is surprisingly heat-stable and difficult to denature. Over time, most of the ovalbumin in freshly laid eggs gradually converts to the S-ovalbumin form. In

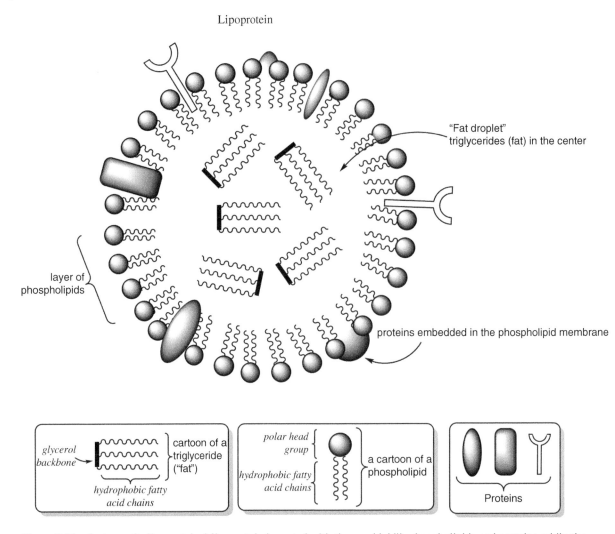

Figure 8.16 Cartoon of a lipoprotein. A lipoprotein is coated with the *amphiphilic* phospholipids and proteins, while the interior is made of hydrophobic fats.

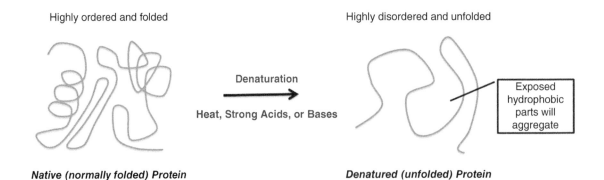

Figure 8.17 Protein denaturation at the molecular level.

Figure 8.18 Egg protein denaturation. Protein denaturation caused by heat as seen with the unaided eye. uckyo/Adobe Stock Photos

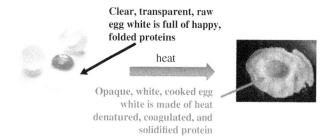

Clear, transparent, raw egg white is full of happy, folded proteins

heat

Opaque, white, cooked egg white is made of heat denatured, coagulated, and solidified protein

fact, by the time an egg reaches the grocery store shelf, approximately 50% of its ovalbumin has converted to the S-ovalbumin form. The accumulation of S-ovalbumin raises the temperature at which the protein denatures. Fresh ovalbumin denatures at 84.5°C, while S-ovalbumin denatures at 92.5°C. The "melting of ovalbumin" is shown in Figure 8.19 and Table 8.3. This change in ovalbumin's three-dimensional structure makes it more challenging to denature and results in "runnier" whites when cooking [5].

All this egg protein coagulation happens well below the boiling point of water (212°F/100°C). Overcooking an egg with too much heat and for too long can evaporate the water molecules that were trapped among the unfolded protein—the result is a rubbery solid. You may have noticed this effect if you have prepared scrambled or fried eggs in a too-hot skillet!

Denatured proteins coagulate into a network stabilized by interactions between hydrophobic regions of the unfolded protein and disulfide bonds that form between the protein changes (Figure 8.20). In the cooking of a whole egg, the white will always set before the yolk. This interesting fact makes possible the soft yolks of delicately poached eggs, sunny-side-up fried eggs, and baked eggs. While egg white protein begins to set at approximately 93°C/145°F with coagulation above 99°C/151°F, egg yolk lipoproteins only begin to denature around 95°C/149°F and finally coagulate at 70°C/158°F. A large amount of fat—relative to protein—in the egg yolk makes it harder for the denatured proteins to find one another and stick their exposed hydrophobic parts together; therefore, more energy in the form of higher temperature is required to accomplish denaturation and coagulation of yolk proteins.

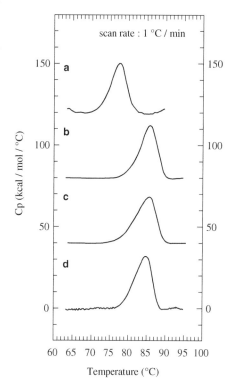

Figure 8.19 Thermostability of native ovalbumin and S-ovalbumin. The stability of ovalbumin and S-ovalbumin to heat. (a) Native ovalbumin, (b) S-ovalbumin from stored egg, (c) S-ovalbumin produced by alkaline treatment, and (d) the solubilized purified S-ovalbumin as analyzed with a differential scanning calorimeter. The temperature was scanned at 1°C/min. The peak for native ovalbumin is at a lower temperature than for S-ovalbumin. *Source:* Yamasaki, JBC 2003, http://www.jbc.org/content/278/37/35524.full / with Permission of Elsevier.

Table 8.3 Protein melt temperatures.[a]

Protein	Temperature of denaturation (°C)
Avidin	85
Egg globulins	92.5
Lysozyme	75
Ovomucoid	79
Ovalbumin	84.5
S-ovalbumin	92.5

a. Adapted from [4].

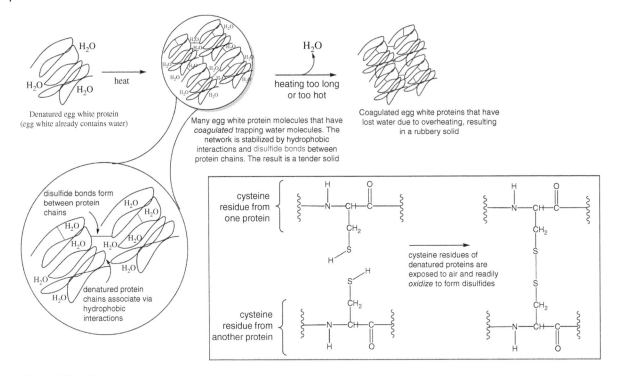

Figure 8.20   Coagulation of egg white protein.

8.8   Custards

Heat denatures (or unfolds) proteins, and when proteins denature, the exposed hydrophobic parts of the protein join together with exposed hydrophobic parts of other proteins, clumping together in coagulation. We observe this process in the cooking of an egg (discussed above in Section 8.7); the clear, runny white becomes a white solid mass of denatured and coagulated protein.

When we add other liquids and/or molecules like starch, sugar, and/or fat (from milk, butter, cheese, etc.) to the eggs, we dilute the protein mixture and raise the temperature at which the proteins begin coagulating. Dilution of the proteins surrounds the protein molecules with much more water (and/or sugar or fat) molecules, which makes it harder for the proteins to find one another and stick their hydrophobic parts together (i.e. coagulate), so we have to raise the temperature, which makes the molecules move around that much more rapidly and find each other. When the coagulated protein network does finally form in these diluted mixtures, the solid matrix of coagulated protein is tender and fragile—the large networks of coagulated proteins are filled with water, sugar, and/or fat molecules—for example, as in a custard.

If you heat a custard preparation too high or fast, the proteins will *curdle* instead of thickening. In curdling, proteins denature and form hard, tight lumps of coagulated protein that exclude the water, sugar, and/or fat molecules (in custard making, the effect is called **syneresis**; Figure 8.21). Because custards are so sensitive to overheating, they are often cooked in an oven while sitting in a water bath. The water bath keeps the temperature constant during cooking. The boiling point of water is 212°F/100°C, while a typical oven temperature is 325–350°F. No matter how hot you make the oven, the water will only ever reach 212°F/100°C, upon which it will boil and eventually evaporate away. Considering that custards are best cooked at approximately 180°F/82°C, the water bath method works particularly well. Additionally, the water bath increases the gaseous water content inside the oven, preventing the surface of the custard from drying out. Pastry cream and crème anglaise are

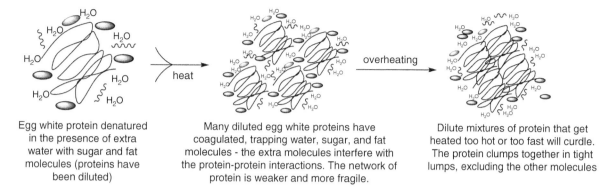

Figure 8.21   Egg protein coagulation diluted with water, sugar, and fat molecules.

both egg-thickened pourable creams and good examples of classic, stirred custards. Making crème anglaise requires the use of a double boiler. The bowl with the egg yolks, vanilla, milk, cream, salt, and sugar sits over another bowl full of boiling water in a variation on the water bath (Figure 8.22). The bowl of ingredients is only ever heated with the steam from the boiling water. This double boiler is essential to successful crème anglaise because the gentle heat from the steam allows the diluted mixture of protein to denature and coagulate while trapping the water, fat, and sugar molecules among the protein matrices.

Pastry cream is a very similar dish, but the addition of cornstarch makes pastry cream more heat stable, that is, it can reach higher temperatures without curdling. Therefore, pastry cream can be prepared without the aid of a double boiler. Examining the recipes reveals that the only difference between crème anglaise and pastry cream is the addition of cornstarch. So, why does adding starch make the denaturation and coagulation of the egg proteins in pastry cream so much easier? The answer lies in the effects of the large starch polymers—amylose and amylopectin. At lower temperatures, the starch polymers interfere with the coagulation of the denatured egg proteins by physically separating the chains of denatured protein. Crème anglaise (made without starch) thickens at 190°F/71°C; the thickening is evidence of protein denaturation and coagulation. If crème anglaise is overheated to 180°F/82°C, the mixture will curdle, that is, you end up with a scrambled egg mixture instead of a smooth cream. In contrast, pastry cream (made with starch) does not begin to thicken until 180°F/82°C and thickens dramatically near the boiling point of water (212°F/100°C). This example shows that the starch is preventing the denatured protein from coagulating into a network. When the temperature is finally high enough, the starch will *gel* to form a three-dimensional network of polymer chains (see Chapter 6—Fruits and Vegetables for a discussion

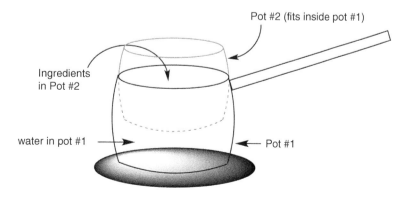

Figure 8.22   Double boiler.

on starch gelation), and this additional polymer network surrounds and stabilizes the denatured protein. Starch is so good at stabilizing the dilute coagulated protein network that stirred or baked custards and creams, including starch molecules, can be heated to much higher temperatures without the risk of curdling. However, the starch does change the final taste and texture.

Baked custards—such as flan or cheesecake—operate on the same principles. Recipes that use no cornstarch must be carefully heated in a water bath to ensure the slow formation of the delicate network of denatured and coagulated egg protein diluted with fat and sugar. Recipes with cornstarch can be baked without a water bath.

If you are using cornstarch in your baked or stirred custard, you should be aware that raw egg yolks contain an enzyme called alpha amylase, an enzyme that breaks starch molecules into individual sugar molecules. When making a custard or cream with added starch, the primary ingredient is egg yolks, but alpha amylase can ruin the product by cleaving the starch molecules. The solution is to heat the raw egg yolks above 180°F/82°C before adding them to the starch; this high temperature denatures and inactivates the alpha amylase enzyme.

8.9 Egg White Foams

Denaturing a protein always uses some energy to break the hydrogen bonds holding the protein together and subsequently unravel its globular structure (Figures 8.23 and 8.24). The unraveled protein now has exposed hydrophobic regions that can stick together or coagulate with exposed hydrophobic regions of other proteins. Agitation can also denature proteins, specifically the aggressive agitation that introduces large amounts of air into the protein mixture, creating a foam. When cooks use a whisk or wire beater to beat egg white, the protein will—over time—stiffen into a white foamy semisolid. This stiffening and change in color from clear to white mirrors what we see in the heat denaturation of egg white protein by energy from heat. When egg white protein is denatured by whipping or beating air into the mixture, the denatured proteins coagulate and trap air bubbles in addition to water. The denatured proteins cluster around and stick their exposed hydrophobic portions into the air bubbles. The large numbers of trapped air bubbles give the coagulated protein matrix a soft, semisolid texture. The semisolid foam of denatured protein and air bubbles can be mixed with other ingredients and then heated. Heating accomplishes additional protein denaturation within the foam, especially of ovalbumin. Ovalbumin is not particularly susceptible to denaturation by agitation, and when the foam is heated, ovalbumin finally denatures, hardening the foam even further. Heating expands the air bubbles in the foam, which causes considerable rise, and it also dries out the foam as water molecules evaporate. When the proteins finally set enough from the heat, the gas bubbles are trapped and can't expand anymore. The result is a puffy/tall, hardened yet airy solid. Meringue cookies like the one shown here are classic examples of cooked egg white foam (Figure 8.25). The structure of the cookie is created almost entirely by egg white protein.

Figure 8.23   An egg white foam (creative commons license). *Source:* katie kuksenok / Flickr / CC BY 2.0.

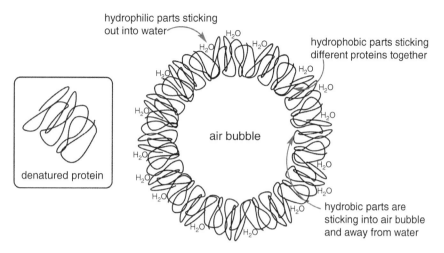

hydrophilic parts sticking out into water

hydrophobic parts sticking different proteins together

air bubble

denatured protein

hydrobic parts are sticking into air bubble and away from water

Figure 8.24 Denatured egg white protein coagulating around an air bubble. This creates a foam.

Figure 8.25 A lemon tart topped with meringue. larik_malasha/Adobe Stock Photos

As we saw with custards, adding other molecules dilutes the proteins, making it more difficult to form the coagulated protein network that stiffens, thickens, or solidifies. In particular, when making an egg white foam, fat molecules are particularly destructive. Adding fat to a foam stabilized by denatured protein is disastrous to the foam. Since fats are very hydrophobic, the denatured protein preferentially clusters around the fat molecules, and the air bubbles escape or never form in the first place (Figure 8.26). Instructions for creating an egg white foam begin with separating the egg yolks from the egg whites. Since the yolk contains all the fat within the egg, removing this fat is essential to creating an egg white foam. This is demonstrated in the preparation of a soufflé. A soufflé is made by first beating egg white into a stiff, glossy foam and then *folding* in a "base" to add flavor to the whites, and the subsequent mixture is then baked. The base typically contains egg yolks, milk, butter, flour, sugar, and flavoring. Because egg white foams are created by trapping air bubbles by denatured protein, the base must be carefully folded in so as not to destroy the foam.

When a soufflé hits the oven, it rises dramatically as the heat expands the air bubbles within the foam. Simultaneously, the heat further denatures the protein, preventing the air bubbles from escaping. The soufflé base typically contains starch and protein to reinforce the walls of the air bubbles in the egg white foam—this gives the soufflé a stronger structure that can stand up straight out of the dish.

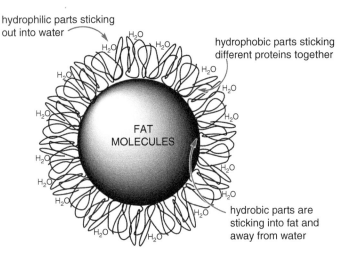

hydrophilic parts sticking out into water

hydrophobic parts sticking different proteins together

FAT MOLECULES

hydrobic parts are sticking into fat and away from water

Figure 8.26 Protein denaturation by agitation in the presence of fat molecules.

Interestingly, it has been known for centuries that egg white proteins form better foams when beaten or whipped in a copper bowl. The secret lies with the protein ovotransferrin (also known as conalbumin). Remember that ovotransferrin is present in egg white to bind iron—that's its job (Figure 8.27). Well, when a cook whips egg whites in a copper bowl, the ovotransferrin binds the copper from the bowl. Copper and iron are both metals, and ovotransferrin will bind copper as easily as iron. When ovotransferrin (a.k.a. conalbumin) binds copper, the structure organizes around the metal ion, forming additional hydrogen bonds and ionic bonds. This increases the stability of ovotransferrin, which makes it denature at a higher temperature. The result is that the egg-white foam is more elastic and stable. When the egg white foam is heated, the air bubbles can expand further before the copper-bound ovotransferrin coagulates; this makes the cooked foam taller or puffier [6].

When ovotransferrin binds the copper from the bowl, the egg-white foam takes on a yellowish-golden hue. The golden yellow copper atoms are trapped by ovotransferrin, and the color is seen in the egg-white foam itself. In the same way, if you crush up an iron supplement tablet and add a pinch to egg whites (the whites should be beaten in a glass or stainless steel bowl), the red color of the iron bound by the ovotransferrin will turn the egg white foam pink.

If you don't have a copper bowl (they are expensive and hard to clean), then the addition of acids like cream of tartar (i.e. tartaric acid) or beating the egg white over a bowl of hot water can improve the formation of an egg white foam by encouraging protein denaturation. In each case, acid or heat disrupts the hydrogen bonds holding the protein structure together, causing denaturation. The acid-induced denaturation of egg protein is similar to the formation of curds in the acidification of milk that occurs when making cheese (see Chapter 5).

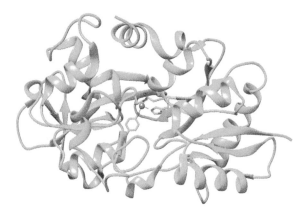

Figure 8.27 The first ~300 amino acid residues of the egg white protein ovotransferrin, represented with a green ribbon. Several amino acid residues are shown as sticks binding a metal ion (gold sphere) at the center. Image created with UCSF Chimera from PDB 1N04.

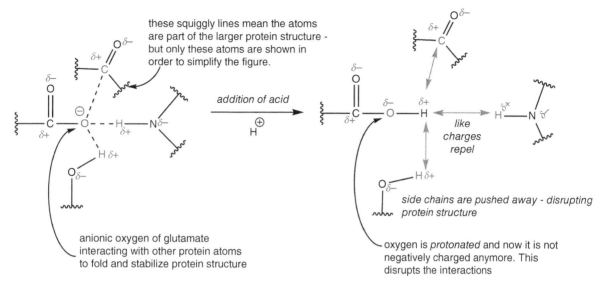

Figure 8.28 Effect of pH on protein coagulation. Several amino acids will ionically interact depending on pH. Tartaric acid is used to reduce the aggregation of egg proteins.

The addition of acid unravels the protein by adding acidic protons (i.e. H$^+$) that change the charges of some amino acid residues—like glutamate, shown in Figure 8.28. The change of charge disrupts the non-covalent interactions that hold the protein structure together. The normal, folded, three-dimensional protein structure has water-loving (hydrophilic) parts on the outside and water-hating (hydrophobic) parts on the inside of the folded protein, but the addition of acid unfolds the protein by disrupting the stabilizing interactions. The acid-denatured protein coagulates with other unraveled proteins by sticking the exposed hydrophobic areas together.

8.10 Egg Pasteurization

Heating eggs at 140°F/60°C for 5 minutes or at 190°F/88°C for 1 minute is essential to kill *Salmonella*. *Salmonella* is a type of bacteria that can be found inside raw eggs. The bacteria come from the chicken that laid the egg—not necessarily from poor handling of eggs in the grocery store or farm. Since the bacteria live inside the egg, washing the surface of the egg has no effect on removing them. Since *Salmonella* bacteria are living organisms, they prefer warm temperatures to grow; bacterial growth is slowed at cold (refrigerator) temperatures, and the bacteria are killed by hot temperatures. There is no way to tell if an egg harbors *Salmonella*—you simply have to prepare egg dishes to kill *Salmonella*, were it present.

Unfortunately, pasteurization does affect the quality of the foam prepared from the pasteurized egg whites, such that whipping a foam takes longer for pasteurized egg whites than for raw egg whites, and pasteurized foam is less stable. The brief increase in temperature during pasteurization denatures ovotransferrin, which has the lowest denaturation temperature (57°C/135°F) of all the egg white proteins and also causes partial denaturation of ovomucin and lysozyme [7]. Lysozyme is a very alkaline protein, so it binds tightly to acidic proteins in egg white, including ovomucin, and it is the ovomucin–lysozyme network that denatures during pasteurization. To improve foaming after pasteurization, metal ions can be added to stabilize the ovotransferrin, and/or the denatured ovomucin–lysozyme network can be removed.

Box 8.2 The roles of eggs in recipes

Eggs can have several roles within a recipe. As we've seen in this chapter, eggs can thicken a custard, emulsify mayonnaise, and provide the structure of a soufflé. But why include eggs in puff pastry and cakes? Why do we find them in the coating on Chicken Française or Chicken Parmesan? In cakes, eggs add leavening, that is, they contribute to the rise of the cake. Beating the cake batter incorporates air bubbles into the egg protein network, generating an egg foam; those air bubbles then expand and are trapped by the solidifying network of protein in the hot oven. In puff pastry, adding egg whites (versus whole eggs) provides additional protein for structure and a drier puff. In Chicken Française and Chicken Parmesan, the raw chicken breast is coated in beaten egg and then dipped in a breading mixture before being placed in the hot skillet. In this case, the denaturation and coagulation of the egg protein help glue the breading to the meat. Nonetheless, eggs are one of the most versatile ingredients in a kitchen.

8.11 Heating Egg Protein Causes Chemical Reactions

What comes to mind when you think of hard-boiled eggs? Perhaps the unpleasant and complex sulfurous odor of cooked eggs? The strangely greenish-yellow yolks that come from overcooking? What about the complex odor of well-cooked eggs? Odor is due to volatile molecules released into the air, and cooking eggs causes the release of a variety of molecules that we see and smell.

Much of the flavor and sensory perception of eggs comes from chemical reactions that produce gaseous or volatile molecules. Cooked whole eggs produce 141 unique, identifiable odor compounds. Most are nitriles (an organic compound with carbon–nitrogen triple bonds) and alkylbenzenes (benzene rings with one or more carbon atoms), and another 33% were either ketones or aldehydes from fatty acids. The Maillard reaction between sugars and amino acids produces pyrazine and thiazole compounds giving another strong egg-like flavor and odor (Figure 8.29).

Heating proteins causes them to denature and coagulate into a solid, but heating also causes chemical reactions within the proteins. In particular, the characteristic "eggy" odor of hard-boiled eggs can be attributed to the degradation of cysteine amino acid residues within the albumen protein to produce H_2S (hydrogen sulfide; Figure 8.30).

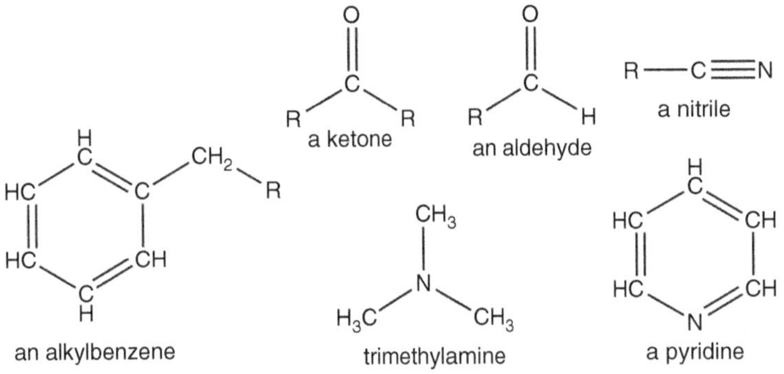

Figure 8.29   Classes of odor molecules produced by cooking eggs.

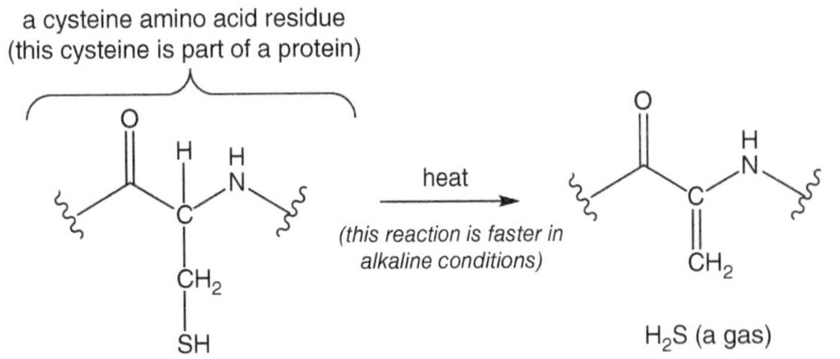

Figure 8.30   The reaction of cysteine and heat to produce H_2S.

The sulfurous odor of H_2S is what we associate with cooked eggs. In high concentrations, H_2S can be objectionable. The cysteines within the egg white albumen protein react more readily to produce H_2S when the protein has unfolded (denatured). This exposes cysteines previously buried inside the protein and allows them to react.

The reaction to produce H_2S also happens more readily in alkaline conditions (pH > 7), and older eggs inevitably turn alkaline due to the loss of carbon dioxide. Some cookbooks recommend adding lemon juice or vinegar to the water when hard-boiling eggs to reduce the "eggy" smell. The acid of the vinegar or lemon juice is supposed to both lower the pH and reduce the formation of H_2S (Figure 8.30).

The breakdown of egg white protein upon heating produces nitrogen-containing compounds, including ammonia, methylamine, and dimethylamine. The latter can be further altered to form the fishy-smelling

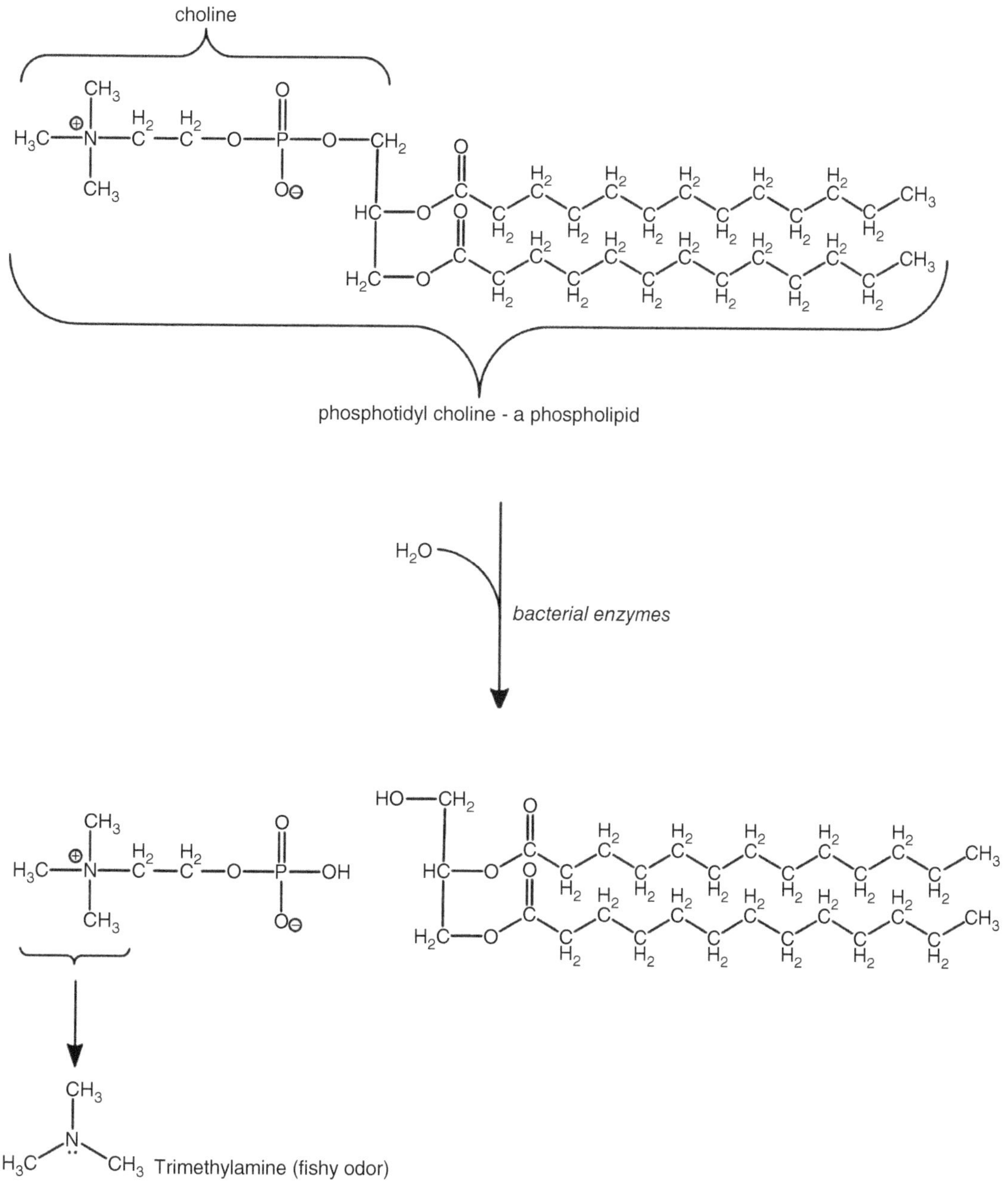

Figure 8.31   Formation of fishy TMA in aged eggs.

trimethylamine (TMA) (see Chapter 7 for more on TMA). When the egg white is more alkaline (pH around 9–10), heating the egg produces several pyridine types of molecules (Figure 8.29). These strong-smelling, nitrogen-containing, cyclic molecules are only partially soluble in water, making the escape of their distinctive, fishy odor slow and continuous. The yolk contains more fats than proteins and, upon cooking, produces odor molecules from the breakdown of fatty acids, including ketones, aldehydes, and short-chain fatty acids. Bacterial decomposition of phospholipids such as phosphatidylcholine inside a contaminated egg or the human gut can produce choline and eventually the characteristic fishy odor of TMA (Figure 8.31).

Figure 8.32   Green eggs and FeS.

Sometimes when hard-boiling eggs, a green-gray discoloration can occur at the interface of the yolk and white and sometimes discolor the entire yolk. This greenish color is caused by the compound iron sulfide (FeS). The reaction to produce FeS happens when the eggs are heated for extended periods. Boiling eggs in water for 15 minutes is a sure way to make FeS. The source of the sulfur (S) in FeS comes from the H_2S released by albumen cysteines, while egg yolk is particularly iron (Fe) rich. The H_2S gas migrates from the white into the yolk, forming the green iron sulfide (FeS; Figure 8.32).

Science for the Chef: Lemon angel food cake with lemon curd

Figure 8.33   Lemon angel food cake topped with lemon curd and raspberries. valentinamaslova/Adobe Stock Photos

Ingredients:

Lemon angel food cake	Lemon curd
140 g (1 ⅛ cups) cake flour, sifted	4 egg yolks, at room temperature
425 g cold egg whites (about 2 cups, from 12 fresh eggs)	Zest of 1 lemon, about 1 tablespoon
425 g (2 cups) granulated sugar	80 ml (1/3 cup) fresh lemon juice (about 2 large lemons)
25 g (or 1 ounce) freshly squeezed lemon juice	134 g (⅔ cup) sugar
1 g (¼ teaspoon) kosher salt	1/8 teaspoon fine salt
1 tablespoon (15 ml) pure vanilla extract	86 g (6 tablespoons) unsalted butter, cut into 6 pieces, at room temperature
1 ½ teaspoons grated lemon zest (two lemons)	
Stand mixer with whisk attachment	Double boiler
Flexible spoonula	Whisk

Instructions:

Part 1: Make the angel food cake

The structure of the angel food cake is made from an egg white foam, egg white protein that has denatured and coagulated around air bubbles due to the energy of whipping and the addition of acid.

Step 1: Separate the eggs and dissolve the sugar

- Separate egg whites from egg yolks. Combine egg whites, sugar, and vanilla extract in the bowl of a stand mixer fitted with the whisk attachment. The recipe calls for fresh eggs because separating the yolk from the white is easier with a fresh egg. The yolk membrane of a fresh egg is tight and intact, and the yolk has not absorbed much water yet because the eggs are fresh. As the chef separates white from yolk, the yolk of a fresh egg is more likely to separate without breaking.
- Whisk on low for 1 minute to just combine and loosen the egg white, then increase speed to medium-low and whip for 3 minutes. At this stage, the sugar will have dissolved in the water of the egg white, but much of the egg white protein is still soluble.

Step 2: Denature egg white protein with acid and whipping

- With the mixer still running, add the lemon juice and salt, then immediately increase the speed to medium and whip for 3 more minutes. Adding citric acid from the lemon juice encourages protein denaturation due to adding H^+ and a reduction in pH. Coupled with the increased whipping speed, the egg white protein begins to denature and coagulate, and the meringue becomes foamy and white in color. Some metallic mixing bowls—like those made of stainless steel, copper, or aluminum—will react with the acid to release metal ions [8,9]. Those metal ions dissolve into the food and give it a metallic taste. Use nonreactive cookware and whisks made of porcelain or silicone to avoid the leaching of metal ions.
- Increase the speed on the mixer again to medium-high and whip until the meringue is glossy white and thick enough that a whisk dragged through the meringue leaves a pattern; this should take 2–4 minutes. The meringue should run off the whisk and make soft mounds in the bowl. The agitation from the whipping continues to denature some of the egg white protein, ovotransferrin, primarily. The trapped air and denatured, coagulated protein are stabilized by the surrounding, hydrogen-bonded, sugar–water matrix. The meringue is not too stiff because it needs to expand easily as the cake rises in Step 4. Because the angel food cake requires an egg white foam, removing the yolk in Step 1 was essential. It would be nearly impossible to form a good egg white foam in the presence of egg yolks because the denatured protein would clump around the molecules of fat from the yolk instead of trapping air bubbles, generating a collapsed foam.

Step 3: Fold in flour

- Remove the whisk attachment. Sprinkle the cake flour gently on top of the meringue and stir loosely to combine with the flexible spoonula, then fold through the center until no pockets of flour remain. The cake flour is a low-protein flour, meaning it has less glutenin and gliadin. Gentle mixing ensures minimal gluten network formation and allows the cake structure to be dominated by the egg protein. Folding is used to maintain the essential air pockets within the foam. On the contrary, aggressive mixing can deflate the foam.

Step 4: Bake and cool

- Scrape the batter into a 10-inch aluminum tube pan with a removable bottom; do not butter or grease the pan. Bake in a preheated 350°F oven until the cake is light golden color, puffed up, and firm to the touch. This should take about 45 minutes or until the cake reaches an internal temperature of 206°F. Fats—such as those in nonstick cooking spray or butter—are disastrous to egg-white foams. Denatured, coagulated protein will preferentially cluster around hydrophobic fat molecules instead of around the air bubbles.

The cake will rise as heat causes the expansion of air bubbles within the foam. There is no additional rising (leavening) agent like yeast or baking powder. As the cake batter heats up in the oven, the increase in temperature will finally denature the remaining egg white protein, mostly ovalbumin, and set the foam into a puffed, airy solid.

- Invert the pan onto stilts or a trio of cans and cool completely while upside down—at least 2 hours. Loosen the cake from the pan using a butter knife or small offset spatula, remove the insert, and flip it onto a serving plate. Slice with a serrated knife using a sawing motion. Cooling upside down ensures the delicate network of set egg protein does not collapse and compress the cake's tender crumb.

Part 2: Lemon curd

Lemon curd is a creamy, rich topping made from egg yolks, lemon juice, sugar, salt, and butter. The thickness of the curd is a function of denatured and coagulated egg yolk protein that is diluted with water, fat, and sugar. The looser curd is pourable due to the weaker network of denatured and coagulated protein.

Step 1: Prepare double boiler

- Place 1–2 inches of water in the bottom pot of your double boiler and bring to a boil, then reduce heat to maintain a simmer. If you don't have a double boiler, place a heat-safe glass bowl over a pot of water, making sure the bowl does not directly contact the boiling water in the lower pot. Double boilers allow the chef to cook indirectly with the heat of steam (212°F/100°C) rather than the direct heat of the stovetop.

Step 2: Dissolve sugar and acid. Heat to denature egg yolk protein

- Whisk together the egg yolks, lemon zest, lemon juice, butter, sugar, and salt in the top of a double boiler. As the mixture heats up, the sugar will dissolve, and the butter will melt. Whisk continuously to prevent the eggs from curdling. Maintain whisking until the mixture thickens. At 10 minutes, the whisk should leave a defined trail in the thickened mixture, and by 13 minutes, the curd will be fully thickened. The thickening of the lemon curd comes from the denaturation and coagulation of the egg yolk protein. Egg yolks contain fat and water in addition to protein, and the recipe also has added sugar. These extra molecules dilute the protein, so the thickened result is looser and less firm.

Step 3: Cool

- Remove from the heat. The curd will thicken further as it cools. Transfer the curd to a heat-proof bowl and cover the surface with plastic wrap to prevent skin from forming. Refrigerate until chilled and firm. Serve with lemon angel food cake and fresh fruit.

Key Concepts

1. An egg is a type of specialized **germ cell** for **sexual reproduction**. An **egg cell** contains *half* the genetic material necessary to make an individual.

2. A chicken egg contains the **germ cell** within the **yolk**. The yolk is surrounded by a **yolk membrane** and suspended by **chalazae** within the egg white or **albumen**. The albumen is contained within **inner** and **outer membranes**, followed by a hard shell coated by a **cuticle**.

3. Eggs are a balanced source of amino acids, and they include a plentiful supply of **linolenic acid**—an essential polyunsaturated omega-3 fatty acid—as well as several minerals and most vitamins.

4. Egg yolk is 50% water; the rest of the yolk is primarily lipids, with some protein, carbohydrates, and the **phospholipid lecithin**.

5. Egg white is 90% water; the remaining 10% is a mixture of proteins called albumen.

6. Egg membranes are made of protein; the porous eggshell is comprised of protein and the ionic compound, calcium carbonate, and the cuticle coating on the eggshell is also made of protein and pigment molecules.

7. Eggs lose carbon dioxide as they age, making the egg white's pH more alkaline. More alkaline egg white is runnier due to changes in the interactions between the egg white proteins.

8. Not only do eggs lose water to evaporation as they age, but water also crosses from the white into the yolk by crossing the yolk membrane. This age-dependent gain of water by the yolk stretches the yolk membrane and indicates egg freshness.

9. Egg proteins (and other water-soluble proteins) fold into three-dimensional **globular structures** made of **α-helices**, **β-sheets**, and loops held together by **hydrogen bonds**. The protein folds to hide **hydrophobic** parts from the water and expose **hydrophilic** parts to water.

10. The egg yolk contains all of the fat within an egg. Since fat is hydrophobic, and water and fat do not mix, the yolk fat is suspended in the water phase of the yolk within emulsifying **lipoproteins**.

11. **Denaturation** is the unraveling of ordered **secondary structure** within a protein due to the addition of energy that breaks non-covalent interactions between amino acid residues. The denaturation process exposes previously buried hydrophobic regions of protein structure. The hydrophobic regions of many denatured proteins stick together in **coagulation**, trapping water and other molecules between the coagulated protein chains.

12. Denaturation and coagulation of egg proteins can be accomplished by heat, as in the cooking of eggs in a hot skillet, or by **agitation**, as in the whipping of egg white proteins to make a meringue.

13. **Dilution** is the addition of extra molecules of water, sugar, fat, air, etc., to the egg protein. More energy is required to denature and coagulate diluted egg protein; the resulting coagulated solid is more fragile and tender.

14. When egg white protein is denatured by whipping or beating air into the mixture, the denatured proteins create a **foam** by coagulating and trapping air bubbles between the denatured and coagulated proteins.

15. Egg white protein denaturation and coagulation are affected by the presence of metal ions (i.e. copper and iron cations), which stabilize the egg white protein **ovotransferrin**, and acid (i.e. H^+), which encourages protein denaturation by lowering the pH and altering the charge state of amino acid residues within egg white proteins.

16. Degradation of cysteine amino acid residues within the albumen protein produces **hydrogen sulfide** or H_2S, the characteristic "eggy" odor of hard-boiled eggs. The H_2S can also react with iron in the yolk to form the greenish-gray **iron sulfide** or FeS.

17. Fishy-smelling amines and pyridines are the byproducts of protein, fat, and phospholipid breakdown in cooked eggs.

References

1 McGee, H. (2004). *On Food and Cooking*. New York: Simon and Schuster Inc.

2 Kenneth, F. K. and Kriemhild Coneè, O., ed. (2000) *The Cambridge World History of Food, Volume I*. New York: Cambridge University Press.

3 Smith, P. & Daniel, C. (1975). *The Chicken Book*. Boston: Little, Brown and Company.

4 Donovan, J. W., Mapes, C. J., Davis, J. G., Garibaldi, J. A. (1975). A differential scanning calorimetric study of the stability of egg white to heat denaturation. *J Sci Food Agric*, *26*, 73–83.

5 Huopalahti, R., López-Fandiño, R., Anton, M. & Schade, R. (2007). *Bioactive Egg Compounds*. New York: Springer Science Business Media.

6 Harold, J. M., Sharon R. L., Winslow, R. B. (1984). Why whip egg whites in copper bowls? *Nature*, *308*, 997–998.

7 Lomakina, K. and Míková, K. (2009). A study of the factors affecting the foaming properties of egg white—a review. *Czech J Food Sci*, *24*, 110–118.

8 Kamerud, Kristin, L., Kevin, A., Hobbie Kim, A. Anderson. (2013). Stainless steel leaches nickel and chromium into foods during cooking. *Journal of agricultural and food chemistry*, *61*(39), 9495–9501.

9 Mazinanian, Neda, Yolanda S., and Hedberg. (2016). Metal release mechanisms for passive stainless steel in citric acid at weakly acidic pH. *Journal of The Electrochemical Society*, *163*(10), C686.

Additional Readings

Bhat, Z. F., Morton, J. D., Bekhit, A. E.-D. A., et al. (2021), Effect of processing technologies on the digestibility of egg proteins. *Compr Rev Food Sci Food Saf*, 20(5), 4703– 4738.

Cardoso, M.J., Nicolau, A.I., Borda, D., Nielsen, L., Maia, RL, Møretrø, T, Ferreira, V, Knøchel, S, Langsrud, S., & Teixeira, P. (2021). *Salmonella* in eggs: From shopping to consumption—A review providing an evidence-based analysis of risk factors. *Compr Rev Food Sci Food Saf*. 2021 20(3), 1– 26.

Croguennec, T., Nau, F. and Brulé, G. (2002), Influence of pH and Salts on Egg White Gelation. *J. Food Sci*, 67(2), 608– 614.

Yao, L., Wang, T., Persia, M., Horst, R.L. and Higgins, M. (2013), Effects of Vitamin D3-Enriched Diet on Egg Yolk Vitamin D3 Content and Yolk Quality. *J. Food Sci*, 78(2), C178– C183.

End of Chapter 8 Questions

1. There are three important proteins in the egg: ovalbumin, ovotransferrin (i.e. conalbumin), and ovomucin. If you were to slowly heat an egg, you would find that ovomucin denatures first at 54°C, followed by conalbumin (at 63°C), then ovalbumin (at 80°C). Which of the following statements most accurately explains the order of protein denaturing?
   a. Ovalbumin has fewer amino acids than the other proteins
   b. The proteins with less intermolecular forces take less energy (heat) to denature the protein
   c. Proteins without disulfide (R-S-S-R) groups will require higher heat to denature, thus ovalbumin must not have disulfide groups
   d. The greater the hydrophobic side chains, the greater the melting (denaturing) point of a protein
   e. None of the above

2. When cooking, heat and/or acid does what to the structure of a protein?
   a. Binds to fats
   b. Always generates curds
   c. Often unfolds and denatures proteins
   d. Leaves the proteins in solutions where they cannot interact with each other

3. Consider the reactions shown below and the questions that follow:
   i. Reaction #1 is an example of a chemical process that takes place inside a raw egg. The carbon dioxide gas and water molecules both escape through the pores of the eggshell.

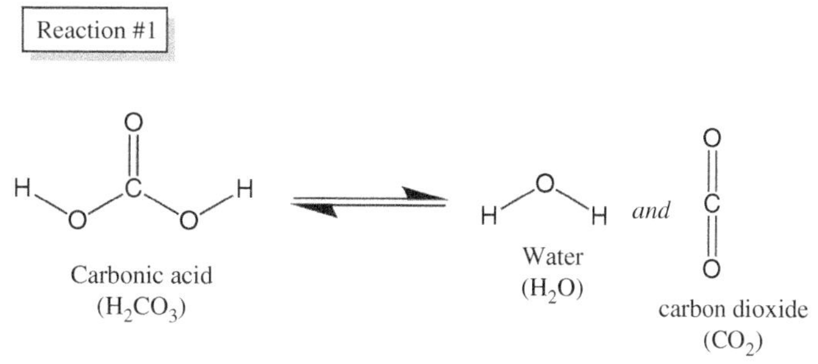

   Reaction #1

   Carbonic acid (H_2CO_3) ⇌ Water (H_2O) and carbon dioxide (CO_2)

   a. Why does the escape of water and carbon dioxide from the egg make the contents of the egg more alkaline? Discuss pH in your answer.
   b. The yolk of Grade B eggs will break immediately upon hitting the pan (in comparison to Grade A or AA eggs). Grade B eggs are older than Grade A or AA eggs
      i. How are Grade B eggs different than Grade A or AA with respect to reaction #1?
      ii. Why does the yolk break so easily in a Grade B egg?
   c. Grade B eggs smell more "eggy" when cooked. What is causing the eggy smell, and why do you smell more of it from a Grade B egg?

4. Overbeaten or overwhipped egg whites are grainy from little clumps of egg protein that have tightly coagulated. Which strategy or strategies can help prevent this problem and help you form perfect, stiff peaks?

   Select all correct answers:
   a. Aging the egg whites
   b. Using a copper bowl
   c. Adding water
   d. Adding a small amount of yolk
   e. Adding acid

5. The binding of a metal cation increases the denaturation temperature of ovotransferrin—an egg white protein. Which of the following statements are true?

    a. Metal-bound ovotransferrin requires more energy to denature because it contains fewer non-covalent interactions between its atoms

    b. Metal-bound ovotransferrin requires more energy to denature because it contains more non-covalent interactions between its atoms

    c. Metal-bound ovotransferrin requires more energy to denature because the number of amino acids in the protein is greater

    d. The number of covalent bonds in the protein doesn't change. The greater number of non-covalent interactions (due to the organizing effect of the metal) means more energy is needed to break them

6. The two parts of Reaction #2 take place in an egg upon hard boiling. As some cooks have observed, older eggs that have been sitting in the refrigerator for a while smell more "eggy" when they are hardboiled.

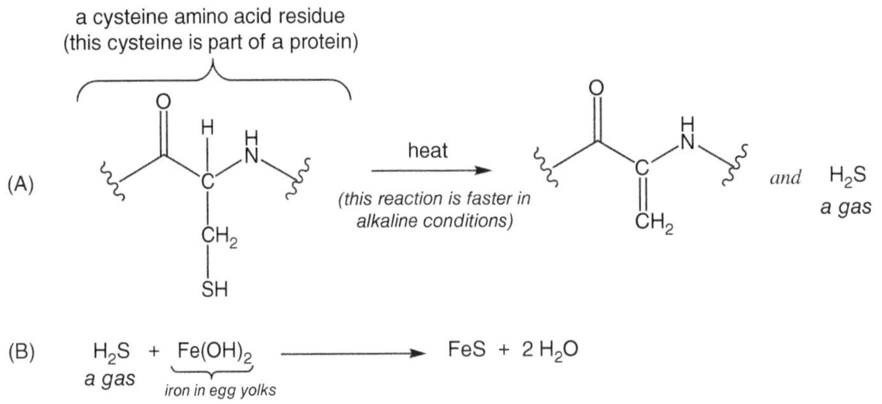

    i. What molecule is creating the eggy smell?

    ii. Reaction (A) is faster under alkaline conditions. How and why does egg white become alkaline?

    iii. Why do you create more of this eggy-smelling molecule when the eggs are older?

    iv. If you hardboiled an egg, then peeled and cut it open, how could you tell if reaction (B) had occurred? Explain the chemical reason behind your observations.

7. Mayonnaise is an emulsion made of three main ingredients: oil, vinegar, and egg yolks (lemon juice and mustard are sometimes added as well). Explain how the three main ingredients work together to make the stable emulsion.

9

Bread, Cakes, and Pastry

Guided Inquiry Activities (Web): 6, Higher Order Protein Structure; 7, Carbohydrates; 9, Fats Structure and Properties; 18, Starch; 27, Bread

Learning Objectives

1. Understand the chemical composition of and the structural components of a grain and their roles in seed protection and nutrition.
2. Recognize the chemical transformations that occur during grain milling processes to produce various types of flour, such as white flour and whole wheat flour.
3. Describe the differences in protein composition between "soft" and "hard" flours and their impact on the texture and baking properties of the flour.
4. Understand the components and molecular structure of starch in flour, including their respective roles in the formation of starch granules.
5. Identify and analyze the impact of protein interactions of the flour proteins, gliadin, and glutenin, on the formation of gluten in dough.
6. Discuss the role of disulfide bonds formed between cysteine residues in the stability and mechanical rigidity of the gluten network in bread dough.
7. Understand the process of yeast fermentation and its role in bread making, including the conversion of starch and sugar into CO_2 and alcohol through yeast metabolism.
8. Know the mechanism behind the expansion of bread during baking by applying the ideal gas law to the behavior of gases in the dough.
9. Explain the chemical processes involved in the formation of the crust and crumb of bread, including the Maillard reactions and their contribution to color, flavor, and aroma.
10. Evaluate the impact of different ingredients, such as salt, fat, and sugar, on the gluten network, dough texture, bread staling, and desired baking outcomes.
11. Explain celiac disease and the challenges of baking high-quality gluten-free baked goods.
12. Relate the role of chemical leavening agents in the production of carbon dioxide gas, and the importance of mixing order and technique in generating a well-raised, light baked good, even in the absence of a leavening agent.
13. Analyze factors contributing to moist and dry cakes through the study between moist and dry cakes based on ingredient ratios and types.
14. Describe the chemical interactions between fat, starch granules, and water that lead to the separation of starch particles and the formation of flaky layers in pie crusts and puff pastries.

The Science of Cooking: Understanding the Biology and Chemistry Behind Food and Cooking, Second Edition. Joseph J. Provost et al.
© 2025 John Wiley & Sons, Inc. Published 2025 by John Wiley & Sons, Inc.
Companion website: www.wiley.com/go/provost/food_science_2e

9.1 Introduction

Few of us can walk by a bakery without being drawn to the aromas and visuals of freshly baked bread, pies, or cakes. Sometimes, you might bite into a chewy, dense wheat bread. You might be drawn to a flaky, buttery croissant on other occasions. On a third occasion, you might choose a light and fluffy cake. Flour and water provide the basis for all of these baked goods. Yet, they have significantly different textures, aromas, and mouthfeel due to additional ingredients, treatment of, and structure of the dough or batter. In this chapter, you will learn about flour and its molecular components, the process of making yeast-based bread, quick bread, sponge cake, and pastry, and variations in other ingredients that help to yield the final baked product with the desired texture, aroma, and flavor. The flour that is used in most baked goods comes from wheat. Not only is wheat readily available, but it produces flour with the components and properties ideal for different types of baked goods, ranging from leavened breads to flaky pie crusts. Thus, most of this chapter will center on wheat-based flour and baked goods. However, because many individuals have celiac disease, we will also discuss some other flour options and characteristics of the disease itself.

9.2 Wheat-based Flour, Where It Comes From and Its Components

All flours come from a grain that is milled or ground into a fine powder; different flours are produced when each component of the grain or various types of wheat are milled. First, let's talk about the structural components of a grain.

A grain (or seed) consists of three major structural components: the germ (or embryo), the endosperm, and the bran (Figure 9.1). The plant sprout emerges from the part of the grain called the germ or embryo. The germ is nutrient-rich, containing fats, vitamins, and mostly soluble proteins. The endosperm makes up most of the bulk of the grain and contains cells that are filled with carbohydrates and proteins. Carbohydrates, in the form of starch, make up about 70–80% of the endosperm. The protein component of the endosperm (about 7–15%) consists of soluble and insoluble proteins. The bran makes up the grain's outer shell; it contains some minerals (~7%), soluble proteins (~16%), and carbohydrates but mostly consists of the structural, insoluble carbohydrate called fiber. In a seed, fiber has the primary role of protecting the endosperm and germ components.

You might be asking, what do the structural components of a seed have to do with flour? Different flours are distinguished by the component of the grain milled during the production process. For example, white flour

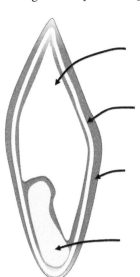

Endosperm
- *Starch (Complex Carbohydrates - flour)*
- *Limited proteins*
- *B vitamins*

Aleurone
- *Outermost layer of endosperm*
- *Lining made of specialized cells*
- *Contains most of nutritional whole grain components*

Bran – Just under the hull/husk
- *Insoluble carbohydrates (Fiber)*
- *Trace minerals, phenolic compounds*
- *B vitamins*

Germ
- *"New plant" embryo*
- *Nutty sweet flavor*
- *Vitamin E, Folic acid, Thiamine, metals*

Figure 9.1 Anatomy of a grain. Flour is primarily composed of starch and proteins found in the endosperm.

Table 9.1   Protein content and use of white, wheat-based flours.

Flour type	Protein (% of total)	Primary use in cooking
Cake	7.5–8.5	Cakes, quick breads, muffins, and pancakes. Produces a tender crumb
Pastry	8–10	Pie crusts and pastries
Instant	9.5–11	Sauces and gravies
Bleached, all-purpose	9.5–12	General use flour for all cooking and baking purposes but contains a little too much protein for the best pie crusts, muffins, and pancakes and too little flour for the best yeast breads.
Bread flour	11.5–12.5	Yeast breads, pasta, pizza
Durum wheat (semolina)	13–13.5	Pasta

comes almost solely from the endosperm portion of a grain of wheat, while whole wheat flour contains the ground bran, germ, and endosperm. You have probably seen or used other types of flour, including all-purpose, bread, cake, semolina/durum, pastry, and instant. These white, wheat-based flours are distinguished by their protein content (Table 9.1).

Wait a minute, protein content! How do you get flours of different protein content when these flours all come from wheat? Each component of a wheat grain contains a different amount and type of protein (soluble and insoluble). Moreover, differences in wheat type, growing climate, and soil conditions will all impact the amount of protein present within a grain of wheat. The endosperm of spring-sown wheat is more brittle, disintegrates readily on milling, and typically has a higher protein content; thus, spring-sown wheat produces "hard" flours. "Soft" flour is usually obtained from winter wheat, as these flours have less than 10% protein content.

Try this at home. Place a few tablespoons of bread, all-purpose, and cake flour in different unlabeled dishes. Feel the texture of each. Can you tell the difference between the softer flours (low-protein content) and the hard flours (high-protein content)?

The "all-purpose" white flour you purchase in the grocery store is usually blended flour to give a more consistent product in terms of protein content. Most flours typically do not contain any other additives, aside from drying agents, to prevent the flour from forming lumps when exposed to moisture in the air unless the flour is designated as "self-rising." Self-rising (self-raising in some countries) flour contains a raising agent like baking powder; this flour is sometimes called for in quick bread, biscuit, or cake recipes.

There is clearly a lot to learn about the protein in flour. However, you can't consider protein in the absence of carbohydrates. Let's learn more about flour's carbohydrate and protein components and how each impacts the characteristics of a loaf of bread, pie crust, or cookie.

9.3   Carbohydrates in Flour

9.3.1   Starch

You already know much about carbohydrates and their importance in cooking, food, and nutrition. To refresh your memory, glucose is a carbohydrate (it contains carbon, hydrogen, and oxygen atoms) in which five of the carbons and oxygen are most commonly arranged in a "ring" or cyclic structure (Figure 9.2). When glucose is not bonded to any other carbohydrate molecules, it is a monosaccharide.

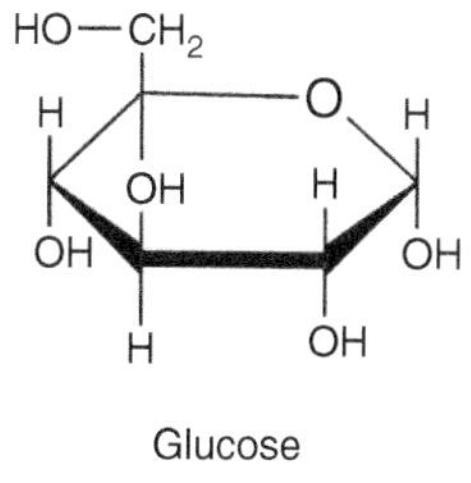

Glucose

Figure 9.2   The structure of glucose.

However, most of the glucose molecules in flour don't exist in this monosaccharide molecular form; they are present in the form of a glucose polysaccharide called starch. You have likely heard of the term "starch," used the term "starch," and maybe even tried reducing "starch" from your diet, but have you ever thought about what starch looks like at the molecular level?

Starch is composed of two different types of glucose polysaccharides, amylose, and amylopectin, both of which consist of glucose molecules that are covalently bonded to one another via glycosidic bonds to make long molecular chains of glucose (Figure 9.3). A single molecule of amylose may contain up to several thousand glucose molecules that are bonded together via $1 \rightarrow 4$ glycosidic links. An amylopectin polysaccharide can contain many more glucose molecules (on the order of a million) bonded via $1 \rightarrow 4$ glycosidic bonds and $1 \rightarrow 6$ glycosidic bonds. The term glycosidic bond is used to describe a covalent bond that connects a saccharide (sugar) molecule to another molecule. The numbering system describes the point of connectivity of the two molecules.

A starch granule in flour, which contains many individual molecules of amylose and amylopectin that comprise 98–99% of the total molecular components, has a size of 2–40 μm and consists of 10–20% amylose and

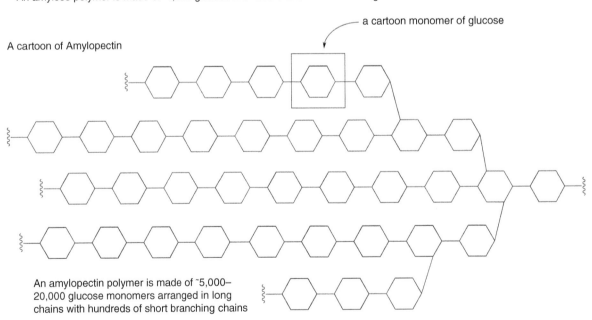

A cartoon of Amylose

An amylose polymer is made of ˜1,000 glucose monomers attached in one long extended chain

a cartoon monomer of glucose

A cartoon of Amylopectin

An amylopectin polymer is made of ˜5,000–20,000 glucose monomers arranged in long chains with hundreds of short branching chains

Figure 9.3 Starches of flour. Amylose and amylopectin are both polymers of glucose monomers. The branched amylopectin has two types of glycosidic bonds ($1 \rightarrow 4$ and $1 \rightarrow 6$). Amylose molecules pack together more tightly due to their more linear structure, while amylopectin packs more loosely and provides many "ends" for the storage and quick release of glucose by digestive enzymes.

80–90% amylopectin. Amylose has a more linear three-dimensional structure, while the $1 \rightarrow 6$ glycosidic bonds in an amylopectin molecule result in a branched structure that looks like a bristled brush. The different polysaccharide molecules interact loosely via intermolecular interactions with themselves, with one another, and with the small amount of protein present within the flour. However, because of the differences in three-dimensional structure, amylose molecules pack more tightly against one another and any interacting protein molecules than amylopectin. Thus, amylose molecules and any interacting proteins present in the starch granule are more resistant to degradation by enzymes since the enzymes have a difficult time accessing the molecules to break them down (Box 9.1).

Box 9.1 Starch, flour, and gravy

The amount and type of protein in a starch granule, along with its location, are critical to cooking. You see this when you make a flour-based gravy or sauce. When cold water is added to flour, it is absorbed by a small amount of water-soluble protein that is present, but it can't penetrate the amylose and amylopectin carbohydrates due to the tightly packed structure of the carbohydrates. Once the proteins absorb water, they become sticky and cause the starch granules to clump together. Can you say "lumpy gravy"? Once a large lump has formed, the molecules near the center of the lump become protected from the outside, so the lump will not break down, even when you add more water or try to whisk it away. What are some possible ways to prevent the problem?

1. Use a low-protein flour (e.g. Wondra™). Less protein decreases the likelihood that the proteins will absorb all of the water, allowing some water to penetrate and interact with the carbohydrates.
2. Initially dissolve a small amount of flour in a large amount of water with rapid mixing. If there is enough water and stirring for the water to penetrate the starch granule, the starch granules will begin to swell and absorb enough water that they will not clump (however, they will separate from the water if you leave the mixture to sit in a bowl for a few minutes).
3. Dissolve the flour in warm water or broth, which will allow for more complete water penetration into the starch granule due to the heat of the water, which will break up some of the weaker intermolecular interactions that hold the granule together. Since the granules can absorb more water, they swell and disrupt the intermolecular interactions of the starch molecules on the inside of the granule, allowing for even more water absorption.

Why doesn't the gravy thicken until you boil it? Around 120°F/49°C, the starch granules become gel-like, as you have an interconnected network of starch and water molecules. We observe this change as a thickened gravy, which also becomes more translucent because the starch molecules are less tightly packed.

Two additional classes of proteins make up approximately 75% of the total protein content of a wheat grain, which is key to the structure of the starch granule and the behavior of flour in cooking and baking. These proteins are water-insoluble, structural proteins known as gliadin and glutenin. You may recognize the name's similarity to "gluten," the term associated with products made from wheat-based flour. In the following section, you will learn that gluten is not a protein but is formed when the two classes of proteins, the gliadins and glutenins, interact with each other (Figure 9.4).

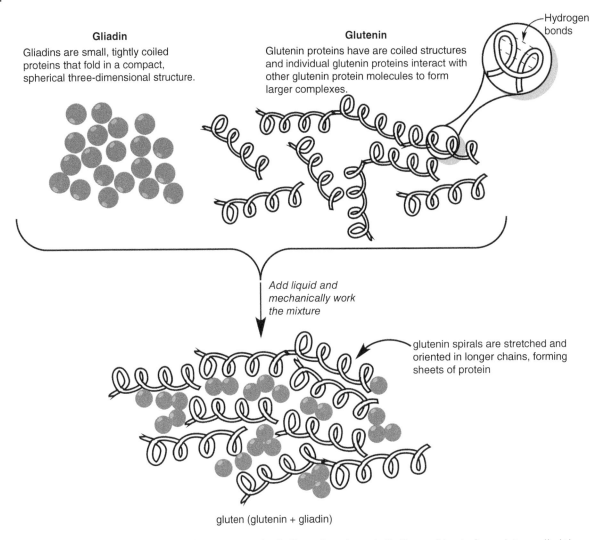

Figure 9.4 The muscle of flour. Two of the key proteins in flour, glutenin, and gliadin, combine to form gluten, called the muscle of flour.

9.4 Wheat Proteins and Gluten Formation

How significant is the impact of gliadin and glutenin proteins in baked goods? Huge! The gliadin and glutenin proteins greatly affect the texture and consistency of the final product, playing a crucial role in whether a pie crust will be flaky and tender or dense and chewy and whether a loaf of bread will have a light, tender crumb or a dense, heavy texture. If you are a baker, it is very useful to have some knowledge of the structure and behavior of these proteins so that you can create the baked result that you desire.

Gliadins are small, tightly coiled proteins forming a compact, spherical shape. Each gliadin protein molecule can fold and maintain stability in a monomeric form due to internal interactions such as hydrogen bonds and electrostatic interactions. There are numerous types of gliadins, with slight variations in sequence and structure, and a single variety of wheat may contain over 40 different types.

Glutenin proteins, on the other hand, have coiled and helical components, thus, their structures are less tightly packed. Individual glutenin molecules interact with each other to form larger protein aggregate complexes. These

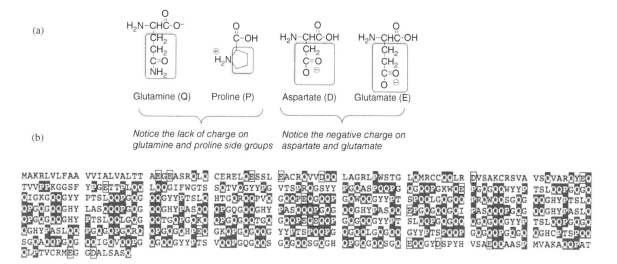

Figure 9.5 The nature of glutenin. (a) The amino acid structures of glutamine and proline possess noncharged side chains (the nitrogen on proline is tied up with the peptide backbone in a protein) and the acidic, negatively charged amino acids of aspartic acid and glutamic acid. (b) The protein sequence in single-letter amino acid abbreviations for glutenin. Notice the high proportion of Q and P (highlighted in blue shading) and the low abundance of charged D and E (yellow shaded). [1]

complexes have distinct structural and protein sequence features containing a central domain of repeated sequences of amino acids, particularly glutamine (Q) and proline (P), with a low proportion of negatively charged amino acids (Aspartate-D and Glutamate-E). See Figure 9.5 to analyze the high quantity of Q and P versus D and E amino acids in the amino acid sequence for glutenin. The glutamine amino acids in glutenin have two important roles. They act as an important nitrogen source for a germinating grain of wheat, and, structurally, the amide side chain of glutamine (R—NH—R′) allows for the formation of hydrogen bonds between different glutenin molecules and between glutenin and water (Figure 9.6). However, even though glutenin can hydrogen-bond with water, the glutenins remain insoluble due to the low relative proportion of charged amino acids within this central domain of the protein complex.

Are there any other structural domains within the glutenin complex that impact protein behavior? Yes, the glutenin central domain is surrounded by N- and C-terminal domains that contain a high proportion of the amino acid, cysteine (Figure 9.7). Remember cysteine? It has a sulfur-containing side chain and readily forms strong, covalent,

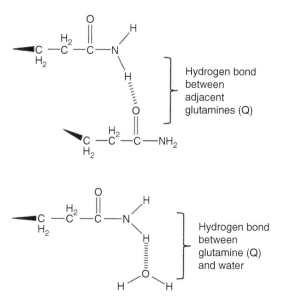

Figure 9.6 Glutamine of glutenin. The amine side group of glutamine can hydrogen-bond with another glutamine forming cross-links between strands, within strands, and with water.

disulfide bonds with other cysteine residues in oxidizing environments (Figure 9.8). These structural domains within the glutenin complex further impact the behavior of the protein.

In a dry flour or a "just-combined" flour/water mixture, the gliadin molecules and glutenin protein complexes do not significantly interact with one another, starch molecules, or water molecules; they have a random and disorganized arrangement. However, once a flour/water mixture begins to be worked through stirring, mixing, or kneading,

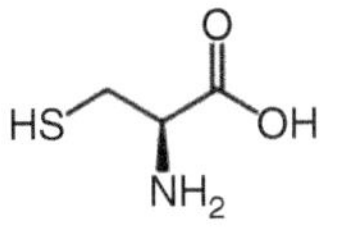

Figure 9.7 Cysteine.

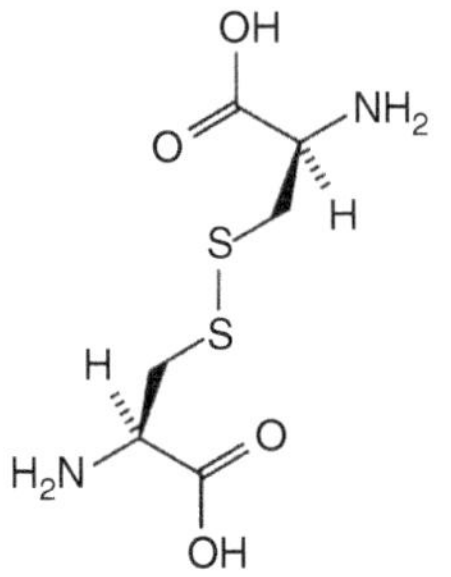

Figure 9.8 Cysteine disulfide. Two cysteine amino acids can bond through the S atoms. In bread, this covalent bond cross-links proteins making them more stable and mechanically rigid, impacting the characteristics of the bread.

the gliadins and glutenin complexes begin to interact via intermolecular interactions. How and why does this happen? As you work with a dough, the gliadin and glutenin proteins are stretched out, literally. In other words, the protein molecules can no longer be balled up in spherical, compact structures because the mechanical action of stirring, mixing, or kneading extends and stretches the proteins out, like a slinky. As they are stretched, the proteins have more opportunity and space to interact with other protein molecules and they do. The protein molecules align against one another in more orderly, stretched, interacting protein sheets that are given a new name: gluten. Thus, gluten is the complex network of interacting gliadin and glutenin proteins that has properties like strength, elasticity, and stretchiness that are distinct from either of the individual gliadin and glutenin proteins. Gluten is of the utmost importance in baked goods. To make good bread, you desire gluten formation. In a pie crust, you want to avoid the formation of gluten. In foods like a tortilla, some of the characteristics of gluten are desired, but not all. Let's review each of these examples to learn more about how a cook or baker can achieve the results that they desire in the final product.

Box 9.2 Thickeners

The ability to thicken a soup, gravy, sauce, or beverage is a critical tool for any cook. The trick to creating a successful thickened sauce, soup, or gravy is through the incorporation of a complex carbohydrate and/or protein that binds water, yielding a more viscous liquid. The feeling that comes with tasting and eating a thick gravy or salad dressing is called mouthfeel. In order to understand how thickening works, you need to learn some physics about liquids. Water is considered a Newtonian fluid because there is a proportional relationship between viscosity and the force applied to the fluid. The more viscous the Newtonian fluid, the more force required to get it to flow. Gravies and other thickened foods are non-Newtonian fluids; these fluids require a larger force to start movement, but once in motion, non-Newtonian fluids move with a greater ease. Most of you have experience with ketchup. What a great example of a fluid that does not easily start to move but, once flowing, flows much faster than you often desire! What does a non-Newtonian fluid have to do with mouthfeel? The stickiness and viscosity of a non-Newtonian fluid are notably detected by the human mouth and give foods containing these fluids a distinctive, pleasurable character. Think of a soup that seems watery versus one that has been thickened with starch or gum. The thickened soup will taste better even when the flavor of the soups is identical.

One cooking technique used to thicken a recipe is called reduction. Heating a mixture provides enough energy to the water molecules to allow them to rotate, vibrate, and break the intermolecular forces that hold the water molecules in place. As the water "boils off," the remaining molecules of complex carbohydrates or proteins are forced to interact with each other, which increases the viscosity of the fluid. Unfortunately, a preparation that requires high heat or long cooking times often alters the flavor and nature of the fluid in unintended ways. The starches found in flour (in the form of a paste) can be used to thicken gravies and stews. While thick, non-lumpy flour pastes are difficult to form, as the starch solidifies in a low-water environment, cooks may use fats to coat and interact with the starch. Such a mixture, called a roux, is the base for pasta gravies and some thick stews.

How do you decide what thickening agent to use—a protein, starch, or gum? It depends upon the flavor and final presentation of the desired food. Is the food to be served hot or cold? Proteins coagulate upon cooling (making a solid mess), while gums and starches are more effective thickeners when cold. Starch thickeners quickly clump when reheated. If you want a clear thickened soup, some gums (like xanthan gum) can be used in a small amount to avoid clumping and cloudiness. Both starch and xanthan gums work well for more viscous preparations. pH can also impact the behavior of a thickening agent. Acidic foods cause some gums to be less effective; thus cornstarch or arrowroot starch is a better choice. Clear Jel is a chemically modified cornstarch that is specially made for baking and freezing acidic foods; it is a good choice for acid-containing fruit pies that need thickening.

9.5   Gluten, Fumaric Acid, and Tortillas

A handmade tortilla is a simple food made from wheat flour, water, shortening, and salt. Although some cooks prepare their own homemade tortillas, many individuals buy a commercially prepared tortilla from their local grocery store or market. Most of these store-purchased tortillas are made using a machine, rather than by hand. Why is this important? As you have just learned, the two proteins that make up gluten, gliadin and glutenin, include large numbers of cysteine (sulfhydryl-containing) amino acids. Through the process of mixing, stirring, and kneading a tortilla dough, the gluten network is formed with disulfide bonds helping to provide an elastic, chewy texture to the tortilla. Unfortunately, machine processing of tortillas creates excess links between the gluten network proteins resulting in a rubbery, less than satisfying tortilla. The solution lies in a molecule called fumaric acid, which contains a carboxylic acid functional group. Fumaric acid, naturally made in tissues of plants and animals, acts as a reducing agent, keeping the —SH groups from forming disulfide bonds (S—S) and decreasing the pH level of the flour dough, therefore defeating the toughening disulfide bonds of gluten (Figure 9.9). This results in softer, more machinable tortilla dough. In addition to limiting the gluten cross-linking, fumaric acid also acts as an antimicrobial agent and does not easily bind to water from the atmosphere, increasing the shelf life of the food from a few days to over two months.

9.6   Yeast-raised Bread

The ingredients and steps of baking a fresh loaf of bread are pretty simple: mix flour, water, yeast, and salt; knead the dough; let the dough rise; and bake. Yet, anyone who has made bread can attest that a perfect loaf is not easily attained. The bread may be dense, flavorless, or have a pale and soft exterior. In addition, there is some variation in ingredients and the specific instructions associated with a basic white bread. Your mother might mix yeast with water first and allow the mixture to sit (or proof) for a few minutes, while your uncle might first combine yeast with the flour and then add water to the yeast/flour mixture. Does it really matter? Yes! Let's begin our discussion about bread by learning more about the organism that is key to it all: yeast.

9.6.1   Yeast

Humans have been eating raised bread for about 6,000 years; however, we only began to understand the process of leavening through the work of Louis Pasteur in the 1850s. Now, we know that the CO_2 and alcohol generated during yeast fermentation are key components to a bread's taste, aroma, and texture. You have heard about yeast, metabolism, and fermentation in an earlier chapter, but let's review some of the basics again.

Yeast is a living organism, a fungus commonly used in baking and the brewing of alcoholic beverages. Its Latin name, *Saccharomyces cerevisiae*, means brewer's sugar. In bread dough, yeast consumes starch and sugar that are

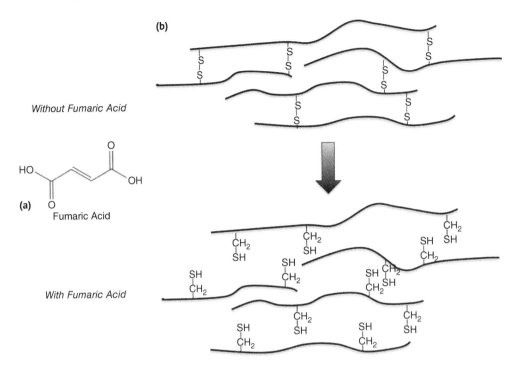

Figure 9.9 Soft tortillas. (a) The molecular structure of fumaric acid. (b) The impact of fumaric acid on disulfide formation in gluten found in tortillas. The addition of the acid keeps the disulfide bonds in the reduced state limiting the cross-linking of gluten for a softer chewier food.

present in flour and converts it to CO_2 and alcohol through fermentation. These yeast products give a baked loaf of bread its texture, flavor, and aroma. How does the yeast accomplish this? Can yeast begin "chewing" up the starches and producing CO_2 immediately when added to flour? Not exactly. Yeast can't break down the starch in its gigantic, polysaccharide form; the starch is too compact and tightly packed for the yeast enzymes to access it for metabolism. Thus, although there is a lot of potential, because the glucose molecules are stored in the form of the starch polysaccharide, the yeast can't initially use it to support its own growth and reproduction. Fortunately, flour contains soluble enzymes called amylases. When activated in the presence of water, the amylases carry out hydrolysis reactions, using water to break down the massive starch molecule into smaller molecular compounds: maltose, glucose, and other dextrins (smaller glucose polysaccharides) (Figure 9.10). These smaller sugars can serve as a food source for the fermenting yeast. Isn't it great that flour contains precisely the enzymes that yeast need to thrive?

Once the yeast has a usable food source, yeast begins to carry out glycolysis, an anaerobic (i.e. no oxygen is required) metabolic process essential for the yeast organism's growth and reproduction. You've learned about the importance of glycolysis to all organisms, as it is the primary metabolic pathway whereby organisms can generate the energy source molecule ATP. After an organism carries out glycolysis, however, it needs to do something with the products of the pathway, by breaking them down further or by repurposing them into molecules that are essential for the organism's function. Yeasts have the enzymes necessary to accomplish both of these tasks in the fermentation process (Figure 9.11). In fermentation, yeast breaks down the pyruvate generated by glycolysis via two reduction reactions, converting it first to acetaldehyde, carbon dioxide, and ethanol. NADH is used as the reducing agent and is converted to NAD^+. Thus, through fermentation, the yeast generates ethanol and carbon dioxide gas and regenerates the NAD^+ that is necessary to allow glycolysis to continue (Figure 9.11). What does this have to do with bread? Pretty much everything! The ethanol and carbon dioxide products are key to a great loaf of bread, as you will find out later in this chapter. But before we move on, we have a couple of other yeasty things to touch upon that may affect how you approach your next loaf of homemade bread: the different types of yeast and the order of addition of the yeast.

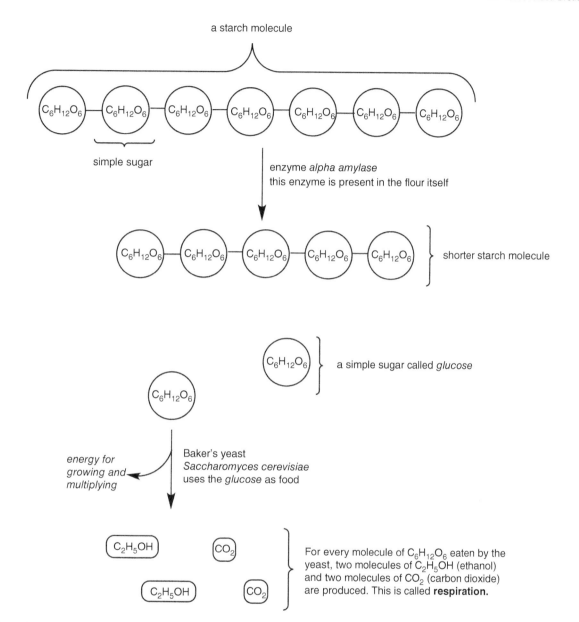

Figure 9.10 Yeast metabolism. Yeast utilizes the enzyme, amylase, which is found in flour, to break down complex starches to the simple sugar, glucose. Glucose is then converted to carbon dioxide and other compounds by the yeast via respiration.

9.6.2 Different Types of Yeast

The three most common forms of baker's yeast used in the home kitchen (and readily available on the grocery store shelf) are active dry, rapid-rise, and instant yeast. All these forms are dried versions; this is useful for the home baker, as they have a refrigerator shelf life of 1–2 years. Commercial bakers tend to use fresh or wet yeast; it is more dependable but has a shelf life of only about 2 weeks, making it less ideal for the home baker. However, to become active, the yeast needs heat and moisture, which occurs through the addition of warm water or milk, as well as the baking process (for heat). The differences between the three types center on particle size, recipe treatment, and the rise the yeast provides.

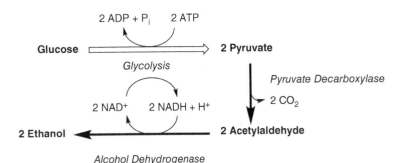

Figure 9.11 Ethanol production in yeast. Continued respiration (glucose metabolism to pyruvate and carbon dioxide) requires a regeneration of NAD^+ from NADH. Yeast uses alcohol dehydrogenase for this step. The by-product (from the yeast's perspective) is ethanol.

Active dry yeast is the yeast that most people think of when they picture "yeast" in their heads. It is dry and granular, having the consistency of cornmeal. Active dry yeast begins as a wet, compressed cake of yeast that is pressed through screens to make yeast strands. The strands are cut into pellets, dried (via exposure to high temperatures), and ground into granules. To utilize active dry yeast, you can dissolve it in a tablespoon of warm water and add it to the rest of the ingredients. This yeast will remain active (i.e. continue producing CO_2 and ethanol) for two rises, so the dough requires a rise immediately after kneading and a second rise in the bread pan.

Rapid-rise yeast is dried at lower temperatures and is milled into smaller particles than the active dry version. Because of this, rapid-rise yeast does not need to be dissolved in water and can be added directly to the dry ingredients. Typically, rapid-rise yeast has other additives (like enzymes) that also help to make the dough rise even faster. In addition, with rapid-rise yeast, you only need to do one rise. In other words, right after kneading, you can shape the dough into the bread pan, give it time to raise, and toss it into the oven for baking. However, in saving time, your bread will lose flavor and structure. A bread prepared from rapid-rise yeast will be fairly bland and commercial tasting. If substituted for either the active dry or instant yeasts, you will see a difference in the final bread product, both in the rise (or fall) of the bread and in its taste/texture.

Instant yeast is also known as bread machine yeast. It is milled into finer particles than active dry yeast, so it doesn't need to be dissolved in water. However, unlike rapid rise, instant yeast will give you two separate rises and can be used interchangeably with active dry yeast.

9.6.3   First Steps: Yeast Proofing

Most home bakers use the active dry version of yeast, thus, "proofing" the yeast by combining it with a warm liquid, along with a little sugar or honey, is a necessary step for a good rise (Figure 9.12). Following a resting period of 5–10 minutes, you should see a foam or bubbly liquid, as the yeast begins to ferment the added sugar if the yeast is active. The added sugar, although not absolutely necessary, jump-starts the activation of the yeast. If there are no bubbles or foam, something has gone wrong—the yeast may have expired, the liquid may be too cool to activate the yeast, or too hot, and it may have killed them. The optimal temperature for the liquid (which is typically water, milk, or some combination of the two) during proofing is 105–110°F. For rapid-rise yeast that is added directly to

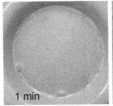
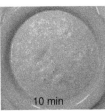
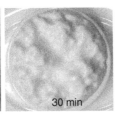

Figure 9.12 Baker's yeast. *Saccharomyces cerevisiae* (baker's yeast) proofing

the flour first, the liquid should be at a slightly higher temperature (120–130°F/49–54°C). Once the yeast is activated, you are ready to add the liquid to the other ingredients (typically flour and salt), which is where the yeast and the molecular components of the flour come together to yield a great loaf of bread.

9.6.4 Next Steps: Mixing, Kneading, and Gluten Formation

Mixing may seem unimportant relative to the aforementioned yeast and kneading processes that will be described next. However, there is more to mixing than simply combining the physical ingredients. Recall that when water is added to flour, the water is absorbed by the starch granule. Thus, the glutenin and gliadin proteins on the exterior of the starch granules generate intermolecular interactions with the water molecules (and vice versa), and the proteins become hydrated. These proteins were already involved in intermolecular interactions with themselves to create their three-dimensional, compact, spherical structures prior to the water addition. By involving water in the intermolecular interactions, the overall three-dimensional structure of the proteins is loosened and less compact. This loosening allows for new intermolecular interactions to form with neighboring protein molecules. As these new interactions are created, so are the first strands of gluten. However, that is not all! As the proteins become hydrated, so do the starch granules that the proteins are a part of. Thus, the individual starch granules that were separate entities in dry flour become sticky and bind together; you have experienced this with your eyes and hands if you have ever mixed bread dough. And that is still not all! As the water gets absorbed by each starch granule, the amylase enzymes (found in flour) now have the water needed to break down starch into smaller molecules, specifically maltose and glucose. The yeasts that were activated during proofing can use these sugars as a food source and begin to carry out the processes of glycolysis and fermentation. Wow! Who would have thought that all of this is happening during the mixing of a bread dough: hydration, enzyme activation, yeast fermentation, and the beginning stages of gluten formation? However, mixing is only the beginning of the process. Let's bring together some concepts as we learn more about kneading.

Have you ever tasted bread that has been kneaded for 1 minute and one that has been kneaded for 10 minutes? There is a difference in how the bread tastes, how it feels to the touch, its mouthfeel, its chewiness, etc. These differences are based on what happens at a molecular level during kneading, which promotes the formation of gluten, impacting dough texture, rising, baking, and taste. As mentioned previously, during mixing, the starch granules bind together and the proteins change structure as they begin to have intermolecular interactions with water and each other.

During kneading, the "bound" starch granules are moved apart, which forces the proteins to unwind and become even more linear (while still kinked and coiled), allowing for interactions with other protein molecules in different and more ways via both intermolecular interactions (like hydrogen bonds) and covalent bonds (like disulfide bonds; Figure 9.13). As the dough continues to be kneaded, these interactions are broken and reformed over and over again, which eventually results in glutenin protein molecules that are interacting with one another end to end, forming long, coiled protein chains that might be a few hundred glutenin molecules in length, like a slinky (Figure 9.14). This arrangement leads to an elastic dough that stretches but doesn't break unless you stretch it apart really far. As the dough is stretched, the proteins uncoil and lengthen; when you release the stretch, the

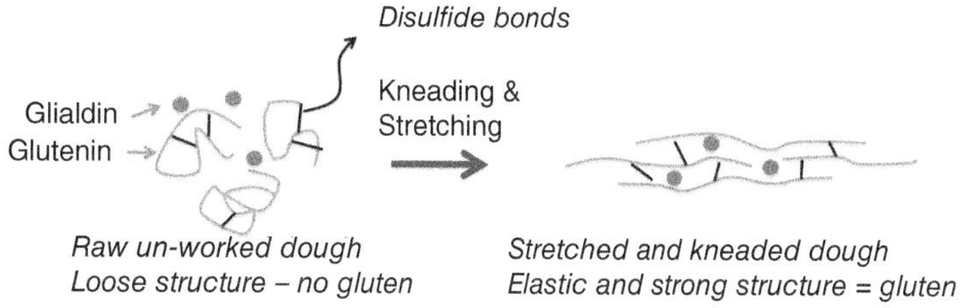

Figure 9.13 The molecular effects of kneading on dough structure.

Glialdin
Glutenin

Disulfide bonds

Kneading & Stretching

Raw un-worked dough
Loose structure – no gluten

Stretched and kneaded dough
Elastic and strong structure = gluten

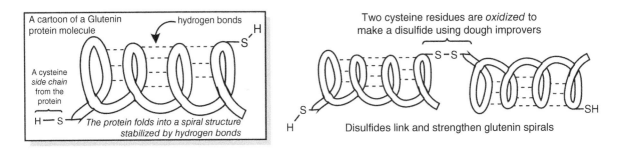

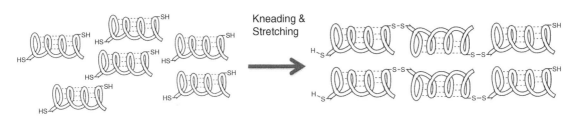

Figure 9.14   Gluten forms during kneading dough. Stretching and kneading flour dough lengthens the proteins where they can then cross-link with each other forming long, stronger gluten complexes. The cross-linking of gluten comes from the disulfide bond between strands of glutenin. The high glutamine (Q) content supports the helix by forming many hydrogen bonds, as shown with the dashed lines.

dough bounces back to its original shape as the proteins recoil and reform the disulfide and hydrogen bonding interactions with the neighboring glutenin proteins. You can recognize when this happens during kneading; the dough becomes less sticky, smoother, and more elastic, yet it remains firm.

You might be thinking, what are the gliadin proteins doing as the glutenin proteins are stretching out into these long coils? The gliadins contribute to a dough property called plasticity. Remember that the gliadins are typically compact in their native structural state, so they actually interfere with or limit the intermolecular interactions between glutenins. Why don't these interruptions reduce the strength of the gluten network? The interrupting gliadins actually do disrupt the network, and this is to our advantage in bread baking. If the gluten network is too tight (i.e. the glutenins are held too strongly or tightly together via intermolecular interactions), then the dough won't stretch, it would not deform under the pressure of your kneading hand, and it won't mold into a ball for rising. Thus, the gliadins help to prevent a gluten network that is too strong.

We have a couple of questions left to address about gluten. What additional factors (besides kneading time) affect the quality of the gluten network? Can you break a gluten network, once it has been created? What happens when the gluten network becomes too strong?

9.7   Control of Gluten Formation

There are a multitude of ways to control gluten formation that do not require an excessive amount of time or money. A relatively easy way to control gluten formation is by adjusting the type of flour that you use for a particular type of baked good. Since bread flour has a higher protein (i.e. more gliadin and glutenin) content than all-purpose flour, a stronger gluten network will be created with the same amount of kneading and ingredients in a dough made from bread flour versus all-purpose flour. Secondly, the presence of oxidizing agents or air enhances gluten formation. This may seem a bit unusual, but there is a scientific basis for this action.

Because disulfide bonds are more likely to form in an oxidizing environment and the formation of gluten involves the creation of disulfide bonds, a stronger gluten network is formed if more air (i.e. oxygen gas) is incorporated into the dough. While on the subject of disulfide bonds, aged flour also makes a better yeast-raised bread than fresh flour, as the thiol groups (i.e. cysteine's sulfide group) are more oxidized in aged flour since aged flour has been exposed to oxygen for a longer period of time than fresh flour. Thus, disulfide bonds are more likely to form in aged flour, leading to a stronger gluten network. The fourth factor has already been addressed: kneading time. The more kneading, the better the gluten network to a certain point. If you are hand kneading, your hands and arms will tire long before you have reached the point of kneading too long. However, over-kneading can happen, especially if kneaded with an electric mixer. When the dough is over-kneaded, the gluten network is so tight, it feels very dense and tough, it tears apart when it is stretched, and it doesn't flatten and fold easily. If you are using unbleached flour, the dough color might even change when you reach the point of over-kneading, as the excessive oxidation causes the flour to bleach. If you are using an electric mixer to knead, pay close attention to the springiness of the dough, typically kneading for 2–4 minutes with a dough hook. If you over-knead, the best option is to start over.

One technique that you can use to prevent over-kneading and reduce kneading time is one that was developed by a French breadmaking authority named Raymond Calvel in the 1970s. In this technique, called autolysis, the flour and water are first mixed and allowed to rest before kneading. As the mixture rests, the protease enzymes in the flour begin to break down/cleave the glutenin and gliadin proteins into smaller pieces. These smaller proteins are more easily straightened and aligned during the kneading process, so an organized gluten network is created more quickly. The autolyzed bread dough requires a much shorter kneading time.

Clearly, gluten formation is important. The question is, why is it important? The formation of a strong, elastic gluten network is key to the leavening or rising step of bread baking; this is where the fermenting efforts of the yeast and your kneading efforts come together.

9.8 The Rising Bread

Any bread baker visually knows what happens when bread rises; it gets bigger, often doubling in size after sitting in a warm location for about 60–90 minutes. You already know that it has something to do with the yeast, but what exactly is happening at the molecular level? During the rise, the yeast continues to metabolize the simple sugars produced from the breakdown or hydrolysis of starch (carried out by the amylase enzymes in the flour), releasing carbon dioxide (CO_2) and ethyl alcohol (ethanol). Both of these molecules (and other alcohol byproducts) are really important to bread; the CO_2 is particularly key to the rising phase. Why? Remember, CO_2 is a gas. Gases can bubble out of a solution, similar to what you see in a carbonated beverage. However, a gas can also be trapped inside a mixture. If your bread dough has a strong gluten network, the sheets of gluten become blown up with the CO_2 gas like a balloon. As the gas molecules are captured, the dough rises. However, in order for the CO_2 to be captured, the gluten sheets need to be robust enough to not break during expansion and plentiful enough to capture the CO_2 gas molecules as very small bubbles. If the gluten sheets are weak, the CO_2 bursts out of the gluten network, rather than being captured, and the dough will not rise properly. If the CO_2 is captured in large, rather than small bubbles, large, gaping holes are present in the final loaf. Through the formation of a strong, elastic, and plastic gluten network, small CO_2 bubbles are captured, thus actually allowing a bread to rise. Furthermore, with the production of a greater amount and smaller CO_2 bubbles, the bread will be more tender due to the CO_2 pocket interruptions in the gluten networks.

Bread rising typically happens in a warm space. Why is there a temperature dependence? Yeast are living organisms and thrive in warmer temperatures (just like some humans). When the yeast are at a cooler temperature, they are not happily duplicating, growing, and fermenting, and less CO_2 production will occur over a period of time. For this reason, refrigerated bread dough requires a rising time of 8–12 hours; it just takes longer for the yeast to produce the same amount of CO_2 at 39°F/4°C relative to 73°F/23°C. The optimal temperature for bread rising is 27°C (81°F). Although yeast actually reproduces more rapidly at 95°F/35°C, the yeast releases bitter metabolites, which may give a bread raised at this temperature a less desirable bitter flavor (Box 9.3).

Box 9.3 Raising your dough

It might be convenient to allow your dough to rise overnight. However, the slower rise time does lead to the production of different flavor molecules than a more typical rise. The stronger, often more sour flavors that develop in an overnight, slow rise are preferred for some breads or in some individuals. A slow rise also encourages the development of a more open, irregular crumb structure in the bread due to the redistribution of the dough gases that occurs when the cooled dough warms.

Some bakers do not disturb a rising bread dough, while others incorporate a turning step into the dough rise. Turning involves delicately folding the dough over several times at one or multiple points during the rise. Turning the dough stretches it and builds strength to the gluten network, as it helps to align any misaligned gluten sheets. In addition, turning removes some CO_2 from the dough, an excessive amount of which inhibits yeast activity, to ensure the maximum rise and production of flavor molecules. Throughout this chapter, reference has been made to flavor molecules, both bitter and favorable flavors. What are these flavor molecules? One product of fermentation is ethanol, but as yeast proliferate, they continue to synthesize proteins, aldehydes, ketones, diacetyl, and other aromatic molecules that contribute to the scent and flavor of a raised bread dough and final product. We will discuss the specifics of the flavor molecules later in the chapter.

Finally, is bread raised enough and is it bigger or better? Most cookbooks tell the baker that the bread has risen enough "when it is doubled," but it is hard to know when enough is enough. Rising is complete when you can poke your finger into the dough and the hole doesn't immediately close (Figure 9.15). At this point in the rise, the gluten has been stretched to the limit of its elasticity. If you try to raise the bread any more, the gluten network will break, and the bread will deflate, releasing all of the captured CO_2. The rising time for a typical dough that sits in front of a sunny, draft-free window is 60–90 minutes.

Figure 9.15 Rising bread dough. The dough rises with the production of CO_2 gas. Vidady/Adobe Stock Photos

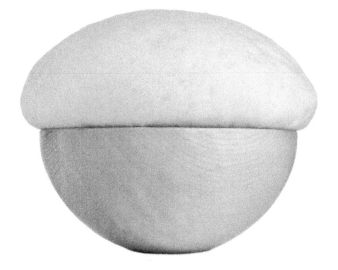

9.9 The Punch and Second Rise

In many bread recipes, after the first rise, the baker knocks down or punches the dough and allows it to rise again. Is this a step needed to relieve human aggression or does it have some purpose? To start, the punch is not actually a punch but more of a pushdown, also known as degassing. As the degassing term implies, when pressure is put onto a raised dough, some of the CO_2 bubbles are expelled from the dough, and those that remain break into smaller gas bubbles. These smaller bubbles can be blown up again during a second rise, giving a final bread product a finer crumb and light texture (with fewer large air pockets). The yeast cells, the sugars, and the products of fermentation are also redistributed throughout the dough during the punch, which leads to better yeast growth and distribution during the second rise. Interestingly, the temperature of the dough is also equalized during the punch. The inside of a raised dough is warmer than the outside due to the activity of the yeast; by equalizing the temperature, the second rise is more evenly distributed across a punched dough relative to one that is not punched. Finally, by punching down the dough, the gluten network has a chance to relax before it is stretched again, which allows for a dough that is easier to shape. Thus, there are scientific and culinary reasons why you should punch down the dough, which requires some practice in proper technique (Figure 9.16). For a final bread product with a fine crumb (like a sandwich bread), first, push your fist quickly but gently into the center of the dough. Then, pull the edges of the dough to the center. Take the dough out of the bowl, pat it, turn it over, and do two or three kneadings to release any additional air bubbles and leave the dough for the second rise. For a more coarse bread that has a more robust chew, turn the dough (gently folding it over) to reactivate and redistribute the yeast and sugars, without pressing out as much air.

Once the dough has been punched accordingly, it is shaped into the form of the final loaves and then left to sit and rise in a warm place. This second rise is sometimes called "finishing" or "appret," as it is the final opportunity

Figure 9.16 Punching down dough.

Figure 9.17 Appret. Slightly slicing the surface of a growing dough to limit tearing by allowing a "pleat" to stretch.

for the yeast to do their work before they succumb to the heat of the oven during baking. Some bakers slit the tops of the shaped dough before or after the second rise, which ensures that the gluten network is not stretched to its limits during baking, reducing the presence of unattractive tears in the crust (Figure 9.17).

9.10 Baking

You may view the baking process as simply "cooking" the dough, but there is so much more to it, including gas expansion and browning reactions, to name a few.

First, let's talk about why the bread expands (even more) during baking. Remember that a raised dough is like a gluten balloon filled with yeast-generated carbon dioxide and air that was introduced into the dough during mixing and kneading (Figure 9.18). Your experiences (in the kitchen and in life) tell you that gases expand when heated: a balloon left in a warm car explodes, and the air pressure in your car tires decreases in the winter. The scientific law that tells us that gases expand when heated is known as the ideal gas law: $PV = nRT$.

The letters in the ideal gas law stand for: pressure (P), volume (V), amount of gas in moles (n), a gas constant (R, which doesn't change in value), and temperature (T). The behavior of most gases is consistent with this equation, including the gases present in the rising dough. Thus, if the amount of gas that you have is the same in the prebaked and the baking dough (i.e. n is constant), then the increase in temperature that occurs during baking causes an increase in the overall value on the right-hand side of the equation. If the right-hand side of the equation increases in value, then the terms on the left-hand side of the equation must, together, account for this temperature increase. Not only does the gas pressure (P) inside the baking loaf increase, but so does the volume of the bread (V). How does this happen? As the gas bubbles expand, the dough, which is still soft during the initial stages

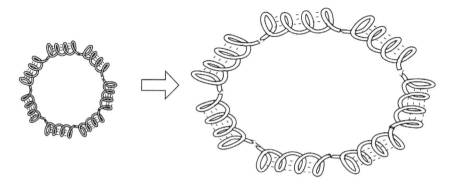

Figure 9.18 Expanding gases are held in place by stretched gluten protein cages.

of baking, is pushed aside and visibly expands. If the starch and gluten networks are too rigid (i.e. an over-kneaded dough), the bread doesn't expand because the gluten network cannot move.

When does the bread become more rigid? In other words, what happens to make the bread fully baked? During baking, the starch granules in the bread's interior absorb water, swell, and create a permanent solid structure that is a network of starch and protein filled with tiny air pockets called alveoli. This sponge-like, soft structure is referred to as the "crumb" by bakers. In contrast, the outer surface (also known as the crust) develops a dry, dense, hard texture. Clearly, different things are happening in the crumb versus the crust; let's learn more about each component.

As the dough heats, it actually becomes more fluid due to the increased movement of the molecules, the gas bubbles expand, the gluten stretches to capture the expanded gas bubbles, and the dough rises. Moreover, at about 60°C/140°F, more gas is generated in the expanding bread due to the vaporization of the yeast-produced ethanol and water; the additional gas molecules and larger gas bubbles expand the dough by as much as half of the initial dough volume. However, bread expansion stops when the crust becomes too firm and stiff to be pushed around anymore, at about 90°C/194°F. Furthermore, at 100°C/212°F, the water in the bread's interior and on the exterior fully evaporates. In the interior, the generated steam (water vapor) is redistributed throughout the bread because it cannot escape. However, on the bread's surface, the water fully evaporates, resulting in a drying and hardening of the bread's crust. As the temperature continues to rise in the bread's interior to about 150–180°C/300–350°F, the starch begins to "gel" into a more solid state due to the absorption of water vapor and resulting swelling of the starch granule. Moreover, the proteins that make up the gluten network, now denatured due to high temperatures, form even stronger intermolecular and disulfide bond interactions with one another. In essence, then, the starch and protein network on the bread's interior becomes much tighter and stronger, and the gas bubbles have no more space for expansion. Pressure builds as the gas bubbles try to expand until the bubbles rupture, turning the structure of the loaf from a network of separate gas bubbles into a more open, porous network. In the end phases of baking, the starch forms an even more solid, gel-like structure. Even after baking and during cooling, the starch granules continue to become more firm, making the bread easier to slice when cool than when piping hot. Simply put, during the process of baking, the interior of a bread dough generates more gas molecules, expanded gas bubbles, starch granules that are swelled with water to the point of being more solid and gel-like, and a tighter molecular network that only allows the bread to expand to a certain point upon baking. All of this chemistry and biology generates the crumb of the bread, so it is easy to see why your bread crumb may vary from loaf to loaf.

What about the crust? The crust's color and aromas are a result of Maillard reactions. You have heard about Maillard reactions in Chapter 3. These reactions, discovered by Louis Camille Maillard in 1911, occur between sugars and amino acids (Figure 9.19). Because there are 20 different amino acids and hundreds of different sugars, the Maillard reactions are best described as a group of reactions rather than a single one. The first step of any

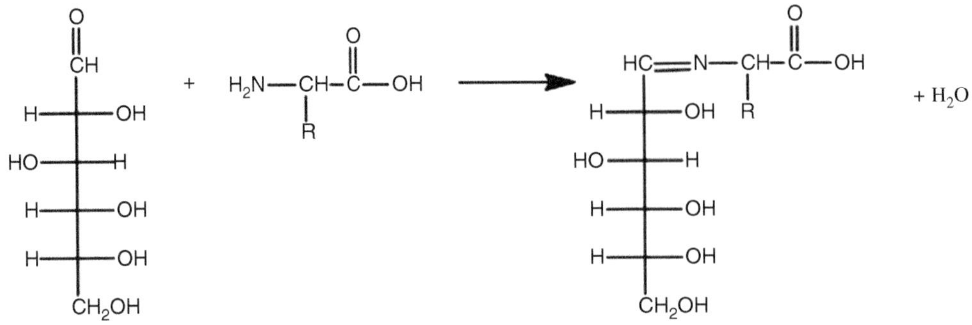

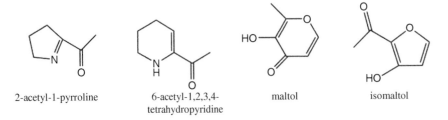

Figure 9.19 Maillard reaction browning bread crusts. A sugar molecule, shown on the left in the straight chain form, will condense with the nitrogen of an amino acid releasing water as the first step of the browning reaction.

Maillard reaction occurs between the carbonyl group of a sugar molecule and an amino group, resulting in the formation of a new covalent double bond between the original molecules and a new molecule of water. The reactive amino group may be part of a free amino acid or an amino acid that is still part of a protein chain.

What types of Maillard reactions occur in bread and what do they do? These reactions certainly contribute to the brown color of a bread crust, but they also contribute substantially to the flavor and aroma of the bread. You already know some of this chemistry from Chapter 3, but let's look at just a few of the hundreds of Maillard reaction products that are generated in the dough, crust, and crumb of a great loaf of bread.

In an early scientific study on the aroma compounds of wheat bread, 372 compounds were identified in the crust and 86 in the crumb of wheat-based white bread. In the crumb, non-Maillard molecules, including aldehydes, alcohols, and ketones predominate, while in the crust, the Maillard products predominate, including pyrazines, pyrroles, and furans. With all of these aromatics (Figure 9.20), it is no wonder that bread smells so good! Two Maillard products associated with bread flavors and aromas are 2-acetylpyrroline (roasty, bread crust-like) and 6-acetyl-1,2,3,4-tetrahydropyridine (roasty, cracker-like flavor), while maltol and isomaltol contribute to the freshness of baked bread.

The chemistry behind the formation of these molecules is shown in Figure 9.21, The aldehyde group (CHO) of the sugar and the amine group of a free amino acid or amino acid that is present within a protein react to form a compound called Schiff's base (it has the C=N group. Schiff's bases are not stable compounds, so this compound rearranges itself to a more stable compound called an Amadori compound, the ketosamine. All of the chemistry up to this point is fully reversible and is the same chemistry for all Maillard reactions. What happens next, however, varies widely depending upon the exact conditions of the reaction. The Amadori compound can rearrange, condense with another compound, isomerize, polymerize with a like compound, cyclize, or undergo degradation reactions. In short, it is the products of the Amadori intermediates that give us the extensive array of color, flavor, and aromatic compounds associated with baking or freshly baked bread. That being said, one of the major pathways produces a compound (the 1,2-eneaminol), followed by 5-hydroxymethyl-2-furaldehyde (labeled as HMF). From HMF, the brown, melatonin pigments associated with a brown crust form (maltol and

| 2-acetyl-1-pyrroline | 6-acetyl-1,2,3,4-tetrahydropyridine | maltol | isomaltol |

Figure 9.20 Aroma compounds from baked bread.

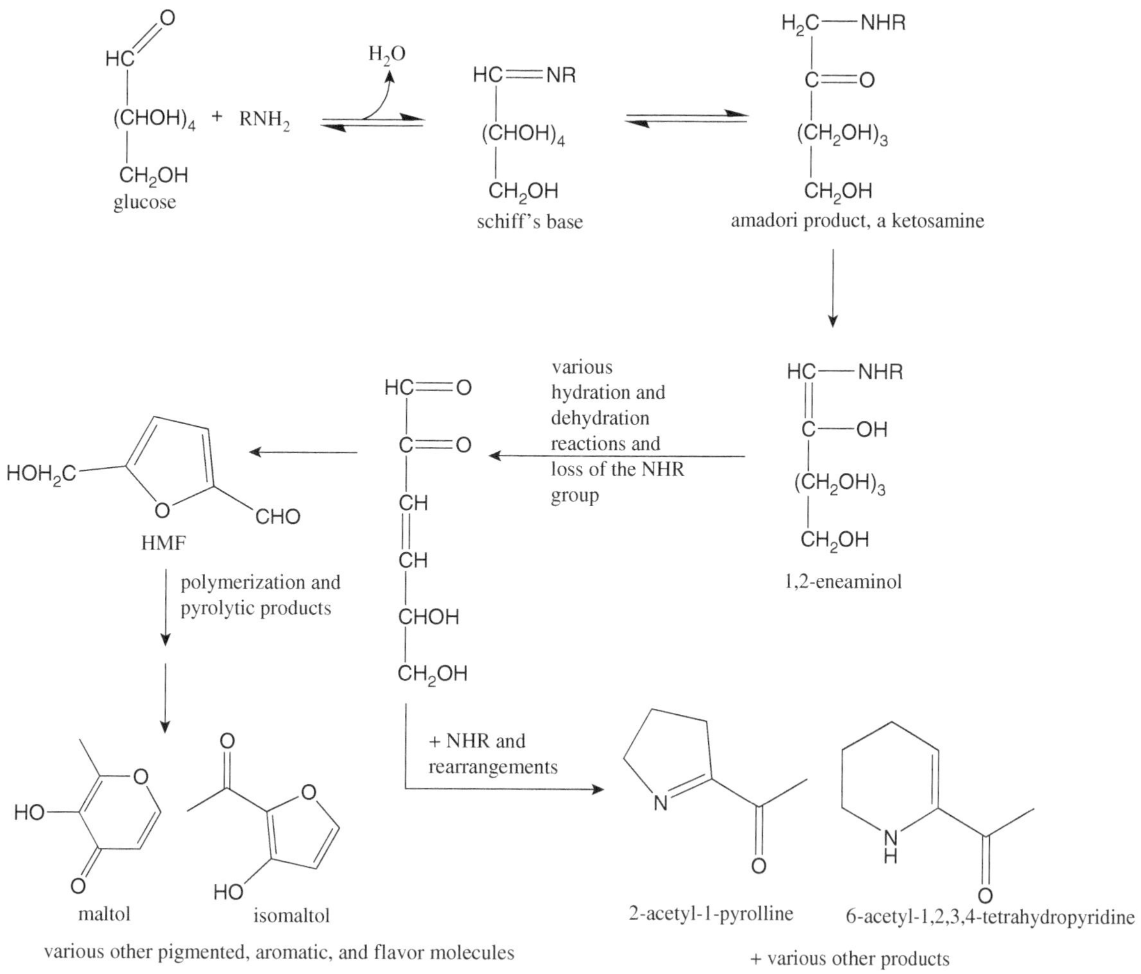

Figure 9.21 The browning reactions forming brown crusts of bread.

isomaltol). A more minor pathway produces a 1,3-enediol to eventually produce various alpha-dicarbonyls. A third pathway is called the Strecker degradation, which produces aldehydes. The complexity and variability of this chemistry is significant, and small changes in temperature and conditions can dramatically impact the resulting molecules and therefore, the bread.

In addition to the flavor molecules produced by the Maillard reaction, the products of yeast and bacterial fermentation also contribute to bread flavor. Sourdoughs and breads derived from "starters" are much more tangy and flavorful than traditional homemade white bread. What is a starter? A starter consists of a small amount of dough that was saved from the last time a loaf of bread was made. The baker adds a little more water and flour to the starter and then allows the mixture to ferment for a longer period of time (i.e. 2–24 hours) before adding the other components of the bread. This fermented starter has garnered its own characteristic yeast and bacteria, depending upon storage conditions. The increased acidity of a sourdough, due to lactic acid production by bacteria and yeast, along with other distinctive flavor components produced in the dough, gives sourdough its tanginess and lack of brown color (because acidity slows browning reactions). Because of the increased resting period and time, the carbohydrates, fats, starches, and other sugars have more time to break down yielding many more and more diverse flavor molecules than those generated during the traditional (shorter) white bread dough preparation and rising.

9.11   Other Ingredients in Bread

Now that we have talked about the aspects of traditional white bread ingredients and baking that are common to most bakers, what about possible variations, and how will that affect a finished loaf?

Most bakers add salt and fat during the preparation of the bread dough. In addition to the added flavor, salt modifies the activity of the yeast and strengthens the gluten network. Yeast requires a certain amount of salt (i.e. Na^+ and Cl^-) for optimal growth. The addition of too much salt will kill the yeast. If the baker adds too little salt, the yeast may multiply more rapidly than normal, giving the final product a strong "yeasty" flavor. However, in addition to allowing for optimal yeast growth, the charged nature of the ions in salt (i.e. Na^+ and Cl^-) allows the ions to interact with and stabilize any charged amino acids that are present within the gliadin and glutenin proteins. This interaction decreases any possible charge repulsion between individual protein molecules (e.g. two positively charged groups repelling one another), allowing the two salt-bound protein molecules to bond (covalently and via intermolecular interactions) more extensively.

If you have ever tried to crunch your way through a two-day-old baguette loaf, you have experienced that the addition of fat prevents bread staling. What does fat do at the molecular level? Fats inhibit the formation of the gluten network, increase the volume and lightness of a dough, and slow the staling process. Fats like to bind to other fats and other hydrophobic species. The glutenin and gliadin proteins do have hydrophobic components, thus, the fat molecules bind to the proteins, preventing them from interacting with one another. However, fat also helps to stabilize the walls surrounding the gas bubbles, preventing them from rupturing, which leads to a lighter, more voluminous bread. Bread staling is actually an issue of water loss due to changes in starch. Over time, the amylose and amylopectin molecules in a starch granule begin to associate more with one another and less with water. This phenomenon is called retrogradation, and it occurs in most any starch-based food, including potatoes and pastas. During the first phase of staling, which begins to take place within a few hours of baking, amylose molecules begin to associate and interact (or crystallize) with one another, while in the second stage of staling, amylopectin molecules retrograde. These phenomena result in a more crystalline and solid structure of the starch-based food and loss of the water molecules that were involved in intermolecular interactions with the starch granule. Fat protects against retrogradation, as it coats the starch granules, preventing the amylose and amylopectin from interacting with one another.

Staling can be reversed (in part anyway) by "melting" the crystals through heat and moisture. Thus, many cooks will place an older loaf of bread into the oven before a meal to soften the bread. Although the softening is short-lived, it does work! Perhaps surprisingly, placing a loaf of bread in the refrigerator expedites retrogradation, while freezing the bread prevents the crystallization of the starch molecular components and water migration.

One additional additive that impacts the taste, structure, and texture of bread is sugar. While a small amount of sugar is important in "jump-starting" the yeast, gluten development is challenged with a large amount of sugar. Because sugars bind to water, water is unable to form intermolecular interactions within the gluten network, yielding a weakened gluten structure. Thus, a rich dough (containing a high sugar and fat content) is quite fragile and relatively soft. Some bakers overcome this difficulty by kneading in the fat and sugar after the gluten network has been established within the dough.

9.12   Gluten and Celiac Disease

Most grocery stores and college cafeterias now have a large number of products and areas that are designated as gluten free. You or your friends may have adopted a gluten-free diet for medical or nonmedical reasons. What is the medical basis behind celiac disease? How does a baker prepare high-quality, tasty baked dough without gluten?

Celiac disease is an autoimmune disease. In any autoimmune disease, the body recognizes a molecule as foreign and utilizes its immunological system to attack the foreign molecule. In most cases, the immune system can recognize harmful foreign species (e.g. bacteria) from nonharmful species (e.g. foods, human biological molecules), thus helping to protect the body from harmful foreign species. However, in the case of celiac and other autoimmune diseases, the immune system triggers an abnormal response against a molecule or organ. In celiac disease, gluten is the recognized foreign molecule. Thus, when a person with celiac disease eats gluten, the lining of his or her intestine becomes inflamed and damaged due to the body's immunological response, leaving the intestine unable to properly digest and absorb nutrients from other foods.

An estimated 1% of people worldwide are thought to have celiac disease; a gluten-free diet is the only treatment for those afflicted, which is problematic given that gluten is present in so many foods. Gluten is found in wheat (of course), barley, rye, bulgur, farro, and possibly oats, all of which are a good source of vitamins, minerals, and fiber in one's diet. Thus, in order to remain healthy, a celiac-afflicted individual must avoid most breads, crackers, cereals, pasta, pastries, and a number of processed foods but can indulge in eggs, meat, fish, fruits, and vegetables.

In addition to those individuals formally diagnosed with celiac disease, you may have friends who are gluten sensitive or have made a dietary decision not to eat gluten. What should you do if you would like to serve bread at your next dinner party? The easy response is to go to the gluten-free aisle in the grocery store. But maybe you would like to experiment with baking gluten-free bread. How can you achieve a chewy, light texture without gluten?

Often gluten-free flours are supplemented with (or you can add) xanthan or guar gum or other emulsifiers. These products can contribute to the elasticity (the gums), stabilization of, and reduction in the diffusion of the gas bubbles (the emulsifiers). It is tricky, but with some experimentation with different flours, gums, and emulsifiers, you can still bake a great loaf of bread.

9.13 Muffins and Batter Breads

Perhaps on a given morning, you desire the tender structure and sweet taste of a blueberry muffin or a lemon poppy seed loaf rather than a piece of chewy, toasted sourdough. Although still "bread," these breads are a lot more like cake; they are quickly prepared (hence the name quick bread), use chemical leavening agents, and have a completely different texture from yeast-based breads. In muffins and batter-based breads, you do not want to generate the chew of a robust gluten network but a nice, tender, porous crumb; both the ingredients and techniques help to accomplish this.

The ingredient list for a muffin is expanded relative to a yeast-based bread and includes more sugar and fat, eggs, baking soda or baking powder, or a combination of the two, and in some recipes buttermilk, sour cream, or milk. In addition to providing richness, sweetness, and body, the "extra" eggs, fats, and sugars all interfere with gluten formation by interacting with the gliadin and glutenin proteins, thereby preventing the protein molecules from interacting with one another to form the gluten network. However, even in the presence of these interfering molecules, simply stirring the mixture can generate a gluten network. Thus, any muffin man knows that you should combine the dry ingredients, add the combined wet ingredients in bulk (often making a "well" in the middle of the dry ingredients), and fold the wet into the dry with a rubber spatula only until the dry ingredients are barely wet and the batter is still lumpy (Figure 9.22).

Figure 9.22   The wet and the dry. Keeping the two sets of components of a cake or muffin mix separate until needed is a secret to avoiding excess gluten formation.

Using this method, your muffins will be tender, with a porous, honeycomb-like interior, but will stale quickly due to the lack of an even distribution of the fat to coat the starch particles. That being said, if you make a good muffin, the entire batch will be eaten within a day, so stalling won't be a problem.

9.14 Chemical Leavening Agents

If you have ever forgotten to add the baking powder to a batch of muffins, the flattened, dense result illustrates the importance of a leavening agent. The unraised muffin probably felt as heavy as a pool ball. Under the right conditions, chemical leavening agents produce carbon dioxide gas, the same gas bubbles that are released by yeast during fermentation. The two most commonly used chemical leavening agents are baking soda and baking powder. Although they both produce carbon dioxide, they do so under different conditions and over a different time frame, thus, if you interchange them in a recipe, you will probably be underwhelmed with the resulting product (Figure 9.23).

9.14.1 Baking Soda

Baking soda consists of a single, basic chemical called sodium bicarbonate ($NaHCO_3$), which produces carbon dioxide, water, and salt in the presence of an acid (H^+) (Figure 9.24). Wait a minute! An acid? It is accurate that many acids are corrosive and should be avoided in the kitchen. However, many foods are, in fact, acidic. Buttermilk and sour cream contain lactic acid; molasses has various acids, including acetic, propionic, aconitic acids; lemon juice has citric acid; and cream of tartar (tartaric acid) are the most common acidic ingredients in a baking soda-leavened quick bread or muffin. The great thing about the chemical reaction between sodium bicarbonate and the acid is that it causes the production of millions of tiny carbon dioxide gas bubbles in the thick muffin batter. If these expanding gas bubbles are captured in the structure of the batter as it becomes firm in the hot oven (transitioning from foam into a sponge), you will have a tender, porous muffin. The problem with this chemical reaction is that it begins immediately when the bicarbonate and acid are exposed to liquid water. Thus, as soon as you add the wet to the dry ingredients, carbon dioxide production begins. If you are a muffin baker, you need to get the batter into the oven as quickly as possible after mixing so that it reaches its firm state before all of the carbon dioxide gas is lost to the air (Box 9.4).

This chemical leavening reaction is an acid–base reaction, with baking soda (sodium bicarbonate) acting as the base. The acidic additive (e.g. buttermilk) not only assists in the production of CO_2 gas molecules but also neutralizes the mixture so that the final product doesn't have the bitter taste associated with alkalinity (basicity). Thus, it is critically important to add the exact amount of baking soda that is called for in a recipe, both for leavening and taste. Baking soda may be the only leavening agent used in a favorite chocolate chip cookie recipe, as it causes the cookies to spread out and brown due to Maillard reactions, while in other recipes (e.g. buttermilk pancakes) baking soda is used in combination with baking powder. Interestingly, it is the basicity of baking soda that causes the reddening of cocoa powder, hence the name devil's food cake.

Figure 9.23 Baking soda versus baking powder. Baking soda is pure sodium bicarbonate, while baking powder contains sodium bicarbonate, along with other additives that help to govern its leavening power.

$$NaHCO_{3\,(s)} + H_{(aq)}^+ \longrightarrow H_2O_{(l)} + CO_{2(g)} + Na^+$$

Figure 9.24 Sodium bicarbonate reaction in acid.

Box 9.4 Tartar as an acid

Cream of tartar is a white powdery substance, whose most common use is in the stabilization of beaten egg whites in a meringue. Given its slightly acidic nature (i.e. a pK_a of 4.3) and solid structure, it reacts relatively slowly when compared with a stronger aqueous acid like lactic acid (pK_a 3.9). In a reaction with baking soda (sodium bicarbonate), carbon dioxide is liberated (Figure 9.25), which contributes to some of the volume of the meringue.

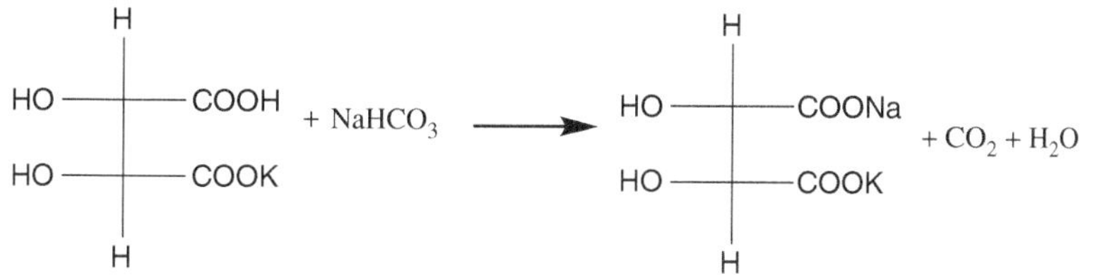

Figure 9.25   Cream of tartar reaction with sodium bicarbonate.

9.14.2 Baking Powders

Baking powder contains a mixture of three components: baking soda (sodium bicarbonate), one or more acids (e.g. tartaric acid, mono- or dicalcium hydrogen phosphate, dicalcium hydrogen phosphate, sodium aluminum sulfate or phosphate), and a filler/diluent. What is a filler? Given that baking powder is primed to produce carbon dioxide as soon as it gets wet (since it contains both the base and acid components), it is prone to lose its effectiveness under warm, humid conditions. A filler, like cornstarch, is added to keep the molecules of acid and base physically separate from one another. However, even with the presence of a filler, baking powders will degrade over time and have an expiration date that a user should pay attention to. You have probably figured out several perks associated with the use of baking powder in a recipe: (i) you don't need to add an acid as part of your ingredients list; (ii) baking powders are preneutralized, so your baked product will not have a bitter taste due to an unneutralized base; (iii) baking powders are double acting.

What is a double-acting baking powder? Double acting baking powders actually have two acid additives, so they act "twice". One acid (e.g. monocalcium dihydrogen phosphate) is soluble in the mixture and immediately reacts with the available sodium bicarbonate to produce carbon dioxide, much like the reaction and behavior that you see with baking soda. However, as discussed earlier, in order for these gas bubbles to be retained within the structure of the baking batter, the batter needs to be set relatively quickly in the oven. In reality, only a portion of the gas bubbles are actually retained to leaven the final product, thus this soluble acid really just lightens the batter a bit before it cooks (through retention of a limited amount of CO_2 bubbles). The second acid (e.g. sodium aluminum sulfate) in a baking powder only becomes reactive in the presence of heat because it is insoluble (or only slightly soluble) in a batter mixture at room temperature. Thus, as the muffins start to bake, with the batter not yet set enough to capture the expanding CO_2 bubbles being produced by the soluble acid one, acid two dissolves and begins to react with the remaining sodium bicarbonate base, producing more CO_2 that will be captured within the muffin structure as it firms and sets. Thus, a recipe that utilizes baking powder will typically yield a larger rise than a recipe that uses baking soda, even if the baking is delayed by 15–20 minutes. Since much of the reactivity of acid, one is complete within the first 2–5 minutes of mixing and initial baking, the final raised product is largely due to the work of acid 2.

9.14.3 Baking Soda versus Baking Powder

Some recipes call for one, some recipes call for the other, and some recipes call for both. How do you decide? The best chemical leavening agent depends upon the desired characteristics of the final product.

Baking powder will give a product that is more raised, while baking soda will provide a flatter or browner structure, as seen in a chocolate chip cookie. The baking soda in cookies causes them to brown (due to the Maillard reaction) and spread out on the pan, rather than up into the air. A baking powder cookie will taste more cakelike than a baking soda cookie because of the rise. If you are a fan of the cake-like cookie and would like to carry out an experiment in the kitchen, remember to add more baking powder to a baking soda-based cookie recipe, given that baking soda comprises only about a third of the same amount of baking powder.

If a recipe calls for both baking powder and soda, the baked good is actually being leavened by the baking powder, which contains the correct proportions for all of the acids to react with the basic bicarbonate, while baking soda is added to neutralize any excess acid (like buttermilk) that is present within the ingredients, preventing a sour overtone. Check out a recipe for buttermilk pancakes relative to milk-based pancakes, and you will see this difference in the leavening agents. What is the point of buttermilk in the first place if you don't need an acid, why not just use milk? In addition to providing the milk proteins that interfere with gluten formation, which are common to both milk and buttermilk, buttermilk is thick. When you add a thick liquid to your batter, then you don't need to add as much flour, also reducing the development of gluten. If you want to do a breakfast experiment, prepare milk-based and buttermilk-based pancakes and see if an eater notices the difference. Buttermilk pancakes tend to be fluffier, lighter, and more distinctive in taste.

9.15 Cakes

Cakes are similar to quick breads and muffins in many ways. They have a similar basic ingredients list, including flour, eggs, sugar, fat, and a chemical leavening agent. Most cakes also have a tender, porous, and moist texture, but are much lighter, finer, and spongier than a muffin (some cakes are formally called sponge cakes). You already know that "lightness" is generated by the retention of gas bubbles within the molecular structure of a food. However, what does this structure look like in a cake, and does the structure come from the ingredients or their treatment? A cake is an egg foam, but you need a lot more than eggs to make a great cake. Let's take a closer look at the ingredients list and cake-making techniques with an eye toward science.

9.15.1 Generation of the Foam

You may recall learning about egg foams in Chapter 8, specifically where you learned how whipping egg whites can make a meringue. The egg foam for a sponge cake needs to be more robust than that of a meringue, as it has to withstand the weight of the cake and any associated fillings In fact, the generation of the foam begins right at the start through the incorporation of air bubbles into the batter. You may have the first step of a cake recipe memorized, "cream butter and sugar together until light and fluffy." Although you might want to skip this step, it is actually one of the most important steps in cake baking and significantly influences the final texture. By beating the butter and sugar for an extended period of time, you incorporate air bubbles into the mixture that will eventually expand in the heat of the oven (along with the steam and baking powder-generated carbon dioxide bubbles) to make a properly risen cake that is light and fluffy. In doing so, you also evenly distribute the sugar, which is a key part of the air bubble incorporation. How? The solid sugar crystals, along with your mixing beaters, carry air molecules into the semisolid (crystalline/liquid) room temperature structure of the butter or fat. The air bubbles are retained in the fat, thus, as the butter/sugar mixture is creamed, the color of the mixture becomes lighter and the volume increases (i.e. it becomes fluffier) (Box 9.5).

The next step of cake batter preparation typically involves addition of the eggs and other liquid-based ingredients (e.g. dairy products, flavorings). The eggs are important to the development of the foam (see Chapter 8), but they also provide moisture, richness, and flavor to the cake. Recall what you have already learned about eggs and egg foams. Eggs are made of protein molecules. As eggs are beaten, they undergo mechanical stress that causes the proteins to denature. The hydrophilic parts of the proteins organize themselves to interact with other water-based molecules, while the hydrophobic parts of the proteins stick to the fats so that they can avoid water. With continuous beating, the incorporated air becomes enclosed in a protein/fat bubble with the hydrophobic parts directly toward the inside where it can interact with the nonpolar air molecules (like oxygen), while the hydrophilic parts are directed toward the outside where they can interact with the hydrophilic sugars. With more beating, air continues to be incorporated, and the air bubbles within the foam get smaller, making the cake more tender (Figure 9.26).

9.15.2 Flour

Although there are recipes for flourless cakes, most cakes contain flour. Flour is very important to a cake, as it helps to stabilize the egg foam, thus preventing the cake from collapsing in the oven. However, in a cake, the baker does not want to generate gluten, pop all of the newly generated air bubbles, or break down the egg foam. Therefore, it is best to gently fold the flour (and other dry ingredients) into the eggy, voluminous batter with a rubber spatula, particularly with an unleavened cake (see the following). As soon as all of the dry ingredients are

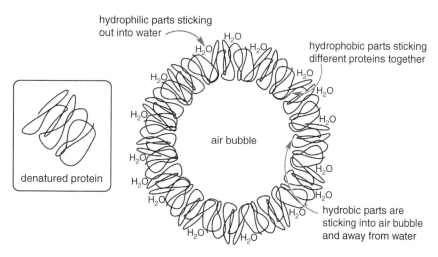

Figure 9.26 Proteins and foam. The denatured proteins will form a cage around air mixed with protein and fat (not shown here).

incorporated, the folding should cease. In a cake that utilizes a leavening agent, you may fold in the flour (as aforementioned) or mix your batter slightly to generate a limited amount of gluten. The gluten development will make the batter a little more capable to the capture of some of the leavening agent-generated carbon bubbles (due to the limited gluten sheets), and the cake will be well-raised and light. The problem with gluten generation is that your cake will not have as tender a texture. If you are interested in conducting an experiment in your home kitchen, prepare a cake batter entirely with an electronic mixer and one in which you fold the dry ingredients into the wet ingredients with a rubber spatula. Make a hypothesis about the texture and density of your two cakes and determine if the resulting products support your hypothesis.

9.15.3   To Use a Leavening Agent or Not?

Many cakes use chemical leavening agents. First, let's give an example of a cake that does not use a chemical leavener: a Genoese sponge cake. The key aspect of a tasty unleavened cake is the air bubbles; they must be small and you must generate a lot of them. In other words, in a non-leavened cake, you need to generate a stiff egg foam that can retain the incorporated air bubbles. However, the egg foam needs sugar because sugar increases the viscosity of the eggs. Without the sugar, a baker would not be able to beat the eggs fast enough to create an egg foam that could withstand the weight of the cake batter without heat to denature the egg proteins. The careful folding of the flour, additional dry ingredients, and melted butter (for a Genoese cake) into the mixture such that the foam is not destroyed is also key, along with baking the cake as quickly as possible since the fat begins to "melt" the egg foam relatively quickly. Even in the absence of a leavening agent, a Genoese cake rises a little during baking due to the expansion of the air bubbles and steam generation that is captured in the denatured egg and flour protein structure. At this point you may be thinking, isn't it simpler to use a leavening agent? Please read on and judge for yourself.

Baking powder (as opposed to baking soda) is the most common leavening agent for a cake because you desire a good rise and do not desire browning reactions. However, even with baking powder, good leavening is not foolproof and still requires some baking technique, particularly since the leavening-generated carbon dioxide bubbles need to be small. Small bubbles can be promoted in a number of ways by the baker: (i) preparing a "thicker" batter (relative to a Genoese sponge cake) and (ii) controlling the baking conditions. With a double-acting baking powder, some small carbon dioxide bubbles will be generated as soon as the "wet" ingredients are added to the "dry" ingredients (as you learned with muffins). A substantial increase in carbon dioxide production (due to the "slower acting" second acid) will occur when the temperature of the batter reaches approximately 120°F/50°C. Also at this temperature, much to the bakers dismay, the previously made CO_2 bubbles will grow larger. Large gas bubbles, particularly when generated in a thin batter, may come to the surface of the cake, be lost like a popped balloon, and the cake will have some bubble pockets dotted across its top. The more gas lost in this way, the fewer carbon dioxide bubbles in the cake, and the cake may not rise adequately. At about 140°F/60°C, protein denaturation occurs, which stabilizes the batter and retains the air bubbles in the batter. These bubbles can expand to a certain point, but if there is still a substantial amount of new carbon dioxide being produced from the baking powder, eventually the bubbles will burst. With all of these competing factors, some bakers think that making a sponge cake without a leavening agent is actually less complicated and more likely to be successful than sponge cakes that contain leaveners. You will have to judge for yourself which type of cake is preferred to bake or eat.

9.15.4   The Woes of a Collapsed Cake

If a cake is undercooked, it will collapse. Most bakers have experienced this woe at one point or another. What does baking do at a molecular level? Baking stabilizes the structure of the cake by increasing protein denaturation and thus the formation of more protein–protein intermolecular interactions (within and between the egg and flour proteins). Recall that as the proteins that surround the air bubbles denature, they unfold and stretch out. When a protein stretches or linearizes, it has more of an opportunity to interact with other protein molecules that

Figure 9.27 Pie crust.
MSPhotographic/Adobe Stock Photos

are also denatured. When the denatured protein molecules form intermolecular interactions and covalent cross-links (e.g. disulfide bonds) with one another, they become much more rigid and less fluid in their behavior. Thus, a baking cake makes a transition from a viscous liquid batter into a more solid, sponge-like structure, and the foam structure that had been holding the air bubbles is strengthened enough to withstand the increasing pressure associated with the expansion of the gas and vaporization of water. If baked properly, the structure of the cake should hold even as the cake cools, the air bubbles contract, and the vapor condenses. If it does, the cake's interior retains small air pockets, making the cake "light." Why does a cake collapse in the middle? Cakes are more cooked along the edges because the cake's edges interact with the metal pan; the pan both conducts heat and structurally supports the edge of the cake. Thus, a cake is gooier in the middle (i.e. it doesn't get as hot or "baked") and more crusty/crisp along the edges (Figure 9.27).

9.15.5 Moist versus Dry Cakes – The Components of a Cake

Cakes provide a good example of the science behind baking. Understanding the chemistry and biochemical mechanisms behind cakes can help bakers create the ideal texture, rise, and moisture, empowering a baker to make informed decisions about ingredient ratios, mixing techniques, and baking methods. By applying this knowledge, bakers can create moist, tender, and delicious cakes that showcase the intricate interplay between ingredients and scientific principles. Let's focus on the key factors that contribute to the moistness and texture of cakes, focusing on the chemical and biochemical mechanisms involved.

One of the critical factors in achieving a moist cake is the balance of ingredients. Cakes can be categorized into either the dry or moist category, depending on the ratios and types of ingredients used. The scientific principles that govern this categorization lie in the interactions between the ingredients and their impact on moisture retention and texture.

Moist cakes are known (by a baker and eater) for their tender and moist texture. They typically contain higher amounts of fats, sugars, and liquids, all of which play a crucial role in moisture retention. Fats, such as butter or oil, are hydrophobic and repel water. During baking, fats create a barrier that slows down moisture loss, trapping interior water molecules and enhancing the cake's moistness. Additionally, fats contribute to the overall richness and mouthfeel of the cake. Hygroscopic ingredients, such as sugars like sucrose, honey, molasses, and certain starches, also play a significant role in moisture retention. Because sugars have an affinity for water (remember that the hydroxyl groups present in sugars form hydrogen bonds with water molecules) and act as humectants, which are substances that attract and hold onto moisture. The effectively trapping it within the cake. Moreover, because of the higher amount of liquid in the batter, there is more moisture to trap. Examples of moist cakes include:

Butter cake: Butter cakes, such as pound cakes, have a relatively high fat content, containing equal parts (by weight) of butter, sugar, eggs, and flour.

Chocolate cake: Moist chocolate cakes also have a high fat content, often from ingredients like butter or vegetable oil. The presence of cocoa powder or melted chocolate adds additional moisture and richness to the cake.

Carrot cake: Carrot cakes are moist due to the inclusion of ingredients like grated carrots, crushed pineapple, and oil, which contribute to the cake's moisture and tenderness.

Dry cakes, known for a more firm and dry texture, are also beloved by bakers and eaters. As you might expect from the previous discussion, dry cakes typically have a lower fat content when compared to other cakes. Examples of dry cakes include:

Sponge cake: Sponge cakes have a lower fat content and a higher proportion of eggs. They are light and airy, with a delicate crumb. The high proportion of eggs contributes to the cake's structure, while the limited fat content leads to a drier texture.

Angel food cake: Angel food cakes are predominantly made with egg whites, sugar, and flour. They contain no added fats. The absence of egg yolks and minimal fat content results in a drier, lighter cake.

Biscuit or shortcake: Biscuit or shortcake recipes typically have a higher proportion of flour and minimal amounts of fats and sugars. These cakes have a crumbly texture and are very dry compared to other types of cakes, so they are often eaten with cream, as in strawberry shortcake.

As discussed earlier in this chapter, in addition to the ingredients, the structural changes that occur during the baking process also impact the moistness of the cake. When cakes are baked, the proteins in the flour and eggs denature and coagulate, forming a network that provides structure to the cake. This protein network acts as a sponge, holding onto water and preventing it from evaporating. Once the denatured proteins are "set" during baking, the moisture is trapped and retained.

The migration of starches and sugars within the cake batter also contributes to its moisture retention. During baking, starches and sugars dissolve in the liquid and migrate towards the outer edges of the cake. This migration helps create a gradient of moisture, with the outer edges being slightly drier than the center. As a result, the center of the cake retains more moisture, contributing to its moist and tender crumb. The movement of the starches and sugars is facilitated by diffusion and the movement of water molecules within the batter. Thus, the difference in taste and mouthfeel of an edge piece versus a center piece of cake is real.

As implied from the previous discussion, the ratios of ingredients used in a cake recipe also influence its moisture retention. Moist cakes often have a higher proportion of fats, sugars, and liquids compared to drier cakes. The fats contribute to moisture retention by coating the flour particles, forming a physical barrier that limits gluten formation and moisture loss. The sugars attract and hold onto moisture due to their hygroscopic nature, preventing the cake from drying out. Increased liquid content provides additional moisture, resulting in a more moist and tender cake.

To bring these differences to light, let's compare two cake recipes to better understand how the incorporation of different ingredients at different ratios impacts the moisture content of the final product (Table 9.2).

Scientific Explanation for Moistness: First, look at the fat content of the two cakes. This chocolate cake recipe contains vegetable oil, which adds moisture and richness to the cake. The fat molecules, being hydrophobic, create a barrier that slows down moisture loss during baking and after the cake is baked. The inclusion of cocoa powder or melted chocolate, which contains additional fat, in the recipe adds and helps to further retain moisture in the cake.

Table 9.2 Comparison of two cake recipes.

Dry sponge cake recipe	Moist chocolate cake recipe
Ingredients	Ingredients
• 1 cup cake flour	• 2 cups all-purpose flour
• 1 cup granulated sugar	• 1 3/4 cups granulated sugar
• 6 large eggs, separated	• 3/4 cup unsweetened cocoa powder
• 1/2 teaspoon cream of tartar	• 1 1/2 teaspoons baking powder
• 1/4 teaspoon salt	• 1 1/2 teaspoons baking soda
• 1 teaspoon vanilla extract	• 1 teaspoon salt
• Zest of 1 lemon	• 2 large eggs
	• 1 cup whole milk
	• 1/2 cup vegetable oil
	• 2 teaspoons vanilla extract
	• 1 cup water

The presence of sugars, such as granulated sugar, acts as humectants in the recipe. Sugars attract and hold onto moisture, preventing the cake from drying out. They form hydrogen bonds with water molecules, effectively trapping moisture within the cake and enhancing its moistness.

The recommended mixing technique of combining the dry and wet ingredients (which typically involves the use of an electric mixer) ensures proper hydration and even distribution of moisture throughout the cake batter. The interaction between the ingredients, such as flour, eggs, and milk, creates a protein network that contributes to the structure of the cake while also helping to retain moisture.

Scientific Explanation for Dryness: The dry sponge cake recipe contains no addition of fat as a stand-alone ingredient, which results in a drier texture due to reduced moisture retention. Moreover, the generation of an egg white foam in the sponge cake recipe contributes to its drier texture. Egg whites contain proteins that denature and coagulate during baking, forming a structure for the cake. However, without the presence of egg yolks, which contain fats and emulsifiers, there is less moisture retention. The lower proportion of sugars in the sponge cake recipe also contributes to its dryness. Given that sugars attract and hold onto moisture, the reduced amount of sugar in this recipe results in less moisture retention. The mixing technique of beating the egg whites with cream of tartar creates a stable foam with trapped air bubbles, with a gentle folding in of the dry ingredients. These air bubbles from the foam expand during baking, giving the sponge cake its light and airy texture.

As stated earlier, factors beyond the ingredients also influence the moisture level of a cake. Proper mixing techniques ensure the ingredients are evenly distributed without overmixing, which causes excessive gluten development, resulting in a drier and tougher crumb. The incorporation of air during mixing is also important, as it contributes to the cake's lightness and moistness. Moreover, reheating the oven, using the correct oven position, following recommended baking times, and choosing the desired cake pan also contribute to even heat distribution and prevent over- or underbaking.

9.16 Pastries: Flaky Pie Crusts and Puff Pastries

What is the common advice given to every early-career pie baker? Handle the dough as little as possible, and now you know the reason why. Mixing, working with, and kneading dough generates gluten, and gluten does not lead to a flaky, tender pie crust. However, there are a few other aspects of preparing a pie crust or pastry that make these baked goods quite distinct from a yeast-based bread, a muffin, a cookie, or a cake. The ingredients for a pie dough are simple: flour, water, and butter. You don't need an electric mixer or even a whisk, just two knives or a pastry cutter. Simple? Hmmm... most bakers would tell you otherwise.

In a typical bread dough, flour and water are the primary ingredients used to make the thick, heavy, nonsticky mass that we call a dough. Fat or any other ingredients are not present at a very high concentration. By contrast, in pastry dough, the high-fat content is necessary for the flaky texture. Why? The fat separates the starch particles present within flour from one another and prevents water from accessing the starch, thereby preventing gluten formation. This separation begins right at the beginning of the preparation of a pie crust. In most recipes, the cold fat is first mixed into a low-protein content flour using two knives, a pastry cutter, or a food processor until the pieces of fat are pea-sized. Pea-sized is actually important in order for the fat to be distributed evenly among all of the starch granules. The smaller the pieces, the more separated the granules, the less likely water will access the granules, and the flakier the crust. You might ask, why can't you just melt the butter and mix it into the flour such that it is evenly distributed? Remember that butter is made of fat, milk proteins, and water. When melted butter is added to flour, the water from the butter penetrates the starch granules, and the granules stick together before the molecules of fat can surround them. Therefore, you end up with water-filled starch aggregates (i.e. gluten alert) rather than individual starch particles. Moreover, when cold butter goes into a hot oven, the water evaporates immediately. When warm, soft butter goes into a hot oven, the butter water seeps into the structure of the crust, resulting in a soggy mess. It is the same rationale for using cold butter to make a crust, rather than butter at room temperature.

Once you have your fat-separated starch granules (which feel and look like a bunch of pea-sized pieces of butter covered in flour), the moisture is added to make a cohesive (but not pretty) mass. The addition of a cold liquid is important to avoid the warm butter phenomenon (see above), but the exact liquid will vary depending on the recipe. It may be icy water, cold buttermilk, or even vodka (see Box 9.6); the key part about the cohesion mass is to not make it homogeneous and pretty. The more moisture, the greater the chance for gluten development, thus a recipe will typically suggest that you sprinkle in the liquid in tablespoon amounts until you have just enough for the dough to coalesce. You basically want just enough moisture so that you can make a disk-shaped mass that can be wrapped and placed into the refrigerator or freezer. Because you have already coated the starch granules with fat, the starch granules will be unable to swell, and the starch proteins will be unable to interact with one another and with water. At colder temperatures, proteins also absorb less water (because solubility is temperature dependent), further reducing the hydration of the granules and preventing gluten development. If you start to knead or work with the dough extensively or the dough begins to warm, the developed gluten will make the crust chewy and tough. Pie crust preparation will take less than 5 minutes for a proficient pie baker.

Box 9.6 Pie crust liquids

When making a pie crust, the goal is to limit gluten formation. You can limit its formation by using a low-protein flour (e.g. pastry flour), not kneading the dough, and carefully selecting the type of liquid that you use to coalesce the dough into a mass. Recall that water promotes gluten formation by entering starch granules and forming intermolecular interactions with the gliadin and glutenin proteins, causing them to unfold and interact with one another. The more water, the better the chance of gluten formation. Eighty-proof vodka consists of 60% ethyl alcohol (CH_3CH_2OH) and 40% water. Interestingly, the ethanol does not readily interact with the gliadin and glutenin proteins, thus it does not promote protein unfolding and the corresponding gluten formation. By using vodka or another alcohol, a baker can add a little more moisture to the dough (making it easier to work with) without worrying about gluten formation due to the extra water. For those vodka fans out there, a vodka-prepared crust does not have any alcohol taste, as the alcohol evaporates out of the crust in the heat of the oven.

In the next step, the pie crust dough (and you) has a chance to relax. The dough goes into the refrigerator; you might go to the couch. In addition to making the dough cold, the rest period gives the protein molecules a chance to relax, such that if any were stretched out during mixing (which is not desired), they have the chance to shrink back without forming intermolecular interactions that stabilize gluten formation.

Once the dough is chilling, you might think that you are home free. All that is left is the simple step of rolling out the dough, placing it into the pan, and baking it. Not so fast! During rolling, the dough is stretched, flattening the fat chunks into sheets. Once again, with any stretching, gluten develops, and the baker needs to keep the gluten at a minimum. How is this accomplished, since the rolling process promotes gluten development? The pie dough needs to be kept cool during rolling, ensuring that the fat does not melt. Some professional baking kitchens have a marble or metal countertop for rolling pastry doughs to help keep the dough cool, with metal being the best surface for conducting heat, followed by marble, and then (in last place) come wood or plastic countertops. Because metal and marble conduct heat, they will stay as cold as the pie dough because both will readily remove the rolling pin-generated heat energy.

During rolling, you quickly learn whether you added the correct amount of water or generated (unwanted) gluten in your pie crust by the elasticity of the dough. Elastic, sticky dough that readily bounces back to its original shape and size when rolled has a high gluten content and excess water. You might recover the dough by adding a little extra flour to the work surface during rolling to achieve the right consistency. Pastry dough that is too dry will be very stiff, won't hold its shape upon rolling, and may even have chunks break off. Try adding a little more water to a dry dough in order to achieve a dough that can be rolled to approximately 3–4 mm thick.

Is the rolled crust now ready for the oven? Not yet. Once rolled, it is important to let the dough stand for another 10–20 minutes, as the dough has just undergone a lot of mechanical stress, which stretches the proteins, promoting gluten formation. As the dough relaxes during the rest period, the rolled sheet will contract, and any stretched proteins that have not yet formed fully into gluten will relax. The amount of contraction will depend upon the elasticity of the dough; the more elastic the dough, the greater the contraction. If the dough is not allowed to contract on the counter, the dough will contract during the baking, potentially spilling your pie filling contents over the edge of the crust. Ideally, the dough will keep its shape when cooked, so that you know what the final shape will be!

9.16.1 Baking the Crust

You have your pie crust rolled out on the counter, and you have a choice of an aluminum foil, glass, or metal pie plate. Which one should be chosen? Any shiny metal pan will reflect heat away from the pie; your pie will take longer to cook. Heat readily passes through a glass dish and your pie crust will cook and brown quite evenly. At all costs, avoid the flimsy aluminum foil pan. Your crust will not cook evenly, and the pans are so thin that they do not retain any heat, and the crust will cook slowly. You know this if you have ever picked up aluminum foil-wrapped bread directly out of the oven with your bare hands. As long as you hold onto the edges of the foil, you will not get burned.

By now you likely know that your crust gets "cooked" during baking and that there must be some science behind the baking process. During the baking of a pie, the desired crust becomes firm, crisp, and brown. This state is achieved by baking at a high enough temperature that the dough heats up quickly, causing the internal moisture (from added water and butter) to be converted to steam and evaporate out of the crust. If this works effectively, as the steam evaporates, it makes layers in the crust, giving the crust a characteristic flakiness. In contrast, if the butter melts slowly, that water seeps into the starch granules, resulting in a soggy crust. The same "soggy crust" problem exists when your crust is cooked with a filling; think about the characteristic soft, soggy bottom crust of Aunt Ruth's pumpkin pie. How do you avoid a soggy crust in a filled pie? Prebake the crust, also known as blind baking. Place the pie crust in the oven, filled with dried beans or pie weights to ensure that the crust doesn't develop air pockets, and bake it until the edges are barely golden (~15–20 minutes), then brush the crust with an egg wash (i.e. a beaten egg with a little water added). Then, you add the filling and finish baking the pie. The prebaking causes the crust to be partially set, while the protein layer of the egg wash prevents the liquid filling from seeping into the crust. You can also make that bottom crust a little more firm and browned by baking your pie on the lowest oven rack.

There you have it, simple as pie! The trouble is that you might make a great pie crust on one occasion with the perfect balance of structure and flakiness, then on another occasion, it doesn't turn out quite right. If you haven't already figured it out from this extensive section about pie crust preparation, pie crusts are finicky things; one extra teaspoon of water, a humid day, or variation in the protein content in your flour can all be the culprits of a crust disaster or the basis for the best crust ever. However, with knowledge of the science and practice, you will be able to make the necessary adjustments to make a great crust every time.

9.16.2 Puff Pastry

A puff pastry is a dough that has flaky layers. Baklava provides the perfect example of the flakiness of a puff pastry; it is so flaky that flakes of it drop onto your plate as you take a bite (Figure 9.28). Interestingly though, although puff pastry is flaky, each layer of the pastry has a high gluten content. How can this happen when you just learned that gluten makes pie crusts chewy, not flaky? In puff pastry, a layer of fat separates all of the gluten-rich pastry layers. Thus, when it is baked, the moisture in each fat layer is converted to steam, causing separation of each individual gluten-heavy layer from the other layers, none of which are connected.

In order to achieve this type of layered structure, you prepare a dough of flour, water, and sometimes a little bit of fat, and mix it minimally. You roll the dough into a square or rectangle and let it rest to allow the gluten to relax. Then, you add the fat by pounding (literally hitting it with your rolling pin) out a large piece of butter (by hitting it with your rolling pin) and placing it on the dough (Figure 9.29). The dough is turned onto itself (over the top of the butter), is rolled out again, and the steps are repeated five more times. Throughout the process, you develop layers of gluten that are separated by layers of fat, 729 and 728 layers to be exact, respectively.

Recall from the discussion about pie crusts, if the butter-generated steam is critical to the preparation of a puff pastry, you need to keep everything cold. Thus, don't try preparing puff pastry on a hot day; you will have better results making it in your friend's kitchen, who doesn't turn the heat on in the winter. Another solution is to store the rolled pastry in the refrigerator during the resting periods. After the final turn, the dough rests for about an hour, then is rolled out to a thickness of about 1/4 inch/6 mm and baked in a very hot oven. The escaping steam causes the layers to puff to a height that is at least four times the starting height of the unbaked pastry.

Figure 9.28 Saint Honore cake with chocolate and raspberry. Notice the flaky layers of cooked dough. The lack of interactions is due to fat that breaks up the cross-linking of proteins in flour between the individual layers. fahrwasser/Adobe Stock Photos

Figure 9.29   Folding puff pastry.

Science for the Chef: Whole wheat bread

One recipe that demonstrates several of the learned concepts and chemical transformations from this chapter is a traditional whole wheat bread recipe.

Ingredients:
- 3 cups whole wheat flour
- 1 cup bread flour
- 2 1/4 teaspoons instant yeast
- 2 tablespoons honey
- 1 1/2 teaspoons salt
- 1 1/2 cups warm water
- 2 tablespoons olive oil

Instructions:
1. In a large mixing bowl, combine the whole wheat flour, bread flour, instant yeast, and salt. Stir to evenly distribute the dry ingredients.
2. In a separate small bowl, mix together the warm water and honey until the honey is dissolved.
3. Make a well in the center of the dry ingredients and pour in the honey-water mixture and olive oil.
4. Stir the mixture with a wooden spoon or a dough whisk until it starts to come together.
5. Transfer the dough onto a clean, lightly floured surface and knead it for about 10 minutes until the dough becomes smooth and elastic. This kneading process helps to develop the gluten network.
6. Place the dough in a greased bowl, cover it with a clean kitchen towel, and let it rise in a warm place for about 1–2 hours until it has doubled in size. During this rise, the yeast ferments the sugars and starches in the dough, producing carbon dioxide gas, which causes the dough to rise.
7. Once the dough has doubled in size, punch it down to release the gases (remember technique here), and transfer it to a lightly floured surface.
8. Shape the dough into a loaf by rolling it tightly from one end and pinching the seams to seal. Place the shaped dough into a greased loaf pan.

9. Cover the loaf pan with a kitchen towel and let the dough rise for another 30–45 minutes until it has risen slightly above the edges of the pan.
10. Preheat the oven to 375°F (190°C) while the dough is rising.
11. Once the dough has risen, place the loaf pan in the preheated oven and bake for about 30–35 minutes, or until the bread is golden brown and sounds hollow when tapped on the bottom.
12. Remove the bread from the oven and let it cool on a wire rack before slicing.

The Science Behind the Recipe

Whole wheat and bread flour contain different levels of gluten-forming proteins, which contribute to the texture and structure of the bread. Whole wheat flour retains the bran and germ from the wheat grain, while bread flour is higher in gluten-forming proteins from the endosperm. When yeast is introduced, it metabolizes the sugars present, converting them into carbon dioxide gas and alcohol through fermentation. This gas production causes the dough to rise, resulting in a lighter and airy texture. Kneading the dough promotes gluten development. Gluten is formed when two proteins, gliadin and glutenin, interact with water and each other, creating a network that provides elasticity and strength to the dough. During baking, the heat causes the starch granules in the flour to absorb water and swell, a process known as starch gelatinization. This gives the bread its structure and contributes to the crumb texture. The browning of the bread crust is a result of Maillard reactions. These reactions occur between amino acids from proteins and reducing sugars in the dough, leading to the development of the desired color, flavor, and aroma. As the bread bakes, the carbon dioxide gas trapped within the dough expands due to the ideal gas law. This expansion pushes against the dough, causing it to rise further, creating a light and fluffy loaf.

The gluten network, created by the gliadin and glutenin proteins, contributes to the bread's strength, elasticity, and stretchiness. The disulfide bonds formed between cysteine residues in glutenin proteins provide stability and mechanical rigidity to the gluten network. The ratio of ingredients (i.e. flour, water, and fat ratio) determines the bread's moisture content and texture. The crust forms on the outer surface due to heat-induced reactions, while the crumb, the interior structure of the bread, is most influenced by gluten development and gas expansion. Ingredients like honey, salt, and olive oil not only add flavor, but also impact the dough's fermentation, gluten development, and final texture.

Key Concepts

1. A grain consists of the **germ**, **endosperm**, and the **bran**. The germ consists of fats, vitamins, and soluble protein. **White flour** is milled from the endosperm, which contains mostly starch (carbohydrates), along with soluble and insoluble proteins. The bran contains mostly insoluble carbohydrates (fiber) and limited proteins and starch. **Whole wheat flour** is milled from the germ, endosperm, and bran. **Soft flours** are flours that have a relatively low protein concentration and are best suited for baked goods with low gluten content, while hard flours that have a high protein concentration are best suited for goods with high gluten content.

2. **Starch** is composed of **amylose** and **amylopectin** carbohydrates. The molecular structure of amylose is a linear chain of glucose molecules, while amylopectin is a branched molecule. Amylose molecules pack together more tightly due to their more linear structure, while amylopectin packs more loosely and provides many "ends" for the storage and quick release of glucose by digestive enzymes. A **starch granule** forms through the intermolecular interactions of individual starch molecules. When heated in the presence of water, hydrogen bonds between the starch molecules break, allowing water to penetrate a starch granule, resulting in the swelling of the granules, solubilization of amylose, formation of hydrogen bonds between

starch molecules and water, and the formation of a viscous **gel**. When the starch gel cools, the disrupted hydrogen bonds between starch molecules start to reform, releasing water. This **retrogradation** process leads to the formation of a more firm structure.

3. **Gliadin**, a water-soluble protein, and **glutenin**, a water-insoluble protein, the two main proteins in wheat flour, interact to form **gluten** when water is added to the flour. Gliaden and glutenin interact through various types of bonds, including hydrogen bonds, hydrophobic interactions, and disulfide bonds. **Disulfide bonds** form from **cysteine** residues in glutenin, under oxidizing conditions, and are rearranged during dough mixing and kneading. Gluten, the term given to the network of interacting gliadin and glutenin proteins, can trap gases produced by yeast and give the dough its elasticity and structure.

4. **Yeast**, a single-celled microorganism, metabolizes sugars and starches present in the dough through a process called **fermentation**. Yeast converts glucose and other simple sugars into **ethanol** (alcohol) and **carbon dioxide** (CO_2) gas. The carbon dioxide gas is trapped within the gluten network, causing the dough to rise and creating air pockets, which result in a light and airy texture in the baked bread. As the bread is baked in a hot oven, gas bubbles expand according to the **ideal gas law**, $PV = nRT$, also contributing to the expansion of the bread.

5. **Maillard reactions** occur between amino acids (such as lysine and asparagine) and reducing sugars (such as glucose and fructose) during baking at high temperatures. This **non-enzymatic browning** reaction leads to the development of a brown, flavorful outer layer. Inside the bread, the distribution of starch, proteins, and water determines the formation of the **crumb**. **Starch gelatinization** contributes to the formation of a moist and tender crumb, while proteins provide structure and texture.

6. Ingredients like salt, fat, and sugar influence the formation and properties of the gluten network. Salt enhances gluten development by strengthening the gluten structure and improving dough **elasticity**. Fat (such as butter or oil) coats the gluten proteins, inhibiting gluten development and creating a **tender** texture. Sugar competes with proteins for water, reducing gluten formation and resulting in a softer texture.

7. **Celiac disease** is diagnosed when an individual has an immunological reaction to gluten. Due to the importance of the gluten network in many breads, in gluten-free baking, alternative ingredients and techniques are used to achieve the desired texture and taste. Ingredients like xanthan gum or guar gum are often added to mimic the structural properties of gluten. Other alternatives include using **gluten-free flours** made from rice, corn, or tapioca. Various chemical and physical adjustments are necessary to create gluten-free bread with acceptable texture and volume.

8. The most common **chemical leavening agents** are baking powder and baking soda, which are not interchangeable. When these chemical leavening agents come into contact with water and heat, a chemical reaction occurs, releasing carbon dioxide gas, causing the baked good to rise (quick bread), spread (cookies), and/or brown (cookies). In goods that do not include a leavening agent, extended mixing of butter and sugar and the development of an **egg foam** to incorporate air into the batter, whose gas bubbles will expand and be trapped during the baking process.

9. **Cakes** are categorized as dry or moist. Moist cakes typically contain higher amounts of fats, sugars, and liquids, which play a crucial role in moisture retention. Fats prevent moisture loss, and sugars and liquids help to retain moisture in the cake through intermolecular interactions. Dry cakes have a low fat content and often a high egg content, yielding a texture that is crumbly (as in a shortcake) or light (as in an angel food cake).

10. A flaky and tender pie crust is one that has as little gluten formation as possible. Gluten formation is avoided in pie crust preparation through the use of cold fat (typically butter), very limited water/moisture, limited kneading and handling of the dough, and allowing the dough to chill and rest to reduce the heat or mechanical stress impact due to rolling.

References

1 Anderson, O.D., Greene, F.C., Yip, R.E., Halford, N.G., Shewry, P.R. & Malpica-Romero, J.M. (1989) Nucleotide sequences of the two high-molecular-weight glutenin genes RT from the D-genome of a hexaploid bread wheat, *Triticum aestivum* L. cv RT Cheyenne. *Nucleic Acids Res*, 17, 461–462.

Additional Readings

Ajamian, M., Rosella, G., Newnham, E.D., Biesiekierski, J.R., Jane G. Muir, J.G., and Gibson, P.R. Mol. Nutr. Food res. Vol 65. 1901275 2021.

Chavan, R.S. and Chavan, S.R. (2011), Sourdough Technology—A Traditional Way for Wholesome Foods: A Review. *Compr Rev Food Sci Food Saf*, 10(3), 169–182.

Effect of Gluten Ingestion and FODMAP Restriction on Intestinal Epithelial Integrity in Patients with Irritable Bowel Syndrome and Self-Reported Non-Coeliac Gluten Sensitivity.

Gao, H., Jorgensen, R., Raghunath, R., Nagisetty, S., Ng, P. K. W., & Gangur, V. (2021). Creating hypo-/nonallergenic wheat products using processing methods: Fact or fiction? *Compr Rev Food Sci Food Saf*, 20(6), 6089–6115.

Gelinas, P., Roy, G., and Guillet. Relative Effects of Ingredients on Cake Staling Based on an Accelerated Shelf-life Test., M. J Food Sci. Vol 64, 937–940. 1999.

Godefroidt, T., Ooms, N., Pareyt, B., Brijs, K. and Delcour, J.A. (2019), Ingredient Functionality During Foam-Type Cake Making: A Review. *Compr Rev Food Sci Food Saf*, 18(5), 1550–1562.

Merlino, M., Arena, E., Cincotta, F., Condurso, C., Brighina, S., Grasso, A., Fallico, B. and Verzera, A. (2022), Fat type and baking conditions for cookies recipe: A sensomic approach. *Int J Food Sci Technol*, 57:5943–5953.

Struyf, N., Van der Maelen, E., Hemdane, S., Verspreet, J., Verstrepen, K.J. and Courtin, C.M. (2017), Bread Dough and Baker's Yeast: An Uplifting Synergy. *Compr Rev Food Sci Food Saf*, 16(5), 850–867.

Uthayakumaran, S., Newberry. M., Keentok, M., Stoddard, F.L, and Bekes, F. Cereal Chem. Basic Rheology of Bread Dough aith Modified Protein Content and Glutenin-to-Gliadin Ratios. Vol 77. pp 744–749. 2000

End of Chapter 9 Questions

1. What are the three main structural components of a grain?
   a. Germ, bran, and starch
   b. Endosperm, germ, and amylose
   c. Bran, endosperm, and gluten
   d. Germ, endosperm, and bran

2. Which part of the grain is nutrient-rich and contains fats, vitamins, and soluble proteins?
   a. Bran
   b. Endosperm
   c. Starch
   d. Germ

3. What is the primary difference between "soft" and "hard" flours?
   a. Soft flour has higher protein content than hard flour.
   b. Hard flour has higher protein content than soft flour.
   c. Soft flour contains more vitamins than hard flour.
   d. Hard flour contains more starch than soft flour.

4. Which two proteins interact to form gluten in dough?
   a. Gliadin and starch
   b. Glutenin and amylopectin
   c. Glutenin and gliadin
   d. Amylose and amylopectin

5. What is the role of disulfide bonds formed between cysteine residues in glutenin proteins?
   a. They contribute to the stability and mechanical rigidity of the gluten network.
   b. They enhance the flavor and aroma of baked goods.
   c. They facilitate the expansion of bread during baking.
   d. They prevent starch from separating in pie crusts and puff pastries.

6. During yeast fermentation of starches and sugar, what end products are formed that provide energy to yeast and contribute to a yeast-raised baked good??
   a. CO_2 and water
   b. CO_2 and alcohol
   c. Alcohol and water
   d. CO_2 and yeast

7. What chemical processes contribute to the color, flavor, and aroma of the crust and crumb of bread?
   a. Fermentation and gluten formation
   b. Amylose and amylopectin interactions
   c. Maillard reactions
   d. Disulfide bond formation

8. In a cake, which ingredient helps to stabilize an egg foam, preventing it from collapsing in the oven?
   a. Sugar
   b. Salt
   c. Flour
   d. Fat

9. Which type of leavening agent is commonly used in cakes?
   a. Baking powder
   b. Yeast
   c. Baking soda
   d. Cream of tartar

10. Describe the process of starch gelatinization and retrogradation. How does changing starch sources impact the type of thickening results while cooking?

11. Explain the role of gluten in baked goods and how it is formed during the preparation of dough.

12. Describe the chemical reactions involved in the Maillard reactions and their contribution to the color, flavor, and aroma of bread crust.

13. How does celiac disease affect the body's immune response to gluten, and why is it challenging to bake high-quality gluten-free bread?

10

Seasonings: Salt, Spices, Herbs, and Hot Peppers

Guided Inquiry Activities (Web): 13, Flavor; 14, Cells and Metabolism; 19, Plants; 20, Plants and Color

Learning Objectives

1. Understand the chemistry and molecular structure of salt and its role as an ionic compound formed by a reaction of an acid and a base.
2. Explain the process of salt dissolution in water and the role of water molecules as a solvent in disrupting the attractive force between salt ions.
3. Recognize the impact of salt on cooking, baking, and human health, including its effect on taste, protein behavior, and boiling point of water.
4. Understand the distinction between salt, herbs, and spices as seasonings.
5. Identify the botanical origins of herbs and spices and their different plant parts used.
6. Explain the role of volatile organic compounds in the aroma and flavor of herbs and spices.
7. Describe the chemical characteristics and biological activities of phenols, esters, terpenes, and pungent compounds found in herbs and spices.
8. Explain the chemical compounds responsible for the flavor and aroma of vanilla and the difference between natural and synthetic vanilla.
9. Explore the genetic basis for the love/hate relationship people have with cilantro and its connection to the perception of flavor.
10. Identify the chemical components and unique characteristics of saffron, nutmeg, mace, and curry and their role in flavor profiles.
11. Understand the chemical composition of chili peppers, including the primary compound responsible for their heat, capsaicin, and its role in activating pain receptors.
12. Explore the various colors exhibited by chili peppers and their underlying pigments, such as carotenoids and chlorophyll.
13. Investigate the factors that contribute to the heat levels of chili peppers, including the different capsaicinoids present and their binding affinity to receptors.
14. Analyze the biological purposes served by capsaicin in chili plants, including its role in chemical defense against herbivores and fungal pathogens, as well as its involvement in seed dispersal.

The Science of Cooking: Understanding the Biology and Chemistry Behind Food and Cooking, Second Edition. Joseph J. Provost et al.
© 2025 John Wiley & Sons, Inc. Published 2025 by John Wiley & Sons, Inc.
Companion website: www.wiley.com/go/provost/food_science_2e

10.1 Introduction

We will learn about the molecular structure and behavior of simple inorganic salts, organic compounds that make up spices and herbs, and the more complex biomolecules found in hot peppers. We will discuss and investigate how these compounds impact food, how they are best prepared and used, and examine the history and use of hot peppers.

As you sprinkle salt on your popcorn, have you ever thought about the historical impact of those tiny grains on humans and our livelihood? Humankind's historical search for salt, spices, and herbs has caused wars, sent people to discover lands that were unknown to them, and encouraged the development of new methods to access and acquire particular spices. Seasonings not only enhance the flavor of food, but they have helped define some cultures and significantly changed how food can be prepared and stored safely. We all have experienced how adding a small amount of a compound like salt can enhance the flavor of food.

Many historical tales describe the use of seasoning to mask the flavor and odor of tainted foods. In 1939, biochemist J.C. Drummond in *The Englishman's Food: Five Centuries of English Diet*, told of the medieval recipes used to mask the flavor and odor of rotten meat [1]. In *Food and Cooking in Roman Britain*, Marian Woodman describes the difficulty in storing food and the use of seasoning to mask the lack of freshness of food [2]. However, food historians will argue these tales as myths and point out that those who could afford spices and herbs were the least likely to eat spoiled meat and food. In fact, some seasonings are now known to reduce the incidence of spoilage and contamination rather than being used to "cover" it up. Salts dry and inhibit bacterial growth in cured meat; some spices and herbs even behave as antibiotics. Thus, it is entirely plausible that people who found certain flavors attractive were more likely to use them in cooking and were less likely to become sick from microbial pathogens. These people would teach the use of spice and pass on their genes for the seasoning taste receptors, along with the heightened desire for spiced food. Hot peppers have grown in popularity to rival most herbs and spices and have a fascinating science, earning their own place in books on cooking. Hot peppers may have been particularly important in hot and humid climates (where they grow easily and well) where refrigeration was historically at a premium. Salt, spices, and herbs all play an important role in the health and taste of our food. Let's start our discussion with the most simple, most utilized, and perhaps most important seasoning of all, salt.

10.2 Salt: Flavor Enhancer and a Driving Force of History

Salt, whether mined from the ground or dried from the sea, is a critical component of human health and has been integral in shaping much of the world's history. Many modern roads were initially paths created by animals to salt licks. The discovery and harvest of salt (and spices) created trade routes, resulting in global power shifts and colonization around the world. The Romans used salt as part of a soldier's pay, in fact, salt is the root of the term "salary" [3]. In order to understand how salt is indispensable in cooking, baking, and human health, it is important to start with the basics: the chemistry and molecular structure of salt.

10.2.1 Chemistry of Water and Salt

The chemical definition of a salt is an ionic compound formed by a reaction of an acid and a base. Salts are compounds composed of cations (positively charged ions) and anions (negatively charged ions) whose charges balance out one another. Common examples of monovalent (singly charged) salts include table salt (i.e. sodium chloride, consisting of an equal number of Na^+ and Cl^- ions), and potassium chloride (consisting of an equal number of K^+ and Cl^- ions). Magnesium chloride is an example of a divalent salt, consisting of one Mg^{2+} for every two Cl^- ions. In some cases, polyatomic ions make up a salt, such as that found in sodium sulfate, which consists of two Na^+ cations and one polyatomic sulfate (i.e. SO_4^{2-}).

Figure 10.1   Ionic lattice of sodium chloride.

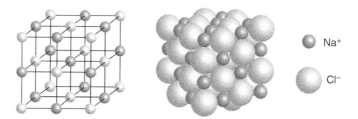

In the solid state, a salt forms a well-organized, three-dimensional network called an ionic or crystal lattice, where each ion is surrounded by ions of opposite charge (Figure 10.1). This ionic attraction holds the cations and anions in place, allowing for the formation of large crystals due to the presence of a repeating geometric pattern.

When table salt or some other salts are dissolved in water, the ions dissociate from one another (Figure 10.2). This means that the cationic and anionic components separate from one another. Why? Remember that water molecules are polar, where the oxygen atom of water has a partial negative charge due to its attraction or affinity for electrons, while the two hydrogen atoms in water have a partial positive charge. In solution, the water molecule surrounds the individual ions of salt (Na^+ and Cl^-) due to its large dielectric constant.

What in the world is a dielectric constant? A dielectric constant is a numerical value that indicates how likely two ions will separate in a particular solvent. Water has a large dielectric constant because water counters the attraction between the cation and anion of the salt since water is polar. In short, water (H_2O) reduces the force (defined by Coulomb's law; Figure 10.3) that holds the ions together because it is highly attracted to and surrounds the cationic and anionic components of the salt. In a salt water solution, shells or cages of water surround the ions, with the negative or positive pole of water molecules orienting themselves to neutralize the ionic compound. This arrangement shields the attraction of positive and negative ions (e.g. Na^+ and Cl^-) from one another, keeping the ions in solution (surrounded by water molecules), and thus the salt dissolves in water.

When water is lost due to evaporation, the ions lose some interactions with water molecules, become more concentrated, and begin to form into ion clusters (consisting of interacting cations and anions). With increased loss of water, these ion clusters get large enough to form crystals and precipitate out of the aqueous solution. Moreover, since every salt has a different solubility in water, every salt precipitates more or less readily at different water concentrations! In water above 104°F/40°C, KCl is more soluble than NaCl. Thus, by carefully heating a

sodium chloride (i.e. table salt) is an ionic compound. It is made of 2 different types of atoms that are held together by a positive-to-negative attraction called an *ionic bond*

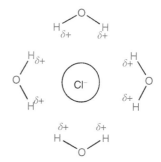

The sodium cation (Na^+) is surrounded by a cloud of water molecules that are oriented to present their *slightly* negative oxygens toward the positively charged sodium

The chloride anion (Cl^-) is surrounded by a cloud of water molecules that are oriented to present their *slightly* positive hydrogens toward the negatively charged chloride.

Figure 10.2   Salt dissolves in water. The polar nature of water helps to disrupt the attractive force between ions in a salt crystal.

solution (like seawater) that contains both KCl and NaCl, you can remove NaCl from the mixture by precipitation while leaving the bulk of the KCl still dissolved in water. The solubility properties of salts are quite important in the kitchen; thus, we will come back to this topic later in the chapter.

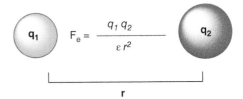

Figure 10.3 Coulomb's law. The magnitude of force between charged particles is described by Coulomb's law. The electrostatic force, F_e, is equal to the quantities of the two charges divided by the square of the distance times the dielectric constant, ε. ε is equal to 80.4 for water and 2.2 for benzene.

10.2.2 Sodium and Health

You have probably seen a "reduced sodium" can of soup in the grocery store or been told to watch your salt intake by your doctor. What do sodium and salt have to do with human health? Everything! Salt is essential for human life. Sodium is critical for function in the blood, lymphatic system, and cells throughout the body. Nerve cells use the concentration/charge gradient of sodium and potassium ions to fire signals throughout the nervous system. Sodium and potassium ions are essential in generating the high-energy molecule, ATP, from ADP. A 180 lb adult has nearly 0.2 pounds of sodium and on average, sodium comprises 0.1–0.2% of an individual human. The recommended daily allowance of sodium for adults is 2,000–2,300 mg or about one-half teaspoon of table salt each day, most of which comes from food and drink. Although this seems like a lot of salt, most of the salt that humans ingest does not come from a salt shaker but from highly processed, canned, frozen, or salty snack foods. Diets high in sodium significantly contribute to high blood pressure, heart disease, and stroke. These diseases kill more people in the United States than all cancers combined. A recent study of over 3,000 participants [4] found that a moderate four-week-long decrease in sodium intake from 9–12 to 5–6 g/day decreased blood pressure in both participants with and without high blood pressure problems. While increased amounts of the kidney enzyme renin also accompanied the decrease in sodium intake and blood pressure, which may contribute to several health issues, including diabetes and vascular and renal disease, the benefit of sodium reduction is thought to outweigh this negative impact. A different recent study of over 3,000 people found that moderate levels of sodium intake did not translate into a greater risk of hypertension or coronary disease. In contrast, in this same 2011 study, lower sodium intake was associated with higher heart disease mortality [5]. Although these studies provide some mixed results, generally speaking, low to moderate salt intake makes for a healthier long-term lifestyle.

10.2.3 Use of Salt in Cooking

The culinary use of salt is vital. Salt improves the way we see, taste, and smell food. Salt in food helps to keep proteins from aggregating into a clotted mess. Sodium chloride can stabilize oil-in-water emulsions, reducing the separation of oil and water. We have already discussed the preservative role that salt plays in food. Although excessive salt intake is detrimental to human health, cooks cannot ignore the benefits and role of salt in cooking.

Over 5,000 years ago, humans began to use salt to enhance flavor and preserve food. How do we taste salt, and how, biologically, does it enhance flavor? Proteins on the surface of taste bud cells transport Na^+ into the taste bud, which signals "salty food" to the brain. However, if you have ever added salt (or forgotten to add salt) to a cookie recipe, you know that salt does more than just cause food to taste salty. Low concentrations of salt suppress bitter flavors, allowing other flavors to come through the palate. At high concentrations, salt can increase umami or savory flavors by decreasing sour and sweet flavors. For example, most tasters found a mixture of table sugar (sweet) and urea (bitter) to be equally bitter and sweet. However, adding sodium to the mixture made the mixture seem overwhelmingly sweet. A salad with a bitter green like spinach or arugula can be made sweeter by adding salt. This amazing property of salt may help you to recover and fix a seemingly bitter and ruined meal!

In addition to its contributions to taste, salt impacts the behavior of proteins during cooking and baking. As you have learned in previous chapters, during cooking, proteins often denature in the presence of acid or heat. The denatured, unraveled protein molecules get tangled up with other proteins, forming large protein aggregates that precipitate. You can recognize this aggregation process as clots or protein curds (as in curdled or sour milk). How does salt impact protein denaturation? In low-salt conditions, salt ions will interact with the protein molecules, preventing their aggregation and the formation of an insoluble (precipitated) complex. This property comes in handy when preparing meat loaf or meatballs. Most meatloaf or meatball recipes include an egg in the ingredient list. The egg works as a binding agent helping the meatloaf to hold its shape; this binding property is due to the egg white proteins. What does this have to do with salt? The meatloaf needs to be baked to a high enough temperature to denature the egg white proteins. However, a meatloaf does not have an eggy-looking exterior or interior like a fried egg. The addition of salt to the meatloaf (containing the egg white proteins) helps the proteins remain in the "solution" (even in acidic or heated conditions) and not aggregate, thus enhancing the ability of the egg white proteins to unfold and form an interconnected network, thus binding the food together.

What about baking? Why does salt-free homemade bread have a horrible taste? In bread, salt plays a different role with proteins. As you learned in Chapter 9, gluten is a complex network of proteins responsible for creating a stretchy bread dough due to the formation of cross-links with other wheat proteins. These cross-links govern the texture of the final bread product and cannot form until water hydrates wheat flour. The proteins that make up gluten have many positively charged amino acids, and in solution, these charges cause the protein molecules to repel each other. A negatively charged chloride ion, provided by table salt, binds to the positive charges in the proteins, allowing them to come close together and to form the cross-links and connections that strengthen the dough (i.e. formation of the gluten network). However, the same interaction (between protein molecules and sodium chloride ions) slows down the hydration of gluten proteins because the interaction of proteins with ions reduces interactions between protein molecules and water (since the protein charges are now neutralized by the salt). Bakers will sometimes begin to mix their dough without salt to reduce this effect and then add the salt following the hydration of the flour. If you want to taste the difference that salt can make in the flavor and texture of bread, try a low- or no-salt bread recipe. The flavor will likely be flat, perhaps unlike any bread you may have tasted. A real-world example of this type of bread is the traditional bread baked in Tuscany/Florence Italy. Due to either the high taxes of salt in early central Umbria (now modern Tuscany, Italy), the papal state taxation, or a regional war versus Pisa, salt is not used in traditional *pane toscano* (Tuscan bread) or *pane sciapo* (bland bread). The result is just that—a bland and flavorless, dense, and chewy bread.

Salt can also act as an emulsifier in cooking. Emulsifiers help two solutions that would normally remain separated (like oil and water) to be one homogeneous mixture or solution. How does this work? The salt ions interact with the water molecules (as described above), which increases the density of the aqueous solution and allows the polar and nonpolar components to more readily mix. While able salt can be used as an emulsifier; however, other more complex salt compounds such as sodium citrate, phosphates, and tartrate are more commonly used to emulsify foods such as cheese and dairy products. In some foods, table salt will enhance the water binding to proteins, forming a gel and keeping phases from separating.

Do you add salt to pasta water to make the water boil faster? While the old tale actually has a bit of scientific truth to it, it is not very accurate. It is accurate that adding salt or other compounds to water will elevate the boiling point of the solution. The boiling or melting point of a solution is called a colligative property. Any substance, whether salt, sugar, or cinnamon, dissolved in water will alter both the boiling and melting points of the solution. The more substance dissolved, the more the melting or boiling point will change. How much is the change, and does it matter?

The equation for boiling point elevation is shown below, where ΔT_b is the change in boiling point, i is the van't Hoff Factor, K_b is a constant for the solvent, and m is the molarity or concentration of the solution.

$$\Delta T_b = i K_b \; x \; m$$

The van't Hoff Factor, i, is equivalent to "2" for table salt since it dissociates into two ions in water (Na^+ and Cl^-). Thus, for a water and salt solution, the only factor that will impact the change in boiling point is the mass of table salt dissolved in the water (m). The more the salt, the greater the change in boiling point. Because m is a positive value, ΔT_b will be positive and the boiling point will increase. What does this mean for the cook? When you add salt, more heat is required to bring the mixture to a boil than pure water. It will also take longer to boil the same volume of salt water versus pure water. However, the boiling saltwater solution will have a higher temperature; therefore, food cooked in salted, boiling water could cook faster than food cooked in unsalted water.

Unfortunately, validation of the old tale ends here. Let's imagine that you add 533 g (about 2.25 cups) of salt to one gallon of water. The boiling point will increase by a whopping 2.5°C. In other words, the teaspoon of salt that you may add to your water does not change the boiling point by a significant amount. However, the addition of a pinch or two of salt to water is important in cooking! Salt increases the volatility of some compounds, making it easier for these compounds to escape the boiling water and enhance the flavor and aroma of the cooked food. Salt water saturates the starches in boiling pasta, enhancing the flavor of the pasta. So, keep adding salt to your boiling water; it does make a difference in the flavor of the final product.

10.2.4 Too Many Kinds of Salts

A trip to the gourmet grocery store or a close inspection of various recipes may lead you to think there are different kinds of salt: smoked salt, sea salt, table salt, flake salt, kosher salt, etc.—the list is long and confusing. However, in chemistry, salt is a general term referring to any salt consisting of anions and cations. Potassium chloride (KCl), sodium chloride ($NaCl$), and magnesium sulfate ($MgSO_4$) are all salts. By contrast, in cooking, salt takes on a more specific definition. The culinary definition of salt is sodium chloride ($NaCl$) or table salt. The various salts that are described in recipes or are present on the grocery store shelf are all sodium chloride. The difference between each version of sodium chloride lies in the preparation and presence of additional components (contaminants).

Around the world, salt is primarily produced by two techniques: mining or evaporation of seawater. Most table salt comes from rock salt that is mined from the earth. Large, ancient underground salt formations can be mechanically removed with mining equipment and explosives, ground into small pieces, and dissolved in water. Alternatively, water can be pumped through mines, which dissolve the salt, making a saturated brine solution that is pumped from the caverns. The salt solution is evaporated to concentrate the salt, and the salt crystallizes and precipitates out of the solution. In windy and warm conditions, the brine is left to concentrate (and the salt to crystallize) in open pans. Modern evaporation techniques use vacuum chambers to reduce the water content. The salt brines contain a number of contaminating minerals and other compounds, which are removed by a process called selective precipitation.

Seawater, which contains approximately 3.5% $NaCl$, provides a nearly inexhaustible source of salt. The simplest method to harvest salt from seawater is to pump the seawater into smaller drying ponds where the water evaporates in the wind and sun. The evaporated seawater is rinsed with a pure, saturated saltwater solution to rinse off impurities without dissolving the salt crystals. This simple process can be easily recreated if you live near a saltwater body. In short, filter a clean, nonpolluted gallon of seawater (to remove rocks and debris), place it in a shallow pan, and evaporate the water using the heat of an oven or the sun. A gallon of seawater will produce a little less than a cup of salt. If you heat the final product in an oven at a high temperature, you can evaporate the last drops of water and kill any microorganisms. Is there any difference in the sun evaporation versus the oven evaporation methods? Maybe. Rapid evaporation and the addition of small salt crystals (which will initiate or start the crystallization process) cause the formation of small granular crystals that we associate with table salt. Slow evaporation allows salt crystals to grow larger, with a more irregular size. These salt "flakes" are sought-after due to their sharp edges, which allow the salt to stick to cooking surfaces, producing a crunchy, salty finish to foods. Crystal size and shape are two ways of distinguishing different types of salts. Let's look at some other differences that lead to the different types of salt found on the grocery store shelf.

10.2.4.1 Table Salt

Table salt, consisting of small, regularly shaped grains, is slow to dissolve because of the shape and purity of each grain or crystal. Iodine is often added to table salt (you may have seen "iodized" and "uniodized" versions) to reduce the incidence of goiter, a thyroid disease associated with problems in mental development, which, prior to iodized salt, plagued the United States and other countries. Glucose is also added in very small amounts to stabilize the iodine for long-term storage. To keep the salt free-flowing, additional additives may be used at low concentrations, such as other sodium, calcium, aluminum, phosphate, and silicon salts. These salts help to absorb any water that may be present in the salt or in the air, thus preventing clumping and caking of the table salt.

10.2.4.2 Kosher Salt

According to Jewish dietary law, meat must be treated with salt to draw out the blood. Given that salt already adheres to the Jewish dietary law, there are no additional dietary or production requirements for Kosher salt. However, most producers either do not include any additives (iodine or other) or just add anticaking agents. Kosher salt is typically produced by slow evaporation techniques, yielding large, thin sharp-edged salt flakes. The thinner and jagged-edged flakes quickly dissolve and more readily stick to foods than the table salt granules. For this reason, some cooks prefer to use Kosher salt in cooking.

10.2.4.3 Canning or Pickling Salt

Canning or pickling salt is produced in the same manner as table salt; however, it doesn't contain any iodine or anticaking agents. As these additives are insoluble at the high salt concentrations required in canning processes, the use of regular table salt in canning will yield a cloudy (and visually unappetizing) canned good. Both canning and pickling salts are produced and milled to form a small, fine-grain crystal that dissolves even more quickly than the larger cubed table or rock salt.

10.2.4.4 Rock Salt

Rock salt is the product of raw, undissolved, crushed, or large crystal sodium chloride formation. Because it doesn't go through a purification process, it contains contaminants of minerals and other compounds. Consequently (as you might predict), rock salt is a less expensive salt that is usually not used for food and cooking, except when making homemade ice cream using a maker that requires the use of a salt/ice mixture for freezing. Rock salt is not used in the salt grinder that you might see on your dinner table; salt grinders contain large crystals of purified salt.

10.2.4.5 Gourmet Salt

The diversity and custom flavors found in gourmet salts are popular with chefs and at-home cooks alike (Figure 10.4). Gourmet salts are flake salts formed from local water sources, which often contain regional or added contaminants (i.e. organisms, other salts, and minerals trapped in the salt flakes) that give the salt a unique color or taste. Some gourmet salts are made with added flavorings. The most famous artisan salt is made by an ancient French technique. Traditional fleur de sel (flower of salt) comes from salt beds in west-central France. Minimally disturbed, the salt flakes grow by dehydration on the surface of the pond and are collected by workers who scrape this layer of salt before the crystals sink to the bottom of the container or ponds. Unrefined and often not washed, these salt flakes contain a range of minerals and even small amounts of algae, all of which contribute to the flavor. Is fleur de sel safe to eat? Yes, this salt is safe for consumption, but given the high cost (approximately $30/lb), it is most frequently used as a garnish or sprinkled to finish a dish.

Himalayan pink salt is crystallized rock salt that comes from the Pakistan Himalayan mountains, where a high mineral content leads to its characteristic color. The lava or coral particles added to make Alaea Hawaiian sea salt give this salt its red or pink-brown color. You can buy or make numerous other flavored salts to provide additional unique and interesting flavor to a dish. Common flavored salts include garlic, celery seed, and lemon, but if you check out the spice section at your local grocery store, you will find a multitude of others.

Figure 10.4   Gourmet salt. Three types of salt with sea salt on the right. tbralnina/Adobe Stock Photos

10.2.4.6 Sea Salt

Like gourmet salt, sea salt is produced from solar heating or thermal evaporation of seawater. Since sea salt is less purified and refined than table salt, it is off-colored (sometimes gray) with large pyramid-shaped crystals and sharp edges. Because of the size and shape of the flakes and their expense, sea salt is best used to stick to the surface of a prepared food as its delicate structure dissolves quickly in the mouth, providing an immediate crunchy, salty sensation. However, there is little evidence to support some claims that sea salt tastes differently or is healthier than other salts. Why might there be a health benefit? The rationale behind this claim is that there is less sodium per tablespoon in sea salt relative to table salt due to the presence of potassium and calcium salts. However, both table and sea salts contain about 40% sodium by weight. Moreover, the additional minerals found in sea salt are often included in our diet from other sources. The true benefit of sea salt is the quick dissolving, crunchy mouthfeel that is present when the salt is used as a finishing salt for a dish, added immediately prior to serving.

10.3   Herbs and Spices

While some argue that salt is a spice, it is not. Although salt, herbs, and spices are indeed all seasonings, herbs and spices are the products of plants (and now you know that salt does not come from a plant and is, in fact, an inorganic mineral). Herbs are derived from the leaves of a plant, while spices are harvested from the rest of the plant (i.e. the root, stem, bark, seed, or plant fruits). Some plants, such as cilantro and dill, produce both spices and herbs, while others, like basil and cinnamon, produce herbs or spices, respectively. Generally, herbs are grown in more temperate climates, while spices grow in warmer or tropical zones.

As with salt, herbs and spices have significantly influenced humankind. The earliest evidence for the use of herbs or spices comes from ancient humans who wrapped their food in leaves, which presumably made for a more flavorful food. Table 10.1 shows some common herb–food pairings, some of which may have been first discovered during these earlier times. However, there is also historical evidence for using spices and herbs to preserve food,

Table 10.1 Herb–food suggested pairings.

Food	Suggested herb or spice combination
Beef	Bay leaf, marjoram, nutmeg, onion, pepper, sage, thyme
Lamb	Curry powder, garlic, rosemary, mint
Pork	Garlic, onion, sage, pepper, oregano
Chicken	Ginger, marjoram, oregano, paprika, poultry seasoning, rosemary, sage, tarragon, thyme
Fish	Curry powder, dill, dry mustard, marjoram, paprika, pepper
Carrots	Cinnamon, cloves, dill, ginger, marjoram, nutmeg, rosemary, sage
Corn	Cumin, curry powder, onion, paprika, parsley
Green beans	Dill, curry powder, marjoram, oregano, tarragon, thyme
Potatoes	Dill, garlic, onion, paprika, parsley, sage
Squash	Cloves, curry powder, marjoram, nutmeg, rosemary, sage, cinnamon, ginger
Rice	Chives, green pepper, onion, paprika, parsley

as perfumes, in religious ceremonies, and for medicinal purposes. Because of the myriad uses, the human desire for access to and control of herbs and spices drove colonization and expanded exploration routes. Historians find links to trade routes, changes in political power, and geopolitical conflict based on access to herbs and spices.

The aroma and flavor associated with most herbs and spices are due to what chemists call volatile organic compounds. Volatile organic compounds are primarily made of carbon atoms (remember that any molecule that is carbon-based is organic). These compounds are not very polar or ionic, they have a high vapor pressure and low water solubility. Simply put, this means that such molecules have few interactive forces with water and have enough energy at relatively low temperatures to escape from a liquid into a gas phase, where the aroma can reach our nose.

Let's look at two herbs as an example: cilantro (coriander) and parsley. If you visually compare these two leafy herbs, they can be easily confused (although the mnemonic "c"ilantro has curved leaves, while "p"arsley has pointy leaves might help you remember). However, by crushing a leaf of either plant, which allows the volatile compounds to escape into the air, will immediately inform you of the herb's identity. Cilantro's aroma and flavor mostly come from a family of carbon compounds called decanals (Figure 10.5). Decanals are 10-carbon-containing molecules with an aldehyde functional group on the first carbon. Some of the decanals also have a double bond within the carbon chain. The compounds associated with the aroma of parsley also lead to a complex scent, but the principal compounds found are 1,3,8-*p*-menthatriene and limonene (Figure 10.5). 1,3,8-*p*-Menthatreine provides parsley with its floral scent, while limonene is the same compound found in oranges and lemon; this mixture contributes to the complex aroma bouquet of parsley.

If you compare all three compounds, you will notice that none are charged and all are hydrophobic and nonpolar. The structures of the two parsley compounds are very similar, with the main difference being the placement of the double bonds in the ring structure, while the shape of decanal is quite distinct. Nevertheless, as you might

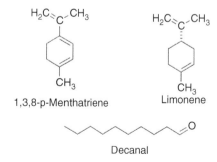

Figure 10.5 Aroma compounds of cilantro and parsley. The two ring compounds 1,3,8-*p*-menthatriene and limonene are responsible for the smell of parsley, while the long carbon chain decanal gives cilantro its odor.

predict, all of these compounds are poorly soluble in water and have a high vapor pressure that is characteristic of volatile aromatic compounds. The compounds smell differently because of the sensitive receptors for taste and smell in the nose that detect these subtle molecular differences with great discrimination and at very low concentration.

The flavorful and aromatic compounds found in herbs and spices are often called essential oils. This is an appropriate term because the organic compounds found in the herb or spice more easily dissolve in oil than water. Some of these organic compounds are less volatile than others, which will lead to a longer-lasting flavor and aroma, as they will remain mostly in the liquid form. What makes a compound more or less volatile? The chemical shape and functional groups contained within the molecule (see Chapter 1). Let's talk more about chemical shapes and functional groups that are important in herbs and spices: aldehydes, ketones, alcohols, amines, esters, ethers, terpenes, and thiols.

10.3.1 Terpenes

The most volatile and aromatic molecules found in herbs and spices fall within a family of compounds called terpenes. Terpenes, found both in plant and animal cells, are a diverse set of carbon-based structures, built from smaller five-carbon units called isoprenes or isoprene units (Figure 10.6). Terpenes are organized by the number of isoprene units combined to make the compound (Table 10.2).

Terpenes are common and have a number of functions that extend beyond cooking. In animal cells, isoprenes are the building blocks for cholesterol, testosterone, estrogen, and sterol hormones, including corticosteroids. In plants, terpenes are very common but play a secondary role. The blue smoky haze of the Appalachian Mountains is formed by terpenes secreted by the pine trees. Monoterpenes also serve as seeds of cloud formation. In cooking and food, terpenes found in herbs and spices are fairly volatile and are the primary component of essential oils. The volatile nature of these compounds is why you immediately smell a strong odor upon heating the herb or spice. Terpenes are also fairly reactive, especially with oxygen. The combination of terpenes with oxygen is a chemical change that creates a new compound that may not be detected by the receptors in your nose. Thus, the aroma may seem to "disappear" after extended heating or cooking or after aging in oxygen-rich environments. Generally speaking, the larger the terpene compounds, the less volatile the compound is. This property is beneficial because larger, less volatile terpenes will remain in the food during cooking, providing a more enduring taste to your food. Rosemary and ginger both contain less volatile, larger terpene molecules, allowing for their lingering taste and smell in a roasted turkey or batch of gingerbread cookies.

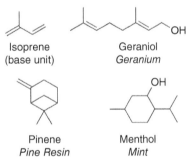

Figure 10.6 Terpenes of spices and herbs. A terpene is a base unit used by the enzymes in plants to produce an amazingly diverse set of compounds including those shown here.

Table 10.2 Isoprenes in herbs and spices.

Isoprene Unit	Name	Formula	Use/Example
2	Monoterpene	$C_{10}H_{16}$	Citral, thymol (mandarin orange), menthol, pine, geraniol
3	Sesquiterpene	$C_{15}H_{24}$	Chamomile, cinnamon, clove, ginger
4	Diterpene	$C_{20}H_{32}$	Vitamin A, rosemary
6	Triterpene	$C_{30}H_{48}$	Lanosterol, cholesterol

10.3.2 Phenols

Thousands of compounds in herbs and spices contain or are derived from a group of molecules called phenols. Phenols are compounds that have a benzene ring (a six-carbon ring system with alternating double bonds) with an attached hydroxyl group (–OH; Figure 10.7). When several phenol groups are bonded together, a polyphenol is created. Polyphenols have diverse biological and chemical uses; they are used as dyes and in the generation of plastics. In herbs and spices, some polyphenols come from a component of the plant cell wall called lignin, while others are used in defense against herbivores or diseases.

One type of polyphenol that is particularly important in food and drink is the tannins (Figure 10.7). Another complex family of flavor compounds, tannins are polyphenols derived from bark, stems, and woody plant material, which provide a pucker-like feeling called astringency. Several spices contain tannins, including tarragon, cumin, vanilla, cinnamon, and cloves.

Oregano, cumin, thyme, bay, and cinnamon are all examples of spices or herbs that contain phenol groups. Extraction and analysis of these five herbs and spices using a sensitive mass spectroscopy analysis found 52 different phenolic compounds [6]. Rosmarinic acid, first found in rosemary plants, is common to all five of these herbs or spices and is found at very high levels in oregano, rosemary, and thyme (Figure 10.8). Caffeic acid, found in coffee, is another polyphenol compound identified in all five herbs or spices, although it is found in lesser amounts in cinnamon, cumin, and bay (Figure 10.8). Caffeic acid is a key intermediate in the production of lignin and is found in nearly all plants. A third phenolic compound, chlorogenic acid, is found at relatively similar levels in each of the five herbs or spices (Figure 10.8). Chlorogenic acid is produced by the modification of caffeic acid and is important in lignin biosynthesis. In addition to its presence as a flavorant and odorant in herbs or spices, it is also found in coffee beans and some fruit.

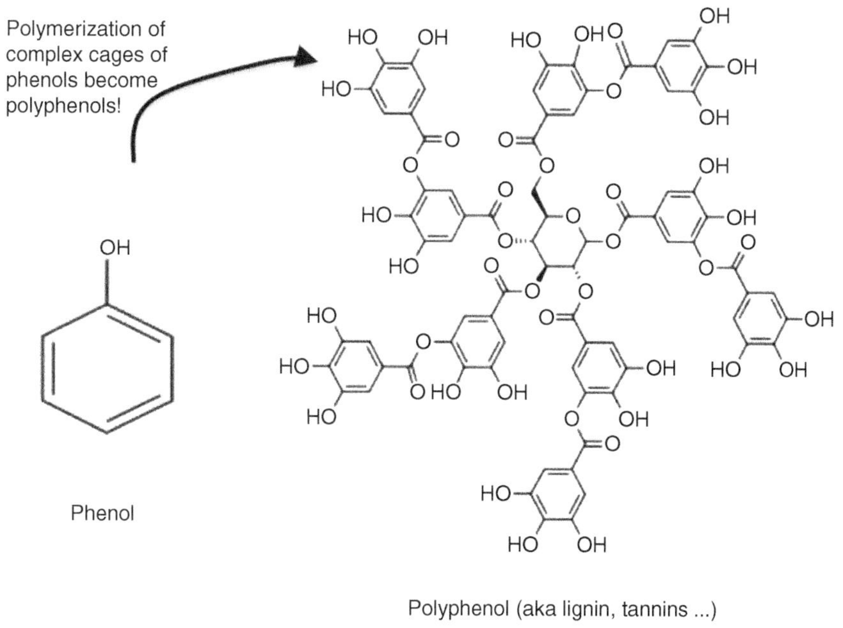

Figure 10.7   The creation of polyphenols. Lignin is shown on the right as an example of a polyphenol.

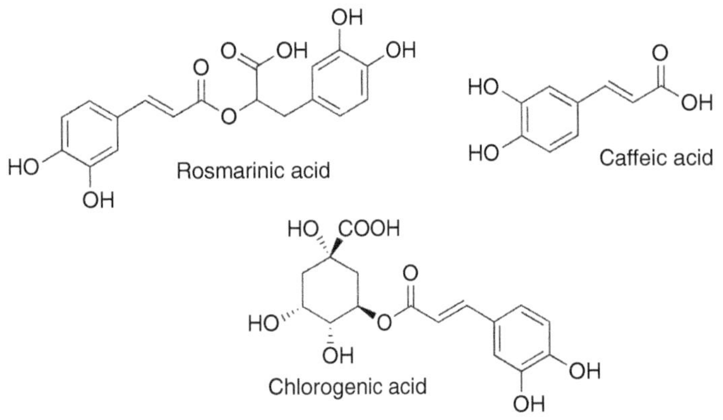

Figure 10.8 Spices as phenols.

Rosmarinic acid

Caffeic acid

Chlorogenic acid

10.3.3 Esters

Remember that you have already learned about several chemical functional groups in Chapter 1, including the esters. The presence of functional groups in a compound leads to different chemical characteristics and unique biological activities. Esters are commonly produced from the reaction of carboxylic acids and alcohols (Figure 10.9). Esters tend to offer a fruity taste and aroma to our food and drink. Pine, cinnamon, and jasmine are a few spices that contain high concentrations of esters.

10.3.4 Pungent

A flavor family not defined by the chemical structure of its flavorants and odorants but by the sensation of heat, "hotness," or unpleasantness that they bring upon us, is appropriately called the pungent family. In general, pungent compounds do not bind and activate food and odor receptors, but compounds in this family interact with receptors that signal pain or thermal events. The formal definition for this type of perception is chemesthesis—the activation of senses in the mouth, nose, or throat for pain, touch, heat, or cold. The cooling sensation of menthol is a chemesthesis event, as well as the heat sensation of wasabi. Horseradish, mustard, wasabi, ginger, pepper, and chilies make up the herbs and spices of this flavor family. Later in this chapter, we will focus on chilies and capsaicin. However, first, we will classify four groups of compounds based on their chemical structure: thiocyanates, alkylamines or alkaloids, and everything else.

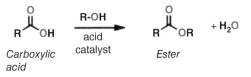

Carboxylic acid → (R-OH, acid catalyst) → *Ester* + H_2O

Figure 10.9 Formation of esters from organic acids. Many odorants are generated as esters by the loss (dehydration) of water from a carboxylic acid and combined with another carbon-containing compound.

Horseradish, cabbage, wasabi, and mustard all contain a chemical functional group called thiocyanate (Figure 10.10). Allyl isothiocyanate is a pungent compound found in horseradish, wasabi, and mustard oil. This compound is produced when each plant's root is crushed; the crushing process releases enzymes that catalyze the breakdown of the sulfur-containing carbohydrates in the cell wall (called glucosinolates) into isothiocyanates. Although horseradish and wasabi have a similar flavor, the various forms and total amounts of the glucosinolates provide the unique flavor profile of wasabi that distinguishes the two. There is almost 10% more allyl isothiocyanate in wasabi than in horseradish. However, because of the similarity, the less expensive and easier-to-obtain horseradish is often tinted green and used as "wasabi." Don't let the color fool you though; 10% more allyl isothiocyanate makes a world of difference to the receptors in your nose, throat, and mouth!

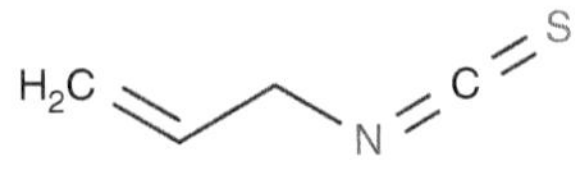

Figure 10.10 Allyl Isothiocyanate. This sulfur compound is one of the pungent classes of odorants responsible for horseradish, wasabi, and other well-known flavors.

Dried mustard seeds or powders may seem less pungent than a liquid or moistened version of the same product. In these dried products, the drying process has halted enzyme activity. However, once hydrated with water, the enzymes are able to produce allyl isothiocyanate, leading to the pungency that you associate with mustard. It can, however, take several minutes to hours for the enzymes to generate enough isothiocyanate for detection. For example, if mustard is mixed with an acidic solution such as citric acid or vinegar, the enzymes will be activated, but at a much slower rate, leading to a less pungent dish. In addition, isothiocyanates are fairly unstable and break down quickly. Interestingly, although the addition of acid reduces the rate at which the enzymes produce isothiocyanate, the lower pH substantially prevents the allyl isothiocyanate from breaking down in your recipe. Extended exposure of the isothiocyanates to heat also increases the breakdown and formation of a nonpungent product. What is to be learned from this discussion? The cook who thrives on the preparation and consumption of pungent dishes should wait until the end of the cooking period to add the mustard or horseradish (Box 10.1).

Box 10.1 Bugs and pungency plants

The use of allyl isothiocyanate is common in the natural world as a chemical defense against bugs and plant-eating animals. In humans, allyl isothiocyanate binds and activates a protein receptor (transient receptor potential cation channel, member A1 [TRPA1]), which signals a pain chemosensor in the body. Interestingly, cinnamaldehyde, one of the compounds found in cinnamon, also binds to the receptor and is one of the reasons why cinnamon is added to spicy dishes of Indian and other cultures.

The second group of chemical structures that define the pungent family of flavors is the alkaloids. Alkaloids are a diverse family of carbon-based compounds that contain a nitrogen base. In plants, alkaloids are important in the development of the plant, fruit, and seed. However, while they have interesting chemistry and biology, most alkaloids are not flavorants. If, however, you enjoy a blackened grilled salmon or spicy salsa, then two alkaloids are critical to enlivening your taste buds. The alkaloids, piperine and capsaicin, are responsible for the pungency of black pepper and chili pepper, respectively.

Piperine, produced by black peppers of the fruit of *Piper nigrum*, acts by binding and exciting the receptors (TRV1) for pain nerve cells (Figure 10.11). If pepper has ever caused you to feel pain, now you know why! Three different piperine isomers are found in the pepper fruit: chavicine, isochavicine, and isopiperine. If you remember, isomers are compounds with the same atomic makeup or molecular formula, but they possess a different organization or structure. Chavicine also has a strong bite and aroma, but it slowly degrades, while piperine remains stable and pungent. The pepper berries also include other aromatic volatile compounds, adding to the aroma and pungency of pepper. Terpene, limonene, and linalool all combine with piperine to give fresh black pepper its woody floral taste with a bite of pain. Piperine acts similar to, but has a greater efficacy than allyl isothiocyanate (spoken about earlier in this section).

Did you ever wonder why freshly cracked black pepper tastes and smells differently than ground pepper, especially ground pepper that has been left in a shaker for a long time? The difference in taste and aroma is all due to the chemistry and biology of the pepper plant and its compounds. The mature fruit or berries from the pepper vine are dark red and contain a single seed. The blanched and dried berries are left to age in the sun, which ruptures the cell wall of the berries, allowing for enzymatic and Maillard browning reactions to take place. These reactions

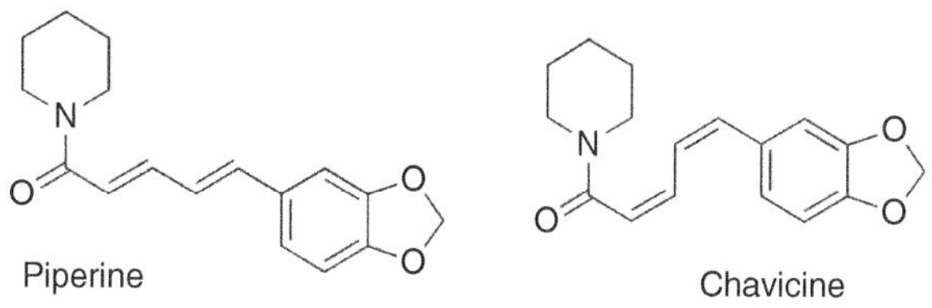

Figure 10.11 Pungent alkaloids. Piperine and chavicine are two alkaloid isomers that each bind and activate the receptors responsible for sending pungent pepper odor to the brain.

produce dark-colored polyphenols and other volatile compounds in the intact berry, now called a peppercorn. When you grind peppercorns in your mill, many volatile compounds are released into the air, resulting in that "peppery" aroma. Although aged ground pepper still contains many piperine and terpene compounds, a significant number of the volatile compounds have already evaporated. This is the reason why whole peppercorns are often used for longer forms of cooking or preserving instead of ground pepper.

Have you ever seen white pepper? White pepper is derived from the same *P. nigrum* vine berries as black pepper; however, the outer fruit layer is removed by bacterial decomposition in water. This process results in the loss of most of the terpene aromatic compounds but allows for the retention of the pungent piperine molecules. This milder preparation of pepper has less aromatic flavor and aroma while still adding pungency to a food. White pepper is often used in foods, salads, and cream sauces in which some of the "pain" of pepper is desired but not the stronger taste or color of black pepper.

Do chilies and black pepper give the same "pain" sensation? Yes and no. This mixed response isn't surprising because the key component in both chilies and pepper binds and activates the TRV1 receptor for pain in humans (Figure 10.12). However, capsaicin (in chilies) does so with an efficacy that is 1,000 times greater than black pepper's piperine. The difference in pungency or hotness in the many varieties of hot chili peppers is primarily due to the level of capsaicin in each pepper. What makes capsaicin so powerful? It is the structure. Capsaicin, like piperine, is a relatively large molecule that has some polar functional groups; it is even able to hydrogen-bond with water! This hydrogen bonding characteristic makes capsaicin much less volatile, which you are likely grateful for if you've ever touched a sliced jalapeno and then your eyes! We will talk more about the characteristics of hot peppers later in this chapter.

There are many other compounds in the "pungent" family; a few more are worth noting. Gingerol (found in ginger) closely resembles capsaicin but does not have the nitrogen base of an alkaloid; rather, it is a modified phenol. While pungent, gingerol is rated less pungent than pepper. However, age and heat cause the degradation of gingerol to another compound called shogaol. Shogaol happens to be twice as pungent as its parent, gingerol. Thus, dried ginger has a more pungent flavor than fresh ginger. A lesser-known spice compound called paradol is found in the seeds of Guinea pepper. The compound is a phenol, is similar in structure to gingerol, and is rated with the same pungency as piperine. Paradol has an interesting property in that it activates a process called thermogenesis, a biochemical metabolism that burns fat to produce energy. The length of the carbon chain of paradol seems to be critical for its fat-burning ability. In a study in which mice were fed a high-fat diet, the shorter the chain on the compound, the lower the weight gain in the mice that were fed the compound (Box 10.2).

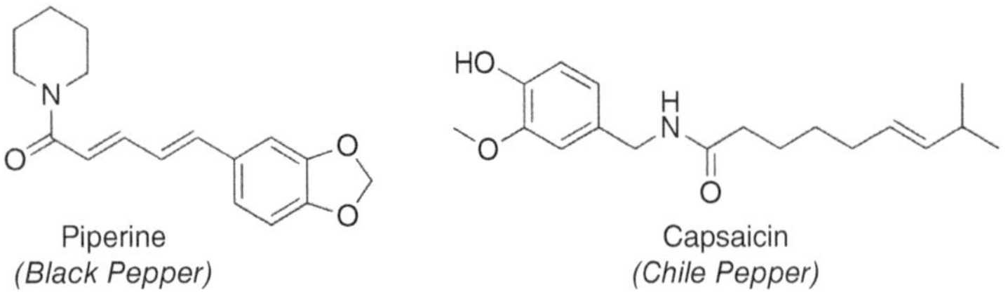

Piperine
(Black Pepper)

Capsaicin
(Chile Pepper)

Figure 10.12 Hot or not? Two pungent alkaloids give food a hot flavor but act by very different receptors. Capsaicin, but not piperine stimulates our pain receptors, giving a hot feeling.

Box 10.2 Cooking with herbs and spices—now it should make sense!

Cookbooks and advice articles recommend a few approaches to using herbs and spices:

- Crush leaves and grind spices immediately before adding to your dish.
- Use whole spices or herbs while cooking when you desire a slow release of flavor (e.g. adding whole cloves or bay leaves early in a recipe).
- Add ground spices no more than 15 minutes prior to the end of cooking time.

- Add more ground, dried, or pre-prepared herbs and spices than fresh.
- Dried herbs and spices are best used when cooking with oil.
- When using fresh herbs and spices, chop and grind the leaves, seed, or root into small pieces to release the flavor.

Now that you understand the basics of the components of herbs and spices, you can recognize the reasons behind each of these hints. More volatile compounds evaporate over time or react with oxygen; thus, fresh herbs and spices have more potency and should be used more sparingly than their dried counterparts. In whole herbs and spices, the cells need to be broken before the enzymes can be released that make the flavorant molecules, so cooking for longer periods of time is effective. Because many of the flavorful compounds in herbs and spices are poorly soluble in water, a better liquid for mixing is oil. Application of these hints and your newly gained knowledge of the science behind the hints will certainly help you better understand what and how to best approach the kitchen pantry as an experimental cook and baker.

10.4 A Closer Look at a Few Herbs and Spices

If you haven't already gathered this, the biology and chemistry of herbs and spices are extensive. The impact of flavor, the evolution of the originating plants, and the biological impact of herbs and spices are the subjects of many fascinating books. Here, let's focus on a handful of herbs and spices that have an interesting scientific story and play a significant role in the kitchen.

10.4.1 Vanilla

Vanilla is one of the world's most popular flavorings, finding its way into food, beverages, perfumes, and even pharmaceuticals! Vanilla, which originated in Mexico and Central America, comes from the vanilla orchid, a vine that produces vanilla beans as a dried seed pod of its fruit. The value of the bean pods was recognized by the Aztecs, who used them to flavor their drinks made with powdered cocoa beans, ground corn, and honey. However, due to the hermaphroditic character of the plant (this means that the plant has both male and female reproductive organs), the plant flower requires pollination to set the fruit. This characteristic was problematic during the first attempts to cultivate vanilla outside of Mexico and Central America, given that the natural pollinator of the flower was not native to other tropical areas where vanilla was first transplanted (Figure 10.13).

Due to the great culinary value of the vanilla, cultivars of the plant were brought around the globe in the early to mid-1800s, including the West Indian island of Réunion, where the vine would grow but the pod would not develop. Although this gave the Central American growers a lock on the much-desired vanilla flavor, scientists continued to strive to find a way to cultivate the plant outside of Central America. In 1836, a Belgian botanist discovered the importance of the *Melipona* bee for pollination; however, it wasn't until 5 years later, in the West Indies, that a 12-year-old slave discovered and developed a hand pollination method for the vanilla flower. This method is still used today. Thanks to 12-year-old Edmond Albius, who won his freedom for the development of this process, vanilla vines can now grow and fruit in many tropical areas around the world.

Madagascar and Indonesia are the world's largest producers of vanilla. Madagascar and the West Indian island of Réunion (previously called Bourbon) produce Bourbon or Madagascar vanilla. These pods produce the rich flavor that you most often think of as "vanilla." Tahitian vanilla is derived from a plant hybrid that is grown in the Philippines. While Tahitian vanilla has a desirable flowery and fruity flavor, it is susceptible to breakdown by heat. The sensitivity of Tahitian vanilla to heat is particularly crucial, as part of the curing process of the vanilla bean is to heat the pod, which promotes the browning reactions necessary to form mature

Figure 10.13 Vanilla bean pod. An open, close-up image of a vanilla pod. Notice the small black seeds held within the pod. These seeds are used to extract vanillin compounds for cooking and baking. thelefty/Adobe Stock Photos

vanilla flavors. Mexican and Indonesian beans also have a more subdued vanilla flavor and smoky or wine-like aroma than do the Madagascar/Bourbon pods.

When you cook with a vanilla pod, rather than vanilla extract or flavoring, the food has a much more interesting and complex flavor. Why? Most of the vanilla flavor resides in the sticky material inside the pod, as well as in the small black bean seeds. How do you work with a vanilla bean pod in the kitchen? Slice down the length of a bean pod, scrape out the sticky black material and seeds, and include the combination of scraped seeds and the bean in the recipe (Figure 10.13). This is particularly delicious when you are making a dish comprised of milk or cream. Because the compounds that provide the flavor and scent of vanilla are more soluble in fat and oil than water, the fats in milk solubilize the vanilla flavor molecules, leading to wonderful concoctions like vanilla milk or vanilla bean ice cream. An interesting additional use of unused sliced pods is to submerge the opened, uncooked pods into a closed container of table sugar; this creates a rich, vanilla-scented sugar that is worthy of baking and candied treats.

Which molecules give vanilla its characteristic vanilla flavor, and how are these flavoring molecules produced? During the aging and browning process of the bean pods, some of the glycoside components of the seed and plant cell walls (remember that glycosides are sugars that are covalently bonded to other sugars or functional groups via a glycosidic bond) are converted to vanillin, the molecule most responsible for vanilla flavor and aroma (Figure 10.14). There are many other compounds detected and responsible for part of the flavor of vanilla, but vanillin is responsible for most of the flavor and aroma. The worldwide demand and expense for vanillin far exceed (by about 10-fold) the capacity of the plant to produce the flavor. Therefore, synthetic vanillin accounts for most of the vanilla flavoring market and is produced at one-hundredth the cost of the natural product (Figure 10.14).

What is the difference between a vanilla that is made naturally and synthetically? Natural vanilla extract is a complex mixture that includes vanillin extracted from alcohol-soaked vanilla beans or processed beans that are repeatedly washed over with alcohol. Pure natural vanilla extract is best characterized by its sweet, fruity, spicy flavor, and aroma. Most vanilla is a synthetic production of vanillin, which contains added sugar and other compounds. This is still pure vanilla; it is just not naturally produced by the plant. However, regardless of whether vanilla is artificial or a pure vanilla extract, the compound is very volatile, so you should add it later in the cooking process to avoid evaporation and loss.

Vanillin

Figure 10.14 Vanilla flavorant. Vanillin is one of the key compounds responsible for the flavor and aroma of vanilla.

10.4.2 Coriander and Cilantro

Part of the carrot family and a native Middle Eastern plant, coriander and cilantro are two widely used herbs grown and utilized around the world. Cilantro and coriander come from the same plant; the leaf is used as an herb, while the spice comes from the fruits or seeds of the plant and is typically ground. Coriander and cilantro provide a great example of how herbs and spices are distinguished. Coriander is a spice, as it comes from the fruit of the plant. Cilantro is an herb since it is the leaf of the plant. If you have ever tasted or smelled cilantro and coriander side by side, you know that there is a distinct difference in flavor and aroma. These differences are due to the distinct molecules that are present in the fruit relative to the leaf of the plant. Cilantro leaves contain 41 different volatile compounds, including decanal and similar isomer compounds [7]. Coriander flavor and aroma come from the terpene flavor molecules, linalool, and pinene, which give the spice a fruity, pine- or sage-like flavor and odor. Linalool is a branched carbon chain with an alcohol (OH) functional group. Pinene contains a complex carbon ring system and is the molecule also found in pine resin, pine oil, and lemon oil. Mixed with cumin, ground coriander seeds provide the base for many Indian culinary dishes (Figure 10.15).

Cilantro deserves some discussion because the flavor of this herb is polarizing. Some people love cilantro and mix it with their homemade salsas and Mexican food. Other people hate cilantro, claiming it tastes like soap. Why is there such a love/hate relationship? Let's take a closer look at the compounds found in the herb and inspect our DNA.

Why do some claim that cilantro has a soapy taste? At the molecular level, some of the compounds in soaps are structurally similar to the decanal-based flavor molecules of cilantro. The aldehyde component of the decanals is also structurally similar to the odor molecules that some bugs produce as a defensive weapon. In a post for The New York Times [8], Harold McGee posits that these similar compounds can remind people of experiences with soap, earth, or even bugs. It is no wonder some people have strong negative feelings about the herb! But why doesn't everyone think that cilantro tastes like soap? This has to do with DNA. Fourteen to twenty-one percent of people with Asian, European, and African ancestry report the detection of a soap-like flavor and a corresponding dislike of cilantro. In contrast, only about 3–7% of people from South Asia, Central America, and the Middle East (where cilantro is heavily used) do not like the taste of cilantro [9]. In a recent study, scientists found a genetic change (a change in a single DNA base) in the chromosomes of about 10% of the population that is linked to a dislike for cilantro. This change in one nucleotide (the chemical building block of DNA) is called a single nucleotide polymorphism (SNP, pronounced "snips").

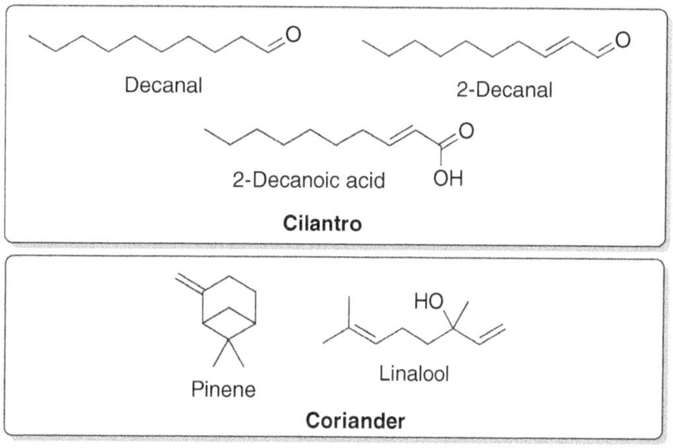

Figure 10.15 Compounds of the coriander plant. Cilantro and coriander, while from the same plant, are an herb and a spice, respectively. The compounds responsible for their unique characteristics are shown here.

SNPs are not uncommon; they are found in approximately every 300 nucleotides on a chromosome. SNPs are part of what brings about genetic diversity in humans, plants, and most organisms. Since the human genome has about three billion nucleotides, our genes have about 10 million SNPs. Although 10 million seems like a lot, most of these single mutations do not affect us, as most nucleotides that make up human chromosomes do not code for proteins. However, when an SNP happens in a part of the chromosome that codes for a protein, the proteins that the gene codes for may have some unique characteristics.

There are two SNP variants linked to the perception of cilantro that were discovered through scientific research. In a study containing over 14,000 people, individuals were asked whether they detected a soapy cilantro taste and had their genes sequenced. Based upon the results, a connection was made between two SNPs and a group of genes on chromosome 11 that had a single mutation that codes for olfactory receptors [10]. The OR6A2 gene was altered in nearly half of the participating European descendants and codes for a receptor highly sensitive to aldehyde-containing compounds. Another study found a link between a dislike for the herb and three genes. What is the take-home message? Small genetic differences in chromosomes can alter the structure and function of proteins. These genetic differences may be why two individuals differ in their perception of flavor and odor.

10.4.3 Cinnamon

One of the most popular spices or herb is cinnamon. Cinnamon is broadly used in the cooking of sweet and savory foods, beverages, and candies or is sprinkled on a piece of toast. Cinnamon is a spice that comes from the inner layer of the bark of tropical evergreen trees and shrubs from the *Cinnamomum* genus, which consists of 250 trees and shrubs. Given the different species used to make the spice, the term "cinnamon" doesn't fully capture the different characteristics of molecules present in each preparation. Consistent among all cinnamons though is the main flavor ingredient, cinnamaldehyde (Figure 10.16), while the minor components of the spice will vary from source to source.

What are the different types of cinnamon? "True cinnamon" is derived from the *Cinnamomum verum* tree that is native to South India and Sri Lanka and has also been transplanted to grow in Madagascar. Because of the historic importance of the spice, true cinnamon is called Sri Lankan (or Ceylon—Sri Lankan's former name) cinnamon. A second species of tree, *Cinnamomum cassia*, used to make the spice called "cassia cinnamon" (also called Chinese, Padang, Saigon, or Batavia cinnamon), grows more widely in Vietnam, India, and Indonesia. The *Cinnamomum burmannii* tree is the source of burmannii cinnamon, which is also called Indonesian or korintje cinnamon. In order to understand what makes these three types of cinnamons different from one another in taste and use, we need to talk about the production of cinnamon.

Cinnamon is obtained by stripping the inner bark from the shoots of 2–3-year-old stems of the tree. The inner bark is dried and curls in a characteristic way that we associate with cinnamon sticks. The dried bark of any form of cinnamon or cassia contains 0.5–3% volatile oils that provide most of the flavor and aroma of the spice. The cinnamon trees are cut or pruned to allow for the growth of new shoots for the next crop of bark. The work is difficult and requires skilled peelers. Sri Lankan or true cinnamon is paper-thin and forms into a single curl (or quill), while cassia and burmannii cinnamons are thick and curl into a double curl/quill. Only true cinnamon will have many thin layers rolled into its single quill. Once ground into powder (in the absence of chemical analysis), it is nearly impossible to distinguish between true and cassia cinnamon. Cinnamon powders typically come from low-grade and chipped bark; the leaves and low-grade bark can also be distilled or solvent extracted to harvest cinnamon oil.

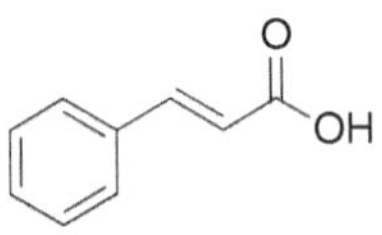

Is there a difference in taste between the different types of cinnamon? Sri Lankan cinnamon, which was used in many early European desserts and Mexican recipes, is slightly sweeter and has a mild flavor. Cassia and burmannii cinnamons have a stronger, almost peppery flavor, are the "cinnamon" that most buy in the grocery store, and comprise the cinna-

Cinnamaldehyde

Figure 10.16 Cinnamon flavorant. Cinnamaldehyde is the main compound responsible for the flavor of cinnamon.

mon spice aromas and flavors found in cinnamon gum and apple pie. In other countries, cassia cinnamon is distinguished from others by labeling the spice as cassia, not cinnamon.

Since there is a difference in flavor in the different types of cinnamon, you can likely surmise that each cinnamon type must have a different chemical composition. The key molecular component, cinnamaldehyde, makes up between 75% and 90% of all the compounds in cinnamon oil (Figure 10.17). Cinnamaldehyde is a modified phenol compound, where a short carbon chain containing an aldehyde has replaced the phenol OH. The compound can be detected at very low concentrations (0.1–0.5 of the total percent of food), and upon binding to its receptors, it provides a pungent sensation in addition to the sweet taste.

While all cinnamons contain cinnamaldehyde, true cinnamon also contains volatile terpene compounds, including the pine-scented pinene and the sweet, floral compounds linalool and eugenol, which are found in many plants that smell of cloves and honey. In contrast, cassia cinnamon includes a small amount of vanillin, higher concentrations of tannins, and only trace amounts of eugenol. These molecular differences partially explain the difference in flavor between the two (with true cinnamon being sweet and mild and cassia being more potent and peppery) (Box 10.3).

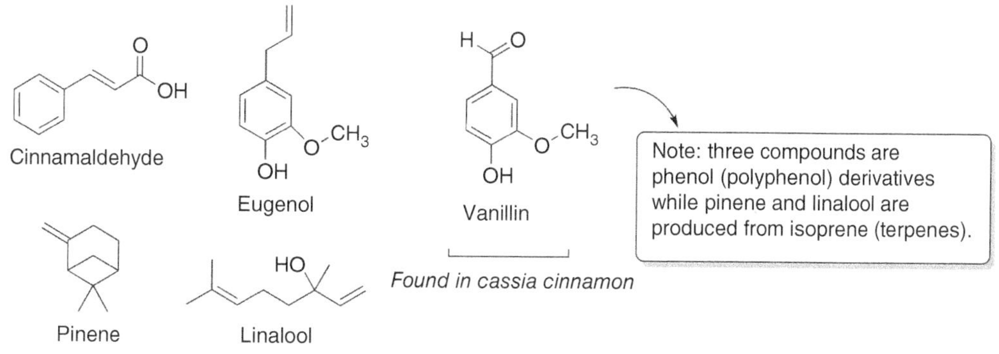

Figure 10.17 Compounds of cinnamon. While both true and cassia cinnamons have cinnamaldehyde, true cinnamon has more eugenol than cassia, while cassia cinnamon has small amounts of vanillin.

Box 10.3 Cinnamon challenge

Coumarin is found in all forms of cinnamon, although in cassia cinnamon the concentration is 60 or more times higher than in true cinnamon. Coumarin, a naturally occurring fragrant compound found in many plants, is used in perfumes and as a precursor compound to the anticoagulant warfarin (coumadin). Coumarin has a modest and incompletely tested antidiabetic property where it may help to lower blood sugar and lipid levels. However, at higher levels, coumarin is toxic; it causes liver failure in a genetic subset of the population and, upon exposure to animals at high concentrations for long periods of time, may have carcinogenic activity. An adult weighing 130 lbs would have to consume 2 g of cassia cinnamon a day to approach the toxic level of coumarin that has been set by the German government. While other countries have not set a minimum or maximum intake standard, the European Food Safety Authority recommends a coumarin daily intake limit of 0–0.1 mg/kg of body weight per day or up to 50 mg/kg of food. As of 2014, the US Food and Drug Administration does not limit the amount of coumarin used in cinnamon-flavored products. However, the USDA does prohibit coumarin as a food additive. Current recommendations are that those who take in high doses of the spice in food or as a health supplement should take caution, as several of the animal studies are difficult to interpret and extrapolate to humans.

10.4.4 Saffron

Saffron is a fascinating spice that, perhaps, you have never used (or even heard of). Why is it interesting? For starters, it is the most expensive known seasoning, selling for $1,500 and $2,500 per pound. Why is it so expensive? The spice comes from a flower, the *Crocus sativus*, which is grown in limited regions of the world. Only three threads of the spice are produced by each flower, and harvesting of the spice is performed by hand! Are you intrigued? Let's learn more about this interesting and expensive spice (Figure 10.18).

Saffron is sold in its native form as thin red threads; the threads are the stigmas of the crocus flower. Given the small number of saffron threads per flower, it takes about 75,000 plants to produce a pound of the spice; one acre of saffron plants yields only about 10 pounds of the spice annually. Iran, Spain, and Portugal are the main producers of the spice; however, the flower is also grown in India. The high cost and low availability of saffron have led to a significant counterfeit market for the spice. True saffron has a unique smell; the threads will turn a cup of water yellow, the resulting water will have a bitter taste, and upon addition of baking soda, the water will remain yellow. Many counterfeit saffron will turn water red or brown with an increase in pH. American or Mexican saffron, which comes from a daisy flower, does not impart the flavor of true saffron. However, turmeric is used in many Indian dishes as a saffron substitute, providing a similar color and flavor to the food.

Like the other spices and herbs we've studied, saffron is a complex mixture of volatile substances; over 150 unique compounds have been identified in the spice. Some of the volatile compounds that contribute to the aroma and flavor are produced while drying the flower stigmas. Analysis of the volatile compounds produced during the aging process identified 23 different compounds. The two key volatile components are safranal (which makes up ~70% of the volatiles) and beta-isophorone (Figure 10.19). Both of these small organic compounds have very little water solubility, leading to good volatility and fragrance.

The deep red color of saffron is due to the presence of the fat-soluble carotenoids, crocin, and crocetin. These strongly colored pigments are formed by the breakdown of a compound called zeaxanthin. A by-product of zeaxanthin degradation is a bitter compound, picrocrosin. Picrocrosin, a glucose derivative and water-soluble flavor molecule, is the telltale compound found in true saffron. Interestingly, the other by-product of this reaction is the aromatic safranal. Why do the stigmas have to be dried to yield these flavor molecules? As we have seen before, the drying process breaks open the cells, releasing the enzymes that catalyze these reactions (Figure 10.20).

Saffron is common in Middle Eastern and Spanish dishes such as rice, risotto, and paella. Fortunately for those cooks who are on a tight budget, a pinch of thread is all that is needed for sufficient flavor and development of the characteristic yellow color. Often, the spice is steeped in warm water or milk for 30 minutes prior to use; this process draws out the color and assists in generating a homogeneous mixture of flavor in the finished dish.

Figure 10.18 The delicate spice saffron. zhekos/ Adobe Stock Photos

10.4.5 Nutmeg and Mace

Originally grown in the Indonesian islands, the tropical evergreen *Myristica fragrans*, commonly known as the nutmeg tree, produces both mace and nutmeg. Nutmeg is made from the seed or pit, while mace comes from the webbing that covers the shell of the pit. Growing nutmeg trees requires great patience. The nutmeg tree is dioecious; this means that there are separate male and female trees. Furthermore, only the female trees can produce fruit, and it takes nearly eight years to identify a tree as male or female. To overcome this complication and reduce the risk of an all-male orchard, cuttings are used to clone the female plants, and 10 female trees are transplanted for each male tree (Figure 10.21).

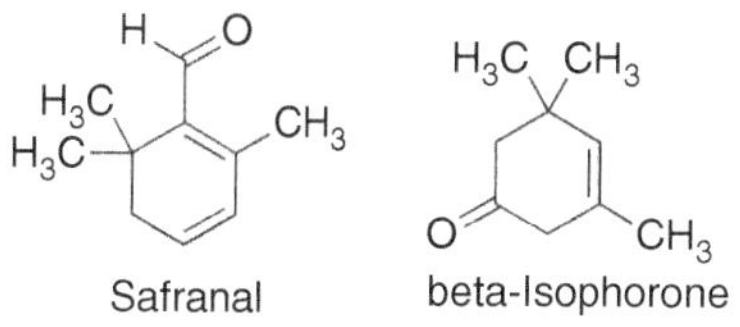

Figure 10.19 Volatile compounds of saffron. The two key components of authentic saffron.

In the United States, nutmeg is used in potato-based dishes, cookies, pastries, sausages, and, during the winter holidays, eggnog. The savory flavor of nutmeg is used in many Indian, Middle Eastern, and Indonesian dishes. If you have never cooked with mace, it has a nutmeg-like aroma, with greater pungent and savory tones. Although it was once a highly sought-after spice, mace is now primarily used as a flavor in spice doughnuts and spice cakes. Both spices are often included with a mix of other spices, including cinnamon, cumin, and vanilla, and the compounds contained in both spices are terpene-based with woody and floral tones (similar to cloves).

Figure 10.20 Saffron color agents.

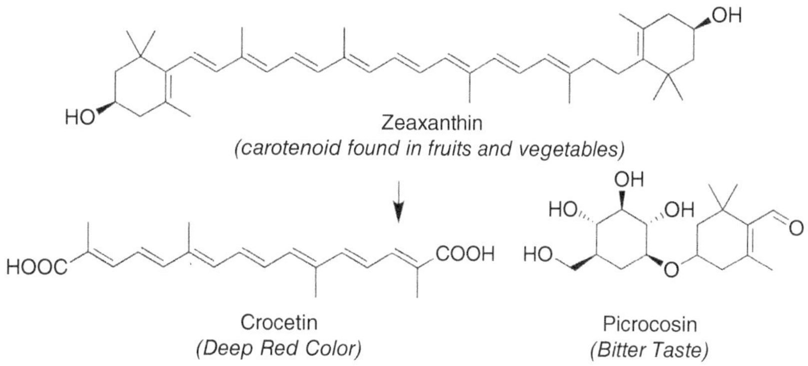

Zeaxanthin
(carotenoid found in fruits and vegetables)

Crocetin
(Deep Red Color)

Picrocosin
(Bitter Taste)

Figure 10.21 The nutmeg fruit. Kenishirotie/ Adobe Stock Photos

Off and on in recent years, the potential for nutmeg to produce drug-like, hallucinogenic high moves through communities. However, consumption of the amount of nutmeg needed to produce the high will result in severe side effects including vomiting, diarrhea, nausea, kidney and central nervous system issues, and irregular cardiac rhythms. The nutmeg molecule that is believed to contribute to the hallucinogenic effect is myristicin. Chemists and biologists believe that the drug is either acting directly in the body or must undergo a chemical change to exhibit its activity. Understanding which of these processes are occurring is called the mechanism. There is scientific evidence that myristicin is chemically converted to a psychedelic amphetamine called MMDA in the liver. However, other studies show that myristicin is chemically converted to nonhallucinogenic compounds in the liver. Additional studies show that myristicin itself binds to and stimulates the receptors that are activated by other psychedelic drugs such as amphetamines, serotonin, and dopamine. Given that all of these studies were conducted in rodents or petri dishes, it is not clear which of these possible mechanisms might be at work in humans. However, either way, the dangers and severe unpleasant reactions to myristicin are nasty and can lead to an emergency room visit.

10.4.6 Curry

Distinct from the seasonings we have discussed thus far, curry powder is a mixture of herbs and spices, most of which do not come from either of the two plants named "curry." Curry leaf comes from the *Murraya koenigii* plant and has a lime–lemony taste with woody overtones. Although most curry powders do not contain curry leaves, Indian curry powder mixtures often include curry leaves, or the leaf is added directly to Southern Indian cuisine. The unrelated curry plant comes from *Helichrysum italicum*, a plant similar to the flowering sage bush that smells of curry. Interestingly (and perhaps confusingly), neither of these "curry" plants is a traditional component of most curry spice mixes.

Curry is a term used for both a dish and the powder that provides the base seasoning. Curry describes a variety of dishes from around the world, including foods from India, Pakistan, Sri Lanka, Singapore, Thailand, and Japan. Like dialect and cultural variation, there are as many kinds of curry spice mixtures as villages. In most cases, the recipes for curry powder start by toasting or browning the spice. The traditional spices in a curry powder include turmeric, coriander, cinnamon, allspice, and cumin, to name a few. Northern curry powder, like the Punjabi style of curry known as garam masala, tends to be sweet and contains black pepper, cardamom, and coriander as its main components.

Thai curries, unlike Southern Indian dishes, do not contain curry leaves, are sold as a powder or paste, and are typically identified by the color of the powder/paste: red, green, and yellow. Thai curry paste tends to be particularly spicy, where the spice level depends on the chili being used. Red Thai curry is made with red chilies, garlic, lemongrass, ginger, and shallots. Green Thai curry paste has a green chili pepper and includes coriander and cumin. Yellow Thai curry gets its distinctive color from adding turmeric and cumin; it is a little sweeter and creamier than the green and red versions due to the addition of coconut milk.

A Northern African/Arabic curry called Ras el hanout is a blend of black pepper, cardamom, sea salt, ginger, cinnamon, mace, turmeric, allspice, nutmeg, and saffron; this blend is common to Moroccan and Arabic cuisines.

Madras curry sauce is a British version of a hot Indian-inspired curry paste containing chili powder, turmeric, cumin, and cinnamon, which is added for its pungent and savory flavor that contributes to the "heat" of a Madras curry. Given that we are discussing the "heat" associated with a curry, let's transition into the chilies and capsaicin that give many other dishes "heat."

10.4.7 Chilies, Capsaicin, and Heat

The fruits of the flowering plant Capsicum include the mild green pepper and the hottest peppers, including ghost pepper or bhut jolokia, Carolina Reaper, and the Trinidad moruga scorpion. The primary compound contained within capsicum plants is capsaicin. However, the name of the fruit that produces the compound varies. At times, the term "chili" is used to describe the pepper and the name for the food, derived from chili con carne, a tomato-based dish made with peppers. Chile, the name of the South American country, and consistent with the Spanish "e" ending, has also been used as a name for hot peppers. However, perhaps the most historically correct way to refer to a pepper is chili. A Spanish physician and botanist, Francisco Hernández de Toledo, in Four Books on the Nature and Virtues of Plants and Animals for Medicinal Purposes in New Spain, used the Aztec native language to describe white habanero peppers as "arbol chili" in 1615. We will use this historical common name, chili or pepper, here.

Chilies belong to a larger family of flowering plants, including tomato, potato, and petunia plants, and over 2700 other species called Solanaceae (nightshade). From this larger family, plants are organized into smaller subsets (or genus). Chili is in the *Capsicum* genus, which includes 22 wild and more than three domestic species. Most hot peppers lie in the *Capsicum annuum* species, including bell, anaheim, banana, jalapenos, cayenne, and some other commonly used peppers. The breadth of taste and heat that lie within the *C. annuum* species is surprising and significant. A few familiar peppers do belong to a separate, unique species: Tabasco and Thai (*Capsicum frutescens*) and habanero and Scotch bonnet (*Capsicum chinense*).

10.4.7.1   Anatomy of a Chili

Have you ever thought about a chili as a hollow container for seeds (Figure 10.22)? That is essentially what it is! Inside the hollow pod, the thin-shelled seeds are attached to the glands and placenta of the fruit. Chilies are mostly water (70% or more); the dry mass consists of fibrous, soluble, and insoluble complex carbohydrates with a significant concentration of glucose and free amino acids that provide a flavor that is hidden behind the heat of a chili. Although there are volatile oils and other fats that also contribute to the flavor and aroma of chili fruit, most of the characteristic colors, flavors, and aromas come from molecules called carotenoids and capsaicinoids.

10.4.7.2   Color

Have you ever seen or tasted an orange bell pepper? The varied colors of raw and powdered peppers are highly valued for the aesthetics that the vivid colors bring to food. The pigments that give plants and select microorganisms yellow and red colors are isoprene and phenol-based compounds called carotenoids. There are over 20 different carotenoids in the fruit of chilies. The carotenoid, beta-carotene, is responsible for much of the yellow-orange color of peppers (Figure 10.23). In contrast, the red color of cayenne, red bells, and even some red spices like paprika come from the less common carotenoids, capsanthin and capsorubin.

What about green peppers and plants? Chlorophyll is the molecule that gives plants their green color. While several different types of chlorophyll (a, b, c, d) exist, all of which have slight structural differences, all forms have a four-ringed structure (Figure 10.24). This structure causes chlorophyll to absorb blue and red light, thus causing the reflection of green light and the light and color that we see (i.e. green). However, there are a variety of colors of green found in different chilies. This variation occurs because

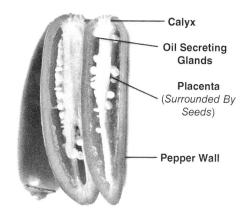

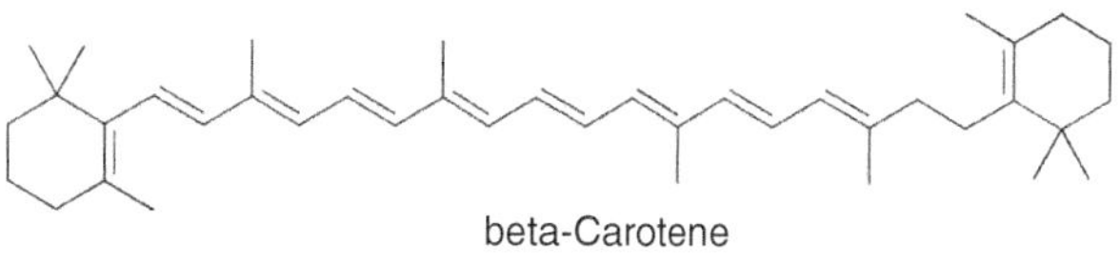

Figure 10.22   Anatomy of a pepper. Most of the "hot" compound, capsaicin, is found in the placenta. The seeds are filled with bitter-tasting cell wall material and are coated by the oil glands with capsaicin.

beta-Carotene

Figure 10.23   Beta-carotene, the color of fruits and vegetables. The compound is responsible for many of the colors in plants and vegetables. The metabolism of this compound generates a variety of colors in vegetables and fruits.

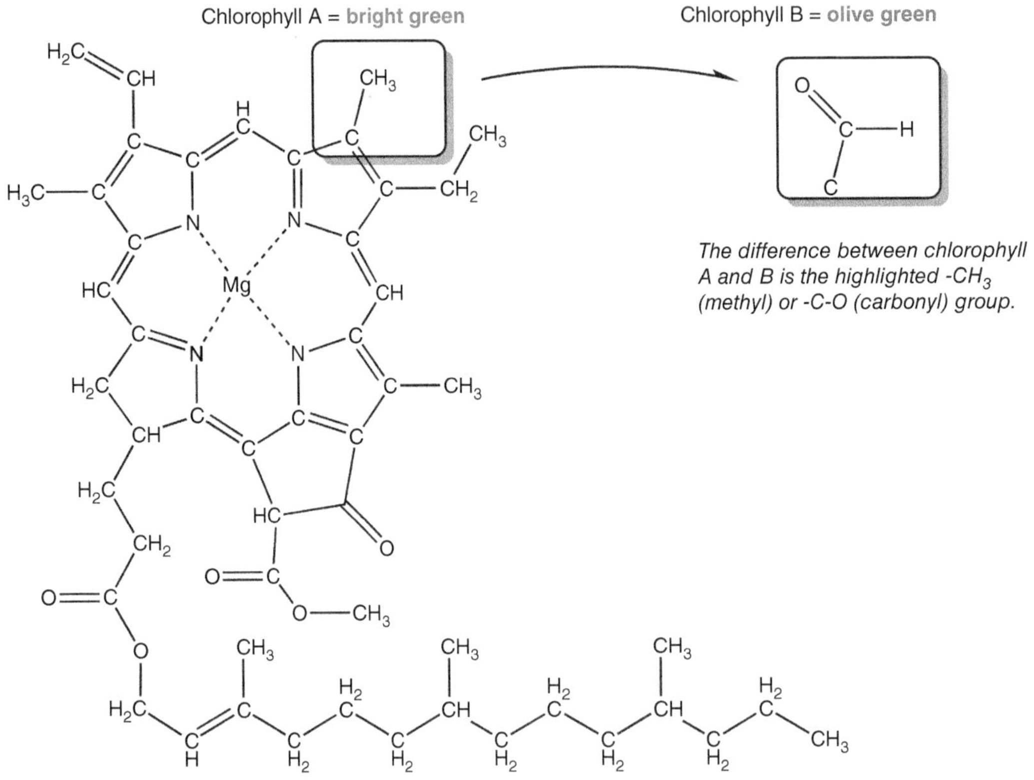

Figure 10.24 Chlorophyll.

green chilies contain different types of chlorophyll (in different amounts) and also contain other carotenoid pigments. When combinations of beta-carotene, chlorophylls, and other carotenoids come together, red, blue-green, and blue lights are absorbed, leaving a spectrum of green light reflected from different types of chili fruits.

Have you ever seen a pepper fruit change color as it matures on the plant in your garden? Chlorophyll is an unstable compound; when a plant stops producing it or when a fruit is harvested, the green pigment quickly decomposes and no longer absorbs light. Carotene, on the other hand, is more stable and degrades more slowly than chlorophyll. Thus, as your green pepper sits on a dying plant, it will turn yellow or red, which is a color consistent with carotene. Likewise, if you place a green pepper on your kitchen counter, it will eventually also turn yellow/orange/red. Moreover, in several types of fruit, red carotenoids are significantly produced at maturity when the plant hormones shut off chlorophyll production. You may be wondering about the type or structure of the mature fruit carotenoids, and the answer is that, it is complicated. Thirty-four different carotenoids were identified in a ripening extract of Hungarian capsicum. Yellow carotenes are less stable than red carotenes, giving aged pepper powder its characteristic red color vs. a yellow or orange color.

Perhaps you have also seen purple, black, or blue-ish peppers at a farmers market. These less typical pepper colors are due to another group of molecules called anthocyanins. These molecules can change color in response to the pH of the fruit (red for acidic and purple in more basic conditions). For more information about anthocyanins, turn to the chapter on wine.

10.4.7.3 Capsaicinoids
What is the star of hot chicken wings, extra spicy salsa, or an eye-tearing Tabasco sauce? The heat, of course! The molecule responsible for the heat is a group of alkaloid molecules called capsaicinoids (Figure 10.25). Often, the capsaicinoids are grouped together and called capsaicin, as this is the single molecule that comprises

the majority (64–72%) of the capsaicinoid compounds in a chili. The capsaicinoid that comprises the second-highest amount, dihydrocapsaicin, accounts for about 22% of the pungent compounds. Five other closely related compounds comprise the remaining 10–20% of the capsaicinoids. A close inspection of the compounds shows that they are similar in molecular structure to other flavorants, such as piperine, gingerol, and vanillin, particularly in the ring component of the molecules. In fact, capsaicinoids are part of a group of molecules called vanilloids (e.g. gingerol, vanillin, and capsaicin) due to the presence of the phenol group (Figure 10.26). However, you know that chilis are very different than vanilla. Why are the two so different? The variation in the carbon "tails" provides great distinction in the signaling of each compound to our brain; the longer, more hydrophobic tail of capsaicin is chemically very different from the OH and straight chain found in gingerol (as one example). Because of these structural differences in particular areas of the molecules, each compound binds to entirely different receptors and sends very different signals to our sensory system.

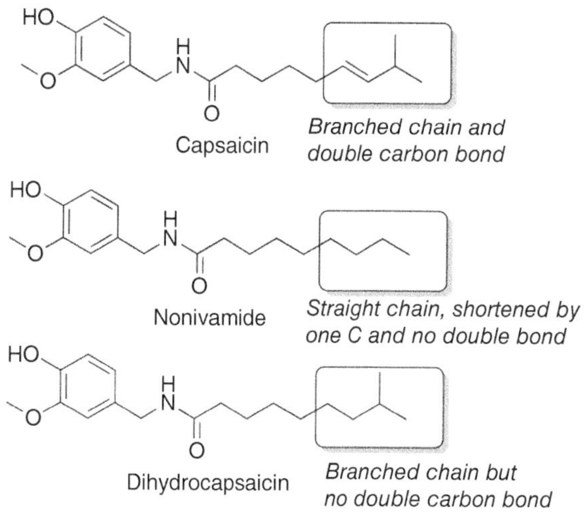

Figure 10.25 Capsaicinoids. Capsaicin is the parent compound of the capsaicinoids. Carefully inspect the differences in the tails; the remaining structures are all identical.

Perhaps you have noticed a difference in taste when you include (or don't include) jalapeno seeds in your homemade salsa. Many people associate the heat of a chili with the seeds. However, remember that capsaicin is produced in the placenta of the fruit. Thus, while some of the "heat" compounds find their way to the thin, delicate seeds, the seeds are not a major source of capsaicin, nor do they make positive contributions to taste (i.e. they have membranes with tannins and polyphenols that make them bitter). Thus, the best way to appreciate the heat, retain the flavor, and reduce the less appealing, bitter overtones of a chili is to remove the seeds and retain the rest of the chili in a dish.

You may have also noticed that the same type of chili (like a jalapeno) may have different heat levels. Additional stressors (from heat or dry conditions) during the growing process increase the capsaicinoid level in the fruit. The fruit also increases the production and secretion of capsaicinoids as the color changes and the fruit withers. Thus, a less mature jalapeno tends to be less hot than a more mature fruit.

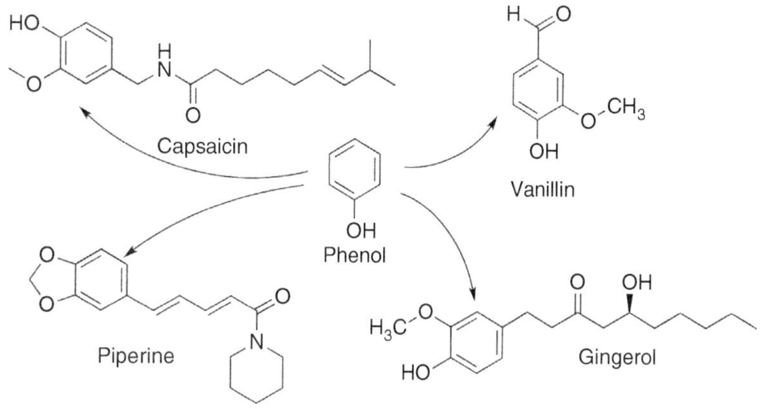

Figure 10.26 The vanilloids. Based on the common phenol head group, the vanilloids are a diverse set of molecules responsible for many odors and flavors. Note the diversity in tails or nonphenol components for each class of compound. Each compound will bind to distinct receptors, eliciting a very different response.

10.4.7.4 How Do Chilies Work?

When you bite into a chili, the sensory impact is obvious: pain. However, there is little to no real biological damage that occurs when you eat chilies. Capsaicin (and related compounds) interact and bind to a pain-sensitive nerve, not a taste receptor! More precisely, capsaicin compounds bind to a class of protein receptors located on the surface of specialized sensory nerves called nociceptors. Nociceptors, found on the skin and mucous membranes, are typically activated by a noxious stimulus that provides a damage signal to the body. Damaged tissue associated with trauma, heat, or chemicals leads to the activation of nociceptor nerve fibers, which provide a signal to the person experiencing the event to withdraw, stop, or get away from the event. Thus, capsaicin does not cause damage; it just sends a signal that pain or heat is being experienced. Let's take a closer look at the biology of this process.

Each capsaicinoid binds to a protein in the nociceptor nerve fibers called the transient receptor potential vanilloid type 1 ion channel. That is a mouthful! Fortunately, the receptor is also known as the capsaicin receptor or TRPV1. TRPV1 responds to various signals, including capsaicin, the pungent compound allyl isothiocyanate, heat above 109°F/43°C, acid conditions below pH 5.3, and electrical current (Figure 10.27). Gingerol also binds to the TRPV1 receptor when it is found in higher concentrations, providing the pungency found in aged ginger powder. Once stimulated, TRPV1 activates dozens of proteins within the cells and opens the membrane for calcium and other ions to enter the nerve cell. The result is a slew of biological signals propagating pain nerve fibers' activation. This pain signal is sent as long as capsaicin binds its receptor, and capsaicin binds the TPRV1 receptor very well with an affinity of less than 700 nM. What does this mean? Humans can detect capsaicin in solutions of 10 parts per million, that is, about 3.5 teaspoons dissolved in an Olympic-sized swimming pool. In other words, there is plenty of capsaicin in most peppers to be noticed by a human.

As you can surmise, the more capsaicin present in the fruit, the more likely the TPRV1 receptor will bind and send the signal of burning pain to the brain. Some individuals will eventually want to stop and limit the pain; what is a solution? The best solution lies in the structure of capsaicin. Because the compound is fat-soluble, not water-soluble, drinking water or soda will only worsen things because water will spread the hydrophobic oil-like capsaicin throughout the mouth to mix and coat other membranes, which will set off more nerve fibers! Soda may worsen things as the acidic drink synergizes the pain signal. The best solution is to remove the unbound molecule from the system and tease the capsaicin from its receptor with a glass of fat-containing milk or oily (fat-based) food that you spit out. If you swallow the fat-based drink or food, you will only deliver the compound to the rest of your digestive tract, providing you with additional pain for a long time.

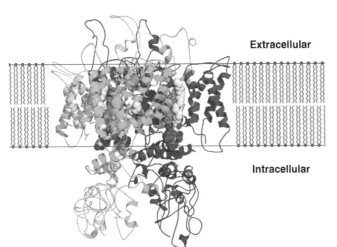

Extracellular

Intracellular

Figure 10.27 Capsaicin activation of pain nerves. Several signaling cues can activate the membrane receptor TRPV1, including capsaicin. This results in an influx of calcium and sodium ions and sets off pain signals to the CNS (central nervous system).

Another piece of advice for working with chilies is don't touch your eyes or any other mucous membrane tissue if you have been cutting a chili with your bare hands! Because capsaicinoids are hydrophobic and primarily nonpolar, the molecules will dissolve in the fat of your mucous membranes, leading to a burning, pain sensation. On the skin, a rub in butter will help solubilize the compound into the fat (butter), where it can be washed or wiped away.

Do you know anyone who doesn't seem to be bothered by spicy, hot foods? The tolerance for chilies or other foods with capsaicinoids is easily explained at the TRPV1 receptor level. If the receptor is activated for long periods of time or repeatedly stimulated due to the presence of capsaicinoids, enzymes modify TPRV1, shutting down the nerve's ability to send signals to the brain. This property, called desensitization, has been used therapeutically for a long time. Ancient Mayans used capsaicin (chilies) to treat sore throats, while the Aztecs used them to treat toothache. The desensitization caused by capsaicin continues to be used by the pharmaceutical industry. Capsaicin is topically used to alleviate pain and is a key ingredient in the creams that are used to treat shingles, muscle aches, and pain patches; it is even sold to reduce the pain of arthritis. A second mode of action utilized by capsaicin to relieve pain is through a peptide called substance P. Substance P is associated with several chronic pains, including arthritis; it is responsible for some of the pain signals of different nociceptor nerve fibers that respond directly to capsaicin. Routine application of capsaicin depletes the amount of substance P available; with less substance P, there is less pain associated with some chronic illnesses. Thus, as a drug, capsaicin is used with modest success in treating the pain associated with shingles, fibromyalgia, osteoarthritis, and some post-cancer surgeries.

10.4.7.5 How Hot Is It?

Inspection of the structures of a few of the more than 20 known capsaicinoids shows seemingly minor changes in double bonds and orientation of the tail or other functional groups (see Figure 10.25 for a simple example). However, these structural differences significantly alter the ability of each compound to bind and send a signal through the TRPV1 receptor. Capsaicin and dihydrocapsaicin, representing 80–90% of the compounds in the family, have the highest binding affinity for the TRPV1 receptor. The other 20 known capsaicinoids each bind to TRPV1 with lower efficiency, with no significant detectable difference. Therefore, if you are looking for the hottest chili, you should focus on the amount of capsaicin. The more molecules of capsaicin, the more receptors are activated, and the more nerves send the painful burning signal to the brain.

The famous Scoville heat unit (SHU) is the unit by which the "heat" of a chili is measured. The SHU is a rather arbitrary and difficult measurement technique, so how did this semi-quantifiable scale come about? Parke–Davis pharmaceutical chemist, Dr. Wilbur Scoville, was following up on some initial studies performed by a Hungarian physician, Hőgyes. In 1878, Hőgyes noted that ingesting capsaicin created a sharp sensation on his tongue and warmth in his gut, followed by belching and flatulence. While Hőgyes' work eventually led to the drug's pharmacological use, Dr. Scoville searched for a way to measure the "hot" compound in ground peppers. Why was this work necessary? Well, chilies vary significantly in heat from plant to plant or season to season. The result of Scoville's work came in 1912 when Dr. Scoville published his one-and-a-half-page paper "Note on Capsicums," in which he described the process of measuring "heat" to further the drug discovery potential for capsaicin [11].

In this historical method, Scoville ground an exact amount of dried chili (a grain or about 64 mg of fruit) in 100 mL of ethanol to dissolve the capsaicin. He then diluted the extract with sugar water. The dilution in which the pungency was no longer detected (by the human tongue) was the measurement of the fruit's capsaicin strength. The heat scale was based on this dilution and later became known as the Scoville heat index. However, this taste-based (organoleptic) test method is problematic, as tasters eventually desensitize their receptors (remember what we discussed earlier in this chapter). Thus another more modern method, utilizing high-performance liquid chromatography (HPLC), has been devised. Similar to Scoville's procedure, a chili mass is ground in 95% ethanol with sodium acetate. The fluid is then run through fine, silt-like beads in a steel column that binds and releases the compounds in the extract differently. Some compounds do not bind with the beads and flow through the column quickly, while other compounds bind tightly and are only released after a long time (or anything in between). This process, called chromatography, separates the compounds in a complex mixture since each individual compound flows through the column beads at a different rate. How are these compounds measured or detected? Each compound can be measured by its absorbance of light as the solution flows through a light detector; the larger the absorbance peak, the greater the compound present. Based on the structure of the compound, the properties of the column beads, and the properties of the fluid (or solvent) that is running through the system, a scientist can predict which peak corresponds to which molecule. Now, scientists can carefully determine the heat of chili using this standardized technique, where SHU (from Scoville's historical method) is the measurement of the heat. Within this HPLC method, the parts of capsaicinoid present per million of solvent (ppm) are multiplied by 15 to obtain SHU.

For years, the hottest pepper crown was given to the Bhut Jolokia from India. This pepper was commonly called the "ghost bhut" or the ghost pepper because of how the pain sneaks up on the consumer. However, the pepper lost its standing as "hottest" in the Guinness Book of World Records in 2012 to the Trinidad moruga scorpion pepper. In the 2012 study, conducted by Dr. Bosland from the New Mexico State University Chile Pepper Institute, the bhut jolokia was found to have an average SHU rating of 1.02 million, with the hottest plant possessing a rating of 1.58 million SHU. The Chinese variety Trinidad averaged 1.2 million SHU, with the hottest individual pepper found to have just over 2.0 million SHU. To ensure the uniqueness of this pepper, the scientists measured several genes with unique SNPs for each. This study showed the ghost bhut was not a variety of the Trinidad pepper. Then, one year later, the top spot in Guinness was taken over by the Carolina Reaper. Measured in an undergraduate analytical chemistry laboratory at Winthrop University, the average pepper was 1.57 million SHU, while the top hot pepper produced a mouth-screaming 2.2 million SHU. To give you some perspective on this kind of heat level, pure capsaicin has a rating of 16 million SHU. For a better perspective, the Carolina Reaper is 440 times hotter than an average jalapeno pepper. Those plugged into the pepper world have heard rumors of new, even hotter variants being developed and tested—Pepper X. It will be interesting to see how much capsaicin a single pepper fruit can produce!

10.4.7.6 Why? Chemical Warfare!

Eating a chili with two or more million SHU is beyond most people; however, the curious (but not crazy) might ask, why would a plant produce such a noxious compound? Every molecule that is produced by a cell requires energy. Thus, if, over time, an organism retains the gene that allows it to produce a molecule, there is usually a biological rationale behind it. Chili plants produce capsaicin for two biological reasons: chemical defense and seed dispersion.

Plants do not have an immune system, nor can they move if attacked or threatened. To defend themselves against being eaten or killed by pathogens, plants have developed an amazing assortment of defense mechanisms to detect and stop invading organisms and animals. The bitter taste of some plants is a deterrent to animals biting and eating the plant. Damage caused by insects opens a wound that allows microbes to enter, damage, and kill the plant. Thus, in response to insect or animal damage, plants release volatile repellant compounds such as terpenes and polyphenols to deter further attacks. In other words, the same volatile oils that provide the flavor and aroma of spices act as insect repellants. In addition to repelling further damage, breaking the cell wall in some plants releases enzymes, such as the polyphenol peroxidases, that chemically transform the lignin carbohydrates in the cell wall. The modified lignins act as a scar and patch the wounded area, slowing the growth of microbial pathogens already infecting the plant.

Capsaicin also plays a very different role in chili plant survival. Most plants need to disperse seeds away from the parent plant to avoid the new plants competing with the parent for sunlight and nutrients. One method that plants use to achieve this is called zoochory. Zoochory occurs when seeds are dispersed through the gut of animals. In fact, zoochory is why the flesh (fruit) surrounding seeds is attractive in color, flavor, and aroma. The fruit is an attractant for an animal to eat and spread the seed through its feces. However, this method of seed dispersal is problematic for chilies because the seeds of most chilies are more fragile than seeds found in other fruits. Yet, the chili fruit has considerable sugar content and an attractive color. Let's consider this problem with an example. Pear, orange, and apple seeds have a tough coating. A herbivore with flat grinding teeth would not destroy an apple seed, but a chili seed would be seriously damaged. The problem is to attract the right animal to eat and disperse the seed and deter those that would damage the seed and block propagation. Fortunately, an evolutionary change generated a molecule (capsaicin) that would specifically bind to a receptor in animals that warns for burning pain, deterring the animals from eating the fruit and destroying the seeds. Birds, however, that do not have the grinding molars of a mammal have TRPV1 receptors, but capsaicin does not bind to these genetically distant receptors. Thus, birds with a more delicate way to pick and swallow the seed without injury can spread and disperse the seeds after traveling through the gut.

An interesting study on this problem tested the consumption of fruit and chilies by mice, rats, and birds. All three animals ate the hackberry fruit, while neither the cactus mouse nor the packrat consumed the capsicum fruit. Yet, thrashers (birds) readily ate the hot pepper fruit. Moreover, the seeds eaten by the birds were passed and germinated the same as nonconsumed seeds. However, the rodents did not pass intact seeds, and no plants grew from the nonintact seeds that passed through either the rat or the mouse [12]. It was also found that the birds spread the seeds under their tree, where shade helped the plant to grow. This study illustrates that capsaicin is an effective deterrent to ineffective seed dispersers, which increases the evolutionary fitness of the plant.

Another set of studies by the same scientist (Dr. Joshua Tewksbury, University of Washington) highlights a second chemical defensive and evolutionary reason for the capsaicinoids [13]. Recall that the placenta produces capsaicin, and some compounds coat seeds attached to the placenta. Several pathogenic microbes, including fungus, can grow inside the pods of punctured or damaged chilies. These fungal plant pathogens produce several toxic compounds in the plant host. Since the bite of a foraging insect will leave a wound within which a fungus can grow and bring disease to the plant, if capsaicin could detract insects from biting or fungus from growing, the compound would serve a second chemical plant warfare purpose.

Looking at plants in Bolivia, Argentina, and Paraguay, Dr. Tewksbury and his colleagues found an interesting directional gradient of capsaicinoid-producing plants. There were more plants in the southwest than the northwest by more than two-thirds. Interestingly, the number of insect bites in both groups of plants was the same, indicating that capsaicin was not a deterrent to the insects. However, plants that contained capsaicinoids were half as likely to be infected and rot due to fungus compared to plants with less capsaicin. This study suggests that capsaicin protects seeds from the pathogenic fungus.

Thus, if you want to know "why heat?" the answer is clearly chemical defense. Capsaicin stops the molared mammals from eating and grinding the seeds into oblivion, and the "heat" slows fungal infection, thus sparing seeds from rot. With some humans, however, this evolutionary protection mechanism seems to be lost. We can postulate that we have adapted to the spice as it inhibits fungus in our food, but the organoleptic pleasure of spices is lost when eating a habanero. In fact, one-third of humans seek out the pain of eating chili peppers daily. This is nothing new, as ancient civilizations in the Old and New World included peppers in food and drink. Our innate aversion to a key biological response to pain has somehow been short-circuited. Psychologists have attempted to understand the reason behind the phenomenon of chili consumption. Two hypotheses have been proffered. The first is the social and learned behavior of cultures that consume peppers.

Simply put, Mom, Dad, and the relatives eat it, and the kids learn to like it too. The second is "benign masochism." Like the thrill of danger and our fondness for roller coasters, chili heat presents a unique human activity, similar to a trip through a haunted house or a scary amusement park ride. Whatever hypothesis is true for the chili lovers of the world, these pepper heads will continue to fight to claim to have eaten the hottest chili, despite the obvious reasons for avoiding it.

10.5 Medical Uses of Herbs and Spices

There are plenty of nonculinary uses of herbs and spices listed in social media outlets, on the internet, or in your great, great grandmother's book of wisdom. Historically, herbs and spices had a distinct antimicrobial activity that kept food from spoiling, which was critical when refrigeration space was nonexistent or at a premium. In addition to fighting microbial growth, several herbs and spices have antioxidant activity, which preserves the food (and the person ingesting the food). There are plenty of medical uses of some compounds to decrease blood pressure, fight cancer, and reduce depression. Let's learn more examples of these medical uses in the following; you may have some medicine in your spice rack without having known it!

10.5.1 Antimicrobial Activity

Before refrigeration, people who lived in warm, tropical climates struggled to maintain safe stores of meat and vegetables, much more than those in the cooler climates. Taking advantage of the natural resources growing in these tropical areas, the herbs and spices added to the foods during curing and cooking provided more than an organoleptic purpose. They were added to ensure that the food was safe to eat! In fact, nearly half of the recipes for meat in tropical climates included spices, while less than 5% of those dishes from cooler climates called for the spice. Does this explain the flavoring of Midwestern foods versus those in the southwest corners of the United States? Maybe! Very low amounts (0.05–0.1%) of some herbs and spices inhibit the microbes that cause salmonella, cholera, and food poisoning. Specifically, linalool, eugenol, myristicin, cinnamic aldehyde, allyl isothiocyanate, and vanillin inhibit 50–75% of bacterial growth compared to nontreated control samples! While the exact mechanism by which the compounds in herbs and spices inhibit bacterial growth is unknown, some evidence points to interference with the bacterial membrane continuity or inactivation of DNA and RNA needed for cell growth and division.

In addition to the bioactive compounds in herbs and spices already discussed, a wealth of scientific literature focuses on the role of herbs and spices in cancer. Some compounds that cause or promote the formation of cancer tumors must first be chemically altered by enzymes in the liver, such as the P450 enzyme system. Compounds in pepper, rosemary, turmeric, and cinnamon stop this process by inhibiting several of the P450 enzymes, thus potentially preventing an accumulation of cancer-causing material. Once formed, tumors begin with uncontrolled growth and spread throughout the body in a process called metastasis. Turmeric decreases the signal from the tumor cell that induces its growth. Curcumin and capsaicin inhibit several activation signals to the tumor cell and inhibit the proteins that support cancer cell metastasis. Many herbs and spices also have components that may compete with estrogen for its receptor in human breast cancer cells. Thus, some herbs and spices may reduce the incidence or progression of estrogen-dependent breast cancers.

Although these important effects imparted by herbs and spices on human health are interesting, they must be treated with a dose of cautious skepticism. Herbs and spices are familiar and considered "natural"; thus, many are drawn to herbal treatment to achieve a positive health outcome for their ailment. Unfortunately, few clinical applications for herbs and spices have been found. Many studies have been performed in culture dishes of bacterial or animal cells, and some work in animal models such as mice or rats show promise; however, the transfer of the cell or animal model effect into humans has largely failed. One example is a study that found that cinnamon extracts suppressed the growth of a bacterium (*Helicobacter pylori*), which is the major cause of stomach cancer and some esophageal cancers. Unfortunately, patients with the bacteria given high doses of cinnamon extract did not experience an altered bacterial growth or cancer outcome. However, not all of the studies have yielded disappointing results. The curcumin compound found in turmeric blocks inflammation and other cancer-related enzymes. Early clinical (phases I and II) trials suggest an anticancer activity to the compound that has successfully translated to the patient.

Another danger is the thinking that "natural" herbs and spices automatically make them safe. We have already seen examples of the strong side effects of nutmeg and the potential health hazard of high doses of cassia cinnamon. Many other stories could fill a book about both the positive and negative roles of herbs and spices. Care and caution with healthy skepticism are always a good spice to include in your decision-making process.

Science for the Chef: Mango pineapple hot sauce

This is one of the author's favorite hot sauce recipes to make, especially with the fresh variety of peppers he grows in his garden. Hot sauce is a popular condiment known for its spicy kick and flavor enhancement. It typically consists of a combination of peppers, fruits, acids, spices, and preservatives. In this recipe for mango pineapple hot sauce, the author combines the heat of habanero peppers and super hot peppers with the sweetness of mangoes and pineapples. The addition of lime and lemon juice, ginger, rice wine vinegar, salt, and other ingredients further enhances the flavor profile. With the perfect balance of heat, tanginess, and sweetness, this hot sauce is a delicious addition to various dishes.

Ingredients:
- 2 cups peeled chopped mangoes
- 2 cups pineapple, crushed, not drained
- 1 can (~14 oz) mango juice (Jumex mango nectar is great)
  - I've played with pineapple juice or even substituted mango for canned peaches and syrup.
    Peppers (will make a pretty hot sauce—adjust as your taste requires)
    - 8–12 habanero peppers (remove seeds)
    - 5–10 super hots (examples include Trinidad Morgua Scorpion, Peach Ghost Scorpion, Bhut Orange Copenhagen: remove seeds to avoid bitter flavors from the phenols of the seed coating)
- 1 tablespoon sugar
- 1 lime juiced
- 2 lemon juiced
- 2 tablespoons ginger, fresh minced
- ½ teaspoon black cumin
- ½ cup rice wine vinegar (white will do—malt vinegar might be interesting)
- 1 teaspoon salt
- 1/8th teaspoon potassium sorbate (optional—also trace metabisulfate)
- 1/8th teaspoon Xanthan gum

Instructions:
1. Add lime juice with fruit. Mix in half of mango juice and blend until smooth.
2. Blend all remaining ingredients until smooth. Filter pulp through a fine strainer, return a bit to the filtered juice but toss most.
3. Deseed the peppers—the seeds are bitter, and most of the heat comes from the ribs (placenta) of the fruit.
4. Roughly cut/dice peppers and place them in the remaining mango juice with lemon juice and vinegar, then blend until smooth.
5. Add salt, sorbate (preservative), and xanthan gum and stir. Allow the gum to swell (10–15 minutes) then blend the entire mixture smoothly. Add peppers as needed.
6. Sterilize bottles in boiling water then transfer sauce and store at 4°C (fridge).

Science Behind the Recipe

Mangoes and pineapple: Mangoes and pineapple contain natural sugars, acids, and enzymes. The enzymes in the fruits can break down proteins, enhancing the texture and flavor of the sauce. The sugars in the fruits provide sweetness, while the acids contribute to the overall tanginess of the sauce.

Habanero peppers and super hots: The habanero peppers and super hot peppers, such as Trinidad Moruga Scorpion, Peach Ghost Scorpion, and Bhut Orange Copenhagen, are rich in capsaicinoids. Capsaicinoids are responsible for the spiciness or heat in peppers. The heat level of the sauce can be adjusted by varying the quantity and type of peppers used. Capsaicinoids stimulate receptors on the tongue and create a sensation of heat.

Lime and lemon juice contain citric acid, which enhances the tangy flavor of the sauce. Citric acid also acts as a natural preservative, helping to inhibit the growth of microorganisms.

Fresh minced ginger adds a warm and slightly spicy flavor to the sauce. Ginger contains compounds called gingerols, which have antioxidant and anti-inflammatory properties.

Rice wine vinegar contributes acidity and adds a unique flavor profile to the sauce. The acetic acid in vinegar acts as a preservative and helps to prevent the growth of bacteria.

Salt not only enhances the overall flavor of the sauce but also acts as a preservative by drawing out moisture from microorganisms, inhibiting their growth.

Potassium sorbate is an optional preservative added to prevent the growth of yeast, mold, and some bacteria. It helps extend the shelf life of the sauce.

Xanthan gum is a thickening agent used to give the sauce a desired texture. It forms a gel-like consistency and helps prevent separation of ingredients. Xanthan gum is also a stabilizer, providing a smooth and consistent sauce.

Key Concepts

1. **Salt** is an example of an ionic compound formed by the reaction of an acid and a base, consisting of positively **charged ions (cations)** and **negatively charged ions (anions).**

2. When salt is dissolved in water, the polar nature of water disrupts the attractive force between salt ions, leading to the **dissociation** of cationic and anionic components.

3. Salt plays a vital role in cooking by improving taste, preserving food, stabilizing emulsions, and influencing protein behavior during cooking and baking.

4. **Volatile** organic compounds are carbon-based, non-polar compounds with high vapor pressure and low water solubility that are responsible for many of the aromas and flavors of herbs and spices.

5. **Phenols**, **esters**, **terpenes**, and **pungent compounds** give herbs and spices their unique blends of aroma and flavor. These different compounds bind to sensitive receptors for taste and smell in the nose that can detect molecular differences with great discrimination and at very low concentration.

6. Differences in DNA sequences called **single-nucleotide-polymorphisms** or **SNPs** can influence an individual's perception of flavor and odor.

7. Pigments present in chili peppers give them their vibrant colors, such as **beta-carotene** for yellow-orange hues, chlorophyll for green, and **capsanthin** and **capsorubin** for red colors.

8.   A group of alkaloid compounds, including capsaicin and dihydrocapsaicin, contribute to the heat of chili peppers by binding to the TRPV1 receptor and activating pain signals. Capsacins are part of a plant's **chemical defense** strategy against microbes and unwanted foragers.
  • The Scoville heat unit (SHU) is the unit by which the "heat" of a chili is measured. Using modern chromatography methods, the parts of capsaicinoid present per million of solvent (ppm) are multiplied by 15 to obtain SHU.

References

1 Drummond, J. C. & Wilbraham, A. (1939). *The Englishman's Food: A History of Five Centuries of English Diet.* London: J. Cape.

2 Woodman, M. (1978). *Food and Cooking in Roman Britain.* Cirencester: Corinium Museum.

3 McGee, H. ed. (2004). *On Food and Cooking.* New York: Simon and Schuster, Inc.

4 He, F. J., Li, J. & Macgregor, G. A. (2013) Effect of longer term modest salt reduction on blood pressure: cochrane systematic review and meta-analysis of randomised trials. *BMJ, 346,* f1325.

5 Stolarz-Skrzpek, K. et al. (2011). Fatal and nonfatal outcomes, incidence of hypertension, and blood pressure changes in relation to urinary sodium excretion. *JAMA, 305,* 1777–1785.

6 Vallverdú-Queralt, A., Regueiro, J., Martínez-Huélamo, M., Rinaldi Alvarenga, J. F., Leal, L. N. & Lamuela-Raventos, R. M. (2014). A comprehensive study on the phenolic profile of widely used culinary herbs and spices: rosemary, thyme, oregano, cinnamon, cumin and bay. *Food Chem, 154,* 299–307.

7 Potter, T. L. & Fagerson, I. S. (1990). Composition of coriander leaf volatiles. *J. Agric. Food Chem, 38*(11), 2054–2056.

8 McGee, H. (2010). Cilantro haters, it's not your fault. *The New York Times.* Available at http://www.nytimes.com/2010/04/14/dining/14curious.html (accessed November 16, 2015).

9 Mauer, L. & El-Sohemy, A. (2012). Prevalence of cilantro (*Coriandrum sativum*) disliking among different ethnocultural groups. *Flavour, 1,* 8.

10 Nicholas, E., Shirley, W., Chuong, B. D., Amy, K. K., Joyce, Y. T., Joanna, L. M., David, A. H. & Uta, F. (2012). A genetic variant near olfactory receptor genes influences cilantro preference. *Flavour, 1,* 22.

11 Scoville, W. L. (1912). Note on capsicums *J. Am. Pharm. Assoc, 1*(5), 453–454.

12 Tewksbury, J. & Nabhan, G. (2001). Seed dispersal: directed deterrence by capsaicin in chillies. *Nature, 412,* 403–402.

13 Tewksbury, J., Reagan, K., Machnicki, N., Carlo, R. T., Haak, D. & Penanaloza, A. (2008). Evolutionary ecology of pungency in wild chilies. *PNAS, 105,* 11808–11811.

Additional Readings

The Colors of Health: Chemistry, Bioactivity, and Market Demand for Colorful Foods and Natural Food Sources of Colorants. de Mejia, E.G. Zhang, Q., Penta, K., Eroglu, A., and Lila M.A. Annual Review of Food Science and Technology Vol. 11, pp 145–182 2020.

Darré, D., and Domene, D. Binding of Capsaicin to the TRPV1 Ion Channel. Molecular Pharmaceutics 12, pp. 4454–4465 2015.

Functional properties of wasabi and horseradish. Kinae, N., Masuda, H., Shin, I.S., Furugori, M., and Shimoi, S. BioFactors Vol 13, pp.265–269, 2000.

Gorgani, L., Mohammadi, M., Najafpour, G.D. and Nikzad, M. (2017), Piperine—The Bioactive Compound of Black Pepper: From Isolation to Medicinal Formulations. *Compr Rev Food Sci Food Saf., 16*(1), 124–140.

Hernández-Pérez, T, Gómez-García, MdelR, Valverde, ME, Paredes-López, O. (2020), Capsicum annuum (hot pepper): an ancient Latin-American crop with outstanding bioactive compounds and nutraceutical potential. A review. *Compr Rev Food Sci Food Saf.*, 19(6), 2972–2993.

Mathot, AG, Postollec, F, Leguerinel, I. (2021) Bacterial spores in spices and dried herbs: the risks for processed food. *Compr Rev Food Sci Food Saf.*, 20(1), 840–862.

Paup, V.D., Barnett, S.M., Diako, C. and Ross, C.F. (2019), Detection of Spicy Compounds Using the Electronic Tongue. *J. Food Sci.*, 84(9), 2619–2627.

Synergistic Anticancer Activity of Capsaicin and 3,3′-Diindolylmethane in Human Colorectal Cancer. Clark, R., Lee, J., and Lee, S.H. J Agricultural Food Chemistry *63* (17), pp 4297–4304, 2015.

End of Chapter 10 Questions

1. What is the chemical definition of salt?
   a. A compound formed by the reaction of an acid and a base
   b. A substance consisting of sodium and chloride ions
   c. Any edible substance used to enhance flavor in food
   d. A mineral mined from underground salt deposits

2. How does salt affect protein behavior during cooking?
   a. It denatures proteins, leading to the formation of protein aggregates.
   b. It prevents protein aggregation and promotes protein solubility.
   c. It enhances the binding properties of proteins, helping them hold food together.
   d. It stabilizes emulsions and prevents separation of oil and water.

3. What distinguishes salt from herbs and spices?
   a. Salt is an inorganic mineral.
   b. Salt comes from plants.
   c. Salt is used for medicinal purposes.
   d. Salt has a pungent flavor.

4. Which part of a plant are herbs derived from?
   a. Leaves
   b. Root
   c. Stem
   d. Fruit

5. What is the main factor contributing to the aroma and flavor of herbs and spices?
   a. Terpenes
   b. Phenols
   c. Esters
   d. Volatile organic compounds

6. What is the significance of pollination in vanilla cultivation?
   a. It helps in the production of vanilla beans.
   b. It enhances the flavor and aroma of vanilla.
   c. It ensures the growth of vanilla vines in different tropical areas.
   d. It prevents the natural pollinators from damaging the vanilla flowers.

7. Which compound is primarily responsible for the flavor of vanilla?
   a. Vanillin
   b. Cinnamaldehyde
   c. Saffron
   d. Myristicin

8. What genetic change is associated with the dislike of cilantro?
   a. Single nucleotide polymorphism (SNP)
   b. Chromosomal mutation
   c. Genetic recombination
   d. Insertion-deletion (Indel)

9. What gives saffron its distinct deep red color?
   a. Cinnamaldehyde
   b. Crocin
   c. Safranal
   d. Myristicin

10. Explain the chemical compounds responsible for the distinct flavor and aroma of curry powder, and how they contribute to its sensory experience.

11. Which pigments are responsible for the yellow-orange color in some chili peppers?
    a. Capsanthin and capsorubin
    b. Carotenoids
    c. Chlorophyll
    d. Anthocyanins

12. How do capsaicinoids contribute to the heat levels of chili peppers?
    a. By binding to taste receptors on the tongue
    b. By activating pain receptors in the nervous system
    c. By increasing the production of capsaicin in the placenta
    d. By interacting with enzymes that modify the heat sensation

13. Explain the significance of the mechanism of action of TRPV receptors in sensory perception and its implications for understanding pain and temperature sensations.

14. Many people feel there are health benefits to using Himalayan salts in food. Similar claims are made for sea salt. What is chemistry or biology support for such claims? How likely are these claims?

15. What are the different kinds of capsaicin found in pepper plants, and do they have different heat levels? If so, how?

11

Beer and Wine

Guided Inquiry Activities (Web): *3, Mixtures and States of Matter; 4, Water; 7, Carbohydrates; 14, Cells and Metabolism; 15, Metabolism, Enzyme, and Cofactors; 29, Alcohol and Beer Brewing; 30, Beer and Wine*

Learning Objectives

1. *Explain the role of oxygen in aerobic yeast metabolism (i.e. respiration) versus anaerobic yeast metabolism (i.e. alcoholic fermentation) of glucose. Identify the substrates and products of yeast metabolism of glucose in the presence and absence of oxygen.*
2. *Describe structural similarities and differences between fusel alcohols and ethanol and understand the conditions that create fusel alcohol production during fermentation.*
3. *Understand the solubility properties of ethanol in water and its ability to dissolve fat-soluble compounds due to the presence of a polar hydroxyl group and a nonpolar carbon chain.*
4. *Comprehend the process of distillation and describe its application in raising ethanol concentration in alcoholic beverages.*
5. *Define the concept of an azeotropic point and explain its significance in determining the maximum achievable ethanol content during distillation.*
6. *Describe the processes of alcohol absorption, distribution, metabolism, and excretion in the body.*
7. *Understand how genetic variations in ADH and ALDH enzymes can impact an individual's response to alcohol consumption, including the risk of alcohol use disorder (AUD), the "Asian Flush" reaction, and fetal alcohol syndrome (FAS). Recognize the interplay of genetic and environmental factors in alcohol dependence.*
8. *Understand the delicate balance between GABA(A) and NMDA receptors in response to low and high levels of ethanol and during acute ethanol exposure, chronic ethanol exposure, and withdrawal, and their role in the development of alcohol tolerance, withdrawal symptoms, and cravings in individuals with AUD.*
9. *Understand the process of beer production, including malting, mashing, fermentation, and conditioning.*
10. *Identify the key enzymes and reactions involved in the conversion of starch to simple sugars during malting and mashing.*
11. *Describe the role of hops in beer production, including their impact on flavor, bitterness, and preservation.*
12. *Differentiate between top-fermenting and bottom-fermenting yeasts and their respective roles in ale and lager beer production.*
13. *Identify the key compounds involved in flavor and aroma in wine, including volatile organic compounds, polyphenols, and anthocyanins.*
14. *Describe the impact of yeast strains on the production of flavor compounds during fermentation.*
15. *Understand the use of additives such as sulfur dioxide, sorbate, and oak in wine production and their influence on flavor development.*
16. *Identify the components of wine and their reactions that produce the flavors and odors of wine.*

The Science of Cooking: Understanding the Biology and Chemistry Behind Food and Cooking, Second Edition. Joseph J. Provost et al.
© 2025 John Wiley & Sons, Inc. Published 2025 by John Wiley & Sons, Inc.
Companion website: www.wiley.com/go/provost/food_science_2e

11.1 Introduction

Beer, wine, whiskey, gin, sake, and tequila: at the heart of these and many more liquors is an essential process that shares its science and technology with baking bread, making cheese, and other microbiological processes. Once upon a time, someone first found wild yeast growing in a watery liquid with a carbohydrate source. Someone tried the resulting ethanol, and the alcoholic drink was born (Figure 11.1). The process has been refined for over 7000 years. Sumerians wrote poems about the effects of wine on cranky teenage princesses, chemists found trace molecules from beer in ancient Chinese containers, and the Code of Hammurabi included punishments for over-charging tavern customers for their drink [1].

For any alcoholic beverage, the basics are simple: combine water, yeast (a microorganism), a source of carbohydrates, and time in a process called alcoholic fermentation. The carbohydrate is either a source of simple sugars—for example, through the harvest of ripe grapes or other fruit—or more complex sugar polymers (i.e. starches) from seeds and cereals that are converted into simple sugars using enzymes. The simple carbohydrates are then metabolized as food by the microscopic yeast, yielding the waste products of ethanol and carbon dioxide. While bakers use carbon dioxide to leaven their dough, when making alcoholic drinks, ethanol is the prized final product of these microbiological factories.

Alcoholic beverages owe their flavor and color to the starting compounds, the yeast strain, and how the fermented mother liquor is processed. Some beverages are aged for more complex flavors, and others are bottled for immediate consumption (Table 11.1). Wine, beer, sake, cider, and mead involve minimal post-fermentation

Figure 11.1   Types of alcoholic beverages. boule1301/Adobe Stock Photos

Table 11.1   Alcoholic beverages sugar sources and processing.

Alcoholic beverage	Carbohydrate starting material	Post-fermentation processing
Wine	Grapes or fruit	Aging for oxygen and tannin reaction
Beer	Barley, wheat, rice, corn	Added hops and adjuncts for flavor and minimal aging
Mead	Honey, some add fruit or spices	Solids are settled by gravity
Cider	Primarily apples, some other fruits	Pectin is removed by precipitation and solids settled
Sake	Polished white rice	Molds digest starches for yeast, additional alcohol added, solids filtered
Vodka	Potatoes, grains (wheat, rye), fruit	Distillation and rectification (repeated distillation for high alcohol content)
Tequila	Agave cactus, sugars, pineapple	Distilled. Silver—bottled after distillation, Anejo/Reposado—aged in barrels
Rum	Sugarcane products, juice, and molasses	Distilled and aged in oak casks for color and flavor
Whisky	Barley, corn, rye, wheat	Distilled using copper to remove sulfur and aged in oak barrels or casks

processing and are not enriched in their alcoholic content. Liquor, hard liquor, spirits, or more formally distilled spirits, begin with the same basic principle of fermented beverages: a sugar source and yeast. As per the name "distilled spirits," the fermented liquid is enriched in its ethanol content by distillation.

11.2 Yeast: Metabolic, Ethanol-producing Factories

Historic fermentation required some luck to produce alcohol from fruits and grains, and that's because alcoholic fermentation requires a microscopic organism, yeast. Early winemakers unknowingly added yeast from the environment; that yeast was already growing on the grapes themselves, the stems of the vine, or on the feet of those stomping the grapes to release juices. Ancient beer production relied on wild yeast blown in from the air or from the barley (and, in some cases, bread) to start the process. So what are yeasts?

11.2.1 Aerobic Respiration versus Anaerobic Fermentation

Yeast are a large class of eukaryotic fungi. They are eukaryotes because their cells—the smallest structural and functional unit of the organism—place genetic material inside a nucleus, just like the cells of humans but unlike the cells of bacteria (see Box 11.1). Species of yeast most often exist as individual single cells, but in some cases, as in the yeasts used to make lager or ale beer, they form long strands of individual cells. Yeast can

---

Box 11.1 Classifying living things

When classifying living things, biologists make groups of organisms using observed similarity based on genetic analysis. Every living thing is made of at least one *cell*, and many organisms are comprised of many cells. The cell is the smallest structural and functional unit of the organism. Cells are defined by an outer boundary called a membrane—depending on the organism, a wall may be added to the membrane—and cells contain the genetic material that codes for the organism. That genetic material is inherited as cells grow and divide. It is possible to analyze a cell's genetic material, much like reading a blueprint that describes the construction of an entire building, and learn about the relationships or similarities between organisms.

The largest categories of living things are called domains [2], and there are three very large groups: archaea, bacteria, and eukarya. You probably recognize the word "bacteria." The domain of bacteria includes the *Lactobacillus* that you find in yogurt and the *Streptococcus* that cause the frequent childhood illness, "strep throat." Together with archaea, bacteria make up the prokaryotic organisms; these are microorganisms consisting of only one cell whose genetic material is loose inside the cell. The third domain, eukarya, includes every animal and plant in our world, from humans to chickens and from banana trees to mushrooms. What do all these "eukaryotic" organisms share in common that makes them different from prokaryotic organisms? The cells of organisms within eukarya have their genetic material contained within a second compartment called a nucleus. The nucleus is also defined by a membrane or boundary, and it is possible to see the nucleus inside a eukaryotic cell when looking through a microscope, as in Figure 11.2.

The domain of eukarya is further divided into five kingdoms, three of which are animalia (i.e. animals), plantae (i.e. plants), and fungi. Those kingdoms are further subdivided into phyla, which are further subdivided into classes. Classes are then subdivided into orders, orders into families, and families into genera. A single genera is called a genus, and each organism within a genus is assigned a species name. When you see the scientific name of an organism, such as *Saccharomyces cerevisiae*—the scientific name for brewers's yeast—that is the genus (i.e. *Saccharomyces*) and species (i.e. *cerevisiae)* for a unique organism. The scientifically named *Saccharomyces bayanus* is a different species of yeast from the genus *Saccharomyces* that is used in making wine and cider.

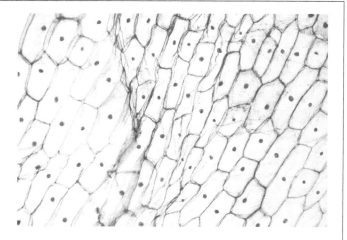

Figure 11.2 A single layer of cells from an onion as viewed under a microscope. There are many cells, each delineated with a boundary, and a nucleus is visible within each cell as a blue circle. Peter Hermes Furian/Adobe Stock Photos

In conclusion, let's review the classification of *Saccharomyces cerevisiae* from kingdom to species.

Domain: Eukarya
Kingdom: Fungi
Phylum: Ascomycota
Class: Saccharomycetes
Order: Saccharomycetales
Family: Saccharomycetaceae
Genus: *Saccharomyces*
Species: *Cerevisiae*

convert food into energy in the presence of oxygen or in the absence of oxygen (Figure 11.3). When oxygen is present, yeast metabolism is aerobic, from the Greek *aēr* for "air" and the Greek word *bios* meaning "life." So aerobic metabolism depends on "the air," or rather the oxygen within that air. Yeast use oxygen (i.e. O_2) in the biochemical process of respiration to convert glucose and oxygen to carbon dioxide (i.e. CO_2) and water. When oxygen is absent, the prefix an- is added to create the word, anaerobic. Without oxygen, yeast convert glucose to ethanol and carbon dioxide through the process of alcoholic fermentation. There are thousands of species of yeast in nature, but the species of yeast within the genus *Saccharomyces* have been grown, or cultured, and domesticated for their ability to metabolize glucose into ethanol during alcoholic fermentation.

Alcoholic fermentation is useful to humans, but it is ultimately toxic to yeast. Ethanol kills yeast when it reaches concentrations of 10–20%. So why do yeast even make it in the first place? Ethanol may be desirable for making beer and wine, but to yeast, ethanol is simply a waste product of energy production. When yeast break down sugar (i.e. glucose), the primary goal is to extract energy in the form of ATP (i.e. adenosine triphosphate). ATP is a way to store chemical energy for use later, almost like an energy currency. Yeast first converts glucose to pyruvate in a process called glycolysis: *glyco* refers to the "sweet" sugar that is the starting material and *-lysis* means "break" or "split" as in the chemical reactions that break apart the glucose molecule. The biochemical steps of glycolysis synthesize two molecules of ATP from two molecules of ADP (i.e. adenosine diphosphate) and two phosphate anions (i.e. two PO_4^{2-}) while breaking the glucose into two molecules of pyruvate and consuming two molecules of NAD^+. NAD^+ is a large, complex molecule called nicotinamide adenine dinucleotide that is essential for the oxidation chemistry that ultimately turns glucose into pyruvate. The NAD^+ gains electrons and two H^+ ions and becomes NAD–H. It is at this point that aerobic respiration parts ways with anaerobic

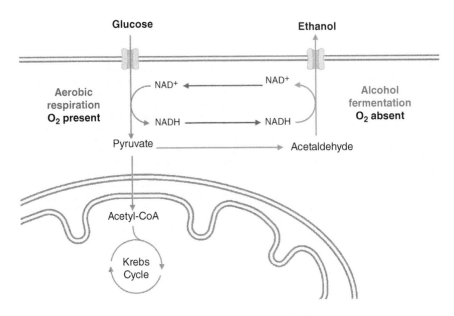

Figure 11.3 The presence of oxygen determines the product of yeast metabolism.

fermentation. For the breakdown of glucose to continuously yield energy, the NAD–H must be recycled to NAD^+ for use in additional "glucose-breaking" reactions. If the yeast have access to oxygen, then the pyruvate molecules are converted ultimately to six molecules of CO_2, and the NAD–H molecules are recycled to NAD^+, both using oxygen-requiring biochemical reactions. If the yeast do *not* have access to oxygen, then each molecule of pyruvate is converted to ethanol (i.e. C_2H_5OH) and CO_2 while recycling NAD–H to NAD^+ as part of the overall process (Figure 11.4). The biochemical processes of glycolysis, aerobic respiration, and anaerobic fermentation are conserved across almost all living things, including humans. You can read more about the details of these essential biochemical processes in Chapter 15.

In the beginning stages of beer and wine production, oxygen is a critical component as respiration fuels the growth of yeast to a high density. Adding oxygen produces more yeast growth and more ATP, but little ethanol. In

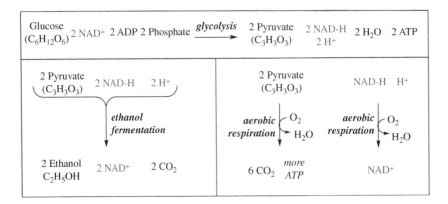

Figure 11.4 The biochemical breakdown of glucose by yeast begins with glycolysis. Glycolysis is followed by aerobic respiration when oxygen is present, but ethanol fermentation in the absence of oxygen.

contrast, the density of yeast cells at the start of fermentation is too low to make more than a percent or so of alcohol. Therefore, before starting fermentation, the brewer gives the yeast plenty of food and oxygen to grow and multiply to a high density.

11.2.2 Making Ethanol

Once the brewer has achieved a satisfactory density of yeast growth, and sufficient yeast cells have converted from aerobic respiration to anaerobic fermentation, sealing the yeast off from any more oxygen exposure will ensure the cells shift to producing ethanol in earnest. Very little cell division occurs at this stage, as the total ATP production is decreased in the absence of oxygen. When oxygen is scarce, ethanol production offers yeast a chance to convert NAD–H back to NAD$^+$ and serves a second purpose. Ethanol serves as a poison to other microorganisms. Three to five percent of ethanol inhibits a range of bacteria and other organisms from growing; higher concentrations will stop most from growing. Yeast growth is also limited by ethanol concentration, although not to near the extent of other microorganisms. Higher levels of ethanol will damage the mitochondrial DNA and inactivate some of the enzymes involved in glycolysis, including hexokinase and NADH-producing dehydrogenases (Box 11.1).

Box 11.2 Yeast and alcohol tolerance

Most yeast in the *Saccharomyces* genus can tolerate up to 10–20% ethanol by volume before growth is inhibited, depending on the specific strain used. That is why most wines or strong beers are limited in alcohol content. However, scientists have been working to generate a new strain of yeast or ways to help yeast tolerate higher alcohol concentrations and temperatures for higher biofuel (ethanol) yield/production [3]. At levels above approximately 20% ethanol, the membranes of yeast cells become porous, and the cells die. The new yeast strain produces a steroid that strengthens the membrane to withstand the effects of ethanol production. Another set of scientists found that adding potassium salts helps membrane pumps compensate for the influence of alcohol [4]. While this was done using yeast strains specific for biofuel, it is interesting that changing the culture conditions of the yeast improved the ethanol yield by almost twice compared to without changing the culture conditions. We don't know how these additions will alter the flavor of fermented beverages or if yeasts growing in such conditions will also generate an additional set of compounds that may impact the final flavor of beer, wine, or distilled spirits.

For those who want more than the ethanol concentration found in beer or wine, further concentration of the alcohol is required. This is most commonly achieved by removing water through distillation. Thus, the total starting amount of simple sugar and the mass of yeast when switching to anaerobic fermentation and strain selection significantly contribute to the final concentration of ethanol in beer or wine. Of course, stopping fermentation early in the process will result in residual sugars providing a sweet taste, while allowing the fermentation to run to completion (sugar depletion) provides for a dry (less sweet) but higher alcohol content beverage.

In addition to ethanol, other higher molecular weight alcohols called fusel alcohols are also produced, although at low levels. Examples include butyl alcohol, isoamyl alcohol (isopentanol), and isopropyl alcohol (Figure 11.5). Several fusel alcohols provide off-flavors and are poisons. The danger in amateur distillation is that these dangerous alcohols can be concentrated into the distillate (moonshine can hurt you!). These are produced if the fermentation temperature is too high, the pH is too low, and when there are not enough nutrients available for the yeast (principally nitrogen). Other compounds adding to the complex flavor and aroma of fermented beverages depend on the chemical nature of the starting material and the strain of yeast used. Other commonly produced molecules include acetaldehyde (green apple aroma),

---

(Continued)

Box 11.2 (Continued)

diacetyl (buttery butterscotch flavor), dimethyl sulfide (sweet corn aroma), sulfur (rotten eggs or burnt matches), and many phenolic compounds discussed in the spice chapter.

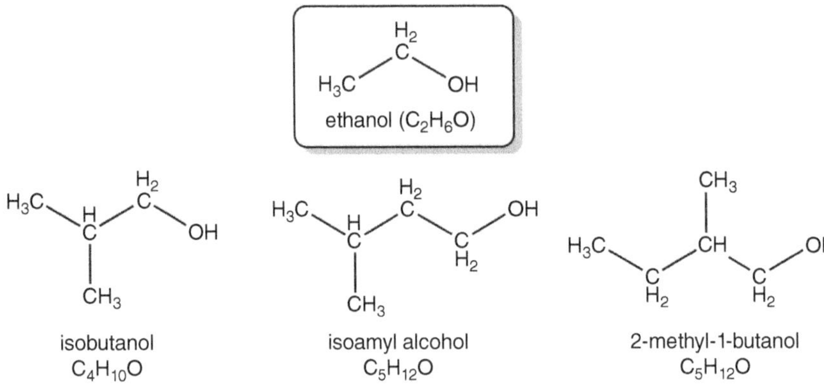

ethanol (C_2H_6O)

isobutanol
$C_4H_{10}O$

isoamyl alcohol
$C_5H_{12}O$

2-methyl-1-butanol
$C_5H_{12}O$

Figure 11.5 Ethanol is the two-carbon alcohol of wine and beer, while isobutanol, isoamyl alcohol, and 2-methyl-1-butanol are amino acid-derived fusel alcohols formed during yeast fermentation under conditions of low ammonia. The fusel alcohols have greater than two carbon atoms, so they are also sometimes called "higher alcohols."

11.2.3 Fusel Alcohols

In order for yeast to grow, they need more than just glucose (i.e. $C_6H_{12}O_6$); yeast also need a source of nitrogen to build more biomass, grow and multiply. Yeast can utilize ammonia (i.e. NH_3) as a nitrogen source, but during fermentation and under conditions of low ammonia, yeast rely on amino acids as a nitrogen source. We first saw amino acids in Chapter 2 (see Figure 2.3), and we learned they are the building blocks of protein. When extracting nitrogen from amino acids, yeast turn the leftover parts of the amino acid skeleton into fusel alcohols, a group of alcohols with more carbon atoms than the basic ethanol. Because of their extra carbon atoms, fusel alcohols are sometimes called "higher alcohols." When talking about "alcohols," chemists refer to any molecule containing an -OH group of atoms. Ethanol (i.e. CH_3CH_2-OH) is the "alcohol" of wine and beer and it is made of two carbon atoms, hydrogen atoms, and the characteristic -OH group of atoms. Fusel alcohols (Figure 11.5) also contain an -OH group of atoms, and they also have greater than two carbon atoms connected in a straight line or branched. Fusel alcohols have a flavor profile that is distinct from ethanol, so the brewer needs to understand what they are and where they come from.

The process of fusel alcohol production by yeast begins with the deamination, that is, amino group removal of specific amino acids, releasing their nitrogen-containing amino groups and leaving behind α-keto acids (Figure 11.6). These α-keto acids are then further reduced to aldehydes and finally transformed into fusel alcohols through additional reduction reactions. Fusel alcohols, such as isobutanol, isoamyl alcohol, and others, contribute to the complex flavor and aroma profiles of fermented beverages like beer and wine. The intricate interplay between yeast, amino acids, and limited ammonia availability during fermentation showcases the remarkable adaptability of yeast and the biochemical versatility of amino acids in facilitating the production of desirable compounds.

Fusel alcohols can positively and negatively impact the flavor of fermented beverages like beer, wine, and distilled spirits. While they contribute to the overall complexity and character of these beverages in small amounts, excessive levels of fusel alcohol can lead to undesirable off-flavors and potential health hazards. High concentrations of fusel

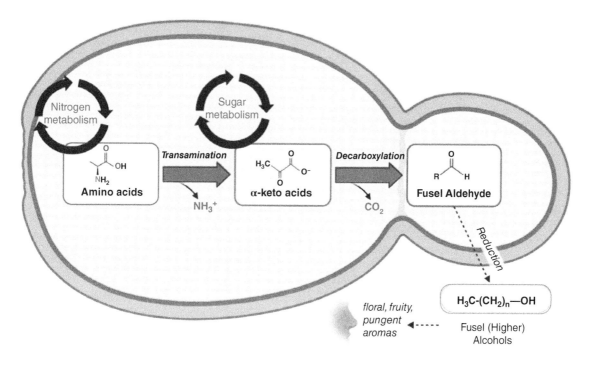

Figure 11.6   Fusel alcohol production in yeast during fermentation.

alcohols, especially isoamyl alcohol and isobutanol, can produce harsh, solvent-like aromas. These off-flavors are generally perceived as undesirable. While other fusel alcohols, particularly those with longer carbon chains, can impart a perceived "hotness" or burning sensation on the palate due to their higher boiling points. This can lead to astringency and a less smooth mouthfeel, which are not preferred qualities in well-balanced beverages. Elevated levels of fusel alcohol may negatively affect the overall drinkability of the beverage. Consumers may find the drink less smooth and enjoyable, leading to a decrease in its appeal.

   Although the exact relationship between fusel alcohol and hangover symptoms is not fully understood, some studies suggest that higher levels of fusel alcohol might contribute to the severity of hangovers, potentially leading to headaches and other discomforts. In distilled spirits, excessive fusel alcohol content may lead to health and regulatory issues. Some countries have strict regulations on the allowable levels of fusel alcohol in spirits to ensure consumer safety and product quality.

   To mitigate the negative impact of fusel alcohols, brewers, winemakers, and distillers carefully control fermentation conditions, including yeast strains, temperature, and nutrient availability. Yeast selection plays a crucial role since different strains may produce varying levels of fusel alcohol. Proper fermentation management and monitoring also help maintain fusel alcohol concentrations within acceptable ranges, allowing for a desirable and balanced flavor profile in the final product.

11.3   Ethanol

A chemist defines an "alcohol" as any molecule containing a carbon bound to an oxygen that is also bonded to a hydrogen (i.e. C—OH); chemists call this arrangement of atoms a hydroxyl group. Ethanol has the chemical formula of C_2H_6O, but those same atoms can be arranged as CH_3CH_2OH (Figure 11.5) to emphasize the presence of the hydroxyl group that makes ethanol an alcohol. When talking about food and cooking, the word "alcohol" is synonymous with ethanol, also known as ethyl alcohol or grain alcohol. The term "proof" is a historical measure of the alcoholic content in a beverage. In the United States, the proof of alcohol content is twice that of ethanol's percent (weight by volume) in solution. A 50-proof alcoholic drink will be 25% (w/v) by these standards. Historically, proofing alcohol was a measure used by sailors in the eighteenth century to test their rum rations to see if it was watered down. Daily rum rations were given to sailors who would mix the rum with gunpowder. If the mixture of rum and gunpowder would ignite, the rum was "proofed." However, diluted rum would contain too much water for the gunpowder to catch fire and be "underproof." The relationship between proof and percent is because gunpowder will not burn unless the rum contains at least 57.15% ethanol and is considered to have "100 degrees of proof." Thus, the historic 7/4 times the alcohol by volume as a proof is why we do not use precisely one-half percent. Eventually, the proof was simplified to twice the percent alcohol content.

Box 11.3 Distillation

The process of making consumable ethanol begins with sugar and water. After fermentation, the liquid ethanol is mixed with the water in the fermentation mixture (sometimes called a "mash"). The concentration of ethanol is approximately 10–12% because that is how much ethanol yeast can tolerate before they die. In order to raise the ethanol concentration and produce such beverages as vodka, tequila, rum, or whiskey, the ethanol must be purified from the mash. Heating a liquid gives some of the molecules enough energy to leave the liquid and boil into a gas or vapor (Figure 11.7). The temperature needed to convert a pure sample of a

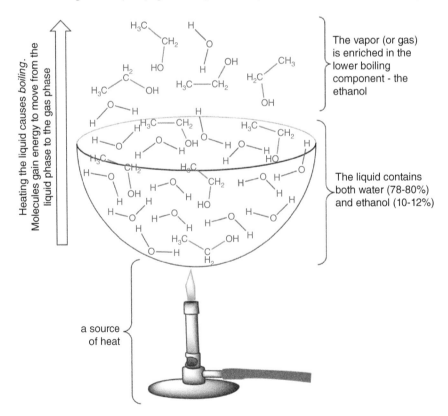

Figure 11.7 Ethanol evaporation. The physical change as an ethanol–water mixture is evaporated. This is the first part of distillation.

liquid into a gas/vapor is called its boiling point. Water boils at 100°C/212°F and ethanol boils at 78.5°C/173°F, but if you heat a mixture of ethanol and water, it will boil somewhere between 78.5°C and 100°C. The vapor of the ethanol–water mixture will contain both water molecules and ethanol molecules, but the vapor of the ethanol–water mixture is more concentrated in the lower boiling component—in this case, the ethanol. If we were to capture this ethanol–water vapor and condense it back into a liquid, the concentration of ethanol would be higher (Figure 11.8 shows a common fractional distillation apparatus to distill liquid mixtures). This is the process of distillation—vaporization (boiling) and condensation (cooling the vapor back to a liquid). If we continue the process of vaporizing and condensing the ethanol–water mixture over many rounds of distillation, we will eventually get a liquid that is highly enriched in ethanol.

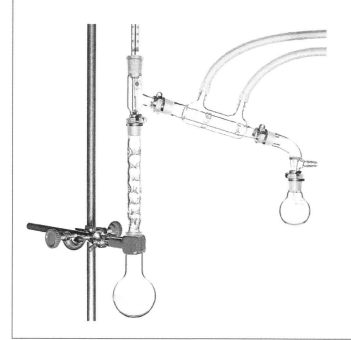

Figure 11.8 Fractional distillation apparatus. The mixture is heated where the vapor is cooled in a condensing tube, cooled with water.

While ethanol is CH_3CH_2OH, there are other alcohols, such as methanol and propanol (1-propanol or 2-propanol; Figure 11.9). Each of these alcohols has that hydroxyl group (i.e. C—OH). The hydroxyl group of atoms is *polar* because both the C—O bond and the O—H bond have an unequal sharing of electrons (see Chapter 1.4.1.2). Since the electron-rich oxygen atom and the polar bonds are at one end of the molecule, alcohols like ethanol, methanol, and propanols are polar molecules (Figure 11.9). This polar nature of small alcohols makes them excellent solvents for a variety of small flavor molecules, and also makes alcohols mix well with water. The C—H and C—C bonds that make up one end of ethanol are hydrophobic and nonpolar, which helps dissolve fat-soluble or hydrophobic molecules if the ethanol is at a high enough concentration. Slightly larger alcohols like isopropanol will dissolve such molecules at even lower concentrations. This is why liquors are used to extract spices and flavor compounds for cooking. For example, vanilla extract is made by soaking vanilla beans

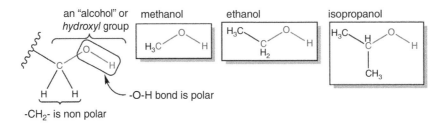

Figure 11.9 Alcohol is a polar molecule.

in ethanol. The characteristic flavors of the vanilla bean—including the molecule vanillin—dissolve into the ethanol. You can make your own vanilla extract at home by using vodka as the source of ethanol. Why is vodka a good choice as a source of ethanol? Because vodka is a distilled liquor with a higher concentration of ethanol (Box 11.2).

The hydrogen bond between water and ethanol also creates a loss of volume when the two solutions are mixed. In pure liquid water, individual molecules form and break hydrogen bonds with other water molecules at a rate of a million times every second; that's not a typo. A million times is a 1 with 12 zeros! This rapid exchange of hydrogen bonds means that pure liquid water molecules are inefficiently packed, with an average of 2.3 hydrogen bonds per water molecule. However, when mixed with water, ethanol (i.e. CH_3CH_2OH) packs more closely with the water molecules, and the more efficient packing results in shrinking the volume of the ethanol : water mixture.

Because of the nonpolar carbons of the molecule, ethanol has a lower boiling point than water. It takes less energy (heat) to disrupt the interactions between ethanol molecules, allowing the individual molecules to escape their intermolecular interactions into a gas. A common misconception when cooking with alcohol is that heating or flaming food, baked goods, or drinks will cause all of the ethanol to "burn off" or evaporate. This doesn't happen. Baked bread retains some of the ethanol for hours even after cooking, something you can smell with freshly baked bread. The interaction between water and ethanol can explain what is going on. Water and ethanol molecules form hydrogen bonds between each other and alter the boiling point for both.

There is a point where heating food will boil off the same amount of ethanol as water, leaving the percent alcohol unchanged (Figure 11.10).

The chemical term for this is an azeotropic point, a specific mixture of two compounds that can't be altered by heating or distillation. As a solution of alcohol and water is heated, ethanol, with its lower boiling point, will evaporate first at its lower boiling point. Thus, the ethanol content does reduce in contact during heating but will not completely "boil off." Part of the reason not all of the ethanol is removed during heating is explained by its azeotropic point. The azeotropic point is the mixture or ratio of ethanol and water where both molecules evaporate simultaneously during a simple distillation. Thus, at the azeotropic point, both ethanol and water will evaporate at the same rate. The azeotropic point for ethanol and water is about 95.6%. Thus, simply boiling will result in a minimum of 5% of the water. The US Department of Agriculture reports that after an hour of simmering, 25% of the initial ethanol remained in solution [5]. Eighty-five percent of the original ethanol was found when alcohol was added to a boiling liquid before being removed from heat. Some of the ethanol remaining behind is likely due to the hydrogen bonding capacity of ethanol. Ethanol loves water!

Figure 11.10 Boiling ethanol: Azeotropic point. The vapor below the boiling point will be enriched in EtOH (lower bp) until the azeotropic point is reached (where both vapor and liquid mole fraction are the same and can no longer be distilled/separated)—EtOH AzoPt = 95.6%.

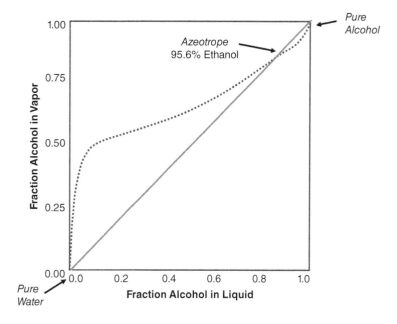

Box 11.4 Distillation and whiskey

Distillation is a fascinating process used in making whiskey, where both science and artistry play roles in creating the beverage's unique flavors and aromas. Whiskey begins with grains like rye, wheat, corn, or barley, or a mix of these, which are ground into a mash, similar to the early stages of beer brewing. For instance, rye or barley may be malted to trigger the germination of seeds, which in turn starts the expression of amylase enzymes. These enzymes break down complex carbohydrates into fermentable sugars, which yeast will later convert into alcohol during fermentation.

The drying method of these grains is crucial to the flavor of whiskey. The Maillard reaction, occurring between reducing sugars and amino acids, can produce hundreds of flavor compounds like esters and lactones. If the grains are smoked over peat, phenolic compounds are infused, adding a smoky flavor to the final spirit.

Once fermentation is complete, the liquid, now called the wash, is distilled. Initial distillation occurs in pot stills, where the wash is heated, vaporizing the alcohol and other volatile compounds. These vapors condense, and the resulting liquid, known as low wine, has an alcohol content of around 25–35%. This low wine is then distilled further in spirit stills to increase the alcohol concentration and refine the flavor profile.

During distillation, different fractions are collected. The first, the "foreshot," contains methanol and other unwanted compounds and is discarded. The "head" follows, enriched in acetone and acetaldehyde, which are also not desired. The "heart" is the main collection, rich in ethanol and the flavors we seek in whiskey. The "tails" come at the end, containing higher boiling point substances, and are sometimes redistilled to recover more ethanol. The chemical compounds that distill over include:

- Acetaldehyde has a fruity odor and is sometimes implicated in hangovers.
- Acetone, the same substance used in nail polish remover, is also found in whiskey.
- Esters provide a fruity aroma and can vary widely in their boiling points.
- Longer chain fusel alcohols and compounds like acetic acid emerge toward the end of distillation, known as the tailing distillation.

(Continued)

Box 11.4 (Continued)

Once distilled, whiskey is aged in wooden casks, often made of American or European oak, which contributes additional flavors through tannins and the effects of charring or toasting the barrels. Aging can also involve casks previously used for other spirits, which can impart unique flavors.

Different types of whiskey include:

- Single malt whiskey: Made at a single distillery from one type of malted grain, usually barley.
- Scotch whisky: Must be distilled, aged, and bottled in Scotland, with specific regional variations that contribute to its smoky character, derived from peat.
- Bourbon whiskey: An American whiskey with at least 51% corn in its mash bill, aged in newly charred oak barrels, known for its nutty flavors and caramel sweetness.
- Tennessee whiskey: Similar to bourbon but filtered through sugar maple charcoal, giving it a distinctive flavor profile.
- Rye whiskey: Contains at least 51% rye and has a spicier taste compared to other whiskeys.

The aging process, known as maturation, further refines the whiskey's flavor. The type of wood, the age of the casks, and whether they've been used before all impact the final product. For example, first-fill casks impart the most robust flavors, while reused barrels offer subtler notes. The interaction between the whiskey and the wood over time leads to a smoother, richer drink.

11.4 Alcohol and the Body

11.4.1 Alcohol: Impact and Distribution

Alcohol, one of the most widely consumed psychoactive substances globally, has long been intertwined with human culture and social rituals. As a central nervous system depressant, its effects on the body are complex and varied, ranging from relaxation and euphoria to impaired judgment and coordination. Despite its widespread acceptance and availability, the dangers associated with alcohol ingestion cannot be overlooked. According to the World Health Organization (WHO), alcohol misuse contributes to a staggering number of deaths and disabilities each year. Globally, an estimated 3.3 million deaths are attributed to alcohol consumption annually, accounting for approximately 5.9% of all deaths. Moreover, excessive alcohol intake poses significant health risks, including liver cirrhosis, cardiovascular diseases, various cancers, mental health disorders, and addiction. As we delve into the intricate relationship between alcohol and the body, it becomes crucial to acknowledge the potential consequences of its misuse and strive for informed and responsible consumption.

At a BAC of 0.08%, which corresponds roughly to 12 mM of ethanol in the bloodstream, individuals are considered legally intoxicated in many countries, and it is the common threshold for impaired driving (Figure 11.11). At this level of intoxication, ethanol affects various molecular targets in the brain and body, leading to a range of effects on humans. A simple explanation of how our neurotransmitter system works is that it involves chemicals in the brain that transmit signals from one neuron to another through a synaptic cleft (Figure 11.12). The chemical signals called neurotransmitters, influence everything from our mood to muscle movement. Enzymes found in the synapse help to remove the neurotransmitters. Ethanol alters the balance of these neurotransmitters, notably affecting those involved in signaling pleasure and pain, as well as those that regulate awakeness and sedation. In the central nervous system, ethanol interacts with neurotransmitter systems, including enhancing the inhibitory effects of gamma-aminobutyric acid (GABA) and inhibiting the excitatory actions of glutamate (GLU).

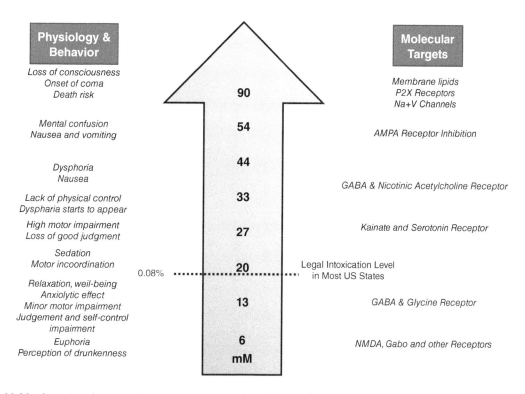

Figure 11.11 Impact and target of increasing concentration of blood alcohol concentration (BAC). The level of blood alcohol is displayed against both the molecular targets to highlight how alcohol impacts neurons as well as the physiology and behavior at increasing BAC.

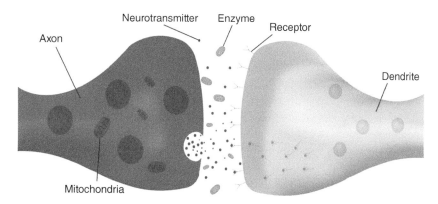

Figure 11.12 Nerve transmission of chemical signals. Neurotransmitters created or recycled in the axon end of one nerve are released into the synapse, the junction between nerves. There the neurotransmitters are removed by enzymes in the synaptic cleft or bound by protein receptors in the receiving dendrite nerve. This allows messages from one neuron to send the message to other neurons. joshya/Adobe Stock Photos

As a result, cognitive functions such as judgment, reasoning, and reaction times are impaired, leading to reduced coordination and increased risk of accidents. Additionally, ethanol at this concentration can affect motor skills and speech, resulting in slurred speech and loss of balance. Furthermore, alcohol impacts the reward pathways in the brain, leading to feelings of euphoria and relaxation, but it can also impair decision-making, increasing the likelihood of engaging in risky behaviors (Figure 11.13).

At 0.08% intoxication, individuals may experience impaired memory and concentration, making it difficult to perform complex tasks effectively. It is essential to recognize the effects of alcohol at this concentration, and it highlights the importance of responsible drinking to ensure the safety and well-being of individuals and others around them.

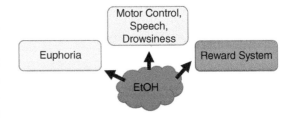

Figure 11.13 Ethanol impact on behavior. Ethanol is considered a dirty drug as it affects many targets in different parts of the brain producing a wide range of effects both emotional and physiological.

BAC is influenced by four key processes: absorption, distribution, metabolism, and excretion. Each of these processes plays a significant role in determining the level of alcohol in the bloodstream and its effects on the body.

- **Absorption:** It refers to the passage of alcohol from the gastrointestinal tract into the bloodstream. When alcoholic beverages are consumed, the alcohol is primarily absorbed in the small intestine. The rate of absorption depends on several factors, including the type and concentration of alcohol consumed, the presence of food in the stomach, and individual differences in metabolism. Carbonated beverages, for example, tend to increase absorption rates due to the gas bubbles pushing alcohol into the stomach lining more rapidly. Conversely, consuming alcohol on an empty stomach leads to faster absorption, as there is no food to slow down the process. Once alcohol enters the bloodstream, it begins to exert its effects on the central nervous system.
- **Distribution:** After absorption, alcohol is distributed throughout the body via the bloodstream. The water content in the body plays a crucial role in alcohol distribution, as alcohol is water-soluble. As a result, organs and tissues with higher water content tend to have higher alcohol concentrations. The brain and liver, being high in water content, are particularly susceptible to alcohol's effects. This distribution process contributes to the initial feelings of euphoria and relaxation associated with alcohol consumption, as the central nervous system experiences a surge in alcohol concentration.
- **Metabolism:** Alcohol metabolism occurs mainly in the liver, where enzymes break down ethanol into acetaldehyde and then further into acetic acid, which is eventually converted into carbon dioxide and water. The primary enzyme involved in alcohol metabolism is alcohol dehydrogenase (ADH), which has a limited capacity to process alcohol. The rate of metabolism is relatively constant, typically eliminating about 0.015% BAC per hour in most individuals. This means that, regardless of how much alcohol is consumed, the body metabolizes alcohol at a relatively steady rate. However, certain factors, such as genetics and liver health, can influence individual variations in alcohol metabolism.
- **Excretion:** Alcohol is excreted from the body through various routes, with the majority being eliminated through the breath, urine, and sweat. The breathalyzer test, which measures alcohol concentration in the breath, is a common method used to estimate BAC for law enforcement purposes. Alcohol excretion rates can also vary based on factors such as hydration levels, kidney function, and individual differences.

Ethanol absorption and distribution in the body are complex processes influenced by various factors. As mentioned earlier, ethanol is primarily absorbed in the small intestine after consumption. It is worth noting that ethanol is highly water-soluble but practically insoluble in fat. This solubility property affects its distribution throughout the body (Figure 11.14). Since the stomach contains a minimal amount of alcohol-degrading enzymes, most absorption occurs in the small intestine, where the larger surface area and higher enzyme activity facilitate more efficient uptake into the bloodstream.

A person's percentage of body fat plays a significant role in altering the effective volume of alcohol concentration in the body. As ethanol is highly water-soluble but practically insoluble in fat, individuals with a higher percentage of body fat will have a smaller volume of distribution for alcohol compared to those with a lower percentage of body fat (Figure 11.14). To understand this concept, consider two individuals who have consumed the same amount of alcohol. The person with a higher percentage of body fat will have less body water available to dilute the alcohol, resulting in a more concentrated alcohol distribution. On the other hand, the person with a lower percentage of body fat will have a higher proportion of body water, leading to more dilution of alcohol and a lower concentration in the bloodstream. As a result, individuals with higher body fat percentages are likely to experience higher

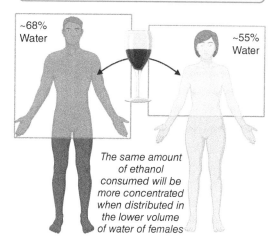

Equilibrium of ethanol in tissues is reached quickly - but not in lipid tissues

~68% Water

~55% Water

The same amount of ethanol consumed will be more concentrated when distributed in the lower volume of water of females

Figure 11.14 Distribution and concentration of alcohol depends on body composition.

BAC for a given amount of alcohol intake compared to individuals with lower body fat percentages. This phenomenon can be particularly significant in women, as they generally have a higher proportion of body fat compared to men, as well as in elderly individuals, whose body fat percentage tends to increase with age. Additionally, women tend to have lower levels of ADH and acetaldehyde dehydrogenase, the enzymes responsible for metabolizing alcohol, compared to men. This means that women may metabolize ethanol at a slower rate, further contributing to higher BAC levels.

After entering the bloodstream, ethanol swiftly permeates throughout the body due to its water-soluble nature. Organs with high blood perfusion, including the brain and liver, experience a rapid surge in ethanol concentration. As the brain is exceptionally sensitive to alcohol's influence, this swift distribution can induce noticeable alterations in behavior and cognition. Conversely, tissues with lower blood flow and lower water content, like bone and fat, will exhibit lower ethanol concentrations. Ethanol achieves equilibrium across the body relatively rapidly, often within minutes of consumption. However, the rate at which equilibrium is attained in the brain and other tissues may differ. The central nervous system benefits from its high perfusion, enabling quick distribution and yielding immediate effects on brain function and behavior. In contrast, fat tissue, characterized by limited blood flow and reduced permeability to ethanol, experiences slower equilibration. This variance in distribution rates may contribute to the diverse time frames of intoxication and the delayed onset of certain alcohol-related effects.

11.4.2 Metabolism of Alcohol in the Body

The body metabolizes alcohol through a two-step process called first and second pass metabolism. First-pass metabolism of alcohol refers to the initial breakdown of a minor fraction (about 10–30%) in the stomach and liver before entering the bloodstream. Second-pass metabolism occurs in various tissues, including the brain, where alcohol undergoes additional metabolism at a slower rate, contributing to its prolonged effects on cognitive function and behavior.

Two important proteins, ADH and aldehyde dehydrogenase (ALDH), are found throughout the body and are involved in the removal of ethanol into acetic acid (Figure 11.15). Later, by different enzymes, acetic acid is further broken down into carbon dioxide and water, allowing the body to fully eliminate alcohol. Reduction of alcohol from the body takes place in two phases: the first pass removal of ethanol happens in the stomach and the second pass of ethanol metabolism takes place throughout the body but mostly in the liver.

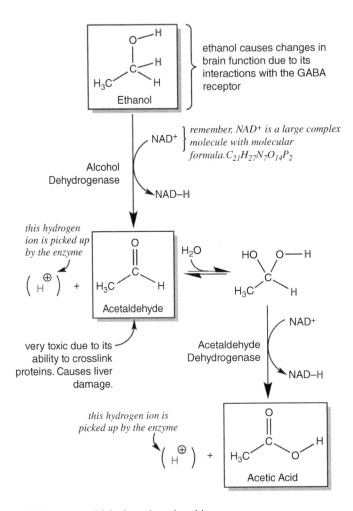

Figure 11.15 Ethanol metabolism to acetaldehyde and acetic acid.

Upon alcohol consumption, ADH produced in small amounts in the stomach contributes to the initial reduction of ethanol in the first-pass metabolism. Once ethanol distributes into the bloodstream, the molecule is transported to the liver where ADH plays a central role in the breakdown of ethanol. The liver's high concentration of ADH makes the liver a crucial site for the second-pass metabolism of alcohol. The product of ADH, acetaldehyde, is a toxic compound that must be removed quickly. This is done by ALDH, which converts the toxic acetaldehyde into acetic acid, a less harmful compound. This essential step occurs primarily in the liver, where abundant liver ALDH facilitates the conversion. A simple version of the actions of ADH and ALDH are shown in Figure 11.16. Once acetic acid is produced, it can be metabolized into carbon dioxide and water, allowing the body to eliminate alcohol efficiently.

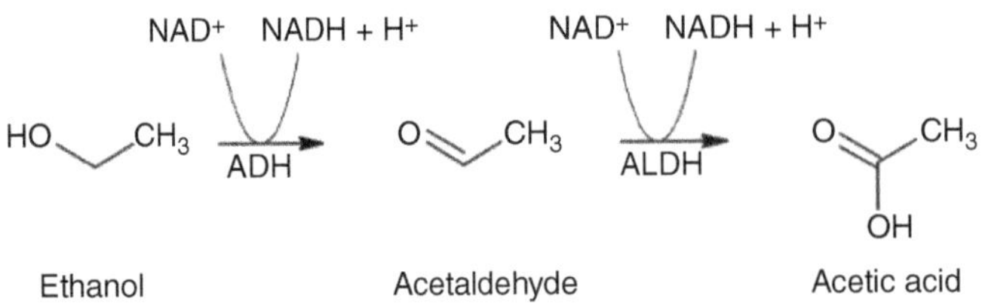

$$NAD^+ \quad NADH + H^+ \qquad NAD^+ \quad NADH + H^+$$

HO—CH₃ $\xrightarrow{ADH}$ O=CH₃ $\xrightarrow{ALDH}$ O=CH₃, OH

Ethanol   Acetaldehyde   Acetic acid

Figure 11.16 Acetaldehyde build-up requires high rates of activity by ADH and slow removal by ALDH.

As mentioned, both ADH and ALDH are found beyond the stomach and liver; in fact, our body produces these enzymes in different parts of the body. ADH is located in places like the stomach lining, pancreas, lungs, and some areas of the brain. ALDH is also spread throughout the body, including the brain, kidneys, and the digestive system, but it's more concentrated in the liver. Both ADH and ALDH come in various versions, called "isoforms." These isoforms are variations of the enzyme that are produced from different copies of the genes that have changed over time. The specific type of isoform of the enzyme present in a person can affect how well the enzyme works. This is because genetic differences can change the enzyme's activity. A common example is with ALDH2, a specific variant of ALDH. Some people have a genetic variant of ALDH2 that makes it less active. This can lead to a buildup of acetaldehyde, a chemical in the body after drinking alcohol (Box 11.3).

In simpler terms, "isoform" refers to different versions of the same protein. These versions are slightly different in their structure due to variations in the genes that produce them. This can lead to differences in how well they perform their functions in the body.

The ADH enzyme has several isoforms, including ADH1, ADH2, ADH3, and ADH4 (Table 11.2). Each of these isoforms works differently, with variations in how effectively and quickly they process alcohol. Scientists measure this effectiveness and speed using two values: K_m and V_{max}. The K_m value indicates an enzyme's affinity, or attraction, to its reactant. In the case of ADH, a lower K_m means the enzyme can work well even at low alcohol levels. ADH1, mainly found in the liver, has the lowest K_m, so it's very good at breaking down alcohol even when there's not much around. This makes ADH1 a key player in initially processing alcohol in the liver. ADH2 and ADH3 need more alcohol to work at their best, with ADH2 located in the liver and stomach and ADH3 spread out in various tissues. ADH4, mostly in the stomach lining, requires the highest alcohol level to function efficiently.

V_{max} represents the maximum rate an enzyme can convert reactants to products. For ADH the product is acetaldehyde, a dangerous and toxic product. ADH1 is the fastest, making it highly efficient. ADH2 and ADH3 work at a moderate pace, while ADH4 is the slowest.

Box 11.5 K_m and V_{max} are used to describe how enzymes function

K_m (Michaelis constant) is the affinity constant of an enzyme for its substrate. This is like the worker's pickiness level. K_m represents the concentration of alcohol at which the worker is half as efficient as it could be. Enzymes with a low K_m are like easygoing workers; they don't need much alcohol to get to work efficiently. On the other hand, enzymes with a high K_m are more selective; they need a higher concentration of alcohol to really get going.

Affinity is essentially how much an enzyme "likes" its substrate, in this case, alcohol. Enzymes with high affinity have a strong attraction to alcohol, even at low concentrations, so they can start working effectively with just a small amount of alcohol present. Enzymes with low affinity need more alcohol to bind to them before they become active. The lower the K_M, the less amount of ethanol needs to be present in the blood before ADH will bind and react. If the ethanol concentration is below the K_m, then very little ethanol will bind and react with the enzyme. Thus, enzymes with a high K_m value will not remove ethanol until the alcohol level is higher than the K_m.

The V_{max} (maximum velocity) is another constant used to measure how an enzyme works. V_{max} is like the worker's maximum speed. It represents the maximum rate at which the worker can process alcohol. Enzymes with a high V_{max} are like the speedy workers; they can process alcohol at a rapid rate once they get going. Enzymes with a low V_{max} are slower; they can only process alcohol at a slower pace, even when there's plenty of it around.

So, when you have different types of workers (i.e. enzymes) with different K_m, affinity, and V_{max} values, they work together as a team to break down alcohol in your body. Some enzymes might have low K_m and high affinity, meaning they can start working efficiently with just a small amount of alcohol, and they can process it quickly once they do. Others might have high K_m and low affinity, so they need more alcohol to become active and work at a slower pace.

This combination of enzymes with varying affinities, K_m, and V_{max} values ensures that your body can efficiently metabolize alcohol, adapting to different levels of intake and processing it at different rates.

Table 11.2 Distribution of ADH and ALDH in the body. When investigating the isoform of ADH and ALDH, it helps to remember that K_m reflects an enzyme's affinity for its substrate, the lower the K_m, the higher the affinity. A low K_m for ADH and ALDH in ethanol metabolism indicates that these enzymes efficiently bind to ethanol and acetaldehyde, respectively. Conversely, V_{max} tells scientists the maximum rate at which ethanol can be broken down. High V_{max} values for ADH and ALDH mean a faster metabolism of ethanol and acetaldehyde, allowing the body to clear alcohol efficiently. This highlights the importance of both K_m and V_{max} in understanding how efficiently the body can metabolize ethanol. A low K_m indicates high enzyme–substrate affinity, ensuring efficient binding and metabolism of ethanol and acetaldehyde. High V_{max} values suggest a rapid rate of enzymatic activity, enabling the body to process alcohol efficiently. Together, these enzymatic characteristics play a significant role in determining an individual's response to alcohol consumption.

Enzyme	Gene	Tissue	K_m (µM)	V_{max} (µmol/min/mg)
ADH1	ADH1B, ADH1C	Liver, Stomach, Pancreas, Lungs, Brain	26–32	56–1800
ADH2	ADH4	Liver, Stomach	73–110	170–660
ADH3	ADH1A, ADH1B, ADH1C	Liver, Lungs, Pancreas, Brain	140–170	12–65
ADH4	ADH1B, ADH1C	Stomach, Small Intestine	320–800	100–200
ALDH1	ALDH1A, ALDH1A2, ALDH1A3	Liver, Brain, GI Tract, Kidneys	50–200	14–91
ALDH2	ALDH2	Liver, Heart, Brain	3–20	1200–6700

The ALDH enzyme takes over the next step of breaking down alcohol via the intermediate, acetaldehyde. ALDH also has different isoforms, mainly ALDH1 and ALDH2 (Table 11.2). ALDH1 is found throughout the body and works best at moderate alcohol levels. ALDH2, primarily in the liver, is the most effective, efficiently handling even low levels of acetaldehyde. Thus ALDH plays an essential role in converting acetaldehyde into acetic acid, a less toxic substance. Each isoform's unique characteristics help our body handle alcohol in various situations, ensuring it's broken down safely and effectively.

Box 11.6 Genetic influences on alcohol metabolism

Alcohol metabolism is a complex process influenced by a variety of factors, including genetic variations in the genes encoding ADH and ALDH. Genetic differences in ADH alleles, alternative forms of the same gene, can significantly impact an individual's response to alcohol consumption.

Fetal alcohol syndrome and ADH genetic variants: One profound consequence of ADH genetic variants occurs during pregnancy. Pregnant women who carry certain ADH alleles may metabolize alcohol differently, leading to higher acetaldehyde levels in their bloodstream. As a result, alcohol exposure during pregnancy can pose a significant risk to the developing fetus. Elevated acetaldehyde levels can adversely affect fetal development, leading to a condition known as fetal alcohol syndrome (FAS). FAS is characterized by a range of physical, behavioral, and cognitive impairments in the child. It highlights the importance of alcohol abstinence during pregnancy and the need for individuals to be aware of their genetic predisposition to alcohol metabolism.

Increased risk to alcohol use disorder and the "Asian Flush" reaction: Genetic variations in ADH alleles can also impact an individual's vulnerability to alcohol use disorder (AUD), a chronic condition characterized by an inability to control alcohol consumption despite negative consequences. Certain ADH alleles, like ADH1B*2 and ADH1C*1, are associated with a reduced ability to break down ethanol into acetaldehyde. Paradoxically, individuals with these alleles may experience a reduced risk of alcohol dependence due to their increased sensitivity to alcohol's effects. On the other hand, individuals with ADH alleles that metabolize alcohol more efficiently may be at a higher risk of developing AUD as they require higher alcohol consumption to achieve the same effects. Furthermore, individuals of Asian descent who carry a specific ALDH2 genetic variant experience an intense facial flushing reaction and other discomforts when drinking alcohol, known as the "Asian Flush." This occurs due to a genetic deficiency in ALDH enzyme activity, which leads to an accumulation of acetaldehyde, causing adverse reactions to alcohol consumption. These genetic insights underscore the importance of personalized approaches to alcohol consumption, taking into account an individual's genetic profile and potential risk factors.

A link to increased alcohol dependence: Certain ADH genetic variants, such as ADH1B*2 and ADH1C*1, result in more efficient alcohol metabolism, leading to a quicker onset of alcohol-induced euphoria. This heightened sensitivity to alcohol's pleasurable effects can drive increased alcohol consumption and contribute to the development of alcohol dependence. Similarly, specific ALDH genetic variants, like ALDH2*2, lead to reduced ALDH enzyme activity, causing an accumulation of acetaldehyde, the toxic metabolite of alcohol. While some individuals may experience aversive reactions to acetaldehyde, others may develop tolerance, leading to increased alcohol intake and an elevated risk of alcohol dependence.

Interplay of genetic and environmental factors: It is essential to emphasize that the development of alcohol dependence is a complex interplay of genetic and environmental factors. While ADH and ALDH genetic variants can contribute to an individual's vulnerability to alcohol dependence, environmental factors such as social influences, stress, and family history also play significant roles. Individuals with a genetic predisposition may be at higher risk of alcohol dependence if exposed to environments that promote heavy or problematic drinking. Identifying individuals with genetic risk factors may help healthcare professionals tailor interventions to address their specific needs. Additionally, increased awareness of genetic influences on alcohol dependence can promote early screening and intervention to prevent the development of problematic drinking behaviors. Education about the potential risks associated with certain genetic profiles may also empower individuals to make informed decisions regarding alcohol consumption and seek support if needed. It is crucial to recognize that alcohol dependence is a multifactorial condition, and genetic factors interact with environmental influences to shape an individual's risk. As such, comprehensive approaches that address both genetic and environmental factors are necessary to effectively combat alcohol dependence and promote healthier drinking behaviors.

11.4.3 Dangerous Metabolism

Alcohol metabolism can be dangerous due to the intermediate product acetaldehyde, which is highly toxic and reactive. Unlike acetate, which is relatively harmless and breaks down into carbon dioxide and water, acetaldehyde can build up in the body, particularly in individuals with certain genetic variations and isoforms of ADH and ALDH enzymes. This buildup can lead to serious health risks.

Acetaldehyde is dangerous because it readily forms stable chemical bonds, known as "adducts," with proteins, DNA, and other cellular components. These adducts can disrupt the normal functioning of cells, leading to widespread damage. For example, when acetaldehyde binds to proteins, it can change their structure and function, impairing vital activities and causing dysfunction. When it binds to DNA, it can cause mutations and chromosomal issues, increasing the risk of cancer. Consequently, acetaldehyde accumulation from alcohol metabolism poses significant health risks, including liver damage, increased cancer risk, and overall cellular toxicity.

The risk of acetaldehyde is particularly concerning when considering its impact on hypoxia and liver damage, as well as its role in the development of fatty liver (steatosis), fibrosis, and cirrhosis. Hypoxia, a condition of reduced oxygen supply to tissues, can be exacerbated by acetaldehyde. It interferes with mitochondria, the cell's energy producers, by disrupting their normal functioning, leading to reduced energy production and oxygen supply. This can cause cellular stress and damage, contributing to liver dysfunction and disease progression.

In fatty liver development, acetaldehyde accumulation promotes fat synthesis and storage within liver cells. It disrupts lipid metabolism, leading to an increased uptake of fatty acids and decreased oxidation. Consequently, fat builds up in the liver, leading to steatosis, which can worsen into more severe liver disorders if alcohol consumption continues.

Acetaldehyde also contributes to fibrosis and cirrhosis, more advanced stages of liver disease. Fibrosis, the scarring of liver tissue due to chronic injury and inflammation, is promoted by acetaldehyde's role in increasing connective tissue proteins like collagen. Over time, this can progress to cirrhosis, a severe condition marked by extensive scarring and loss of liver function, leading to potential liver failure and other life-threatening issues.

Furthermore, alcohol metabolism is associated with an increased risk of certain cancers, such as those of the liver, mouth, throat, larynx, and breast. Acetaldehyde contributes to cancer development by forming DNA adducts that interfere with DNA replication and repair, leading to mutations of DNA causing the cell with the altered DNA to become cancerous. This uncontrolled cell growth, along with acetaldehyde-induced inflammation and oxidative stress, contributes to a cancer-promoting environment. Additionally, alcohol metabolism produces reactive oxygen species, which exacerbate cellular damage and increase cancer risk.

11.4.4 Alcohol and the Brain

The neurobiology of ethanol involves intricate interactions that are not confined to a selected few targets but instead involve a network of interactions that give rise to its diverse behavioral and cognitive effects. To start with, one must understand the basics of neurotransmitters and receptors.

Receptors: In biochemistry, receptors are protein molecules located on the surface of or within cells. They are designed to recognize and bind to specific molecules (like neurotransmitters, hormones, or drugs) in a lock-and-key manner. When a receptor binds to its corresponding molecule, it triggers a specific response inside the cell, which can alter how the cell behaves or communicates with other cells. This response is crucial in transmitting signals across the nervous system and throughout the body.

While the biological impacts of ethanol are complex, our focus is on two types of protein molecules called receptors, specifically GABA(A) and one of the GLU receptors, N-methyl-D-aspartate (NMDA), which are responsible for some of ethanol's effects. These receptors recognize and respond to chemical messengers, including the neurotransmitter molecules affected by ethanol. The signaling of GABA(A) and NMDA receptors in response to low and high levels of ethanol involves a delicate balance between excitatory and inhibitory neurotransmission (see Figure 11.17). This balance is crucial because neurotransmitters, the chemical messengers in the brain, can either activate (excitatory) or calm (inhibitory) the brain's activity. The complex interactions between these receptors and neurotransmitters contribute to the wide range of effects observed with alcohol consumption, from relaxation and euphoria at moderate levels to sedation, cognitive impairment, and potential toxicity at higher levels.

11.4.4.1 Neurotransmitters

Neurotransmitters are chemical messengers that neurons (nerve cells) use to communicate with each other and with other types of cells. They are released from one neuron at a junction called a synapse and travel across a tiny

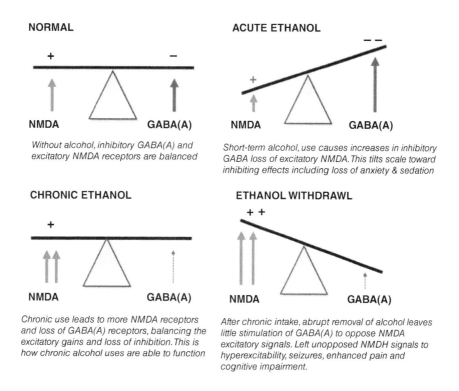

Figure 11.17 Balance of excitatory and inhibitory signaling in AUD.

gap to bind to receptors on the next neuron. This binding can either stimulate (excitatory neurotransmitters) or inhibit (inhibitory neurotransmitters) the receiving neuron, influencing many bodily functions and emotions.

Several key neurotransmitter systems are central to the brain's response to ethanol, including serotonin (5-HT), dopamine (DA), GABA, GLU, acetylcholine (ACH), NMDA receptors, and kainate receptors (Table 11.3).

Serotonin: Ethanol influences the serotonergic system by modulating serotonin release, synthesis, and receptor sensitivity. Chronic alcohol consumption can lead to altered serotonin levels, which may contribute to mood disorders and psychiatric symptoms associated with alcohol use. The serotonergic system plays a crucial role in mood regulation, impulse control, and emotional responses, and its dysregulation can influence an individual's susceptibility to alcohol dependence and other addictive behaviors.

Dopamine: The dopaminergic system is a primary target of ethanol's rewarding effects. Ethanol increases dopamine release in specific brain regions, including the nucleus accumbens, a key component of the brain's reward pathway. This heightened DA signaling is linked to feelings of pleasure and reinforcement, driving individuals to seek and continue alcohol consumption. Activation of the brain's reward system is a critical factor in the development of alcohol dependence and addiction.

Gamma-aminobutyric acid: It is the brain's major inhibitory neurotransmitter, reducing neuronal activity and promoting relaxation. Ethanol enhances the inhibitory effects of GABA by interacting with specific GABA receptors, known as GABA-A receptors. This interaction leads to increased GABAergic signaling, resulting in sedative and anxiolytic (anxiety-reducing) effects. Ethanol's actions on GABA-A receptors contribute to its ability to induce relaxation and reduce anxiety, but they can also lead to impairments in coordination, memory, and cognitive functions.

Table 11.3 Neurotransmitter systems that are central to the brain's response to ethanol.

Neurotransmitter	Role	
Acetylcholine *Choline esterified acetic acid*	Acetylcholine is a very widely distributed excitatory neurotransmitter that triggers voluntary muscle contraction and stimulates the excretion of certain hormones. It is involved in wakefulness, attentiveness, learning, memory, sleep, anger, aggression, sexuality, and thirst.	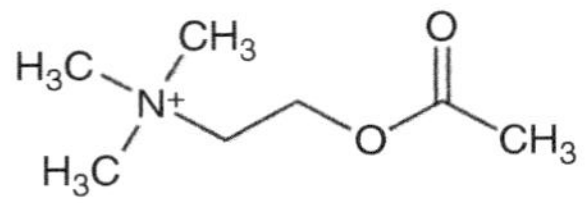
Dopamine *Monoamine amino acid derivative (Tyr)*	Dopamine correlates with movement, attention, and learning. Dopamine is involved in controlling movement and posture. It also modulates mood and plays a central role in positive reinforcement and dependency.	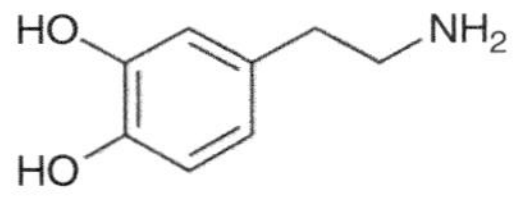
Norepinephrine *Monoamine amino acid derivative (Phe/Tyr)*	Norepinephrine is associated with alertness. neurotransmitter that is important for attentiveness, emotions, sleeping, dreaming, and learning. Norepinephrine is also associated with the "fight or flight" response.	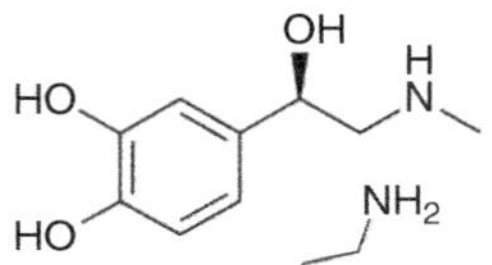
Serotonin *Monoamine amino acid derivative (Trp)*	Serotonin plays a role in mood, sleep, appetite, and impulsive and aggressive behavior.	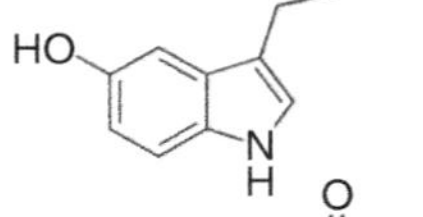
GABA (Gamma- Amino Butyric Acid)	GABA is the major inhibitory neurotransmitter in the CNS, contributing to motor control, anxiety regulation, vision, and many other cortical functions.	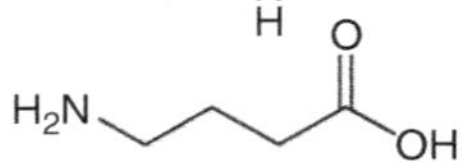
Opioid System - peptides	Opioids effect is to alter other neurotransmitter release. Resulting in pain relief, euphoria and other behaviors including alcohol consumption. This is key to craving alcohol in addiction/withdraw. Met-enkephalin is one example of an opioid	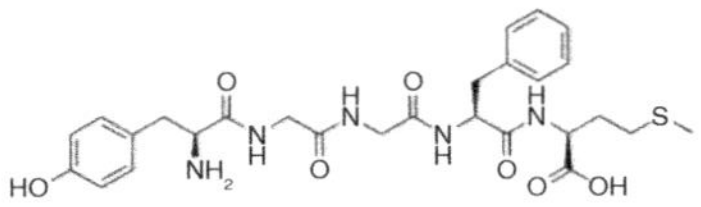

Glutamate: It is the brain's primary excitatory neurotransmitter, promoting neuronal activity. Ethanol exerts complex effects on GLU signaling. It can suppress the release of GLU, leading to its depressant effects on the central nervous system. This inhibition of GLU activity is associated with alcohol-induced cognitive impairments, sedation, and the potential for alcohol-related blackout episodes. Moreover, ethanol also interacts with GLU receptors, including NMDA and kainate receptors, contributing to its broader effects on brain function.

Acetylcholine: Ethanol's actions on ACH are diverse and region-specific within the brain. It can both inhibit and enhance ACH function, depending on the brain region and receptor subtype involved. Ethanol's effects on ACH have implications for cognition, attention, and memory. For example, alcohol's inhibition of ACH activity in certain brain regions may contribute to impaired learning and memory formation, while its enhancement of ACH signaling in other areas may influence attention and arousal.

11.4.4.2   More on Receptors

The NMDA receptor is a type of ionotropic GLU receptor found in the central nervous system (CNS). It's essential for excitatory signals in the brain and is involved in several vital brain functions, including learning, memory, and synaptic plasticity. Synaptic plasticity is the ability of synapses (the connections between neurons) to strengthen or weaken over time, and it's crucial for learning and memory. The NMDA receptor is particularly important for a process called long-term potentiation (LTP), which strengthens the connections between neurons as you learn and remember.

On the other hand, the GABA(A) receptor is also an ionotropic receptor in the CNS, but it works quite differently. It's crucial for inhibitory signals, helping to calm down brain activity and maintain balance. It's named after its primary neurotransmitter, GABA. GABA reduces neuronal activity, making neurons less likely to fire off signals. This calming effect is crucial for preventing excessive brain activity and keeping the brain stable. The GABA(A) receptor is involved in many physiological processes like motor control, anxiety regulation, sleep, and sedation.

When it comes to alcohol (ethanol) exposure, the balance between GABA(A) and NMDA receptors is quite complex. At low alcohol levels, GABA(A) receptors are more sensitive because of a process called allosteric modulation. Ethanol binds to specific sites on the GABA(A) receptor, making it more responsive to GABA. This leads to more chloride ions entering the neurons, hyperpolarizing them and making them less likely to fire. As a result, GABA's calming effects are stronger, leading to relaxation and reduced anxiety—the familiar effects when you first start drinking.

NMDA receptors, however, are inhibited by low levels of ethanol. This inhibition reduces the influx of calcium ions into neurons, decreasing excitatory signals and contributing to the overall calming effect of alcohol.

However, as alcohol levels rise, the increased sensitivity of GABA(A) receptors can lead to too much inhibition. This means even more suppression of brain activity, resulting in sedation, impaired coordination, and cognitive problems. At very high alcohol levels, excessive inhibition can lead to dangerous effects like coma or loss of consciousness. So, the interaction between GABA(A) and NMDA receptors and alcohol is crucial in determining the effects of drinking, from the initial relaxation to the more severe consequences of heavy consumption.

11.4.4.3   Balance of GABA and GLU Signaling in AUD: Involvement of GABA(A) and NMDA Receptors

In a healthy brain, there exists a delicate balance between GABA and GLU signaling (Figure 11.17), two critical neurotransmitters with opposing effects. GABA, mediated by GABA(A) receptors, exerts inhibitory actions, promoting relaxation and reducing anxiety. GLU, acting through NMDA receptors, plays a pivotal role in excitatory transmission, promoting alertness, learning, and memory formation. AUD involves intricate changes in the balance of GABA and GLU signaling, with key involvement of GABA(A) and NMDA receptors. Acute ethanol exposure enhances GABAergic signaling and inhibits NMDA receptor activity, leading to sedation and euphoria. With chronic ethanol exposure, the balance shifts, with decreased GABA(A) receptor sensitivity and increased NMDA receptor activity, resulting in tolerance. Withdrawal from ethanol accentuates these receptor changes, leading to hyperexcitability, withdrawal symptoms, and intense cravings, perpetuating the cycle of AUD.

During acute ethanol exposure, GABA(A) receptors are markedly enhanced by ethanol's allosteric modulation, allowing more chloride ions to enter neurons. This hyperpolarizes the cell membrane, intensifying the inhibitory effects of GABA and leading to the characteristic sedative and anxiolytic properties of alcohol. Additionally, ethanol inhibits NMDA receptors, dampening the excitatory effects of GLU. This dual action results in decreased overall brain excitability, contributing to the initial calming and euphoric effects experienced after alcohol consumption.

With chronic ethanol exposure, the brain adapts to counteract the inhibitory effects of alcohol on GABA(A) receptors. It does so by reducing the number and sensitivity of GABA(A) receptors, resulting in a decreased response to GABA signaling. At the same time, chronic ethanol exposure can upregulate NMDA receptors, increasing their sensitivity to GLU. This shift in the balance between inhibitory and excitatory signaling contributes to the development of alcohol tolerance, where higher amounts of alcohol are required to achieve the same effects.

Withdrawal from ethanol is characterized by a pronounced impact on GABA(A) and NMDA receptors. Upon abrupt cessation of alcohol, the decreased sensitivity of GABA(A) receptors and increased NMDA receptor activity lead to a state of hyperexcitability in the brain. This hyperexcitability contributes to the onset of withdrawal symptoms, such as anxiety, irritability, tremors, and seizures. The heightened activity of GLU through NMDA receptors during withdrawal can be neurotoxic and may play a role in the development of withdrawal-induced seizures and excitotoxic damage in the brain.

The changes in GABA(A) and NMDA receptors during withdrawal also contribute to the intense cravings for alcohol, as the brain seeks to restore the disrupted balance and alleviate the hyperexcitability. Seeking alcohol becomes a coping mechanism to mitigate withdrawal symptoms and reinforces the addictive cycle of AUD.

Box 11.7 AUD and the importance of language

AUD is a medical condition characterized by problematic patterns of alcohol consumption, leading to significant impairment or distress. It involves a complex interplay within the brain's reward pathway, which comprises four stages:

Try (initiation): Some individuals may initially try alcohol experimentally, but it is essential to remember that occasional or social drinking does not equate to AUD. Describing this stage as "alcohol experimentation" or "occasional use" avoids stigmatizing language and acknowledges that many people try alcohol without developing a disorder.

Like (risky use): As alcohol use progresses, some individuals may engage in risky patterns of consumption. Describing this stage as "risky alcohol use" focuses on behavior rather than labeling individuals, fostering empathy and understanding.

Crave (disorder): When alcohol use becomes compulsive, and individuals experience difficulty controlling their consumption, they may enter the stage of AUD characterized by cravings and loss of control. Referring to this stage as "Alcohol Use Disorder" or "Problematic Alcohol Use" emphasizes the medical nature of the condition, promoting a compassionate and supportive approach to treatment.

Need (dependence): At this stage, individuals may develop physical and psychological dependence on alcohol. They may experience withdrawal symptoms and an overpowering urge to continue using alcohol to avoid discomfort. Using phrases such as "alcohol dependence" or "alcohol addiction" recognizes the severity of the condition while avoiding stigmatizing labels.

Choosing non-stigmatizing language is crucial in promoting empathy and understanding towards individuals with AUD. It emphasizes that AUD is a health condition that involves changes in brain function within the reward pathway. By using appropriate language, we can help reduce the shame and self-blame that individuals with AUD may experience and encourage them to seek the support and treatment they need. Person-first language is especially relevant when discussing AUD within the context of the reward pathway. Emphasizing that individuals are not defined by their disorder helps humanize their experience. For example, using phrases like "a person with alcohol use disorder" rather than "an alcoholic" ensures that the focus remains on the individual's identity beyond the condition. Using non-stigmatizing language and highlighting the availability of evidence-based treatment and support services can encourage individuals with AUD to seek help. By conveying that AUD is a treatable condition, we create an environment where individuals feel empowered to take steps toward recovery.

11.5 Beer

11.5.1 The Art and Science of Beer

The process of making beer is fairly simple. Start with a source of complex carbohydrate (e.g. starch) from a seed. Convert the starch to simple sugars using enzymes and then add yeast to ferment the sugar to alcohol. Adding spice, fruits, and hops can create flavored beer. Ancient civilizations made beer in an amazing variety of ways. The Babylonian Code of Hammurabi proscribed a "fair" price for beer and outlined the punishment for overcharging (drowning the innkeeper) [1]. An ancient recipe found in a Sumerian poem to the goddess of beer described using barley seed to make bread, which was then soaked in water where wild, environmental yeast fermented to produce beer. Incans chewed corn, where digestive enzymes broke down the starches for yeast, producing a beer that is still available in Peru called chicha beer. Egyptians and Mesopotamians began a technique still used to use grains as the source of sugars. The process stimulates the seeds to germinate (grow) to produce their own starch-digesting enzymes. Modern crafters and commercial brewers use a mix of art and science to create an array of types and flavors of beer.

The entire process used by modern brewers is shown in Figures 11.18 and 11.19 and Table 11.4. The starting ingredient for beer is the complex starches found in grain seeds (Figure 11.20). The grains of rice, corn, or other crops have been used, but barley seed corn is the chief source of sugar for beer production. Barley is a seed grain from the grass family whose seed is arranged in two or six rows at the end of each stem. Like wheat, each seed is covered in a woody husk, lined with living cells (aleurone layer cells), filled with starch-containing endosperm and the plant embryo.

The first challenge is the conversion of the complex starches located in barley endosperm into smaller mono- and disaccharides. Amylose is an unbranched glucose polymer made of thousands of glucose monomers and comprises 20–40% of total barley starch. Amylopectin is a branched polymer of up to 250,000 glucose units and makes up 60–80% of total barley starch. Yeast enzymes digest neither glucose polymer. Instead, these complex starches must be separated from the grain, hydrated, and digested into the smaller mono- and disaccharides, glucose, and maltose. The entire process is completed as part of the malting step.

11.5.2 Malting

The goal of the malting step is to produce enzymes from the grain to digest starch and proteins into smaller components. Malting consists of three steps: steeping, germination, and kilning. Cracked and milled barley seed is soaked (steeped) in enough water to wake up the cells in the aleurone and embryo. The process takes up to 40 hours when the embryo cells begin to produce plant hormones and will start to sprout (germinate). At this point, water is removed, and the seeds are thinly spread to absorb moist air and incubated at 15.5°C/60°F for several days to allow the embryo to begin to develop. During this germinating phase of malting, the embryo will produce hormones such as gibberellin to stimulate aleurone layer cells to produce many different enzymes used to metabolize and digest starch and proteins in the endosperm and the rest of the barley grain. Timing is critical

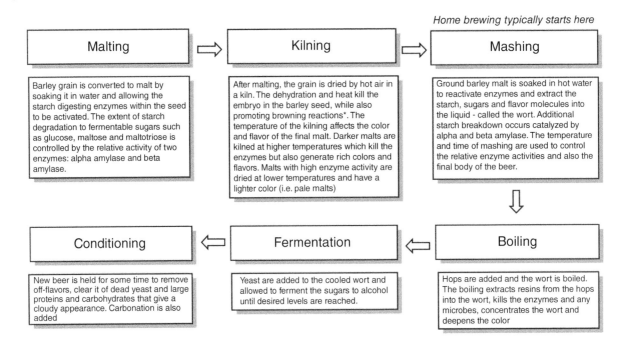

Malting

Barley grain is converted to malt by soaking it in water and allowing the starch digesting enzymes within the seed to be activated. The extent of starch degradation to fermentable sugars such as glucose, maltose and maltotriose is controlled by the relative activity of two enzymes: alpha amylase and beta amylase.

Kilning

After malting, the grain is dried by hot air in a kiln. The dehydration and heat kill the embryo in the barley seed, while also promoting browning reactions*. The temperature of the kilning affects the color and flavor of the final malt. Darker malts are kilned at higher temperatures which kill the enzymes but also generate rich colors and flavors. Malts with high enzyme activity are dried at lower temperatures and have a lighter color (i.e. pale malts)

Mashing

Ground barley malt is soaked in hot water to reactivate enzymes and extract the starch, sugars and flavor molecules into the liquid - called the wort. Additional starch breakdown occurs catalyzed by alpha and beta amylase. The temperature and time of mashing are used to control the relative enzyme activities and also the final body of the beer.

Conditioning

New beer is held for some time to remove off-flavors, clear it of dead yeast and large proteins and carbohydrates that give a cloudy appearance. Carbonation is also added

Fermentation

Yeast are added to the cooled wort and allowed to ferment the sugars to alcohol until desired levels are reached.

Boiling

Hops are added and the wort is boiled. The boiling extracts resins from the hops into the wort, kills the enzymes and any microbes, concentrates the wort and deepens the color

Figure 11.18 The process of brewing.

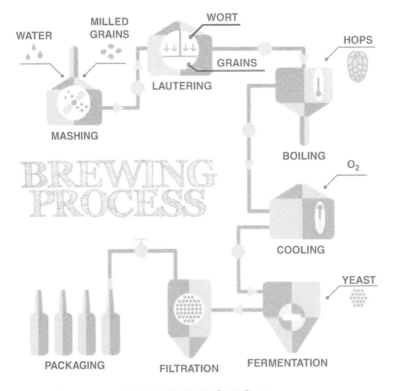

Figure 11.19 A simple view of the brewery process. fosin/Adobe Stock Photos

Table 11.4 Overview of beer brewing.

Milling	Cracking and grinding grains for dissolving gels in water
Malting	Conversion of grain starch into usable form for yeast metabolism, some protein hydrolysis
Mashing	Converting water-soaked malt into sweet liquid wort
Fermentation	Metabolism of carbohydrates to ethanol
Maturation	Final production of flavor compounds and precipitation of proteins and cell debris
Finishing	Filtering, carbonation, and storage

Figure 11.20 Anatomy of a wheat kernel. The sugar for beer comes from the endosperm of a wheat kernel. Other cells including the aleurone produced enzymes and hormones to break down the complex carbohydrate found in the endosperm.

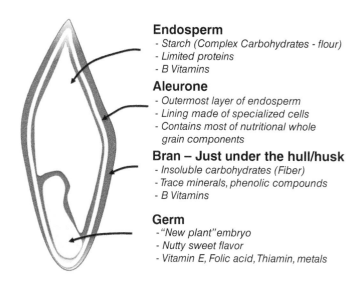

Endosperm
- *Starch (Complex Carbohydrates - flour)*
- *Limited proteins*
- *B Vitamins*

Aleurone
- *Outermost layer of endosperm*
- *Lining made of specialized cells*
- *Contains most of nutritional whole grain components*

Bran – Just under the hull/husk
- *Insoluble carbohydrates (Fiber)*
- *Trace minerals, phenolic compounds*
- *B Vitamins*

Germ
- *"New plant" embryo*
- *Nutty sweet flavor*
- *Vitamin E, Folic acid, Thiamin, metals*

for this stage. Some enzymes are lost late in germination, and the balance of total protein production versus the types and amounts of each enzyme can change from batch to batch of malt.

Converting the milled grain to a "malt" results in the production of enzymes that break down starch and protein along with a host of other compounds. During this process, the sprouting cells use some of the starch and other nutrients of the grain. This process typically takes 4–7 days. Shoots and the beginning of roots are produced while the enzymes begin to digest endosperm components, and the grain itself begins to swell with water content and softens. The process must be stopped at a point where enough enzymes are made to digest the remaining starches and proteins but not left too long, where the growing embryo digests too much of the starch.

Once the signs of a growing shoot and root tip from the embryo are observed (commercial malting professionals will also measure the enzyme production in the grain), the process is stopped by heating and air-drying in a kiln. At this point, the grains will be dried to store the barley malt for the next step. Drying lasts for 30 or more hours at 80–100°C. Darker beers are heated longer and at higher temperatures to allow Maillard and caramelization reactions to occur. Heating and drying is a delicate step that requires careful testing by the malter. The darker the malt, the more digestive enzymes the heat destroys. If too much moisture remains in the malt, germination will continue, and mold may contaminate the malt during storage. The last step is to remove the shoot tips and roots. At this point, the malt will appear like a swollen barley seed with a lighter color.

11.5.3 Mashing

Mixing ground malt with hot water will begin to reactivate the enzymes produced during malting and dissolve most of the starches into solution. Starch processing during mashing can be divided into three phases: gelatinization, liquefaction, and saccharification.

The process of dissolving the long tangles of starches into water is complicated. Gelatinization is the process whereby water begins to dissolve starch by interacting via hydrogen bonds with the —OH groups of the starch molecules. Water forms organized shells around the starch, causing the starch to swell in size. At some point, amylose leaks out of the granules of dried starch and bursts the granules. Barley is an excellent choice for beer as its starches absorb water at a lower temperature than corn or rice starches; these grains must be boiled for gelatinization. Thus, it takes less heat to dissolve the barley starch, avoiding further denaturation of the digestive enzymes produced during malting. The result is a gel-like solution thick with swelled starches. This step is critical for starch digestion to smaller, simpler sugars. Enzymes need starch to be soluble for proper digestion.

Once the large starch chains are suspended in the water (they have gelled), liquefaction can occur. During liquefaction, enzymes break the large chains of starch into much smaller, simpler sugars. Aleurone cells produce two major classes of digestive proteins: starch-digesting and protein-digesting enzymes. The enzymes that degrade starch are similar to those found in saliva and the digestive system. There are three types of starch-degrading enzymes: alpha amylase, beta amylase, and limit dextrinase. Each enzyme uses water to cleave (hydrolyze) starch into smaller components (Figure 11.21). Alpha amylase will cut the 1,4 glycosidic bond between glucose units on amylose and amylopectin. Alpha amylase randomly binds and cleaves starches producing much shorter chunks of starch. However, these shortened starches are still too large to be used as food for yeast. Beta amylase also cleaves 1,4 glycosidic bonds, but unlike alpha amylase, beta amylase binds at the nonreducing end of the polymer, cutting off two glucose units at a time. Beta amylase thus produces the disaccharide maltose. Partially digested starches include short runs of glucose collectively called dextrin and the trisaccharide maltotriose. Amylopectin contains 1–6 glycosidic branch points not digested by either amylase. Another enzyme called limit dextrinase (sometimes called "debranching enzyme") is needed to cleave the 1,6 branches allowing the amylases to continue to digest the starch into maltose and glucose. The result of this complex digestion is a mixed population of glucose, maltose, maltotriose, and shorter chains of starch called oligosaccharide dextrins.

Alpha amylase and beta amylase are key enzymes in beer brewing, breaking down starches from grains into fermentable sugars at different temperature ranges and with distinct results. Alpha amylase operates optimally at a higher temperature range, typically between 154°F and 162°F (68–72°C), and functions by randomly cutting the long starch chains into smaller sections, targeting the internal α-1,4 glycosidic bonds in the starch molecules. This action produces larger, more diverse sugars such as maltose, maltotriose, and limit dextrins. These larger molecules are less fermentable by yeast, which contributes to the beer's fullness and body, leaving a residual sweetness and complexity in the final product, making it desirable for styles where a fuller body is appreciated.

Conversely, beta amylase works best at lower temperatures, around 131–150°F (55–65°C). It methodically cleaves off units of maltose from the nonreducing end of the starch molecule. Maltose, a disaccharide consisting of two glucose molecules, is the primary sugar produced by beta amylase. It is smaller and more fermentable than the sugars produced by alpha amylase, leading to a lighter beer with potentially higher alcohol content. This results in a drier, crisper finish, often desired in lighter beers where a clean, refreshing taste is the goal.

The molecular nature of the sugars produced during brewing is crucial. Starches are long chains of glucose molecules, and the manner in which these chains are broken down determines the type of sugar produced and, subsequently, the characteristics of the beer. Larger sugars from alpha amylase activity are not fully fermentable by typical brewing yeast, meaning they remain in the beer, adding sweetness and body. In contrast, the smaller, more fermentable sugars from beta amylase lead to higher alcohol content and a drier finish. By controlling the

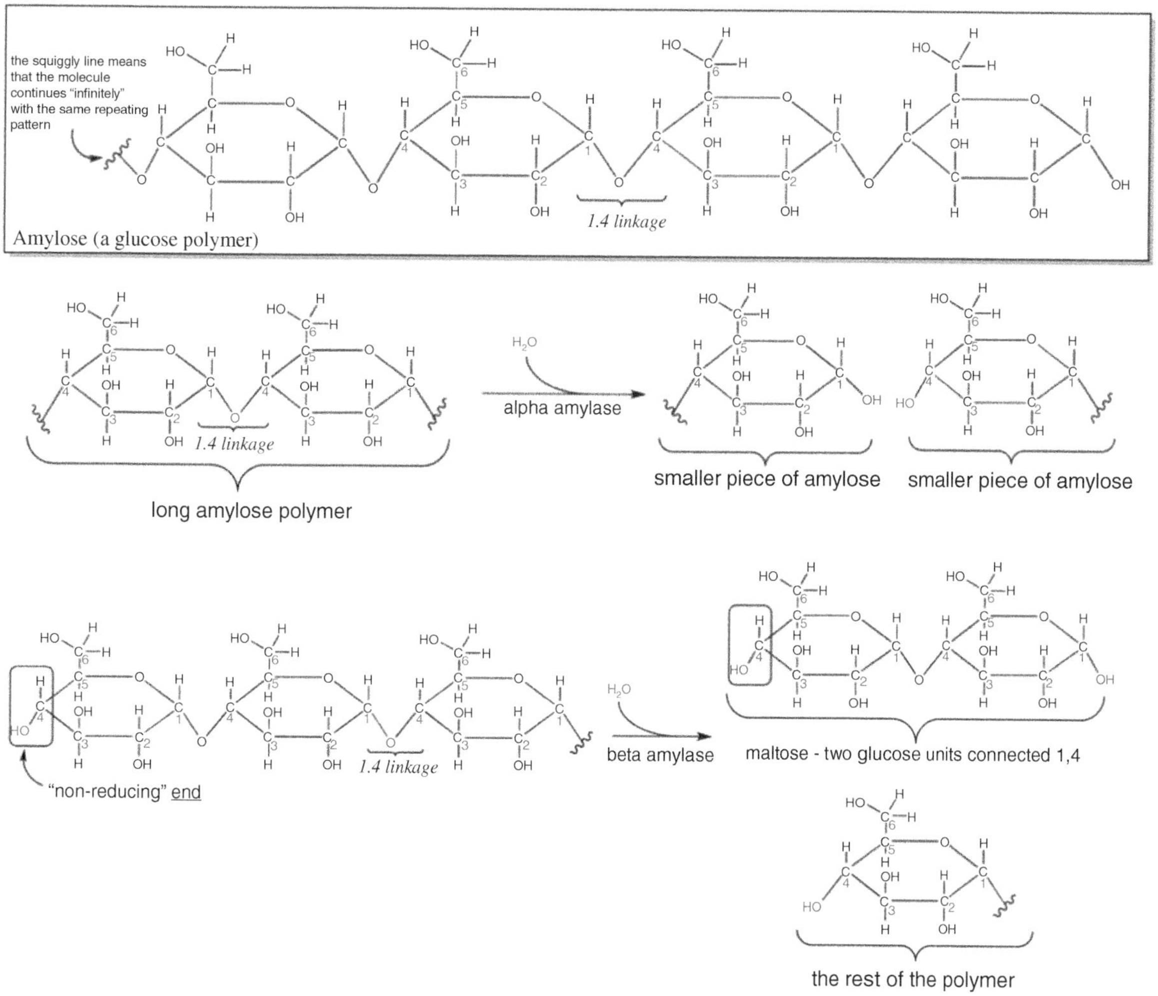

Figure 11.21   Alpha and beta amylase reactions.

mash temperature, brewers can influence which enzyme is more active and thus the sugar profile of the wort. This control allows for fine-tuning of the beer's flavor, alcohol content, and body to achieve the desired characteristics of the final product. Understanding and manipulating the actions of alpha and beta amylase is a vital part of the brewing process, enabling brewers to craft a wide range of beers, each with its unique balance and taste.

Protein-digesting enzymes called proteases are also in malt and activated during mashing. These enzymes use water to break the peptide bond of proteins. These proteases, called peptidases, provide smaller protein fragments called peptides and amino acids from protein degradation. These amino acids provide the nitrogen needed for robust and healthy yeast growth during fermentation. Dextrins, amino acids, and peptides contribute to the mouthfeel and taste of beer and support a stable foam during drinking. Large, undigested proteins will aggregate and cause a cloudy look or haze to the finished product.

When the large chains of starch have been broken down into tiny fragments (i.e. glucose, maltose, and malto-triose), saccharification occurs (Figure 11.22). The level of starch digestion into the smaller carbohydrates is the fermentability of the mash. The more glucose and maltose produced, the more food there will be for yeast to produce ethanol. A more robust beer will have more body provided by fewer shorter and more of the longer starch

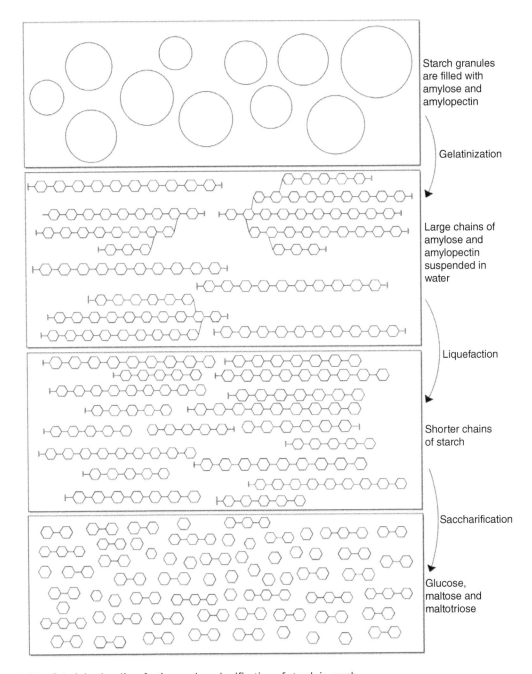

Figure 11.22 Gelatinization, liquefaction, and saccharification of starch in mash.

intermediate compounds. Light beer will have a high-fermentability mash where nearly all the starch has been digested into glucose and maltose.

During the mashing process, a mixture of malt and water is placed in a metal container called a mash tun. The tun is heated and held at specific temperatures to allow various classes of enzymes to function. Each temperature stop is called a rest. The mash rest is first held at 40–45°C/104–113°F to allow cell walls to be digested by beta-glucanase enzymes. These enzymes break down the cell wall components breaking the cells containing the starches and reducing the insoluble cell wall material that can impart poor flavor and haze if left intact. The protease rest allows for the digestion of proteins into peptides and amino acids. Increasing the heat to a final amylase rest can be done at two different temperatures or one for both alpha and beta amylases. A final mashout step will raise the temperature of the mash to slightly above 172°F/78°C. This will denature and inactivate most of the enzymes from the malt and most microbial contaminants remaining in the mash.

Darker beers may include removing a fraction of the mash and boiling it to caramelize and add flavor to the mash. Adding a fraction of boiled mash back to the tun is called decoction. Another form of decoction is to boil unmalted adjunct cereals (rice, oats, wheat, sorghum, or corn) to gelatinize these starches. The boiled adjunct starch is then added to the barley mash to allow the enzymes to saccharify the adjunct starch. The result is a low-protein, highly fermentable mash that yields a different taste and higher alcohol content in the finished product. Other adjuncts can also be added at this time to produce different flavors (spice, chocolate, etc.) and increase sugar content (kettle sugars including sucrose or brown sugar).

11.5.4 Wort Processing

Separation of mash solids (grist) from the liquid (wort) happens by filtering through a lauter tun, creating a brown, sweet liquid called the wort. The wort is mixed with hops and often boiled in a copper kettle for an hour. Boiling kills any residual yeast, bacteria, and enzymes not already lost in the mashing process. Boiling the wort in this fashion will also help remove volatile compounds such as dimethyl sulfide (made from amino acid breakdown during mash heating) that may impart bad flavor to the final product. Wort boiling will also help to convert some of the compounds from hops, making them more soluble. Denatured protein will aggregate and mix with some of the cell wall compounds, forming a scum called a hot break. The wort is quickly cooled to limit further sulfide products and helps to further denature and aggregate proteins, preventing haze.

Heating wort also concentrates the sugars and increases the overall density. The remaining compounds from the hops add a required bitterness, after which the wort is called a "hopped" wort. A typical mixture of sugars in wort is as follows: maltose (~50%) > dextrins (25–30%) > maltotriose and glucose (10–20%) > sucrose or fructose (1–2%). Maltose and maltotriose are taken into yeast cells using ATP to transport the disaccharides across the membrane. In contrast, glucose and fructose are passively transported into the yeast cells without expending energy, which impacts the potential ethanol output of the yeast during fermentation. Yeast will grow more slowly due to the increased transport needs of maltose. Once inside the cell, the oligosaccharides will be converted to glucose and used to produce ethanol. Overboiling wort has an additional impact on the final quality of beer; concentrated, highly boiled wort has higher quantities of ethyl acetate and diacetyl. Both compounds produce off-flavors in beer and are considered highly undesirable.

11.5.5 Hops

Resembling a pine cone, hops are the female flowers of the hop vine (*Humulus lupulus*). Hops are relative newcomers to beer brewing. For the first few thousand years, beer makers used herbs or spices to provide flavor and what turned out to be a preservative effect for the fermented beverage. Around 820 AD monks in Eastern Europe began to add the oil-containing cone into their beer. Unspiced or unhopped beer quickly spoils with bacterial or wild yeast growth. Some of the oils from hops inhibit microbial growth and provide a unique flavor to the drink. Original ale was unhopped while beer was the term for hopped brews. By the 1600s, nearly all beer was "hopped." Modern commercial production of hops takes place in Germany and in Washington State in the United States (Figure 11.23).

Like wine grapes, there are a dizzying number of strains of hops whose content is influenced by the environment and earth in which the vine is grown. Hops provide much of the aroma and most of the desired bitter flavor of beer. Located in the lupulin glands of the hop is a yellow powder containing resins. Hops are a complex mixture

Figure 11.23   Cones of hops on a vine. toomler/Adobe Stock Photos

of compounds, including acidic resin oils, tannins, pectins, proteins, waxes, and carbohydrates. The sticky resin of the hop includes α-acids and β-acids. Unaltered versions of both acids dissolve poorly in water. In fact, less than one part per million (PPM) of either acid is usually detected in most beers. Depending on the hop variety, there are low levels of either hop (Noble hops, such as the German Hallertau Hersbrucker variety, ~3–4% for each acid) or a 3.5 ratio of α-acids to β-acids (14–16 to 4% α-acids for the North American Nugget and Zeus hop varieties). The α-acids are slightly less soluble in water than the β-acids leaving the β-acids to have a larger impact on the flavor and taste of beer. Boiling hop acids catalyze the isomerization of α-acids to isoacids (Figure 11.24). Iso-α-acids are much more soluble in beer than their non-converted α-acid forms. Slightly less than half of the α-acids are converted to the iso-α-acid form and remain in beer after cooling.

There are three main α-acid forms: humulone, cohumulone, and adhumulone. Humulone is the predominant type (48–75%) of α-acid, and its isomerized form, isohumulone, contributes to the desirable soft bitterness of the beer. The vitamin riboflavin will absorb light in the green wavelength, which then transfers the captured energy via a reactive oxygen species to isohumulone. The reaction continues and further alters the iso-α-acid. The reaction of light with isohumulone creates a "skunky" or "light-struck" foul odor to the beer. Breaking the bond by light creates an intermediate compound with a single electron. Such single, unshared electron-containing compounds are called free radicals and are very reactive. The iso-α-acid free radical will react with the amino acid

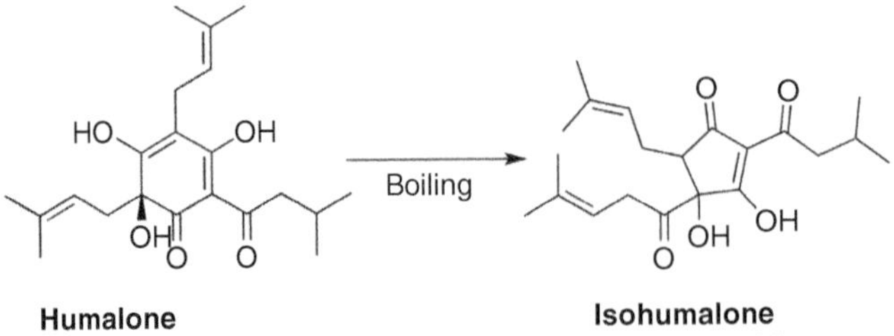

Humalone
(alpha acid)

Boiling

Isohumalone
(Iso-alpha acid)

Figure 11.24   The conversion of alpha acids by heat. Boiling hops converts the alpha acid humulone to the less bitter isohumulone.

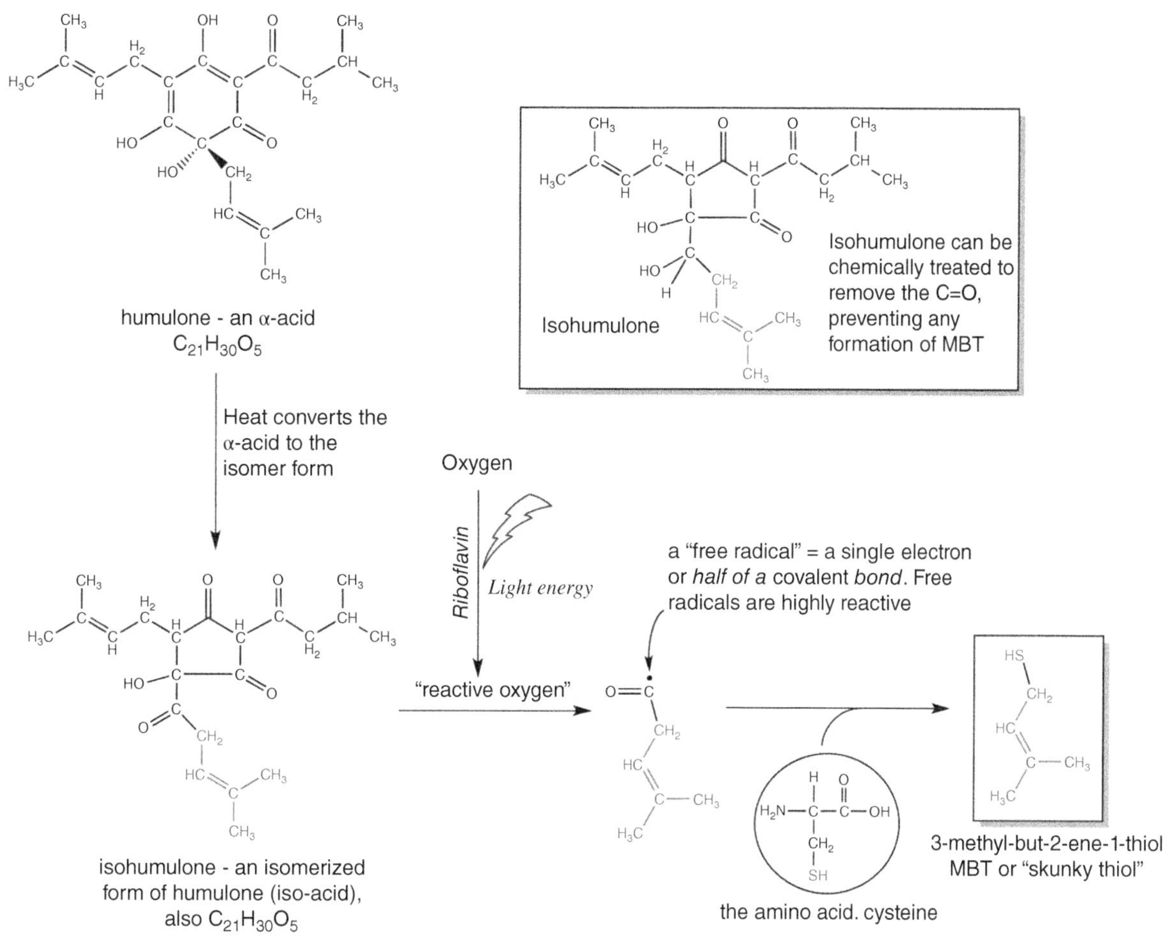

Figure 11.25 Skunk-proofing beer. Conversion of isohumulone followed by light-induced oxidation (free radicals) results in foul-smelling sulfur compounds.

cysteine found in the beer to create 3-methylbut-2-ene-1-thiol (MBT), which smells like (but is not) skunk spray. Brown or green beer bottles absorb green light inhibiting the reaction while on the shelf. Removing one of the iso-acid's carbon–oxygen double bonds by adding hydrogen across the bond is called skunk-proofing (Figure 11.25).

While boiling converts some of the hop acids to a bitter, soluble isoacid, much of the aroma of the volatile hop compounds is lost during boiling. Brewers will often place hops, pelleted hops, or extract of hop cone back into the wort for the last few minutes of boiling or to the finished, cooled wort. This second hopped or "dry hopped" wort will have a much stronger aromatic component from the hop resins and oils that don't evaporate. Oils include some of the same aroma and flavor compounds found in spices, including the terpenes linalool, geraniol, pinene, limonene, and citral, providing fruit, citrus, and pine aromas to beer.

11.5.6 Fermentation of Beer

The finished wort is a rich, nutrient-dense solution ready to turn into beer. Two different types of yeast strains perform the fermentation of wort into beer. There are top fermenters and bottom fermenters. This explains how the two basic categories of beers are organized. Top-fermenting yeasts are used to make ale, and bottom-fermenting yeasts produce a lager beer style. Top-fermenting yeasts prefer warmer temperatures (12°C/55°F) and culture faster, often finishing the job in under 8 days. Bottom-fermenting yeasts are much slower fermenters and are best cultured at 4°C/40°F. Of course, the type of starting materials and how the grains are processed, how the wort is produced, and the infusion of adjuncts and hops, together with various strains of each type of yeast, create a rich and diverse beer spectrum for both ales and lagers.

There are hundreds of ale-producing yeasts (top fermenting), most of which originate from the parent yeast strain *Saccharomyces cerevisiae* but may also share genetic identity with other yeasts, including strains used for winemaking. Most top-fermenting strains will clump (flocculate) and rise to the surface as they trap CO_2 gas within the flocculated cells at the end of the culture/fermentation. These strains will also produce a stronger set of flavor molecules, including the esters (Figure 11.26): isoamyl acetate (banana aroma and flavors), ethyl hexanoate (red apple), ethyl acetate (a flowery aroma), and ethyl caproate (a fruity, wine-like aroma). The level of these and other flavor-producing molecules depends on the fermentation strain and finely tuned conditions. Ester production takes place inside the yeast. The primary reaction is a condensation reaction between a carboxylic acid and an alcohol. Acetic acid and acetyl-CoA both provide the carboxylic acid for the reaction, and ethanol or other longer carbon chain alcohols (butanol, propanol, etc.) react to form many final floral and fruity-smelling ester products.

Careful brewers will select strains that produce enzymes involved in making acetyl-CoA. To get a larger bouquet of esters, higher yeast growth rates will increase the longer (often called fusel) chain alcohols. Increasing the temperature to encourage faster yeast growth can help this along. After yeast begins to grow at a sufficient rate, lower oxygen levels induce ethanol and fusel alcohol production. This shifts the metabolism from fatty acid production to alcohol leaving some of the acetyl-CoA for the ester reactions. Commercial brewers do this in a number of ways, including changing the pressure and temperature of the fermentation, reducing the aeration of wort or extended boiling of wort for a higher-density starting solution. Home brewers can achieve similar effects by ensuring the starter yeast culture will produce the necessary components.

Lagers are often described as clear and crisp, while ales are fruity and complex. The key difference, after the production of wort, is in the species of yeast used for fermentation. Lagers are produced by bottom-fermenting yeast that grow slower and at much lower temperatures. Like the top fermenters, bottom fermenters aggregate or flocculate but do not trap CO_2 and instead sink to the bottom of the fermenter near the end of culturing. The big difference is the rate and metabolism of the two types of yeasts. Bottom-fermenting yeast is more of a misnomer as much of the ethanol production is taking place throughout the vessel. It is not until the yeast population is high and flocculation occurs that the organism settles to the bottom of the fermenter.

As with ale yeast, there are hundreds of variants of bottom-fermenting yeast providing a diverse set of flavors and aromas for the beer. Lager yeasts will produce pilsners, bocks, and many of the more mild beers produced in the United States. Historically, these bottom-fermenting yeast strains were thought to be *Saccharomyces uvarum*

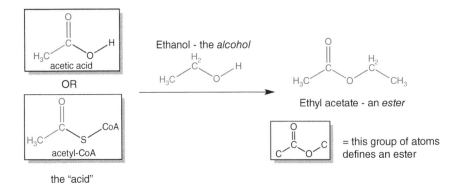

Figure 11.26 Formation of esters from acids and alcohol.

or Saccharomyces pastorianus/Saccharomyces carlsbergensis. The latter two are the same strain and were isolated by the Carlsberg brewery research group, the very same institution that defined the pH scale to describe acidity for fermentation. However, genetic analysis has identified that many of the genes in what was thought to be a different strain of bottom-fermenting yeast are identical to genes found in the *S. cerevisiae* top fermenters. In fact, "lager yeasts" are limited to breweries; they are not found in the environment as are other fermenting yeasts. What likely happened is *S. cerevisiae* was crossed and genetically fused with a cold-tolerant yeast sometime before lager beers were brewed in the fifteenth century. A close investigation into the genetic sequence of three unique bacterial strains, *S. uvarum, S. eubayanus*, and *S. cerevisiae*, found that the genes of all three were fused (most of the genes coming from the parent strain, *S. cerevisiae*) to form what was thought to be a novel species, *S. carlsbergensis.* One of the interesting changes was that specific genes involved in sugar metabolism and how the yeast cell used sulfite compounds were transferred from *S. eubayanus* to *S. cerevisiae.* These and other traits from *S. eubayanus* and *S. uvarum* allow the once top-fermenting *S. cerevisiae* to grow at colder temperatures and change the flocculation rate and abilities of the cells [6].

Most likely the strains were cultured together allowing fermentation to take place in the colder, dark caves of Bavaria, the home of lager brewing. The German use of "lager" means "to store" and reflects the longer fermentation time of these beers compared with the short time needed for ale fermentation. The result of "bottom fermenting" is a mild beer without the esters of the ale beers. The mild sulfur odor is due to the sulfite metabolism produced by lager yeast and is appropriate for lager beers.

11.5.7 Conditioning

After the fermentation is over, the beer is considered "green," not carbonated, and unfinished. Conditioning beer requires transferring the young beer from the dead yeast, wort, and hop debris. Yeast is cleared from top-fermenting beer and transferred to a new container. Lagers are traditionally kept at very cool temperatures to stop further yeast fermentation and encourage flocculation and precipitation of fermenting debris. Some add stabilizers or powdered crustacean exoskeleton (chitin) or polyvinylpolypyrrolidone (PVPP) to keep proteins and polyphenols in solution during chilling. Carbonation can be done by directly adding liquefied carbon dioxide, pressurizing a keg with carbon dioxide gas, where the gas will eventually dissolve into the liquid, or by inducing a second fermentation. Here, a small amount of yeast is inoculated to clarify green beer with sugar or reserved wort. The mixture is promptly bottled, and the continued metabolism of the remaining sugar will create the dissolved carbon dioxide gas.

11.6 Oenology: The Science of Wine and Winemaking

11.6.1 Grapes and Juice

Like beer, the process of making wine requires a source of sugar (grapes or other fruit), water, yeast, time for fermentation to take place, and some post-fermentation handling. Unlike beer, the carbohydrate needed for fermentation does not need enzymes or other biological processes to convert complex sugar polymers into glucose and other mono- and disaccharides. Well-grown, mature wine grapes will have plenty of glucose and fructose. However, the handling of wine before, during, and after fermentation is quite a bit different and can be more complex than maturation of most beers (Figure 11.27).

Figure 11.27 Oenology, the science of wine. indigolotos/ Adobe Stock Photos

A simple description of making wine starts with grapes harvested when the sugar content is highest. The grapes are then crushed with the stems and seeds and mixed into a must. Depending on the type of wine, the skins will remain in contact with the juice, and yeast is added to begin fermentation. After most of the sugar has been metabolized to ethanol and other products, including CO_2, solids are precipitated, and the clarified wine is aged until consumption.

11.6.2 From Grapes to Must, Preparation of Sugars for Metabolism

Grapes are the most common but not only source of sugar for winemaking. Many fruits, including apples, pears, and even non-fruit plants, such as dandelions, can be used to make wine. Using fruit requires additional resources, including enzymes to break down the pectin from cell walls. For this chapter, we will focus on grapes.

Like beer, winemaking is an ancient discovery. There is abundant evidence for wine production throughout the ancient world, 3000–4100 BCE in China and the Middle East areas. The simplest and most likely origins of grape wine were from environmental yeast landing and metabolizing the sugars from the fruit. Wine as a preserved source of social lubrication has been in favor since.

The formal term for crushing the grapes, stems, and seeds is *maceration*. At their peak, grapes will consist of 20–30% sugar. Most of the sugar comprises the monosaccharide glucose with a significant portion of the sugar being fructose. Glucose is transported into the yeast and metabolized at a higher rate than fructose, but both provide the metabolic carbon needed for fermentation. Grapes also provide micro- and macronutrients essential for yeast growth. Macronutrients such as lipids, proteins, and complex carbohydrates are needed to build new yeast cell material and micronutrients such as vitamins and minerals, which are crucial for enzyme activity inside the yeast.

Wines are often termed "varietal," which means that a particular wine is made from one type of grape variety rather than a mixture of grapes. The genetic makeup of the grape heavily influences the nature of the finished wine. The grape can be simply divided into the exocarp (skin) and mesocarp (flesh) (Figure 11.28). The berry, its skin, seeds, and stems are all crushed together. The initial crushing of the grapes produces water, sugar, and organic acids in a clear juice called the *free run*. The pulp is what remains after the free run is removed by crushing. The juice remaining in the pulp is found in the skin, seed, and stems but lacks many of the flavor compounds. Further crushing or "pressing" of the pulp results in greater extraction of terpenes, tannins, thiols, and other more bitter-tasting components. The combination of the free run and any other liquid pressings is the "must," which is Latin for young wine. The crushed solid portion is called the pomace. The longer the pomace remains mixed with the must, the more of the color and other compounds like tannins that will be extracted into the liquid must.

The grape varietal, location and environment of the vine, condition of the fruit, and maturation of the grape berry are all key to making a good wine. Prior to maturation, there is less glucose and fructose; instead the fruit produces organic acids, tartrate, and malate. These two bitter-tasting acids are produced as a defense mechanism against foraging animals and birds until the seed develops. As the fruit reaches maturity and the seed is ready, plant hormones switch metabolism to stimulate the production of pigments (anthocyanins in red grapes) and decrease glycolysis to allow the buildup of glucose and fructose. Malate is used to produce other compounds and several volatile organic compounds are created. This results in a sweet, tasty, and attractive-smelling berry for animals to eat and spread the seeds (Figure 11.29).

In an introductory biology course discussing the maturation of wine, it's important to understand how the development of sugars, acids, and aromatics in the grapes contributes to the final product (Figure 11.30). From the time of fruit set, when the flowers of the grapevine are fertilized and tiny grapes begin to form, a timeline of development unfolds. Sugars, primarily glucose and fructose, start accumulating as the grapes grow

Exocarp (Skin)

Terpenes
- Geraniol
- Terpineol
- Nerolidol
- Linalool

Norisoprenoids
- beta-damascenone
- beta-ionone

Thiols
S-3-(hexanol-Cysteine)

Mesocarp (Fleshy Fruit)

Organic Acids
- Malic acid
- Tartaric acid

Sugars
- Glucose
- Fructose

Figure 11.28 Anatomy of a grape. The various compounds in wine as found in grapes. New Africa/Adobe Stock Photos

Figure 11.29   Red and white grapes for a Shiraz or Chardonnay. Red or white wine has the color due to the time the skin color is extracted from the wine. Pink Zebra/Adobe Stock Photos, leungchopan/Adobe Stock Photos

and photosynthesize. These sugars are crucial as they will later be fermented into alcohol. Acids, mainly tartaric and malic, develop alongside the sugars. Tartaric acid is stable and plays a significant role in the wine's taste and structure, while malic acid, which is less stable, can soften during the ripening process and through malolactic fermentation (MLF) postharvest, affecting the wine's acidity and flavor. Tannins, which contribute to the bitterness, astringency, and complexity of wine, also develop in the skins, seeds, and stems of the grapes; their evolution is crucial for the aging potential and structure of the wine.

The term "véraison" is a key milestone in this timeline. It refers to the onset of grape ripening, a phase when the grapes change color, sugars increase rapidly, and acids begin to decrease. This stage marks the transition from berry growth to berry ripening, and it's a critical period where the balance of sugars, acids, and aromatics in the grape begins to take shape. During véraison, the flavors and aromas of the grapes become

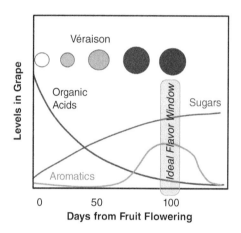

Figure 11.30   Grape development. Development of sugars, acids, aromics, and color during ripening.

more complex, setting the stage for the quality and characteristics of the wine to come. The exact timing of véraison and the subsequent ripening period can vary based on the grape variety and environmental conditions, but understanding this process is essential for predicting the quality and style of the wine produced.

The maceration of grapes and stems is the beginning of wine. The distinction between a white, rose, or red wine is in large part due to the color compounds in the pomace where the skin and crushed seeds lie. When must and pomace are in contact, tannins and other compounds are extracted from the solid pomace into the liquid must. These compounds impart color and give body and flavor to the wine. The pale color of white wines are a product of a short time mixing the must and the pomace, while heavy-bodied reds like the Italian Amarone and Chianti will be fermented in the presence of the pomace. In this case the heat from fermenting yeast and the ethanol will help to extract the flavor and color compounds into the water (aqueous) phase. Of course, red grapes will provide more color than white grapes, but much depends on the temperature and time of maceration.

An important factor in growing wine grapes is "terroir," a French concept widely embraced by viticulturists. Terroir is a concept that encompasses the myriad ways in which a particular region's environment shapes the characteristics of the wine produced there. The scientific principles behind it involve the complex interaction of environmental factors with the grapevine's biology, influencing the molecular composition of the grapes and the sensory attributes of the wine. More simply, terroir is the effect of local soil and climate on the development of the complex nature of flavor and aroma molecules in grapes. The basic idea is that exact clones of a grape will produce different-tasting wine depending on the environment, geography, and local climate of the vineyard.

While the concept of terroir is broadly discussed in the context of the sensory characteristics of wine, the scientific principles behind it relate to how these environmental factors influence the molecular composition of the grapes and, consequently, the wine. These include the following:

Soil composition and water access: Different soil types can affect the water and nutrient availability to the grapevines. For instance, deep, rich soils might produce grapes with lower concentrations of flavor compounds, whereas poorer soils might stress the vines, leading to grapes with more concentrated flavors. Minerals and compounds in the soil can also influence the levels of certain elements in the grape, affecting flavor and growth.

Climate and weather: Temperature, sunlight, and rainfall during the growing season influence the ripening of grapes and the development of sugars, acids, and various flavor compounds like esters and phenols. Cooler climates might produce grapes with higher acidity and lower sugar levels, leading to wines with more tartness and less alcohol.

Topography: The slope, elevation, and aspect of the vineyard can affect how much sunlight and heat the grapes receive, which in turn affects the synthesis of sugars and the evaporation of water, concentrating the juices.

Local yeast populations: Wild yeasts present in the environment can contribute to the fermentation process, potentially adding unique flavors to the wine. Different regions have different yeast species and strains, which can influence the fermentation process and the resulting wine's character.

These factors are suggested to impact the metabolic composition of grapes, including the levels of flavonoids, polyphenols, sugars, organic acids, and amino acids. One in-depth chemical study using a technique called mass spectrometry focused on four different vineyards located in the Burgundy region in France. In this work, the chemists found a significant, discernible difference between the Pinot Noir grapes before and after maceration. They also found that after several years in the bottle, several flavorants were at different levels even though the vineyards were only 40 km from each other [7]. Polyphenolic compounds seemed to have the greatest differences in these studies. A few molecular changes include:

Sugars: The synthesis of sugars like glucose and fructose is influenced by sunlight and temperature. These sugars are crucial for fermentation, where yeast converts them into alcohol and carbon dioxide.

Acids: Organic acids, such as tartaric and malic acid, contribute to the wine's flavor and stability. Their concentration is affected by climate, with cooler regions typically producing grapes with higher acidity.

Phenolic compounds: These include tannins and color pigments (anthocyanins). They are responsible for the color, astringency, and bitterness of the wine and are influenced by factors such as sunlight exposure and water stress.

Aroma compounds: Terpenes, esters, and thiols are among the molecules that contribute to the aroma and flavor profile of the wine. Their development is influenced by the grape's genetic makeup and the environmental conditions, including the soil type, climate, and vineyard practices.

11.6.3 Fermentation of Wine

At this point, as the must is mixed with pomace or separated from the solids, wine-making has begun. Environmental (or more excitingly named "wild") yeast and bacteria have already been introduced to the fruit juice and begun to ferment. As in beer and rising bread, alcoholic fermentation is conducted by the yeast genus *Saccharomyces*. Over 3,000 yeast species and strains are available for biomedical research that are nearly as complex as the hundreds of yeast variants available from wine and beer supply houses. The two most common species are *S. cerevisiae* (although there are many strains of this yeast) and *S. bayanus*. A strain is made from small differences within a species of an organism. For yeast, changes in the genetic material (i.e. DNA) lead to altered proteins, metabolism, and characteristics of a new strain. Over time, both species of yeast have mutated slightly to produce subtle but very important differences in growth rate and optimal growth temperature, metabolism, and the production of important flavor compounds. For example, there are two closely related strains of *S. cerevisiae*, but one strain produces a low concentration of isoamyl acetate (responsible for a pear fruit taste) of 1.2–3.5 mg/l of wine, while a second closely related *S. cerevisiae* strain produces 9.0–16 mg of the ester per liter of wine. To fully appreciate the different characteristics of wine yeast strains, there are several prominent and trusted suppliers of yeast strains for beer and wine production.

There are key characteristics for any yeast strain to be considered for winemaking. First is the ability to metabolize most of the sugar into ethanol and the other products. A yeast strain inhibited by lower levels of ethanol (< 10–14%) will leave considerable sugar in the end product, resulting in a very sweet wine. Lack of sugar is what makes a dry wine. A good strain should also ferment at a reasonable and predictable rate to compete with other environmental microbes. However, a high fermentation rate will result in higher temperatures, which can cause the evaporation of desirable volatile esters and other volatile aroma compounds. Strains of yeast for white wines are fermented at lower temperatures (12.14°C) while red wines are fermented at higher temperatures to assist in the extraction of color and aroma flavonoids (35–42°C). Strains that do not produce "off-flavors" are also an important consideration. For example, hydrogen sulfide (H_2S) produces a rotten egg smell, while some strains may produce acetic acid giving a vinegar smell. Like fish in an aquarium, if the wine-maker mixes yeast strains or bacteria, the cells must play together nicely. Some yeast produce toxic proteins or peptides called "killer factors," which are lethal to other strains of yeast or bacteria. If an advanced winemaker plans to use a mixture of yeast or bacteria (see MLF in the following), it is important to use a strain of yeast resistant to the "killer yeast" toxins. Many wild yeasts produce toxins and are resistant to the killer toxins, while *Saccharomyces* does not produce toxins yet is resistant to their effects.

Wild yeasts (indigenous to the grapes versus added or inoculated yeast) are found near the pedicels and stomata of grapes and flourish in sites on damaged or broken skin. *Saccharomyces* is not typically found on grapes in the field; rather, the two classes, Ascomycetes and Basidiomycetes are mostly commonly found on grapes. Molds, such as *Aspergillus, Penicillium, Rhizopus*, and *Mucor*, and bacteria (*Bacillus, Pseudomonas, Micrococcus*, and *Acetobacter*) are also part of the wild flora of grapes. *Saccharomyces* does not tolerate the harsh conditions of the vineyard and is considered a domesticated strain of yeast.

The question of whether the winemaker should produce wine from the existing "wild" yeast or instead add a precise strain of *Saccharomyces* remains a topic of discussion. The predominant argument for using wild yeast to ferment the wine is that these unique strains of yeast create a complex and unique character in the finished wine, not repeatable with *Saccharomyces*; however, these strains can also produce a diverse set of flavor molecules, some of which may not be acceptable in the wine, for example, higher amounts of acetic acid and ethyl acetate. Depending on the number of yeast starting on the grape, the initial cell seeding may be too low to ferment before bacteria or other microbes take over the fermenting wine. Because most wild yeasts cannot tolerate high levels of ethanol, winemakers may mix their favorite strain of *Saccharomyces* to the culture or add back a small reserved volume of wine fermented with *Saccharomyces*. However, like handcrafted unique beer brewers, some winemakers stay with the tradition to accept the benefit of native yeast fermentation to provide notes and flavors difficult to achieve and unique from mass-produced wine.

A typical yeast growth curve has three or four phases (Figure 11.31). The culture begins with adding a portion of living yeast cells—called the inoculum—to the larger culture. The original inoculum grows, divides, and multiplies (Figure 11.32). This is a crucial step for the production of wine. The yeast cells must have sufficient oxygen and micro- and macronutrients to build sufficient biomass before converting their metabolism over to ethanol production. Winemakers also carefully measure the sugar concentration of the must (in winemaking, units of

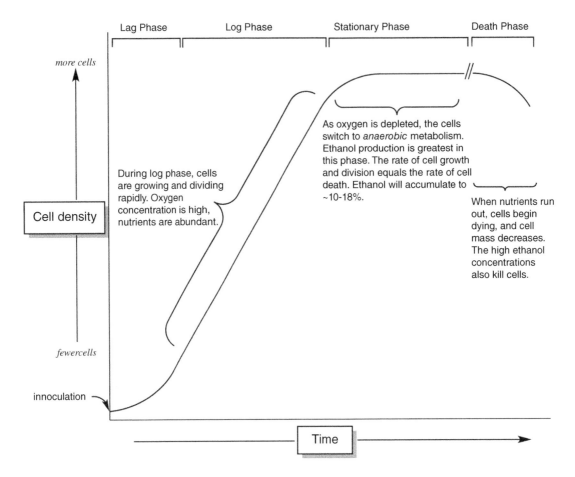

Figure 11.31   Cell growth of wine yeast.

sugar concentration are measured in Brix, where 1°Bx is 1 g of sucrose in 100 g of the liquid), because high sugar concentrations are toxic to the yeast cells. In this beginning phase, the wine-maker will add either dried yeast, rehydrate the yeast with nutrients, or, if available, add an active liquid culture purchased from a supplier. As mentioned, if the starting inoculum is small, bacteria and molds may dominate the fermentation. A typical starting concentration of yeast is 10^8–10^6 yeast cells per liter of juice. Dry active yeasts typically have 25×10^9 cells/g. Starting with higher numbers of cells in the starting inoculum can produce more fusel alcohols and esters, which, depending on the wine and grapes, may be desirable.

After the lag phase, where cells adapt to the new environment and begin to grow and divide into new daughter cells, they begin to grow at a higher rate called the log phase. Temperature and oxygen are critical in the early phases of cell growth. The number of cells will double depending on temperature.

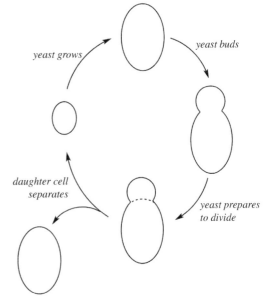

Figure 11.32   Growth cycle of a yeast cell.

At 10°C, cells double every 12 hours, while at 20°C they divide every 5 hours and every 3 hours at 30°C. White wines fermented at lower temperatures produce fewer colloids (particles of insoluble protein and carbohydrate), which cloud during cold temperatures. Furthermore, cold growth temperatures produce a brighter, fruity bouquet as more esters are retained in the wine (the colder temperatures prevent evaporation of volatile compounds), whereas the tannic and heavy-bodied red wines need the higher temperatures to extract the color and flavor compounds from the pomace.

In both lag and log phases, cells are using nutrients to create the biological building blocks (proteins, lipids, DNA, RNA, and complex carbohydrates) needed to make new cells. A small amount of ethanol is produced as the must contains dissolved oxygen. Most of the sugar is metabolized using oxygen to produce CO_2, water, and ATP via glycolysis, the tricarboxylic acid cycle, and the electron transfer chain in mitochondria. As long as oxygen is present, most of the sugars will be metabolized to CO_2, producing up to 38 ATP molecules per molecule of glucose, or less for fructose. Remember that yeasts can grow with or without oxygen, and in the absence of oxygen, they will produce ethanol as a way to regenerate NAD^+ from NAD–H to allow continued glycolysis and ATP production; the coupling of glycolysis to ethanol production ultimately produces two ATP and two ethanol molecules for every molecule of glucose. While ethanol production produces fewer ATP molecules per glucose, it is *faster* than the aerobic version, which produces 38 ATP. So in the early stages of fermentation when sugar concentration (Brix) is high, yeast will use a mix of anaerobic and aerobic metabolism and produce low levels of ethanol.

In the stationary phase, cells are no longer actively growing and dividing; they are stationary. Yeast cells also die and divide at the same rate, so there is no net accumulation of cell mass. It is in this phase that the bulk of the ethanol is produced. After the flurry of metabolism that took place in the log phase, there is little dissolved oxygen left—the conditions are becoming increasingly anaerobic. If the must is left still, escaped carbon dioxide gas will produce a barrier layer of gas at the surface of the liquid, reducing further oxygenation of the liquid. The result is anaerobic metabolism as cells convert pyruvate to acetaldehyde and ethanol. Of course, not all of the acetaldehyde is converted to ethanol, and some are shuttled to make other compounds (aldehydes, ketones, and esters), which we will learn about later in this chapter. As long as sugar and other nutrients (nitrogen from proteins, phosphate, and vitamins) are available, viable cells will continue to metabolize sugar into pyruvate and ultimately ethanol.

Eventually, waste products overcome the fermentation, and cells will no longer grow and continue to die. Ethanol inhibits its own production from acetaldehyde in a term called feedback inhibition. Fermentations that are held too long will begin to produce more and more acetaldehyde once ethanol concentrations rise. In addition, ethanol will disrupt the way proteins interact with cell wall lipids, and the proteins themselves will unravel and denature in higher concentrations of ethanol. An interesting note is that ethanol allows H^+ (protons) to enter through the membrane acidifying the cell and causing cell death.

The final result of fermentation will be from 10% to 16% ethanol depending on the wine, yeast strain, and fermentation conditions. The carbons from glucose and fructose will have mostly been converted (95%) to carbon dioxide and ethanol. Other final products of wine include pyruvate (organic acid), acetate, acetaldehyde, esters, ketones, glycerol, phenols and other volatiles, tartaric acid, and lactate (Figure 11.33).

11.6.4   Sulfur, Sorbitol, and Oaking: Additives in Winemaking

Both the home and larger professional winemakers use several additives to ensure a safe, noncontaminated final product and to produce the full spectrum of finished

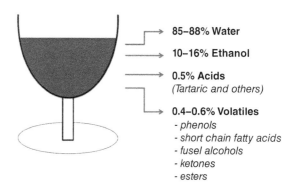

85–88% Water

10–16% Ethanol

0.5% Acids
(Tartaric and others)

0.4–0.6% Volatiles
- *phenols*
- *short chain fatty acids*
- *fusel alcohols*
- *ketones*
- *esters*

Figure 11.33   Chemical composition of a typical wine.

wine flavor. Sulfur dioxide (SO_2) is a strong antioxidant (reducing unwanted reactions between oxygen and polyphenols and other compounds) and microbial growth inhibitor. This explains why Romans burned candles (producing sulfur compounds) inside wooden casks to prevent a vinegar smell. The sulfur compounds would inhibit the vinegar produced by bacteria contaminating the wood; thus the ancients practiced a form of sterilization without realizing it.

The SO_2 produced by burning candles or matches has a sharp smell, easily detected but different from another sulfur compound, H_2S. To prevent mildew growth, winemakers can dust their grapes with sulfur to inhibit growth of the mold. However, when yeasts work on these sulfur-dusted grapes, the product H_2S smells like rotten eggs. SO_2 easily dissolves in water where it reacts to form bisulfite (HSO_3^-), which will further react to form sulfite (SO_3^{2-}):

$$H_2O + SO_2 \rightarrow H^+ + HSO_3^{-1} \rightarrow 2H^+ + SO_3^{2-}$$

Sulfites can react irreversibly with other compounds in *the* wine by forming covalent bonds or reversibly by binding non-covalently to other molecules—in either case, the sulfite is considered "bound." The remaining unbound and unreacted sulfite molecules are considered free sulfites. The free sulfite has antioxidant and antimicrobial properties. Many winemakers measure both bound and free sulfites, which sum to total sulfite.

In red wines, polyphenolic compounds can react with oxygen reducing the color and astringency of the wine. Phenolic compounds in white wines will react with oxygen to form darker, polyphenolic pigments. The addition of sulfite to wine will compete for such reactions. SO_2 is added to must to inhibit the browning enzymes released during the grape crush. SO_2 also reacts with the products of polyphenol oxidation, producing acetaldehyde as a side reaction. Unfortunately, SO_2 gas is most soluble at very low pH (pH < 1). At the pH of wine, approximately 3–4 pH units, the SO_2 converts to HSO_3^- from reaction with water. Therefore, in must, SO_2 is added as a liquefied gas or added as potassium metabisulfite that will convert back to SO_2 in the water.

Wild yeasts are typically inhibited by low concentrations (20–50 ppm free SO_2) of SO_2, while *Saccharomyces* growth rate is only slightly inhibited at these levels of sulfide dioxide. Resistant yeasts express higher levels of protein pumps in the cell membrane (SSU1), which help to transport the sulfur compound out of the cell where it will not impact cell growth. Wild yeasts do not produce these transporters and are more susceptible to SO_2. Bacteria are also very susceptible to sulfur inhibition, and if a winemaker plans to perform MLF, they should avoid the compound altogether. Because of the sensitivity of bacteria to sulfites, it is used to inhibit most non-yeast microorganisms. By the end of fermentation, most of the added sulfite will be in the bound form and not effective in inhibiting bacterial growth, and a second batch is often added for longer-term storage. This helps to block unwanted contaminant bacteria from turning wine to vinegar. Sodium metabisulfite is often used at higher concentrations (50 g/l of water ~1250 ppm) as a sanitizing agent. At these levels, a small amount will revert to SO_2 gas where it will act as a strong antimicrobial and can sanitize equipment. However, since SO_2 is most soluble at acidic pH, some will add citric acid to decrease the pH and shift the equation to maximize the concentration of the SO_2 gas form. This can be dangerous in confined conditions, where the gas can be harmful, and short-term exposures induce asthma and other lung problems.

Some are sensitive to sulfites and avoid wine for fear of headaches or inducing asthma attacks. The FDA estimates that 1 in 100 people have some sort of sulfite sensitivity and within this group 5% have asthma. Most of these reactions are the result of breathing in SO_2 gas, which is very low in wine. Yeast will produce small amounts of sulfites, and the USDA allows up to 350 ppm (Table 11.5).

11.6.4.1 Sorbate

Once the wine has finished fermenting, some residual yeast can continue to grow. If the wine is not dry—that is, it still contains sugars—existing or wild yeast can begin to grow and ferment as the wine ages. If the wine has already been bottled, this is particularly troublesome. Carbon dioxide gas will build up as the yeasts metabolize residual sugars, eventually bursting the glass bottles. Potassium sorbate is a simple organic acid (a short-chain fatty acid originally isolated from the Mountain Ash Tree) commonly added to wines to inhibit remaining yeast.

Table 11.5 Sulfur compounds in wine.

Molecular SO$_2$	Bisulfite HSO$_3^-$	Sulfite SO$_3^{2-}$	Bound versus free sulfur
Antioxidant and inhibitor of browning oxidizing enzymes Gas, soluble in water Only found in appreciable levels in pH < 1	More prevalent at the low pH of wine Binds aldehydes, sugars, and anthocyanins (bound SO$_2$) Prevents browning enzymes (quinones) and chemical oxidation (reacts with R—C=O groups)	Major form at pH greater than 7.5 Strong antioxidant Slowly reacts with oxygen and is one of the effects of aging	Bound, sulfite bonds or interacts with carbohydrates, polyphenols, and aldehydes Free, unbound sulfur, in wine bisulfite, in more pH neutral water a mix of sulfite and bisulfite. Only free forms of sulfites act as an antimicrobial

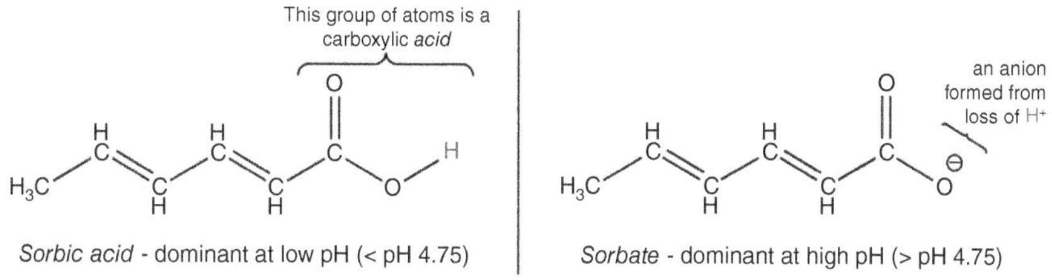

Figure 11.34 Sorbic acid is the dominant form at the low pH of wine.

Sorbate is also often used as a food preservative to inhibit mold and yeast growth. While not typically required for dry, red wines, sweeter reds, fruit, and white wines will have enough sugar to support yeast growth—these wines benefit from the addition of sorbate. In the acidic pH of the wine, sorbate is present as the acid form—sorbic acid (pK_a of 4.75; Figure 11.34), which is 20 times more inhibitory to yeast than the ionized sorbate form.

11.6.4.2 Oak
Traditional aging of wine in oak barrels is not common in modern winemaking, as industry home winemakers most often use stainless steel. However, the impact of oak on wine aroma and flavor is significant, and wood is often included in the fermenting process. There are a number of compounds that are extracted from wood that positively impact the flavor and aroma of wine. Vanilla-like compounds and specialized esters called lactones are a few examples of important compounds brought to wine by oak (Figure 11.35). These molecules will also react with the grape tannins and anthocyanins and other phenolics, providing an additional spectrum of flavors while reducing some of the astringency of a young wine. If barrels are not used for aging, adding wood to the fermentation is a critical step for achieving depth and complexity of flavor.

For wine production, oak is divided into French and American oak (see Table 11.6). French oak is found in Limousin Burgundy, the Central and Vosges regions of France, and is now also grown in Eastern Europe (*Quercus robur* and *Quercus sessilis*). American white oak is grown in Kentucky, Missouri, Arkansas, and Michigan (*Quercus alba*). In general, the French oaks have a higher polyphenol (tannin) content, while American oaks have a lower phenol, higher lactone, and a very high aromatic content—the lactones are responsible for the oak and coconut aromas. Both "wine lactone" and "whiskey lactone" have intense coconut aromas that are potent odorants; humans can smell these molecules in *tiny* amounts (10–100 ng/l). While some lactones, such as the "wine lactone" (Figure 11.35), are produced by grapes and during fermentation, the woody, coconut-like fragrance of oak is due primarily to β-methyl-γ-octalone—the "whiskey lactone" produced by the oak tree. Vanillin and similar molecules

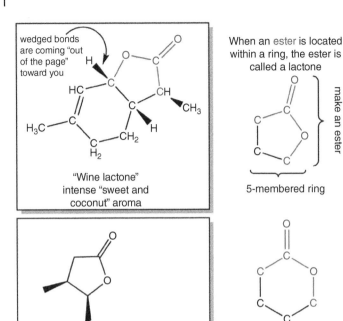

When an ester is located within a ring, the ester is called a lactone

The red atoms make an ester

5-membered ring

6-membered ring

Figure 11.35 Lactones provide some of the flavors of wine and whiskey. A compound with a ring structure containing a carbon double-bonded oxygen (i.e. ester) is a lactone.

wedged bonds are coming "out of the page" toward you

"Wine lactone" intense "sweet and coconut" aroma

β-methyl-γ-octalactone "whiskey lactone" intense coconut aroma

Table 11.6 Types of oaks and their aromas used in winemaking.

Vanillin	Vanilla aroma
Fresh oak	*trans*-Lactones
Coconut	*cis*-Lactones
Spice and clove	Eugenol and isoeugenol
Caramel, butterscotch, and sweet	Furfural and 5-methylfurfural
Cinnamon and spice	Coumarin
Charred and smoky aromas	Guaiacol and 4-methylguaiacol

will be drawn from the wood into the ethanol during fermentation. Some of the tannins in the wine are absorbed and modified by oak; however, toasting decreases the ability of oak to absorb the tannins and thus produce a more astringent wine.

Toasting or slightly heating the wood will alter the compounds in the wood and produce another set of flavor and aroma options for the winemaker. A "heavy toast" or high-temperature toasting breaks down wood carbohydrates to make caramel and butterscotch flavors that can be extracted into the wine. Heavy toasts also promote Maillard reactions, creating new "browned" molecules from the reaction of amino acids and simple sugars. Toasting will reduce vanilla compounds but produce a smoother tannin profile.

Alternatives to using expensive barrels are wood chips, sticks of oak, and liquefied oak extract. Choosing the oak type, amount, and level of toast provides an impressive array of diversity of wine beyond grapes and yeast strains.

11.6.5 Post-fermentation Clarification

Now that the fermentation is done, the dead cells (called lees) and grape solids need to be left still to settle to the bottom of the fermentation vessel. Transferring the remaining liquid is a process called racking. The process of racking the young wine into a new vessel every few months is another important step. Not only does this remove unpleasant particles from the wine, but the process exposes the wine to oxygen, which can lead to unwanted oxygenation reactions with flavor molecules. Red wines with more polyphenols to react with the oxygen (and are typically treated with higher sulfites) have a greater capacity to react with the oxygen introduced during racking. Whereas white wines with less phenols will be more affected by oxygen. An additional trick winemakers use to reduce cloudiness is to rack at lower temperatures. This decreases the solubility of several components and reduces the need for multiple racking transfers.

Some wines, like champagne or Chardonnays, are left to "sit on the lees" (sur lie in French) for weeks with periodic stirring. The bready and smooth mouthfeel of these wines is due to the aging of the dead yeast (lees), and the process involves gentle stirring to mix the lee with the wine without adding oxygen. The entire process is called battonage and results in the release of mannoproteins—cell wall glycoproteins that are a combination of protein (10%) and carbohydrate (90%). During battonage, these glycoproteins (the classification of protein–carbohydrate hybrids) are released from dead, broken yeast debris. The mannoproteins stabilize the wine, keeping proteins and tartaric acid in solution, but they also interact with other flavor compounds, including tannins. Through an unknown mechanism, these complex carbohydrates and protein–carbohydrates increase the perception of a creamy denseness to the wine that is attractive to some.

Once the cell debris has been removed by racking, there may remain an unattractive haze or cloud in the wine. Over time, the compounds causing the haze will precipitate as crystals or a fluffy solid at the bottom of your wine bottle that will alter the aging and flavor of the wine. Hazing is due to either complex polysaccharides or proteins remaining partially in solution. Proteins from the grapes or broken cells may polymerize into a semisoluble cloudy solution. Polysaccharide hazes are the result of cell wall carbohydrates of the grape and yeast. This is a significant issue when using fruit to produce wine and is solved by adding the enzyme pectinase to degrade and remove the haze.

Bentonite is a fining agent used to bind proteins and precipitate to the bottom of the vessel. Bentonite is a clay that readily absorbs water and binds positively charged proteins. At the low pH of the wine, most proteins will be positively charged with the $R-NH_3^+$ groups and a significant fraction of the carboxylic groups neutral $R-COOH$. The balance will be a positively charged protein, which will bind to the bentonite particles and through coprecipitation will be removed from the solution. The clay will also remove flavorants, and in one study, up to 13% of the volatile compounds were removed by bentonite. Polysaccharides also cause hazing but will not be removed by clay. Enzyme digestion of this haze is often the first choice producing smaller, more soluble pieces of carbohydrate.

Red wine will produce a phenol polymer haze that will leave a fine solid precipitate in aged bottles of wine. Proteins that bind to the polymerized phenols creating a larger insoluble mixture often remove such particles. Isinglass is a collagen protein made from swim bladders of fish often used in white wines. Gelatin is another type of collagen derived from bone that is used to remove tannins and phenol hazes in red wine. Two commercial products, Sparkolloid and Kieselsol, are also used. Sparkolloid is a mix of algae extract with positively charged components. Kieselsol is a suspension of silicon dioxide prepared as a side product in glass production. These are negatively charged particles that will bind to positively charged proteins and tannins and encourage settling. Chitosan is the ground exoskeleton of crustaceans and is positively charged.

Combinations of charged fining agents with an understanding of the source of haze are a good application of chemistry and biology.

After racking and fining, wine is often given one more treatment to stabilize it and avoid precipitation or haze during aging. Cold stabilization is achieved by cooling wine to 81°F/27°C for 2 or more weeks until acids like tartrate precipitate. Heat stabilization is used to remove undefined proteins.

11.6.6 Flavor and Aroma

The flavor and aroma profile of the wine is established during maceration, but it continues to evolve and complexify post-maceration due to the myriad of compounds involved. Over 1,000 distinct aroma and flavor compounds have been identified in various grapes and wines. The volatile compounds, which become airborne and are detectable at concentrations as low as 2 ng/l (akin to two drops in a swimming pool), significantly contribute to the sensory experience of wine. Notably, flavor compounds in wine differ from those in living plants or plant material, where many flavorant molecules are covalently bonded to sugars or amino acids, enhancing their solubility and reducing volatility. In the winemaking process, enzymes from the grape or yeast cleave these bonds, releasing the aroma compounds as the wine matures.

In the context of red wines, they are often described as bitter and astringent, whereas white wines may exhibit bitterness but rarely astringency. Astringency refers to the sensation that causes mouth puckering or dryness post-consumption, attributable to the presence and level of tannins in the wine. Tannins with lower molecular weight, which are both bitter and astringent, are prevalent in young wines lacking in body and flavor.

The bouquet of flavor and aroma in wines is intricate and multifaceted. For instance, Sauvignon Blanc, a white wine, is characterized by high levels of 2-methoxy-3-isobutylpyrazine, a compound typically produced by plants as a deterrent to herbivores. In grapes, this methoxypyrazine compound is synthesized in young, immature berries and gradually metabolized as the fruit matures. The residual level of methoxypyrazine influenced by sun exposure and local climate is critical; while a hint enhances the wine, an excess can be detrimental.

In addition to these aspects, glycosides play a significant role in the development of wine's flavor and aroma. These compounds consist of a sugar molecule bonded to another functional molecule, often an aromatic compound. During wine maturation, enzymes can break these bonds, releasing small, aromatic molecules like tannins, which contribute to the wine's complexity and sensory attributes. The enzymatic release of these aromatic compounds from glycosides is a crucial factor in the development of a wine's final bouquet.

Furthermore, thousands of flavor and aroma compounds, ranging from small organic molecules metabolically produced by the grape and yeast during fermentation to polyphenols and anthocyanins originating from the grape's skin, seeds, and stems, as well as from oak used in storage or during fermentation, contribute to a wine's character. The diversity of these compounds, their extraction during maceration, the yeast strain used, and the interactions between these molecules during aging add layers of complexity to the wine's flavor and aroma profile. Moreover, individual taste perception depends on the variation and abundance of receptors in one's mouth and nose.

Grasping the nuances of the compounds influencing wine's flavor and aroma extends beyond scholarly pursuits; it is an essential asset for creators, enthusiasts, and aficionados of wine. Such an understanding empowers both winemakers and wine lovers to traverse the complex world of viticulture and oenology, discerning the delicate molecular dance that weaves the elaborate sensory fabric of wine. As we transition from this exploration of wine's bouquet to a more focused look at the chemistry of small organic reactions and aroma compounds, we take a deeper dive into the reactions and interactions at the molecular level that are fundamental to crafting the wine's unique character.

11.6.7 Small Organic Flavor and Aroma Compounds

11.6.7.1 Organic Acids

Wine is basically an acidic aqueous (water) solution with ethanol. The pH of wine is fairly acidic with a pH between 3.3 and 3.6 for red wines and 3.1 and 3.4 for white wines. Acids are organized into soluble or fixed acids that are responsible for most of the acid content of wine (tartaric, malic, and several other acids) and volatile acids (including acetic, butyric, and propionic acids).

The bulk of the acid in wine is tartaric acid, which imparts a sour flavor. In aged wine, tartaric acid reacts with oxygen in the presence of Cu^{2+} or Fe^{2+} to produce glyoxylic acid. This new product binds phenolics like tannins resulting in overall less tannins and a mature, less astringent taste. Tartaric acid is less soluble at cold temperatures; the acid will precipitate as crystals with calcium or potassium ions. Some winemakers take advantage of this chemistry by cold treating the wine prior to bottling and removing any precipitated tartaric acid.

There are two forms of tartaric acid, D- and L-tartaric acid (Figure 11.36). These two similar forms of the same chemical are called enantiomers. Enantiomers are molecules that are mirror images of each other, much like how your left hand is a mirror image of your right hand but cannot be perfectly superimposed onto it. A racemic form of both forms of a chemical compound, like tartaric acid, thus consists of a mixture of two enantiomers. In the natural context of grapes, the racemic form of tartaric acid is the most common. This racemic mixture spontaneously and slowly changes, or isomerizes, primarily into the D-form over time. Isomerization is the process where molecules with the same chemical composition rearrange their structure without changing the types or numbers of atoms involved. A notable characteristic of racemic mixtures like tartaric acid is their solubility. When D-tartaric acid and L-tartaric acid are present in equal amounts, the resulting mixture is much less soluble than either of the pure enantiomers alone. This reduced solubility is a critical aspect of racemic mixtures.

Malic acid in mature grapes is much less than half of the total acid content, but in immature grapes it can be equal to tartaric acid. Grapes grown in warm temperatures metabolize (or remove) much of this acid, but grapes picked early or grown in cool conditions will have much higher malic acid. The sour, green apple flavor from malic acid is undesirable. One way to remove the offending acid is to initiate a second fermentation using bacteria to convert malic acid to lactic acid. This process called MLF is carried out by *Oenococcus oeni*, a lactic acid bacterium that can grow in the acidic conditions of wine. Bacteria convert malic acid to lactic acid and carbon dioxide, thus reducing the overall acidity (lactic acid has one less H^+ producing carboxylic group) and help create a smoother, creamy mouthfeel due to the lactic acid (Figure 11.37).

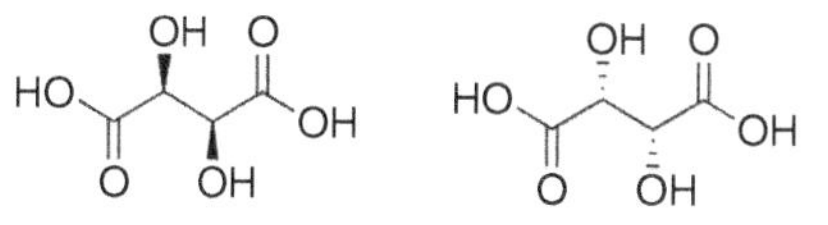

D-(-)-tartaric acid   L-(-)-tartaric acid

Figure 11.36 Stereoisomers of tartaric acid. Notice the difference between d and l stereoisomers is the spatial arrangement of the two —OH groups. The darkened dart indicates the OH is coming out of the page and the dashed darts show the OH is behind the page.

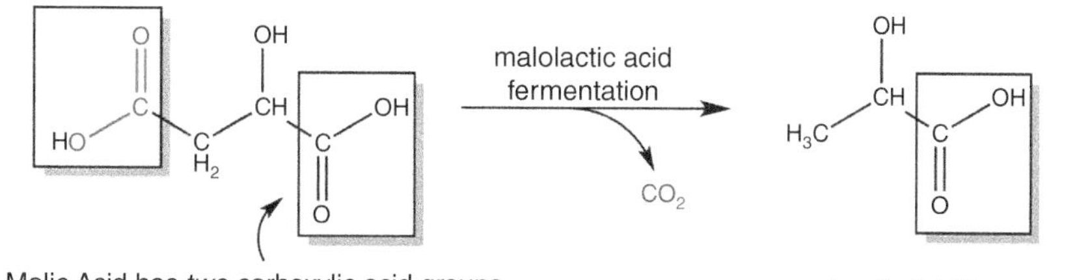

Malic Acid has *two* carboxylic acid groups

Lactic Acid has one carboxylic acid group

Figure 11.37 Malolactic fermentation. A secondary fermentation by bacteria is performed after yeast fermentation to produce lactic acid from malic acid.

Fermenting yeasts produce several volatile short-chain fatty acids. The principal volatile acid is acetic acid, which provides a vinegary flavor and smell. Smaller levels of propionic acid give wine a fatty taste, while butyric acid has a rancid butter aroma. Typical levels of acetic acid are low; however, bacterial contamination of wine will convert ethanol into acetic acid, and this is how wine vinegars are produced.

11.6.7.2 Higher, Fusel Alcohols

While ethanol is the primary alcohol produced by yeast, there are several other longer carbon chain alcohols produced along the way (Figure 11.38). The most important of these are propanol, isobutanol, and isoamyl alcohol, which account for nearly 50% of the aromatic constituents of wine after ethanol but only about 0.2% of the total mass of wine. In addition to the direct impact of these higher alcohols on the aroma of wine, they play an important indirect role in developing aged wine bouquets. Alcohols will react with organic acids found in wine to produce a new class of volatile compounds, the esters.

11.6.7.3 Esters

Esters are the result of a reaction in the acidic solution between an alcohol and an organic acid (Figure 11.39). These reactions take place slowly and generate a wide range of fruity aromas during fermentation and aging. Over 300 different esters are created between acids and alcohols in the winemaking process. These products are the compounds you sniff as a warmed wine is swirled in a glass. For example, acetate esters give wine its characteristic "wine-like" or vinous scent. Isoamyl acetate has a banana flavor, while ethyl butyrate smells like apples and ethyl hexanoate has a pineapple odor. As the carbon chain of the ester gets longer, the ester becomes less volatile, and the flavor turns soapy and lard-like [8]. Because the alcohols, acids, and enzymes that react to make these esters are specific to the type and strain of yeast and the fermentation conditions, the bouquet of esters can vary drastically from wine to wine, and the ester concentrations and ratios can change during wine aging.

11.6.7.4 Aldehydes and Ketones

Aldehydes and ketones are both types of molecules that contain a C=O double bond; they only differ in where the C=O is placed within the molecule. Aldehydes have a C=O on the terminal carbon, whereas ketones have the C=O in the "middle carbon." Acetaldehyde is the dominant aldehyde in wine. At lower levels, this aldehyde gives a fresh-cut apple aroma. At higher levels (100–125 mg/l), acetaldehyde is an off-odor and flavor (also known as a wine fault or defect) and is considered to be a contributing factor in hangovers. There are few ketones in grapes. β-Ionone and β-damascenone are two examples of grape-derived ketones that survive fermentation and have powerful scents (Figure 11.40). Both β-ionone and β-damascenone can be detected by the human nose at very low levels, and due to this fact, they contribute to the wine bouquet even in very small amounts. The diketone diacetyl is most commonly produced during MLF, where higher levels can even give a caramel-like, buttery aroma to the wine.

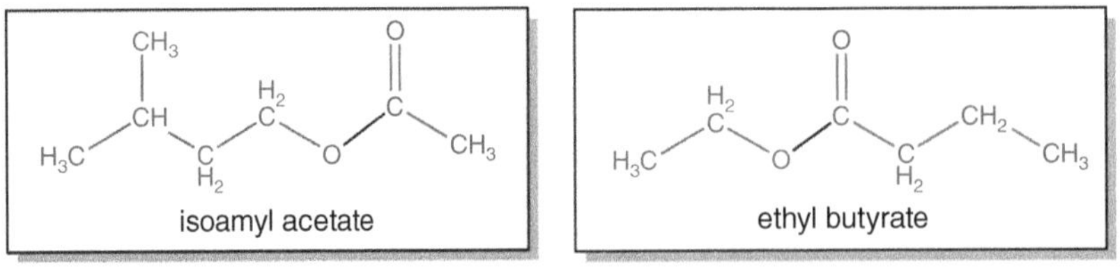

Figure 11.38 Fusel alcohols.

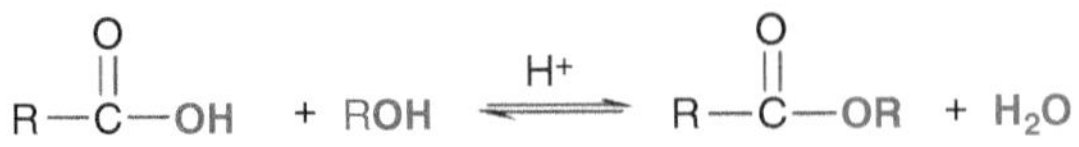

Figure 11.39 Conversion of an acid and alcohol to an ester.

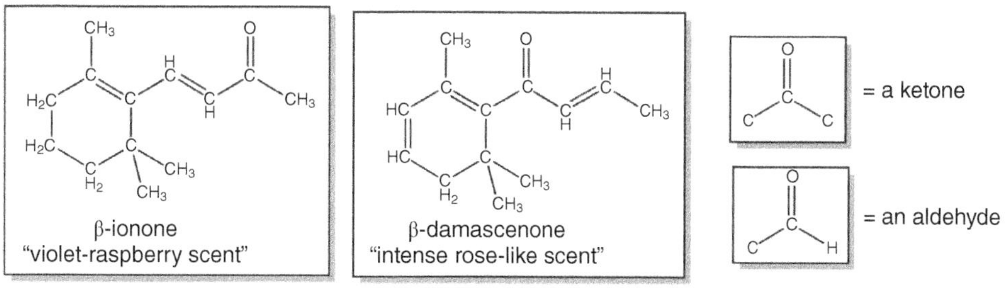

Figure 11.40   A ketone (Ionone) and an aldehyde (damascenone).

11.6.7.5 Thiols/Mercaptans

Organic compounds that have a sulfur–hydrogen group (R–SH) are collectively termed thiols and mercaptans. Thiols, which may bind mercury, are given the term mercaptan. There are several sulfur-containing compounds produced by the grape or by yeast. Many of these thiols are chemically bound to grape proteins through the cysteine amino acid. During fermentation, yeast enzymes cleave the bond producing a volatile thiol. The flavor of blackcurrant and grapefruit in Sauvignon Blanc and Semillon wines is due to such thiols.

11.6.8 Large Organic Polyphenol Molecules

Flavor and aroma compounds mostly originate from grapes, seeds, stems, and, to a lesser extent, the yeast. In addition to the smaller organic molecules described previously are much larger, more complex carbon compounds. A few classes of these compounds include polyphenols (tannins, flavonols, anthocyanins, stilbenoids, and phenolic acids), terpenes, carotenoids, and alkaloids.

11.6.8.1 Flavonoid Polyphenols

The largest class of such compounds is the group known as polyphenols. Phenols are benzene rings with an alcohol functional group. Polyphenols are large polymers of phenols that are often covalently bonded to carbohydrates and play an important part in maintaining the structure and color of cell walls. Like spices, the grape skin, woody parts of the stem, and seed all possess a complex range of different, large phenol compounds, which provide many flavors, aromas, and colors to the wine. Flavonoids are a large class of polyphenols to which tannins, anthocyanins, and flavonols belong. The types of polyphenols found in skin versus stem or seeds are different. Remember that red wine pomace is left in contact with the must so that more polyphenols are found in red than in white wine.

   The astringent and some of the bitter characteristics of wine are due to a class of polyphenols called tannins. When these molecules hit the tongue, they give a sensation of dryness (i.e. the opposite of salivation), and at high concentrations, tannins taste bitter. These substances are found bound to proteins, carbohydrates, and free in solution, and as polymers with other phenolic compounds. Grape and oak tannins are placed in two categories: condensed (also called proanthocyanidin) and hydrolyzable. The bonding of catechin and/or epicatechin monomers together forms condensed tannins. As shown in Figure 11.30, catechin and epicatechin are stereoisomers of one another. When part of a condensed tannin polymer, the A ring of one monomer is joined to the C ring of another by covalent bonds between the 4 and 8 positions (numbers from standard flavonoid numbering, also shown in Figure 11.41). Condensed tannin polymers can also release an anthocyanidin pigment molecule from the end of the polymer during an oxidative cleavage reaction, and it is this reaction that gives condensed tannin polymers the name "proanthocyanidin" (Figure 11.42). Condensed tannins polymerize with anthocyanins and are mostly found in seeds. These tannins are derivatives of the flavonoid catechin or its stereoisomer, epicatechin. Generally, the smaller tannin monomers have more flavor than the larger tannin polymer, since the larger polymers are simply too big to fit into a taste receptor protein. But the larger tannin polymers are responsible for the astringency

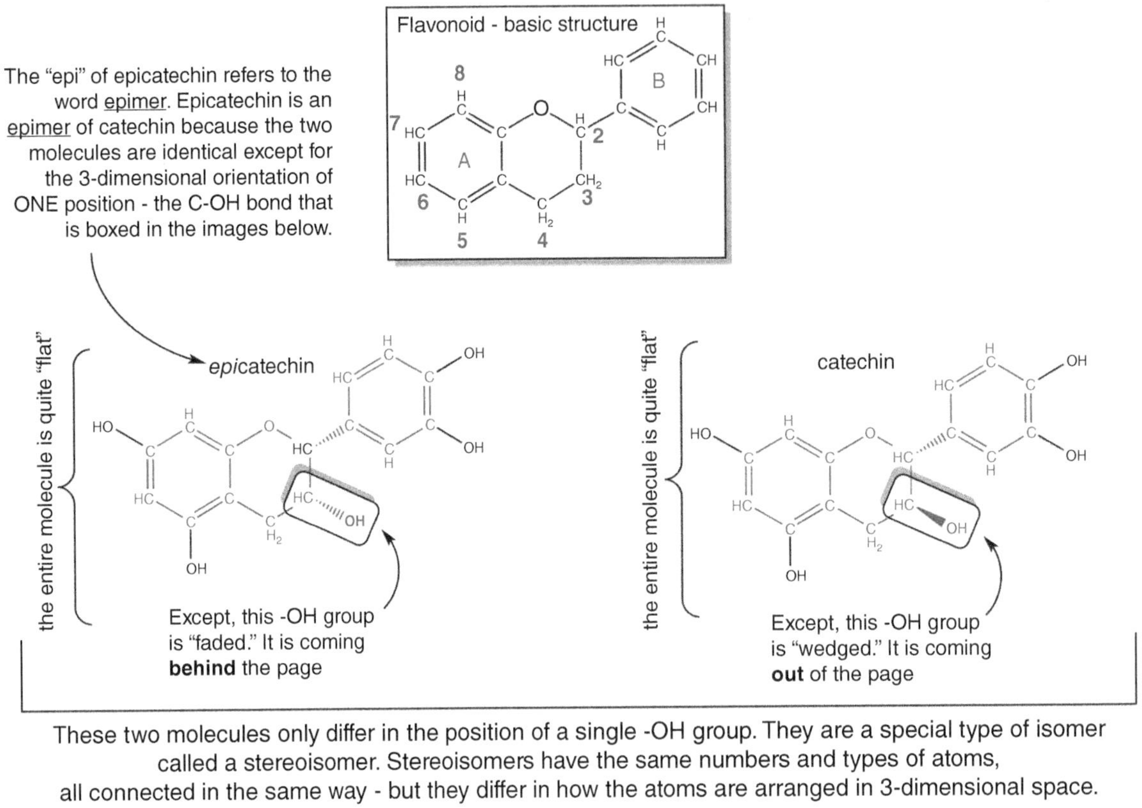

The "epi" of epicatechin refers to the word epimer. Epicatechin is an epimer of catechin because the two molecules are identical except for the 3-dimensional orientation of ONE position - the C-OH bond that is boxed in the images below.

Flavonoid - basic structure

epicatechin

the entire molecule is quite "flat"

Except, this -OH group is "faded." It is coming **behind** the page

catechin

the entire molecule is quite "flat"

Except, this -OH group is "wedged." It is coming **out** of the page

These two molecules only differ in the position of a single -OH group. They are a special type of isomer called a stereoisomer. Stereoisomers have the same numbers and types of atoms, all connected in the same way - but they differ in how the atoms are arranged in 3-dimensional space.

Figure 11.41 Epicatechin and catechin. Basic structures of flavonoid polyphenols.

or "dryness" that wine creates in the mouth. It has been proposed that the polymeric tannins bind to proteins in saliva, causing them to denature and "precipitate" or become insoluble. While this has not been conclusively demonstrated, it would explain the "dry" or "rough" feeling that red wines have in the mouth and on the tongue.

Condensed tannins can undergo an oxidative cleavage reaction to produce anthocyanidin molecules (Figure 11.43). Anthocyanidin pigments are known for their purple and reddish hues. If the anthocyanidin is modified by a sugar (sometimes called a glycone), the molecule is called an anthocyanin. If the anthocyanidin reacts with the end of the condensed tannin polymer, it terminates the polymer. The carbon 4 of the anthocyanidin is doubly bonded to another carbon, this prevents the addition of any other monomers.

The anthocyanins are considered a different class of flavonoid polyphenols, and they are largely responsible for the color and organoleptic character of wine. Anthocyanins (and anthocyanidins) change color with pH and react with oxygen to lose their color. They are also antioxidants (by their nature of reacting with dissolved oxygen sparing the reaction from compounds) and have antimicrobial and reported anticarcinogenic activity.

Hydrolyzable tannins are derivatives of gallic acid and are found bound to carbohydrates (Figure 11.44). Thus, the term "hydrolyzable tannins" comes from the enzymatic removal of tannins from the core sugar by using a water molecule to cleave the connecting bond. There are over a thousand variants of these tannins. The two core forms are the gallotannins and ellagitannins. Both are larger water-soluble tannins that are found in spices, oak, as well as wine grapes. Grapes have low quantities of hydrolyzable tannins. However, oaking a wine (or aging it in a wooden barrel) will extract some tannins into the wine. These tannins are considered soft as they have an astringent feel but are not bitter. This differs from the smaller molecular weight tannins, which are called hard tannins because they are both bitter and astringent (Figures 11.43 and 11.44).

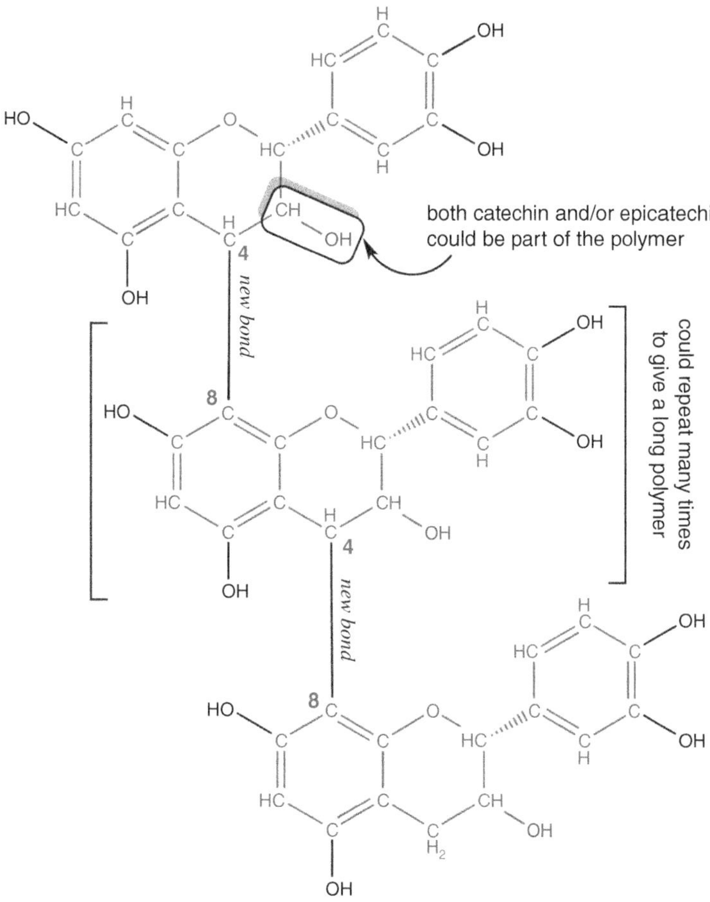

Figure 11.42 An example of a condensed tannin. Notice the building blocks of tannin and the numbering system.

11.6.8.2 Non-flavonoid Phenols

Stilbenes represent a predominant class of non-flavonoid phenol compounds found in wine; the more simple is resveratrol (Figure 11.45). These compounds originate from the stems and skin rather than the pulp or seed of the grapevine. Several studies indicate a significant role for stilbenes in cardiovascular disease and some types of cancer [9]. Caftaric acid is an example of a second non-flavonoid phenol class known as hydroxycinnamates. These compounds are easily found in most plants and react with tartaric acid. Caftaric acids, like stilbenes, have little or no flavor component.

11.6.9 Terpenes in Wine

Terpenes are a large class of compounds created by building polymers of isoprene units (Figure 11.46). These are not considered polyphenols. Several terpenes found in wine are the same as those found in spices and, along with esters, provide spicy notes in wine. Terpenes accumulate in the skin and pulp of the berry and are, therefore, prevalent in white and red wines. Most terpenes are modified with an alcohol making the compound volatile and, therefore, part of the wine fragrance. The most common terpenes in wine give the aromas of floral, rose-like (geraniol and nerol), coriander (linalool), and citrus (citronellol and limonene).

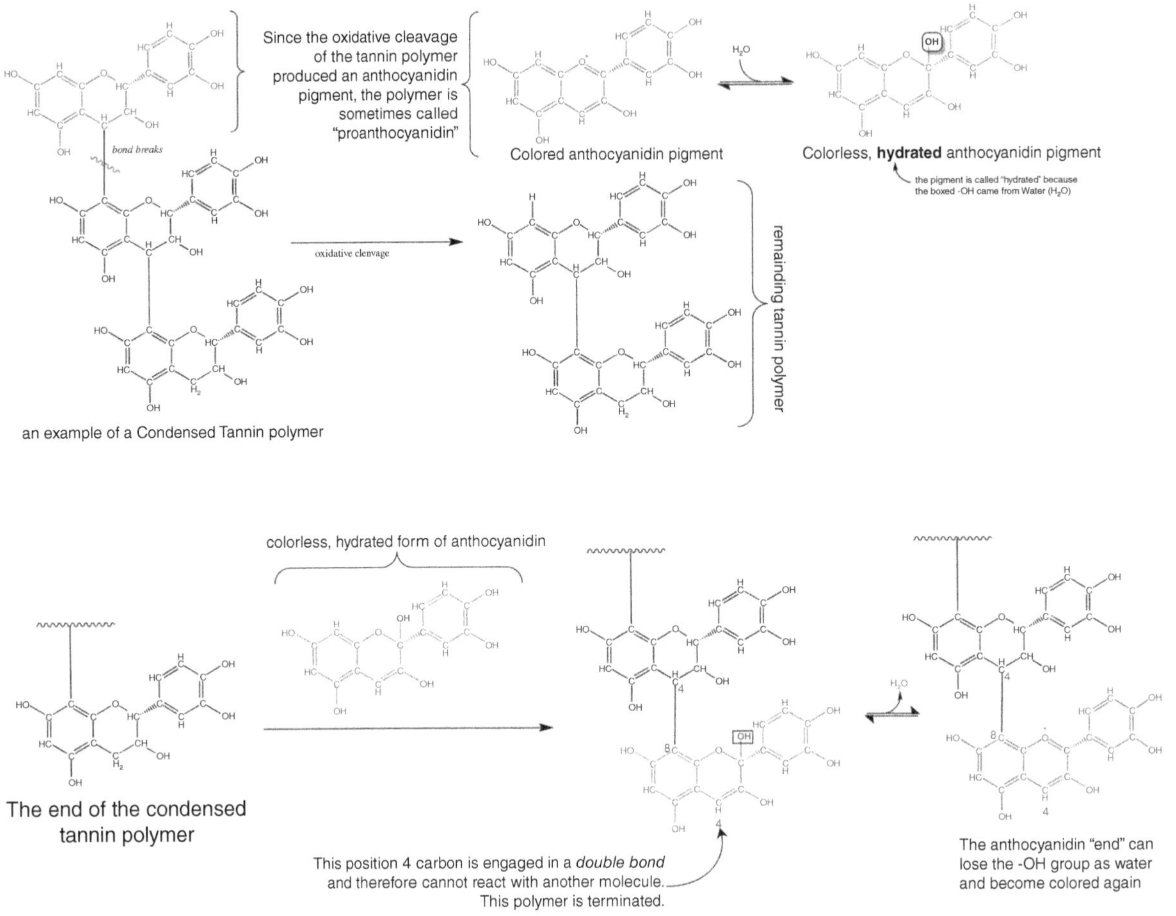

Since the oxidative cleavage of the tannin polymer produced an anthocyanidin pigment, the polymer is sometimes called "proanthocyanidin"

Colored anthocyanidin pigment

Colorless, **hydrated** anthocyanidin pigment

the pigment is called "hydrated" because the boxed -OH came from Water (H₂O)

oxidative cleavage

an example of a Condensed Tannin polymer

remainding tannin polymer

colorless, hydrated form of anthocyanidin

The end of the condensed tannin polymer

This position 4 carbon is engaged in a *double bond* and therefore cannot react with another molecule. This polymer is terminated.

The anthocyanidin "end" can lose the -OH group as water and become colored again

Figure 11.43 The complex reaction of tannin.

11.6.10 Wine Flavors: Aging and Reactions

Dr. Ann C. Noble, a professor at the University of California Davis Viticulture and Enology Department, created the aroma wheel to characterize and appreciate the complexity of flavor and aroma in white and red wines [10]. This wheel is a fantastic demonstration of the diversity of flavorants in wine. While many (but, by far, not all) of the important components have been described in this chapter, we have yet to account for reactions between many of these molecules as a wine ages. The interaction of wine with proteins, fining agents, and oxygen will result in a loss or change in the flavor and aroma molecules.

Consider the impact of adding bentonite to remove fines. Tannins will bind to proteins and grow to a large insoluble polymer. Acetic acid and acetaldehyde will react with several large molecules, enhancing and changing their impact on wine flavor and aroma (Figure 11.47).

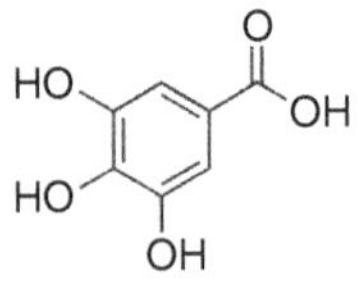

Young red wine is often very astringent with excess tannins and bitter compounds. Aging will lead to loss of some of these and other aromatic compounds, mellowing out the wine. Tannins will react with sugars in the wine reducing their bitterness and softening the wine's astringency. Esters of fatty acids change during aging, diminishing the volatile notes. Young white wines will lose a few types of esters while other more stable esters

Gallic Acid

Figure 11.44 Gallic acid is used to form hydrolyzable tannins.

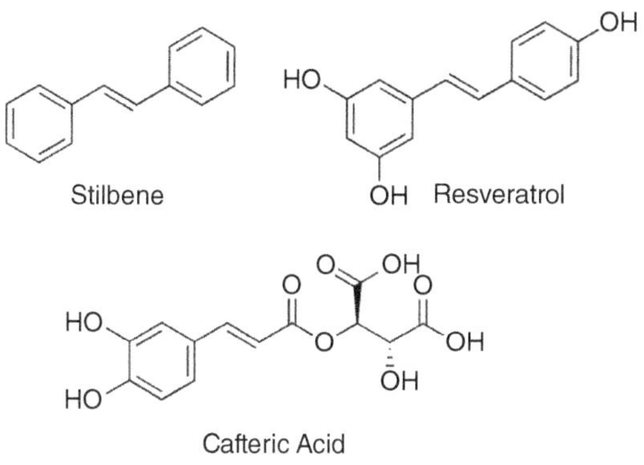

Figure 11.45   Non-flavonoids.

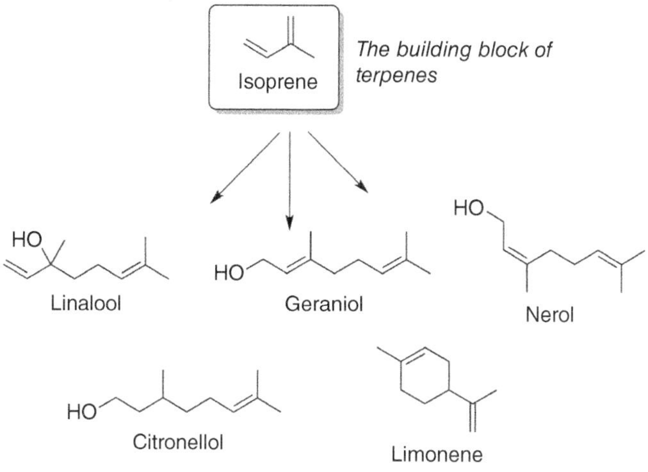

Stilbene

Resveratrol

Cafteric Acid

Figure 11.46   Common terpenes created from units of isoprenes.

Isoprene

The building block of terpenes

Linalool

Geraniol

Nerol

Citronellol

Limonene

remain. This can explain why a Chardonnay has a strong pear flavor, but, after several years, the wine loses those notes and gains the notes of buttery diacetyl (diacetyl is higher in content and more stable). The reaction of oxygen with esters and tannins will remove the stronger flavor and aroma compounds allowing the oak vanillins and softer tannins to remain and dominate in the aged wine.

"Wine varieties" typically refer to the different types of wines that are classified based on the grape species used, the region where the grapes are grown, and the methods employed during fermentation and aging. The chemical makeup of these wines is complex and significantly influences their flavor profiles. Here's an overview of how the chemical composition affects the characteristics of some common wine varieties:

Cabernet Sauvignon
- Tannins: High tannin content contributes to a robust structure and the aging potential.
- Anthocyanins: These pigments give the wine a deep red color.
- Pyrazines: Compounds that provide the distinctive green bell pepper notes.
- Flavor profile: Dark fruits like blackcurrant, with hints of cedar and tobacco, especially when aged in oak.

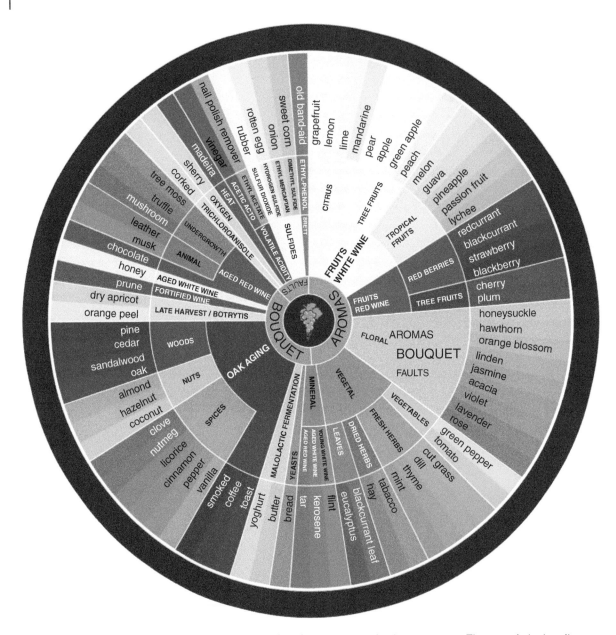

Figure 11.47 Wine aroma wheel. The smallest inner circle is the general wine bouquet terms. The next circle describes the simple terms, while the outer circle groups the individual flavors of that category. *Source:* Aromaster / Wikimedia Commons / CC BY SA 3.0

Pinot Noir

- Tannins: Lower in tannins compared to Cabernet Sauvignon, which makes for a softer mouthfeel.
- Esters: These contribute to the fruity aromas, such as cherry and raspberry.
- Terpenoids: Responsible for the floral aromas in some Pinot Noir wines.
- Flavor profile: Red fruits, floral notes, and an earthy undertone.

Chardonnay
- Acidity: Typically has a higher acidity, lending to a crisp flavor profile.
- Diacetyl: A byproduct of MLF that gives a buttery flavor.
- Sulfur compounds: Can contribute to a wide range of aromas from tropical fruit to struck match.
- Flavor profile: From apple and citrus in cooler regions to tropical fruits in warmer areas, often with a buttery or oaky note if aged in barrels.

Sauvignon Blanc
- Acidity: High acidity providing a sharp, refreshing taste.
- Thiols: These sulfur-containing compounds can give flavors of grapefruit and gooseberry.
- Pyrazines: Contribute to the green, herbaceous characteristics.
- Flavor profile: Citrus and tropical fruits, with grassy and herbaceous notes.

Merlot
- Tannins: Moderate tannin levels, contributing to a smooth texture.
- Anthocyanins: Lower than in Cabernet Sauvignon, leading to a lighter color.
- Alcohol: Generally high, giving a sense of sweetness and body to the wine.
- Flavor profile: Plums and black cherries, with chocolate and herbal notes when aged.

Riesling
- Acidity: High acidity balanced by natural grape sugars.
- Monoterpenes: Aromatic compounds that impart floral and fruity notes.
- Flavor profile: Green apple, beeswax, and honeycomb, with petroleum notes developing with age.

The chemical makeup of wines is also influenced by factors such as soil type, climate, winemaking techniques (like oak barrel aging, which imparts vanillin and toastiness), and the presence of yeast strains during fermentation. These components work together to produce the myriad of possible flavors, aromas, and textures found in wine. Each variety has its unique profile, which can be further influenced by terroir—the complete natural environment in which a particular wine is produced, including factors, such as the soil, topography, and climate. Two of the most important factors are acids and tannins.

Acids and wine flavor: In the world of winemaking, acidity plays a pivotal role in the taste and quality of the final product. Acidity in wines originates primarily from the grapes themselves. Young, green grapes are typically high in acidity, which diminishes gradually as the grapes ripen. The total acidity in wine refers to the concentration of all acids present, and this is a crucial factor in the wine's profile. The sensation of acidity in wine is not solely dependent on the total acidity but also on the pH level, which measures the intensity of the acid taste. Wines with higher acidity tend to have a lighter body and a less sweet flavor, often described as crisp or tart. Conversely, wines with lower acidity usually have a fuller body and a sweeter taste, which can be perceived as more rounded or rich. A wine with excessively low acidity can taste flat, dull, or "flabby," lacking the vivacity or freshness that acidity imparts. The key acids in wine are tartaric, malic, citric, and succinic acids. These acids come mostly from the grapes but can also be influenced by the fermentation process and other winemaking practices. Tartaric and malic acids are the most prominent acids in grapes, contributing to the wine's structure and stability. Citric acid, although present in smaller quantities, plays a role in the wine's freshness. Succinic acid, produced during fermentation, adds complexity to the wine's flavor profile. Understanding the balance and interaction of these acids is crucial for winemakers aiming to craft wines with desired taste characteristics and structural integrity.

Tannins and wine flavor: As we've learned, tannins are a group of complex organic compounds that play a significant role in wine's taste, texture, and aging process. They can bind to proteins and react with oxygen, serving as a defense mechanism in plants. This property gives plants a bitter taste, deterring insects and animals from eating them. Moreover, tannins act as inhibitors against microbial and fungal contamination, especially when the plant's skin is damaged. In the context of wine, tannins are found in higher concentrations in red wines compared to white wines. This is because red wine is typically made with skin contact, where many of the tannins are located. Tannins can precipitate with proteins through hydrophobic interactions. The proteins in our saliva, which are rich in proline amino acids, bind well with tannins, leading to the removal of lubricating proteins from the mouth. This interaction is responsible for the rough, dry sensation known as astringency. The astringency of tannins contributes to the overall flavor profile of the wine. When wine ages, tannins react with oxygen and ethanol to produce acetaldehyde, which then reacts with flavonols (tannin monomers), tannins, and anthocyanins, causing them to precipitate out of the wine. This precipitation process is why aged wines tend to be smoother and have less astringency; the tannins that cause the rough sensation diminish over time.

Examples of wines with varying levels of tannins include:

- Low tannin levels: Pinot Noir and Gamay, which tend to be lighter and less astringent.
- Medium tannin levels: Merlot and Sangiovese, offering a balance between smoothness and the characteristic dryness of tannins.
- High tannin levels: Cabernet Sauvignon and Nebbiolo, which are known for their higher astringency and are often better after aging, which mellows the tannins.

11.7 Sake Rice Wine

Sake, often referred to as rice wine, is a traditional Japanese beverage that undergoes a unique fermentation process distinct from that of grape wine or beer. The journey to crafting sake begins with the harvesting of rice. Once harvested, the outer husk of the rice grains is removed, revealing the starchy core. This rice is then steamed until the kernels soften, preparing them for the next crucial step.

In making beer, a process called malting and mashing is used to convert the starches in grains to sugars, which yeast can then ferment into alcohol. However, sake production employs a different technique involving a special mold called koji (*Aspergillus oryzae*). Koji mold spores are sprinkled onto the steamed rice, initiating the fermentation process. Koji is a cultured mold spore used not only in sake but also in other fermented products like soy sauce and miso (soybean cheese). This mold is vital because it produces enzymes such as alpha-amylase and proteases, which break down the rice's starches into sugars and amino acids necessary for yeast fermentation. It also releases vitamins that contribute to the fermentation process.

The koji mold spores are thoroughly mixed and kneaded into the rice, and this mixture is stored to allow the enzymes to work, breaking down the grain's structure and releasing the components needed for fermentation. During this period, the koji converts the rice starches into fermentable sugars.

After the fermentation is complete, the mash—which now contains alcohol—is carefully separated, preserving the liquid. This liquid is what we know as sake, and it is then bottled for consumption. Sake offers a complex flavor profile, often depending on the specific strains of koji and yeast used, as well as the type of rice and water. The science of sake is a beautiful blend of tradition and biochemistry, where microorganisms transform simple rice into a beverage enjoyed by many around the world.

Science for the Chef: Dragon's blood (raspberry, blackberry, strawberry wine)

Dragon's Blood Wine, with its distinctive bright red color and robust fruity flavor, is a creation from the home winemaking community. The vivid hue and taste profile of this wine come from the specific selection and combination of fruits used in its production. The science behind the flavors of Dragon's Blood Wine involves both the natural compounds found in the fruits and the chemical transformations that occur during fermentation. A few interesting points include:

Fruit selection: The base fruits for Dragon's Blood Wine often include a mix of red and dark berries like raspberries, blackberries, and blueberries. Each of these contributes its unique flavor profile: Raspberries contain ketones, particularly raspberry ketone, which gives a characteristic ripe berry aroma. Blackberries have high levels of esters and alcohols that contribute to their sweet, yet slightly tart, aroma. Blueberries bring in delicate floral notes and a slight tartness from their anthocyanins and phenolic compounds.

Anthocyanins: These are the pigments responsible for the red, purple, and blue colors of many fruits, including the berries used in Dragon's Blood Wine. They can also act as pH indicators, changing color with the acidity of the wine, which affects its final appearance.

Yeast: EC-1118 yeast is a robust choice for fruit winemaking, known for its vigorous fermentation and ability to work in a wide temperature range (10–30°C) and high-alcohol environments (up to 18%). Its strong fermentation ensures complete sugar conversion and is particularly suited for sparkling wines due to its pressure tolerance. EC-1118 preserves the fruit's natural flavors with its neutral profile and contributes to a clearer, more stable wine, reducing the need for fining. Its reliability and versatility make it a preferred strain for producing consistent, flavorful fruit wines.

Fermentation process: During fermentation, yeast metabolizes the sugars in the fruit mixture to produce alcohol (ethanol) and carbon dioxide. Alongside these primary products, a host of secondary metabolites are produced, such as esters, higher alcohols, and organic acids such as malic acid and citric acid.

Tannins: Berries contain tannins, which can give wine astringency and structure. To add more bite to the wine, our recipe includes adding more tannins. The level of tannins extracted during the winemaking process will influence the mouthfeel of the wine.

Clarification process: High pectin levels found in these fruits (a complex cell-wall carbohydrate) will remain suspended in the wine. To counter this, the enzyme pectinase is added to shorten these carbohydrates into soluble molecules, preventing a hazy appearance. To further refine clarity, Sparkolloid, a fining agent composed of diatomaceous earth and polysaccharides, is used. It operates by attracting and binding the suspended particles in the wine, which then settle at the bottom, allowing for easy removal.

Yeast energizer: Yeast energizer is a blend of nutrients and vitamins, including diammonium phosphate (DAP), yeast hulls, B-group vitamins, and trace minerals, used in winemaking to support healthy yeast fermentation. DAP, a key component, provides nitrogen and helps limit the formation of off-flavors like diacetyl and acetone. Yeast energizers are crucial for stimulating sluggish fermentations or ensuring robust activity from the start, particularly in nutrient-deficient musts or under stressful conditions like high sugar or alcohol levels.

Maturation: If Dragon's Blood Wine is aged, even for a short period, interactions between the wine's acids, alcohols, and any residual sugars can lead to the formation of new flavor and aroma compounds, rounding out the flavor profile and potentially softening any harsh edges.

The blend of these components and the artistry of the winemaker in balancing them determines the final flavor profile of Dragon's Blood Wine. It is a personalized wine, with each batch reflecting the choices made in fruit selection, fermentation, and aging.

Dragon's Blood Wine Recipe from DangerDave's Easy Peesy (Skeeter Pee) Recipe

Ingredients:

- 2 × 48 oz bottles of 100% Lemon Juice
- 6 lbs of Frozen Fruit - Triple Berry Blend (raspberry/blackberry/blueberry)
- 6 gallons of Spring water (not distilled)
- 20 cups of white granulated sugar (adjust for desired ABV)
- 1 teaspoon tannin
- 4 teaspoon yeast nutrient
- 2 teaspoon yeast energizer
- 3 teaspoon pectic enzyme
- Nylon fine mesh bag
- 1 packet EC-1118 Yeast
- 1/4 teaspoon Potassium Metabisulfite
- 3 teaspoon Potassium Sorbate
- 1 tablespoon Sparkolloid

Instructions for a 6-gallon (30 bottles) batch:

Instructions:

Step 1: Prepare primary fermenter

1. Clean and sanitize a seven-gallon primary fermenter.
2. Add two bottles of lemon juice. Adjust quantity for acidity preference.
3. Add water to about five gallons.
4. Dissolve 20 cups of sugar, aiming for a specific gravity (SG) between 1.085 and 1.09 after filling to six gallons.
5. Stir in 1 teaspoon of tannin, 4 teaspoon of yeast nutrient, 2 teaspoon of yeast energizer, and 3 teaspoon of pectic enzyme.
6. Top up with water to six gallons, aiming for the desired SG.
7. Add 6 lbs. of Triple Berry Blend in a nylon bag to the primary fermenter. Squeeze the bag to work in the pectic enzyme.
8. Let sit undisturbed for 12–24 hours.

Step 2: Initiate fermentation

1. Prepare one packet of EC-1118 yeast as per instructions and add to the primary.
2. Stir the primary vigorously to introduce oxygen.

Step 3: Daily care

1. Check temperature and SG.
2. Squeeze juices from the fruit pack into the fermenter and temporarily remove the pack.
3. Stir primary vigorously to introduce oxygen. Replace fruit pack.

Step 4: Post-fermentation

1. Once SG is < 1.000, squeeze and remove the fruit pack.
2. Rack to a clean and sanitized six-gallon carboy.
3. Add 1/4 tsp. potassium metabisulfite and 3 teaspoons potassium sorbate. Stir well.
4. Degas thoroughly.
5. Add Sparkolloid prepared as instructed.
6. Allow to clear undisturbed for at least 1 week.

Step 5: Pre-bottling

1. Once clear, rack off lees into another clean carboy.
2. Dissolve four to five cups of sugar to taste, ensuring it's fully dissolved.
3. Allow the wine to clear completely, which may require additional racking.

Step 6: Bottling

1. Once the wine is completely clear, bottle it in clear bottles.
   Enjoy crafting your batch of Dragon's Blood wine, respecting the craft and the legend behind the name!

Key Concepts

1. **Alcoholic fermentation** is a fundamental process utilized in the production of various alcoholic beverages, involving the conversion of carbohydrates, such as sugars from fruits, grains, or honey, into ethanol and carbon dioxide by yeast in the absence of oxygen.

2. During fermentation, **yeast** produces **fusel alcohols** and higher alcohols like isobutanol and isoamyl alcohol, which significantly contribute to the flavor and aroma profiles of fermented beverages like beer, wine, and distilled spirits. While fusel alcohols can enhance complexity in small amounts, excessive levels can lead to undesirable off-flavors, harsh aromas, and potential health hazards.

3. Ethanol's **polar** nature, attributed to the hydroxyl group, allows it to form **hydrogen bonds** with water, making it miscible in water and a suitable solvent for both polar and **nonpolar** compounds.

4. **Distillation** is a technique used to separate components of a liquid mixture based on differences in their boiling points. By heating an ethanol-water mixture and condensing the vapor back into a liquid, the concentration of ethanol can be enriched.

5. At the **azeotropic point**, a specific mixture of ethanol and water evaporates at the same rate during simple distillation, preventing further enrichment of ethanol content beyond a certain limit.

6. The processes of alcohol absorption from the gastrointestinal tract, its distribution throughout the body via the bloodstream, its metabolism in the liver by **ADH** and **ALDH** enzymes, and its excretion through breath, urine, and sweat play crucial roles in determining alcohol's effects on the body.

7. Different ADH and ALDH isoforms, encoded by various genes, have varying affinities (K_m **values**) and capacities (V_{max} **values**) for ethanol and acetaldehyde metabolism. Genetic variations in these enzymes can influence individual responses to alcohol consumption and the risk of alcohol-related conditions like AUD and FAS.

8. Ethanol interacts with various neurotransmitter systems in the brain, including **serotonin** (5-HT), **dopamine** (DA), **gamma-aminobutyric acid** (GABA), **glutamate** (GLU), and **acetylcholine** (ACH). These interactions underlie the diverse effects of ethanol on behavior, mood, cognition, and addiction.

9. The **GABA(A) receptor** is an inhibitory receptor that promotes relaxation and reduces anxiety, while the **NMDA receptor** is an excitatory receptor involved in learning and memory. Ethanol's effects on these receptors contribute to its range of effects, from sedation to cognitive impairments.

10. The enzyme **amylase** is key in breaking down complex starches into simple sugars during malting and mashing.

11. Different compounds present in **hops** contribute to beer flavor and bitterness. Wine flavor and aroma are influenced by a wide range of compounds, including volatile organic compounds, **polyphenols**, and **anthocyanins**, originating from the grape and yeast during **fermentation**.

12. **Maceration**, the process of soaking the grape solids in the liquid must, plays a crucial role in extracting flavor compounds, color, and tannins from the skins, seeds, and stems of the grape.

13. Different yeast strains used during wine fermentation can produce varying levels of flavor compounds, such as **esters** and aromatics, which contribute to the unique character of the wine.

14. **Additives** like sulfur dioxide and sorbate are used to inhibit microbial growth and preserve wine quality, while **oak aging** can introduce additional flavors, including vanillin and lactones, to enhance complexity.

References

1 *The Code of Hammurabi*. Translated by L. W. King with Commentary from C. F. Horne. The Avalon Project. Yale University. Retrieved July 12, 2015. http://avalon.law.yale.edu/subject_menus/hammenu.asp (accessed on November 17, 2015).

2 Woese, C., Kandler, O., & Wheelis, M. (1990). Towards a natural system of organisms: proposal for the domains Archaea, Bacteria, and Eucarya. *Proc Natl Acad Sci USA*, *87*(12), 4576–4579.

3 Caspeta, L., Chen Y., Ghiaci, P., Feizi, A., Buskov, S., Hallstrom, B.M., Petranovic, D., & Nielsen, J. (2014). Altered sterol composition renders yeast thermotolerant. *Science*, *346*(6205), 75–78.

4 Lam, F. H., Ghaderi, A., Fink, G. R. & Stephanopoulos, G. (2014). Engineering alcohol tolerance in yeast. *Science*, *346*, 71–75.

5 USDA Table of Nutrient Retention Factors. Release 6. Retrieved July 12, 2015, http://www.ars.usda.gov/SP2User Files/Place/80400525/Data/retn/retn06.pdf (accessed on November 17, 2015).

6 Libkind, D., Hittinger, C. T., Valerio, E., Goncalves, C., Dover, J., Johnston, M., Goncalves, P. & Sampaio, J.P. (2012). Microbe domestication and the identification of the wild genetic stock of lager-brewing yeast. *Proc Natl Acad Sci*, *108*(35), 14539–14544.

7 Roullier-Gall, C., Boutegrabet, L., Gougeon, R. D. & Schmitt-Kopplin, P. (2014) A grape and wine chemodiversity comparison of different appellations in Burgundy: vintage vs terroir effects. *Food Chem*, *152*, 100–107.

8 Jackson, R. (2008). *Wine Science: Principles and Applications*. 4th edn. Elsevier.

9 Tome-Carneiro, J., Larrosa, M., Gonzales-Sarrias, A., Tomas-Barberan, F. A., Garcia-Conesa, M. T. & Espin, J. C. (2013). Resveratrol and clinical trials: the crossroad from in vitro studies to human evidence. *Curr Pharm Des*, *19* (34), 6064–6093.

10 Ann Nobles and the Aroma Wheel. Retrieved July 12, 2015, http://winearomawheel.com (accessed on November 17, 2015). [New Ref 1] https://www.ncbi.nlm.nih.gov/pmc/articles/PMC2293160/

Additional Readings

Chen, L., & Darriet, P. (2021) Strategies for the identification and sensory evaluation of volatile constituents in wine. *Compr Rev Food Sci Food Saf*, 20(5), 4549–4583.

Gilpin, N.W. & Koop, G.F. (2008) Neurobiology of Alcohol Dependence: Focus on Motivational Mechanisms. *Alcohol Research & Health*. 31(2), 185–195.

He, N.X. & Bayen, S. (2020) An overview of chemical contaminants and other undesirable chemicals in alcoholic beverages and strategies for analysis. *Compr Rev Food Sci Food Saf*, 19(6), 3916–3950.

Piornos, J. A., Koussissi, E., Balagiannis, D. P., Brouwer, E., & Parker, J. K. (2023). Alcohol-free and low-alcohol beers: Aroma chemistry and sensory characteristics. *Compr Rev Food Sci Food Saf*, 22(1), 233–259.

Solovyev, PA, Fauhl-Hassek, C, Riedl, J, et al. (2021) NMR spectroscopy in wine authentication: an official control perspective. *Compr Rev Food Sci Food Saf*, 20(2), 2040–2062.

Volkow, N.D., Gordon, J.A, & Koob, G.F. (2021) Choosing appropriate language to reduce the stigma around mental illness and substance use disorders. Neurophyschopharmacology. 46, 2230–2232

Wang, L., Chen, S., & Xu, Y. (2023). Distilled beverage aging: a review on aroma characteristics, maturation mechanisms, and artificial aging techniques. *Compr Rev Food Sci Food Saf*, 22(1), 502–534.

Wilson, D.F., and Matschinsky, F.M. (2020) Ethanol metabolism: the good, the bad, and the ugly. Medical Hypothesis 140, 109638.

End of Chapter 11 Questions

1. What is the primary goal of alcoholic fermentation?
   a. To produce carbon dioxide and ethanol
   b. To produce carbon dioxide and water
   c. To produce ethanol and water
   d. To produce glucose and ethanol

2. Which group of higher alcohols contributes to the complex flavor and aroma profiles of fermented beverages?
   a. Ethanol and methanol
   b. Isobutanol and isoamyl alcohol
   c. Ethanol and propanol
   d. Methanol and butanol

3. What property of ethanol allows it to mix well with water and dissolve fat-soluble compounds?
   a. Nonpolar carbon chain
   b. Hydroxyl group
   c. Presence of oxygen
   d. Long carbon chain

4. What is an azeotropic point in the context of ethanol-water mixtures during distillation?
   a. The point where the liquid mixture becomes completely immiscible
   b. The point where ethanol evaporates faster than water
   c. The point where both ethanol and water evaporate at the same rate
   d. The point where water evaporates faster than ethanol

5. Which enzyme primarily metabolizes ethanol into acetaldehyde in the liver?
   a. Aldehyde dehydrogenase (ALDH)
   b. Alcohol dehydrogenase (ADH)
   c. Glutamate dehydrogenase
   d. Acetylcholinesterase

6. Explain the process of alcohol metabolism in the body.

7. How do genetic variations in ADH and ALDH enzymes influence alcohol metabolism and alcohol-related conditions?

8. How does our brain adapt as alcohol intake shifts from acute to chronic alcohol use and then during withdrawal? How does this change the way we may act?

9. Which step in beer production involves the enzymatic conversion of complex starches into simple sugars?
   a. Boiling
   b. Fermentation
   c. Mashing
   d. Conditioning

10. Which process plays a crucial role in extracting flavor compounds, color, and tannins from the grape solids?
    a. Fermentation
    b. Clarification
    c. Maceration
    d. Aging

11. What compounds are responsible for the bitter and astringent taste often found in red wines?
    a. Volatile organic compounds
    b. Polyphenols
    c. Anthocyanins
    d. Sorbates

12. What role do yeast strains play in wine fermentation?
    a. They produce sugars from grape juice.
    b. They inhibit the growth of bacteria.
    c. They metabolize sugar into ethanol and other flavor compounds.
    d. They remove haze and sediments from wine.

13. What is the primary purpose of adding sulfur dioxide to wine?
    a. To enhance color extraction during maceration.
    b. To stabilize the wine and prevent haze formation.
    c. To add desirable flavors to the wine.
    d. To increase the alcohol content of the wine.

14. Explain the significance of yeast strains in wine fermentation and their impact on flavor development.

15. Explain the role of bottom-fermenting yeasts in lager production, including the fermentation temperature, their position during fermentation, and the impact on the beer's flavor profile. Evaluate why this might be preferred for creating certain types of beer.

12

Nonalcoholic Beverages

Learning Objectives

1. *Explain the purpose of caffeine in plants.*
2. *Describe the molecular structure of caffeine and the biochemical process by which it acts as a stimulant and feel-good molecule in humans.*
3. *Explain the concept of half-life and why it is important to caffeine metabolism.*
4. *Explain how the caffeine, chlorogenic acid, fat, and sucrose content differences between Arabica and Robusta coffees impact their taste and aroma profiles.*
5. *Describe the anatomy of a coffee cherry and initial steps by which it is processed to yield a "green" coffee bean.*
6. *Understand the concepts of roast profile, first crack and second crack, and the corresponding Maillard, caramelization, and pyrolysis reactions that occur during coffee bean roasting.*
7. *Understand why there is no change in overall caffeine content during roasting, but that there are changes in caffeine-to-mass and caffeine-to-volume ratios.*
8. *Explain how/why the brewing process is a type of solid–liquid extraction and the factors that impact solubility.*
9. *Describe coffee brewing variables that impact the flavor and quality of the product.*
10. *Describe the health benefits of coffee and tea.*
11. *Describe the three major bioactive compounds in tea and their biological function.*
12. *Describe the five steps in tea processing and the most distinguishing processing step in the production of the six major tea types: green, yellow, white, oolong, black, and dark.*
13. *Explain fermentation in tea leaves as it relates to levels of catechins and different tea types.*
14. *Describe tea brewing variables that impact flavor and quality of the product.*

12.1   Caffeine

What do coffee, tea, kombucha, and energy drinks all have in common? The answer is caffeine, the most popular psychoactive drug that is universally accepted and the only psychoactive drug we routinely give children. It has been estimated that approximately 90% of humans consume caffeine regularly. Also known as 1,3,7-trimethylpurine-2,6-dione, caffeine can change our consciousness by making us more alert, focused, energetic, and even happy. Some scholars believe that the psychoactive properties of caffeine have even transformed the course of history by revolutionizing the workforce. Before caffeinated beverages like coffee and tea became popular in Europe, Europeans drank diluted beer and wine throughout the day due to unsanitary water conditions. You see, alcohol acts as a natural disinfectant by reducing the number of pathogenic bacteria in water. However, drinking alcoholic beverages naturally lowered the workforce's productivity throughout the workday. Drinking caffeinated beverages had the

The Science of Cooking: Understanding the Biology and Chemistry Behind Food and Cooking, Second Edition. Joseph J. Provost et al.
© 2025 John Wiley & Sons, Inc. Published 2025 by John Wiley & Sons, Inc.
Companion website: www.wiley.com/go/provost/food_science_2e

opposite effect of beer and wine and increased workers' productivity. Caffeine sharpened the minds of workers who were dampened from drinking alcohol. Additionally, since preparing coffee or tea relies on boiling water to extract aromatic compounds, waterborne pathogens are killed in the preparation. Therefore, these new caffeinated beverages were safer to drink than diluted beer and wine. Eventually, coffeehouses allowed people to gather and foster discussions on philosophy, politics, and scientific discoveries. Some historians even argue that caffeine is responsible for the Age of Enlightenment and the Industrial Revolution. Without our favorite caffeinated beverages, what would civilization look like today?

12.1.1 Nature's Stimulant

Like many psychoactive drugs, caffeine evolved in plants as a natural defense mechanism to discourage herbivores from chewing their leaves. As a result, caffeine is found in the leaves, seeds, and fruits of over 60 plants, including Cacao (chocolate), Yerba mate, Kola, and Guarana (Table 12.1). But the most well-known plants to produce the simulant are *Coffea* (coffee) and *Camellia sinensis* (tea) (Figure 12.1). Due to their high caffeine content, *Coffea* and *Camellia sinensis* have become two of the most prosperous species on the planet. Caffeine is characterized as an alkaloid compound and is thus bitter. The bitterness of caffeine discourages herbivorous predators and is even toxic to insects in high concentrations. However, do not be alarmed that caffeine is a toxin! The toxicity of a compound comes in its dosage, and what may be poisonous to insects is harmless to humans.

Although plants use caffeine as chemical warfare to discourage pests, certain species like *Coffea* and *Citrus* use caffeine in the nectar to attract pollinators. Interestingly, scientists have found the levels of caffeine in nectar are below the threshold to which pollinators detect a bitter taste. So then, what's the purpose of caffeinated nectar? Like humans, caffeine also produces a notable psychoactive response in bees: it helps bees learn and recall a particular scent allowing bees to return to flowers offering caffeinated nectar [1]. By enhancing pollinators' memory of reward (glucose-rich nectar), plants use caffeine to improve their reproductive success by having greater success in spreading their genetic material. Singlehandedly, caffeinated plants have evolved to use a single psychoactive drug to ward off predation and entice certain species (pollinators and humans) to disperse their seeds for proliferation. Our morning craving for caffeine highlights one of the important reciprocal relationships we have with plants.

12.1.2 Discovery of Caffeine

While bees and other pollinating insects have been enchanted by caffeinated pollen for ages, how and when did humans stumble upon caffeine? As with many discoveries, a human's first encounter with caffeine was likely accidental. For example, an ancient Chinese legend dating back to 3,000 BCE states that Emperor Shen Nung

Table 12.1 Caffeine content in plant material.

Species	Caffeine % Mass (dry mass)
Cacao (beans)	0.05–0.3%
Yerba mate (leaves)	1–2%
Coffea (beans)	1.2–2.2%
Kola (seeds)	2–3.5%
Camellia sinensis (leaves)	3%
Guarana (seeds)	3.6–5.8%

A caffeine content of 1% means 1 g of caffeine per 100 g of dry plant matter.

Figure 12.1 Coffee and tea. Coffee is the second most popular beverage in the world, only second to water. Coffee is produced from the extraction of *Coffea robusta* or *Coffea arabica* beans, while tea is made from the leaves of *Camellia sinensis*. yaisirichai / Adobe Stock Photos

haphazardly discovered tea after leaves of *Camellia sinensis* blew into his boiling pot of water. However, the discovery of coffee occurred much later. Lore says that Kaldi, a ninth century CE Ethiopian herder, noticed how his goats would stay awake all night after consuming the berries of the *Coffea arabica* plant. So Kaldi brought the bright red berries to a local monastery; after grinding up the beans and dissolving them in hot water, the world's first cup of coffee was born [2]. Yet, it wouldn't be until the early 1800s that the molecule responsible for alertness and overall heightened brain activity be characterized and named.

Caffeine was first isolated and purified from coffee beans in 1819 by German chemist Friedlieb Ferdinand Runge. At the time, coffee beans were a delicacy, so how did Runge acquire these precious beans? Interestingly, a famous poet, Johann Wolfgang Von Goethe, was impressed with Runge's previous experiments, particularly a demonstration performed for Goethe. So, as a gift of appreciation, Goethe handed Runge a carton of Arabian mocha coffee beans and requested he analyze the beans. Soon after receiving the beans, Runge isolated a white, bitter powder he named Kaffebase, German for "coffee base." After caffeine's isolation, Runge and other chemists quickly worked to identify its physical and chemical properties. Later in 1895, German chemist Hermann Emil Fischer determined the chemical structure of caffeine and was the first to synthesize the compound. Fischer's work with caffeine, in addition to his work with sugar and purines, was awarded the Nobel Prize for Chemistry in 1902. Recently, on February 8, 2019, Google paid homage to Runge's caffeine discovery by celebrating his 225th birthday with the day's Google Doodle, making caffeine's chemical structure a part of the iconic Google logo.

12.1.3 Caffeine and the Mind

Soon after that first cup of a hot caffeinated beverage in the morning, our minds awaken from the presence of caffeine. Caffeine is a stimulant for the central nervous system and thus causes heightened brain activity. But how does this small molecule increase our focus, memory, and alertness? Due to caffeine's structure (Figure 12.2), the molecule is water- and fat-soluble, allowing it to cross the blood–brain barrier readily. Once in the brain, caffeine blocks or competes with the ability of adenosine to bind to A_1 and A_{2A} adenosine receptors. When

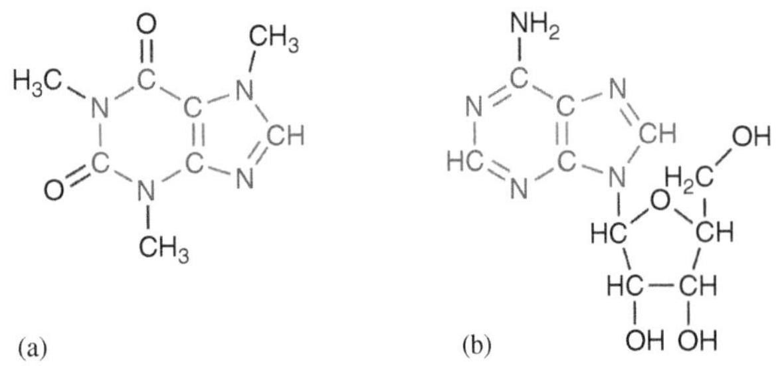

(a)

(b)

Figure 12.2 Caffeine and adenosine. Chemical structures of caffeine (a) and adenosine (b). The xanthine structure is colored red, highlighting the similarities between the two molecules and their ability to interact with the adenosine receptor.

adenosine binds to these receptors, the receptors release a cascade of signals that produce a sleep-inducing effect and muscle relaxation. Specifically, the binding promotes the release of GABA, a neurotransmitter that inhibits neurons responsible for arousal and wakefulness, thus promoting sleep. In looking at the chemical structure of caffeine, we can see that it is remarkably similar to adenosine. Both caffeine and adenosine are xanthines, which are structures composed of two concentric rings (one made of six atoms, the other made of five atoms), with a total of four nitrogen atoms within the rings (Figure 12.2). Because of the structural similarity, caffeine fits into the adenosine receptors without activating them and thus acts as a competitive inhibitor (Figure 12.3). While caffeine promotes alertness, this feeling is only transient. In the meantime, adenosine accumulates, and its concentration rises in the brain as caffeine postpones our sleepiness.

Now, this begs the question, what is causing the buildup of adenosine? Remember ATP, adenosine triphosphate? This high-energy molecule is responsible for driving essential biochemical reactions in the body. The phosphodiester bonds of ATP are cleaved to free the chemical energy, producing adenosine and phosphate as byproducts in this energy-harnessing process. High levels of adenosine signify that our body has depleted ATP, our body has been especially active, and rest is required to restore ATP. Thus, adenosine receptors serve as a mechanism to carefully monitor the levels of ATP in our body. Caffeine blocks our body's surveillance of ATP, deceiving it to believe our body can fire on all cylinders—hello, heightened brain activity! Rest, who needs that when you have caffeine playing tricks with your brain? The ability of caffeine to trick our body into a state of alertness is unquestionably the main reason why caffeine is the most popular psychoactive drug on the planet.

Not only does caffeine bolster wakefulness, but it also improves our mood. So how does caffeine influence our mood? The answer is dopamine—the "feel-good" molecule that makes us sense joy and happiness. Dopamine is part of our brain's reward system, and when we are flooded with these feel-good sensations, we seek out activities that cause dopamine release in our bodies. Thus, we reach for that second or third cup of coffee in the morning as our bodies crave that minor dopamine release. In our brain, adenosine receptors are linked to dopamine receptors.

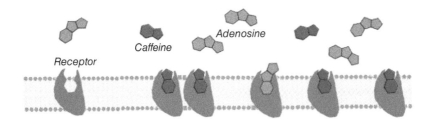

Figure 12.3 Adenosine receptors. Schematic representation of A_1 and A_{2A} adenosine receptors (blue) embedded in a lipid membrane. Caffeine (red) docks in the receptors blocking the activation of the adenosine receptors. Meanwhile, concentrations of adenosine accumulate in the brain.

Adenosine binding reduces the activity of dopamine by inhibiting its release and reducing the affinity of dopamine to bind to its receptor. In essence, dopamine binds less to the dopamine receptor when adenosine is attached to its receptor. Therefore, when caffeine blocks the adenosine receptor, it improves the ability of the dopamine receptor to bind dopamine, increases dopamine signaling, and improves our mood [3].

If you are a habitual user of caffeine, at some point, you may notice that your go-to volume of a caffeinated beverage no longer gives you that same energetic surge. There is a reason for this. Our brains adapt to the over-consumption of caffeine by creating more adenosine receptors. Thus our tolerance for caffeine increases, meaning the same dose of caffeine no longer delivers the same level of alertness. So, what are we to do? Innately, we consume more caffeine to produce similar effects! Whether you have intentionally or inadvertently removed caffeine from your diet for several days, you are probably familiar with caffeine withdrawal symptoms. The symptoms of caffeine withdrawal have been well documented for over 170 years in medical publications and include the following: headaches, tiredness, depressed mood, irritability, and mental fog [4]. When caffeine is absent, the adenosine receptors work overtime, but thankfully, the withdrawal symptoms are short-lived. The additional adenosine receptors vanish within a week, and the withdrawal symptoms subside.

So, you may wonder, between caffeine tolerance, withdrawal, and the fact that caffeine can stimulate dopamine activity in the brain, is caffeine addictive? This question is complicated. Caffeine produces similar behavioral effects comparable to other addictive drugs; it has reinforcing effects that make some caffeine users dependent. To categorize a substance as addictive, it needs to increase dopamine concentrations in the brain. While caffeine increases dopamine signaling in the brain, recent research shows that our favorite psychoactive drug does not increase the amount of dopamine in our brain [3]. Therefore, caffeine is not addictive in the traditional sense. However, the 5th edition of the *Diagnostic and Statistical Manual of Mental Disorders* (DSM-5) proposes adding "Caffeine-use Disorder" as a condition that needs further study [5]. A recent survey of the prevalence of caffeine-use disorder found that 8% of sampled caffeinated Americans qualified for the disorder and suffered from caffeine impairment, which included poor sleep, anxiety, and increased rates of depression [6].

12.1.4 Metabolism and Health Effects

After caffeine plays tricks on the adenosine receptors, the liver begins metabolizing it into three specific metabolites: paraxanthine, theobromine, and theophylline, each of which has its unique physiological response in the body (Table 12.2). The half-life of caffeine, the time our liver needs to process 50% of the caffeine consumed, varies among individuals due to differences in liver health and genetics. However, scientists have found the half-life of caffeine to be a minimum of 2 hours to a maximum of 12 hours [7]. Let's consider what this means. So, at 7 a.m., you consume a cup of coffee containing 160 mg of caffeine. Within 60 minutes, all 160 mg of caffeine is absorbed by your small intestines and quickly transported throughout your body by the circulatory system. For example, if your half-life of caffeine is 6 hours, by 1 p.m., you have 80 mg of caffeine circulating in your system, and by 7 p.m., only 40 mg of caffeine remains. In the meantime, the three metabolites are further altered in the liver and eventually excreted in the urine. So, before reaching for that after-lunch pick-me-up, remember that you likely have plenty of caffeine in your body from your morning caffeinated beverage due to the lengthy half-life of caffeine. Because caffeine delays your body's natural circadian rhythm, the American Academy of Sleep Medicine suggests consuming your final caffeinated drink at least 6 hours before bedtime.

In addition to being a stimulant of the mind, caffeine imparts a wide range of effects on your body. Notably, it has been well documented that caffeine consumption via pre-workout supplements increases athletes' aerobic and anaerobic endurance. Caffeine ingestion is most effective in 3–6 mg/kg body weight doses, 60 minutes before exercise. Due to the increased performance of athletes, caffeine was included in the list of prohibited substances by the International Olympic Committee (IOC) in 1984 and later by the World Anti-Doping Agency (WADA) in 2000. However, both the IOC and WADA removed caffeine from the list of controlled substances

Table 12.2   Metabolites of caffeine.

Metabolite	Structure	Physiological response
Paraxanthine (84%)		Increased lipolysis (triglyceride breakdown to free fatty acids and glycerol) provides fuel for muscle cells and improves athletic performance.
Theobromine (12%)		Acts as a vasodilator to dilate blood vessels and encourages blood flow. Results in increase flow of oxygen and nutrients throughout the body.
Theophylline (4%)		Acts as a bronchodilator by opening up airways in the lungs.

in 2004 since caffeine enhances athletic performance at levels that are impossible to differentiate from background levels caused by daily ingestion of caffeinated foods and beverages. The National Collegiate Athletic Association (NCAA) continues to classify caffeine as a controlled substance. Athletes must have urine caffeine concentrations less than 15 ug/ml—equivalent to ingesting approximately 500 mg of caffeine. While most of us are not professional or collegiate athletes, we can all reap caffeine's benefits when used before our daily workouts.

12.1.5   Overconsumption of Caffeine

How much caffeine is too much? The FDA declares that 400 mg of caffeine daily is generally recognized as safe (GRAS) for healthy adults without any negative health consequences, which is equivalent to 22 cups of green tea, 9 cups of black tea, 5 Rockstar energy drinks, 3–4 cups of coffee, 2 Celsius energy drinks, or 1.33 Bang energy drinks (Table 12.3). For teenagers, the FDA states that 200–300 mg of caffeine per day is generally recognized as safe, and pediatricians advise against caffeine for children under 12 years. High doses of caffeine, 500–600 mg, have been linked to insomnia, anxiousness, rapid heart rate, headache, and nausea. The FDA reports that toxic effects, like seizures, can begin at 1200 mg of caffeine consumption with lethal doses of 10 000 mg (10 g). However, death from caffeine intoxication is infrequent, and most deaths have been attributed to caffeine consumption in over-the-counter dietary supplements, not from caffeinated food and beverages. Pure caffeine powder is marketed and sold as a dietary supplement within the United States with little to no regulation. Accurately measuring a safe serving size of pure caffeine (100–300 mg) is nearly impossible at home and has resulted in accidental deaths. In 2021, a British personal trainer succumbed to cardiac arrest from caffeine toxicity after mistakenly weighing out the supplement on a kitchen scale that could only accurately weigh items greater than 2,000 mg (2 g).

Table 12.3   Caffeine content of common beverages.

Beverage	Total caffeine	Caffeine concentration (mg/oz)
Coffee	100–160 mg in 8 oz	12.5–20.0
Green tea	18 mg in 8 oz	2.2
Black tea	42 mg in 8 oz	5.2
Soda–Coca-Cola	34 mg in 12 oz	2.8
Soda–Mountain Dew	54 mg in 12 oz	4.5
Energy–Red Bull	80 mg in 8.5 oz	9.5
Energy–Monster	160 mg in 16 oz	10.0
Energy–Celsius	200 mg in 12 oz	16.7
Energy–Bang	300 mg in 16 oz	18.8
Shot–5 Hour Energy	200 mg in 1.93 oz	103.6
Shot–Bang	300 mg in 3 oz	100

Source: Adapted from https://www.caffeineinformer.com/the-caffeine-database. Caffeine concentrations in coffee vary due to roast type (blonde vs. dark), brewing style, water temperature, and grind size.

12.2   Coffee

Coffee is one of the most popular beverages in the world, with over 2 billion cups consumed every day, only behind water and tea. Daily, we drink over 450 million cups of coffee in the United States alone! So how did coffee, originating in Ethiopia, take over the world? While the legend of the first cup of coffee dates back to Ethiopia in 850 CE, written evidence shows the first coffee consumption in Yemen at the port city of Mocha. By the 1500s, coffee was extensively consumed in Yemen and played a unique role in Islamic traditions. For example, coffee assisted in periods of fasting during Ramadan and aided Muslims in staying awake during overnight ceremonies and times of prayer. By the end of the fifteenth century, coffee had disseminated throughout the Arabic world. In Yemen, the drink was known as "Qahwa," Arabic for wine, and pronounced "Kahveh" in Turkish, which led to the pronunciation of "Caffe" in Italian and finally "Coffee" in English. By the 1600s, coffee had spread throughout Europe. Coffee houses were trendy in both the Arabian Peninsula and Europe. They served as social spaces outside the home where people gathered to discuss politics, business, and religion. Contrary to European pubs, where people were usually intoxicated and violence ensued, coffeehouses were considered a place for intellectuals. In Great Britain, they were known as "Penny Universities," where a person could buy a cup of coffee for a penny and gain knowledge from the daily intellectual discussions. It's believed that Parisian coffeehouses encouraged the Age of Enlightenment, as Jean-Jacques Rousseau, Voltaire, and Benjamin Franklin were frequent patrons.

   With the popularity of coffee houses sweeping across the continent, Europeans were determined to remove their dependency on Yemen exports and cultivate their own beans. However, doing so proved to be more difficult as Yemen boiled or roasted their beans before export, preventing the germination of the seeds. In 1616, Dutchman Pieter van den Broecke smuggled live coffee plants out of Mocha, Yemen, and took them to Amsterdam, where he struggled to keep them alive. Native to Ethiopia, the coffee plants did not thrive in the northern climate. So the Dutch East India Company transported the contraband plants to more tropical climates and began plantations in India, Sri Lanka, Java, and Sumatra to satisfy Europe's coffee bean demand. The cultivation of the coffee plant eventually circumnavigated the globe. Known as the "*Bean Belt*," the geographic region between the Tropics of

Cancer and Capricorn provides a climate ideal for coffee farming. How coffee made it to the Americas is yet another tale of smuggling. In 1714 King Louis XIV received two coffee plants from the Mayor of Amsterdam, which were grown in the Royal Botanical Garden in Paris. With a desire to take the coffee plant to the Americas, Frenchman Gabriel Mathieu de Clieu stole a cutting from the king's trees. After a grueling voyage, de Clieu started a prosperous coffee plantation on the island of Martinique. With this single coffee seedling, coffee cultivation spread to the Caribbean, Central, and South America, with Brazil becoming the world's leading coffee producer for the past 150 years.

12.2.1 The Coffee Bean: *Coffea arabica* versus *Coffea canephora*

There are over 100 species in the genus Coffea, but only two species are commercially produced: *Coffea arabica* (Arabica) and *Coffea canephora* (Robusta). Coffee branches contain ripe and unripe cherries, thus the need for handpicking, making coffee a labor-intensive crop (Figure 12.4). Yearly over 20 billion pounds of coffee are produced in the Bean Belt. Brazil is the top producer, growing roughly 25% of the world's coffee, followed by Vietnam, Colombia, and Indonesia. Of the two species, Arabica is preferred by consumers for its smoother taste and therefore contains the larger share of the global coffee market (60–70%). In addition to taste, growing conditions, caffeine content, and appearance differ between Arabica and Robusta (Table 12.4).

Arabica originated in Ethiopia and is believed to be the species of coffee that Kaldi's fidgety goats found in the desert. The Arabica coffee shrub grows in a narrow temperature zone, 59–75°F (15–24°C), and requires 47–86" (1,200–2,200 mm) of rain annually. Meanwhile, the Robusta coffee shrub is hardier and can tolerate warmer temperatures, 64–97°F (18–36°C), but requires more rainfall, 86–118" (2,200–3,000 mm). Arabica is the more challenging coffee to grow of the two species, as it is more susceptible to disease and pests, slower to mature, requires specific growing conditions, and produces a lower yield than Robusta. With these limitations to cultivation, Arabica beans are more expensive to produce than Robusta beans. So then, why do consumers favor the pricier Arabica bean? The taste and sensory profile of Arabica are far superior to Robusta.

Figure 12.4 Coffee berries. A cup of coffee begins with ripe, handpicked berries from the *Coffea* plant. The berries are dried, roasted, and eventually brewed. SHUTTER DIN/Adobe Stock Photos

Table 12.4 Comparison of Arabica versus Robusta.

	Arabica	Robusta
World production	60–70%	30–40%
Top producers	Brazil, Colombia, Ethiopia, Honduras, Peru	Vietnam, Brazil, Indonesia, Uganda, India
Caffeine content	1.2–1.5%	2.2–2.7%
Chlorogenic acid content	5.5–8.0%	7–10%
Fat content	15–17%	10.5–11%
Sucrose content	6–9%	3–7%
Flavor profile	Smoother and sweeter with caramel, floral, berry, and chocolate notes	Harsher and bitter with oatmeal, peanuts, and earthy undertones

One of the primary reasons Arabica is preferred is due to the differences in aroma compounds produced during roasting. Arabica contains higher concentrations of furaneol, sotolon, abhexon, and 2,3-pentanedione, responsible for the sweet-caramel, spicy, buttery flavors, generating a nuanced flavor profile. Meanwhile, Robusta is known for its robust, spicy, earthy tones from high concentrations of guaiacol, 4-ethylguaiacol, 2,3-diethyl-5-methylpyrazine, and 3-ethyl-2,5-dimethylpyrazine (Figure 12.5). The aroma compounds in Robusta are known to leave behind a burnt, rubbery aftertaste. The more delicate, mild aromas of Arabica are greatly preferred.

The caffeine content of the beans also influences consumer flavor preference. Robusta contains up to twice as much caffeine (2.2–2.7% by mass) as Arabica (1.2–1.5%). More caffeine warrants a more bitter taste in Robusta coffee. Caffeine is an alkaloid, a class of basic organic compounds that contain at least one nitrogen atom (Figure 12.2). Basic compounds bind to our bitter taste receptors, triggering a neurochemical response that registers as bitter in our brain. Evolutionarily, humans have evolved to detect and avoid bitter substances, as many bitter substances in nature are poisonous. After all, many plants produce caffeine for its insecticidal properties. However, caffeine is not toxic for humans in the quantities found in a cup of coffee. For those wanting an enhanced caffeine boost, Robusta may be for you, but expect a more bitter coffee than Arabica.

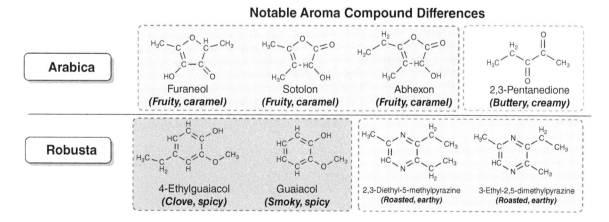

Notable Aroma Compound Differences

Arabica

Furaneol (Fruity, caramel)

Sotolon (Fruity, caramel)

Abhexon (Fruity, caramel)

2,3-Pentanedione (Buttery, creamy)

Robusta

4-Ethylguaiacol (Clove, spicy)

Guaiacol (Smoky, spicy)

2,3-Diethyl-5-methylpyrazine (Roasted, earthy)

3-Ethyl-2,5-dimethylpyrazine (Roasted, earthy)

Figure 12.5 Notable aroma compound differences. Scientists have identified the main aroma compounds responsible for the distinctive flavor profiles between Arabica and Robusta beans.

Higher levels of chlorogenic acids also contribute to the less preferred astringent and bitter taste in Robusta coffee. Arabica beans can contain half as much chlorogenic acid (5.5–8.0% by mass) compared to Robusta (7–10%). You may be wondering why chlorogenic acids contribute to astringent/bitter flavors, considering acids are known to be sour. In roasting, chlorogenic acids inside green coffee beans are converted to chlorogenic lactones and phenylindanes, both classes of these compounds are very bitter. Additionally, higher levels of sucrose (6–9% by mass) in Arabica contribute to the sweetness and provide a source of sugar for the Maillard Reaction during roasting. Arabica also contains roughly twice as much fat (15–17% by mass) than Robusta (10.5–11%), producing a tastier cup of coffee. Fat plays a vital role in brewed coffee's overall flavor and experience. First, hydrophobic fat helps bind and carry volatile, hydrophobic aroma compounds in brewed coffee. Second, fat contributes to the mouthfeel or body of brewed coffee and gives it a smoother texture. Lastly, in espresso brewing, fat contributes to the generation of crema—the thin layer of frothy cream on the top of an espresso shot. Ultimately, lower chlorogenic acid and caffeine levels and higher sugar and fat content make Arabica the preferred bean to brew.

Since Arabica and Robusta coffee beans have distinct characteristics, many coffee roasters generate *blends.* By combining Arabica and Robusta beans, coffee manufacturers can produce better-tasting coffee while cutting the cost of the more expensive Arabica bean. Roasters blend the two beans to balance the coffee's overall aromas, caffeine content, and body. The goal is to produce a superior-tasting coffee.

12.2.2   Bean Processing—Roasting

Before we can enjoy a cup of piping hot coffee, coffee cherries are first handpicked from the coffee shrub, then are pitted, dried, and processed. Once picked, the coffee cherries are pitted, also known as depulping. This process removes the fruit's outer skin (i.e. pulp), leaving behind two green coffee beans (i.e. seeds). The coffee cherry fruit is discarded once the beans are released. The wet beans are covered in a thin layer of pectin (fiber) called mucilage. The damp coffee beans are submerged in water for 12–72 hours and undergo fermentation, which helps remove the mucilage covering the beans. After fermentation, the beans are washed to remove any residual mucilage. The beans are then dried to reduce the moisture content by mass to 10–12%. This drying process can quickly be accomplished with mechanical dryers or can take up to 3 weeks if dried outside in the sun. Finally, the beans undergo hulling, where the parchment layer surrounding the beans is removed (Figure 12.6). This thin, tough, papery layer

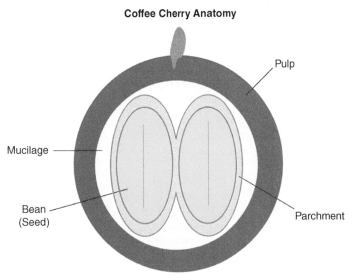

Coffee Cherry Anatomy

Figure 12.6   Coffee cherry anatomy. Processing the coffee cherry aims to extract the internal coffee beans (seeds) after several steps: depulping, fermentation, washing, drying, and hulling.

protects the beans from environmental temperature, humidity, and light fluctuations. Now the tasteless green coffee beans are ready to undergo the final process step, roasting, and develop much-needed flavor.

Roasting is an intricate process where green coffee beans are heated to temperatures between 350°F and 450°F (176–232°C) for 12–15 minutes. During this time, the beans undergo complex chemical and physical changes, producing volatile aroma compounds inside the bean thanks to the Maillard and caramelization reactions, while the beans darken in color. In addition to the browning reactions, chlorogenic acids, lipids, proteins, and polysaccharides inside the bean undergo pyrolysis—thermal decomposition of organic material. Pyrolysis aids in the development of unique flavors inside the roasted bean. A typical cup of brewed coffee contains over 1000 aroma compounds, all thanks to the chemical reactions that occur during roasting. Roasting is a critical step in coffee production as it determines the final flavor of the beans before they are sold to consumers. Roasters carefully control the chemical reactions by monitoring the beans' temperature, heating time, airflow, and cooling times. Given the nuanced differences in the quality of green coffee beans, batches of roasted coffee will be slightly different. Coffee manufacturers rely on professional coffee tasters (i.e. coffee cuppers) who are palate-trained to evaluate and assess the flavor and quality of the roasted beans and brewed coffee. A coffee taster ensures roast consistency between batches by assessing and developing roast profiles.

Green beans are first placed inside a roaster—a rotating heated drum—and the roasters begin to control the temperature inside the drum carefully. The meticulous temperature profile (i.e. heating rate) during roasting is known as the roast profile. Skillful roasters develop the roast profile to enhance specific products (aroma compounds) of the Maillard and caramelization reactions while minimizing the production of other compounds. In doing so, roasters develop highly specialized flavor profiles. During the beginning of the roast, the green coffee beans undergo a drying phase where they release moisture as the beans heat up. The green beans then transition to a yellow color, signifying the start of the Maillard reaction around 250°F (121°C). Reducing sugars and amino acids inside the bean begin to react, producing hundreds of compounds that ultimately create roasted coffee's aroma, taste, and color.

As the temperature rises, the Maillard reaction continues, and the beans darken to a tan color (i.e. light roast). As the beans approach 300°F (150°C), sucrose (disaccharide) inside the bean decomposes into its monosaccharides (glucose and fructose) and begins to caramelize. The caramelization of the bean's sugars produces an additional array of aroma compounds and brown pigments. These reactions are responsible for the toasty, nutty, smoky, earthy, and caramel notes in coffee's distinguished aroma. (See Chapter 3 for a detailed discussion of the Maillard and caramelization reactions).

Once the roasted bean approaches 386–392°F (197–200°C), a process known as the "*first crack*" occurs—signified by a sound similar to the sound of popping popcorn. At this temperature, all liquid water inside of the bean has *vaporized*. The buildup of gaseous water inside the bean expands the bean's volume, and pressure builds. Eventually, the cellulose walls of the bean succumb to the increase in pressure and force the surface to crack. During this time, water escapes from the bean, decreasing its mass. Those wishing to produce a *light roast* coffee halt the roasting process at the first crack (Figure 12.7).

With the removal of water during the "first crack," the Maillard and caramelization reaction rates are greatly enhanced, and the beans quickly darken. This period is known as the "development phase," and coffee roasters closely monitor the temperature and the rate of pigment development. As the bean reaches 437–446°F (225–230°C), the "second crack" occurs. At this temperature, cellulose (fiber) inside the bean's cell walls begins to undergo thermal decomposition, and the structure of the bean starts to collapse. During the second crack, carbon dioxide built up inside the bean from pyrolysis is released, and oils within the bean are pushed to the surface. The characteristic sound of the second crack is similar to oil sizzling in a frying pan. If the beans are pulled immediately once the second crack occurs, roasters develop a medium roast coffee. Further roasting allows the browning reactions and pyrolysis to continue, generating medium-dark (445°F/230°C) and dark roast (475°F/246°C) coffees (Figure 12.7). The beans are rapidly cooled once the desired roast type is produced to stop the high-heat reactions.

Figure 12.7 Roasted coffee beans. The roasting of green coffee beans changes the bean's chemical structure and creates fundamental flavors that will be extracted into a cup of coffee during the brewing process. From left to right: unroasted beans, light roast, medium roast, medium-dark roast, and dark roast. Rixie / Adobe Stock Photos

As the roasting process advances from a light to a dark roast bean, the flavor profile of the coffee and chemical complexity heavily evolve (Table 12.5). Coffee consumers know the differences between the roast types, likely preferring one of the four roast types due to the variances in flavor, acidity, and mouthfeel. Light roasts' floral and fruity flavors surrender to the darker roasts' earthy, toasty, spicy, and bitter flavors. The flavors arrive from the extent of the Maillard and caramelization reactions. Although over a thousand aroma compounds have been identified from the brewing process, only a handful of compounds, roughly 5%, are responsible for

Table 12.5 Types of coffee roasts and their properties.

Roast type	Time	Flavor	Acidity	Body (mouthfeel)	Common roast names
Light	At first crack	Floral and fruity notes, with hints of tea and chocolate	Highly acidic	Light-bodied texture	Cinnamon, New England Roast
Medium	End of first crack - start of second crack	Pronounced roasted flavor, with undertones of fruit, caramel, and nuts	Less acidic than light roasts	Balanced and smooth, not light, nor heavy	American, City Roast
Medium -Dark	Roughly 30% into second crack	Strong caramelization flavors with bittersweet aftertaste	Less acidic than medium roasts	Fuller bodied, with a more pronounced mouthfeel	Full City, Vienna Roast
Dark	Roughly 50–75% into second crack	Bitter, smoky, roasted, burnt taste	Least acidic of all roasts	Full, heavy bodied, bold and oily	Italian, French Roast

the overall flavor profile of a brewed cup of coffee [8]. Two main factors determine which compounds will be aromatically perceived by humans: (i) the concentration of the aroma compound and (ii) the compound's odor threshold—how much of the compound is needed before human detection. Ultimately, the ratio of the compound's concentration to the odor threshold (odor activity value) provides a relative measure of its overall aromatic importance. During roasting, the Maillard reaction produces volatile furans, pyrazines, pyridines, pyrroles, and thiols, with furans being the most prominent product—followed by pyrazines. Notable Maillard products with high odor activity values include 2-furfurylthiol (roasted coffee), 2-isobutyl-3-methoxypyrazine (paprika), 2-Ethyl-3,5-dimethylpyrazine (roasted, earthy) (Figure 12.8). Meanwhile, the caramelization of sugars produces furanones, furans, alcohols, and ketones. Important caramelization aromas contributing to the smell of coffee include diacetyl (buttery, butterscotch), furaneol (caramel, burnt sugar), and hydroxymethylfurfural (buttery, caramel) (Figure 12.8). Roasting is the science of controlling chemical reactions to create a delicate balance in the composition of aromatic compounds inside the coffee bean.

In addition to the formation of aroma compounds, the extent of roasting also impacts the acidity of the bean (Table 12.5). Unroasted beans are most acidic due to high levels of organic acids, such as chlorogenic, citric, malic, and quinic. Although acids are generally known for imparting a tart flavor, chlorogenic acids are associated with bitterness due to a series of chemical reactions during roasting. Chlorogenic acids are polyphenolic compounds that participate in important biological activities inside plants, including defense against pathogens and pests and protection from environmental stressors. Coffee contains one of the highest concentrations of chlorogenic acids, accounting for 5.5–10% of the bean's unroasted mass, with higher levels found in Robusta (Table 12.4). Chlorogenic acids are a group of compounds where types of caffeic acids and quinic acids are joined via an ester bond. While over 40 types of chlorogenic acids have been identified in coffee, 5-O-caffeoylquinic acid (5-CQA) is the most abundant (Figure 12.9).

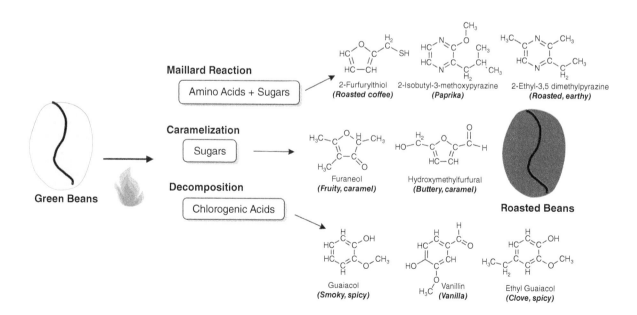

Figure 12.8 Aroma compounds from roasting. The roasting of coffee beans produces thousands of complex flavors from multiple reactions, such as the Maillard and caramelization reactions and the decomposition of chlorogenic acids.

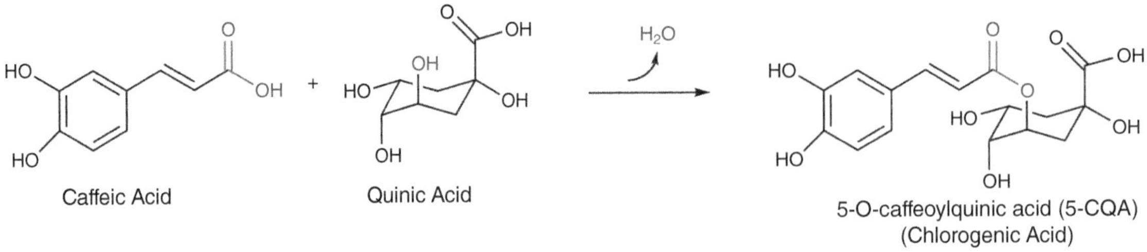

Figure 12.9 Formation of chlorogenic acid. Caffeic and quinic acids undergo a condensation reaction and are joined via an ester bond (red) to form 5-O-caffeoylquinic acid (5-CQA), the most abundant chlorogenic acid in coffee beans.

Upon roasting, two important reactions with chlorogenic acids occur at high temperatures. First, chlorogenic acids, like 5-CQA, decompose into their respective caffeic and quinic acids. These acids further degrade into several phenolic compounds, such as guaiacol, ethyl guaiacol, phenol, and vanillin—generating spicy, smoky aromas in the coffee (Figure 12.10). The concentration of these phenolics increases as the roast darkens. Additionally, chlorogenic acids rearrange to form an additional ester bond—developing a chlorogenic acid lactone (Figure 12.10). A lactone is a functional group (i.e. alcohol, amine, carboxylic acid, thiol, ester) where the ester is part of a ring (cyclic) structure. Approximately 10 chlorogenic acid lactones have been identified as sources of bitterness in roasted coffee. As green beans transition to a light or medium roast, about 50% of the chlorogenic acids are degraded to spicy, smoky phenolic compounds or converted to pleasantly bitter chlorogenic acid lactones. The reduction in acidity and increase in bitterness help balance the flavor profile of the coffee. As the beans continue heating to a medium-dark or dark roast, chlorogenic acid lactones undergo thermal decomposition and form phenylindanes (Figure 12.10). Unlike chlorogenic acid lactones, phenylindanes are potently bitter and linger in one's mouth. Ultimately, the darker the roast, the more bitter the coffee will become, as most chlorogenic acids are decomposed [9].

While the concentration of chlorogenic acids decreases upon roasting, the amount of other organic acids (acetic and formic) increases due to the decomposition of sugars [9]. Therefore, the pH of brewed coffee remains acidic, ranging between a pH of 4.85 and 5.1, and the pH difference between light and dark roasts is negligible. However, the overall sour taste profile decreases as the roast level increases, thanks to the decomposition of chlorogenic acids and the formation of bitter compounds. The acidity does an excellent job of balancing the sweet and bitter flavors in a cup of coffee.

If the flavor and acidity profiles change as the roast transitions from light to dark, what happens to the caffeine molecules during the roasting process? Does caffeine decompose just like chlorogenic acids do as the temperature of the bean increases? It turns out that the caffeine content inside of a coffee bean is relatively stable during roasting (phew!). The stability of caffeine results from its high melting point (460°F/238°C) and resistance to thermal degradation. Therefore, a bean will have the same amount of caffeine before and after roasting. However, you may have heard that darker roasts have more caffeine. If you measure your coffee beans by mass instead of volume, this statement has some truth. During roasting, the bean loses a substantial amount of water due to evaporation at the first crack and nearly 90% of its water content by the second crack. The mass of a bean can decrease between 10% and 20% during the roasting process, but the mass of caffeine stays the same (Figure 12.11). As a result, darker beans have a higher caffeine-to-mass ratio than lighter beans. If you weighed 25 g of a dark French roast and compared its caffeine content to 25 g of a light New England roast, the French roast would contain more caffeine. However, the bean's volume increases due to thermal expansion during roasting. Dark beans have a larger volume than light beans, meaning their caffeine-to-volume ratio (i.e. density) will be lower. If you measure your coffee

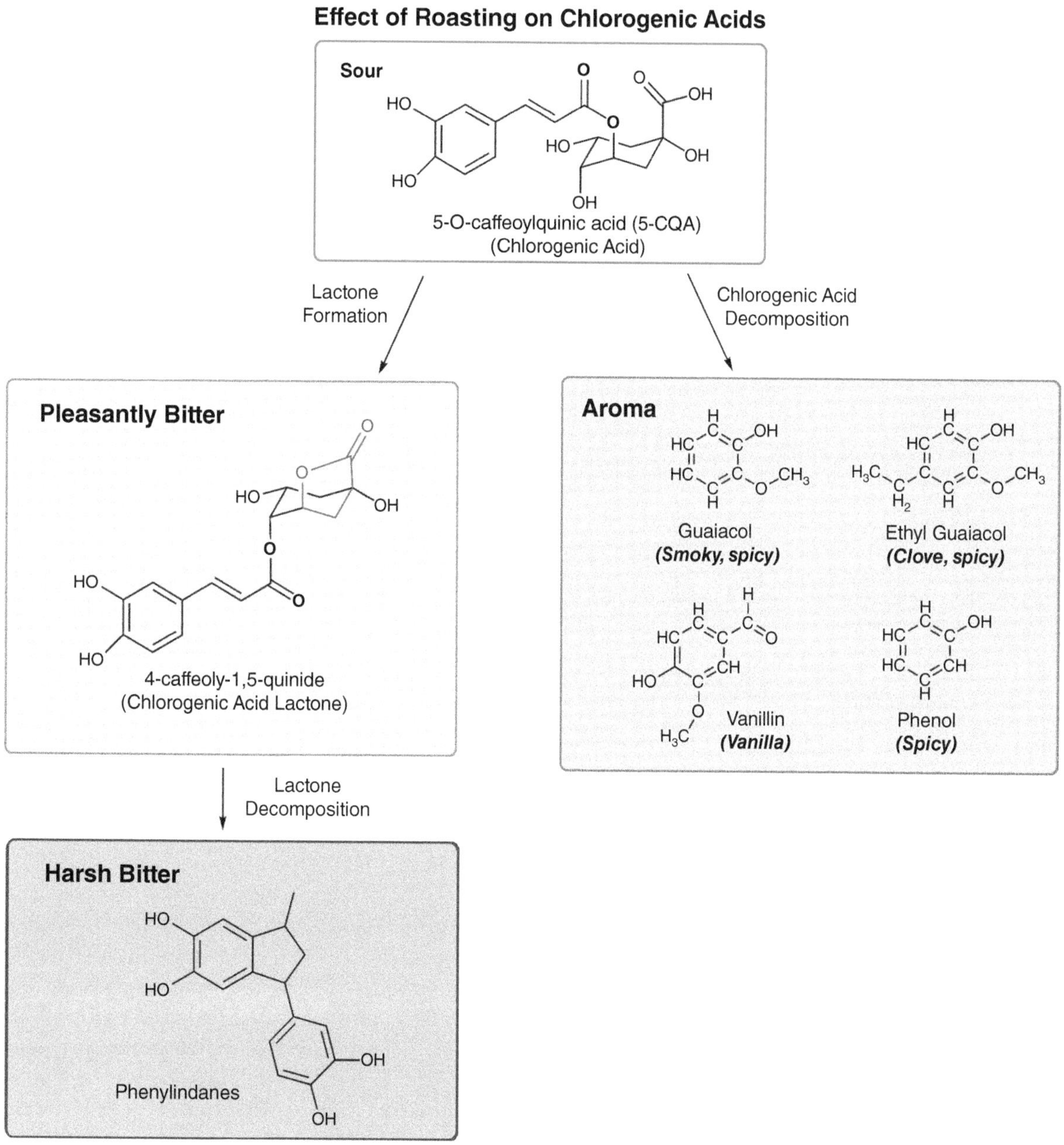

Effect of Roasting on Chlorogenic Acids

Sour

5-O-caffeoylquinic acid (5-CQA)
(Chlorogenic Acid)

Lactone
Formation

Chlorogenic Acid
Decomposition

Pleasantly Bitter

4-caffeoly-1,5-quinide
(Chlorogenic Acid Lactone)

Aroma

Guaiacol
(Smoky, spicy)

Ethyl Guaiacol
(Clove, spicy)

Vanillin
(Vanilla)

Phenol
(Spicy)

Lactone
Decomposition

Harsh Bitter

Phenylindanes

Figure 12.10 Effect of roasting on chlorogenic acid. Decomposition of sour chlorogenic acids leads to the formation of phenolic aroma compounds, bitter chlorogenic acid lactones, and further roasting generates harshly bitter phenylindanes. The lactone functional group is highlighted in red.

beans by volume, dark roasts contain less caffeine than light roasts (Figure 12.11). If you are concerned about maximizing your caffeine content in a dark roast, measure by mass. Or better yet, look for a dark roast coffee blend that contains a higher percentage of Robusta! Remember that Robusta contains roughly twice as much caffeine as Arabica beans.

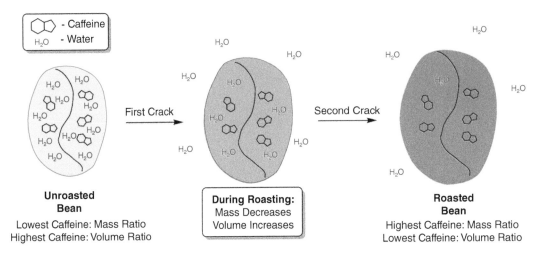

Figure 12.11 Caffeine content during roasting. As the green coffee bean is roasted, the mass of the bean decreases as steam is released once the bean cracks. During this time, the volume of the bean increases due to thermal expansion. Meanwhile, the caffeine content inside a bean remains constant, but the caffeine-to-mass and caffeine-to-volume ratios change as the bean roasts.

12.2.3 Brewing Coffee

While the roasting process develops a myriad of flavors essential in creating a delectable cup of coffee, the brewing process determines which flavor molecules and how many of them reach the final cup. Brewing is a type of solid–liquid extraction, a purification technique. This extraction technique allows one to dissolve desirable compounds (solutes) in a liquid (solvent) while leaving undesirable molecules behind in the solid material. In brewing coffee, water is the solvent of choice and must first be absorbed by the coffee grounds. Water-soluble compounds, such as caffeine, sugar, chlorogenic and organic acids, and aroma compounds, transfer from the coffee ground into the water. How well different molecules can be extracted in a cup of coffee depends on their water solubility, where the polarity of compounds determines how water-soluble a compound will be. The more places water can interact on a compound, the more polar and water-soluble it will be. This explains why caffeine is one of the first molecules to extract during brewing, while the less polar and harshly bitter phenylindanes are one of the last compounds to extract (Figure 12.12). Finally, this solute-solvent mixture must be separated from the remaining coffee grounds—typically through filtration.

In brewing, variables such as grind size, water temperature, extraction time, water quality, pressure, and the coffee-to-water ratio can be manipulated to change the flavor and quality of a cup of coffee. Altering these variables allows us to perfect the extraction yield of the brew—the mass ratio of extracted coffee solubles to the mass of coffee used. The extraction yield determines the flavor of a brew. The Specialty Coffee Association states that a well-balanced brew will contain an 18–22% extraction yield. Yields under 18% are classified as under-extracted and have a sweet, acidic flavor profile. While coffee brews with extraction yields more significant than 22% are said to be over-extracted with bitter flavor profiles. Achieving the best extraction yield requires careful control of the brewing parameters.

The grind size of the coffee beans affects the extraction rate—the speed at which soluble compounds are removed from the coffee bean and extracted into water. Finer grinds have smaller particles and a larger surface area, which holds water well. This allows for a better opportunity to remove more solute molecules and results in

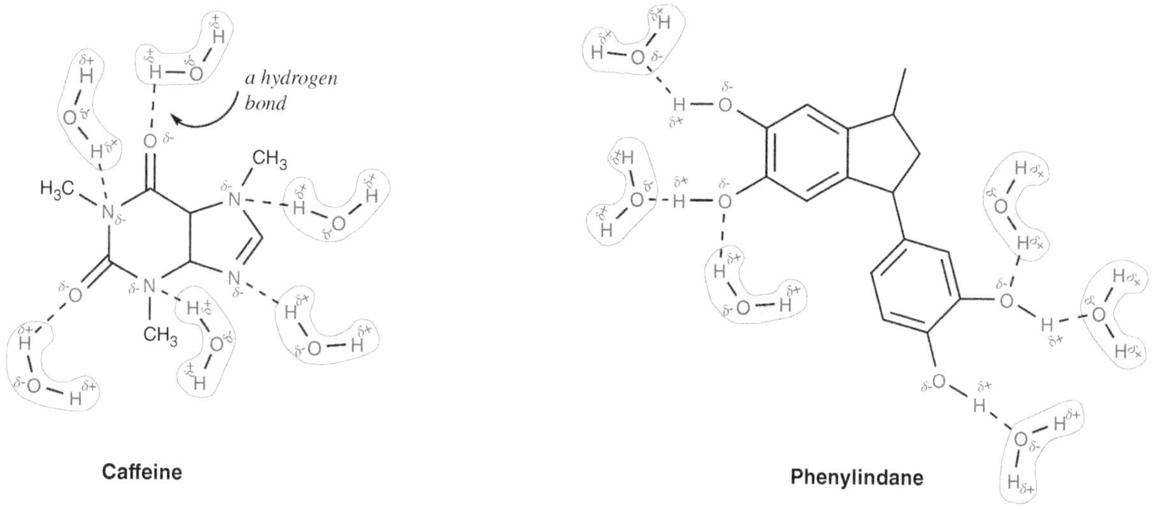

Caffeine

Phenylindane

Figure 12.12 Polarity differences between caffeine and phenylindane. Water surrounds most of the caffeine molecule and forms hydrogen bonds with its partially negative oxygen and nitrogen atoms; this makes caffeine highly water soluble. With a higher percentage of nonpolar bonds (C—C, C—H), water surrounds only a portion of phenylindane, making it less water soluble than caffeine.

a faster extraction. Coarser grinds have larger particles, which allows water to pass quicker through the coffee solids causing a slower extraction. Typically, finer grinds yield more robust, stronger coffees since more coffee solids will be extracted in a shorter period, while coarser grinds produce milder, weaker coffee. It's important to note that different brewing methods (i.e. French press, filter, espresso, percolator) require specific grind sizes (Figure 12.13).

The grind's freshness also impacts the brewed coffee's overall flavor. The roasting process generated hundreds of *volatile* aroma compounds from the Maillard and caramelization reactions, decomposition of chlorogenic acids, and pyrolysis of organic matter. These aroma compounds are trapped inside the bean, but once ground, they

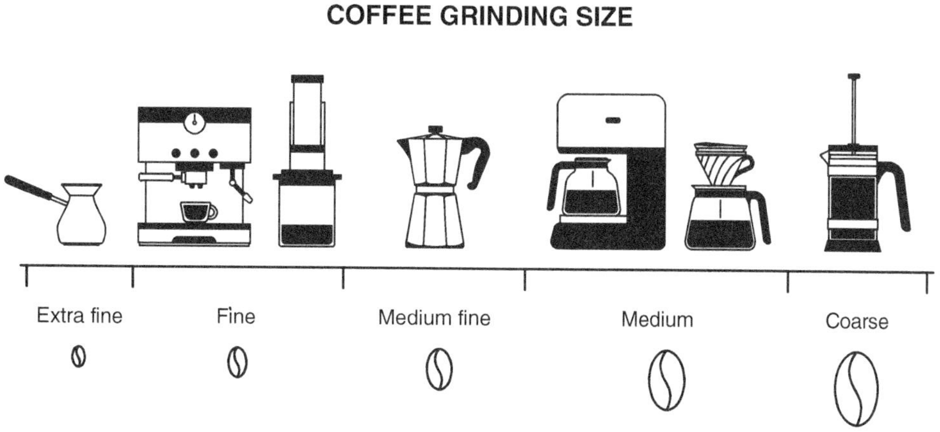

COFFEE GRINDING SIZE

| Extra fine | Fine | Medium fine | Medium | Coarse |

Figure 12.13 Brewing methods and grind size. From left to right: Turkish (extra fine), Espresso (fine), AeroPress (fine), Moka Pot (medium fine), Drip (medium), Pour-Over (medium) and French Press (coarse). WinWin/Adobe Stock Photos

quickly evaporate and escape into the atmosphere. The longer the ground coffee is exposed to air, the more aroma compounds will be lost. Fewer volatile aroma compounds mean a less palatable cup of coffee. This is why it is best to grind coffee immediately before brewing. In addition to losing aroma compounds to the atmosphere, grinding increases the aroma compounds' oxidation rate. Oxidation is adding oxygen or removing hydrogen in a compound by transferring electrons from one substance to another. By increasing the surface area of the coffee grounds, oxygen can interact with more molecules and increase the rate of oxidation of acids, oils, and fragrant aroma compounds. Highly oxidized coffee tastes stale. Exposure to air, heat, and sunlight speeds up the oxidation process of coffee. Storing beans in an airtight container in the freezer can lengthen the freshness of your coffee beans by limiting oxidation and staling.

The water temperature of the brew impacts the solubility of the compounds inside the bean. By manipulating the temperature of the solvent, we can alter how much solute can be extracted from the bean. Typically, increasing the temperature of the solvent increases the solubility of the solute. Higher temperatures give solvent molecules enough energy to break intermolecular forces that hold solute molecules together. Increasing the water temperature means more delicious compounds (chlorogenic and organic acids, caffeine, and aroma compounds) will be extracted inside the bean. However, higher temperatures also increase the solubility of the lesser water-soluble, bitter compounds like chlorogenic acid lactones and phenylindanes. A good brewing guideline is to use water between 198°F and 205°F (92–96°C) to produce a well-balanced cup of coffee. When water is near its boiling point (212°F/100°C), coffee can become over-extracted and very bitter. Another downside of boiling water is that the delicate aroma compounds responsible for most of the brewed coffee's flavor will evaporate due to their low boiling points. However, roast types may impact the temperature of the brew. Darker roasts, with higher concentrations of bitter compounds, benefit from brewing in slightly cooler water (195°F/91°C) to prevent the extraction of these compounds. Likewise, lighter roasts that contain fewer bitter compounds can withstand brewing near the boiling point of water (212°F/100°C).

The extraction time of the coffee beans is the most critical factor in determining the overall flavor profile of the brew. This refers to the amount of time the solvent (water) is in contact with the solid phase (coffee grounds). The longer the extraction time, the higher the extraction yield and the greater the concentration of soluble compounds. Short extraction times may yield an acidic, under-extracted coffee, while long extraction times will result in a bitter, over-extraction. Different brewing methods require specific extraction times, with espresso machines requiring a 30-second extraction, while the French press requires a 46-minute extraction time.

Since water acts as the solvent in extracting coffee's soluble compounds, water quality plays a vital role in coffee brewing. The composition of the water aids in the extraction of acids, bases, sugars, and starches inside the coffee grinds. In particular, the mineral content and water hardness are essential factors in determining the quality of brewed coffee. Minerals, particularly Mg^{2+} and Ca^{2+} cations, favor the extraction of soluble compounds due to their high charge magnitude. Water hardness—specifically carbonate hardness—is the measure of bicarbonate (HCO_3^-) and carbonate (CO_3^{2-}) ions in water (also referred to as the alkalinity of water). These basic ions can accept acidic protons (H^+) in an acid–base reaction. Given that coffee is weakly acidic, the presence or absence of bicarbonates and carbonates will alter the sour/bitter profile of the coffee. Low bicarbonate levels (i.e. soft water) will generate an acidic, tart cup of coffee. High levels of bicarbonate (i.e. hard water) will neutralize the organic and chlorogenic acids in the coffee yielding a flat, chalky cup of coffee. Given that the mineral content and carbonate hardness vary by geographic region, brewing the same bag of beans in two different cities will result in different-tasting cups of coffee. To account for differences in water quality, many coffee shops use filtered water and mineral blends to generate the perfect ratio of Mg^{2+}, Ca^{2+}, and HCO_3^-. Home brewers can rely on bottled water to have maximum control over the water quality in the brew. However, you will want to experiment with different bottled water brands, as the mineral content and carbonate hardness vary.

The coffee-to-water ratio is another driving force in determining a cup of coffee's taste profile and overall strength. The ratio refers to the amount (mass) of coffee grounds used in relation to the amount (mass or volume) of water used in the brew. A higher ratio produces a more robust cup of coffee with a faster extraction rate, while a lower ratio results in a faint cup of coffee with a slower extraction time. The goal is an extraction yield of 20% of the soluble solids inside the bean. According to the Specialty Coffee Association, a general guide uses 1–2 tablespoons of ground coffee for every 6 ounces of water (55 g coffee per 1000 g water, 1 : 18 ratio), referred to as the *Golden Cup Standard*. However, specific coffee-to-water ratios are recommended for the various brewing methods (Table 12.6). Espresso requires a high coffee-to-water ratio of 1 : 2 due to the short brew time, while the typical drip coffee makers need a lower ratio of 1 : 18.

12.2.4 Health Benefits

In addition to providing a jolt of alertness from caffeine and a plethora of delicious aroma compounds, your morning cup of coffee is also overflowing with potential health benefits. Increasing scientific evidence shows that moderate coffee consumption (2–5 cups/day) is linked to a reduced risk of developing several chronic

Table 12.6 Summary of brewing methods.

Brewing type	Brewing method	Description	Coffee-to-water ratio	Brew (extraction) time (min)
Immersion Brewing	Turkish	Water, coffee grounds, and sometimes sugar are gently boiled to produce a sweet, thick coffee. The coffee grounds are not removed from the brewed coffee.	1 : 10	2–2.5
	French Press	Coarse grounds are immersed in hot water, then a fine mesh slowly segregates the grounds.	1 : 14	4–6
Infusion/drip brewing	Filtered/drip	Gravity pulls hot water through coffee grounds, and the brew passes through a filter. The filter holds coffee oils back.	1 : 18	4–6
	Pour over	Hot water is poured in a spiral pattern over coffee grounds held in a conical filter. Brewed coffee is collected in the cup below (manual drip brewing).	1 : 18	3–4
Pressure brewing	Moka	A stove-top coffee maker where boiling water pressurized by steam passes through coffee grounds.	1 : 10	3–4
	Espresso	Pressurized water is forced through a "cake" of coffee grounds. The espresso shot contains an oily foam topper known as "crema." The pressure forces oils and CO_2(g) out of the coffee grounds creating a colloid. The crema binds nonpolar aroma compounds via hydrophobic interactions.	1 : 2 to 1 : 3	0.5
	AeroPress	A manual brewing system that incorporates immersion and pressurized techniques. Coffee grounds are submerged in a chamber for a short brew time. As the plunger is pushed through the chamber, the coffee passes through a microfilter.	1 : 10 to 1 : 15	2–3

diseases: cancer, cardiovascular disease, liver disease, and type 2 diabetes [10]. Why does coffee help prevent these chronic diseases? Coffee is a rich source of bioactive compounds, such as polyphenols (chlorogenic acid and quinic acid) and diterpenes (cafestol and kahweol). These bioactive compounds serve as antioxidants in the human body, which help reduce inflammation and protect cells from detrimental oxidative stress from free radicals.

Coffee consumption has also been linked to reduced risk of neurodegenerative disorders like Parkinson's and Alzheimer's disease. Interestingly, Parkinson's disease is caused by low dopamine levels. Caffeine binds to (and blocks) adenosine receptors in the brain. This action is known to increase dopamine release; therefore, higher levels of caffeine consumption are connected to a lower risk of developing Parkinson's disease. Antioxidants, such as chlorogenic acid, help remove harmful free radicals in the body and reduce oxidative stress, which has also been linked to the development of Parkinson's disease. Coffee's combination of caffeine and antioxidants is the primary means of reducing the risk of Parkinson's disease. The link between coffee consumption and reduced risk of Alzheimer's disease is not fully understood. Alzheimer's disease is depicted by the brain's buildup of amyloid-beta and tau proteins. Research has suggested that caffeine and other bioactive compounds in coffee hinder the production of the protein plaques in the brain. Chlorogenic acid contains neuroprotective properties that protect brain cells from damage. Caffeine's cognitive-boosting effects may also protect against cognitive decline from dementia and Alzheimer's disease.

Coffee's health benefits weren't always touted. In 1991, the World Health Organization (WHO) classified coffee as a "possibly carcinogenic" substance. Specifically, a limited number of studies showed a causal link between coffee consumption and bladder cancer. However, subsequent research repeatedly showed a decreased risk of cancer once studies adequately accounted for confounding variables like smoking history and sedentary lifestyles. Due to the insufficient evidence that coffee consumption is associated with an increased risk of cancer, the WHO officially removed coffee from its list of possible carcinogens in 2016, and coffee lovers rejoiced!

Box 12.1 World's most expensive coffee – Kopi Luwak

Kopi Luwak coffee, also known as civet coffee, is the most expensive coffee in the world, fetching up to US$600 per pound (US$1,300 per kg). Harvested primarily in Indonesia, the steep price tag of Kopi Luwak reigns from the unorthodox production method—coffee cherries are digested and excreted by the Indonesian palm civet, a small cat-like mammal. The civet is known to eat only the highest-quality coffee cherries; thus, the coffee beans extracted from the stool are considered superior quality. Once consumed by the palm civet, the coffee cherry is digested, but the coffee bean (seed) is not. The coffee beans are believed to undergo fermentation during digestion as they pass through the civet's digestive tract. This fermentation gives civet coffee its claim to having enhanced flavor with a smoother, less bitter taste profile. After roughly 24 hours, the civet defecates the fermented coffee beans. The coffee farmers collect the feces, and the coffee beans are extracted from the stool. The extracted coffee beans are washed, dried, and the outer layer of the bean is removed before ultimately being roasted (Figure 12.14). Many people wonder if Kopi Luwak coffee is safe to drink, considering the beans are extracted from civet stools. Remember that coffee beans are roasted for 12–15 minutes to temperatures between 350°F and 450°F, so any bacteria on the recovered beans will be killed.

The limited supply and high demand for the sought-after fermentation-enhanced coffee have led to the controversial harvesting of Kopi Luwak in less-than-ideal conditions. The production method has resulted in plantations farm-raising palm civets in appalling conditions, similar to factory farming. If you try this coffee delicacy, ensure your Kopi Luwak has been ethically sourced.

Figure 12.14 Kopi Luwak coffee is collected from the feces of the Asian palm civet. The coffee beans allegedly have enhanced flavors due to enzymatic digestion in the civet's digestive tract. pooretat/Adobe Stock Photos

12.3 Tea

After water, tea is the second most popular beverage consumed on the planet. Worldwide, an estimated 1.5 trillion cups of tea are consumed every year. But tea's impact extends far beyond being a popular beverage. Tea has shaped trade routes, drove colonialism and imperialism, and started wars. Still, it has also been used for its medicinal properties and created social rituals through distinct tea cultures across the globe. Cultural tea rituals include the Chinese and Japanese tea ceremonies, British afternoon tea, and Indian chai, to name a few.

So, how did tea become one of the world's most influential beverages? The discovery of tea and its medicinal properties started with the Legend of Shennong in 2737 BCE. In one version of the legend, a tea leaf blew into Shennong's pot of boiling water, and he discovered the unique taste of tea-infused water. In a second version of the legend, Shennong found tea after testing hundreds of herbs for their medicinal properties. While many herbs poisoned him, tea leaves cured him of his poisoning. Archaeological data suggests tea has been cultivated in China's Yunnan and Sichuan provinces for over 6,000 years. Then, tea leaves were likely consumed as a vegetable or a bitter medicinal beverage, known for easing stomach aches—a stark contrast to how tea is consumed today. Over time, tea leaves transitioned into an enjoyable drink where steam-dried tea leaves were ground into a fine powder and heated with water—Matcha. The consumption of matcha became popular with Buddhist monks, as tea helped them stay alert during long meditations. As Buddhism spread across China, so did tea consumption. By the time of the Tang Dynasty (seventh to tenth century), tea was a popular drink for commoners, and teahouses spanned throughout China and became a part of everyday life. Before the Tang Dynasty, tea consumption was limited to Buddhist monks and aristocrats. Tea spread to Japan in the eighth century when Buddhist monk Saisho brought tea leaves back from China for religious ceremonies. However, the Japanese tea culture wouldn't become popular until the twelfth century.

Tea wouldn't be imported to the West until the Age of Exploration when Dutch traders brought tea to Europe in the early 1600s. Eventually, tea was introduced to England in 1645. However, coffee-drinking Brits did not favor the bitter beverage. Tea remained unpopular in England until 1662 when the Portuguese tea-lover Catherine of Braganza married Charles II. Catherine's love for tea spread amongst the royals. Eventually, the palate for tea trickled down the socioeconomic ladder to England's middle and working classes. The Brits' passion for tea spread further west to the American colonies, but the colonies would soon boycott tea. In 1773, the British Parliament passed the Tea Act, which permitted the British East India Company a monopoly on selling tea in the American colonies. The Tea Act came after a series of taxes were imposed on the American colonies. It was quickly seen as unfair British control and further taxation without representation. To protest the Tea Act, a group of colonists boarded the British East India Company's ships and threw 342 chests (90,000 lbs.) of tea into Boston Harbor on December 17, 1773. The Boston Tea Party escalated rising tensions between the American colonists and the British government, eventually leading to the American Revolutionary War in 1775. After the Boston Tea Party, tea became a symbol of British tyranny and was quickly refused by the American colonists. To this day, tea is not as popular in the United States of America as it is in other countries, being only the sixth most popular drink after water, coffee, soda, and alcohol.

The British East India Company found a need to increase the import of tea to England in the 1700s due to high demand from British commoners. However, China held all the power in the tea export and demanded silver in exchange for tea. To counter the trade imbalance, the British East India Company started to export opium to China in trade for tea. An opioid crisis in China led to the Opium Wars in 1839 and 1856, which resulted in more Chinese ports opening for tea trade with the West. While more trading ports helped supply the growing demand for tea in England, China was still the sole exporter of tea. The British East India Company planned a tea heist to counter the trade imbalance. At the center of the Great British tea heist was Robert Fortune, a Scottish botanist commissioned by the British East India Company in 1848 to steal tea plants and secrets of the horticultural and manufacturing processes. Robert Fortune disguised himself as a Chinese plantation worker and, for years, learned how to cultivate, manufacture, and process tea. In 1851, he stole thousands of tea plants and convinced several tea masters to travel with him to Assam and Darjeeling, India, where they started British tea plantations. The new supply allowed tea to become more readily available across the world.

12.3.1 Tea: *Camellia sinensis*

All tea varieties are derived from *Camellia sinensis,* a small, perennial shrub or evergreen tree that grows in tropical to subtropical, high-precipitation environments from 16°S to 30°N of the equator. *Camellia sinensis* grows best in high altitudes (4,000 ft. 1,500 m) and mountainous terrain (Figure 12.15). The unique growing conditions limit tea production. Tea plants are indigenous to China, India, and Thailand, and their origin is traced back to the Yunnan and Sichuan provinces of China. As of 2021, global tea production is highest in China (40%), India (21%), Kenya (7%), and Sri Lanka (6%), with smaller contributions from Turkey, Indonesia, Vietnam, Japan, Iran, and Argentina. The tea leaves and buds must be hand-picked. Tea plants sprout tender leaf buds in a series of stages throughout the year, resulting in the tea being harvested between 3 and 4 times a year.

Like coffee, there are two primary varieties of tea: *Camellia sinensis var. sinensis* and *Camellia sinensis var. assamica.* The *sinensis* variety is native to China, with smaller leaves, and thrives in higher altitudes. In contrast, the *assamica* variety is native to India, with large leaves, and tolerates warmer, humid environments. Both tea varieties produce six different types of tea: white, green, yellow, oolong, black, and dark tea, but *assamica* is preferred in black tea production. Under these two varieties are hundreds of subvarieties known as cultivars. These cultivars, or cultivated varieties, are developed by tea farmers through selective breeding of tea plants with unique properties, such as high polyphenol concentrations, ample aroma compounds, or the ability to thrive in specific growing conditions.

Tea is consumed primarily for its flavor, therapeutic, and energizing properties resulting from distinctive bioactive compounds: polyphenols, caffeine, and L-theanine. These bioactive compounds allow tea plants to grow in warm, humid, mountainous regions. Remember, caffeine is an insecticide, a beneficial compound to have in the tropics! Tea plants use polyphenols to provide UV protection and act as antimicrobial

Figure 12.15   Tea plantation. A cup of tea begins with handpicked leaves of *Camellia sinensis*. The leaves are dehydrated and undergo specific processing steps to produce either white, green, yellow, oolong, black, or dark tea. Pavel Timofeev/ Adobe Stock Photos

and insecticidal compounds. These polyphenols undergo chemical reactions during tea leaf processing, generating the exclusive flavors of tea. Tea plants synthesize L-theanine in their roots, a non-proteinaceous amino acid used to transport and store nitrogen within the plant. For tea drinkers, the presence of L-theanine is critical. L-theanine is a taste molecule and gives tea its umami flavor profile. The concentrations of these bioactive compounds are greatly affected by growing conditions such as the microclimate, soil type, altitude, and the amount of shading the tea plant endures. Tea cultivators have determined that shading and excess fertilizer usage reduce the production of polyphenols. Meanwhile, shading increases the synthesis of L-theanine and caffeine in tea leaves. Deliberately shading tea plants is an easy way for tea cultivators to alter the concentrations of bioactive compounds and flavor in tea leaves.

12.3.2   Tea Leaf Processing

Historically, tea leaves were boiled or consumed raw, but fresh tea leaves are bitter and astringent! For the past 1200 years, tea leaves have been processed to a certain degree to make tea leaves more palatable by decreasing the water content, concentrating enzymes or deactivating enzymes, altering the concentration of taste molecules, and producing aroma compounds. Processing involves carefully timed events of withering (wilting), rolling, bruising, fixing, fermenting, and drying tea leaves (Figure 12.16). Traditional processing methods can be done by hand, without electricity, and without adding other compounds to the harvested tea leaves. The chemistry in the tea leaves during processing sets tea apart from one another by playing a pivotal role in the development of flavor compounds and color. The difference between the six major tea types—green, yellow, white, oolong, black, and dark (pu'er)—comes down to which processing steps are used and the specific order of those processing steps.

One of the first processing steps is typically withering, also known as wilting, a method used to improve the flavor quality of the tea. Freshly picked tea leaves lie flat in a well-ventilated area between 5 and 48 hours to allow the leaves to wilt. During this time, the moisture level drops from 80% to as low as 30%, and natural enzymes within the leaves are concentrated. Postharvest, the metabolic processes inside the leaves are still active, and the leaves quickly respond to the human-induced drought stress. One way in which tea leaves react to the

Figure 12.16 Tea processing. Tea leaf processing is the series of methods to transform freshly harvested tea leaves of *Camellia sinensis* into green, yellow, white, oolong, black, and dark tea types. vladayoung/Adobe Stock Photos

environmental stress of withering is through the production of terpenoids. This class of organic compounds helps alleviate drought stress by reducing water loss from plant tissues by creating waxy barriers on leaves. Terpenoids also scavenge harmful reactive oxygen species (ROS) that are formed by the tea leaf during drought stress events. Tea drinkers benefit from the production of terpenoids, as these are also aroma compounds. Withering has been found to increase the floral aroma profile by creating terpenoids such as geraniol, linalool, and linalool oxide (Figure 12.17). The drought stress response has also been shown to increase protease and amylase activity, producing free amino acids, some of which serve as taste molecules, and hydrolyzing starch, a complex carbohydrate, into maltose and glucose to increase the sweetness of the tea leaves. As the tea leaves dehydrate, the cell walls begin to break down. Cell wall destruction increases aroma compounds when hydrolyzing enzymes free glycosidically bound volatiles from their sugar moieties (Figure 12.17). The release of these volatile aroma compounds is believed to be a form of plant defense after wounding and may serve as a means of plant-plant communication. Arguably, the most crucial chemical transformation during withering is the release of phenols, particularly catechin—a bitter, astringent taste molecule, from the vacuole of the tea leaf as the cell wall disintegrates. This allows the mixing of phenols with polyphenol oxidase (PPO), triggering the enzymatic fruit browning reaction. This reaction decreases the bitterness and astringency of the tea as catechins are reacted away (see Section 12.3.3). Withering is an indispensable yet simple process that makes tea leaves palatable as bitterness and astringency reduce and sweet floral flavors develop in the wilted leaves.

Certain teas undergo bruising, where tea leaves are tossed for 1–2 minutes and then allowed to sit undisturbed for an hour. This process physically wounds the leaves, enabling further cell damage and oxygen to enter the plant cells. Tea leaves undergo multiple rounds of stress-induced wounding during bruising. PPO enzymes cause catechin oxidation during this time via the enzymatic fruit browning reaction. As time progresses, green tea leaves transition to red-pigmented leaves, indicating the presence of catechin oxidation products (see Section 12.3.3).

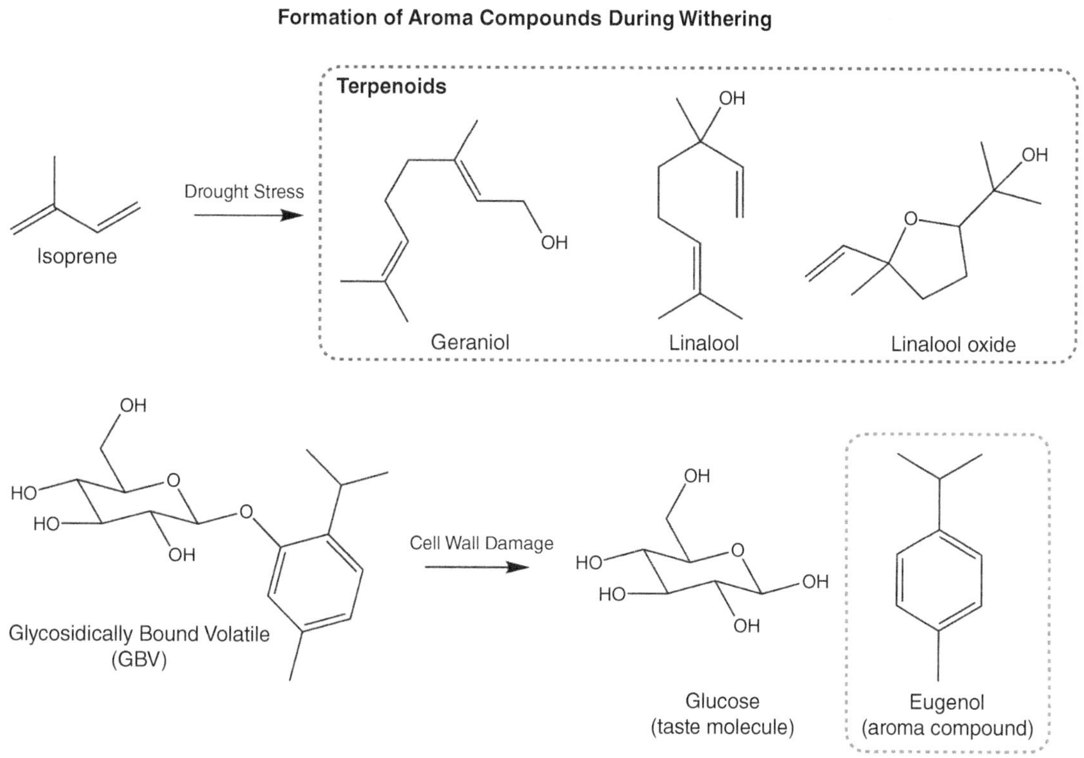

Formation of Aroma Compounds During Withering

Figure 12.17 Chemical changes during withering. During withering, drought stress results in many chemical changes inside the tea leaf, which change the flavor profile. Aroma compounds are formed when plants synthesize terpenoids from isoprene. In another pathway, volatiles (aroma compounds) are liberated from glycosidically bound volatiles when cell walls are destroyed during dehydration.

Similar to bruising, rolling is a process where tea leaves are rolled by hand or mechanically pressed to tear open cell walls. This process is more forceful than bruising. Depending on the timing of this stage, rolling may enhance the enzymatic oxidation of catechins by further releasing PPO enzymes and entering oxygen into the cells. Rolling also plays a pivotal step in helping break down cellulose within the leaf; this allows the release of flavor compounds during the preparation of a cup of tea.

Most teas undergo fixing, a high-temperature heating process to deactivate PPO enzymes, halting or drastically slowing the oxidation of catechins. Two different methods are used to fix tea: (i) dry heat of pan firing—popular in China and responsible for producing a toasty, nutty aroma profile and (ii) wet heat of *steaming*—popular in Japan and produces a grassy, herbaceous aroma profile. Different teas require different fixing temperatures. Green and oolong teas are made when tea is fixed at 428°F/220°C, which destroys all PPO enzymes. Meanwhile, yellow and dark teas are fixed at 392°F/200°C, denaturing most PPO enzymes.

Some teas undergo a process called fermentation (oxidation). During this time, black tea leaves are stored in high-humidity environments and exposed to oxygen to encourage enzymatic PPO activity. *Black tea* fermentation typically lasts for 4 hours. This causes changes in the chemical makeup, specifically the formation of aromas and pigments. In yellow tea production, tea leaves are heaped in piles for 6–8 hours in a process called yellowing. Like fermentation, access to oxygen allows catechins to oxidize (nonenzymatic catechin oxidation) spontaneously, and microbes from the air aid in the decomposition of carbohydrates and lipids in the tea leaves, creating a unique flavor profile. Post-fermentation of dark tea leaves is a prolonged form of fermentation, where leaves sit for

48 hours up to 2 months to allow for nonenzymatic oxidation of catechins, enzymatic oxidation of catechins, and production of flavor and aroma compounds from microbial activities. These leaves may undergo long-term storage, where microbial activity is enhanced.

The final manufacturing step for all tea types is drying (roasting). To promote a long shelf life of tea, the water content needs to be reduced from ~70% to 2–5% (by mass). Water content < 5% will prevent microbial growth in the packaged tea and increase the shelf life. Drying also promotes the Maillard reaction, where dry heat allows free amino acids like L-theanine to react with reducing sugars like glucose and produce savory aroma compounds (see Section 12.3.3).

Tea varieties are formed when *Camellia sinensis* is processed postharvest. The specific processing steps and processing order determine the developed tea variety, as flavor-enhancing enzymes are activated or deactivated in the tea leaves (Figure 12.18).

1. **Green tea**: Mature leaves undergo fixation to deactivate all PPO enzymes. The minimally processed tea preserves the herbaceous, grassy aroma and green pigmentation of green tea. Green tea undergoes the least amount of catechin oxidation, resulting in a bitter, astringent tea. Green tea is the second most popular tea type, accounting for 15% of global consumption.
2. **Yellow tea**: A rare but increasingly popular tea that relies on fixation to deactivate most PPO enzymes. The deactivation of PPO preserves the grassy smell of the tea. After rolling, the yellow tea undergoes enhanced fermentation, called yellowing, to develop a more complex flavor than green tea.

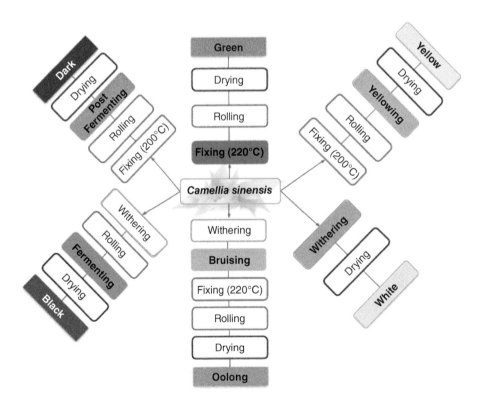

Figure 12.18   The order of manufacturing processes for the six main types of tea. The processing times for each process vary depending on the tea type. The most distinguished processing step in the six tea types is colored (i.e. fermenting for black). watkung/Adobe stock

3. **White tea**: Young tea leaves undergo extended withering for 48 hours. Through enhanced withering, chemical changes reduce astringency and build a fruity, floral tea with enhanced umami and sweetness. Like green tea, white tea is minimally processed.

4. **Oolong tea**: After withering, tea leaves undergo bruising, which increases the release of PPO enzymes and enhances the oxidization of polyphenols. The leaves then undergo fixation once the desired flavor profile is reached, leaving a partially oxidized tea leaf with reduced astringency.

5. **Black tea**: The lack of fixation allows for substantial oxidation of polyphenols in black tea. Black tea undergoes significant rolling to release PPO enzymes, allowing oxygen into the cells. Rolling encourages mixing the enzyme (PPO) and substrate (catechins). Fermentation allows for prolonged oxidation, resulting in fully oxidized tea with robust flavor and red pigmentation. Black tea is the most popular tea type, accounting for 75% of global consumption.

6. **Dark tea**: Dark teas undergo fixation, deactivating most PPO enzymes. However, prolonged fermentation with long-term storage, known as post-fermentation, significantly reduces the astringency of the tea leaves and develops a rich, complex flavor profile over time and brown pigmentation. Dark teas undergo the most oxidation of all the tea types.

12.3.3 Tea Leaf Chemistry

Tea leaves contain several bioactive compounds that undergo chemical changes during tea leaf processing. The chemistry of tea processing plays a vital role in developing flavors, aromas, and colors of each of the six tea types. Two compounds in particular, polyphenols and L-theanine, are used to judge tea quality due to their close association with tea flavor and aroma.

Tea leaves are rich in hundreds of polyphenols, accounting for 20–30% of the dry weight of tea leaves. Tea accounts for the highest concentration of polyphenols, powerful antioxidants, in the human diet. Tea is rich in one class of polyphenols known as catechins. Four catechins in particular are most concentrated: epigallocatechin gallate (EGCG) at ~10% of the dry leaf weight, epicatechin gallate (ECG) at 2.8%, epigallocatechin (EGC) at 1.7%, and finally epicatechin (EC) at 0.8% [11]. Catechins are easily oxidized into several products when exposed to oxygen and the PPO enzyme via the fruit browning reaction. These bitter, astringent catechins give green tea its characteristic taste yet are nearly absent in black and dark teas (Table 12.7). The varying concentrations of catechins are explained by categorizing tea types into unfermented (green and white), partially fermented (yellow and oolong), and fermented (black and dark). In tea processing, fermentation is used

Table 12.7   Average chemical profiles of common bioactive compounds found in processed tea leaves. Compared to unfermented teas, fermented teas like black and dark teas show a distinguished catechin reduction and increased gallic acid during processing. Unfermented white tea has lower levels of catechins than green tea due to the use of young tea leaves which naturally possess lower levels of catechins than mature leaves used for green tea. Adapted from [12].

	Green	White	Yellow	Oolong	Black	Dark
	Unfermented	Unfermented (young leaves)	Partially fermented	Partially fermented	Fermented	Fermented
Catechins (Total) (mg)	113	55	119	75	7	6
Caffeine (mg)	35	27	33	20	29	32
Theobromine (mg)	1.3	0.5	0.9	0.4	1.1	1.4
Gallic acid (mg)	2.0	2.3	1.6	0.7	4.4	3.1

interchangeably with oxidation (fruit browning reaction) and refers to the chemical changes that occur when tea leaves are exposed to oxygen. In tea processing, fermentation does not involve the action of microorganisms such as bacteria and yeast—a point of confusion. Green tea, a type of unfermented tea, undergoes a high-heat process called fixation; this step denatures PPO enzymes, preventing fermentation in green tea. Without fermentation, green tea retains high levels of catechins.

Meanwhile, fermented teas undergo processes such as rolling and bruising, allowing for cell degradation that releases catechins from the vacuole and oxygen to enter the damaged cells. The mixing of the catechins and PPO enzymes causes catechins to polymerize, forming colored polymers of theaflavins (golden-yellow), thearubigins (red), theabrownins (brown) (Figure 12.19), reducing the catechin level to nearly zero (Table 12.7). This causes a substantial difference in the flavor profile of the teas. The high catechin concentrations of green tea are the cause of its bitter, astringent flavor, which has vegetal aromas and a notably green color. Black and dark teas are known for high levels of theaflavins and thearubigins. Theaflavins and thearubigins are less astringent and contribute to fermented teas' full-bodied flavors and added complexity. Partially fermented teas are partially oxidized before the process is stopped by fixation, leaving moderate levels of unreacted catechins in the tea leaves.

Camellia sinensis is the only species of plant to produce L-theanine, a nonprotein amino acid responsible for upwards of 1.5% of the dry weight of tea leaves. L-theanine is a tastant molecule that provides umami and sweet notes that help balance the taste of tea's bitter, astringent catechins. L-theanine participates in the Maillard reaction when exposed to high temperatures above 150°C and reducing sugars, producing a series of aroma compounds with roasted and caramel flavors (Figure 12.19). The Maillard reaction occurs primarily during the drying step of tea processing; however, due to differences in drying times and temperatures, the extent of the Maillard reaction is quite different between tea types. The lower temperatures in drying limit the Maillard reaction in green and white teas, while higher temperatures enhance the Maillard reaction in black, dark, and oolong teas. Therefore, L-theanine concentrations remain elevated in unfermented teas due to the limited Maillard reaction.

12.3.4   Brewing the Perfect Cup of Tea

Like brewing coffee, tea infusion is the process of extracting flavors and aroma compounds from tea leaves. It is heavily influenced by the size of the tea leaves, water temperature, extraction time, and the leaf-to-water ratio. In brewing the perfect cup of tea, one must focus on obtaining good proportions of the important bioactive compounds—catechins, caffeine, and L-theanine, which are known to impact taste. Due to their high surface area to volume ratio, small tea leaves and high water temperatures will produce astringent, harsh cups of tea. Meanwhile, large tea leaves and low temperatures will underextract the bioactive compounds, making an underwhelming cup. It's important to note that inexpensive tea bags, commonly found in grocery stores, frequently contain small, fragmented tea leaves that become easily overextrated in as little as a minute.

One of the most critical factors in brewing tea is the infusion time—how long the tea leaves are steeped in hot water. For hot teas, infusion times can vary from as little as 1 minute to over 15 minutes, depending on the tea type and quality. Due to the water solubility of the bioactive compounds, L-theanine is first extracted from the tea leaves, followed by caffeine and catechins. The taste profiles of these three bioactive compounds are quite different: catechins—astringent, caffeine—bitter, L-theanine—umami and sweet. Thus, the infusion time is critical in producing a balanced cup of tea. A 3–4-minute infusion time generally makes a well-rounded cup of tea with balanced proportions of L-theanine, caffeine, and catechins. Longer extraction times, 10–15 minutes, will be dominated by astringent catechins.

Second to infusion time is the importance of water temperature during brewing. As the water temperature increases, the solubility of caffeine and catechins increases. Therefore, caffeine and catechins are best extracted at 100°C (212°F), producing an astringent, bitter cup of tea. At the same time, 80°C (176°F) optimizes the extraction of L-theanine, contributing to the savory-sweet taste profile and smooth mouthfeel. For the beginner tea enthusiast, a general rule of thumb is to infuse at 90°C (194°F) for a balanced cup of tea. However, due to the vast difference in catechin concentrations in different tea types (Table 12.7), teas with high catechin levels, like green,

Bioactive Compounds of Tea and Changes During Processing

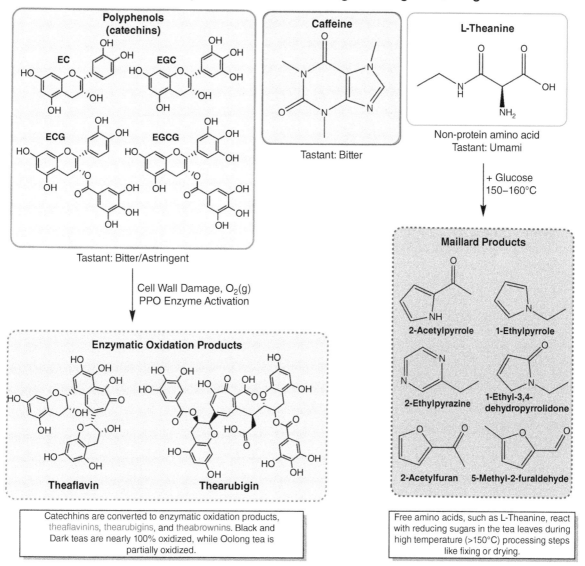

Figure 12.19 Bioactive compounds of tea leaves and their changes during processing. Tea leaves contain multiple bioactive compounds, some of which undergo chemical reactions during tea leaf processing. Catechins (epicatechin [EC], epigallocatechin [EGC], epicatechin gallate [ECG], and epigallocatechin gallate [EGCG]) react with the PPO enzyme when cells are injured and oxidize to theaflavins, thearubigins, and theabrownins. Processing techniques of rolling, bruising, and fermentation promote these reactions in black, dark, and oolong teas. Free amino acids like L-theanine react with reducing sugars at high-temperature processing steps like fixing and drying to generate Maillard products. While all tea types produce Maillard aroma compounds, only certain tea types oxidize catechins to colorful theaflavins, thearubigins, and theabrownins.

are best extracted at 70–85°C (158–185°F). In contrast, low catechin teas, like black and dark, benefit from higher temperature extractions at 90–100°C (194–212°F) (Table 12.8). When brewing tea, a temperature-controlled tea kettle gives you the most accurate water temperature control. It's important to note that individual preferences vary drastically. Experimenting with brewing times and temperatures will help you find the perfect catechin to L-theanine ratio that best suits your taste pallet.

Table 12.8   Recommended water temperature and infusion times for balanced tea.

Tea type	Leaf-to-water ratio	Water temperature	Infusion time (min)
Green	1 tsp/8 oz	70–85°C (158–185°F)	1–2
White	2 tsp/8 oz	80–85°C (176–185°F)	2–3
Yellow	1 tsp/8 oz	85–95°C (185–203°F)	2–3
Oolong	1.5 tsp/8 oz	85–95°C (185–203°F)	2–4
Black	1 tsp/8 oz	90–100°C (194–212°F)	>3–5
Dark	1 tsp/8 oz	90–100°C (194–212°F)	>3–5

12.3.5   Health Benefits

Tea's health benefits result from its unique bioactive compounds, such as polyphenols like catechins, in addition to L-theanine and caffeine. These compounds give tea its distinct health-boosting properties. The bioavailability of tea's bioactive compounds varies. While caffeine and L-theanine are highly bioavailable and readily enter the bloodstream, catechins have lower bioavailability due to their metabolism by gut microbiota. However, the metabolites, or fragments, of catechins formed after microbial breakdown are more bioavailable and can have significant health impacts. After accounting for the bioavailability of catechin metabolites, tea catechins are at least 10 times more bioavailable than previously thought. This new understanding highlights the potential for tea catechins to significantly impact human health more than previously realized, reinforcing the benefits of regular tea consumption. Here's a closer look at how these compounds can enhance your health.

Antioxidant properties: Tea is abundant in polyphenols, particularly catechins in green tea and theaflavins and thearubigins in black tea. These polyphenols are potent antioxidants that help neutralize free radicals in our body that can damage our DNA, thereby reducing oxidative stress and inflammation. This antioxidant action can mitigate cellular damage and lower the risk of chronic diseases related to oxidative stress.

Cardiovascular health: Regular tea consumption has been linked to several cardiovascular benefits. Tea polyphenols can positively impact heart health by improving cholesterol levels, lowering blood pressure, and enhancing blood vessel function. Flavonoids are believed to keep the lining of blood vessels smooth and elastic. These effects contribute to a reduced risk of developing cardiovascular diseases, including heart disease and stroke.

Weight management: Green tea catechins, particularly EGCG, have been shown to support weight management. These compounds can enhance metabolism and promote fat oxidation, which may aid in weight loss. Additionally, tea polyphenols can inhibit digestive enzymes like lipase and alpha-amylase by binding into the active sites of these enzymes. This reduces the breakdown of fats and carbohydrates, which may help manage obesity and type 2 diabetes. Although the impact on body weight can vary among individuals, regular consumption of green tea can be a beneficial part of maintaining a healthy weight.

Gut health: Tea pigments and polyphenols have prebiotic effects that support a healthy gut microbiome. These compounds promote the growth of beneficial gut microorganisms, which are crucial for digestive health. A healthy gut microbiome is associated with improved digestion, immune function, and mental health.

Brain health: Tea's bioactive compounds, including catechins and L-theanine, play a significant role in brain health. L-theanine, an amino acid found exclusively in tea, has calming effects that can improve cognitive function and mood. It acts on the brain by binding to AMPA receptors, increasing the release of inhibitory neurotransmitters like glycine, which ultimately causes the release of dopamine. Additionally, tea catechins and their metabolites can cross the blood-brain barrier, offering neuroprotective benefits, such as reducing oxidative stress and promoting neurogenesis, which may help manage neurodegenerative diseases.

A large prospective study in the United Kingdom, involving half a million tea drinkers, found that consuming two or more cups of black tea daily was associated with a 9–13% lower risk of death from any cause, including cardiovascular diseases and stroke, compared to non-tea drinkers [13]. The study suggests that higher tea intake may offer mortality benefits, regardless of how the tea is consumed or individual genetic differences amongst tea drinkers. Beyond these mortality benefits, tea provides a broad range of health advantages, from antioxidant protection and cardiovascular support to weight management, gut health, and improved brain function. Incorporating tea into your daily routine not only adds flavor but also contributes to a healthier lifestyle and long-term wellness. Whether for its taste or its extensive health-promoting effects, drinking tea regularly can be a valuable component of a balanced diet.

Box 12.2 Kombucha: fermented black tea

Kombucha is a fermented tea beverage with a rich history that dates back nearly 2,000 years. Originating in China, it spread to Japan, Russia, and Europe before finally gaining popularity in the United States during the 1990s. Kombucha is made by fermenting black or green tea with sugar using a symbiotic culture of bacteria and yeast (SCOBY) (Figure 12.20). Various bioactive compounds, including probiotics, B vitamins, and organic acids, are produced during fermentation.

The fermentation of kombucha involves a complex interaction between yeast and bacteria. The yeast ferments the added sugar in the tea, producing trace amounts of ethanol and carbon dioxide gas. At the same time, the bacteria convert the ethanol into acetic acid, giving kombucha its distinctive tangy flavor. This process also generates a range of organic acids, such as glucuronic acid and D-saccharic acid, which contribute to kombucha's potential health benefits and flavor profile. The kombucha fermentation occurs at room temperature for 5–10 days before bottling.

Figure 12.20   Kombucha SCOBY. "Symbiotic Culture of Bacteria and Yeast" is a critical component in the fermentation process of kombucha. The SCOBY is a thick, rubbery biofilm that contains a community of microorganisms that work together to ferment the tea and produce kombucha. The SCOBY usually floats to the top of a brew, forming a seal to prevent contaminants from entering the liquid. SCOBY can be reused after fermentation is complete. flyalone/Adobe Stock Photos

(Continued)

Box 12.2 (Continued)

In addition to its high levels of antioxidants, kombucha is rich in probiotics, beneficial bacteria that can support gut health. These probiotics, mainly lactic acid bacteria, are believed to aid digestion, strengthen the immune system, and may help manage gastrointestinal issues like irritable bowel syndrome (IBS) and diarrhea. By supporting a balanced gut microbiota, the probiotics in kombucha may contribute to overall digestive health.

While kombucha offers potential health benefits, it poses risks, especially if not prepared under strict hygienic conditions. If the kombucha is contaminated or over-fermented, the fermentation process can lead to the growth of harmful microorganisms or mold. There have been reported cases of illness related to homemade kombucha, including liver problems and lactic acidosis. To mitigate these risks, it is advisable to purchase commercially prepared kombucha, which is subject to safety regulations and ensures the alcohol content stays below 0.5%.

Box 12.3 Energy drinks: concentrated caffeine

Energy drinks are popular beverages consumed for their ability to increase alertness and physical performance. They typically contain high levels of caffeine (75–300 mg per can), sugars, and various additives such as guarana, taurine, and L-carnitine. These legal stimulants enhance alertness, attention, and energy while raising blood pressure, heart rate, and respiration. Many students use these drinks to boost energy, though the stimulants can adversely affect the nervous system. Health dangers associated with energy drinks include increased risk of heart problems, sleep disturbances, and potential addiction, with particularly adverse health effects in children and adolescents.

The use of natural stimulants dates back centuries. Tea, which contains caffeine, was first consumed in China around 2737 BC, and coffee was popularized in the Middle East during the fifteenth century. Indigenous South American cultures have long used yerba mate and guarana for their stimulating properties. The modern energy drink industry began in the 20th century. In 1929, *Lucozade* was developed by British chemist William Walker Hunter as an energy drink for recovering hospitalized patients. In 1949, *Dr. Enuf* was launched in the United States. as one of the first vitamin-fortified energy drinks. In 1962, *Lipovitan*, containing taurine and B vitamins, was introduced in Japan and became popular amongst workers and students.

The 1980s and 1990s marked a period of global expansion for energy drinks. This era saw the introduction of Red Bull in 1987 by Austrian entrepreneur Dietrich Mateschitz, a brand that revolutionized the industry with its aggressive marketing and association with extreme sports. Following Red Bull's success, competitors like Monster Energy emerged in 1995, leading to a diverse market. The 2000s witnessed an explosion of brands such as Rockstar, AMP, NOS, and Full Throttle. This global expansion made energy drinks a significant part of our culture, associated with high-performance, adrenaline-inducing activities and a multi-billion-dollar industry.

Common ingredients in energy drinks include caffeine, taurine, L-carnitine, B vitamins, sugars, sweeteners, and herbal extracts (Figure 12.21). Caffeine ($C_8H_{10}N_4O_2$) stimulates the central nervous system, enhancing alertness and reducing fatigue. Energy drinks often contain caffeine levels comparable to or exceeding those found in coffee and tea. Taurine ($C_2H_7NO_3S$), a sulfur-containing compound found in the brain, heart, retina, and skeletal muscle, supports muscle function and mental performance. It synergizes with caffeine, reducing muscle fatigue and enhancing endurance. L-carnitine ($C_7H_{15}NO_3$), a derivative of the lysine amino acid, promotes the conversion of fat into energy by increasing the transport of long-chain fatty acids to mitochondria. B vitamins, such as B3, B6, and B12, play important roles in energy metabolism and nervous system function. They support energy production, cognitive function, and reducing fatigue. For instance, B12 helps convert

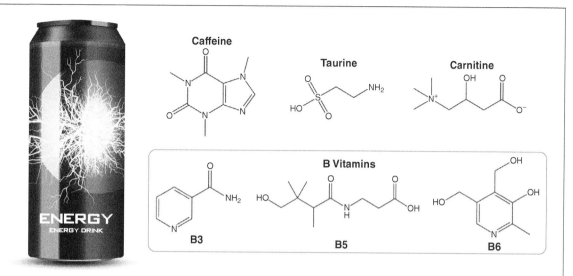

Figure 12.21 Common energy drink ingredients known for their stimulating and energy boosting effects. ZinetroN/ Adobe Stock Photos

food into energy, B6 helps process and store energy from protein and carbohydrates, and B3 aids in releasing energy from food. These vitamins are essential for maintaining overall health and well-being. Energy drinks use simple sugars (sucrose, glucose, and fructose) and artificial sweeteners (aspartame, sucralose, and stevia) to enhance taste and control caloric content.

Herbal extracts like guarana and ginseng are also standard in energy drinks. Guarana, native to the Amazon, is rich in caffeine and other stimulants, providing sustained energy release and synergistic effects with other stimulants. Ginseng, used in traditional Eastern medicine, enhances energy, stamina, cognitive function, and immune support. Since the FDA does not regulate ginseng, it is difficult to know how much is being consumed in energy drinks.

Energy drinks can increase alertness and physical performance due to their stimulant content, but potential side effects include jitters, increased heart rate, and dehydration. Long-term consumption poses significant health risks. These include cardiovascular issues such as elevated blood pressure, irregular heart rhythms, and a higher risk of heart attacks. The high sugar content can lead to weight gain and an increased risk of type 2 diabetes. Frequent consumption can also cause sleep disturbances, anxiety, and gastrointestinal issues. Additionally, combining energy drinks with alcohol can be particularly dangerous. The stimulant effects of the energy drink can mask the depressant effects of alcohol, leading to a higher risk of alcohol poisoning and risky behaviors. This is because the person may feel less intoxicated than they are, leading to excessive alcohol consumption.

Understanding the regulatory guidelines that govern energy drinks is crucial for consumers. Long-term consumption of these beverages can pose health risks, including impacts on heart health, sleep disturbances, and addiction potential. Regulatory bodies like the FDA have set limits on caffeine and other ingredients in energy drinks to ensure their safety. For instance, the FDA limits the amount of caffeine in a single serving of an energy drink to 400 mg and requires that the total caffeine content be listed on the label. It's important to note that the American Academy of Pediatrics recommends children under 12 not consume caffeinated beverages. In contrast, it is recommended that teenagers limit their caffeine intake to 100 mg per day. By being aware of these guidelines, consumers can make informed choices and enjoy energy drinks safely and responsibly.

Science for the Chef: Cold brew coffee versus hot brew coffee (extraction and solubility)

A compelling example to illustrate the underlying scientific principles is comparing cold and hot brew coffee. This offers a clear exploration of how temperature influences the extraction of key compounds such as caffeine, acids, and aromatic molecules, ultimately shaping the beverage's flavor profile. By examining the differences between these two brewing methods, one can gain an enjoyable tasting experience and a deeper understanding of the chemistry behind extraction, solubility, and the role of temperature in developing flavors. This hands-on approach allows for a practical application of chemical principles to everyday activities like brewing coffee, enriching their appreciation of the science in our daily lives.

Cold brew coffee

Ingredients:
- 1 cup coarsely ground coffee beans
- 4 cups cold water

Instructions:
1. Combine the ground coffee and cold water in a large jar or French press.
2. Stir gently to ensure all the coffee grounds are saturated.
3. Cover and let it steep in the refrigerator for 12–24 hours.
4. Strain the coffee through a fine mesh sieve or a coffee filter to remove the grounds.
5. Serve over ice, diluted with water or milk to taste.

Hot brew coffee

Ingredients:
- 1 cup medium-ground coffee beans
- 4 cups hot water (around 200°F or 93°C)

Instructions:
1. Place the ground coffee in a coffee maker or French press.
2. Pour hot water over the coffee grounds, ensuring even saturation.
3. Allow it to brew for 3–5 minutes.
4. Press or strain the coffee to separate the grounds from the liquid.
5. Serve immediately, optionally with milk or sweetener.

Science Behind the Recipe

Temperature and extraction: The pivotal role of temperature in the extraction process is the key distinction between cold and hot brew coffee. This factor significantly influences the solubility and extraction rate of various compounds in coffee, thereby shaping the flavor profile of the beverage.

In cold brew coffee, the lower temperature slows down the extraction process. This results in a beverage that is smoother and less acidic. While compounds like caffeine, oils, and some acids are still extracted, the lower temperature means fewer bitter compounds and acids (like chlorogenic acids) are dissolved. The outcome is a coffee that is typically less bitter and has a mellower flavor profile.

The higher temperature (around 200°F or 93°C) in hot brew coffee accelerates the extraction process. The heat increases the solubility of various compounds, including aromatic oils, caffeine, and acids. This leads to a more complex and robust flavor profile with noticeable acidity and bitterness and the characteristic aromas that are quickly released at higher temperatures.

Coffee's acidity is primarily due to organic acids such as citric, malic, and chlorogenic acids. Hot brewing extracts these acids more efficiently, contributing to its bright and sometimes sharp flavors. With its slower extraction, cold brewing results in lower acidity, making the coffee taste smoother and less astringent.

The volatile aromatic compounds responsible for coffee's rich aroma are more readily released in hot water, which is why the aroma of hot coffee is more intense and immediate. Cold brew coffee, on the other hand, has a subtler aroma because these compounds are less volatile at lower temperatures.

Key Concepts

1. In plants, **caffeine** acts as a chemical toxin (a natural insecticide) to discourage insects from eating the plant leaves and produces a psychoactive response in bees, causing them to more likely return to a plant with caffeine-containing nectar flowers, thus promoting plant proliferation.

2. Caffeine is a **xanthine**. It binds to A_1 and A_2 adenosine receptors due to its structural similarity to adenosine. By blocking adenosine binding, caffeine has a stimulatory effect. Caffeine binding to adenosine receptors also keeps dopamine levels high, enhancing dopamine binding to the dopamine receptor, resulting in an improved or feel-good mood.

3. The **half-life** is the amount of time needed to reduce the amount of a molecule by 50%. The half-life of caffeine in humans ranges from 2 to 12 hours. This variability results in the differing impact of caffeine on individuals.

4. Two species of *Coffea* are utilized for coffee production; roasters often blend beans to balance a coffee's aroma, caffeine content and body, and manage expense. Arabica's aroma is more delicate and mild versus the harsh and robust aromas of Robusta. Robusta contains almost twice as much caffeine as Arabica, resulting in greater bitterness. Higher levels of **chlorogenic acids** (which are converted to bitter chlorogenic lactones and phenylindanes during roasting) also contribute to greater astringency and bitterness in **Robusta** coffee. Higher sucrose in **Arabica** contributes to its sweetness and enhanced **Maillard reactions** during roasting. Arabica beans also contain higher fat content; the fats bind the volatile and aromatic molecules of freshly brewed coffee and contribute to a smooth mouthfeel.

5. A **coffee cherry** consists of the pulp, mucilage, parchment, and bean (or seed). The coffee cherry is processed through **depulping** (which removes the pulp), **fermentation** and **washing** (removing the mucilage), and **drying** and **hulling** (which removes the pulp). The "green bean" that remains is roasted to produce a coffee bean ready for grinding and brewing.

6. The **roast profile**, the temperature profile, and heating rate that take place during coffee bean roasting, significantly impact the compounds produced from **Maillard** reactions (between an amino acid and a sugar), **caramelization** reactions (between two sugars), and **pyrolysis** reactions (thermal decomposition of organic materials) and the flavor profile of the beans. The **first crack** stage of roasting occurs when all of the liquid water in the bean has vaporized. In the **second crack**, the cellulose inside the bean undergoes thermal decomposition. Further roasting leads to the production of medium-dark and dark-roast coffees, which have enhanced flavor and aromatic profiles, reduced acidity, and increased bitterness.

7. Caffeine does not decompose during the roasting process due to its high **melting point** and resistance to **thermal degradation**. However, since the mass of a coffee bean decreases during roasting due to water evaporation, and its volume increases, the caffeine : mass ratio is highest in a dark-roast bean, and the caffeine : volume ratio is lowest in a dark roast bean relative to light-roast beans.

8. Coffee brewing is a form of **solid–liquid extraction**, where the solutes (i.e. caffeine + other coffee molecules) are dissolved in a solvent (i.e. hot water), and are then filtered to yield the final liquid (i.e. coffee). The more hydrophilic or polar the solutes, the more easily they will dissolve in the polar solvent. The higher the temperature of the polar solvent, the greater the solubility of all compounds, both polar and nonpolar.

9. The coffee brewing variables that can impact the flavor and quality of coffee include grind size, freshness of the coffee grounds, water quality, and water temperature. Grind size affects the **extraction rate**, the speed at which the soluble compounds are extracted out of the coffee bean into the water. Coffee beans that have been freshly ground will yield a greater aromatic and flavor profile in the final product. Using filtered water at a temperature just below boiling is ideal for coffee brewing to reduce the negative impact of hard or soft water and to minimize the extraction of unwanted organic molecules. The ideal coffee-to-water ratio depends on the brewing method and time.

10. Coffee is a rich source of bioactive compounds such as **polyphenols** and **diterpenes** that serve as antioxidants. Coffee consumption is also linked to a reduced risk of neurodegenerative disorders due to its ability to bind to adenosine receptors, impacting dopamine levels. Tea has a number of health benefits due to the presence of caffeine, **L-theanine**, and catechins/polyphenols. Polyphenols have antioxidant characteristics, with evidence of improved cholesterol levels, lower blood pressure, and reduced risk of cardiovascular disease. The **catechins** present in green teas support positive weight management and a healthy gut biome. L-theanine and catechin positively impact cognitive function and mood, along with neuroprotection (due to antioxidant characteristics).

11. Tea contains three major **bioactive molecules** that serve flavor and biological properties. In addition to the effects of caffeine, the tea polyphenols provide UV protection and act as antimicrobials and insecticides. L-theanine is a nonprotein amino acid that is used to transport and store nitrogen in plants; it gives the tea its umami or savory taste, along with sweet tones to balance the astringency of catechins (polyphenols). L-theanine reacts with sugars during fixing or drying steps (Maillard reactions) to produce tea flavor molecules.

12. The steps of tea processing include **withering, rolling, bruising, fixing, fermenting, and drying**. During withering, the tea leaves wilt, the plant cell wall begins to break down, and natural plant enzymes are concentrated, resulting in the production of terpenoids, free amino acids, starch hydrolysis products (maltose and glucose), and phenols (i.e. catechin), a bitter, astringent taste molecule. **Enzymatic fruit browning reactions** occur between **polyphenol oxidase (PPO)** and catechin, reducing catechin levels. The rolling and bruising steps, both of which damage plant cell walls, allow for additional **oxidation** reactions (including enzymatic fruit browning). Fixation is a high-temperature heating process that deactivates PPO enzymes. During fermentation or oxidation, tea is stored in high humidity and high oxygen conditions to encourage PPO activity. During drying, water content is reduced to less than 5% and Maillard reactions occur. Production of the six major types of tea varies these steps to yield the distinguishing characteristics found in each tea type. The distinguishing processing steps for each tea types are: **Green** → fixing, **Yellow** → yellowing, **White** → withering, **Oolong** → bruising, **Black** → fermenting, and **Dark** → post-fermentation.

13. Tea naturally has a high content of one class of polyphenols, the **catechins**. The higher the concentration of catechins in the tea, the greater the **bitterness** and **astringency**. Green and white teas have a high concentration of catechins since they are unfermented and, in the case of green tea, utilize fixation to denature all PPO enzymes. Yellow and oolong teas are partially fermented (i.e. partial oxidation, then fixation), while black and dark teas are fully fermented and undergo rolling, where catechins are released from the vacuole and nearly all are converted to theaflavins and thearubigins.

14. **Infusion time**, the length of time that tea leaves are steeped in hot water, is a critical factor in tea brewing. L-theanine is extracted from the tea leaves first, followed by caffeine and the catechins. The longer the extraction time, the more bitter the tea. With a higher water temperature, the less polar caffeine and catechins are more likely extracted into the cup of tea, increasing bitterness.

References

1 Wright, G. A., Baker, D. D., Palmer, M. J., Stabler, D., Mustard, J. A., Power, E. F., Borland, A. M., & Stevenson, P. C. (2013). Caffeine in floral nectar enhances a pollinator's memory of reward. *Science*, *339*(6124), 1202–1204. doi:10.1126/science.1228806.

2 Weinberg, B. A., & Bealer, B. K. (2001). *The world of caffeine: The Science and Culture of the World's Most Popular Drug.* (pp. 3–4). New York: Routledge.

3 Volkow, N. D., Wang, G. J., Logan, J., Alexoff, D., Fowler, J. S., Thanos, P. K., Wong, C., Casado, V., Ferre, S., & Tomasi, D. (2015). Caffeine increases striatal dopamine D2/D3 receptor availability in the human brain. *Transl. Psychiatry*, *5*(4), e549. https://doi.org/10.1038/tp.2015.46.

4 Juliano, L. M., & Griffiths, R. R. (2004). A critical review of caffeine withdrawal: empirical validation of symptoms and signs, incidence, severity, and associated features. *Psychopharmacology*, *176*, 1–29. https://doi.org/10.1007/s00213-004-2000-x.

5 American Psychiatric Association, (2013). *Diagnostic and statistical manual of mental disorders*, 5e. https://doi.org/10.1176/appi.books.9780890425596.

6 Sweeney, M. M., Weaver, D. C., Vincent, K. B., Arria, A. M., & Griffiths, R. R. (2020). Prevalence and correlates of caffeine use disorder symptoms among a United States sample. *J Caffeine Adenosine Res.*, 10(1), 4–11. https://doi.org/10.1089/caff.2019.0020.

7 Daly, J. W., & Garattini, S. (1993). Caffeine, coffee, and health. *Diagnostic and statistical manual of mental disorders*, 5e, 97–150.

8 Yeretzian, C., Jordan, A., & Lindinger, W. (2003). Analysing the headspace of coffee by proton-transfer-reaction mass-spectrometry. *Int. J. Mass Spectrom*, 223, 115–39.

9 Yeager, S. E., Batali, M. E., Guinard, J. X., & Ristenpart, W. D. (2023). Acids in coffee: a review of sensory measurements and meta-analysis of chemical composition. *Crit. Rev. Food Sci. Nutr.*, *63*(8), 1010–1036.

10 van Dam, R. M., Hu, F. B., & Willett, W. C. (2020). Coffee, Caffeine, and Health. *N. Engl. J. Med.*, *383*(4), 369–378.

11 Zhang, L., Ho, C., Santos, J. S. Armstrong, L., & Granato, D. (2019). Chemistry and biological activities of processed *camellia sinensis* teas: a comprehensive review. *Compr. Rev. Food Sci. Food Saf.*, *18*(1), 1474–1495.

12 Yi, T., Zhu, L., Peng, W. L., He, X. C., Chen, H. L., Li, J., & Chen, H. B. (2015). Comparison of ten major constituents in seven types of processed tea using HPLC-DAD-MS followed by principal component and hierarchical cluster analysis. *LWT - Food Sci. & Technol.*, *62*(1), 194–201.

13 Inoue-Choi, M., Ramirez, Y., Cornelis, M. C., Berrington de González, A., Freedman, N. D., & Loftfield, E. (2022). Tea consumption and all-cause and cause-specific mortality in the UK biobank: a prospective cohort study. *Ann. Intern. Med.*, doi: 10.7326/M22-0041.

Additional Readings

Arıkan, M., Mitchell, A.L., Finn, R.D. and Gürel, F. (2020), Microbial composition of Kombucha determined using amplicon sequencing and shotgun metagenomics. *J. Food Sci.*, 85(2), 455–464.

Chindapan, N., Puangngoen, C. and Devahastin, S. (2021), Profiles of volatile compounds and sensory characteristics of Robusta coffee beans roasted by hot air and superheated steam. *Int. J. Food Sci. Technol.*, 56(8), 3814–3825.

Chou, K.-.-H. and Bell, L.N. (2007), Caffeine Content of Prepackaged National-Brand and Private-Label Carbonated Beverages. *J. Food Sci.*, 72(6), C337–C342.

Dai, Y.-H., Wei, J. R., & Chen, X.-Q. (2023). Interactions between tea polyphenols and nutrients in food. *Compr. Rev. Food Sci.Food Saf.*, 00, 1–21. https://doi-org.sandiego.idm.oclc.org/10.1111/1541-4337.13178.

Peixoto, J. A. B., Silva, J. F., Oliveira, M. B. P. P., & Alves, R. C. (2023). Sustainability issues along the coffee chain: From the field to the cup. *Compr. Rev Food Sci. Food Saf.*, 22 (1), 287–332.

End of Chapter 12 Questions

1. What is the primary role of caffeine in plants?
   a. Increase seed germination
   b. Improve nutrient absorption
   c. Act as a natural insecticide
   d. Enhance photosynthesis

2. Which species of Coffea has a more delicate and mild aroma?
   a. Liberica
   b. Robusta
   c. Excelsa
   d. Arabica

3. During which stage of roasting does the cellulose inside the coffee bean undergo thermal decomposition?
   a. First Crack
   b. Second Crack
   c. Caramelization
   d. Pyrolysis

4. Which type of tea is most associated with high levels of catechins?
   a. Black Tea
   b. White Tea
   c. Oolong Tea
   d. Green Tea

5. What process is primarily responsible for the development of the unique flavors in black tea?
   a. Oxidation
   b. Fermentation
   c. Pyrolysis
   d. Caramelization

6. L-theanine in tea is known for its effect on:
   a. Improving cardiovascular health
   b. Enhancing mood and cognitive function
   c. Increasing physical endurance
   d. Promoting cellular regeneration

7. List two health benefits of coffee and explain the compounds responsible for these benefits.

8. How do grind size and water temperature affect the coffee brewing process in terms of flavor?

9. Why is white tea considered to have the highest level of antioxidants among teas?

10. What happens during the fixation step in tea processing, and why is it important?

11. Why are Maillard reactions important in tea processing and what molecules are responsible for these reactions?

12. The concentration of diterpenes, a type of antioxidant, highly depends upon the brewing style of coffee. Unfiltered coffee, like French press, contains the highest levels of diterpenes, while filtered coffee, like a drip coffee maker, contains virtually no diterpenes. Using the chemical structure of Cafestol, a type of diterpene, explain why filter paper removes these antioxidants in filtered coffee.

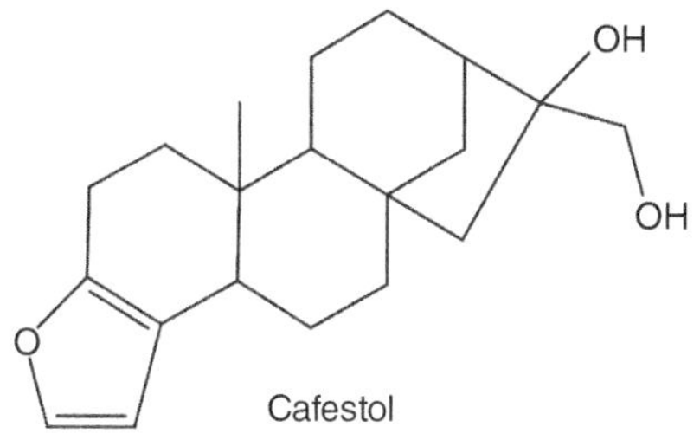

Cafestol

13

Sweets: Chocolates and Candies

Guided Inquiry Activities (Web): *1, Elements, Compounds, and Molecules; 3, Mixtures and States of Matter; 7, Carbohydrates; 12, Emulsion and Emulsifiers; 13, Flavor; 30, Chocolate Properties; 31, Chocolate Tempering; 32 Sugar*

Learning Objectives

1. *Understand the molecular structure of sucrose.*
2. *Compare and contrast the properties and components of white, brown, and powdered sugars.*
3. *Understand the intermolecular interactions between sugars and water and how this contributes to moisture retention and solubility.*
4. *Understand invert sugar and explain its impact on the physical properties in foods and cooking.*
5. *Explain the physical and molecular characteristics of liquid sweeteners and their production.*
6. *Explain the fermentation processes that occur and their importance in the earliest stages of chocolate production.*
7. *Describe molecules that are produced during the roasting of cacao beans or additives that contribute to the flavor, aroma, and distinguishing characteristics of chocolate.*
8. *Explain the purpose of grinding, milling, and conching in the process of chocolate production.*
9. *Understand how an emulsion works and how this impacts chocolate production, taste, and texture.*
10. *Explain the process and importance of chocolate tempering, including the relationship to fat crystallization.*
11. *Describe fat migration and how it changes the properties of chocolate in baked goods.*
12. *Understand the molecular basis behind chocolate seizing.*
13. *Compare and contrast different types of chocolate, based upon their ingredients.*
14. *Understand the correlation between sugar concentration, temperature of cooking, and properties of the resulting non-chocolate candy.*
15. *Describe the scientific principles behind sugar crystallization, including the impact of temperature and stirring on crystal size and the characteristics of the resulting candy.*
16. *Compare and contrast a hard noncrystalline candy, a caramel, and fudge, in terms of preparation, ingredients, and characteristics of the candy.*

The Science of Cooking: Understanding the Biology and Chemistry Behind Food and Cooking, Second Edition. Joseph J. Provost et al.
© 2025 John Wiley & Sons, Inc. Published 2025 by John Wiley & Sons, Inc.
Companion website: www.wiley.com/go/provost/food_science_2e

13.1   Introduction

Few people can walk by a candy shop without being drawn in by the sight of brightly colored rock candies and lollipops and the aromas of chocolate and caramel. And then, once you try a sample of smooth, creamy fudge or a piece of chewy, sweet salt-water taffy, you might yearn for more, want to prepare some at home, or wonder about the history and science behind candy making. For those with a craving for sweet foods, desserts, or other treats, this is the chapter for you! This chapter will cover some of the history and science behind different sweeteners, chocolate, and candy.

13.2   Sugars and Sweeteners

A variety of sweeteners are used in baking and food preparation: white granulated sugar sprinkled on strawberries, molasses in gingersnap cookies, honey on a piece of cornbread, and powdered sugar in a frosting. All of these sweeteners fulfill the primary criterion for a sweetener: they taste sweet. However, if you have ever replaced brown sugar with white sugar in a cookie recipe, you know that this small change makes a big difference in the final product. A cookie that contains white sugar typically has a crisp and crunchy texture, while a brown sugar-based cookie is chewy and flexible. What is the difference in the sugars and sweeteners that we use in cooking? Why might you choose to use one to serve a particular culinary purpose but not another? The answer lies in understanding the structure and chemistry of carbohydrates (Figure 13.1).

13.2.1   Sugars

13.2.1.1   Granulated Sugar

Ah … granulated white sugar. We use it for making a pitcher of cold lemonade in the summer, we sprinkle it on our morning cereal, and we use it in our favorite brownie recipe. When someone refers to "sugar," this is the white crystalline solid that we visualize. Granulated sugar is produced from sugarcane, a member of the tallgrass family

Figure 13.1   Granulated sugar. Table sugar (sucrose) in granular form.

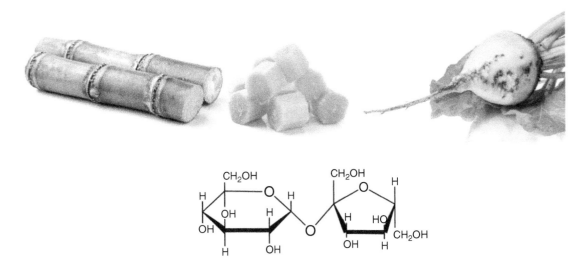

Figure 13.2 Sucrose. Two of the most common sources of sucrose are sugarcane (top left, and top middle) and sugar beet (top right). The molecular structure of sucrose, a disaccharide of glucose and fructose (bottom). sommai/Adobe Stock Photos, ra3rn/Adobe Stock Photos

that grows in tropical regions, and sugar beets, a parsnip-looking vegetable that grows in the more temperate, cooler northern climates (Figure 13.2). About 75% of the world's production of sugar comes from sugarcane. In 2020, the world produced 1.9 billion tons of sugar cane, with Brazil and India as the largest and second largest producers at 760 million tons and 348 million tons, respectively. These two countries will account for about 21% and 18% of the world's total sugar output by 2030 [1]. Sugar beets produce 55–60% of the sugar used in the United States, with the rest produced from sugarcane. (Figure 13.2).

At this point, you already know about the molecular structure of granulated white sugar; it is over 99% pure sucrose, the disaccharide that contains one molecule of glucose linked to a molecule of fructose via an oligosaccharide bond (Figure 13.2). The crystalline structure and molecular components of sucrose give it some really interesting properties in the kitchen. Let's learn more about table sugar.

First, how does granulated sugar become a crystal? The crystalline structure of sugar comes from its production process. To produce sugar, juice that contains sugar is extracted from the cane or beet. Some of the water in the liquid juice is evaporated and impurities removed; the liquid is boiled further to make a supersaturated sugar syrup. Then, as the supersaturated sugar syrup cools (we will discuss this later in this chapter when we look at chocolate), the individual sucrose molecules align themselves to interact with one another via *intermolecular interactions*. The intermolecular forces are noncovalent hydrogen bonds, formed between —OH groups of different sucrose molecules. Through this slow process, the sugar crystal, comprised of many interacting sucrose molecules, develops.

The size of the crystals that form gives us different types of granulated sugars. Typical granulated sugar crystals are 400–600 μm in size, superfine or "caster sugar" crystals are 200–450 μm, and sanding sugar granules that you might use on top of a muffin are 600–800 μm in size. However, even though the crystal size varies, all of these types of sugars are identical in a molecular sense.

13.2.1.2 Brown Sugar

Brown sugar is also a sucrose-based sugar, just like white granulated sugar, but it has a distinctive brown color and texture due to the presence of molasses. Brown sugar contains from 3.5% (light brown sugar) to 6.5% (dark brown sugar) molasses. Its soft, moist texture is due to the hygroscopic nature of molasses. The hygroscopic character means that it attracts water more readily than sucrose causing baked goods to retain moisture and chewiness that is not matched by granulated sugar. Molasses is a thick solution prepared by heating syrup of crushed sugarcane or sugar beets. The liquid is heated to remove water, which thickens the liquid. However, it also promotes Maillard reactions (between proteins and sugars) and caramelization reactions (between sugars) to produce the browned/blackened components of molasses. Crystallized sugars that may result from this process are removed, and the resulting liquid is molasses. The characteristic flavor and aroma (i.e. rich, caramel-like) present in molasses are due to these Maillard and caramelization reactions. Thus, if you replace brown sugar with white sugar in a recipe or vice versa, you will notice a difference. And now you likely know that the myth that brown sugar is healthier than white sugar is not true.

13.2.1.3 Powdered Sugar

The texture and use of powdered sugar are so distinct that you may think that it has a different molecular structure than brown and white sugar. However, powdered sugar (also known as confectioner's sugar) is also sucrose. Powdered sugar is essentially white granulated sugar that has been ground or milled into very small particles that are 10–15 μm in size. Are you surprised? You can try this at home by putting normal sugar into a coffee grinder or pulverizing it with a mortar and pestle. When you try this, your small sugar particles may clump together. Therefore, a small amount of an anti-caking agent, cornstarch, is added during the manufacturing process. The very small particle size of sucrose crystals in powdered sugar results in its wonderful attribute of quickly dissolving into almost anything. Thus, even in the absence of heat, powdered sugar yields a smooth cake frosting or sweetened whipped cream foam lending to its use in sweetening uncooked foods. When heated, the 2–3% cornstarch present in powdered sugar may thicken sauces (which may or may not be helpful in a fruit sauce or syrup). Moreover, it is about twice as expensive by weight as granulated sugar, providing another limitation to its common use.

13.3 Properties of Sucrose-based Sugars and Use in the Kitchen

What are some properties of sucrose that are important in cooking? You have learned that "like" dissolves "like." A molecule of sucrose contains many polar, hydroxyl (—OH) groups, and so does water (Figure 13.3). Thus, when you add a teaspoon of sugar to a cup of water, the sugar crystals dissolve. Why? Each molecule of sucrose disperses away from its sucrose partners within the crystal becoming surrounded by and forming new intermolecular interactions with water because there are so many more molecules of water present than molecules of sucrose. Because of these interactions, sugar easily dissolves in a glass of iced tea, and, as long as you stir it and maintain a 2:1 ratio of water to sugar, it will not sit on the bottom of the glass.

The intermolecular interactions between sucrose and water are not only important for *solubility* but also in the preparation of baked goods and candy. Have you ever thought about how you can bake a pan of brownies at 350°F/177°C and not lose all of the moisture? How do brownies remain soft and chewy when water evaporates at 212°F/100°C? The intermolecular interactions between the sucrose and water help to retain the moisture in brownies and other baked goods. These interactions are so strong that some of the molecules of water remain bound to the sucrose and do not evaporate away, even when baked at this high temperature.

However, you may have also realized that sucrose is not the only thing that is important to the retention of moisture in and appearance of baked goods. Remember the differences between a brown sugar cookie and a white sugar cookie? A brown sugar cookie is soft and chewy but kind of dull looking. A white sugar cookie is dry and crisp, but has a glossy or shiny appearance. There must be something else about these two sugars that explains why you use white sugar in a brownie and brown sugar in a chocolate chip cookie. The "something else" is something called an invert sugar. And, no, there is not an extravert sugar.

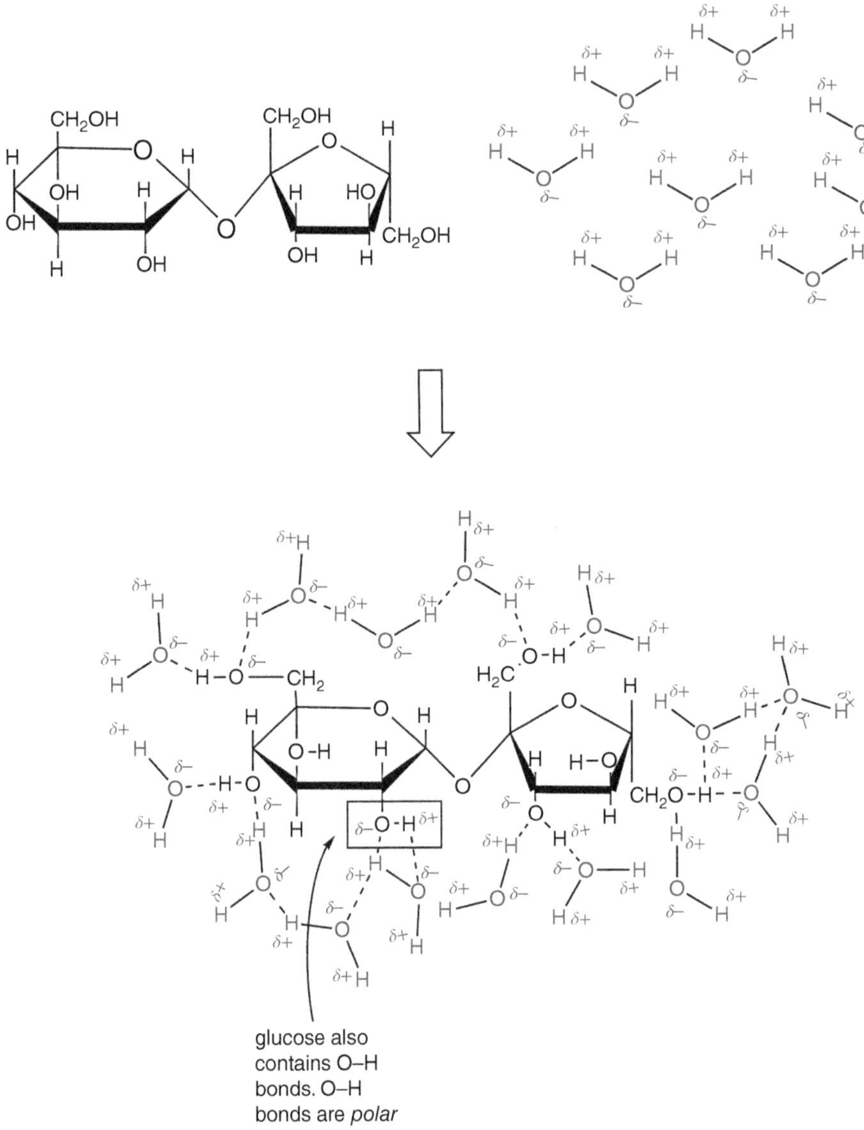

Figure 13.3 Like dissolves like. The polar characteristic of water allows for it to form hydrogen bonds with the polar —OH bonds of sucrose as water dissolves the sugar.

13.4 Inverted Sugars

When sucrose is heated in the presence of acid (like the citric acid from a lemon) or an enzyme called invertase, its disaccharide bond is broken, and the free glucose and fructose molecules are released (Figure 13.4). This process is called inversion; the resulting mixture, which contains a mixture of unbroken sucrose, glucose, and fructose is called an invert sugar or invert syrup. Invert sugars are composed of approximately 75% glucose and fructose and 25% sucrose. Because they contain a mixture of different types of sugars, specifically fructose in the presence of glucose and sucrose, the properties of an invert sugar are different than the properties of sugar. One property of an invert

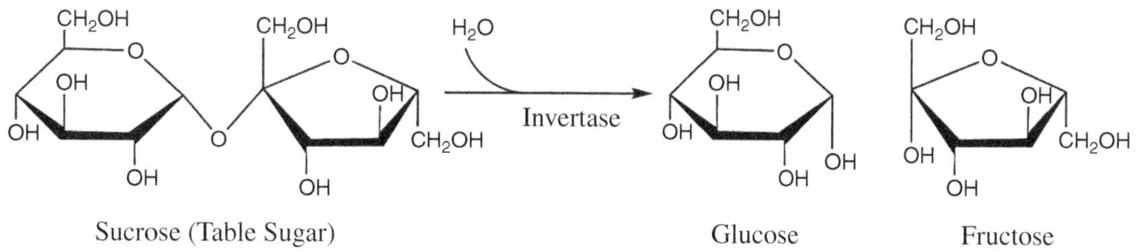

Figure 13.4 Inverted sugar. Heat and acid, or the actions of the enzyme invertase, will hydrolyze (break) the glycosidic bond connecting glucose and fructose as sucrose. Some fruits and vegetables produce invertase as they ripen, increasing the availability of glucose and making the fruit sweeter (due to fructose).

sugar that impacts food and cooking is it doesn't crystallize easily. Why not? The shapes of the disaccharide and the two monosaccharides (glucose and fructose) do not "fit" together in a crystal lattice, thereby reducing the crystallization. This property leads to another physical property change; the saccharides in inverted sugar remain in a viscous, syrupy liquid state, rather than forming a crystalline solid. Moreover, an invert sugar is even more hygroscopic than sucrose because of the additional water molecules that can surround and interact with two sugar molecules (glucose and fructose) relative to one molecule (sucrose) (Box 13.1).

The presence of molasses in brown sugar naturally means that brown sugar contains some invert sugar. Consequently, brown sugar cookies are chewy as the invert sugar attracts and retains moisture, even drawing from the air, during and after baking. However, a glossy, crisp, crystalline top on a sugar cookie prepared solely with brown sugar is not possible, since invert sugar doesn't crystallize. In order to have the glossy top on a cookie, it needs to be prepared with white sugar. If you want both a chewy cookie and a glossy top, then use a combination of white and brown sugar in the recipe.

Box 13.1 Inverting sugar for jam

The inversion of sucrose is an essential part of the process of making jams and preserves. During the jam making process, a fruit and a large amount of added sugar are boiled together for a period of time. During boiling, the fruit softens, its pectin (plus any added pectin) is dissolved into the mixture, and much of the fruit's water is evaporated away. The mild acidity of the fruit promotes the hydrolysis of up to half of the added sucrose to glucose and fructose (i.e. inversion). This chemistry has various beneficial effects in jam making. First, the total number of sugar molecules is increased by up to 50%. This boost gives a modest increase in the sweetness of the mixture, but, most importantly, it increases the proportion of water that is bound to sugar molecules; this water is therefore no longer available to support the growth of microorganisms. Moreover, the invert sugar is much more soluble in water than the original sucrose, repressing crystallization. Because of these characteristics, it is easy to achieve very high sugar levels and low free-water levels that are essential for long-term stability of a jam.

13.5 Liquid Syrup Sweeteners

13.5.1 Honey

Honey is a sweet syrup made by bees using the nectar of flowers. It was the most important sweetener in Europe until the sixteenth century when cane sugar became more widely available. The sweetening power of honey comes from the presence of fructose (25–44%), glucose (24–36%), small amounts of sucrose (0.5–2%), and trace amounts of approximately 25 different *oligosaccharides*. However, because honey is the product of bee-collected pollen, there are also a number of other non-oligosaccharide trace-level components, including proteins, enzymes, amino acids, trace minerals, and polyphenol compounds (that provide aroma). The source of pollen, climate, and type of bee change the ratio and type of the sugars and these important trace components, thus, the aroma, taste, and sweetness of honey vary widely. Given the high concentration of fructose within honey, it is very sweet and does not crystallize easily (remember that fructose doesn't crystallize in the presence of glucose). In cooking, honey is often used as a spread on bread and in teas or other hot drinks where its sweetness and nuanced floral flavors and aromas can be recognized and enjoyed. In baking, its hygroscopic property effectively retains moisture in breads and cakes, and its acidic property (pH between 3.1 and 6.1, depending upon variety) contributes to the leavening of quick breads and baked goods that use baking soda as a leavening agent. The sugars in honey facilitate caramelization and Maillard browning reactions (with proteins) in the crusts of baked goods and meat glazes, such as a honey-baked ham (Box 13.2).

Box 13.2 The health benefits of honey

The antibacterial properties of honey are fairly well known. The low water content of honey makes the sugar inhospitable for microbial growth. In fact, there is so little free-water in honey, the sugar solution will cause water to move out of most microbes into the sugar solution (via osmosis) effectively dehydrating and killing the bacteria. In addition to the low water content, the growth of over 60 different species of bacteria has been found to be partially or fully inhibited by honey. Ancient Sumerians used honey as a drug on wounds, and Aristotle told of the healing power of honey on sore eyes and wounds. What are the molecular components of honey that provide this antimicrobial effect? The enzyme glucose oxidase converts glucose to hydrogen peroxide (H_2O_2) and gluconolactone, which is further transformed to gluconic acid. Hydrogen peroxide modifies the proteins and lipids of bacteria, stopping their growth, while gluconic acid acts as an antifungal and antibacterial agent. Several other compounds, including methylglyoxal and methyl syringate, which are produced by the partial digestion of plant material by the bee, also block microbial growth.

13.5.2 Corn Syrup and High-fructose Corn Syrup

Corn syrup and other starch-based syrups were not developed until the early 1800s but are now major players within the food industry due to their inexpensive production costs and useful culinary characteristics. As you know, all fruits and vegetables contain starch, which are long chains of glucose molecules (see Chapter 1, Figure 1.17).

The long starch molecule can be broken into smaller molecular components by acid or enzymes, which catalyze cleavage at the glycosidic bond between glucose molecules. As this degradation occurs, some smaller chains of starch and some individual glucose molecules remain, impacting the food or cooking process. If you have eaten a "starchy" russet potato, you know that it does not taste particularly sweet. However, when starch is broken down into individual molecules of glucose, it becomes sweeter and takes on a thick, viscous consistency. A corn syrup is produced if the original starch comes from corn, a potato syrup is produced if the original starch comes from potatoes. Although you may have never heard of potato syrup, a syrup can be generated from any high-starch fruit or vegetable. However, although the resulting syrup is sweet, it only has 30–40% of the sweetness of sucrose.

You may not recall how sweetness is measured. Remember from Chapter 2 that the sweetness of a particular molecule or food depends on the type of sugar and how the sugar molecule binds to the taste receptors. Understanding the differences in how sugars taste and their sweetness can help a candymaker create new confections. Lactose, the sugar found in milk, has the lowest sweetness value, while table sugar, high-fructose (45–55%) corn syrup, and fructose all have similar sweet intensity. Glucose is about 30% as sweet as table sugar (sucrose). Clearly, there is a difference in sweetness between corn syrup (see previous paragraph) and high-fructose corn syrup that lends the high-fructose type to be common in many goods found in the local grocery store. In the 1960s, there was a breakthrough in the starch syrup industry that led to the discovery of high-fructose starch syrups. Starch syrup is typically prepared by extracting the starch out of a fruit or vegetable (often-times from "field" or "dent" corn, which is high in starch and low in glucose, making it unpalatable for humans). The starch is then broken into smaller starch fragments and individual glucose molecules using enzymes called amylose (found in your saliva to break down starch into smaller chunks). A second enzyme called a xylose isomerase is then added, which converts (isomerizes) some of the glucose into fructose. Since fructose is sweeter than glucose, the resulting high-fructose corn syrup has greater sweetening power than corn syrup. In fact, a high-fructose corn syrup that contains approximately 53% glucose and 42% fructose will provide the same sweetness as the equivalent weight amount of sucrose. If more of the glucose is isomerized to fructose to reach a 53% fructose and 42% glucose ratio, you have a syrup that is as sweet as honey. As described, all of the starch molecules are not broken down into monomer units, and the longer polysaccharides that remain serve a purpose. These longer molecules tangle up with one another, interfering with crystallization of the individual glucose and fructose molecules and slowing down the motion of all molecules; this results in a high-fructose corn syrup with a much thicker, more viscous consistency than any sucrose syrup.

13.5.3 Molasses

You already know how molasses contributes to the properties of brown sugar, but molasses, by itself, can also be used as a sweetener. Molasses is a by-product of extracting sugar from sugarcane. After the juice is extracted from the sugarcane (by crushing or mashing the cane), the juice is boiled to concentrate the sugar. A raw form of granulated sugar crystallizes out of this boiled solution. However, the remaining liquid syrup (called cane syrup or first molasses) still has a very high sugar content. The cane syrup is mixed with some uncrystallized sugar syrup and boiled again; while more raw sugar crystallizes, a sweet syrup remains. Molasses is derived from this syrup. This process can continue a third time or beyond. With each boiling, the liquid syrup gets darker and darker due to the caramelization of the sugars that occurs during the repeated boiling process. The darker the molasses, the more its sugars have been transformed by caramelization and Maillard browning reactions.

Blackstrap molasses is the syrup that is derived from the 3rd boiling. It is very bitter and not particularly sweet, while molasses made from the first and second boiling are readily used in cooking and baking when a complexity of flavors, including hints of caramel, butter, green tones, and sweetness, are necessary. Not only does molasses provide a sweetness and moisture retention to foods, but it also adds a rich flavor to baked goods such as gingerbread and spice cakes and a savory appeal to barbeque sauces and baked beans.

13.6 Chocolate

Chocolate is arguably one of the world's most loved foods. Over 600 different types of molecules contribute to its acidity, bitterness, astringency, sweetness, creaminess, flavor, and aroma, which makes chocolate an extremely complex and flavorful food. Interestingly, though, chocolate comes from a bean that is itself quite unpalatable; it is crunchy, astringent, bitter, and essentially aromaless. Thus, the realization of the bean's potential as a delectable food has taken hundreds of years and has been refined by numerous people. The details of the very early history of chocolate are not well established. Current thinking supports that the cocoa tree, *Theobroma cacao*, first appeared in South American tropical rain forests, and its potential as a food was realized by the Mayan, Inca, and Aztec civilizations (Figure 13.5).

Around 700 AD, the cocoa tree was carried northward (toward Mexico) by the Mayans and was cultivated and utilized as a food source throughout that region, as the beans contained fat, starch, and protein. Notably, the tree only flourishes in areas that are 20° north and south of the equator, so cocoa tree cultivation was (and still is) quite limited. However, during this time, the beans were exported northward into what is now the United States and were so highly valued that they were used as a form of currency. The word chocolate is derived from the Aztecs as "xocolatl"; xocolatl was the name of a drink in which roasted beans were simmered in hot water, flavored with red pepper and vanilla, and thickened with ground corn. The beans were also used as a spice to flavor meat dishes in the Aztec civilization, similar to moles that are used in Mexican cuisine today.

The detailed history of chocolate became more documented when the first cocoa beans were brought to Europe in 1502. The beans were seized during the fourth and last voyage of Christopher Columbus near an island off of the coast of what is now Honduras, unimpressively mistaken for almonds. However, the Europeans had little understanding of how to use the beans and used them to prepare a drink similar to that of the Aztecs, with less red pepper. Eventually, the chocolate drink was modified and sweetened by mixing the cocoa beans with milk, sugar, and eggs in Spain. It gradually spread throughout Europe, arriving in England in 1650, where it became particularly popular and was the basis for the famous chocolate houses of London. Chocolate drinks continued to be developed and adored throughout the seventeenth and eighteenth centuries in Europe, however, there was an undesirable layer of fat within the drink due to the high-fat content of the beans.

Figure 13.5 Early use of chocolate. The cover of a famous book, *Traités nouveaux & curieux du café du thé et du chocolate* [2] / Philippe Sylvestre Dufour / Public Domain. Dufour first described the beneficial pharmacological effects of chocolate, including counteracting drunkenness, upset stomach, and menstrual disorders.

The first report of the use of a press to remove fat from the beans occurs in a French treatise, published in 1685 [2]. However, the credit for the invention of the process for extracting cocoa butter from the cocoa bean to make cocoa powder was given to the Dutchman, Coennraad van Houten, in 1828. The ability to separate the cocoa butter from the cocoa solids was a breakthrough in the culinary world and allowed the production of solid chocolate by an English company in 1847; 2 years later, milk chocolate was developed by a Swiss firm. Switzerland remains the top chocolate-consuming country at 10.1 kg/year per capita, with the US as the second largest consumer of chocolate (about 9 kg/year, which averages to about three bars of chocolate per week) [4]. Clearly, chocolate is loved throughout the world with an average consumption of one kg/year per capita worldwide (Figure 13.6).

Chocolate is not only delicious to eat, but there is a myriad of science that is fundamental to its flavor, aroma, and behavior. In this chapter, you will learn about the steps and science involved in the production and manufacture of chocolate from cacao beans, the favorable or unfavorable transformations that occur when you work with chocolate in the kitchen, and the different types of chocolate. Let's take a walk into Willy Wonka's factory to learn more about the science of chocolate.

Figure 13.6 Global chocolate consumption. The top 23 chocolate-consuming countries (i.e. chocolate consumed per capita) Adapted from [3] expressed as the amount of chocolate consumed by continent. The population of Europe consumes more chocolate per person than any other continent.

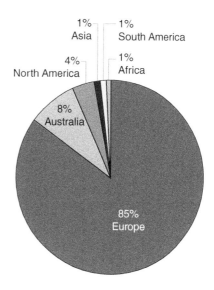

13.7 Chocolate Production

Chocolate is a natural product of the cacao tree, *Theobroma cacao*, which now grows within a narrow range of conditions throughout the wet, lowland tropics of Southeast Asia, South America, and West Africa but was native to South America. The Swedish botanist, Carl Linnaeus, classified the tree as *T. cacao* in 1728. The name was, perhaps appropriately given, as *Theobroma* is Greek for "food of the gods," although, at the time, its use was still largely limited largely to hot drinks. *T. cacao* is a broad-leafed evergreen tree that flourishes in damp places, growing up to 7.5 m tall. When the tree is three to four years old, it begins to produce white flowers; once a tree flowers, it flowers year round, undergoing pollination by tiny flies called midges. Pollinated cacao flowers produce fruit (i.e. seed pods) in approximately 40 days, which becomes melon-shaped with a leathery shell, approximately 20 cm in length and 10 cm in diameter, in another five to six months. Each pod contains 20–40 seeds (i.e. cacao beans; Figure 13.7). Each bean consists of two cotyledons and an embryo surrounded by a seed coat (called a testa) and is enveloped in a sweet, slightly acidic, juicy white pulp that comprises approximately 40% of the seed flesh weight.

Figure 13.7 Cocoa pod. volff / Adobe Stock Photos

Both the bean and the pulp are key components to obtaining the chocolate that we love to eat, although at this early stage, the bean neither tastes, smells, nor looks like chocolate. Chocolate production requires a few steps, some of which are familiar to you from other sections of this book: (i) fermentation of the beans and pulp, (ii) roasting of the fermented beans, (iii) grinding and mixing, (iv) conching, and (v) tempering.

13.8 Fermentation

Fermentation? You might think that this is a mistake, as chocolate doesn't taste or smell like cheese, alcohol, or soy sauce. However, fermentation is a key step in chocolate production and is carried out by similar yeast and microbes that have been previously discussed. The fermentation process removes the fleshy, sweet pulp that envelops the beans and develops the precursor molecules that are essential for chocolate flavor. When and where does fermentation occur in chocolate production? Fermentation occurs on the farms and plantations where the cacao bean seed pods are harvested. Let's take a closer look at the process and science of cacao bean fermentation.

After harvest of the cacao pods, which may occur by a machine, a human with a machete or less ideally via an animal with claws and teeth, the pods are placed in piles and left to ferment in the heat and humidity of the tropics for a few days. During this resting period, the seed pod shell weakens, and, after a few days, workers break open the pods and pile the beans and the pulp together in heaps, covering them with leaves. Exposing the nutrient- and sugar-rich pulp to the warm and humid temperatures of the environment supports the growth of a succession of microbes, including yeasts and bacteria. These microbes are in their element; they have food (i.e. the pod flesh), moisture, and tropical temperatures to grow and flourish for two to eight days. What types of microbes are involved in this activity? The sugar-rich and acidic pulp first favors growth by yeast. The yeasts utilize the sugars (i.e. primarily glucose and fructose) from the pulp to make ethanol and CO_2 (Figure 13.8), generate heat, and metabolize some of the naturally occurring citric acid in the pulp (this citric acid is what makes the pulp acidic). During this process, expectedly, the pH of the pulp increases and the amount of available oxygen decreases, which favors the growth of our microbe friends from cheese making, the lactic acid bacteria (lactobacillus).

As the lactic acid bacteria take over the fermenting mass, they utilize glucose to generate lactic acid (first) and citric acid (second) and then further metabolize the citric acid to reduce acidity, further increasing the pH of the mass. In order to allow for continued fermentation, the farm/plantation workers turn the pile at least once a day. As its temperature rises above 99°F/37°C, acetic acid bacteria become dominant, oxidizing ethanol to acetic acid and then to CO_2 and water (Figure 13.9), this reduces the pH (due to acetic acid production) and further increases the temperature of the fermenting mass to 122°F/50°C or higher due to the exothermic nature of the reactions.

You might be thinking, all of this fermentation is happening within the pulp, but what about the bean? As expected, cacao beans are also affected by fermentation. After all, the acetic acid bacteria essentially create a vat of hot vinegar (i.e. acetic acid), which penetrates and produces holes in the beans. This acidity kills the seed embryo as the compromised cells soak up some of the acids, sugars, and ethanol of the fermenting pulp. At this stage, some of the flavor and aroma of chocolate begin to be generated, as the contents of the cells mix and start reacting with one another due to the acidic conditions. For example, the beans' digestive enzymes mix with proteins and sucrose to degrade them into amino acids and simple sugars (like glucose and fructose), and other

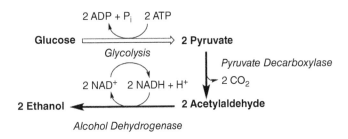

Figure 13.8 Ethanolic fermentation. In the absence of oxygen, yeast will metabolize glucose to pyruvate and ultimately produce ethanol to replace NAD for continued metabolism.

Figure 13.9   Ethanol metabolism. Bacteria will convert ethanol produced by yeast to make ATP under higher temperature conditions.

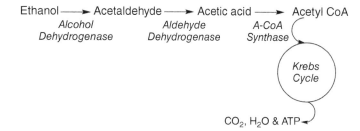

compounds react with proteins and oxygen; these reactions produce some chocolate flavor precursors. Spore-forming bacteria eventually take over fermentation, followed by filamentous fungi that complete the metabolism of acetic acids in the beans (Figure 13.10).

Although fermentation and metabolism of citric, lactic, and acetic acids involve an extensive amount of chemistry and biochemistry, fermentation is largely a preparatory process for the beans, providing them with the molecular precursors that will later develop into molecules associated with the character, flavor, and aromatics of chocolate. The presence (or absence) of these acids also modulates bean pH, affecting the activity of bean enzymes involved in the production of free amino acids, peptides, and reducing sugars, and Maillard reaction chemistry, all of which impact the final chocolate flavor. See Figure 13.10 for an overview of fermentation of cocoa [5].

Once fermentation is complete, farmers dry the beans, often in the sun for one to four weeks. Once the beans are dried to 7% moisture content, they are resistant to spoilage by microbes (as there is not enough water present for a microbe to thrive) and are shipped to manufacturers for further processing.

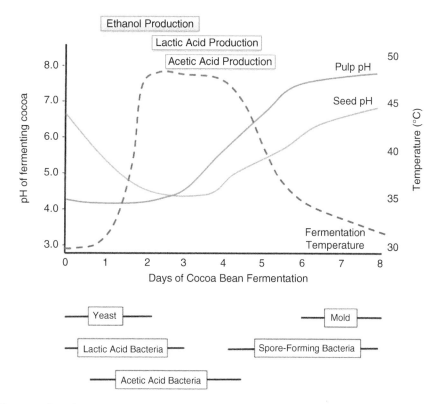

Figure 13.10   Fermentation of cocoa. A series of five different microbiological organisms are responsible for producing conditions to extract and begin the production of chocolate from cocoa beans.

13.9   Cacao Bean Roasting: The Process

Dried, fermented cacao beans are less astringent (due to lower polyphenol content) and more flavorful than unfermented beans, but their flavor has yet to fully develop. In fact, at this point, the cacao bean flavor is dominated by a vinegar aroma and taste due to high levels of acetic acid. By roasting the beans for 30–60 minutes at 250–350°F/ 121–177°C, the glue that holds the cells together (i.e. the cell wall) breaks down, and the reported 200–600 molecules associated with chocolate flavor are produced (Figure 13.11).

It may be surprising that a relatively short-lived and mild heating process can transform a vinegar-tasting cacao bean into something that smells, tastes, and looks more like chocolate. What is happening at a chemical level? This gentle heat drives off the sour-tasting acetic acid and ethanol, while Maillard reactions of sugars, free amino acids, and oligopeptides yield a complex array of molecules that we associate with the flavor and aroma of chocolate.

13.10   Flavors of Chocolate

The cacao bean contains many natural molecules that contribute to its flavor, which are enhanced and generated during the roasting process. If you have ever eaten unsweetened chocolate or cocoa powder and had your mouth go into a sour pucker, you have tasted the molecules that contribute to the bitterness and astringency of a cacao bean, specifically theobromine and caffeine (found in tea) and polyphenols such as catechins, flavan-3-ols, anthocyanins, and proanthocyanidins (Figure 13.12, Box 13.3). Polyphenols undergo oxidation by enzymes

Figure 13.11   Roasted cocoa beans. Popova Olga/Adobe Stock Photos

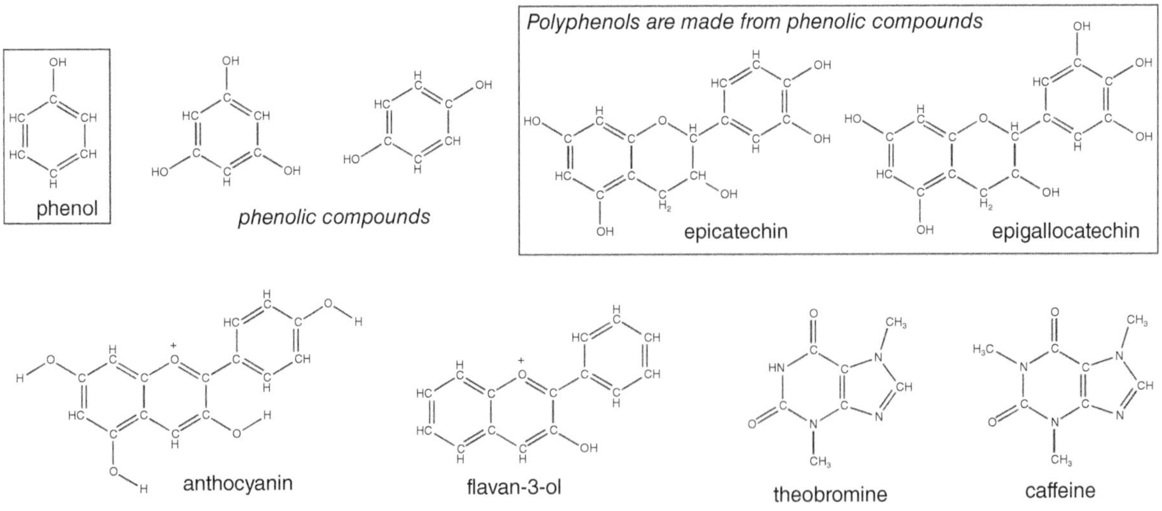

Figure 13.12   Compounds of cocoa bean. There is more than just caffeine in chocolate.

called polyphenol oxidases, producing a range of tannin compounds found in wine, vegetables, and spices. Like in wine, tannin-derived polyphenols play an important role in the flavor and aroma of chocolate. Another source of aromatics and volatiles found in roasted beans comes from the Maillard reactions that occur between the sugars (i.e. sucrose, glucose, and fructose) and free amino acids or small peptides that were produced during fermentation via acid hydrolysis reactions.

The fermented pulp also provides flavors and aromas of fruits, sherry, and vinegar to chocolate in the form of acids, esters, alcohols, and acetaldehyde, the most notable being phenylacetaldehyde, benzaldehyde, phenylethanol, 3-methyl-1-butanol, phenylethylacetate, and 2-heptanone. One class of molecules, pyrazines, is responsible for a roasted, nutty, chocolate note. Dutch processing, which decreases the acidity of chocolate, increases the concentration of the pyrazines. You will notice a difference in color, taste, and recipes that use Dutch-processed cocoa versus regular cocoa powder. Some chocolate manufacturers will add butyric acid during the chocolate-making process, adding a tangy taste to the chocolate. Butyric acid, a common additive to American-made chocolate, is easily detected by humans and is one of the distinguishing characteristics between most European chocolate and American chocolate.

Box 13.3 Is chocolate safe for a human or a pet?

Were you ever told to not eat chocolate at night due to the presence of caffeine? One cannot understand the flavor and taste of chocolate without thinking about theobromine and caffeine (Figure 13.12).

Theobromine and caffeine are nitrogen-containing molecules called alkaloids that are similar in structure to molecules found in DNA and RNA (i.e. the nitrogenous bases G and A). While most chocolates have significantly more theobromine than caffeine (containing 2–10% theobromine in most cocoa solids), the two have similar pharmacological functions and can stimulate heart muscle and nervous signaling. Caffeine and to a lesser extent theobromine block the degradation of cyclic AMP (cAMP) by inhibiting the enzyme responsible for breaking down cAMP. By blocking the breakdown of cAMP, a critical signaling molecule responsible for the control of vasodilatation, stimulation of heart rate, breakdown of glycogen in muscle, and a number of other biological functions, you may have a sense of "stimulation" upon eating a bar of chocolate. Fortunately, humans can metabolize caffeine and theobromine to remove them and their effects from our physiological system, so, any stimulating effects are fairly short-lived in most humans. However, dogs and cats cannot easily break down these molecules, making chocolate is a very dangerous food for a pet due to the long-term impact of increased heart rate, blood pressure changes, and metabolism of glucose. Eating 3 oz of dark chocolate will put a miniature poodle (i.e. a dog of 15–17 pounds) at severe risk of a medical emergency or death. A 7 oz chocolate bar would put a 190 lb. large dog at risk of death. By contrast, a human would be able to eat several large candy bars in one sitting (far more than 7 oz). Although you may develop an expanded waistline, you certainly wouldn't die from the indulgence.

13.11 Grinding and Milling: Cocoa Butter and Cocoa Powder

If you bit into a roasted cacao bean, it would resemble the taste and aroma of chocolate, but would be hard and crunchy due to the presence of the bean shell, perhaps like eating a chocolate-flavored pistachio nut shell. Thus, after roasting, the shells need to be removed and the chocolate ground into small pieces and further milled into tiny particles. The cracking and separation of the shells from the chocolatey product is a logical, necessary step of the chocolate-making process. However, the "why" behind grinding and milling might be less obvious to the chocolate novice.

Cacao beans contain between 50% and 55% fat (i.e. cocoa butter—a triglyceride). As the beans grow, the fat is sequestered in very small pockets, surrounded by rigid carbohydrates and a small quantity of protein. In order to make a smooth chocolate, the rigid carbohydrate structure needs to be broken down. In the first stage of grinding, the beans are ground into smaller pieces called cacao nibs, during which the carbohydrate walls surrounding the pockets of fat are broken down. As the nibs are ground further, the ground-up cocoa solids (mainly protein and carbohydrates) become suspended in the cocoa butter (i.e. fat) to yield a thick, dark chocolate paste called the chocolate liquor. Pure cocoa liquor has a very concentrated chocolate taste, but its flavor is still bitter, astringent (due to polyphenols), and acidic (due to the acetic, lactic, and citric acids).

To turn the bitter, pasty chocolate liquor into something more edible and even delicious, a manufacturer adds ingredients, such as sugar, milk solids (fats, proteins, and carbohydrates from dehydrated milk), vanilla, and an emulsifier. The addition of sugar is obvious to anyone who has tasted unsweetened chocolate; the vanilla adds flavor to the final product. However, what is the "why" behind the addition of more cocoa butter and an emulsifier? These additions do contribute flavor, but also (and more importantly) keep various molecular components of the chocolate mixture in suspension—or emulsified. Let's explain this concept a bit more.

The chocolate liquor contains cocoa solids suspended in melted fat. In the chocolate liquor, there is enough cocoa butterfat to completely surround all of the more polar protein and carbohydrate cocoa solids; you can imagine a sea of butter that contains a limited amount of protein and carbohydrate molecules. However, as sugar is added to the chocolate liquor, the proportion of the fat in the liquor drops, and there is no longer enough fat to coat both the cocoa solids and sugars. The resulting effect is a heterogeneous, pasty mixture, due to lack of fat coating on the sugar, carbohydrate, and protein molecules, rather than a smooth, flowing liquid. In order to obtain a smooth texture, additional cocoa butter is added in an amount that is proportional to the amount of added sugar. This extra fat ensures that the cocoa solids and finely ground sugar remain suspended in the liquid fat, allowing the liquor to flow. How much fat is necessary to create this liquid suspension? Generally speaking, a smooth chocolate requires a minimum total fat content of 30%. However, by also adding an emulsifier that has amphipathic properties (e.g. lecithin, a phospholipid also known as phosphatidylcholine), the cocoa solids and sugar particles can be coated with the hydrophobic components of the amphipathic molecule, which binds all of the particles together and allows the chocolate to have good flowing properties. Thus, even if a manufacturer is not striving to make a lower-calorie, lower-fat chocolate, lecithin is often added to help the hydrophilic sugars and proteins adhere to the hydrophobic fat molecules (Box 13.4).

13.12 Conching

Conching may seem like an unusual name for any step in chocolate production, but its name derives from the shell-like shape of the original machines that carried out this step. In conching, the chocolate liquor is kneaded with the other added ingredients against a solid surface at temperatures of 115–190°F/46–88°C for a period of time between 8 h and 5 days. Although this step sounds a lot like grinding and milling, conching is necessary to create a stable, smooth chocolate emulsion. By breaking the solid particles (i.e. cocoa proteins and carbohydrates) and sugar crystals into even smaller, fine particles during the conching process, the emulsion is no longer gritty. Moreover, conching ensures that the hydrophilic molecules are separate from one another and are coated with cocoa butter particles such that when the finished chocolate melts, it flows smoothly. Finally, due to the presence of heat and mixing, conching mellows the flavor of the chocolate. Aeration and heat cause approximately 80% of the volatile aromatic compounds, including acetic acid and some water molecules, to evaporate. The resulting chocolate product becomes more basic or alkali, and importantly from a consumer standpoint, the chocolate no longer smells or tastes like vinegar. Additionally, a number of other distinctive aromatic and flavorant molecules associated with chocolate are generated, specifically pyrazines, furaneol, and maltol, yielding roasted, caramel, and malty aroma and taste.

Box 13.4 Chocolate and emulsifiers

Many liquids contain a suspension of small solid particles; these mixtures are called **emulsions**. As the solid particles have a tendency to stick to one another and "fall out" of the liquid solution, an amphipathic molecule, called an emulsifier, is added to ensure that the solids remain suspended in the liquid solution. How do emulsifiers work? One end of an emulsifier molecule has favorable interactions with and coats the surface of the solid particle, while the other end of the emulsifier has favorable intermolecular interactions with the molecules that comprise the liquid solution. This interaction with both components of the mixture (solid particles and liquid solution) allows the solid to remain suspended in the liquid. The viscosity (or thickness) of the liquid, as well as the amount and size of the solid particles, governs the flow properties of an emulsion; the smaller the particles, the thicker the emulsion. In chocolate, the properties of the emulsion lead to the different textures found in the myriad of chocolate varieties and types. Generally speaking, the solid particles found in English chocolate are larger; these chocolates flow more quickly around the mouth (i.e. are less viscous), and the consumer detects the feel and taste of the chocolate quickly. American chocolates are typically a thicker emulsion (i.e. smaller solid particles); this chocolate lingers in the mouth for a longer period of time. Try an experiment at home with two different types of chocolate and predict the size of the solid particles based upon how the chocolate "lingers" in your mouth.

Is conching the last step in chocolate production? Yes and no. Liquid dark chocolate, a warm liquid of cocoa butter, molecules of the original cacao beans, the flavorful and aromatic molecules produced during fermentation and roasting, and sugar is the result of conching. Production of milk chocolate requires further addition of dehydrated milk (i.e. milk solids including fat, casein and whey proteins, and lactose), which adds body, emulsification, and smoothness to the mixture. However, it is difficult to broadly sell and eat liquid chocolate of either the dark or milk variety. Thus, the last step of the chocolate-making process is to cool the liquid to room temperature and allow it to solidify in the desired form (e.g. chocolate bars, star-shaped chocolates, chocolate-covered cherries). On the surface, tempering, a term that describes cooling down a material to harden it, seems pretty straightforward; cool the chocolate, it hardens, and you eat it. However, in chocolate-making, this cooling step is the trickiest step of the entire process. To obtain a chocolate bar that looks appealing, snaps when you break it, and melts with a smooth creaminess in your mouth, a manufacturer must pay very close attention to the chocolate liquid as it is cooled and rewarmed. The care required in chocolate tempering is necessary because of the crystallization properties of cocoa butter.

13.13 Tempering

13.13.1 Crystallization

Whether you realize it or not, you have extensive personal experience with crystals in the kitchen in the form of salts, sugars, and maybe even crystallized ginger. Each of these crystalline forms has particular physical properties and characteristics that lend themselves for certain purposes but not others. First, let's start with the basics. What is a crystal? Crystals are solids that are characterized by and form a regular, repeated pattern of connected building blocks (either molecules or atoms). In some solids, the arrangements of the atomic or molecular building blocks are random or very different throughout the material. In crystals, however, the structure of the atoms/molecules is repeated in exactly the same arrangement over and over throughout the entire material. Any variation in the crystalline arrangement leads to different physical properties of the substance.

What does this have to do with chocolate? In order to obtain desirable physical properties of chocolate (i.e. a crisp snap when you break it and a smooth texture upon melting), the fat molecules need to crystallize in a particular manner during tempering. Specifically, the crystals need to be stable enough to hold the chocolate together in a solid bar form, allowing for the characteristic snap when you break it. However, the melting point of the crystals needs to be above room temperature, but below body temperature, so that the chocolate truly melts in your mouth (but not in your hand). Finally, the crystals need to be small, so that the chocolate has a smooth, glossy finish that doesn't look (or taste) gritty. Thus, crystallization is very important in having a chocolate bar with appealing characteristics that we associate with chocolate.

Cacao fat molecules are unusually regular, monounsaturated triglycerides (Figure 13.13). The uniformity of the molecules allows them to pack together tightly and form a dense network of stable crystals involving all of the fat molecules. However, this packing organization is only created when the crystallization of the fat is carefully controlled. Cocoa butter can actually solidify into six different kinds of fat crystals, I–VI called polymorphs (Table 13.1), depending upon the crystallization temperature. A crystallization temperature is the temperature below which the fat molecules (or other types of molecules of interest) crystallize into a particular structure and above which melt into a liquid. Cacao fats (i.e. monounsaturated triglycerides) can stack in various ways, forming double or triple fatty acid chain overlaps. The more stable and tighter the contact, the higher the melting point (Figures 13.14 and 13.15). Of the six forms, polymorph V has the favorable visual and textural characteristics that consumers associate with a great bar of chocolate. Polymorphs I–IV have an unstable, less organized, and looser network of fat molecules. In these polymorphs, some of the fat molecules are included within the crystalline structure, while others remain in a liquid, unpacked form. These "liquid" fats ooze away from the solid, yielding a greasy and soft, less appealing chocolate. Thus, how does a manufacturer or home cook get chocolate to crystallize as the desired polymorph V? By controlling the temperature of fat crystallization. The temperature at which chocolate begins to crystallize primarily governs the particular crystalline form of cocoa butter. Once a crystal starts growing, it continues to grow in the same form. Moreover, if crystals are to be kept small, which is desirable to maintain

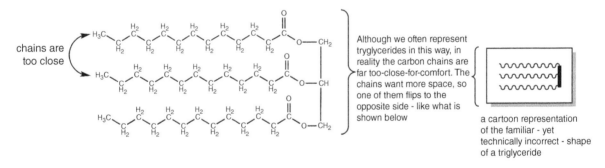

Figure 13.13 Fats of chocolate. Triacylglycerol (backbone glycerol bound to three fatty acids or acyl chains) is a major component of chocolate.

Table 13.1 Chocolate polymorphic forms and melting point.

Form	Melting point [°C]	Chain packing
I	16–18	Double
II	21–22	Double
III	25.5	Double
IV	27–29	Double
V	34–35	Triple
VI	36	Triple

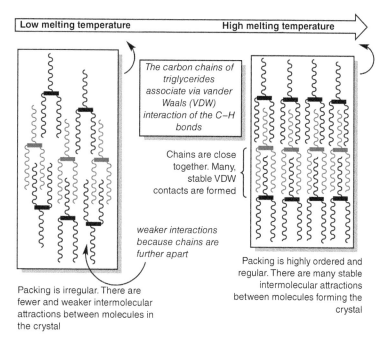

Low melting temperature → High melting temperature

The carbon chains of triglycerides associate via vander Waals (VDW) interaction of the C–H bonds

Chains are close together. Many, stable VDW contacts are formed

weaker interactions because chains are further apart

Packing is irregular. There are fewer and weaker intermolecular attractions between molecules in the crystal

Packing is highly ordered and regular. There are many stable intermolecular attractions between molecules forming the crystal

Figure 13.14 Fatty acid packing and melting point. The more contact the higher the melting point.

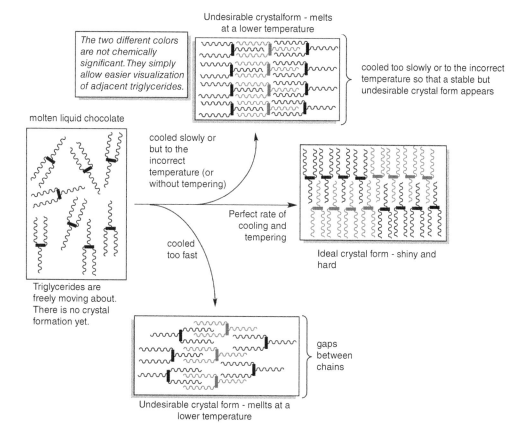

The two different colors are not chemically significant. They simply allow easier visualization of adjacent triglycerides.

Undesirable crystalform - melts at a lower temperature

cooled too slowly or to the incorrect temperature so that a stable but undesirable crystal form appears

molten liquid chocolate

cooled slowly or but to the incorrect temperature (or without tempering)

Perfect rate of cooling and tempering

Triglycerides are freely moving about. There is no crystal formation yet.

cooled too fast

Ideal crystal form - shiny and hard

gaps between chains

Undesirable crystal form - mellts at a lower temperature

Figure 13.15 Packing polymorphs of fat in chocolate.

a smooth texture in chocolate, then many nuclei (i.e. crystal starter molecules) need to be formed within this ideal, desirable temperature range. Further growth or expansion of the crystals outside of the ideal, desirable temperature range leads to tough, chewy chocolate. Conceptually, this may make sense, but practically, how does a manufacturer or home cook prevent formation of the undesirable fat crystals and promote the formation of favorable fat crystals? The answer is tempering.

13.13.2   Tempering Chocolate

Tempering consists of three steps. Step one involves heating the chocolate to approximately 120°F or 49°C. At this temperature, all of the fat crystals melt, but the chocolate emulsion that was created during conching is not destroyed. In step two, the chocolate is cooled to approximately 80°F (27°C) and stirred. The unstable, less organized, and less dense crystals (i.e. polymorphs I–III) will remain "melted" at this lower temperature due to the lack of organization of the individual fat molecules within these crystalline forms. However, polymorph forms IV, V, and VI will begin to crystallize at 80°F/27°C. When combined with stirring, a set of small, starter crystals, primarily of polymorph forms IV and V, are generated. Why not polymorph VI? Polymorph VI forms/crystallizes on a much slower timescale (i.e. days or weeks) than the other crystalline forms, thus very little polymorph VI is formed within the relatively short time frame of chocolate production. In the final step of tempering, the undesired crystals of polymorph IV need to be removed from the mixture. By heating the chocolate to approximately 90°F (~32°C), the polymorph IV crystals melt, while polymorph crystals V remain intact. When tempered correctly and carefully, the desirable starter crystals of polymorph V direct the development of the chocolate crystalline network as the chocolate is cooled and solidifies into its final form, with fat crystals of about 40 millionths of an inch in size.

Have you ever noticed the difference in melting properties and chocolate "snap" in dark chocolate versus milk chocolate? These differences are due to a difference in the crystal structures of milk and dark chocolate. At a crystalline level, the higher relative fat content of dark chocolate (i.e. the lack of milk solids and less sugar) allows the fat to crystallize in a more organized network than found in milk chocolate. The stronger crystalline network means that more heat will be required to "break" the network or melt the chocolate. You can hear and feel this when you break a bar of dark chocolate relative to a milk chocolate bar. You can also do a simple experiment of placing a bar of dark chocolate versus a bar of milk chocolate in your hand. Because milk chocolate contains a more diverse mixture of fats (i.e. the butterfats from the milk), it softens or melts over a wider temperature range than dark chocolate. The result: more chocolate on your hand with the milk chocolate bar. The solution: eat milk chocolate as quickly as possible. These crystallization properties are noticed when you eat a bar of chocolate, but the exact tempering temperatures that a cook uses when working with chocolate will vary depending upon the fat content and type of chocolate that is being prepared or worked with.

13.14   Chocolate Bloom

Have you left an unopened chocolate bar in a car, where the chocolate melted and resolidified in the sealed package? Once opened, you may have noticed that the chocolate developed some white or gray streaks, had a softer than normal texture (even in the solid form), and tasted gritty (if you ate the bar at all). Although your immediate reaction might be that the chocolate spoiled or acquired a mold, in fact, the chocolate experienced a condition called "chocolate bloom." Chocolate bloom provides an excellent example of what occurs when chocolate is not tempered properly (Figure 13.16). While bloomed chocolate is not harmful, it looks unappetizing and loses the expected texture and flavor release of a chocolate bar. From a scientific standpoint, the ideal fat crystals found in the original chocolate bar melted in the heat of the car and were replaced with the less desirable fat crystals. Let's think more about what happened in the car example.

Figure 13.16   Chocolate bloom. Marcpablo8 / Wikimedia Commons / CC BY-SA 3.0.

The desirable form of chocolate has fat particles that crystallize in the 29–32°C (84.2–89.6°F) temperature range (depending upon the chocolate type). During the summer, the temperature in a car can reach temperatures as high or higher than 49°C/120°F. As expected, all of the fat within the chocolate melts at these temperatures. The melting of the chocolate isn't problematic, as this is the first step of chocolate tempering. The problem is the cooling phase, since (in this example), the cooling is happening in the completely uncontrolled condition of a car. In short, the chocolate is cooled to the weather-dictated car temperature, which is typically less than 82°F/28°C. Thus, the chocolate is cooled at a temperature that favors the formation of fat crystals that do not have the desirable properties of chocolate. Specifically, the less stable polymorph forms crystallize first, and then the more stable polymorph crystals develop over a period of hours to days. Remember that polymorph V has a more ordered and stable crystalline structure than polymorphs I–IV. Thus, over time, the crystal matrix becomes more compact (more polymorph V), which pushes the cocoa solids and sugar particles to the surface. It is these particles that lead to the whitish/grayish surface appearance of bloomed chocolate. Moreover, the formed fat crystals are less organized and packed, due to the presence of different fat crystalline states, leading to a "softer" solid state. Thus, although the same molecules that were present in the original chocolate bar exist, the molecules are organized in a manner that results in less desirable physical properties. Can the chocolate be converted back to its original, appealing form? Yes, however, the chocolate needs to be subjected to the entire proper tempering process.

You may be wondering if one type of chocolate blooms more readily than another type of chocolate. As alluded to above, the type of chocolate does impact the exact temperature for chocolate tempering. Thus, the home cook needs to pay close attention to chocolate type and temperature if they are trying to temper chocolate, melt chocolate, or re-temper chocolate to eliminate bloom. In addition, when two fats are mixed, the resulting fat mixture has a different crystallization behavior than either of the two individual fats. The change in fat crystallization behavior will also change the bloom tendency of chocolate. Milk fat (found in butter) and cocoa butter are considered compatible fats. When the two are mixed, the milk fat prevents the bloom of cocoa butter, thus, milk chocolate is less prone to bloom than dark chocolate. However, the fat mixture of milk chocolate also makes the chocolate less snappy. In short, chocolate bloom, fat crystallization, and chocolate tempering are complicated!

13.15 Chocolate Bloom and Chocolate Chip Cookies

Chocolate chips are found in many baked goods, including cookies, brownies, and cakes. When these baked goods come out of the oven, cool (in the uncontrolled environment of your kitchen), and the chocolate chips re-solidify, they do not have a white surface, nor do they taste gritty. Why doesn't bloom happen to chocolate chips in cookies or other baked goods? Your experience and scientific research show that if you have a piece of chocolate that is embedded in a fat-containing cookie or baked good, the chocolate does not undergo bloom. However, if a chocolate chip sits at the very edge of or is not deeply embedded within the cookie dough, it will experience bloom. Thus, the presence of or exposure of the chocolate in a dough or batter reduces the instance of bloom; the key factor in the dough or batter is fat. The phenomenon that is governing the system is called fat migration.

Fat migration can occur when there is a different and higher concentration of fat in one environment (cookie dough) vs. another environment (chocolate chip). If heat is imparted into this system, the heat provides the energy source to drive fat from the higher concentration environment to the lower concentration; this is fat migration. During the baking of a chocolate chip cookie, the milk fat, found in butter-based cookies, migrates from the cookie dough into the chocolate chip during baking. The inflow of fat into the chocolate chip changes the fat crystallization properties of the chocolate, essentially protecting the chips from bloom. Fat migration happens readily between a butter-based cookie dough and a milk chocolate chip because the two are compatible fats. Interestingly, however, vegetable fats present in shortenings are typically not compatible with cocoa butter and may promote or accelerate bloom. Thus, the surface of a peanut butter cup is very prone to bloom. One way to make a shortening-based cookie less prone to bloom is to increase the fat content of the cookie dough. For example, in cookies containing 20% palm oil, bloom did not occur, but it did occur in cookies with 14% palm oil.

13.16 Cooking with Chocolate

As you have probably surmised from the previous discussion about chocolate bloom and tempering, chocolate can be difficult to work with in the kitchen. Another chocolate cooking challenge that cooks may encounter is a phenomenon called seizing. When a chocolate *seizes*, a beautiful, smooth liquid chocolate turns into a grainy, solid mass within seconds. The most common cause of seizing chocolate is a small amount of water. However, a large amount of water can be incorporated into liquid chocolate without any issue. Why the difference?

Let's think about well-tempered chocolate at a molecular level. Tempered chocolate contains millions of microscopic, mostly hydrophilic, cocoa and sugar particles surrounded by a sea of fat molecules. This smooth mixture was created during conching when additional cocoa butter was added to the chocolate liquor to ensure separation of the hydrophilic cocoa and sugar molecules. When a small amount of water is added to this mixture, the cocoa and sugar particles are naturally more attracted to the water (i.e. like is attracted to like; hydrophilic molecules are attracted to hydrophilic molecules) than to the cocoa butter. These hydrophilic molecules begin to clump together in the presence of water, and eventually, there is a clumpy, grainy chocolate ball sitting in cocoa butter. A little bit of water affects many molecules; the more that you mix, the greater the seizing.

By contrast, chocolate containing a large amount of water can maintain a smooth, silky liquid texture and flow. If a large amount of water (or other hydrophilic liquid like alcohol or vanilla extract) is added to a chocolate liquid, there is enough water to keep the sugar and cocoa particles separate from one another and surrounded by water molecules; the resulting chocolate has a smooth, silky texture. Can you recover seized chocolate? Yes, you can continue to add (with heat and stirring) water to a seized chocolate mixture until it is smooth, however, the chocolate will have different properties due to the large amount of water and is likely best used as a chocolate sauce or syrup, rather than in baked goods.

13.17 Chocolate-coated, Filled Candies

The preparation of chocolate-coated candies might seem simple after learning about the complexities of chocolate in the previous sections, but this too, is not trivial and involves some interesting science. Many chocolate-coated candies have a soft center filling that needs to be kept "separate" from the chocolate coating. How is this accomplished? The answer is an enzyme called invertase. Invertase, an enzyme obtained from yeast, catalyzes the hydrolysis or breakdown of sucrose into glucose and fructose. The soft centers of many filled chocolates, which contain a high concentration of sucrose suspended in glucose syrup, flavorings, colorings, and invertase, have a moldable, stiff paste-like texture when first prepared by the candymaker. While this texture is necessary so that the center can be pressed or stamped into a shape, it is not appealing for the consumer. After the center filling is stamped, it is coated with melted chocolate and is then allowed to rest. During the resting period of days or weeks, the invertase enzyme hydrolyzes a proportion of the sucrose to its monosaccharide components, glucose and fructose. Since glucose and fructose are more water soluble than sucrose, they dissolve a bit creating a softer, creamy textured center. This process might take up to 2 months for the desired texture to be achieved. Once achieved, the candies can be sold to consumers for consumption.

13.18 Different Types of Chocolate and Chocolate-like Products

13.18.1 The World of Chocolate

There is a great diversity in the types of chocolate that one might see on a grocery or candy store shelf: dark, milk, bitter, semisweet, or unsweetened chocolate bars, regular or Dutch-processed cocoa powder to name a few. In addition, there are chocolates from a myriad of different countries with different percentage numbers advertised on the wrapper. What are the types of chocolate, what do these numbers mean, and does the type of chocolate matter in cooking?

Each country has its own definition of chocolate. The different components, along with requirements for the chocolate-making process (i.e. all Swiss chocolate must be made in Switzerland), lead to different textures and tastes that a chocolate connoisseur might be able to readily attribute to one country, ingredient, or process. In the United States, the FDA regulates the naming and ingredients of cocoa products based on the percent of chocolate liquor and milk solids.

13.18.2 Different Types of Chocolate

While there are many products that are called "chocolate," what distinguishes one form of chocolate from another are the ingredients, the preparation of the cacao beans, the presence of other additives, and the final fat crystalline structure. For most consumers, different chocolates are distinguished by the sweetness and darkness of the chocolate. Remember that chocolate liquor is the cocoa butter and cocoa solids that result from the fine grinding of cocoa nibs. In terms of its components, the liquor is comprised of approximately 55% cocoa butter, 17% sugars, 10% protein, a few percent of tannins, and a small amount of theobromine. Generally speaking, the more chocolate liquor, the less sugar and fat, and the more bitter the chocolate will taste. Although some chocolates are interchangeable, a cook or baker should be mindful that there is a difference, which will impact the final product.

Unsweetened chocolate, also called or labeled as baking chocolate, is used almost exclusively for baking since it does not have a desirable flavor when eaten by itself. Unsweetened chocolate is pure (or nearly pure) chocolate liquor, but may contain a very small amount of fat (i.e. lecithin) or added sugar. It has the most intense chocolate flavor, since it is almost exclusively chocolate liquor, but again does not have a desirable taste in itself due to bitterness and astringency. In most recipes that contain unsweetened chocolate, sugar or another sweetener is an accompanying ingredient.

Bittersweet and semisweet chocolates are also readily used in cooking and baking but can also be eaten alone, depending upon one's taste preference. Bittersweet chocolate contains between 35% and 50% chocolate liquor, with the remainder composed of fat and sugar. Semisweet chocolate generally has more sugar than bittersweet, but the definition of semisweet chocolate is not regulated, so the sweetness of one type versus another will depend upon the manufacturer. As bittersweet and semisweet chocolates do not contain milk solids, they are types of dark chocolate and may be labeled as such. Dark chocolates can be unsweetened and sweetened chocolate types, never contain milk solids, and are described by the amount of chocolate liquor in the bar. With a higher percentage of cocoa/cacao/chocolate liquor, the chocolate is more bitter, less smooth, harder, and more snappy when broken into pieces. A bitter (but edible) dark chocolate bar will be above 80% cocoa/cacao. Sweet dark chocolate will contain 30% cocoa/cacao, where the added sugar and fat will give the dark chocolate a sweeter taste and softer texture.

Milk chocolate is distinct from dark chocolate in its preparation, look, texture, and taste. Since milk chocolate is prepared by adding milk solids to 10–20% chocolate liquor, milk chocolate is a sweet, creamy, lighter-colored chocolate with hints of caramel and butterscotch flavors. You may recall that milk solids include the sugars (mostly lactose) and proteins (casein and whey) from milk, while milk fat is just that, the fat from milk. The US FDA defines milk chocolate as containing no less than 10% of chocolate liquor, 3.39% by weight of milk fat, and 12% by weight of milk solids. In fact, chocolates made with other fats, such as vegetable oils, cannot be called chocolate or milk chocolate. If you see a chocolatey product that is not clearly labeled as chocolate or milk chocolate, read the label. If the product is made with a plant oil, rather than milk fat, the product label will likely read "made with chocolate" or "chocolate-flavored."

You may see products in a grocery store that contain "white chocolate." White chocolate is, in fact, not chocolate at all, as it doesn't contain any cocoa solids. Instead, white chocolate contains at least 20% cocoa butter, sugar (up to 55%), milk solids, fats, and flavorings including vanilla. Cocoa butter, the fat isolated from the cacao bean, is what gives white chocolate its smooth, melting texture. Why? Cocoa butterfat is a mixture of saturated and unsaturated fats from 20 to 16 carbons long, so the fats melt just below body temperature. These fantastic melting qualities make white chocolate a favorite for cookies and dessert bars. If cocoa butter is replaced with palm or vegetable oil during production, then the product is no longer called white chocolate but white chips or vanilla chips. The taste and properties of these chips may be indistinguishable from most white chocolates since most of the flavor is derived mostly from milk and sugars.

Cocoa powder is another unsweetened form of chocolate, used for baking and for drinking, with added milk and sugar. To make cocoa, unsweetened chocolate (i.e. solidified chocolate liquor) is pressed to squeeze out most of the fat, and the resulting cacao cake is ground into an intense chocolatey powder that contains approximately 80% cocoa solids and between 10 and 22% fat. Cocoa powder is quite acidic (remember the low pH of the chocolate liquor due to acetic and lactic acids), which impacts its taste, characteristics, and reactivity. To counter this acidity, a basic or alkaline solution can be added to the cocoa powder, creating "Dutch" cocoa powder. Dutched cocoa powder has a darker brown appearance than natural cocoa powder due to the neutralization of naturally acidic cocoa powder. In other words, Dutch-processed cocoa powder has a more neutral pH than regular cocoa powder. In addition, Dutch-processed cocoa powder has a milder and detectably more complex flavor. In a comparison of recipes, a recipe containing Dutch-processed cocoa powder (remember neutral pH) will more often be paired with baking powder. By contrast, regular cocoa (i.e. an acid) is more often paired with baking soda (i.e. a base). In addition, the taste of natural/regular cocoa powder can be astringent and harsh, which overwhelms the presence of other flavor molecules; thus, a recipe might also call for Dutch-processed cocoa powder for a milder chocolate taste.

13.19 Candy

All sugar candy, whether creamy, crystalline, soft, or brittle, is made from sugar and water. It sounds pretty simple, doesn't it? If the ingredients are the same, what is the difference between a hard, round peppermint and a soft peppermint-flavored piece of taffy? A cook (or confectioner, in the world of candy) creates different textures by varying the relative proportions of sugar and water, which changes the intermolecular interactions between sugar molecules and sugar and water molecules. Additional variation is imparted by cooking and cooling temperatures, the amount of time spent at a given temperature, the rate of temperature changes, and the presence and timing of stirring or lack thereof. Let's start with the first factor: the relative concentrations of sugar and water.

The first step in the confectionery process is the preparation of sugar syrup. In general terms, some amount of sugar and water are combined, and the solution is heated to or allowed to boil at a particular temperature. As the sugar solution boils, water evaporates, which increases the amount of sugar in the mixture relative to water. Correspondingly, the solution thickens and becomes more viscous and syrup-like; the more water the syrup contains, the less thick the final product. Let's talk more about the science behind this seemingly simple process, specifically phase diagrams.

Figure 13.17 shows the phase diagram for water. A phase diagram is a graph that shows the state of matter of a particular substance under different pressure and temperature conditions. At an atmospheric pressure of 1.0 atm and a temperature of 212°F/100°C, water changes between a liquid and gas phase. Although this sounds very scientific, it is something that you know well: the boiling point of water. As atmospheric pressure decreases, the boiling point of water also decreases. Because of this phenomenon, if you are cooking pasta in boiling water at a high altitude (where there is a lower atmospheric pressure), you will need to boil your pasta for more time to achieve the desired texture, relative to sea level conditions.

What does this have to do with candymaking? A phase diagram for a sugar/water solution, with temperature and sugar concentration as the variables (see Figure 13.17 for reference), is very valuable for the confectioner. The higher the sugar concentration (measured as a weight percent sugar), the higher the boiling point of the syrup. Thus, the value of the boiling point is indicative of the amount of dissolved sugar found in the syrup. Interestingly, though, as a sugar/water solution boils, water molecules evaporate from the solution, and the sugar molecules account for a larger proportion of all of the molecules in the solution. Thus, as the syrup gets more concentrated with sugar, the boiling point rises. A good confectioner needs to boil the syrup while closely monitoring the temperature with an accurate candy thermometer, as this will indicate the exact sugar concentration of the syrup solution. The optimal sugar concentration for fudge occurs at 235°F/113°C, while 270°F/132°C is the best temperature for a taffy maker, and 300°F/149°C is ideal for the preparation of hard candies.

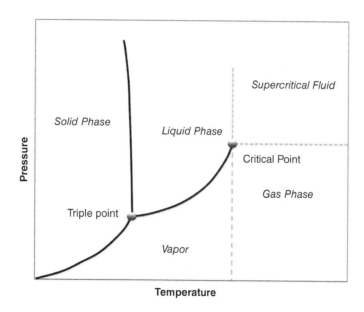

Figure 13.17 Phase diagram for water. The diagram describes the relationship between pressure and temperature for water. Water has a critical point where liquid and gas are indistinguishable and forms a supercritical fluid. At this triple point, all three phases coexist.

In the absence of a candy thermometer, there are other visual and tactile kitchen tests for knowing when a cook has achieved the proper temperature of the desired candy type. Some common terms (e.g. soft ball stage or hard-crack stage) describe the behavior of the syrup by itself or when a drop of it is placed into ice water. Whether you use a thermometer or conduct a kitchen test, the boiling temperature is key to achieving the necessary sugar concentration for the desired candy variety (Table 13.2).

Upon review of Table 13.2, you may identify a correlation between cooking temperature and candy type/texture. Specifically, as the cooking temperature increases, the candy gets harder. What is the scientific basis of this correlation? A candy's texture is determined by the organization of the sugar molecules in the final solid structure, which is governed by the cooling of the liquid syrup. If you think that this sounds like chocolate tempering, you are correct. If the syrup cools and forms a few large sugar crystals, the candy texture will be coarse and grainy. If millions of microscopic sugar crystals develop within a sea of syrup during the cooling process, the candy will be smooth and creamy. If the sugar forms no crystals at all during cooling, you will have a big candy mass. As with chocolate, the trickiest stage of candymaking comes as the syrup cools to room temperature. The rate of cooling, movement of the syrup, and presence of the smallest particles of dust or sugar can have dramatic effects on the structure and texture of a candy.

Before discussing sugar crystal formation in candy, fundamentally, your experience with glistening sugar crystals in a sugar bowl or sanding sugar granules atop a muffin tells you that the molecular structure of sugar favors crystal formation. From the discussion about chocolate, you also know that more stirring leads to smaller crystals, and crystals develop from seeds of the most stable crystal present within the mixture. Let's bring these concepts together to better understand sugar crystallization in candymaking.

Table 13.2 Candy temperature.

Syrup's boiling point and sugar concentration	Type of candy	Syrup's behavior
215–234°F 101–112°C Sugar concentration: 80%	Sugar syrup, fruit liqueur, some icings	Thread stage: The liquid sugar can be pulled into brittle threads between the fingers
234–240°F 112–115°C Sugar concentration: 85%	Fudge, pralines	Soft ball stage: A small amount of syrup dropped into chilled water forms a soft, flexible ball but flattens after a minute in one's hand
242–248°F 116–120°C Sugar concentration: 87%	Caramels	Firm ball stage: Forms a firm ball in ice water that does not flatten when removed from water but remains malleable and flattens when squeezed
250–268°F 121–131°C Sugar concentration: 92%	Nougat, marshmallows, toffee, gummy candy	Hard-ball stage: Forms a hard ball in ice water that holds its shape upon removal
270–290°F 132–143°C Sugar concentration: 95%	Taffy	Soft-crack stage: When dropped into ice water, the syrup separates into hard but pliable threads that bend slightly before breaking
300–310°F 148–154°C Sugar concentration: 99%	Lollipops and other hard candies	Hard-crack stage: The highest possible cooking temperature for a candy. Because there is almost no water left in the syrup, syrup dropped into ice water separates into hard, brittle threads that break when bent

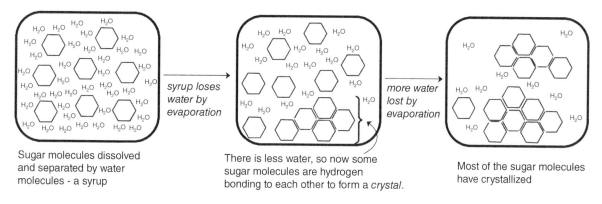

Figure 13.18 Crystallization of sugar. Evaporation of water results in the formation of sugar crystal lattice.

Prior to heating a sugar/water syrup mixture, each molecule of sucrose is completely surrounded by water molecules because there is so much water relative to sucrose (Figure 13.18). However, during heating, the water begins to evaporate and the sucrose molecules have a much greater tendency to bump into and interact with one another. When they do, the interacting sucrose molecules can form a seed crystal. As the water continues to evaporate, the syrup becomes more concentrated, increasing the likelihood of molecular interactions between sucrose molecules. A saturated solution occurs when the tendency of a dissolved substance to bond to itself is exactly balanced by the water's ability to prevent this bonding. The moment at which saturation is reached depends upon temperature and the molecules present. When a sugar/water mixture is boiling, all the molecules (both water and sucrose) are rapidly moving, and intermolecular interactions are constantly being broken and reformed. In this state, the likelihood of crystal formation is low because the molecules are moving so rapidly, even in instances when a large amount of sugar is present relative to water. This type of solution is referred to as a supersaturated solution. However, as soon as the mixture begins to cool (i.e. the molecules stop moving rapidly), the solution contains more sugar than it normally can dissolve at that particular temperature. In this state, even a small disturbance (like a dust particle) will induce the formation and propagation of sugar crystals. As the crystals form, the surrounding solution becomes less concentrated with sugar molecules. When the "new" solution reaches a saturated sugar concentration at the new, cooled temperature, the crystal stops growing until the temperature decreases further. The interplay between sugar molecules that are in solution (surrounded by water) and in the crystalline form (sitting in the sugar/water syrup) continues as the conditions change due to temperature and concentration changes. If this seems complicated, it is! How do these crystals form and grow?

A *crystal seed* is a surface to which other sugar molecules can attach, increasing the size of the crystal. A seed can be a few sugar molecules that come together during the movement of the syrup, which will be facilitated by stirring since the molecules will be more likely to bump into one another. Other events may also induce crystallization, such as the syrup splattering on the side of the pot (which may quickly harden or crystallize). If stirred back into the pot, those crystals may act as a seed! Dust particles or air bubbles may act as seeds. A sitting, metal spoon acts as a seed due to its ability to conduct heat; it can cool a small area of a syrup, leading to a super-supersaturated solution that will quickly crystallize. Thus, the candymaker must know when and how much to stir, as well as which type of utensil (typically a wooden spoon is best) should be used for stirring.

What about the effects of temperature? As with chocolate, the temperature at which crystallization is initiated is very important. Generally, hot syrups produce large, coarse crystals, while cool syrups produce fine crystals. Why? Remember that molecules move faster in a hot solution; once a crystal starts to form, more sugar molecules run into and attach to the *crystal seed*, and the crystals grow rapidly. At the same time, stable *crystal seeds* are less likely to form, as a small complex of molecules is more likely to be broken when the other molecules hit it.

Therefore, crystals made from a hot syrup are larger but smaller in number. In contrast, in a cooler syrup, more *crystal seeds* are present, but slower molecular movement results in fewer molecules attaching to each crystal and slower crystal formation overall. The result: candies that crystallize from hot solutions have a coarse texture, while recipes that require a smooth, creamy texture (i.e. fudge or fondant) call for the syrup to be cooled dramatically before initiating crystallization by stirring (Box 13.5).

How does stirring induce crystallization? Stirring favors the formation of *crystal seeds* by pushing sugar molecules into one another. A syrup that is not stirred will develop only a few crystals, while one that undergoes constant stirring will have a large number of crystals. The more crystals there are in the syrup, the fewer free-floating molecules and the smaller the average size of each crystal. In short, the more a syrup is stirred, the finer the consistency of the final candy. For a smooth and creamy fudge, stir constantly until the process is finished.

In some candies (like transparent, hard, butterscotch disks), a candymaker does not want crystals to form at all. Crystallization is prevented by cooling the syrup very rapidly, such that the sugar molecules stop moving before they have a chance to form any crystals. In these candies, the water content is only 1–2% (making the syrup very viscous); if cooled quickly, the sucrose molecules never have a chance to order themselves into crystals and settle into a disorganized mass called a glass. Just like window glass, sugar glass is brittle and transparent. Transparency is the result of individual sugar molecules being too small to deflect light when they are randomly arranged. Crystalline solids appear opaque because the surfaces of even tiny crystals are big enough to deflect light.

What if the candymaker wants to stop crystallization at a certain point? This can be accomplished by the addition of other ingredients because a new, unique molecule that is unlike the molecules that make up the crystal will effectively block crystallization. These other ingredients, such as corn syrup, milk products, fats, or acid, may also be used to add flavoring or other textural components to the candy (Figure 13.20).

A variety of candies are made using this general method and are consumed and adored by the masses, including non-crystalline candies, crystalline candies, and candies whose texture is modified with other agents (like gums and gels). The best way to make this process come to life is by looking at some specific examples.

Box 13.5 Science of the fondant

Without starting a confectionery war, both buttercream and fondant are great icings for a cake or other baked goods. Fondant is either a soft spreadable icing (often used as filling for pasties) or a more dense, doughlike icing that can be rolled out and shaped to cover cakes and baked goods (Figure 13.19). Made from a sugar syrup using more than one form of sugar (which limits crystallization), fondant provides a pliable and active way to decorate a cake. A basic recipe for rolled fondant includes confectioner's sugar (finely ground table sugar/sucrose), gelatin (a protein that holds water to act as a gelling agent and also interferes with crystalline sugar formation), glycerin (also known as the three-carbon sugar, glycerol), and/or glucose (used to limit the crystal sucrose seeds from forming and to maintain the pliability of the final fondant). Other ingredients can also be added like shortening (fat) to limit the crystallization and add creaminess to the final product and acids (i.e. vinegar, citric, or tartaric) which increase the inversion of sugar. If you recall a discussion from earlier in the chapter, inversion occurs when some of the sucrose disaccharide is cleaved or hydrolyzed to the individual building blocks of glucose and fructose. A very thorough study published by Mary Stephens Carrick in the American Chemical Society *Journal of Physical Chemistry* investigated the volume of water, amount of sugar, and tartaric acid needed to make a consistent fondant. She also found that how the fondant was boiled and treated is critical for a pliable and well-prepared fondant. The final recipe for her fondant is 1 cup of sugar, 3/4 cup of water, and 1/8th teaspoon of cream of tartar. Melt all components and carefully remove the crystals that develop on the pan at the water level, as the crystals will give the fondant an unwanted crunch. Continue to heat for approximately 15 minutes allowing the solution to reach 240°F/116°C. (The next few steps are critical to avoiding the formation of crystals.) Remove from the stovetop without jarring and let stand, allowing all remaining bubbles to rise to the surface. Slowly pour the solution into a shallow platter and cool to 104°F/40°C. Beat the cooled mixture for exactly 3 minutes (she tested the time!) using a wooden spoon with a circular motion (not a cut or folding method), followed by kneading into a soft, moist mass. Kneading, Dr. Carrick found, reduced the stickiness of the fondant. Dr. Carrick's article is a great read, a meticulous approach and an interesting set of experiments for a new connectionist/baker to carry out their craft [6].

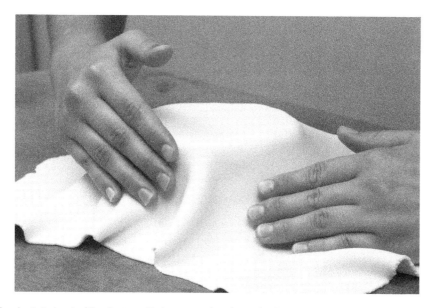

Figure 13.19   Fondant. A sheet of fondant applied over a cake. photopitu/Adobe Stock Photos

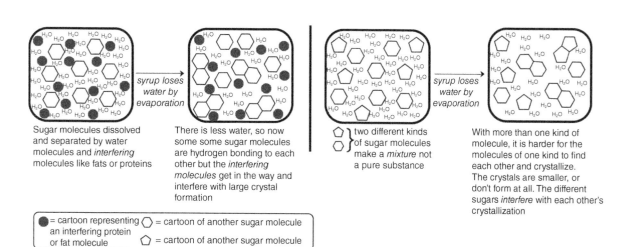

syrup loses water by evaporation

Sugar molecules dissolved and separated by water molecules and *interfering* molecules like fats or proteins

There is less water, so now some some sugar molecules are hydrogen bonding to each other but the *interfering molecules* get in the way and interfere with large crystal formation

two different kinds of sugar molecules make a *mixture* not a pure substance

syrup loses water by evaporation

With more than one kind of molecule, it is harder for the molecules of one kind to find each other and crystallize. The crystals are smaller, or don't form at all. The different sugars *interfere* with each other's crystallization

= cartoon representing an interfering protein or fat molecule

= cartoon of another sugar molecule

= cartoon of another sugar molecule

Figure 13.20   Impact of fat and protein on sugar crystal formation.

13.20 Non-crystalline Candies: Hard Candies and Caramels

13.20.1 Hard Candies

Hard candies, such as butterscotch, clear mints, and lollipops, are the simplest non-crystalline candies in terms of preparation (Figure 13.21). These types of candies have what is called a glass structure because the molecules are not arranged in a stable, ordered, regular pattern of a crystal. If you are thinking that disordered molecules normally lead to a liquid state, you are correct. However, because hard candies have really low water content, they cannot remain in a liquid form. How is this glassy, disorganized structure achieved?

Figure 13.21 Hard candies. Made by boiling sugar water and flavoring. Hayati Kayhan/Adobe Stock Photos

Hard candies are made by boiling sugar, water, and any other ingredients to the hard-crack stage (300–310°F/149–154°C) where the syrup contains only 1–2% moisture. Corn syrup and acids (i.e. tartaric and citric) may be added (as they prevent formation of crystals); notably, crystal formation would lead to an opaque candy with a gritty texture. How does corn syrup or acid interfere with crystallization? In corn syrup, the long glucose polymers are so large, relative to a single molecule of glucose, that they, in effect, interfere with individual glucose molecules from finding each other to form a crystal. Tartaric and citric acids add flavor and induce the chemical reaction that breaks sucrose into its two monosaccharide components, fructose and glucose. Having two dissimilar simple sugars prevents the ordering necessary to form a crystal. Given that any crystallization will ruin the batch of hard candy, once the syrup comes to a boil, no further stirring is done. Moreover, any sugar crystals that form on the side of the cooking vessel are washed from the walls of the pan to prevent the formation of "seed crystals" that would make the whole batch of candy suddenly crystallize.

After cooking the syrup to the hard-crack (310°F/154°C), the low moisture content stage is cooled to 275°F, and color and flavoring are kneaded in while the mixture is still malleable. It is then poured onto a baking pan or marble slab, where it cools quickly to "freeze" the molecules in place so that they do not rearrange themselves into a crystalline form.

13.20.2 Caramels

The chewy, rich, mouthwatering caramel is another example of a non-crystalline candy (Figure 13.22). This may be a surprising revelation given the dissimilarities between caramels and lollipops. However, if you have ever tasted gritty caramel, you have experienced crystallized caramel. The chewy texture of caramel comes from the relatively low temperature at which the syrup is cooked (a mere 245°F), which allows for moisture retention. The expansive ingredient list that includes milk fats and solids gives the caramel richness and its characteristic brown color. Depending on how you make the caramel, you will get the color and flavor from just caramelization reactions (if no protein is present) or a combination of the two reactions (if protein is available) (See Chapter 3 for more details). Let's talk about the process of making caramels first and then get into some of the science.

If you look through 10 different recipes for caramel candy, you will find 10 recipes that vary in the milk products used and the exact temperature at which the syrup is boiled. These differences will result in great variation in the softness, color, and richness of the final product. Generally, though, many caramels are made by cooking sugar, cream, butter, and a small amount of corn syrup to a temperature between 242 and 248°F/117–120°C. As the syrup is heated to this "firmball" stage, it is stirred, but once the desired temperature is reached, stirring ceases to minimize crystallization. The thick syrup is poured into a metal pan, allowed to cool, and cut into bite-sized pieces of caramel candy for eating. This sounds very similar to the preparation of hard non-crystalline candies. What about the differences?

The lower boiling temperature used in caramels increases moisture content, resulting in a softer, more chewy texture. The chewiness also comes from the corn syrup and milk proteins. Not only do the corn syrup and milk interfere with crystallization but also with the formation of a glass structure. The characteristic caramel flavor and brown color come from the Maillard reaction that occurs between the sugar and milk proteins, since we have the three necessary ingredients: heat, sugar, and protein. The aldehyde group of a sugar molecule reacts with the amine group on a protein, resulting in the characteristic brown caramel color and flavor.

Perhaps as you are reading this section, your mouth is watering, just like it does when you eat a caramel. Caramels are mouthwatering because the process of chewing them releases butterfat from the mass of gooey, chewy sugar goodness.

Figure 13.22 Caramels. A different non-crystalline candy form. Brad Pict / Adobe Stock Photos

13.21 Crystalline Candies: Rock Candy and Fudge

13.21.1 Rock Candy

As a child, you may have enjoyed a bag of large, coarse, colored sugar crystals, also known as rock candy (Figure 13.23). Rock candy is one of the easiest candies to prepare and provides a great example of large crystal formation from a sugar syrup. In short, you boil sugar and water to the hard ball stage (250–268°F) and then pour the syrup into a small container (like a glass or jar) that contains a toothpick or *crystal seed* around which the crystals can form. That's it! After a few days of undisturbed sitting, beautiful and large sugar crystals will have formed that will impress your friends. You can extend your experiment by stirring or disturbing the solution, adding flavoring or coloring, or boiling to different temperatures and noting the differences in crystal size. Rock candy is (by far) the easiest type of candy to prepare.

13.21.2 Fudge

Crystalline may not be a word that immediately comes to mind when you think about fudge. After all, good fudge has a creamy and smooth consistency, just like caramel. However, fudge is a crystalline candy, and the microcrystals that are present in fudge are key to making it … fudge (Figure 13.24). These tiny, very fine crystals give the

Figure 13.23   Brown sugar rock candy. eillen1981/Adobe Stock Photos

Figure 13.24   Fudge. Kimberly Reinick/Adobe Stock Photos

fudge its firm texture and are small enough to not feel gritty on the tongue. Thus, the formation of the right type of crystal is key to fudge making. How does the proper crystallization occur? In fudge, it is all about the cooling, not the cooking. Cooking fudge is similar to the preparation of caramel in terms of some added ingredients (i.e. milk solids, milk fat, corn syrup), along with chocolate solids. These ingredients are boiled to the soft ball stage (234–240°F/112–116°C), which is slightly lower than the boiling temperature required for most caramels. The exact boiling temperature for fudge is dependent upon the ratio of other ingredients (e.g. cream to sugar), as well as altitude and humidity. If the boiling temperature is too low, the fudge will be runny, and if the boiling temperature is too high, the fudge will be hard. Thus, the fudge maker may have to experiment (or fudge) with temperature to achieve the right consistency. The next point of distinction between fudge and caramel comes after boiling. The fudgy hot syrup is allowed to cool undisturbed to a temperature of 100–130°F/38–54°C, without any stirring (which would promote crystal formation). If crystals form at this stage, the crystals will be too large, and the fudge will have a gritty, grainy texture. You may be wondering when and how the fudge microcrystals form. In fudge, crystallization is induced after the fudge has cooled, at which time the fudge is stirred/beaten continuously for about 15 minutes until the mixture is thick. The more stirring, the larger the number of small crystals that will form, yielding a smoother candy.

13.22   Aerated Candies: Marshmallows

Jumbo-sized, miniature, bunny-shaped, and the most beloved Peeps™. You can't really talk about the science of candy (or have a campfire) without talking about soft, sticky, and spongy marshmallows (Figure 13.25). Before we discuss the science, let's talk about the name.

The marshmallow confection first came from the mallow plant (*Althaea officinalis*), a weedy relative of the hollyhock that grows in marshes, whose roots contain mucilage. Mucilage is a thick, gluey sap produced by some plants and microscopic animals to help with food storage and seed germination. People of ancient cultures used the plant for both culinary and medicinal purposes. For example, the ancient Egyptians dried the root and mixed it with honey to make mallow treats, while the French, who were introduced to it in the early to mid-1800s, experimented using its gummy juice to soothe sore throats. The French were the first to make "marshmallows" that resemble the confection that we enjoy today by mixing sap from the marshmallow plant with eggs and sugar and beating the mixture into a foam. Marshmallows were so popular in the late nineteenth century that candymakers were unable to keep up with the demand. In response, the modern version of the marshmallow was born, which does not rely on the mallow plant at all. Today, manufactured marshmallows are made by combining gelatin (a protein solution) with a sugar syrup (i.e. sugar and corn syrup) and cooking the mixture to 240°F/115°C. The mixture is whipped into a foam that is two to three times the original volume. The marshmallow foam develops from millions of air bubbles that are stabilized and trapped by the protein molecules as the mixture cools and the gelatin sets. A marshmallow is, in essence, a solid foam that is only 35–45% as dense as water.

Where does the difference in texture and shape of various marshmallow products, such as marshmallow fluff, a Jet-Puffed™ marshmallow, or Peeps come from? Common types of marshmallows have a 1 × 1-inch cylinder shape because in marshmallow factories, the liquid foam is piped through a long, 1-inch diameter tube as it cools. The marshmallow "rope" that emerges from the tube is chopped into pieces that are approximately 1-inch in length. The texture of a marshmallow product can be controlled by adjusting the proportions of ingredients and the amount of whipping. With more whipping, more air bubbles will be incorporated into the foam and a softer marshmallow will result. These soft marshmallows might be great for eating but don't have a firm enough texture

Figure 13.25   Marshmallows. The familiar combustion of sugar and oxygen catalyzed by the heat of a campfire.

to be covered in chocolate or another candy coating, which requires a more firm marshmallow. What happens when a marshmallow is roasted? The fire's heat both melts the gelatin and caramelizes the sugar (Maillard and caramelization reactions), producing a hot, gooey, caramel-flavored concoction.

Science for the Chef: Phase transitions in candy making: chocolate-covered honeycomb

Sponge toffee, also known as honeycomb candy, is a delightful treat that beautifully demonstrates the principles of chemistry in candy making. This airy confection is created by carefully controlling the phase transitions of sugar and utilizing the production of carbon dioxide to form its characteristic bubbly structure. By adding baking soda to a hot sugar syrup, a dramatic foaming reaction occurs, trapping air within the mixture as it cools rapidly into a brittle, glassy solid. When coated in smooth, tempered chocolate, this candy not only provides a delicious contrast of textures but also highlights the intricate science of crystallization and gas production in confectionery. This dessert beautifully illustrates the principles of phase transitions, gas production, and crystallization in candy making. By understanding the chemistry behind these processes, one can create perfect honeycomb toffee with a light, airy texture and a delicious chocolate coating, making it both a scientific and culinary delight.

Ingredients:
- 1 cup granulated sugar
- 1/4 cup honey or corn syrup
- 1/4 cup water
- 1 1/2 teaspoons baking soda, sifted
- 8 ounces tempered dark chocolate (for coating)

Instructions:
1. **Prepare the sugar syrup:**
   - Line a baking sheet with parchment paper or lightly grease it to prevent sticking.
   - In a medium-sized, heavy-bottomed saucepan, combine the sugar, honey (or corn syrup), and water. Stir gently to combine.
   - Heat the mixture over medium-high heat without stirring. Allow the sugar to dissolve and the mixture to come to a boil. If needed, gently swirl the pan to ensure even heating.
2. **Cook the syrup:**
   - Continue to cook the sugar syrup until it reaches 300°F (150°C) on a candy thermometer, also known as the hard-crack stage. At this temperature, the sugar mixture will be clear and have a slightly amber color.
   - Remove the pan from heat immediately to prevent the sugar from burning.
3. **Create the honeycomb structure:**
   - Quickly and carefully sift the baking soda into the hot sugar syrup. Stir rapidly but gently with a wooden spoon or heatproof spatula to incorporate the baking soda.
   - The mixture will foam up dramatically as carbon dioxide is released, creating the characteristic airy, honeycomb structure of the sponge toffee.
4. **Set the toffee:**
   - Pour the foamed mixture onto the prepared baking sheet, spreading it out slightly but not flattening it. The goal is to retain as much air as possible.
   - Allow the toffee to cool completely at room temperature. It will harden and become brittle as it cools.

5. **Coat with chocolate:**
   - Once the toffee has fully hardened, break it into pieces of the desired size.
   - Dip each piece into the tempered dark chocolate, ensuring it is evenly coated. Place the chocolate-covered toffee on a parchment-lined tray to set.
   - Allow the chocolate to harden at room temperature or in the refrigerator before serving.

Science Behind the Recipe

Phase transitions: The key to making honeycomb (sponge toffee) lies in understanding the phase transitions of sugar during cooking. When sugar is heated, it first dissolves in water, creating a syrup. As the temperature increases, water evaporates, and the sugar concentration increases, leading to the different stages of sugar cooking (thread, soft ball, firm ball, etc.). For honeycomb, the syrup is cooked to the hard-crack stage, where the sugar reaches 300°F (150°C). At this stage, the sugar molecules are close to the point of caramelization and will form a glassy, brittle structure upon cooling.

Carbon dioxide production: The addition of baking soda (sodium bicarbonate) to the hot sugar syrup is the critical step in creating the honeycomb structure. The heat causes the baking soda to decompose, releasing carbon dioxide gas (CO_2). This gas expands within the hot, viscous syrup, creating bubbles that are trapped as the sugar rapidly cools and solidifies. The result is the characteristic airy, porous structure of honeycomb toffee.

Sugar crystallization: Once the sugar syrup has reached the desired temperature and the honeycomb structure is formed, the cooling process begins. As the syrup cools, it transitions from a liquid to a solid, but in the case of honeycomb, it solidifies into an amorphous, glass-like structure rather than a crystalline one. This is due to the rapid cooling and the presence of other ingredients like honey or corn syrup, which interfere with sugar crystallization.

Chocolate coating: The final step involves coating the honeycomb in tempered chocolate. Tempering chocolate ensures that the cocoa butter crystallizes in a stable form (specifically the βV crystal structure), which gives the chocolate a glossy finish, a crisp snap, and a smooth mouthfeel. The tempered chocolate also provides a protective layer around the honeycomb, adding a contrasting texture and preventing the toffee from becoming sticky or absorbing moisture from the air.

Key Concepts

1. Sugar is a **disaccharide**, made of one **glucose** molecule and one **fructose** molecule, connected by an **oligosaccharide** bond. Because **sucrose** has many —OH functional groups, it can have many **intermolecular interactions** with water, contributing to its **solubility** in water and moisture retention in baked goods. The sugar particles in granulated, white sugar are in a crystalline form.

2. Powdered sugar is very finely ground white sugar. Brown sugar is a sucrose-based sugar that also contains molasses. The soft, moist texture of brown sugar is due to the **hygroscopic** nature of the molasses.

3. *Inversion* occurs when sucrose is heated in the presence of acid or the enzyme **invertase**, which breaks the **oligosaccharide bond** to release free glucose and fructose. Invert sugars are more **hygroscopic** than sucrose because of additional water that can surround the sugar molecules through intermolecular interactions.

4. **Honey** is a highly viscous liquid sweetener that contains fructose, along with glucose and sucrose. Fructose contributes to the very sweet taste of honey and honey-based products. **High-fructose corn syrup** is produced from the starch in corn, which is first broken into smaller molecular components by the enzyme amylase. Some of the smaller units of glucose are then enzymatically **isomerized** by xylose isomerase to fructose, yielding the very sweet, viscous liquid. **Molasses** is made from the by-product juice left from extracting sugar from sugarcane. Following several boilings of this juice extract, molasses is produced through the concentration and caramelization of the sugars that occur during the repeated boiling process.

5. Chocolate production involves five steps: **fermentation**, **roasting**, **grinding/milling**, **conching**, and **tempering**. During the fermentation of cacao beans, yeast and bacteria ferment the sugars in the cacao bean pulp, yielding a variety of metabolic products such as CO_2, ethanol, lactic, citric, and acetic acids. During conching, the **chocolate liquor** and additives (i.e. sugar, fats, and flavorings) are mixed and kneaded under heat conditions, allowing for the hydrophilic to separate from one another and be coated with cocoa butter particles so that the chocolate flows smoothly. Tempering involves careful heating and cooling of chocolate liquor to ensure that the chocolate solidifies in the ideal crystal structure for a good snap and visibly appealing chocolate. At all stages of chocolate production, the molecules that are associated with the aroma and taste of chocolate are produced.

6. **Chocolate bloom** occurs when chocolate is not tempered properly. In bloomed chocolate, a compact **crystalline** form of chocolate is produced, which pushes the **cocoa solids** and sugar particles to the surface, leaving a visibly white/grayish film on the surface. Chocolate bloom is less likely to occur in baked goods due to fat migration.

7. Mixing a very small amount of water with liquid chocolate will cause the chocolate to **seize** due to the disruption of the proper intermolecular interactions necessary for smooth, flowing chocolate.

8. Different types of chocolate are identifiable by the amount (%) of chocolate liquor/solids, sugar, and milk solids. **Cocoa powder** is an unsweetened form of chocolate that can be used in baking and in drinks. **Dutch-processed cocoa powder** is less acidic and astringent than natural cocoa powder.

9. Non-chocolate candies are prepared from a sugar/water solution or **syrup**. The sugar syrup is boiled to a particular temperature or until a behavior is exhibited by the syrup. The value of the boiling point of the sugar syrup is indicative of the amount of dissolved sugar found in the syrup.

10. The formation of **sugar crystals** is dependent upon the temperature and conditions of the syrup. Crystallization can be induced by a **crystal seed**, the syrup splattering on the side of the pot, dust particles or air bubbles, or a sitting metal spoon. Hot syrups produce large, coarse crystals, while cool syrups produce fine crystals. Candies that crystallize from hot solutions generally have a coarse texture, while candies with a smooth, creamy texture are produced with crystals that were generated under cooler conditions.

11. Caramels and hard, glass-structured, candies are examples of **non-crystalline** candies. Both candy types do not involve extensive stirring and include a variety of ingredients that inhibit the development of a regular crystalline pattern while exhibiting great temperature variation in their preparation. Fudge and rock candy are examples of crystalline candies, where fudge crystals are produced through extensive stirring under low temperature conditions and rock candy crystals are large crystals produced under high temperature conditions.

References

1 OECD (Organisation for Economic Co-operation Development)/FAO (Food and Agricultural Organization of the United Nations). (2021). *OECD-FAO Agricultural Outlook 2021–2030*. Paris: OECD Publishing. https://doi.org/10.1787/19428846-en.

2 Philippe Sylvestre, D. (1685). *Traités nouveaux & curieux du café du thé et du chocolate*, Lyon. Dufour, Philippe Sylvestre, Adriaen Moetjens, Alexandre Toussaint de Limojon De Saint-Didier, Estavayer.

3 Which Country Eats the Most Chocolate in 2024? World Population Review. https://worldpopulationreview.com/country-rankings/which-country-eats-the-most-chocolate (accessed July 25, 2024).

4 Rational Stat. (2022). *How Much Chocolate is Consumed Per Year*? https://rationalstat.com/chocolate-consumption-per-year/.

5 Schwan, R. F. & Wheals, A. E. (2004). The microbiology of cocoa fermentation and its role in chocolate quality. *Crit Rev Food Sci Nutr.* 44(4), 205–221.

6 Carrick, M. S. (1919). Some studies in fondant making. *J. Phys. Chem.* 23(9), 589–602.

Additional Readings

Aprotosoaie, A.C., Luca, S.V. & Miron, A. (2016), Flavor Chemistry of Cocoa and Cocoa Products—An Overview. *Compr Rev Food Sci Food Saf*, 15(1), 73–91.

Hartel, R.W., Ergun, R. and Vogel, S. (2011), Phase/State Transitions of Confectionery Sweeteners: Thermodynamic and Kinetic Aspects. *Compr Rev Food Sci Food Saf*, 10(1), 17–32.

McGill, J. & Hartel, R.W. (2018), Investigation into the Microstructure, Texture and Rheological Properties of Chocolate Ganache. *J. Food Sci*, 83(3), 689–699.

Quelal-Vásconez, MA, Lerma-García, MJ, Pérez-Esteve, É, Talens, P, & Barat, JM. (2020) Roadmap of cocoa quality and authenticity control in the industry: A review of conventional and alternative methods. *Comp Rev Food Sci Food Saf*. 19(2), 448–478.

Valverde García, D, Pérez Esteve, É, & Barat Baviera, JM. (2020) Changes in cocoa properties induced by the alkalization process: A review. *Compr Rev Food Sci Food Saf*, 19(4), 2200–2221.

Wang, Y., Luo, X., Chen, L., Mustapha, A. T., Yu, X., Zhou, C., & Okonkwo, C. E. (2023). Natural and low-caloric rebaudioside A as a substitute for dietary sugars: A comprehensive review. *Compr Rev Food Sci Food Saf*, 22(1), 615–642.

End of Chapter 13 Questions

1. What are the two monosaccharide components of sucrose?
   a. Glucose and fructose
   b. Fructose and galactose
   c. Maltose and glucose
   d. Lactose and fructose

2. Which of the following are necessary for a Maillard reaction to occur?
   a. sugar
   b. heat
   c. protein
   d. a and b
   e. a, b, and c

3. What is one difference between white sugar and powdered sugar?
   a. The size of the sugar particles/crystals
   b. White sugar is sucrose sugar, powdered sugar is fructose sugar
   c. White sugar dissolves more easily in a cold liquid
   d. Powdered sugar contains white flour, which is why it is white.

4. What types of reactions are involved in molasses production?
   a. Caramelization
   b. Maillard
   c. Hydrolysis
   d. a and b
   e. a and c

5. What are the five steps of chocolate production?
   a. Acidification, fermentation, roasting, grinding, heating
   b. Fermentation, roasting, grinding, conching, tempering.
   c. Fermentation, roasting, conching, heating, cooling
   d. Acidification, grinding, fermentation, tempering, cooling

6. What occurs during the fermentation stage of chocolate production
   a. Lactobacilli metabolize glucose to generate lactic acid and citric acid
   b. Yeast metabolize sugar to produce ethanol and CO_2
   c. Acetic acid bacterial oxidize ethanol to water and CO_2
   d. The acidity of the fermenting pulp produces holes in the cacao beans
   e. a and c
   f. All of the above

7. Which of the following statements about chocolate are not true.
   a. White chocolate is not actual chocolate
   b. Chocolate has six different crystalline forms that it can transition between depending upon heating and cooling temperatures
   c. Cacao beans contain high amounts of fat and sugars, so additional sugar does not need to be added to most chocolates
   d. Chocolate bloom is not harmful to humans.

8. Which of the following is NOT an example of a crystalline candy.
   a. Hard, butterscotch disk candy
   b. Fudge
   c. Rock Candy
   d. Chocolate

9. Define invert sugar and how invert sugar results in a chewy cookie.

10. Why is cocoa butter added to a chocolate liquor?

11. Describe why chocolate bloom occurs and how a candymaker can recover their chocolate, after it has undergone blooming.

12. What is the difference between natural and Dutch-processed cocoa?

14

The Science of Taste, Smell, and Flavor

Guided Inquiry Activities (Web): 13, Flavor

Learning Objectives

1. *Understand the relationship between taste, smell, and flavor.*
2. *Define the terms gustation, olfaction, and chemoreception.*
3. *Identify the five basic taste classes: sweet, bitter, sour, salty, and umami, and recognize the structures of ions or molecules responsible for each taste.*
4. *List the steps involved in the sensory transduction of taste and smell.*
5. *Explain how taste and smell receptors interact with tastants and odorants.*
6. *Describe the physiological process by which taste receptor cells detect tastants and transmit signals to the brain.*
7. *Discuss the evolutionary significance of taste as a survival mechanism.*
8. *Analyze the role of membrane proteins and ion channels in the sensory transduction of salty and sour tastes.*
9. *Compare the signal transduction mechanisms of sweet, bitter, and umami tastes using G protein-coupled receptors (GPCRs).*
10. *Evaluate the impact of genetic variation in taste receptors on individual differences in taste perception.*
11. *Illustrate how retronasal olfaction contributes to the overall experience of flavor.*
12. *Summarize the impact of texture, temperature, and pain on the perception of food and beverages.*
13. *Assess the effects of aging on the senses of taste and smell, and how this influences dietary habits.*
14. *Use the principles of chemoreception to predict the response of sensory systems to various food substances.*

14.1 Introduction

Our senses of taste and smell dictate how we perceive the flavor of our food and drink. This chapter will help you understand the molecules and the processes by which we use our senses of taste and smell in cooking and eating.

You've probably experienced the sensation of familiar food smells. Perhaps you walked through the door of a family member's home at holiday time, or perhaps you passed a favorite street vendor or restaurant. Whatever the circumstance, you not only recognized the food smell, but your body also responded with physical changes. Your mouth began creating saliva, and your stomach rumbled. The smell of the food not only allowed you to recognize it was time to eat, but it signaled your body to prepare for digestion. Now, let's imagine you followed up the experience of smelling your favorite food with the next best thing—eating it! Your tongue detects the texture and basic

The Science of Cooking: Understanding the Biology and Chemistry Behind Food and Cooking, Second Edition. Joseph J. Provost et al.
© 2025 John Wiley & Sons, Inc. Published 2025 by John Wiley & Sons, Inc.
Companion website: www.wiley.com/go/provost/food_science_2e

taste as it works with your teeth to crush the food, pushing airborne molecules into your nasal cavity where you smell the food again. Ultimately, the flavor experience is a combination of taste and smell. Jean Anthelme Brillat-Savarin, an eighteenth-century gastronomist, wrote, "smell and taste are in fact but a single sense, whose laboratory is both and whose chimney is the nose." How we perceive food and drink is much more complicated than what food hits our tongue. Let's investigate these senses further!

14.2 The Physiology of Taste, Smell, and Flavor

Smell or olfaction and taste or gustation are part of the human body's sensory system. Specifically, they function through chemoreception, which is the ability of a body to respond to molecules in the immediate environment. In gustation, your body is responding to molecules referred to as tastants, which are molecules that stimulate special proteins called taste receptors. Similarly, in olfaction, receptors in your nasal cavity are responding to molecules called odorants that stimulate other types of proteins called smell receptors. When we comment on how great a specific food or beverage "tastes," we are really commenting on the flavor of that item as perceived by two sensory systems: the olfactory system and the gustatory system.

Sensory systems are specialized tissues that allow you to respond to your environment. Each sensory system sends information to the brain via electrical signals. We have a variety of senses, and each is tuned to different environmental stimuli: touch, temperature, taste, smell, sight, and sound—just to name a few. Each of the sensory systems responds to a specific sensory input and sends electrical impulses to the brain in the process of sensory transduction. When you drink hot coffee, your touch sensor informs you that there is a liquid in your mouth, while your temperature sensors tell you whether that liquid is hot or cold; finally, your taste and smell sensors interact with the molecules (also called odorants and tastants) that make up the coffee to communicate a message of flavor to your brain.

To best understand the sensory experience of flavor, we must first follow the path food takes when it enters the mouth, and how that food engages the two primary sensory systems of gustation and olfaction. First, an initial sniffing of food brings the odorant and tastant molecules directly to the olfactory nerves via the orthonasal route, where small volatile components of food and drink will bind to and activate receptors within olfactory nerve fibers in the nasal cavity (Figure 14.1). Second, the process of chewing or masticating food will break open the plant or animal cells and add to the mixture of volatile compounds filling the mouth and nasal cavity. The third pathway of flavor perception is caused by tastant molecules binding to specific taste receptors in the taste receptor cells (or TRCs) found in the taste buds of our tongue (Figure 14.2). As the food is swallowed, the mouth is closed and exhaled air brings the new mix of volatile molecules back through the nasal cavity, providing a second round of smells and flavors to be experienced in what is called retronasal olfaction (Figure 14.3).

Flavor molecules—odorants and tastants—that bind protein receptors found in receptor cell membranes initiate a cascade of signaling events within the receptor cell that continue to nerve cells, and eventually alert the brain to the presence of a flavor (Figure 14.4). For example, an odorant molecule traveling in the nasal cavity first encounters protein receptors in the membranes of odorant receptor cells. The signal generated by the odorant molecule when it binds to the receptor is carried to olfactory nerves embedded in the olfactory mucosa. These olfactory neurons will then send their signal to the specialized cells of the olfactory bulb, where a second set of nerves then carry the signal through the thalamus and on to the neocortex, where sensory perception occurs. Similarly, the nerves associated with taste receptor cells transmit their signal to the brain using one of three different nerves: the chorda tympani, glossopharyngeal, or vagus nerve. These nerves converge in the brain stem where they signal to the thalamus. From the thalamus, the taste signal is carried to the neocortex for sensory perception. Ultimately, the perception of flavor by the brain requires receiving and distinguishing signals from tens of thousands of compounds [1].

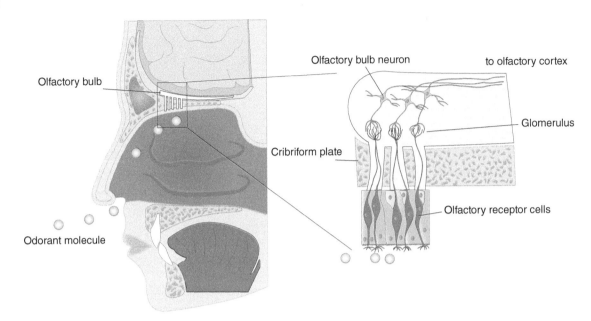

Figure 14.1 Tasting by smell. The olfactory system is able to identify a diverse range of odors and through this system many tastants. Located at the posterior of the nasal cavity, the olfactory bulb is connected to receptor cells exposed to receive volatile compounds. ellepigrafica / Adobe Stock Photos

Anatomy of a taste bud

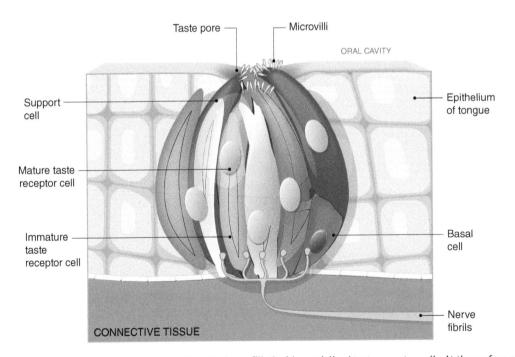

Figure 14.2 Taste bud and taste receptor cells. Taste buds are filled with specialized taste receptor cells. At the surface adjacent to the taste bud, the membrane of the taste receptor cell possesses proteins that bind the flavor molecules and transmit the signal to afferent nerves located deep within the bottom of the taste bud. This is how you taste via the gustatory system. designua/Adobe Stock Photos

Figure 14.3 How we taste our food. Food and drink are sensed in the mouth by smell and through the retronasal pathway odors released by chewing. *Source:* Patrick J. Lynch /Wikimedia Commons/Public domain

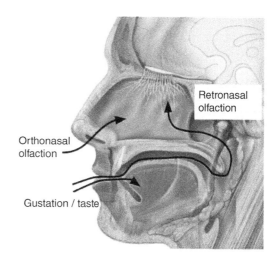

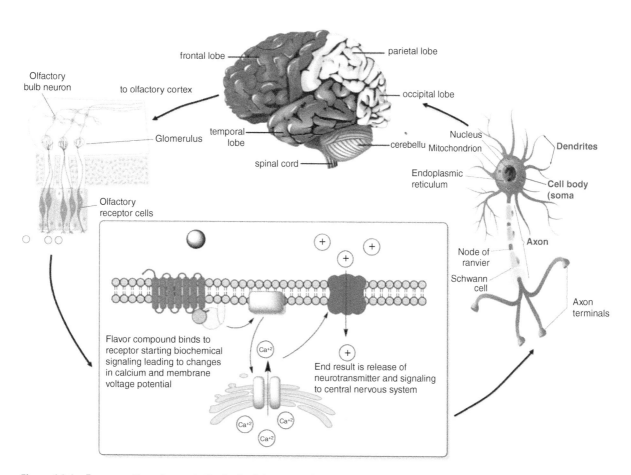

Figure 14.4 From mouth and nose to the brain. Odorants and tastants stimulate receptors in the olfactory and taste bud receptor cells. Depending on the type of receptor activated, biochemical changes lead to membrane depolarization and release of neurotransmitters to the sensory nerves and central nervous system. These signals are then carried to the brain where flavor and smell are perceived. GraphicsRF/Adobe Stock Photos

14.3   Gustation: The Basics of Taste

Of the thousands of different flavors a human can perceive, the gustatory system is responsible for five distinct classes of tastants. To better understand taste distinctions, Tables 14.1–14.5 identify different tastants by the sensory system they activate and the intensity of that activation; this is called a taste index. For each set of tastants, one molecule is arbitrarily assigned a value of 1, and the other taste molecules are compared to it.

14.3.1   Sweet Taste

Sweet is commonly described as having the pleasant taste characteristic of sugar or honey. This sweet taste is not initiated by a single classification of chemicals. The chemicals that stimulate sweet tastes include sugars, glycols, aldehydes, esters, some amino acids, and some small proteins (see Table 14.1). Note that most of the chemicals that stimulate sweet taste are organic molecules.

   Sucrose is a common table sugar and the sweetener we most frequently use in baking. The taste index sets the sweetness level of sucrose at 1. Sucrose is a disaccharide composed of two monosaccharides, glucose and fructose. Notice that in terms of sweetness, glucose tastes less sweet than sucrose while fructose tastes sweeter than sucrose. This is one reason that high fructose corn syrup is used as a commercial sweetener. You get the same level of sweetness with a lower amount of sugar. On the sweetness scale, however, these natural sugars are not leaders. For example, the artificial sweeteners, Equal (aspartame), Splenda (sucralose), and saccharine taste 200, 600, and 675 times as sweet as sucrose. Artificial sweeteners are designed to trick our taste buds into believing we are eating high-calorie sweet food without the calories. This, in theory, allows people to enjoy their favorite sweet-tasting beverages without the additional calories provided by natural sugars. Experiments have found that sweet taste, regardless of its caloric content, enhances your appetite. This means that drinking diet beverages containing artificial sweeteners can lead to an increased food intake, thus undermining a person's effort to reduce calorie intake or lose weight. Aspartame has been found to have the most pronounced effect in this area, but the same applies for other artificial sweeteners, such as acesulfame potassium and saccharin [2].

14.3.2   Bitter Taste

Bitter relates to having a sharp, pungent taste. Similar to sweet taste, bitterness is not caused by a single class of chemicals. Two particular classes of substances typically induce a bitter taste. The first class of substances is the alkaloids, which are a group of naturally occurring chemical compounds containing basic nitrogen atoms (see Table 14.2). Alkaloids such as quinine, caffeine, strychnine, and nicotine are commonly used in medicinal drugs. The second class is made of long-chain organic substances containing nitrogen. It should be noted that relatively small chemical modifications can change a chemical substance from sweet to bitter.

   The taste intensity of different bitter substances does not vary as widely as they do for sweet tastes, but the response to these bitter substances can be quite dramatic (Table 14.2). In this category, quinine is used to set the baseline of one. Quinine is a natural, white-colored crystalline alkaloid with a variety of therapeutic properties. Quinine has been used as a fever reducer (i.e. antipyretic) and an anti-inflammatory agent. The most common use for quinine, however, is in the prevention and treatment of malaria. The Union army alone purchased 595 544 four-ounce tins of quinine sulfate (Figure 14.5) during the American Civil War [3].

   You may have also heard of the presence of quinine in tonic water. As initially developed in India, tonic water was a carbonated soft drink with a significant amount of quinine dissolved in it. The original purpose was as a prophylactic preventative of malaria. Tonic water today still contains quinine but at a much lower concentration. The US Food and Drug Administration limits the amount of quinine in tonic water to 83 mg/L. This is substantially lower than the therapeutic range for quinine, which is 167–333 mg daily. Many of the medicines we take are accompanied by a bitter taste similar to that of quinine.

Table 14.1 Taste classifications, tastant structure, and taste index for sweet compounds.

Taste classification	Substance	Taste index	Structure
Sweet	Sucrose	1	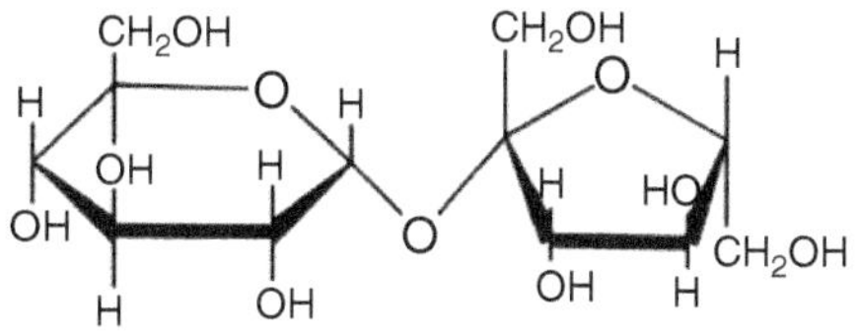
	Glucose	0.8	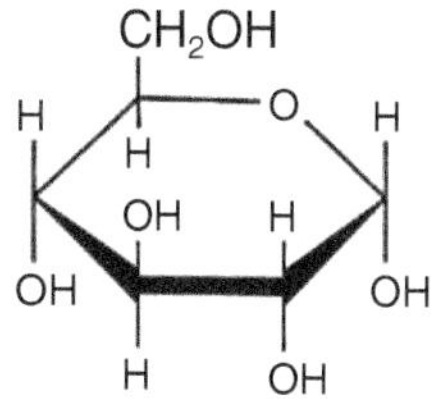
	Fructose	1.7	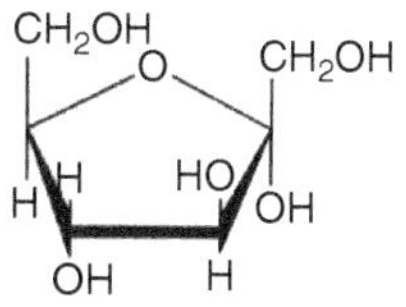
	Chloroform	40	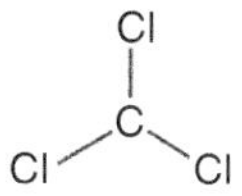
	Aspartame (Equal)	200	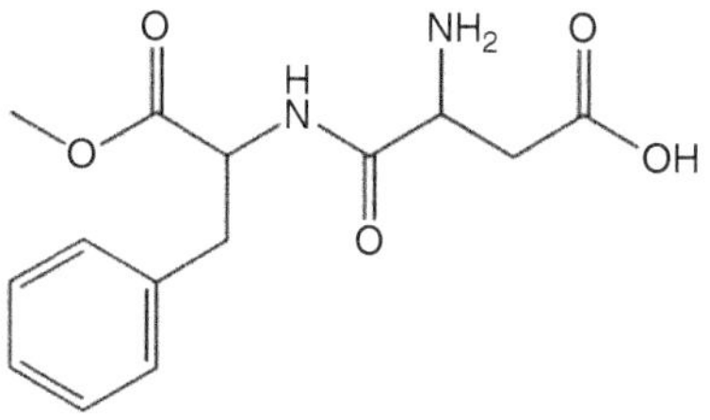
	Sucralose (Splenda)	600	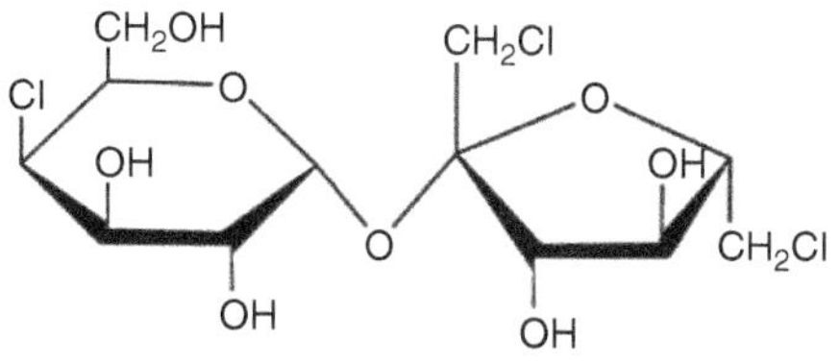
	Saccharin	675	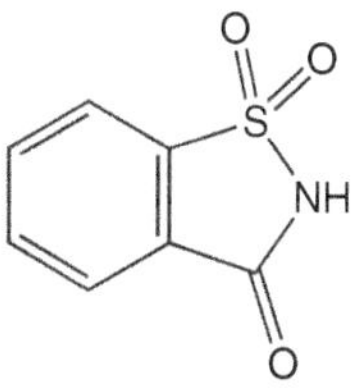

Table 14.2 Taste classifications, tastant structure, and taste index for bitter compounds.

Taste classification	Substance	Taste index	Structure
Bitter	Quinine	1	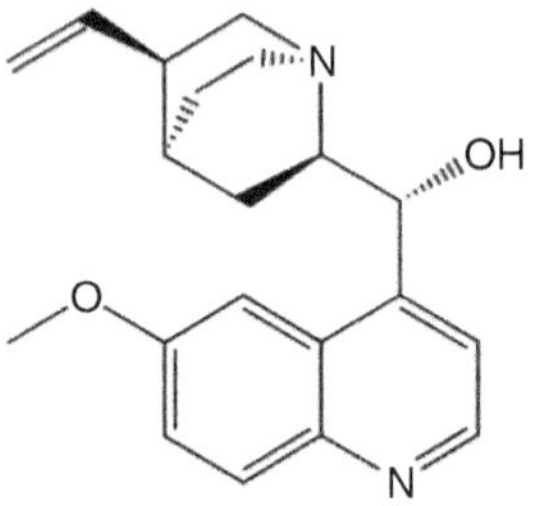
	Caffeine	0.4	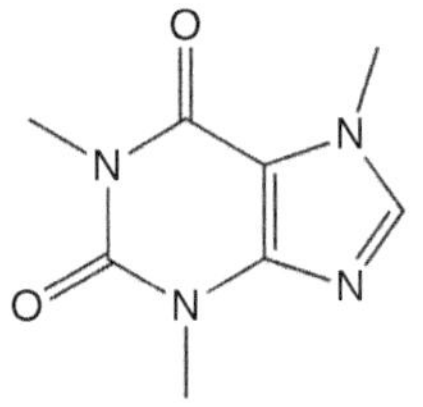
	Nicotine	1.3	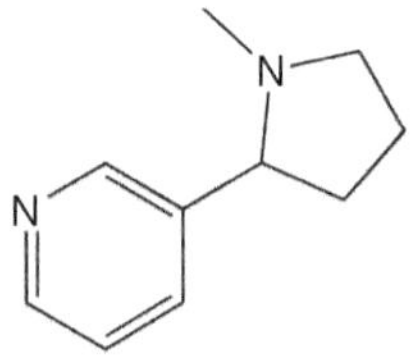
	Strychnine	3.1	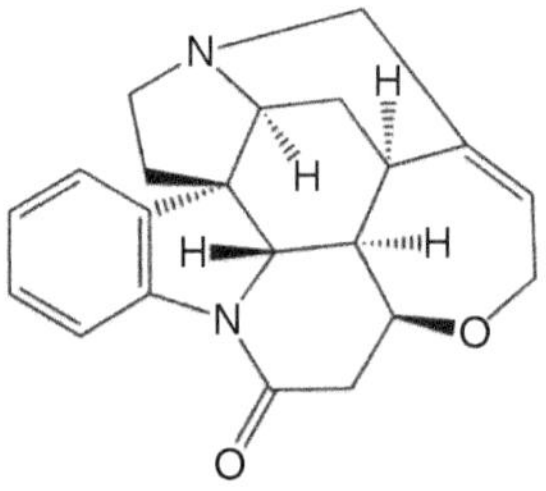
	Brucine	11	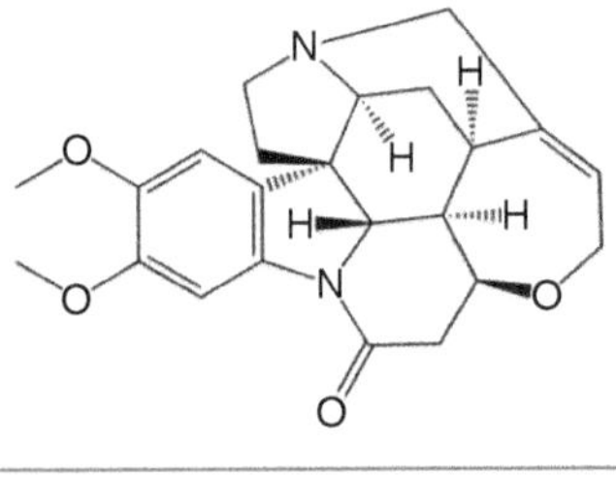

Strychnine is also a crystalline alkaloid. These colorless crystals are highly toxic to humans and were first used as poisons for killing small vertebrates such as mice and rats. Strychnine poisoning causes muscular convulsions and eventually death through asphyxia or lack of oxygen from not being able to breathe. Brucine is a white alkaloid that is closely related to strychnine. While it is poisonous to humans, it requires a substantially higher dose than strychnine does. Medicinally, brucine has been used to treat high blood pressure. Brucine in a concentrated sulfuric acid solution can be used to test for nitrates or nitric acid as the mixture gives off a red color.

14.3.3 Sour Taste

Sour is recognized as having an acidic taste like lemon or vinegar. A sour taste sensation is due to the presence of weak organic acids that will partially dissociate into its conjugate base and proton. The actual signaling tastant molecule is caused by both the hydrogen ion concentration and the type of organic acid, for example, citric acid (citrus fruit sour), malic acid (green apple), acetic acid (vinegar sour), or other weak acids, in the food or drink being ingested (see Table 14.3). The intensity of the sour taste is proportional to the hydrogen ion concentration; thus, the more acidic the food or drink, the stronger the sour sensation will be.

The presentation of sour and salty tastes is much more direct. A simple but not totally complete understanding of sour taste is directly related to the acid content of the food or drink you are ingesting (Table 14.3). The intensity and a main portion of the sour taste come from the presence of the concentration of H^+ in the food or drink. The functional groups discussed in Chapter 1 play a role in the concentration of hydrogen ions in solutions. Most weak acids found in food and drink have one or more carboxyl groups that depending on the acid will produce various amounts of H^+ in our food depending on its pK_a. We use the pH scale to measure proton concentration in solution (Figure 14.6). An increase in the proton concentration in the solution makes the solution more acidic, which is a decrease in pH. Solutions with a pH less than 7 will have a sour taste. The lower the pH, the more H^+ in the solution and the more intense the sourness of the food. Sour candies are often coated with citric acid, and as the solid acid dissolves in our mouth, a high local concentration gives a strong sour taste.

On the taste index, hydrochloric acid (HCl) is used to set the baseline of 1. Hydrochloric acid is a strong acid and is the most common acid in stomach acid, which is very acidic. The other acids on the list are organic acids, which contain a carboxylic acid group. Typically, organic acids are weaker acids and therefore do not release as many protons into solution as HCl and thus have a lower sour taste rating. Formic acid is one organic acid that has a higher taste index than HCl.

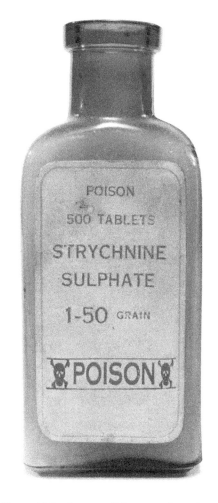

Figure 14.5 From medicines to poisons. The Bitter Truth, early 1800s strychnine bottle. aSculptor/Adobe Stock Photos

14.3.4 Salty Taste

Salty is the basic taste of seawater, or more commonly table salt (sodium chloride, NaCl). Salty taste is stimulated by ionized salts such as Na^+, K^+, and Li^+ (see Table 14.4). The quality of the salty taste varies with the cation present, but sodium cations give the strongest response. The anionic partners to these cations also contribute to the salty taste but to a lesser extent.

Salty taste is primarily controlled by cations in salts. Sodium chloride, or common table salt, is used to set the taste index (Table 14.4). The most commonly occurring natural salts, NaCl, KCl, and $CaCl_2$, give relatively similar taste sensations. In cooking, you will occasionally see recipes that call for sea salt, which is simply salt produced by evaporating seawater. Some cooks believe that it tastes better and since taste is a personal opinion that is justified. There are cooks who talk about sea salt and use the term low sodium. This is inaccurate and misleading. While sea salt is a mixture of salts, it is still approximately 98% sodium chloride by weight. Thus, it does have lower sodium, but certainly not low enough to help lower sodium levels in your body.

Table 14.3 Taste classifications, tastant structure, and taste index for sour compounds.

Taste classification	Substance	Taste index	Structure
Sour	Hydrochloric acid	1	H — Cl
	Carbonic acid	0.06	
	Citric acid	0.46	
	Acetic acid	0.55	
	Formic acid	1.1	

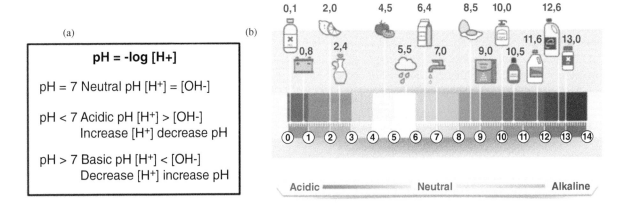

(a)

pH = -log [H+]

pH = 7 Neutral pH [H⁺] = [OH-]

pH < 7 Acidic pH [H⁺] > [OH-]
Increase [H⁺] decrease pH

pH > 7 Basic pH [H⁺] < [OH-]
Decrease [H⁺] increase pH

(b)

0,1   2,0   4,5   6,4   8,5   10,0   12,6
0,8   2,4   5,5   7,0   9,0   10,5   11,6   13,0

0 1 2 3 4 5 6 7 8 9 10 11 12 13 14

Acidic ◄■■■■ Neutral ■■■■► Alkaline

Figure 14.6 The pH scale. (a) Equation for calculating pH and pH concepts. (b) The pH scale and the pH of some different common household items, foods, and drinks. elenabsl/Adobe Stock Photos

Table 14.4 Taste classifications, tastant structure, and taste index for salty compounds.

Taste classification	Substance	Taste index	Structure
Salty	NaCl	1	Na⁺ Cl⁻
	KCl	0.6	K⁺ Cl⁻
	CaCl₂	1	Ca²⁺ 2Cl⁻
	NaF	2	Na⁺ F⁻
	NH₄Cl	2.5	NH₄⁺ Cl⁻

Table 14.5 Taste classifications and tastant structure for umami compounds.

Taste classification	Substance	Structure
Umami	L-Glutamine	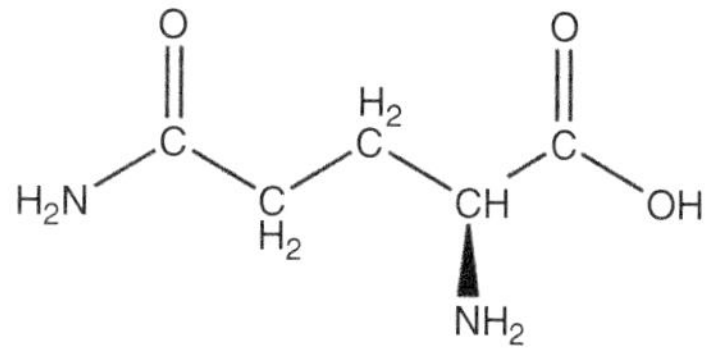
	L-Glutamic acid	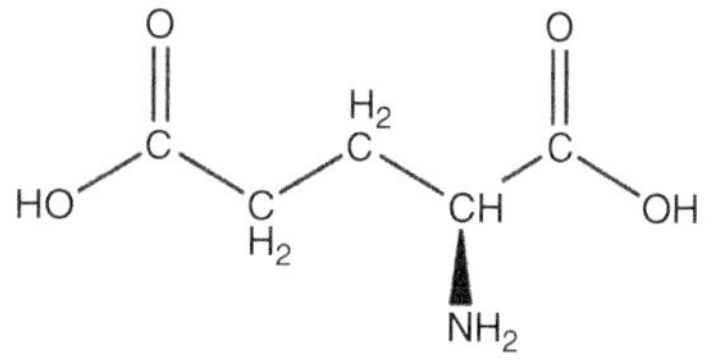
	Monosodium glutamate	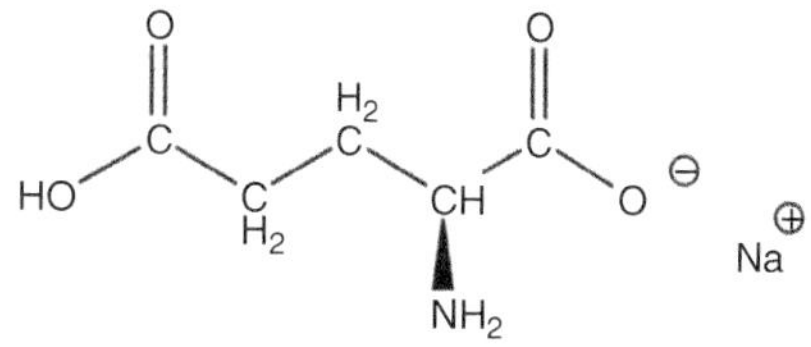
	Inositol monophosphate	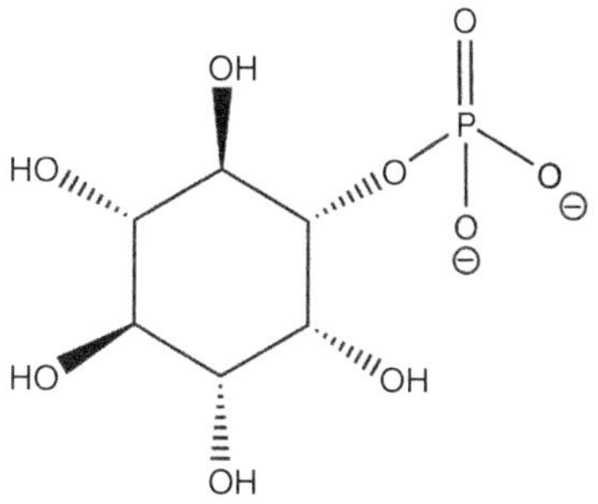

14.3.5 Umami Taste

Umami is the meaty or savory taste of glutamate proteins easily recognized in monosodium glutamate or Parmesan cheese. Umami is a Japanese word meaning delicious. Most people describe it as a pleasant taste that is qualitatively distinct from the other four taste sensations. Umami is the predominant taste of the amino acid L-glutamine (see Table 14.5).

Umami taste is defined by the savory taste of the naturally occurring amino acid L-glutamine. In this case, glutamine has a very similar structure to another amino acid, glutamic acid. Table 14.5 shows that glutamic acid has the same structure as glutamine with the terminal amino group replaced with a hydroxyl group. Another common descriptor for umami taste is monosodium glutamate, the sodium salt of glutamic acid.

14.4 Why Do We Taste?

The value of this system of five basic tastes comes in part through its use as a basic survival mechanism. Taste allows us to evaluate the nutritious content of food while preventing us from eating toxic substances. Salty taste allows us to identify foods and liquids that contain essential salts necessary for the maintenance of electrolyte balance. Sweet taste allows us to identify food that is high in energy-rich nutrients. Umami allows the recognition of amino acids and proteins necessary as building blocks for proteins in our bodies. On the other hand, bitter and sour tastes give us a warning against the intake of potentially noxious or poisonous substances. This taste discrimination becomes even more important during pregnancy. To ensure both the growth of the fetus and the well-being of the mother, specific metabolic adaptations occur. The most predominant change is a need for additional food calories to account for the energy needs of both the mother and the developing fetus. In terms of taste sensation, the two most consistent changes identified in pregnant women come in salty and bitter tastes. Pregnant women have a decreased threshold for salty taste, which means that they show an increased preference for salty food as compared with women who are not pregnant. This increased preference for salty food coincides with an increase in salt requirements that a pregnant woman faces as the fetus develops. Pregnant women also display an increased sensitivity to bitter tastes. Since one role for bitter taste is to prevent ingestion of potentially toxic substances, an increased sensitivity to bitter tastes in pregnancy would protect the developing fetus, which would have a much lower tolerance for these same noxious or poisonous substances than the mother [4].

14.5 Gustation: Signaling Receptors, Cells, and Tissue

14.5.1 The Tongue

Each tastant will diffuse through the fluids of the mouth and bind to protein receptors embedded in the surface of the taste receptor cells clustered in our taste buds. Taste buds are small groups of taste receptor cells found throughout the tongue. The taste receptor cells are located primarily on the top or dorsal surface of the tongue. The taste receptors are localized into tissue projections called papillae. The surface of the tongue has three distinct types of papillae that are located in three different areas. The papillae are small pimple-, ridge-, or mushroom-shaped structures, each a couple of millimeters in size. The three types of papillae are directly involved in taste sensations (Figure 14.7). Fungiform papillae are pimple-shaped structures located on the surface primarily down the sides of the tongue. As you move down the side of the tongue, the next set of papillae is foliate papillae, which are a series of ridges with one distinct set on each side of the tongue. Finally, in the back center of the tongue are the circumvallate papillae that present functionally as a row of mushroom-shaped structures at the back of the tongue. Each of these types of papillae contains multiple taste buds (Table 14.6).

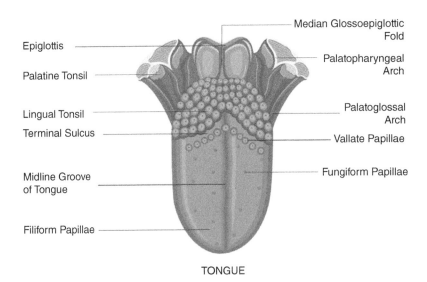

Figure 14.7 Anatomy of the tongue. stockshoppe/Adobe Stock Photos

Table 14.6   Numbers of papillae and taste buds per papillae.

Papillae	Papillae/tongue (average)	Taste buds/papillae	Total taste buds
Fungiform papillae	200	1–18 Front of tongue 1–9 Middle of tongue	1120
Foliate papillae	11	117	1280
Circumvallate papillae	8	252	2200

14.5.2   Taste Buds

Each taste bud is composed of 50–150 taste receptor cells, a substantial series of basal and support cells that surround the receptor cells, and a set of sensory afferent neurons that carry sensory inputs from the receptor to the brain. The number of each type of papillae and the number of taste buds in each type of papillae vary. Table 14.6 gives average values for these numbers. A fourth type of papillae called the filiform papillae, which are long cone-shaped structures, are present on the tongue. The filiform papillae are the most numerous papillae on the tongue. The filiform papillae don't participate in gustation and serve a mechanical function.

On average, people have 2,000–5,000 taste buds, but some can have as few as 500 or as many as 20,000! Someone with 20,000 taste buds would be a supertaster! It is estimated that 35% of women and 15% of men could be categorized as supertasters. The evolutionary value of being a supertaster is unclear. Supertasters would have an advantage in avoiding noxious or toxic substances because of the increased ability to taste them in potential food items. It is difficult to identify other evolutionary advantages. In fact, many supertasters are rather picky eaters as the tastes of many foods and beverages can be overwhelming to them. A few bites of a rich dessert or a salty main dish are typically sufficient to satiate the desire for that taste. But that does not mean that all picky eaters are supertasters or that all supertasters are picky eaters.

The structure of the taste bud lies beneath the surface of the tongue (Figure 14.7). Thus, contrary to popular belief, you cannot see your taste buds. What you see are the papillae on the surface of your tongue. The opening between the surface of the tongue and the taste bud is called the taste pore. The taste pores are not located directly on the top of the gustatory papillae but typically located near the sides and in the crevasses of the papillae. The taste receptor cells are modified epithelial cells with microvilli that project into the taste pore. The microvilli contain the membrane receptors that bind the tastants to initiate the taste process. The taste cells are surrounded by basal cells that support the structure of the taste bud and taste cells but do not directly participate in gustation. The taste cells form a chemical synapse with the sensory neurons leaving the taste buds. When activated, the taste cells release neurotransmitters activating the sensory neuron, which then carries the electrical impulses into the brain.

14.5.3   Taste Receptor Cells

As scientists learn more about the nature of the taste receptors, taste receptor cells, and the anatomy of taste buds, they have found the old "taste map" model of the tongue, where different regions of the tongue were thought to be responsible for a specific taste, is wrong (Figure 14.8). This map did not include umami or

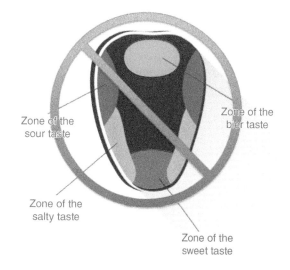

Zone of the
sour taste

Zone of the
bier taste

Zone of the
salty taste

Zone of the
sweet taste

Figure 14.8   The mythical tongue map. luna2631/Adobe Stock Photos

account for all sorts of types of taste receptor cells for each of the flavor types. Each taste bud contains 50–150 taste receptor cells, and each taste receptor cell is unique. A taste bud is comprised of a diverse collection of taste receptor cells with several kinds of taste receptors. Each taste receptor cell will only have one kind of taste receptor and is associated with its own afferent nerve that signals to the thalamus and neocortex. This means that each taste bud is wired and can recognize a variety of different tastes and flavors. The makeup of each bud and the density of the buds can vary, which means a taste map of the tongue can't be right!

Complicating matters is sweet taste receptor cells express only one type of taste receptor, while bitter taste receptor cells have many different receptors produced on a single taste receptor cell. Thus for each cell with only one type of "sweet" receptor, those cells and their associated nerve signal to the neocortex can distinguish between different sweet flavors. Several bitter compounds will each bind and activate the same bitter taste cell brain pathways. The bottom line is that we can tell the difference between different sweet flavors, but, even though there are thousands of different bitter compounds, we cannot tell the differences between them (Box 14.1). They are all signaling through the same taste receptor cell and nerve to the same place in our brain!

Box 14.1 Are you a supertaster?

The first step to determining whether you are a supertaster is to evaluate the number of fungiform papillae present on your tongue. The fungiform papillae are small pink bumps on the surface of the tongue. The following process will make it easier to count the papillae and determine whether you are a super taster.

Supplies
- Vial of blue food coloring
- 2 paper towels per person
- 2 cotton swabs per person
- 2 cotton balls per person
- A magnifying glass
- 2 sticky notebook hole reinforcement circles per student
- A ruler with a millimeter scale

Procedure
- Work with a partner.
- Use the cotton swab as an applicator to paint the surface of the tip of your partner's tongue with the blue dye.
- Put a few drops of the food coloring on the cotton swab.
- Paint the tip of the tongue with the dyed cotton swab.
- Move your tongue around in your mouth to make sure the dye covers the entire tongue. The dye is safe to swallow.
- Pat your tongue dry with one paper towel.
- Place the reinforcement circle on the tip of your partner's blue-dyed tongue. This circle defines the sample area on the tongue.
- Count the fungiform papillae (the pink bumps) inside the sample area.
- Record your data.
- Measure the size of the person's tongue (measure the length and width of the front 1/3 of the tongue).
- Record your data.
- Repeat the experiment for the partner.

Partner	Papillae counted (P)	Diameter of circle (mm)	Area of circle $A_C = \pi(0.5\ D)^2$	Papillae per mm^2 (PPM = P/A)
1				
2				

Partner	Dimensions of tongue	Area of tongue A_T (mm^2)	Papillae per tongue (PPT = PPM $\times$ A_T)	Fungiform papillae taste buds per tongue (PPT $\times$ 8)
1				
2				

Questions
- How does the number of fungiform papillae on your tongue compare with the average in your data table?
- From the calculation, how does your total number of fungiform papillae taste buds compare with the average in your data table?
- Are your numbers high enough to be considered a supertaster? Explain.

14.6 Gustation: Membrane Proteins, Membrane Potential, and Sensory Transduction

As we have seen, tastes are categorized as five distinct sensations detected by different taste receptor cells located within taste buds on our tongues. With these five distinct tastes, humans can distinguish between 4,000 and 10,000 different chemical sensations. Once a taste molecule binds to the receptor in the membrane of a taste receptor cell, there are two classes of receptor proteins and accompanying signal transduction that permit those thousands of taste sensations to reach our brains. Salty and sour tastes function through the direct use of proteins called ion channels, while sweet, umami, and bitter tastes use a family of membrane-embedded proteins called G protein-coupled receptors (GPCRs).

The basic order of taste or gustatory signaling is for the tastant to bind to its specific receptor on one of the clusters of taste receptor cells found in the taste bud (Figure 14.9). Regardless of the type of tastant, the taste receptor cell will cause the release of a neurotransmitter stored in the taste cell. The neurotransmitter will bind to the afferent nerve and start a signal that ultimately ends with nerves in the brain.

14.6.1 Ion Channels and GPCRs Are Transmembrane Proteins

Cells are the structural, functional, and reproductive units of living systems. All cells are enclosed by a plasma membrane that separates the cell from its surrounding environment. The plasma membrane is composed of a bilayer of amphipathic phospholipids and associated proteins (Figure 14.10). Refer back to Chapter 2 and Figure 2.30 for the structure of a phospholipid. To make the membrane, two layers of phospholipids align with their hydrophobic tails forming a nonpolar center of the bilayer, while the polar head groups face the aqueous environment inside and outside the cell.

Proteins are associated with the membrane in one of two ways. Integral membrane proteins reside within the membrane and possess hydrophobic amino acid residues that come in contact with the hydrophobic core of the

Figure 14.9 Tastant signal transmission. The basic flow of information from a taste receptor to the central nervous system begins with a tastant binding to its specific receptor.

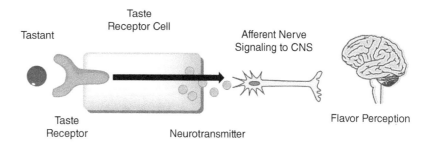

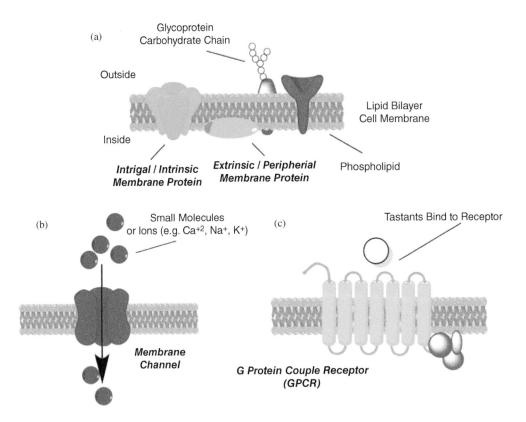

Figure 14.10 Plasma membrane structure and function. (a) A model of the cell membrane. (b) Transmembrane integral protein that functions as a channel. (c) Transmembrane integral protein that functions as a G-protein coupled receptor (GCPR).

membrane. Peripheral proteins are associated with the exterior edges of the membrane. One common type of integral protein that is important for our understanding of taste and smell is transmembrane proteins. These proteins span the entire membrane having part of their structure in the extracellular fluid, part contained in the hydrophobic core of the membrane, and part of their structure in the intracellular fluid. Figure 14.10 shows the basic structure of a plasma membrane with integral and peripheral proteins.

There are two types of integral proteins present in the plasma membrane essential to taste perception: receptor proteins and channel proteins (Figures 14.10 and 14.11). Channel proteins are transmembrane ion pores (Figure 14.10b). When the channels are open, they allow ions to move into or out of cells changing the electrical potential difference across the membrane. The second type of integral transmembrane protein involved with taste reception is the G-protein coupled receptor protein, or GPCR (Figure 14.10c). GPCRs are transmembrane proteins that bind to specific molecules in the extracellular fluid and transmit the taste information into the cell without allowing the tastant molecule to cross the membrane (Figures 14.10c and 14.11).

Once activated by tastant, both transporter/channel and GPCR signaling lead to the release of neurotransmitters, which initiate a cascade of events. Neurotransmitters activate the afferent nerve signaling to the brain stem through the thalamus and finally end in the cerebral cortex. There are three main cranial nerves that bundle all the afferent nerves into the brain stem. The signal is carried from the dendrites through the cell body and axon ending at the synapse. Membrane potential is responsible for the release of neurotransmitters in the receptor cell and for the propagation of signal through neurons.

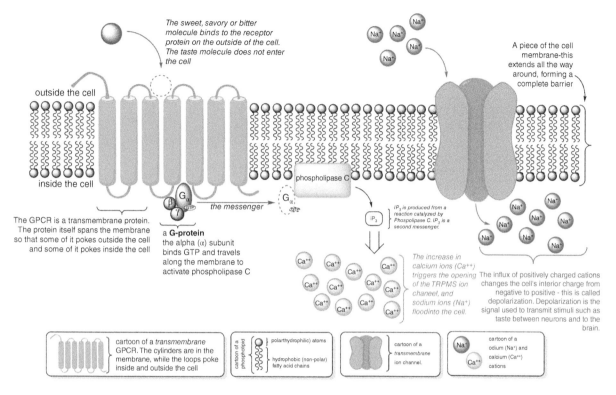

Figure 14.11 Transmembrane proteins that bind and signal the presence of taste and flavor. Ion channel proteins are transmembrane ion pores that allow ions to move into or out of cells. G-protein coupled receptors, or GPCRs, are transmembrane proteins that bind to specific molecules in the extracellular fluid and transmit the taste information into the cell without allowing the tastant molecule to cross the membrane.

14.6.2   An In-depth Look at Membrane Potential and Neuronal Transmission

The basis for the membrane potential is the fact that both the intracellular and extracellular fluids contain cations and anions. When the distribution of anions and cations across the plasma membrane is not equal, there is a difference in net charge across the membrane, an electrical potential difference, which is referred to as the membrane potential (Vm). For most cells in the human body, including sensory receptors and neurons, the inside of the cell is negatively charged relative to the outside of the cell. The resting membrane potential for neurons and sensory receptor cells is around −70 mV (Figure 14.12). That means that there is both an electrical gradient and a concentration gradient across the plasma membrane of these cells (Figure 14.12). This combined gradient is referred to as an electrochemical gradient.

When ion channels in the plasma membrane open, ions can enter or leave the cells, moving down their electrochemical gradient, thus changing the membrane potential. Changes in the membrane potential of sensory cells initiate the electrical events that lead to action potential in the sensory neurons that provide sensory information to the CNS.

14.6.3   Neuron Structure, Action Potentials, and Chemical Synapses

Nerve cells, or neurons, are cells specifically designed to transmit electrical signals from one location to another in the body. All neurons consist of a soma or cell body, dendrites, which are structures that receive neural inputs, and an axon, which sends neural inputs to the next location in the body (Figure 14.13). The junction

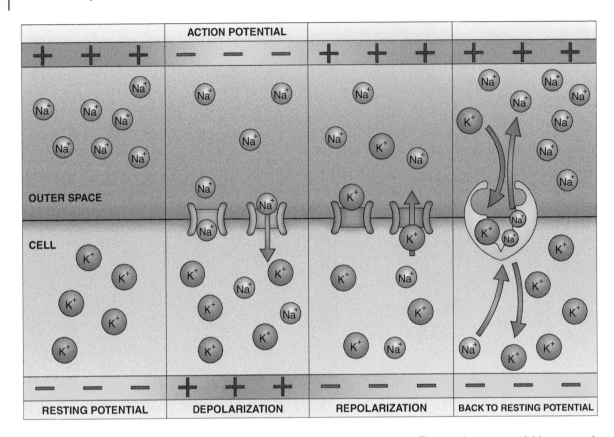

Figure 14.12 Membrane potential and ion gradients across the plasma membrane. The membrane potential is measured as a difference in voltage between the inside and outside of the cell. Differences in concentrations of ions across the plasma membrane establish an electrochemical gradient. extender_01/AdobeStock

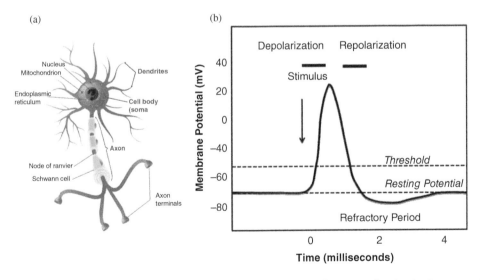

Figure 14.13 Neuron structure and action potentials. (a) The basic structure of neurons showing both presynaptic and postsynaptic neurons. (b) An action potential showing the three distinct phases. designua / Adobe Stock Photos

between two neurons or between a receptor cell and a neuron is called a synapse. There are electrical synapses where the two cells are in direct contact with each other and there are chemical synapses where the presynaptic cell releases a chemical called a neurotransmitter that stimulates the postsynaptic cell (Figure 14.14). The synapses involved in gustation are chemical synapses. The chemical synapse is activated when an action potential arrives at the axon terminal. Figure 14.13 shows the basic structure of the action potential. Neurons have a resting membrane potential of −70 mV. When signals enter the dendrites of a neuron, the membrane potential becomes more positive until it reaches the threshold. Once the membrane reaches the threshold, an all-or-none event known as an action potential is triggered. An action potential involves a rapid depolarization phase when the Vm becomes positive, a repolarization phase when the Vm returns toward normal, a hyperpolarization where Vm drops below normal, and finally a return to resting. Functionally, action potentials move down the axon delivering an electrical signal to another part of the body.

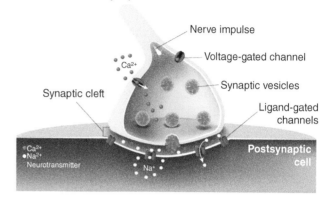

Signal transmission at a chemical synapse

Figure 14.14 Signal transmission at a chemical synapse. designua/ Adobe Stock Photos

When an action potential reaches the axon terminal at a chemical synapse, a distinct series of events occurs that ultimately leads to the initiation of an action potential in the postsynaptic neuron. These events are:

1. Voltage-gated Ca^{2+} channels open.
2. Ca^{2+} enters the axon terminal increasing the Ca^{2+} concentration inside the cell.
3. The Ca^{2+} stimulates the binding of vesicles containing neurotransmitter to the plasma membrane.
4. Neurotransmitter is released into the chemical synapse.
5. The neurotransmitter binds to receptors on the postsynaptic cell.
6. Ion channels in the postsynaptic cell plasma membrane open, causing an increase in membrane potential to threshold.
7. An action potential is initiated in the postsynaptic cell.

14.7 Tasting Through Receptors or Transporters

Sweet, bitter, and umami tastants signal through GPCRs, while sour and salty tastants signal through ion channels. Sweet, bitter, and umami signals trigger a cellular response without ever entering the cell. The tastants are small molecules that bind to the GPCR proteins found in the membrane of the taste receptor cell. The tastant molecules bind the protein receptors in a very specific "lock and key" fashion. The lock (receptor protein) and key (tastant molecule) are very specific in the shape and set of interactions for the two to bind. In fact, most GPCRs will only bind one or a few very similarly shaped tastants. Salty and sour tastes signal to their receptor cells in a very different fashion. In this case, ions and other compounds can bind and open transporters or channels allowing ions to enter into the taste receptor cell. Like receptor signaling, neurotransmitters are ultimately released by the sour and salty receptor cells activating afferent nerves to the brain. Now we will focus on the specifics for each taste receptor cell and tastant.

14.7.1 Sensory Transduction of Salty Tastes

The primary salty taste in our diets comes from Na^+. We have evolved to enjoy the taste, and this helps ensure that we take this essential component into our diet. The Na^+ channel found in taste receptors is called the epithelial sodium channel (ENaC). As sodium channels function, ENaC is somewhat unique in that it is open to conduct sodium ions most of the time. This means that when Na^+ is ingested, whether in food or drinks, dissolved sodium ions in the fluids of your mouth diffuse to taste receptor cells. When the Na^+ containing fluid surrounds the taste receptor cells, the extracellular concentration of Na^+ increases allowing more sodium to move down its electrochemical gradient and enter the cells. The process does not require additional energy and is very efficient. The electrochemical gradient is the combination of electrical and chemical forces that cause ions to move across a membrane. In the case of Na^+, both forces drive the ion to enter the cell. The inside of the cell is negatively charged (Vm = −70 mV), which attracts cations like Na^+ into the cell. Simultaneously, the Na^+ concentration outside the cell is higher than inside the cell. Thus, the chemical gradient uses entropy to drive Na^+ into the cell. The combined impact of the two components of the electrochemical gradient drive Na^+ into the cell and triggers a series of events that leads to action potentials being sent to the brain indicating you are ingesting salty food. Figure 14.15 shows the events involved in the signal transduction process, which are:

1. Na^+ increases in the extracellular fluid around the taste receptor cells.
2. The increase in Na^+ concentration outside the cells increases the electrochemical gradient across the membrane for Na^+.
3. Na^+ enters the taste receptor cell.
4. The membrane potential of the receptor cell depolarizes (becomes less negative). No action potential is formed in the receptor cell.
5. The depolarization of Vm stimulates the opening of voltage-gated Ca^{2+} channels allowing Ca^{2+} to enter the cells.

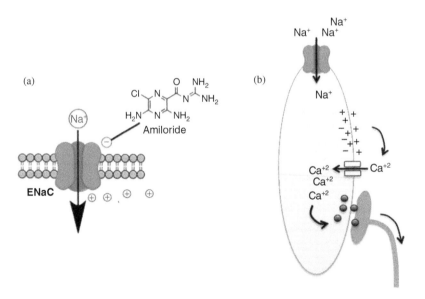

Figure 14.15 Salty taste transmission. (a) The epithelial sodium channel (ENaC) transports sodium into the taste receptor cells and is inhibited by amiloride. (b) Influx of sodium leads to a series of signaling events involving calcium and results in neurotransmitter release.

6. The increase in Ca^{2+} concentration inside the cell stimulates the release of neurotransmitter from the taste receptor cell.
7. The neurotransmitter stimulates the depolarization of the afferent sensory neuron.
8. The afferent sensory neuron sends an action potential to the brain indicating you have ingested something salty (Box 14.2).

Box 14.2 Why do we like salt on popcorn but do not drink ocean water?

While the previous information is our most current understanding of how salt is "tasted," there are some discrepancies. Amiloride is a drug that binds very effectively to ENaC channels and blocks the sodium taste receptor (Figure 14.15a). In fact, amiloride is commonly used to block sodium channels in biology research and is a useful tool to tell a sodium channel from other channels. However, amiloride does not block all of the sodium tastant signaling in taste cells. Other salt ions and high levels of sodium are not affected by amiloride. This indicates that there must be another receptor–channel for salty tastants. At low concentrations (< 100 mM NaCl) the ENaC receptor is activated in an amiloride-sensitive manner and stimulates a positive taste or appetite for salty foods. At higher concentrations of sodium, typically greater than 300 mM NaCl, a yet undetermined receptor is stimulated. Scientists working at the University of California, San Diego, and Columbia University found that high levels of salt activate both the bitter and sour tastant receptor cells, helping to create an aversion for high salt concentrations. In fact, in mice where bitter and sour taste receptor cells have been genetically knocked out, the mice showed a much higher level of salt intake than the control mice. These mice did not avoid high salt solutions, even salt found at ocean water levels. Such results suggest that humans have evolved a trigger to control salt intake. Low levels of salt bind and activate one receptor (ENaC) that is attractive to our senses, and another undiscovered receptor–channel seems to lead to the activation of a sour and bitter reaction helping us recognize and avoid potentially dangerous high salt solutions. It is too early to draw too many conclusions, but it is possible that a better understanding of our aversion to high salt concentrations would help modulate and control our high salt intake, which leads to hypertension and heart health issues. Adapted from [5]

14.7.2 Sensory Transduction Sour Tastes

As we discussed earlier, sour tastes are induced by changes in proton concentration. The protons themselves do not move through the ion channel to change membrane potential as is the case with Na^+ ions in salty taste. For sour tastes, the hydrogen ions appear to alter membrane potential in one of two ways. The protons can bind to and activate the TRPP3 Na^+ channel or bind to and inactivate K^+ channels. The activation of Na^+ channels causes a depolarization due to the increase in positively charged sodium ions entering the cytoplasm of the sensory cell. Blocking the K^+ channel also leads to a depolarization of the membrane. Under resting conditions in the sensory cells, K^+ channels are typically open, allowing positively charged potassium ions to leave the cell. Positive charges leaving the cell help keep the membrane potential more negative. Blocking the efflux of K^+ through the potassium channels retains the positive charge causing a depolarization in the sensory cell. Whether the depolarization is caused by opening Na^+ channels or closing K^+ channels, the end result is the same as we saw with salty taste. Depolarization leads to the opening of Ca^{2+} channels causing an increase in calcium concentration in the cytoplasm and the release of neurotransmitter stimulating the postsynaptic sensory neuron.

As with our understanding of salt gustation, the complete story of how we perceive sour taste is only partly complete. Protons have a major role in activating channels to drive sour signals to the brain. However, it seems that each acid is not the same. Organic acids such as acetic acid and citric acid have a more sour taste than HCl at the same pH and therefore at the number of protons in solution. Because of the number of total molecules needed to achieve the same pH, there are fewer total HCl molecules in a solution with the same pH as acetic or other organic acids. Over the past 30 years, many scientists have tried to understand the role of both the proton and the anion (i.e. the protein donor) with many different and confusing results. One current hypothesis is that there is likely more than one receptor–channel and that both the proton and the anion have a role in triggering the sour taste. This could explain why at equal concentrations weak acids have different perceived sourness.

There are several new and interesting receptor–channels possibly responsible for sour taste. One study found a proton-stimulated (H^+ gated) channel involved in sour detection is inhibited by amiloride, the same compound that inhibits the sodium channel. This observation indicates that sodium is involved in a new taste receptor–channel, and H^+ alone is not enough to stimulate sour taste. Further showing the role for sodium is that this channel protein, called the acid-sensing ion channel 2a (ASIC2a), is produced in sour taste receptor cells [6]. While we have yet to understand the role of sodium in sour taste, there is a second very interesting question about how we sense sourness. The organic acid that produces H^+ may also play a role. Evidence for the organic acid anion was found when comparing various H^+-producing organic acids. When cells were exposed to the same H^+ concentrations and were held at the same concentration, acetic acid produced a larger signal (taste receptor depolarization) than HCl, indicating a role for both the H^+ and the anion of the organic acid, acetate, in sour taste. As both solutions were held to the same hydrogen ion concentration, the two solutions should have given the same result, but did not. The answer may lie in the discovery of ion channel proteins, called polycystic kidney disease proteins (PKDs), in taste receptor cells [7]. PKDs are a family of related proteins (i.e. PKD1L1, PKD2L1, and PKD1L3) where two of the three proteins form integral membrane complexes that allow ions into the taste receptor cells. Mice that were unable to make PKDs could detect sweet and bitter tastes but not sour. While the complete mechanism describing how these PKD-based receptors function in taste cells has yet to be elucidated, it is interesting that several of the PKD acid-sensitive channels are also found in other tissues. Perhaps taste receptors are produced in other tissues that have nothing to do with taste in order to allow those cells to respond to sweet, bitter, or acidic conditions.

14.7.3 Sweet, Umami, and Bitter Tastes Use G-Protein Coupled Receptors

As we briefly presented earlier, the remaining three taste sensations use a family of receptors called GCPRs. GCPRs are involved in the regulation of a large array of body functions. These receptors are involved in the regulation of heart rate and strength of cardiac contraction, stimulate smooth muscle contraction and regulate blood pressure, and function as the light receptors in the eye that allow vision, and these are only a few examples. As the name implies all GPCRs are coupled to an intracellular signaling molecule called a heterotrimeric G protein. These G proteins have three distinct subunits α, β, and γ (Figure 14.16). The G proteins get their name because the α subunit is a guanosine triphosphate (GTP)-binding protein. In the inactive state, the α subunit is bound to guanosine diphosphate (GDP) and is associated with the β and γ subunits. For G proteins to become active, they must bind to a molecule called a guanine nucleotide exchange factor that allows them to release GDP and bind GTP. When the α subunit is bound to GTP, it dissociates from the βγ subunits. The α-GTP can then bind to other proteins activating them and leading to cell-specific responses in the stimulated cells.

Two families of GPCRs—termed taste receptor proteins called the T1R and T2R families—account for sweet, umami, and bitter tastes. There are multiple members of each family that bind to different classes of tastants. See Table 14.7 for a listing of variants in these receptors. For example, the T1R receptor family has three family members named T1R1, T1R2, and T1R3. The G protein involved in this system is named gustducin because of its role in modulating gustation. For each type of taste perception, there is a different combination of T1R and T2R receptor subunits involved. The taste receptor cells for umami and sweet taste present specific receptor

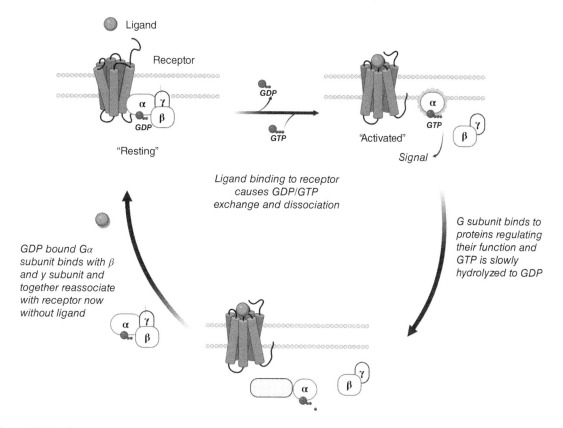

Figure 14.16   G-protein activation and inactivation of GPCRs.

Table 14.7   GPCR and gustation perception.

Tastant	Receptor type	Genetic variation	Misc
Umami	T1R1 and T1R3	T1R1—six different genes with 30 different effective known SNPs T1R3—442 unknown SNPs* Three known mutations increase umami sensitivity twofold or more	Mouse receptors bind most amino acids and nucleotides—human receptor mostly nucleotides and glutamate
Sweet	T1R2 and T1R3	706 SNPs with unknown function	
Bitter	T2R	Over 30 different T2R genes • Highly evolving with many mutations • Some receptors can bind multiple bitter compounds	Many receptors are expressed on a single cell

*SNPs—Single nucleotide polymorphism: single base changes in the gene of a protein. Some changes may alter the coded protein. SNPs provide genetic variation among people. Known SNP effects are genetic changes that lead to a protein (receptor in this example) with a different function. Such changes could increase or decrease a receptor's sensitivity for its tastant.

combinations and thus only respond to one specific taste. The sweet receptors are T1R2/T1R3 dimers. The umami receptors are T1R1/T1R3 dimers. This means that both umami and sweet receptors have T1R3 subunits in their receptors. The taste of amino acids such as glutamate appears to depend upon the T1R1 subunit. As you recall, bitterness usually warns us of noxious or poisonous substances. The receptor family for bitter taste is the T2R receptor. Humans have about 30 different T2R receptor subunits, but there are thousands of different bitter compounds! With these kinds of numbers, it is obvious that we cannot distinguish between different bitter tastes as we can with other tastes. This is because many, if not all, of the 30 different T2R bitter taste receptors are produced on *each* bitter receptor cell. Thus if one or 20 bitter compounds are in a food, the same signal will be sent to the brain: "stop, don't eat!"

The diversity of how we taste and respond to thousands of different tastants is due to variations in the gene coding for each receptor (Table 14.7). Single amino acid changes can enhance how a tastant binds to a receptor, increasing or blocking binding altogether. There are three different amino acid variations in the T1R3 gene that increase the sensitivity for umami flavors 2- to 10-fold. One of many ways in which domesticated cats differ from dogs is in their lack of interest in sweets. Cats do not express the gene (i.e. make protein) for the T1R2 receptor and thus have no ability to distinguish sweet from other tastes. A quick survey of people around you will find a great variation of what is thought to be sweet, bitter, or savory. The T1R receptors (involved with both sweet and umami tastes) have 47 known variants. One bitter receptor gene for the T2R38 receptor (the 38th different bitter receptor) has three different variants. People with different amino acids at position 49-valine, 262-alanine, or 296-valine will not taste the highly bitter phenylthiocarbamide. Such mutations are inheritable and thus families, and, over long periods of time, regions may lose sensitivity to some bitter compounds or tastants. Similar mutations in other receptors are implicated in altering food and affinity for alcohol intake.

GPCR signaling for taste involves the use of second messengers. Second messengers are intracellular signaling molecules produced when a receptor is activated by a primary messenger, in this case, a tastant or an odorant. The role of the second messenger is to trigger a series of intracellular events that ultimately lead to the cell changing its function in response to the primary messenger. For taste, we will see the second messengers inositol trisphosphate (IP_3) and diacylglycerol (DAG).

Once the appropriate tastant is bound to its receptor on the surface of a taste receptor cell the activation pathway that stimulates the afferent sensory neuron is the same for each of the sensory receptor cell types. It is the receptors that distinguish the type of signal the sensory receptor cell will send to the brain.

For example, sweet taste perception includes the following events (Figure 14.17):

1. A sugar compound binds to the sweet T1R GPCR receptor, leading to its activation.
2. The activated receptor binds to an inactive G protein.
3. The inactive G protein releases GDP and binds GTP to become active.
4. The active α-GTP dissociates from the βγ subunits and binds to an enzyme called phospholipase C (PLC).
5. PLC catalyzes the formation of two second messengers, IP_3 and diacylglycerol.
6. The IP_3 stimulates the release of Ca^{2+} from intracellular stores.
7. This increase in calcium ions leads to the activation of another type of Na^+ channel called TRPM5.
8. TRPM5 activation causes a depolarization of the Vm.

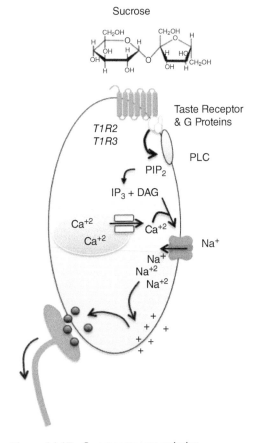

Figure 14.17 Sweet taste transmission.

9. This depolarization opens voltage-gated Ca^{2+} causing a further increase in the intracellular calcium ion concentration.
10. Neurotransmitter is released, stimulating the afferent sensory neuron that sends action potentials to your brain indicating you have eaten something sweet.

The exact same intracellular pathway activating PLC and forming IP_3 is used by both umami and bitter taste receptor cells. The only difference is the GPCR dimers these taste receptor cells express on their surface. Remember that each taste receptor cell contains one type of receptor that defines the adequate stimulus for that receptor cell.

To get a true sense of the flavor of food and drink, you need to have a combination of taste and smell receptors working together. Detecting odor molecules in food also uses GPCR-based signaling and a second messenger called cyclic AMP, which we will investigate in Section 14.8. As Dwayne Johnson used to say on WWE, "Can you smell what the Rock is cooking?"

14.7.4 Flavor Intensity

There is a significant diversity to how we perceive, distinguish between, and recognize the intensity of tastants. The answer to how we achieve this lies in the manner in which each tastant binds and activates its taste receptor(s) on the taste receptor cell. How the tastant fits (lock and key fashion) into the receptor and small amino acid changes to the receptor both can influence the ability of how the tastant binds and sets the signal to

---

Box 14.3 A sixth taste? It may be for fat

Scientific and culinary circles are becoming increasingly more accepting of the idea of a sixth taste—fat. Traditionally, we think of sensing fat in terms of the texture and smell of lipids or fats. However, people who produce a protein called CD36 are more likely to pick out fat in taste samples than those without the protein. This protein, first found in rodents, seems to confer a fat-seeking behavior that encourages rats to seek a high-calorie diet. When this gene was knocked out in rats, the preference for fatty foods was lost. Thus far, three receptors for free fatty acids (see Chapter 1 for a discussion of free fatty acids) have been identified, CD36, GPR120, and GPR40. Each binds and is activated by different sizes, or chain lengths, of fatty acid and signals in a fashion similar to other taste receptors: activating calcium ions and TRPM3 in taste receptor cells.

CD36 is more specific to bind for long-chain fatty acids than GPR120 and GPR40. The latter two receptors are produced in rodents but seem to be at very low levels in the human tongue. Thus, so far in humans, CD36 is produced in larger quantities in taste receptor cells of the circumvallate papillae and is implicated as the leading candidate for taste receptor signaling in humans. Like other receptors, there are several SNP variations reported. The impact of these genetic variations and the overall level of CD36 production in obesity are currently under intense investigation. The level of CD36 produced in the taste receptor cells changes over time. During the nighttime, when food intake is typically low, CD36 levels decrease. Ever feel like avoiding fatty food after eating fat? A high-fat diet also seems to decrease the amount of CD36 produced in the taste buds of humans suggesting that a chronic high-fat intake decreases our appetite for fatty foods. Variants that do not change the production of fatty acid receptors could be responsible for a poor satiation of fat intake. Like sweet taste receptors, hormones that impact appetite regulate CD36. Initial studies using hormones that provide a "full feeling" (i.e. satiation) decreased CD36 and fat intake in mice. However, this had not been determined in humans and care must be taken not to extend the results too quickly. How this knowledge can impact cooking is yet to be truly investigated but has many interesting possibilities. Adapted from [8]

the brain. The better the fit, the more tastant molecules will bind to the receptor for longer times, and the greater the signal of that flavor is sent to the brain. The number of tastant molecules binding to receptors is one of the ways we distinguish intensity. More tastant molecules lead to a larger number of receptors binding to tastants and the greater the signal that will be sent through the taste buds to the neocortex. However, there is a threshold or a minimal concentration of tastant that will trigger a response. The big question that remains unsolved is how we distinguish between types, levels, and mixtures of tastants. How can we tell the difference between different peaches or chocolates? Each taste receptor has a different story providing for its unique ability to respond to the thousands of different tastant molecules we recognize.

14.8   Olfaction, The Other Way To Taste: Basics of Signal Transduction

Most of us remember a time when we had a serious head cold or sinus infection. The familiar taste of everything from our coffee to our favorite comfort food was wrong or simply missing. This is not because colds and sinus infections alter our taste buds; this is due to the fact that much of what we consider our sense of taste actually comes from smell or olfaction.

14.8.1   Olfactory Receptor Cells

A human's ability to smell chemicals is better developed than our ability to taste chemicals. It is estimated that humans can smell more than 400,000 different substances with the great majority being unpleasant smells (80%). We smell with odorant receptor cells found in the olfactory epithelium of the superior portion of the nasal cavity (Figure 14.1). The olfactory receptor cells both structurally and functionally are different from taste receptor cells. Olfactory receptor cells are modified neurons, not modified epithelial cells. The key difference is that olfactory receptor cells send their own electrical signals to the brain; they do not have to release neurotransmitters to stimulate the nerves to the brain like taste receptor cells do.

The olfactory epithelium consists of olfactory receptor cells that have cilia, which is the location of the receptor proteins for smell. These cilia protrude into a mucus layer covering the epithelium. Support cells are also present that help maintain an appropriate environment for the receptor cells, including the production of mucus. Finally, there are basal cells that are responsible for the production of new receptor cells. Basal cells are needed because the olfactory epithelial cells have a life span of 4–8 weeks and thus need to be replaced regularly.

The surface area of the human olfactory epithelium is approximately 10 cm^2 and, while not very large, it is sufficient to allow us to detect odorants at concentrations of only a few parts per trillion. This is equivalent to being able to find one specific drop of water in an Olympic-size swimming pool! It is the size of the olfactory epithelium and the number of receptor cells that determine our olfactory acuity. For example, we all have heard of tracking dogs that can detect the scent of someone who had passed by a location hours earlier. The surface area of the olfactory epithelium of most dogs is greater than 170 cm^2, more than 17 times larger than that of a human. Additionally, dogs have approximately 100 times more receptor cells per cm^2 than humans. For dogs with the highest sense of olfactory acuity like the bloodhound, this means a 10,000,000-fold greater olfactory acuity than an average human.

14.8.2   Olfactory Signal Transduction

With the human's ability to smell over 400,000 chemical compounds, you might think that signal transduction in the olfactory system would be quite complicated. It is not. The variation in olfactory sensory transduction comes in the number of odorant receptors and not the signaling process used. Humans have over 350 genes that code for odorant receptor proteins. The extracellular binding domains of these odorant receptor proteins each have slightly different odorant binding sites. Each odorant receptor cell expresses a single type of odorant receptor protein, and each type of receptor protein binds to a specific type of odorant molecule. This means that the activation of each type of receptor cell corresponds to a specific odor.

Odorant receptor proteins are G-protein coupled receptors or GPCRs—similar to those involved with sweet, umami, and bitter tastes. When you inhale, odorant molecules dissolve in the mucus of the olfactory epithelium. The odorant diffuses through the mucus to the surface of the olfactory receptor cilia and finds an odorant receptor. The sequence of events from that point on is the same for all odorants (Figure 14.18).

The steps involved in olfactory sensory transduction are:

1. The odorant binds to a specific olfactory receptor protein in the plasma membrane of the cilia of an olfactory epithelial cell.
2. The receptor that is a GPCR stimulates the activation of G_{Olf}, the heterotrimeric G protein that functions in the olfactory system.
3. The G_{Olf} α subunit releases GDP and binds to GTP in the activation process.
4. α-GTP dissociates from the βγ subunits.
5. α-GTP binds to and activates the enzyme adenylate cyclase.
6. Adenylate cyclase converts ATP to the second messenger cAMP.
7. cAMP binds to the cAMP-dependent cation channel causing the channel to open.
8. The cAMP-dependent cation channel is a nonspecific channel that changes membrane permeability to Na^+ and Ca^{2+}.
9. The net current flow of Na^+ and Ca^{2+} entering the cell leads to membrane depolarization and an increase in Ca^{2+} concentration.
10. The increase in Ca^{2+} opens a Ca^{2+}-dependent Cl^- channel. This allows Cl^- to leave the cell causing an even greater depolarization.
11. The olfactory receptor cells are modified neurons. Thus when the membrane depolarization reaches threshold, the receptor cells fire an action potential that runs into the olfactory bulb and then to the brain.
12. Olfactory transduction ends when the odorant diffuses away from the protein receptor and is broken down by enzymes in the mucus. The cAMP production in the cell stops, and the existing cAMP is broken down.

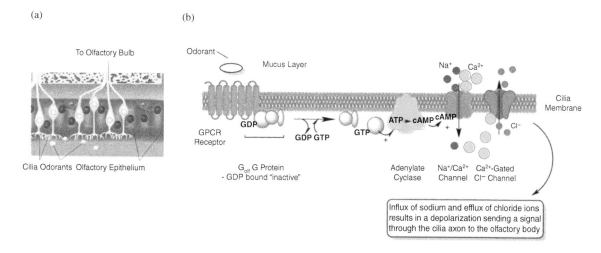

Figure 14.18 Sensory signaling in olfactory epithelium. (a) Odorant compounds bind to receptors in the nasal passage activating neural signals to the olfactory bulb. (b) Odorants bind to specific classes of GPCRs initiating a signaling cascade leading to a depolarization of membrane potential.

Box 14.4 Can you taste that smell?

As presented earlier, the entire experience most of us refer to as taste is better described as flavor. We also know that flavor preferences are very personal and diverse. I know several people to whom having chocolate or strawberry ice cream instead of vanilla is a walk on the wild side. I also know many people who "use both spices on their food, salt and pepper." On the other end of the spectrum, I know many people who cannot eat chili without adding extra Tabasco sauce first, and of course, there are those people who stand in line for the opportunity to sign a waiver to try the newest super atomic hot sauce as the local barbecue rib festival. The flavors we like are very personalized.

To demonstrate the relationship between taste and smell, you can do a simple experiment using hard candies. Can you tell whether a candy is sweet or sour when your eyes are closed and nose plugged? Can you determine whether a candy is lemon, lime, or apple with your eyes closed and nose plugged? What happens if you hold your breath instead of plugging your nose?

Materials
- Five pieces of hard candy that have the same shape and size. Jolly Ranchers are one example.
- A partner.

Process
- The two partners will take turns being the test subject and the experimenter.
- Each partner will taste two candies.

Experiment 1
- The test subject will keep their eyes closed throughout the experimental process.
- The test subject will have to plug their nose for this first experiment.
- The experimenter will open a piece of candy, making sure that the test subject does not see the color of the candy or label of the wrapper.
- The experimenter will give the candy to the test subject.
- The test subject puts the candy in their mouth, closes their mouth, and tastes the candy with their nose plugged.
- The experimenter should ask whether the subject can tell: (yes, this happens with your eyes closed and nose plugged).
- Is the candy sweet or sour?
- What is the candy flavor?
- Next the test subject should stop plugging their nose but keep their mouth closed.
- What is the flavor of the candy?
- How were you able to tell the flavor?
- What happens if you plug your nose again?

Experiment 2
- Once again the test subject will keep their eyes closed throughout the experimental process.
- Instead of plugging their nose, this time the test subject will hold their breath while tasting the candy.
- Repeat the same process as before.
- The experimenter should ask whether the subject can tell: (yes, this happens with your eyes closed and nose plugged).
- Is the candy sweet or sour?
- What is the candy flavor?

- Next the test subject should stop holding their breath but keep their mouth closed.
- What is the flavor of the candy?
- How were you able to tell the flavor?
- What is the difference between the two experiments?
- Record your observations and repeat the process with the other partner being the test subject.

14.9 Texture, Temperature, and Pain

In addition to taste and smell, the texture of our food, the temperature of our food, and the activation of pain receptors also contribute to our eating and drinking experience. Touch and pressure receptors exist in our mouth and tongue just as they do in other surfaces lining our body. We use these to recognize the crispness of a fresh garden salad or the smooth texture of whipped cream. We recognize the change in the texture of foods as they are prepared or eaten in different ways. For example, the texture of a crisp, tart apple compared with the chewier, sweeter apples in our favorite apple pie. Both are pleasing to eat but have different texture qualities. Likewise, temperature receptors exist throughout our bodies, including on the tongue and in the mouth. We have learned to enjoy a steaming bowl of soup on a cold, rainy day, and most of us enjoy the treat of some ice cream during the heat of summer. Throughout our lives, we learn how certain foods are prepared, and we identify a proper temperature for serving and eating these foods. For example, most people in our parents' generation only drank coffee hot. They related the odor and taste of coffee to a hot beverage. For them, the idea of iced coffee sounds foolish, and many of them will never develop a taste for drinking cold coffee. Inversely, many of us have picked up our bottle of soda after it has sat in a car on a warm day, and as soon as we take a drink, we realize that we have made a mistake. The beverage just isn't as refreshing warm as it is cold. During our lives, we learn to have preferred temperatures for specific food items, and for many of us it is difficult to change those preferences.

The sensation of temperature contributes to foods in ways other than the food being truly hot or cold. There are specific food and beverage additives that give a false sense of hot or cold. For example, people will comment on a cool or fresh sensation when consuming food or beverages containing spearmint or menthol. This happens because the cold receptor cells contain a cation channel, TRPM8, which is activated by these compounds. Conversely, we have all eaten spicy food that we identify as "hot" whether the temperature of the food was hot or cold. The spiciness in hot peppers, for example, is caused by the chemical capsaicin. The capsaicin gives us the sensation of heat by directly activating the TRPV1 ion channel that is present in pain receptors that are normally activated by temperatures that could cause damage to our bodies; temperatures we identify as too hot.

14.10 The Absence of Taste and Smell

The loss of the ability to taste and smell can be devastating. Losing your sense of smell can decrease your ability to identify foods by as much as 90%. For many people this takes away the pleasure of the food they eat. Our ability to taste and smell varies by individual. The National Institutes of Health indicates that 25% of Americans are non- or minimal tasters, 50% are medium tasters, and 25% are supertasters. In general, women are more accurate than men in identifying odors. The sense of smell is most acute in humans between 30 and 60 years of age and then it begins to decline. One of the challenges as many people grow older is the loss of sense of smell that leads them to lose interest in their favorite foods, and in some cases, eating in general. At this point, food supplements are typically needed. Those people identified as nontasters typically have receptor deficits for both taste and smell, and they describe the sensation of eating very differently from others.

14.11 Conclusion

As we look at the science of food and cooking you will read many descriptions of food items. The descriptions will talk about the appearance of certain foods and the textures of other food items or ingredients. The most important descriptions will be based upon how the food tastes and smells. As you do different experiments and try different foods, you will gain a greater understanding of these descriptions. You will also learn that flavor preferences of foods vary greatly from one person to another. Your best friend will make their favorite new drink sound absolutely amazing, and when you taste it, your response may be blah! No problem. Everyone's tastes are different. The joy of food and cooking comes from learning from others and discovering exactly what your favorite tastes, smells, and flavors are so you can describe them to your friends and family.

Science for the Chef: A taste and smell experience

This 6-course menu not only provides a delightful culinary experience but also serves as a hands-on exploration of the scientific principles of taste and smell, bringing the chapter's concepts to life in a practical and enjoyable way. Each course is carefully crafted to highlight specific taste sensations—sweet, sour, salty, bitter, umami—and the complex interplay between taste and smell. Through this culinary journey, you will gain a deeper understanding of how different molecules stimulate receptors in your taste buds and olfactory system, and how these signals are integrated in your brain to create the perception of flavor.

Course 1: Sweet and sour - citrus sorbet

Ingredients:
- 1 cup freshly squeezed orange juice
- 1/2 cup freshly squeezed lemon juice
- 1/2 cup water
- 1/2 cup granulated sugar
- 1 tablespoon orange zest
- Pinch of salt

Instructions:
1. Combine the water and sugar in a saucepan over medium heat. Stir until the sugar is completely dissolved to make a simple syrup. Remove from heat and let cool.
2. Mix the orange juice, lemon juice, orange zest, and a pinch of salt in a bowl.
3. Stir in the cooled simple syrup until well combined.
4. Pour the mixture into an ice cream maker and churn according to the manufacturer's instructions.
5. Transfer the sorbet to a container and freeze for at least 2 hours before serving.

Science Behind the Recipe

The sorbet's sweetness comes from the sugar (sucrose) in the recipe, which binds to sweet taste receptors on your tongue. The sourness is due to citric acid in the orange and lemon juice, which releases hydrogen ions that activate sour taste receptors. The combination of these two taste sensations demonstrates how different molecules interact with specific receptors, leading to a complex flavor experience. The cold temperature of the sorbet also engages thermoreceptors in your mouth, adding another layer to the sensory experience.

Course 2: Salty umami - sea-salted roasted vegetables with truffle oil

Ingredients:
- 1 large carrot, sliced
- 1 zucchini, sliced
- 1 red bell pepper, sliced
- 2 tablespoons olive oil
- 1 teaspoon sea salt
- 1 tablespoon truffle oil

Instructions:
1. Preheat the oven to 400°F (200°C).
2. Toss the sliced vegetables in olive oil and spread them on a baking sheet.
3. Sprinkle with sea salt and roast in the oven for 20–25 minutes until tender and slightly caramelized.
4. Drizzle the roasted vegetables with truffle oil before serving.

Science Behind the Recipe

The sea salt provides sodium ions, which are detected by sodium channels in your taste buds, leading to the perception of saltiness. Sodium ions are essential for maintaining the electrochemical gradients necessary for nerve function, highlighting the importance of salt in diet and taste. Truffle oil contains compounds like glutamate, which bind to umami receptors, a type of G protein-coupled receptor. This course illustrates the concept of umami, which enhances the depth and savoriness of foods, making them more flavorful.

Course 3: Umami - miso soup with shiitake mushrooms

Ingredients:
- 4 cups dashi (Japanese soup stock)
- 1/4 cup miso paste
- 1 cup sliced shiitake mushrooms
- 1/2 block tofu, cubed
- 2 green onions, chopped

Instructions:
1. Bring the dashi to a simmer in a pot.
2. Add the sliced shiitake mushrooms and tofu. Cook until the mushrooms are tender, about 5 minutes.
3. Turn off the heat. Dissolve the miso paste in a small bowl with some of the hot dashi, then add it back to the pot.
4. Stir in the chopped green onions and serve immediately.

Science Behind the Recipe

Miso and shiitake mushrooms are rich in glutamate, an amino acid that activates umami receptors (T1R1 and T1R3) in your taste buds. Umami is recognized as a distinct taste sensation, often described as savory or meaty. This course emphasizes the role of amino acids in taste and the importance of umami in enhancing the overall flavor of dishes.

Course 4: Bitter and sour - endive and arugula salad with grapefruit vinaigrette

Ingredients:
- 1 head of endive, chopped
- 2 cups arugula
- 1 grapefruit, segmented
- 1/4 cup olive oil
- 1 tablespoon grapefruit juice
- 1 teaspoon honey
- Salt and pepper to taste

Instructions:
1. In a large bowl, combine the endive, arugula, and grapefruit segments.
2. In a small bowl, whisk together the olive oil, grapefruit juice, honey, salt, and pepper to make the vinaigrette.
3. Drizzle the vinaigrette over the salad and toss to combine.

Science Behind the Recipe

The bitter compounds in endive and arugula, such as sesquiterpene lactones, bind to bitter taste receptors (T2Rs) in your taste buds. Bitter taste perception serves as a protective mechanism, often signaling the presence of potentially harmful substances. The grapefruit adds a sour element, with its citric acid contributing to the sour taste by increasing the hydrogen ion concentration, which activates sour taste receptors. This course highlights the contrast and balance between bitter and sour tastes.

Course 5: Flavor complexity - gourmet cheese platter

Ingredients:
- 2 ounces blue cheese
- 2 ounces aged cheddar
- 2 ounces brie
- 2 tablespoons honey
- A handful of walnuts
- A handful of dried apricots

Instructions:
1. Arrange the cheeses on a platter.
2. Drizzle the honey over the blue cheese.
3. Add the walnuts and dried apricots as accompaniments.

Science Behind the Recipe

This course explores how taste and smell work together to create complex flavors. Each cheese provides a unique combination of tastes (umami from the blue cheese, sharpness from the cheddar, and creaminess from the brie) and smells that are perceived through both orthonasal and retronasal olfaction. The honey and dried fruits add sweetness, while the walnuts contribute a contrasting texture, demonstrating the interplay between different sensory inputs in flavor perception.

Course 6: Interactive taste and smell experiment - chocolate and olfaction

Ingredients:
- High-quality dark chocolate (one piece per participant)

Instructions:
1. Give each participant a piece of chocolate.
2. Instruct them to hold their nose while placing the chocolate in their mouth.
3. After a few seconds of chewing, have them release their nose and observe the difference in flavor perception.

Science Behind the Recipe

This experiment demonstrates the critical role of smell in flavor perception. Initially, holding the nose blocks the olfactory input, so the participant primarily perceives the basic tastes (sweet, bitter) through gustation. Once they release their nose, the retronasal olfaction kicks in, significantly enhancing the flavor experience. This illustrates how the combination of taste and smell creates the full perception of flavor.

Key Concepts

1. **Chemoreception** is the process by which sensory systems detect chemical **stimuli**, which is crucial for the senses of taste (**gustation**) and smell (**olfaction**).

2. Taste **receptors** are specialized proteins located on taste receptor cells that bind to specific **tastants**, such as those responsible for **sweet**, **sour**, **salty**, **bitter**, and **umami** tastes, initiating the perception of taste.

3. **Signal transduction mechanisms** involve the biochemical pathways that convert the binding of tastants or **odorants** to their respective receptors into electrical signals that the brain interprets as taste or smell.

4. **G protein-coupled receptors** (GPCRs) play a key role in the perception of sweet, bitter, and umami tastes by triggering complex intracellular signaling cascades when they bind to their respective tastants.

5. The **membrane potential** refers to the electrical charge difference across the cell membrane, which is essential for generating action potentials in response to taste stimuli, leading to the perception of taste.

6. **Olfactory receptors** are specialized GPCRs found in the olfactory epithelium that bind odorant molecules, initiating the biochemical processes that lead to the perception of smell.

7. **Flavor perception** is the result of the brain integrating signals from both the taste and smell sensory systems, with retronasal olfaction playing a significant role in enhancing the overall sensation of flavor.

8. The **structure-activity relationship** describes how the specific molecular structure of tastants influences their ability to bind to taste receptors, thereby determining the intensity and quality of the taste experienced.

9. **Neurotransmitter** release is the final step in the signal transduction process for taste, where taste receptor cells release neurotransmitters that transmit the taste signal to the brain.

10. Genetic variability in taste perception is due to differences in the genes coding for taste and olfactory receptors, which result in varying levels of sensitivity to taste and smell among individuals.

References

1 Bushdid, C, Magnasco, M. O, Vosshall, L. B., & Keller, A. (2014). Humans can discriminate more than 1 trillion olfactory stimuli. *Science*, *343*(6177), 1370–1372. doi: 10.1126/science.1249168. PMID: 24653035.
2 Yang, Q. (2010). Gain weight by "going diet?" Artificial sweeteners and the neurobiology of sugar cravings. *Yale J. Biol. Med.*, 83(2), 101–108.
3 Maguder, T. (2006). Quinine was a lifesaver during the civil war. Available at: http://fredericksburg.com/News/FLS/2006/082006/08122006/212900 accessed on November 2, 2015.
4 Marijke, M. F., Barbro, N. M., & Paul de Vos. (2010). A brief review on how pregnancy and sex hormones interfere with taste and food intake. *Chemosens. Percept.*, *3*, 51–56.
5 Oka, Y., Butnaru, M., von Buchholz, L., Ryba, N.J., & Zuker, C. S. (2013). High salt recruits aversive taste pathways. *Nature*, *494*(7438), 472–475.
6 Shimada, S., Ueda, T., Ishida, Y., Yamamoto, T., & Ugawa, S. (2006). Acid-sensing ion channels in taste buds. *Arch. Histol. Cytol.*, *69* (4), 227–231.
7 Lopez-Jimenez, N. D., Cavenagh, M. M., Sainz, E., Cruz-Ithier, M. A., Battey, J. F., & Sullivan. S. L. (2006). Two members of the TRPP family of ion channels, Pkd1l3 and Pkd2l1, are co-expressed in a subset of taste receptor cells. *J. Neurochem.*, *98*(1), 68–77.
8 Besnard, P., Passilly-Degrace, P., & Khan, N. A. (2016). Taste of fat: a sixth taste modality? *Physiol. Rev.*, *96*(1), 151–176.

Additional Readings

Ehrlich, MRM, Dando, R. (2021), The sensory properties and metabolic impact of natural and synthetic sweeteners. *Compr Rev Food Sci Food Saf.*, 20(2), 1554–1583.
Lima, R. S., Ramos, L. M., de Medeiros Sousa, V., Tonucci, L. B., Pereira, C. T. M., Pereira, D. M., de Medeiros, A. C., & Bolini, H. M. A. (2023). Temporal sweet taste dominance according to adult body mass index classification. *J. Food Sci*, 88(5), 2191–2202.
Lioe, H.N., Selamat, J., & Yasuda, M. (2010), Soy sauce and its umami taste: a link from the past to current situation. *J. Food Sci*, 75(3): R71–R76.
Mastinu, M., Melis, M., Yousaf, N. Y., Barbarossa, I. T., & Tepper, B. J. (2023), Emotional responses to taste and smell stimuli: self-reports, physiological measures, and a potential role for individual and genetic factors. *J. Food Sci*, 88(S1), A65–A90.
Peters, J.C., Marker, R., Pan, Z., Breen, J.A., & Hill, J.O. (2018), The influence of adding spices to reduced sugar foods on overall liking. *J. Food Sci*, 83(3), 814–821.
Spence, C. (2023). On the manipulation, and meaning(s), of color in food: A historical perspective. *J. Food Sci*, 88(S1), A5–A20.

End of Chapter 14 Questions

1. Which of the following is a component of taste, but *not* odor perception?
   a. gustation
   b. orthonasal olfaction
   c. retronasal olfaction
   d. taste buds
   e. olfactory receptor cells

2. Fill in the blanks:
   a. Something that tastes *sweet*
   b. Something that tastes *sour/tart*
   c. Something that tastes *savory*

3. Coffee is a highly aromatic drink that is both loved and hated. Coffee has a complex aroma, but a fairly bitter taste. Why can coffee smell wonderful, but taste awful (to some people ...)?

4. Including Lime or Lemon in avocado dip can provide crisp tartness. What class of flavor molecules are responsible?
   a. Acid
   b. Sugar
   c. Salt
   d. amino acids
   e. All of the above

5. Which class of transmembrane proteins are responsible for olfaction, or communicating smell?
   a. Ion channels
   b. GPCRs
   c. PDKs
   d. phospholipids
   e. cAMP

6. Which of the following taste receptors open and allow ions to flow into the cell when a tastant binds?
   a. Sour and Sweet
   b. Salty and Umani
   c. Bitter and Sweet
   d. Salty and Sour
   e. None of the above

7. Which receptor allows the hydrogen ion (H^+) to enter into the taste receptor cell resulting in the activation of a nerve?
   a. Sweet
   b. Sour
   c. Bitter
   d. Umami
   e. Fat

8. Umami taste is elicited by_____
   a. Na^+ ions
   b. Sugars and carbohydrates
   c. Ingestion of poisons
   d. Amino acids and nucleotides
   e. Bacteria found in savory meats

9. If someone says you are awfully T2R receptor like, you are most likely _____
   a. Pretty bitter
   b. Very sweet
   c. Using a salty language
   d. Acidic in your approach to people
   e. Just finished a savory move on the dance floor

10. MSG is ...
    a. Mono Sodium Glutamate—a natural amino acid
    b. An excitatory neurotransmitter
    c. Safe in small doses
    d. Synergizes with salt flavors
    e. All of the above

15

Metabolism of Food: Microorganisms and Beyond

Guided Inquiry Activities (Web): *7, Carbohydrates; 14, Cells and Metabolism; 15, Metabolism, Enzyme, and Cofactors; 18, Starch*

Learning Objectives

1. *Define metabolism and describe its importance in cellular and organism function.*
2. *Describe the basic components of a cell and the differences between prokaryotes and eukaryotes.*
3. *Explain the electron transfer that occurs in an oxidation/reduction (i.e. redox) reaction, including the purpose of $NAD^+/NADH$ and $FAD/FADH_2$.*
4. *Describe glycolysis, including the preparatory and payoff phase, and its function in energy production in nearly all cells under aerobic and anaerobic conditions.*
5. *Explain the difference between aerobic and anaerobic metabolism.*
6. *Compare and contrast alcohol and lactate fermentation as two types of anaerobic respiration, including the use of coenzymes and energy production.*
7. *Describe how the three pathways necessary in aerobic respiration: pyruvate dehydrogenase (pre-TCA), the Krebs of citric acid cycle (TCA), and electron transport chain, yield energy in the form of adenosine triphosphate (ATP) generation.*
8. *Provide an overview of fat metabolism, including why fat is a more effective long-term energy storage molecule vs. carbohydrates.*
9. *Describe how the metabolism of proteins and amino acids contributes to food and cooking, including characteristics that distinguish them from carbohydrates and fats.*

15.1 Introduction

Metabolism is the collective process that cells use to transform matter for energy, to create and store new molecules, and to respond to their environment. Understanding cell theory, the diversity of cells, and how different organisms use food from the environment to make new molecules that can impact our cooking and baking is a fascinating and important part of the food or cooking. Realizing how metabolism impacts plant and animal cells helps a cook or chef make informed choices in selecting food and preparing their dish. Focusing on the metabolism of microorganisms is crucial for preparing cheeses, wine, beer, bread, etc.—the list is endless.

If someone asked you to describe how microorganisms relate to food, what would you say? Most people immediately think about meats, fresh fruits, or vegetables that are contaminated with *Escherichia coli* or *Salmonella*,

The Science of Cooking: Understanding the Biology and Chemistry Behind Food and Cooking, Second Edition. Joseph J. Provost et al.
© 2025 John Wiley & Sons, Inc. Published 2025 by John Wiley & Sons, Inc.
Companion website: www.wiley.com/go/provost/food_science_2e

which causes foods to be pulled from grocery store shelves or people to get ill from eating a very rare hamburger. The relationship between microorganisms and food is typically thought of as detrimental, as in these situations the microorganism is unwanted, harmful to human health, and grows in an uncontrolled manner.

Many foods, however, are dependent upon microorganisms for their preparation and production. Fungi (e.g. yeast) are important in the production of breads and alcoholic beverages. The presence of protists (e.g. molds) leads to a tangy, pungent, and flavorful blue cheese. Different species of bacteria are essential to the production of sourdough breads, soy sauce, yogurt, and pepperoni, to name a few. In contrast to the harmful bacteria that cause human illness, some microorganisms are purposefully added to and proliferate in a highly controlled environment, leading to a product that is safe and tasty to eat or drink. In this chapter, you will learn about various classes of multicellular organisms and microorganisms that are important to the chemistry and biology of food and how these organisms generate the energy resources required for them to survive and proliferate and some of the molecules that microorganisms produce in these processes that are important to food and cooking (Box 15.1).

Box 15.1 Not all microorganisms are good for you

Not all cells and microorganisms are helpful to humankind. Many foodborne illnesses are caused by the ingestion of harmful microbes. According to the Centers for Disease Control, one in six Americans (nearly 48 million people) will get sick from food poisoning in their lifetime, and 3,000 will die of foodborne diseases. Although molds, viruses, and parasites all pose a potential risk of food contamination, some strains of bacteria are the most common culprits. One strain of bacteria, *Salmonella*, is responsible for a quarter of foodborne illnesses and impacts 1.2 million cases annually in the United States. Most of the transmission comes from food, water, or contact with infected poultry. Raw or unpasteurized milk and milk products pose a special risk of *Salmonella* illness. *E. coli O157* is a strain of bacteria that lives in the intestines of some cattle, swine, and deer. Improperly cooked ground meat that is contaminated by *O157* causes hemorrhagic diarrhea and intense abdominal cramps, even resulting in death in extreme cases. Heating milk to 161°F/72°C for 20 seconds is enough to kill most harmful bacteria, including *Salmonella* and *E. coli O157* often found in milk. Another common form of food poisoning is from the bacterium *Clostridium botulinum*. This bacterium, commonly found in soil, produces several neurotoxins that act to stop the nerves from signaling, causing paralysis. Although botulism food poisoning is rare, the proteins that produce the neurotoxins are commonly used for cosmetic reasons to paralyze nerves (Botox injections). How can you reduce your chances of getting food poisoning? Wash fresh fruits and vegetables (even the outer parts of the food that you do not eat, like watermelon rinds and orange peels). Keep your food preparation and eating areas clean. Cook foods, as most cells and microbes do not survive in high cooking temperatures due to disruption of the cell membrane and protein denaturation.

15.2 The Basics of the Cell

The smallest unit able to sustain life is called a cell. A cell can be simply defined as a container of small and large molecules that are essential for the survival of an organism. Unicellular organisms, such as algae, bacteria, and protozoa, contain all the molecular machinery within a single cell that is necessary for their survival. When cells evolved to live as a collection or group, some organisms became, by definition, multicellular and developed specialized types of cells with distinct functions. Although the behavior and biology of the cells in uni- and multicellular organisms vary dramatically, several characteristics are common among cells. The interior of a cell is organized into specialized compartments called organelles. Chloroplasts and mitochondria are examples of organelles. Every cell needs a barrier called a membrane, a cell wall may also be added, which forms a barrier between the cell and the outside world, and allows specific molecules into or out of the cell. For example, a cell might need to import food for energy or raw materials in order to make its own food or for the creation of new molecules. Common to both plant and animal cells is the thin outer membrane called the plasma membrane. Made of a combination of lipids (phospholipids and cholesterol) and proteins (associated at the surface and those that go all the way through the membrane—transmembrane proteins), the plasma membrane regulates the traffic of molecules entering and leaving the cell. Flavor molecules bind to transmembrane proteins found in specialized taste bud cells, signaling to other integral membrane proteins that a flavor molecule is in the food. The cytoplasm occupies the region of space between the nucleus, or the center of the cell, and the plasma membrane, or the edge of the cell. Within the cytoplasm are various organelles and proteins. Enzymes, the cell's catalysts, are responsible for the breakdown and synthesis of molecules in a cell. A diverse and dynamic range of other molecules, including carbohydrates, small molecules, nucleotides (DNA and RNA), and 20,000–30,000 different proteins, also play important roles in the function of the cell. Importantly, residing within DNA is the genetic blueprint or hereditary information of the cell, which is passed from cell to cell during cell division.

Organisms are classified into two groups, prokaryotes and eukaryotes, based upon the characteristics of their cell or cells. The earliest forms of life were the unicellular prokaryote organisms. The distinguishing cellular feature of prokaryotes is the absence of a nucleus, or an organized compartment (i.e. organelle) that stores the cell's DNA. Bacteria are a species of prokaryotes. Plants, fungi, and animal cells contain a nucleus and are thus eukaryotes. Not only do eukaryotes possess a nucleus, but they also contain several types of specialized organelles or structures, including mitochondria, chloroplasts, nucleus, and endoplasmic reticulum.

15.2.1 Plant Cells

Plants are multicellular organisms with specialized cells (Figure 15.1). These have both mitochondria and chloroplasts to process energy for the plant cell. In addition to nuclei, endoplasmic reticulum, and other eukaryotic organelles, plant cells have a rigid, structurally complex cell wall made of a number of large and small molecules. The cell wall is a particularly important part of food and cooking. The plant cell wall is a mixture of complicated carbohydrates and proteins giving plants rigid support. Specific to food and cooking are some very interesting components. Cellulose, hemicellulose, and pectin are complex carbohydrate polymers that add fiber to food. Cellulose and hemicellulose serve as thickeners and emulsifiers in some recipes. The pectin in fruit is used to solidify jellies under acidic conditions. About 10% of the plant cell wall is composed of polyphenolic compounds called lignins. Phenols are organic ring structures with the —OH (i.e. hydroxyl) functional group attached. Lignins are a complex set of diverse compounds found in plants and algae (Figure 15.2). A special subfamily of polyphenols called lignin (i.e. a mixture of polyphenols) is responsible for astringency in food and drink. When oxidized by plant cell enzymes, some polyphenols become brown, taste bitter, and help plants fight infection.

15.2.2 Animal Cells

Animal cells do not have a plant cell wall or chloroplasts and are highly diverse depending on the tissue each cell comes from. Most animal cells will have a nucleus, mitochondrion, Golgi, and other components (Figure 15.3). For food and cooking, we focus on meat, and meat is most often the muscle tissue of an animal. Muscle tissue is comprised of muscle cells, which have a slightly different nomenclature that is described in Chapter 7.

Perhaps you are thinking, how do these cells come together to make a functioning organism or a plant or animal tissue that might be cooked or eaten? Cells do not naturally "stick" together. Groups of cells organized into plant and animal tissue are embedded in a jelly-like substance that acts as "cellular glue." The material surrounding the cells is a mixture of carbohydrates and proteins that are collectively called the extracellular matrix. This matrix is the mortar that holds the cells together into a cohesive mass. As in a tissue, some of these extracellular matrix

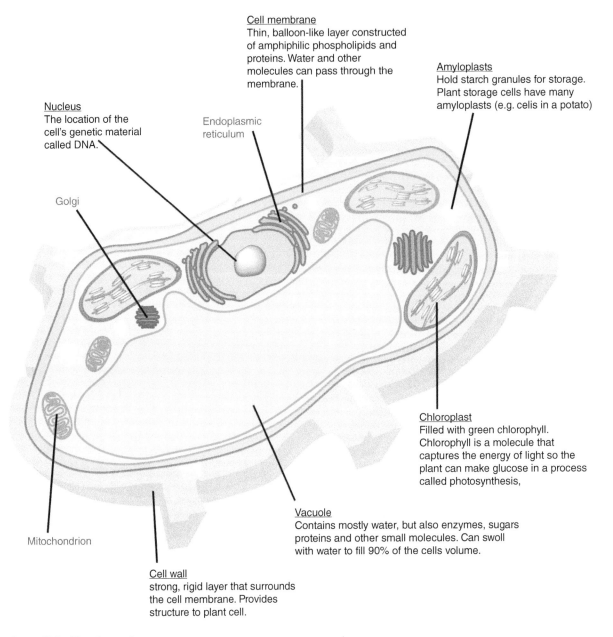

Cell membrane
Thin, balloon-like layer constructed of amphiphilic phospholipids and proteins. Water and other molecules can pass through the membrane.

Amyloplasts
Hold starch granules for storage. Plant storage cells have many amyloplasts (e.g. cells in a potato)

Nucleus
The location of the cell's genetic material called DNA.

Endoplasmic reticulum

Golgi

Chloroplast
Filled with green chlorophyll. Chlorophyll is a molecule that captures the energy of light so the plant can make glucose in a process called photosynthesis,

Vacuole
Contains mostly water, but also enzymes, sugars proteins and other small molecules. Can swoll with water to fill 90% of the cells volume.

Mitochondrion

Cell wall
strong, rigid layer that surrounds the cell membrane. Provides structure to plant cell.

Figure 15.1 The plant cell.

components are useful in a variety of foods. Collagen, a protein found in the extracellular matrix, is used as a thickener in sauces and is the main component of gelatin and the gelatin dessert, JELL-O®. Plant extracellular matrix components, such as pectin, are used in cooking fruit pies.

15.2.3 Yeast Cells

Yeast cells are single-cell eukaryote cells in the fungi kingdom and contain organelles, such as mitochondria, an enclosed nucleus, and a cell wall. Found in nearly every environment, yeast is very diverse with many strains. Yeast is terribly important in food, beverages, baking, and cooking (Figure 15.4). Baking bread requires

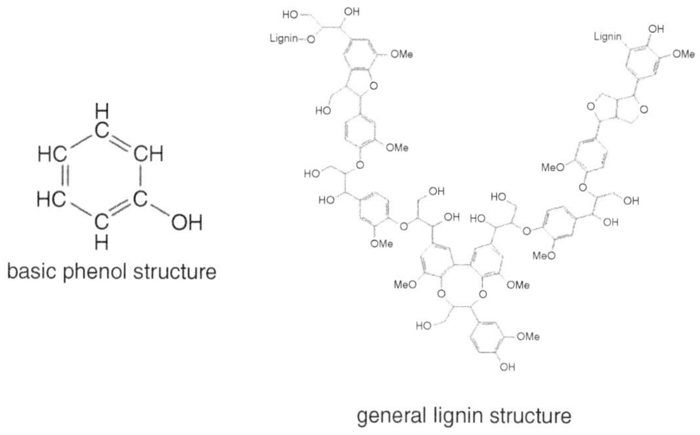

basic phenol structure

general lignin structure

Figure 15.2 Polyphenols. Lignin is a highly diverse and modified polymer of phenol. Lignin is a major component of plant cell walls and in our food and drink.

ANIMAL CELL

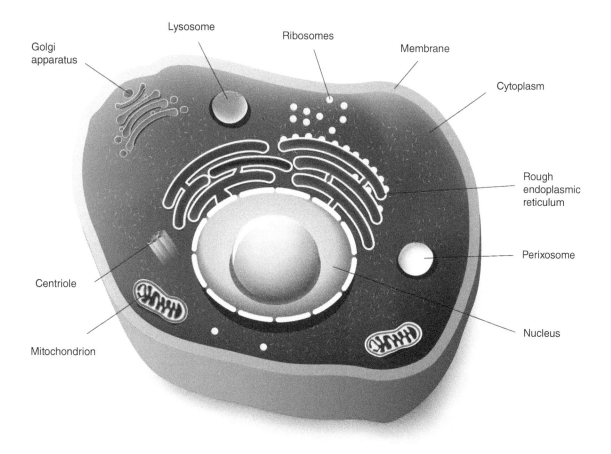

Figure 15.3 The animal cell. designua/Adobe Stock Photos

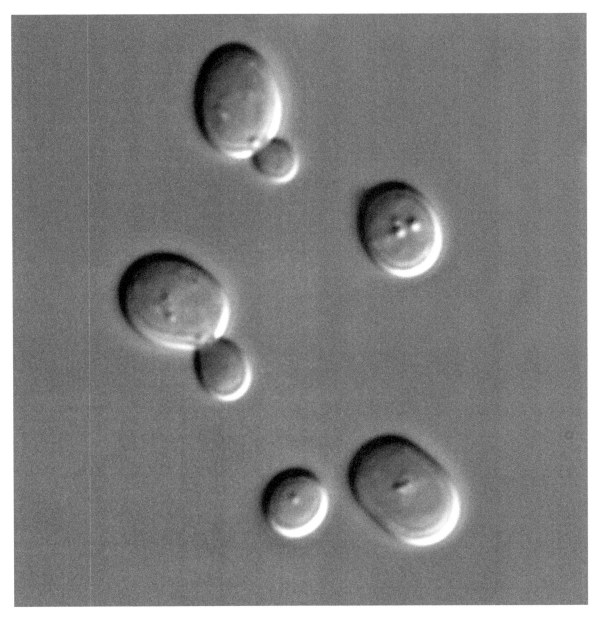

Figure 15.4 Yeast cells. Budding yeast cells (*Saccharomyces cerevisiae*) are used for baking and brewing. *Source:* Masur / Wikimedia Commons / Public domain.

the metabolism of sugars into CO_2 gas to give bread its rise. Because of its mitochondria, yeast can use a number of different food sources and convert carbohydrates, fats, and proteins to a range of final products, including carbon dioxide, ethanol, and acetic acid. Under oxygen-rich conditions, yeast will produce a mixture mostly of CO_2 and some ethanol. Grown in an oxygen-depleted environment, yeast will switch to producing primarily ethanol.

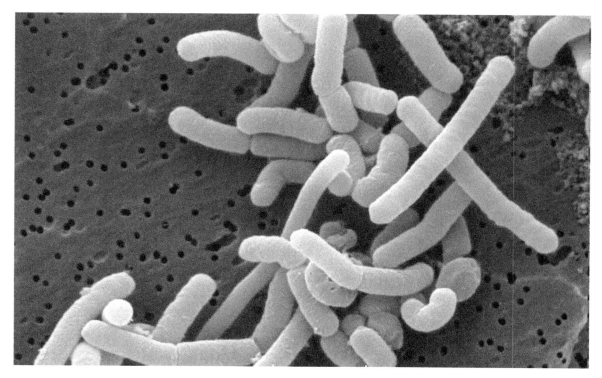

Figure 15.5 Colorized scanning electron micrograph of *Lactobacillus paracasei*. *Lactobacillus* bacteria are the lactic acid-producing bacteria used for yogurt, cheese, and other fermented products. *Source:* Dr. Horst Neve / Max Rubner-Institute/ CC BY-SA 3.0

15.2.4   Bacteria Cells

Bacteria comprise a broad class of prokaryotic single-cell organisms without a nuclear envelope and many of the organelles of eukaryotes (Figure 15.5). Bacteria are an amazingly diverse class of microorganisms, some of which are involved in human health and disease. Throughout this chapter, we discuss bacteria as part of the preparation of food. Bacteria that produce and can live in acidic conditions are used to make cheese, and some of the same bacteria are encouraged to grow on meat as it cures and ages. In this case, the bacterium helps produce acid that inhibits unhealthy bacterial growth. Lactic acid-producing bacteria are used to produce yogurt, sauerkraut, and fermented meats and sausages. Some winemakers will use bacteria to encourage new flavors, including a green apple malonate flavor (malolactic fermentation) in wines. Yet others will encourage bacterial growth in wine to produce wine vinegar.

Having a better understanding of the nature of a cell and how a cell is organized will help you better understand ingredients and steps in cooking and baking.

15.3   Introduction to Basic Metabolism

In order for a cell to survive, grow, and divide, it must have a source of energy. The mechanism by which any cell or organism breaks down and builds the molecules necessary for its function and proliferation is called metabolism. A cell's or organism's metabolism comprises all the chemical reactions that are essential to its survival; thus it is a vast, complicated series of reactions that are specific to a particular organism or cell type.

These metabolic processes that are critical to the life of the organism are also critical to food and cooking. In order to understand this connection, we need to discuss the details of some metabolic pathways and why metabolism is necessary for a cell.

15.3.1 Building Up and Breaking Down

Cellular metabolism consists of many specific pathways, each of which accomplishes a particular task. Some of these pathways are responsible for synthesizing the molecules that are important to the function or structure of the cell. These pathways are called anabolic pathways, from the Greek prefix *ana* meaning "up." Other pathways are important in the breakdown of biomolecules; these reactions are termed catabolism, from the Greek word *kata* meaning "down" (Figure 15.6)

15.3.2 Anabolism Is Endergonic and Catabolism Is Exergonic

The role of the anabolic pathways or anabolism might be obvious; a cell needs to generate biological macromolecules, such as proteins and nucleic acids, which are necessary for its function. Just as it requires energy (in the form of effort and time) to make a cake, building molecules also requires energy. When a chemical

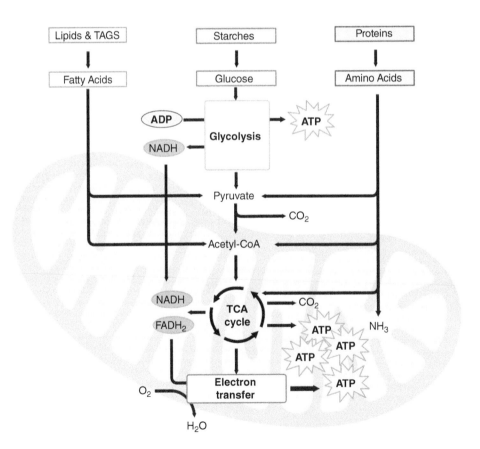

Figure 15.6 Overview of catabolism. Several metabolic pathways, including glycolysis, are involved in using and generating molecules necessary to meet the cell's needs.

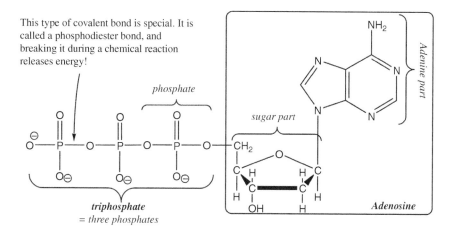

Figure 15.7 Adenosine triphosphate.

reaction requires an input of energy to occur, you have an endergonic reaction. Thus, anabolic reactions are endergonic. Catabolic pathways play two roles within a cell. In catabolism, large molecules are broken down into smaller molecules called metabolites, and there is a net release of energy (i.e. exergonic) to carry out the work of the cell. Here, you might think about eating the cake as a catabolic process. You have to make a minimal effort, moving a bite of cake to your mouth, to yield a significant product, a delicious bite of light, fluffy cake in your mouth. Thus, anabolic and catabolic processes are reciprocal; the energy and metabolites generated through catabolism are used to make new molecules (i.e. anabolism) that are essential for the survival of the organism. The energy source that connects anabolism and catabolism is a molecule called adenosine triphosphate (ATP) (Figure 15.7).

15.3.3   ATP and Its Hydrolysis

As the name implies, ATP consists of an adenosine covalently linked to a triphosphate (Figure 15.7). The adenosine portion of the molecule is made up of an adenine base and ribose sugar. The triphosphate consists of three phosphate groups (PO_4^{3-}) that are bonded to one another via covalent bonds called phosphodiester bonds. ATP is the energy "currency" of the cell. It is generated in catabolic reactions and it is used up in anabolic reactions. How does this work, and where does the energy in ATP come from? ATP's energy comes from the energy contained within the phosphodiester bonds. When the phosphodiester bond of ATP is broken in a reaction with water, called a hydrolysis reaction, about 30 kilojoules/mole (kJ/mol) worth of energy is given off under standard biological conditions. When a reaction, like the hydrolysis of ATP, produces a net energy release, it is called exergonic. When this reaction is linked/coupled to another reaction that requires energy to occur (like an anabolic reaction), that 30 kJ/mol worth of energy can be used to drive the unfavorable reaction forward (Figure 15.8).

15.3.4   Redox: Oxidation and Reduction

In the process of metabolism, the carbons of glucose (i.e. $C_6H_{12}O_6$) are partially or fully oxidized to generate energy in the form of important metabolites that the cell needs. ATP is one of those metabolites. We've learned that ATP is a high-energy compound that can drive the work of the cell. NAD or nicotinamide adenine dinucleotide is another important metabolite.

NAD$^+$ serves as an electron carrier by accepting two electrons and two hydrogen cations or protons from the molecule that is being oxidized, thereby generating NADH plus a proton. The nicotinamide part of NAD$^+$ is largely derived from niacin, a B vitamin that is essential for the survival of cells and organisms. Good sources of niacin include animal products (i.e. meat, eggs, seeds, and legumes). However, only small amounts of NAD$^+$ are present in a cell at any given time. In order for metabolic processes and oxidation reactions to occur within a cell or microorganism, recycling of NADH to NAD$^+$ is a key part of metabolism and how microorganisms meet their energy needs. If you enjoy eating a crusty piece of bread with cheese or drinking a beer, you have benefited from the recycling of NADH to NAD$^+$ by yeast and bacteria.

Nicotinamide adenine dinucleotide comes in two forms: oxidized or NAD$^+$ and reduced or NADH. In Figure 15.9, we can see where these names come from. NAD$^+$ and NADH take part in redox reactions, or reactions that involve the reciprocal processes of oxidation and reduction. Oxidation is the net loss of electrons from an atom or molecule. A molecule that

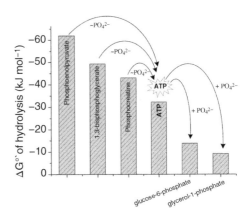

Figure 15.8 Energy of ATP. ATP has an inherent energy called Gibbs free energy (ΔG). While ATP is considered the energy currency of the cell, there are higher and lower energy compounds in the cell. Phosphoenolpyruvate, 1,3-bisphosphoglycerate, and phosphocreatine all have a ΔG of phosphate hydrolysis that is greater than -40 kJ·mol^{-1}, which means these compounds can donate phosphates to make other lower energy molecules. ATP has a ΔG of ~30 kJ·mol^{-1} and can donate a phosphate to make glucose-6-phosphate and glycerol-1-phosphate, both with ΔG's less than -20 kJ·mol^{-1} [1] / Taylor & Francis Group.

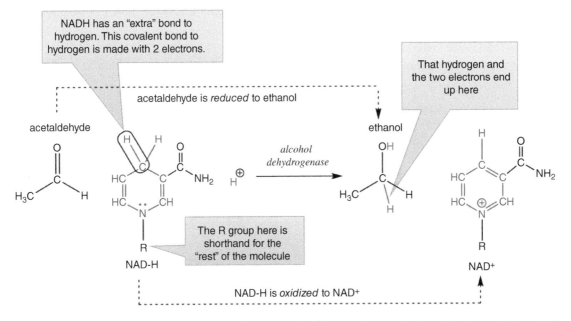

Figure 15.9 Oxidation and reduction are reciprocal processes. In this example reaction found during yeast fermentation, two electrons and a hydrogen (in blue) from the nicotinamide ring of NADH (in red) are transferred to acetaldehyde. The gain of electrons by acetaldehyde is a reduction and the molecule is transformed into ethanol. The loss of electrons from NADH is called oxidation and the molecule is transformed into NAD$^+$. If electrons are donated or lost from one molecule, they must be gained by another molecule—this is the reciprocal nature of redox (*red-* = reduction; *-ox* = oxidation) reactions.

gets oxidized, loses electrons and usually has more bonds to oxygen and fewer bonds to hydrogen when the redox reaction is complete. Reduction is the net gain of electrons to an atom or molecule. A molecule that is reduced gains electrons and usually has more bonds to hydrogen and fewer bonds to oxygen at the conclusion of the redox process. When redox reactions occur in the cell, reduction and oxidation happen over the course of the same reaction. The electrons move from one molecule to another. In Figure 15.9, we can see that NADH gives electrons to acetaldehyde. NADH is therefore losing electrons or becoming oxidized to NAD^+ over the course of the reaction, while acetaldehyde is gaining electrons and getting reduced to ethanol. Notice how ethanol is the product of the reaction with more bonds to hydrogen, which is characteristic of reduction, and NAD^+ is the reaction product that has fewer bonds to hydrogen, which is characteristic of oxidation. NAD^+ and NADH participate in many biological reactions, so much so that the pair, NAD^+ and NADH, are called the "electron carriers" of the cell.

15.4 Catabolism of Glucose (Glycolysis or Fermentation): Glucose To Pyruvate

To launch our discussion of metabolism, we will first concentrate on the breakdown of the molecule glucose (Figures 15.10 and 15.11). Not only is glucose a main energy source in most organisms and cells, but the breakdown of most other energy-rich substances (like fats, complex carbohydrates, and proteins) begins with their conversion into one of the intermediates in the pathway for glucose catabolism. Glucose catabolism can occur under aerobic or anaerobic conditions. In aerobic respiration, the complete breakdown of glucose to carbon dioxide and water occurs in the presence of oxygen. The word "aerobic" is derived from the roots aero-meaning "air" and bios meaning "life." Air contains about ~21% oxygen, and it is that oxygen that determines how a cell will break down glucose. Aerobic respiration is a complex, multi-pathway process that utilizes oxygen as the final electron acceptor for the electrons that are lost from the glucose molecule (Figure 15.6). Because the carbons of glucose are fully oxidized to carbon dioxide, aerobic respiration of glucose yields the greatest amount of energy for the organism.

Most organisms are also able to carry out anaerobic respiration or fermentation, which doesn't require the presence of oxygen. Adding the root an- to "aerobic" gives us a word that directly translates to "not" (an-) + "air" (-aero-) + "life" (-bios); therefore, anaerobic does not require "air," which means it does not require the oxygen that is in air. Anaerobic respiration generates less energy for the organism, as there is no net oxidation of the glucose molecule; however, it is an essential process for organisms to survive in situations where there is an absence of or limited amounts of oxygen. Some organisms even prefer to carry out fermentation! Although there are major differences in the energy generated and oxidation processes for an organism under aerobic versus anaerobic conditions, both processes initially break down glucose using the same pathway called glycolysis.

Glycolysis, derived from the Greek roots *glykos*, meaning sweet, and *lysis*, meaning "loosening" or "splitting," involves a sequence of 10 enzyme-catalyzed reactions that converts one six-carbon molecule of glucose (i.e. $C_6H_{12}O_6$) into two three-carbon molecules of pyruvate (i.e. $C_3H_3O_3$) (Box 15.2). The first five steps of the pathway are called the "investment" or the "preparatory" phase of glycolysis because energy is consumed (see the use of an ATP molecule in step one) to break the six-carbon glucose molecule into two three-carbon units. The next few steps are referred to as the payoff phase of glycolysis, as they include an oxidation reaction that generates the molecule reduced nicotinamide adenine dinucleotide, or NADH, and that generates ATP through direct addition of high-energy phosphate from phosphoenolpyruvate (i.e. PEP) to ATP in a process called substrate-level phosphorylation (Figure 15.10). The final product of glycolysis is pyruvate; from a single molecule of glucose, two molecules of pyruvate are generated.

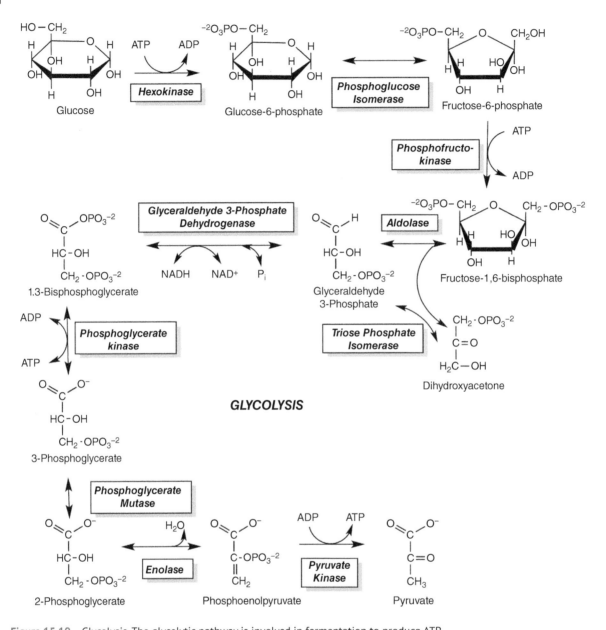

Figure 15.10 Glycolysis. The glycolytic pathway is involved in fermentation to produce ATP.

Any metabolic pathway, including glycolysis, is often viewed as a complicated series of enzymes, substrates, and products. However, all metabolic pathways do have some common themes. The name of the enzyme often describes the features of the reaction it catalyzes. For example, step one of glycolysis is catalyzed by hexokinase. *Hex* is the prefix that is associated with six (in this case, the six-carbon glucose), and *kinase* is the enzyme name for a reaction that involves the addition of a phosphate group. The oxidation/reduction reaction of step six is catalyzed by glyceraldehyde 3-phosphate dehydrogenase, providing a hint as to the substrate of the reaction (i.e. glyceraldehyde 3-phosphate) and the type of reaction that occurs (i.e. a dehydrogenation or oxidation/reduction reaction).

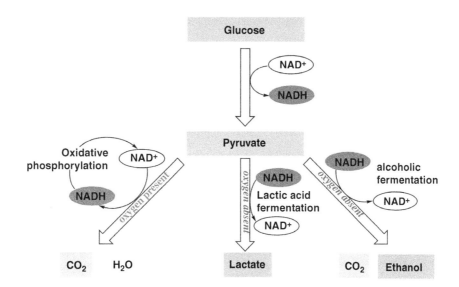

Figure 15.11 The fates of pyruvate. Depending on the organism, cell type, and availability of oxygen, pyruvate can be metabolized to carbon dioxide (i.e. CO_2) and water (i.e. H_2O), lactate, or carbon dioxide and ethanol. When oxygen is present, conditions are aerobic, and glucose is ultimately oxidized to carbon dioxide and water. When oxygen is absent, conditions are anaerobic, and—depending on the organism—glucose is oxidized to lactate or ethanol and carbon dioxide.

Box 15.2 Enzymes are proteins that do work

As we saw in Chapter 2, protein is one of the three biological macromolecules that contribute to nutrition. Cells are full of many different kinds of proteins. Some proteins are for storage, some are for structure, and some proteins do work. Enzymes are proteins that do the work of the cell. Enzymes catalyze chemical reactions; that is, for a given chemical reaction, an enzyme makes it possible for bonds to be broken and formed under the conditions that allow for life to exist. Chemical reactions happen around us all the time, but for chemistry to happen inside a living thing, that chemistry must proceed under the conditions in which the organism lives. While in the laboratory, chemists may use high heat, pressure, strong acids, or heavy metals to make and break bonds, inside living things enzyme-catalyzed reactions make it possible to break and form bonds without those extreme conditions.

15.5 Fates of Pyruvate: Now What?

As mentioned earlier, glycolysis is used by almost all organisms under aerobic and anaerobic conditions. One of the most important features of glycolysis is the final product of the pathway, pyruvate. Nearly every cell type, from simple single-cell organisms to higher-order multicellular eukaryotes, has the capacity to generate ATP from glucose via glycolysis. What happens to the pyruvate produced by glycolysis depends on the type of cell or organism, the nutritional state of the environment, and the access to oxygen (Figure 15.11). Each cell has a limited number of NAD^+/NADH molecules, which means the NAD^+ required for glycolysis can be depleted, and if all the NAD^+ were to be converted to NADH, glycolysis would shut down and the cell would die! This is why cells must have a way to recover the NAD^+ consumed in glycolysis.

 Cells with a functional Krebs cycle and in oxygen-rich environments are aerobic and will continue to metabolize the pyruvate to fats through oxidative phosphorylation (Section 15.6) and the electron transport chain

(Section 15.7) regenerating NAD^+ needed for glycolysis, and making new cellular materials from the pyruvate, including short- and long-chain fats, amino acids, CO_2 and ATP. Cells without mitochondria or those with mitochondria but exposed to low-oxygen environments are functionally anaerobic and must regenerate the nicotinamide adenine dinucleotide (NAD^+) to continue glucose metabolism another way. Anaerobic means of regenerating NAD^+ are typically called "fermentation." Ethanol fermentation and lactate fermentation are two well-known and well-studied forms of anaerobic fermentation (Figure 15.11). Ethanol fermentation takes place in yeast and is the means by which bread rises (Chapter 10) and the ethanol of beer and wine is made (Chapter 11) and is discussed further in Section 15.8.2. Lactate fermentation takes place in human muscles under conditions when physical exertion is strong, for example—running a marathon—and the tissues of the runner are depleted of oxygen, which is discussed further in Section 15.8.1. There are other forms of fermentation (Figure 15.12) common to different species of bacteria in which pyruvate, the product of glycolysis, is converted to a variety of two-, three-, or four-carbon molecules, such as acetate, propionate, or butyrate in an effort to recycle the NAD^+ necessary to fuel glycolysis; this is discussed further in Section 15.8.3.

15.5.1 Bacteria

The continued metabolism of pyruvate in bacteria is highly varied depending on the cell type (Figure 15.12). Bacteria do not have mitochondria, and some bacteria strains do not have the enzymes of the Krebs cycle, and thus must convert NADH back to NAD^+ to continue to use glycolysis for ATP production. *E. coli* is a common bacteria found in the gut of humans; there are several enzymes in *E. coli* that will react with pyruvate. Some produce lactate, and others ethanol and acetate, among others. The bacterial species used to make Swiss cheese,

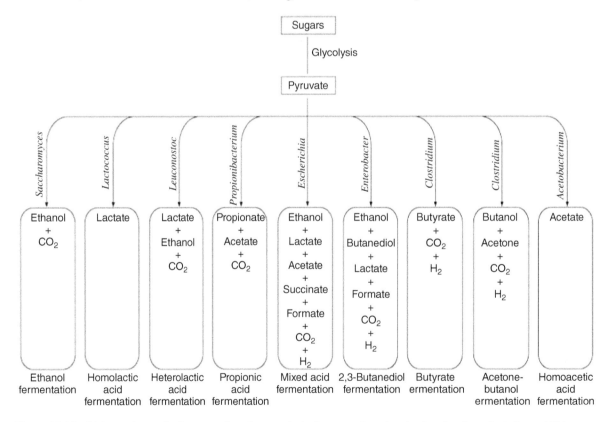

Figure 15.12 Major pathways for fermentation of sugars including organisms involved and end products formed. Taken with permission from Ref. [2] / Taylor & Francis Group.

Propionibacterium, converts pyruvate to propionate, acetate, and CO_2 responsible for the taste and gas holes of the cheese. *Lactococcus* bacterium produces primarily lactate in anaerobic conditions using the enzyme lactate dehydrogenase. This is how the dairy industry produces acidic conditions in fermented milk products and cheese. The major fermentation endpoints of yeast and bacteria are detailed later in section 15.8 [1].

15.5.2 Yeast

Yeast metabolism of pyruvate has two metabolic fates. The yeast cell can use pyruvate to generate CO_2 and ethanol producing two total ATPs (Figure 15.11). A second pathway is found in the mitochondria where pyruvate is further oxidized to CO_2, water, and much more ATP. While both pathways can metabolize pyruvate, oxygen depletion controls the switch from respiration (CO_2 and ATP in the mitochondria) to fermentation (production of CO_2 and ethanol). Depending on the conditions, yeast will slowly convert ethanol to short-chain aldehydes and ketones, acetaldehyde, acetate, and acetyl-CoA.

15.5.3 Mammalian

Animal mammalian cells have two options. Like yeast, the presence of oxygen allows the cells with mitochondria to metabolize pyruvate to CO_2, producing the maximum amount of ATP. However, if the oxygen levels fall due to exercise or other factors, the anaerobic metabolism of pyruvate shifts to produce lactate. This has an important impact on meat before harvest.

Box 15.3 Fermentation versus glycolysis

Let's consider the definitions of glycolysis and fermentation. Glycolysis, as you've learned, is the oxidation of glucose to pyruvate. Fermentation is a broader term that includes glycolysis but includes whatever pathway is used to oxidize NAD^+ back to NADH in anaerobic conditions. Thus fermentation will include glucose to pyruvate and then the continued metabolism to lactate, ethanol, acetate, carbon dioxide, or other compounds depending on the organism and conditions. Fermentation has a more loose historical definition that involves making ethanol in wine and beer. Therefore, fermentation is a combination of different metabolisms to produce ATP from glucose and regenerate NAD^+ from NADH.

To review, in the presence of oxygen, pyruvate undergoes oxidative phosphorylation (further discussion in Section 15.6) to the molecule acetyl-CoA, which can be completely oxidized to CO_2. This process comprises aerobic respiration, which generates a large amount of ATP, H_2O, and CO_2. Under anaerobic conditions or conditions where fermentation is favored, the energy needs of the organism are met by the modest ATP yield of glycolysis, as no additional ATP is generated in the further breakdown of pyruvate. In these cases, however, an organism still requires a means to regenerate NAD^+ so that glycolysis can continue as the organism's primary means to make ATP. The metabolic processes that occur to regenerate NAD^+ that do not require oxygen are called fermentation. As we have begun to learn, fermentation is very important in cooking and foods, such as yogurt, cheese, beer, wine, soy sauce, and pepperoni (Box 15.3 and Section 15.8).

15.6 Aerobic Respiration: The Tricarboxylic Acid Cycle and Oxidative Phosphorylation

In the presence of oxygen, the pyruvate generated in glycolysis is fully oxidized to CO_2, the energy of which is used to make ATP. Three different pathways are necessary to carry out this complete oxidation. The pyruvate is first processed by a large enzyme complex called the pyruvate dehydrogenase (PDH) complex. The PDH complex, as its name implies, oxidizes and decarboxylates the pyruvate into two molecules: a two-carbon molecule called acetyl-CoA and CO_2 (Figure 15.13). The acetyl-CoA can then enter the second pathway, called the tricarboxylic acid (TCA) cycle, which allows for further oxidation of the molecule (Figure 15.14).

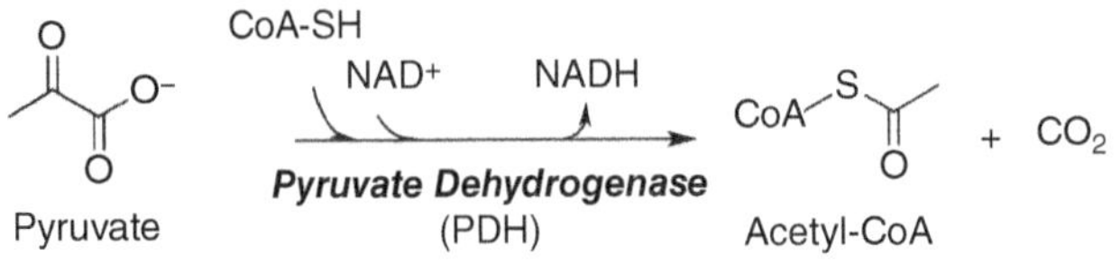

Figure 15.13 Pyruvate dehydrogenase. The enzyme is important for the entry of pyruvate into the Krebs cycle.

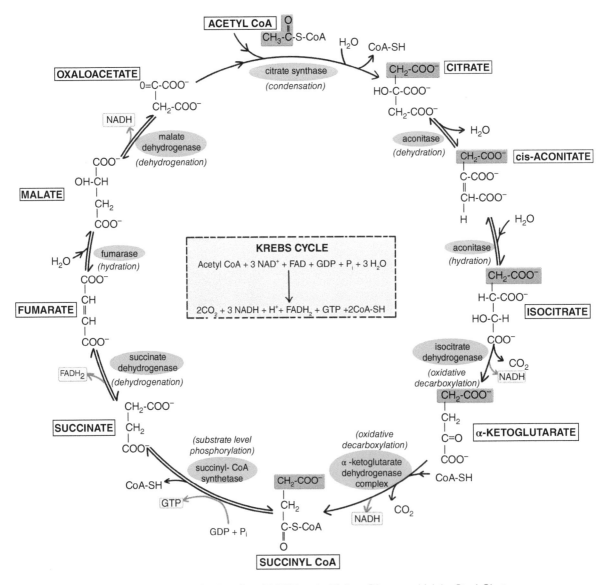

KREBS CYCLE/CITRIC ACID CYCLE/
TRICARBOXYLIC ACID CYCLE

Figure 15.14 The Krebs cycle or the tricarboxylic acid (TCA) cycle. Biology Diagrams / Adobe Stock Photos

The TCA or Krebs cycle (named after Hans Krebs, who first elucidated the pathway), conjugates or attaches acetyl-CoA to oxaloacetate to make a molecule called citrate. Citrate is a TCA, hence the name of the pathway. Although the TCA cycle is complicated, there are a few features of the pathway that are important to our discussion of food and cooking. The first relates to the pathway's namesake, citrate. Have you ever heard of citrate or citric acid? Interestingly, citrate or citric acid is a molecule that gives many drinks and foods a sour taste. It is naturally present at a relatively high concentration in limes, lemons, and other citrus fruits and is added to enhance the tangy, sour flavor of citrus-flavored soft drinks. The citric acid also helps to give the drink an acidic pH (between 2.5 and 4), which prevents microbial growth and emulsifies some of the added flavor molecules to keep them in solution. Interestingly, the citric acid used in the food industry is made by a mold called *Aspergillus niger*. The mold is allowed to metabolize various sugars, and the citrate produced is purified and utilized within the food and other industries.

The second important feature of the pathway is this: for every acetyl-CoA that enters the pathway, two additional CO_2 molecules are generated. Thus, upon completion of glycolysis, the pre-TCA cycle, and the TCA cycle by a six-carbon glucose molecule, six carbons have been fully oxidized to CO_2. It is still not immediately obvious, however, how this oxidation yields energy (and ATP) for the organism. Remember that oxidation must be paired with a reduction (section 15.3.2), so when carrying out an oxidation reaction, a molecule needs to be present to accept electrons. From the six oxidation steps present in glycolysis, the pre-TCA cycle, and the TCA cycle, 12 electrons are transferred to and held by the coenzymes NADH (which we have already talked about) and $FADH_2$, another important electron carrier. If NADH and $FADH_2$ do not get rid of these electrons (become oxidized themselves), then there are two problems. The first is that the cell will soon run out of NAD^+ and FAD, so these metabolic processes will cease and the organism will die. Although we would like disease-causing microorganisms to die, we would not want the yeast in our bread dough to die prematurely. The second is that we haven't yet yielded a large amount of ATP from these oxidative steps. In the last pathway associated with aerobic respiration, called the electron transport chain or oxidative phosphorylation, NADH and $FADH_2$ are oxidized back to NAD^+ and FAD, and the energy associated with the oxidation of these molecules is used to drive the synthesis of ATP.

15.7 The Electron Transport Chain

The electron transport chain is the last pathway associated with aerobic respiration of a cell. It is in this pathway that we finally observe the oxygen dependence of aerobic respiration. How does this happen? The electron transport process requires molecular oxygen as the final acceptor of the electrons from NADH and $FADH_2$. The summarized reactions associated with the electron transfer are shown below:

$$NADH + H^+ + \tfrac{1}{2}O_2 \rightarrow NAD^+ + H_2O$$
$$FADH_2 + \tfrac{1}{2}O_2 \rightarrow FAD + H_2O$$

As you can see, in the process of transferring electrons to oxygen to generate water, the coenzymes (NADH and $FADH_2$) are reoxidized, which is essential for the survival of the organism or cell. These reactions make the electron transport process seem very simple, but they do not show how these electron transfers allow for the generation of ATP.

The electron transport chain comprises a multistep process that involves a series of electron carriers that are embedded within the mitochondrial membrane of the cell. As the electrons are transferred from the NADH or $FADH_2$ through a series of electron carriers that are present within the membrane (who will eventually transfer the electrons to the ultimate acceptor, molecular oxygen), protons (positively charged hydrogen ions) are pumped from the mitochondrial membrane into the intermembrane space of the mitochondrion (Figure 15.15).

MITOCHONDRION

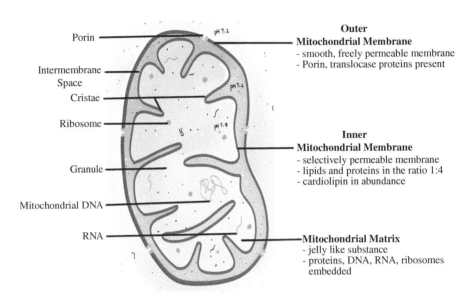

Porin

Intermembrane Space

Cristae

Ribosome

Granule

Mitochondrial DNA

RNA

Outer Mitochondrial Membrane
- smooth, freely permeable membrane
- Porin, translocase proteins present

Inner Mitochondrial Membrane
- selectively permeable membrane
- lipids and proteins in the ratio 1:4
- cardiolipin in abundance

Mitochondrial Matrix
- jelly like substance
- proteins, DNA, RNA, ribosomes embedded

Figure 15.15 Mitochondria. Biology Diagrams/Adobe Stock Photos

As you learned in Chapter 1, when you increase H^+ within a solution, the pH decreases and the solution becomes more acidic. A similar phenomenon occurs within the mitochondrion. As H^+ is pumped into the intermembrane space, the pH of that space decreases relative to the rest of the mitochondrion, creating a "proton gradient" since there is a difference in the number of H^+ ions in the intermembrane space relative to the rest of the mitochondrion. The system wants to equilibrate or "even out" the concentration of H^+ ions on the inside and outside of the membrane, similar to what happens when you put a turkey into a salt brine. The difference in water concentration between the turkey and the brine draws moisture from the interior of the bird to the surface, where it combines with the salt and other seasonings, eventually seasoning the entire bird. In the mitochondrion, the protons cannot pass through the hydrophobic mitochondrial membrane without the assistance of a protein. The protein that recognizes and allows the protons to flow back into the mitochondrial matrix is called ATP synthase. The energy associated with the difference in proton concentration and separation of charge across the inner mitochondrial membrane drives the synthesis of ATP as proteins flow back into the matrix and the concentration and charge difference is reduced (Figure 15.16).

Due to this process, for every NADH or $FADH_2$ that enters the electron transport chain, approximately 2.5 and 1.5 ATP molecules are generated, respectively. Thus, a single glucose molecule that is completely oxidized via aerobic respiration generates 30–32 molecules of ATP (Table 15.1).

Aerobic respiration gives an organism or cell a large energy yield so that it can carry out other processes required for its survival and re-oxidizes coenzymes so that aerobic respiration can continue. Thus, this is the most efficient way for an organism to generate the energy necessary for survival.

ELECTRON TRANSPORT CHAIN

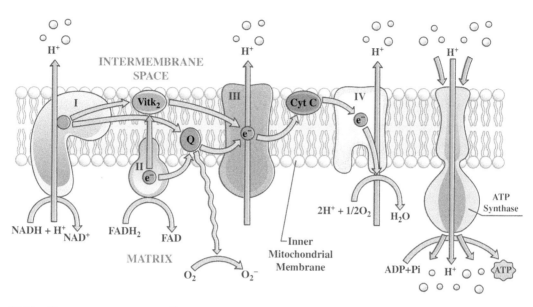

Figure 15.16 Electron transport system. Illustration of inner mitochondrial membrane reactions. From the NADH dehydrogenase complex, electrons are transported along a chain of proton pumping cytochrome electron carrier protein complexes. Reactions involve NAD and flavin adenine dinucleotide (FAD). Oxygen is involved in the step leading to the ATP synthase complex. The final stage involves an ATP/ADP transporter that channels the energy-carrying molecules ATP and ADP. VectorMine/Adobe Stock Photos

Table 15.1 Energy produced from glucose metabolism.

Pathway	Consumed	Produced
Glycolysis	Glucose ($C_{14}H_{12}O_{14}$) 2 NAD$^+$ 2 ADP + 2 phosphates	2 pyruvate ($C_3H_3O_3$) 2 NADH and 2 H$^+$ 2 ATP and 2 H$_2$O
Pyruvate oxidation (PDH complex)	2 pyruvate ($C_3H_3O_3$) 2 NAD$^+$ + 2CoA-SH	2 NADH and 2 acetyl-CoA
Ethanol fermentation	2 pyruvate ($C_3H_3O_3$) 2 NADH and 2 H$^+$	2 ethanol ($C_2H_{14}O$) and 2 CO$_2$ 2 NAD$^+$
Krebs (citric acid) cycle	2 acetyl-CoA 6 NAD$^+$ and 2 FAD 2 GDP and PO$_4^{3-}$ 2 CoA-SH	2 CoA-SH 6 NADH and 2 FADH$_2$ 2 GTP (later converted to 2 ATP) 2 CO$_2$
Total ATP yield per glucose		**ATP**
Glycolysis only		**2**
Glycolysis and ethanol fermentation		**2**
Glycolysis and Krebs (with PDH and electron transport)		**30–32**

The table is calculated based on *one* glucose metabolized. Remember that each NADH can produce approximately 2.5 ATP if mitochondria and oxygen are present and FADH$_2$ can produce approximately 1.5 ATP.

15.8 Additional Metabolic Fates of Pyruvate: Fermentation

The role of a living organism in fermentation was discovered by Louis Pasteur in the late 1850s due to bacterial contamination within fermentation vats within an alcohol distillery. However, long before Pasteur's discovery, humans and other animals enjoyed food and beverage products of the metabolic, fermentative activity of a microorganism. Remember, in both fermentation and aerobic respiration, glucose is first broken down to pyruvate via glycolysis. However, the two pathways then split. In the presence of oxygen, the pyruvate continues to be oxidized through oxidative phosphorylation and the electron transport chain, yielding high amounts of ATP for the organism and products, like citric acid, that are important in foods. In the absence of oxygen or mitochondria, pyruvate isn't fully oxidized to CO_2. Rather, pyruvate is converted to another molecule that allows for the regeneration of NAD^+ and organism survival. The two major fermentation pathways are distinguished by the end product of the pathway. In some animals and many bacteria, the final product is lactate; thus the process of anaerobic glucose catabolism is called lactate fermentation. In most plant cells and yeast, the process is termed alcohol fermentation because the final product is ethanol. Both lactate fermentation and alcohol fermentation are important in food production and cooking; the specific pathway utilized by the organism depends upon the enzymes and conditions available to the organism.

15.8.1 Lactate Fermentation

Lactate fermentation is the anaerobic process of glucose catabolism that results in the production of lactate or lactic acid (Figure 15.17a). The pyruvate generated from glycolysis is reduced (it gains electrons and forms more bonds to hydrogens), and NADH is oxidized to NAD^+ (it loses electrons and has fewer bonds to hydrogen) in a single reaction step catalyzed by an enzyme called lactate dehydrogenase. That's it! The process of lactate fermentation is much simpler than the TCA cycle.

When you consider the reaction in the context of glycolysis, you can write the overall reaction and conversion of glucose to lactate as shown in Figure 15.17b. Again, lactate fermentation is not particularly efficient in terms of ATP production as only two ATPs are generated relative to the 30–32 ATPs obtained from aerobic respiration; however, it allows for the regeneration of NAD^+, which is critical for the organism such that it can continue to carry out glycolysis to obtain some energy from glucose and its intermediates.

For people interested in the chemistry and biology of food and cooking, however, lactate fermentation does more than allow the organism to survive. As its name implies, lactic acid is an acid! Its production changes the pH of the environment in which it exists, and when pH changes, the behavior of molecules changes too. The lactic acid that is produced through the fermentation of bacteria causes milk to curdle, which leads to various dairy products such as yogurt, sour cream, and cheese, and preserves the flavorful fermented meats such as pepperoni and salami.

15.8.2 Alcohol Fermentation

Alcohol fermentation is the anaerobic process of glucose catabolism that results in the production of ethanol. In this two-step process, the pyruvate generated from glycolysis is decarboxylated (i.e. it loses a carbon atom as carbon dioxide) and is converted to acetaldehyde. Acetaldehyde is reduced (it gains electrons and hydrogen) by NADH to yield ethanol (Figure 15.18).

(a)

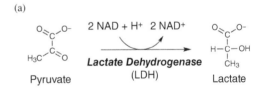

(b)

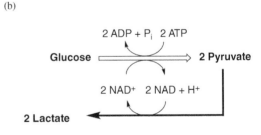

Figure 15.17 Lactate fermentation. To resupply the consumed NAD^+, under anaerobic conditions, cells can utilize the lactate dehydrogenase (LDH) enzyme. (a) LDH-catalyzed reaction. (b) Overall metabolism of glycolysis and LDH.

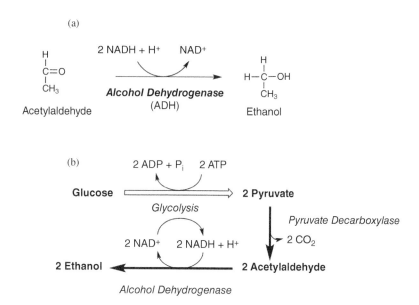

Figure 15.18 Alcohol fermentation. (a) The reaction is catalyzed by alcohol dehydrogenase and (b) integration of ADH supporting anaerobic fermentation in yeast and a few novel bacterial strains.

The enzyme that catalyzes the conversion of pyruvate to acetaldehyde is called pyruvate decarboxylase because a carbon is removed from pyruvate in the form of carbon dioxide gas. The second enzyme is called alcohol dehydrogenase. When you consider the breakdown of glucose via glycolysis and alcohol fermentation, the full reaction is this:

$$\text{Glucose} + 2\,\text{ADP} + 2\,\text{P}_i \rightarrow 2\,\text{CO}_2 + 2\,\text{ATP} + 2\,\text{ethanol}$$

As with lactate fermentation, this process does not have a high energy yield, but it allows for the regeneration of NAD^+ (which is critical for the organism) and the production of alcohol and carbon dioxide, both of which are important to the chemistry and biology of food and cooking. Alcohol fermentation by yeast is a key process in the production of beer, wine, and various baked goods. In yeast breads, the carbon dioxide produced through fermentation is what causes the dough to rise. The ethanol becomes part of the bread's aroma when it vaporizes during baking. For the brewer, ethanol makes the beverage alcoholic, and the carbon dioxide contributes to the carbonation (i.e. gas bubbles).

15.8.3 Other Products of Fermentation

Although lactate and ethanol are the most common fermentation products for organisms in general, many other molecules can be produced through fermentation pathways that are microorganism specific. In the production of Swiss cheese, the hole-making bacterium *Propionibacterium shermanii* converts the lactic acid product of fermentation to propionic and acetic acids and carbon dioxide. The carbon dioxide gives the cheese its distinctive holes, and the propionic and acetic acids provide the distinct aroma and nutty, sweet flavor of Swiss. Other fermentation processes yield acetone, isopropyl alcohol, and butyrate, which are components of the smells of rotten food (Figure 15.19).

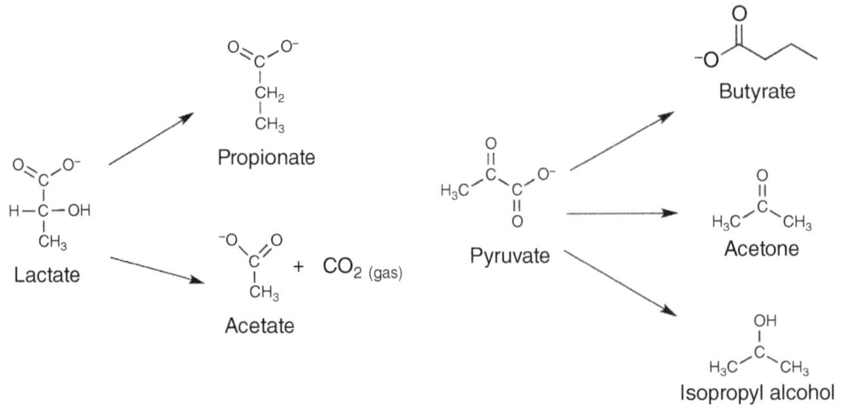

Figure 15.19 Additional fermentation products. Several alternative routes of fermentation will produce unique and interesting flavor and aroma compounds in fermented foods and drinks.

15.9 Metabolism of Other Sugars

So far, we have assumed glucose to be the starting point for metabolism and the most important source of energy for an organism. Glucose is a very important energy source for aerobic and anaerobic respiration; however, it is not the only available energy source for these processes. In fact, in many organisms, cells, and food-related processes, glucose is not the starting point. How do these other sources provide us with energy? In the same way as glucose! Most monosaccharides or simple sugars are structurally similar. When these nonglucose monosaccharides encounter the right enzyme, they are converted to a glycolysis intermediate. For example, the fructose in the honey on your toast can enter the glycolytic pathway at the product of step 2, fructose-6-phosphate. Complex carbohydrates such as lactose, a disaccharide, and starch, an oligosaccharide, are first broken down into monosaccharide units and then enter glycolysis.

Why do organisms utilize the same primary pathway for the breakdown of so many of their carbohydrates? Because essentially all carbohydrates feed into the same pathway, this reduces the need to have completely independent pathways for the breakdown of all the different carbohydrates that may be available to an organism. Thus, it is biologically efficient for an organism to have a common metabolic pathway for the catabolism of carbohydrates. In food production, this metabolic efficiency is also advantageous, as it allows for diversity in the types of carbohydrate sources that the microorganism uses in the food or cooking process. In yeast-based bread doughs, honey (fructose), table sugar (sucrose), or the starch found in the flour can be used as a carbohydrate source for fermenting yeast.

15.10 Metabolism and Degradation of Fats

Sugars are not the only source of energy for cells and the only molecules whose breakdown is important to food and cooking. The metabolism and degradation of fats are also very important, as fats are an important energy source for organisms, and fat metabolism and degradation yield many molecules that contribute to the taste and aroma of foods. In Chapter 1, you learned about the hydrocarbon molecular structure of fat molecules (Figure 15.20).

In this chapter, you have learned that catabolism involves the oxidative breakdown of larger molecules into smaller ones. If more oxidation can occur, more ATP can be generated, and the better the energy source for the organism. Fat molecules contain many carbons that are highly reduced (i.e. the carbons have many hydrogens bound to them); thus they have great potential to undergo oxidation (Box 15.4). This "oxidative potential" makes a fat the most important long-term energy storage molecule for many organisms. However, when you look at the molecular structure of a fatty acid, it looks nothing like glucose or any other sugar molecule! How are these molecules metabolized? How does the breakdown of fat impact food and cooking?

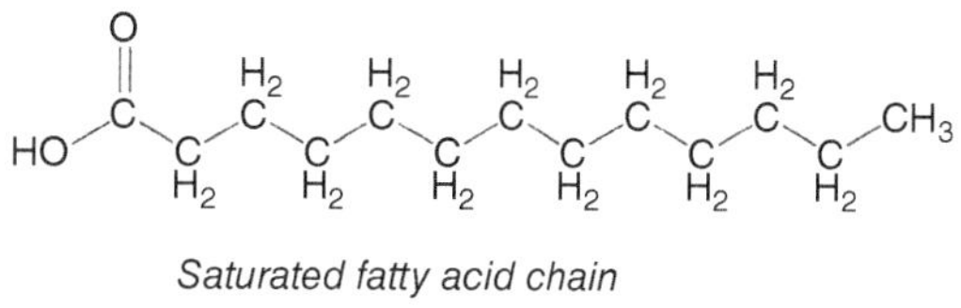

Saturated fatty acid chain

Figure 15.20 Fatty acid.

Most fats are stored as triacylglycerides (Figure 15.21), which are comprised of glycerol and three fatty acids (see Chapter 2). In order to break down the triacylglycerol, the fatty acids are first cleaved away from the glycerol in a hydrolysis reaction to yield a molecule of glycerol and three "free" fatty acids. The glycerol, which is structurally similar to intermediates within the glycolysis pathway, can enter glycolysis following a two-step conversion to dihydroxyacetone phosphate.

In contrast, fatty acids are structurally distinct from carbohydrates; they hardly contain any oxygen at all! Thus, many more oxidation reactions need to be carried out to convert all the carbons in a fatty acid to CO_2. Fatty acids are broken down in a stepwise process called beta-oxidation that involves oxidation and removal of successive two-carbon units. This process generates acetyl-CoA, NADH, and $FADH_2$, which can enter the TCA cycle and electron transport chain to yield energy in the form of ATP. The longer the fatty acid chain, the greater the amount of energy yielded for the organism (Figure 15.22).

The breakdown of fats and fatty acids is also key to the flavor, aroma, and properties of many foods, in positive and sometimes not so positive ways. The molds found in blue cheese, such as *Penicillium roqueforti* transform fatty acids into methyl ketone molecules, such as 2-pentanone, 2-heptanone, and 2-nonanone, which create the cheese's characteristic and pungent aroma. However, oxidation of fatty acids by oxygen in the air leads to some undesirable products. Over time, the oxygen reacts with fats, breaking them down into smaller, volatile fragments, which have an odor that we often characterize as rancid. An expired box of crackers or a bag of tortilla chips may have this rancid scent. A triglyceride of butyric acid, found in butter, has a pleasant taste and aroma. However, as butter ages, the butyric acid is cleaved from the glycerol, leading to an unpleasant odor that some would characterize as vomit-like (Figure 15.19). When you smell an old bottle of olive oil, your nose knows if its fat molecules have been oxidized in detrimental ways. To help prevent this from happening, you can keep oils in the dark (the energy from the light catalyzes these negative reactions), or you can slow the reaction by placing the oil in the refrigerator.

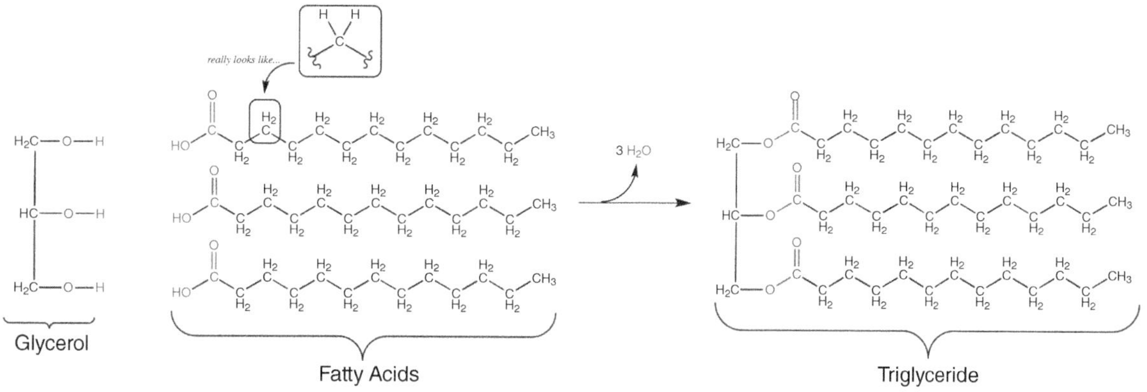

Figure 15.21 Formation of triacylglycerol.

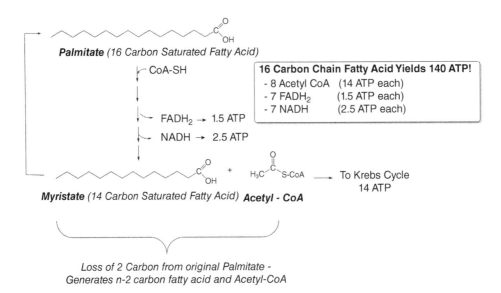

Loss of 2 Carbon from original Palmitate -
Generates n-2 carbon fatty acid and Acetyl-CoA

Figure 15.22   Beta-oxidation of fatty acid provides a high ATP yield.

15.11   Metabolism of Proteins and Amino Acids

Proteins are metabolized for energy and to generate the molecular building blocks for an organism, as was described for carbohydrates and fats. However, their breakdown also makes a notable contribution to the flavors and aromas of uncooked and cooked foods. Some single amino acids and short peptides have tastes of their own. Glutamic acid, the amino acid from which the food additive monosodium glutamate (MSG) is derived, has a taste designated as "savory" or "umami." The breakdown of sulfur-containing amino acids such as cysteine during cooking may lead to an "eggy" aroma. These amino acids and other flavorful small molecules are derived from the breakdown of proteins during cooking and food preparation. Degradation of the milk protein, casein, leads to two distinct events important in the process of making cheese. In the first step of cheese making, the breakdown of casein causes the milk to curdle. Further degradation of the protein into amino acids and small molecules like trimethylamine and ammonia gives different varieties of cheese their distinct flavors ranging from sweet to savory. When foods are seared or grilled at high temperatures, the food browns and generates savory aromas and flavors due to protein degradation to amino acids and their further reaction with sugars to produce molecules such as bis-2-methyl-3-furyl-disulfide, a molecule associated with a "meaty" odor (Figure 15.23).

Figure Earlier in this chapter, you learned that both fatty acid and carbohydrate metabolisms converge on the central catabolic pathways of glycolysis and the citric acid cycle. You see a similar phenomenon in the metabolism of protein and amino acids; following the cleavage of a protein into its amino acid components, the carbons of the amino acids find their way into the citric acid cycle where energy is generated in the form of ATP (Figure 15.24).

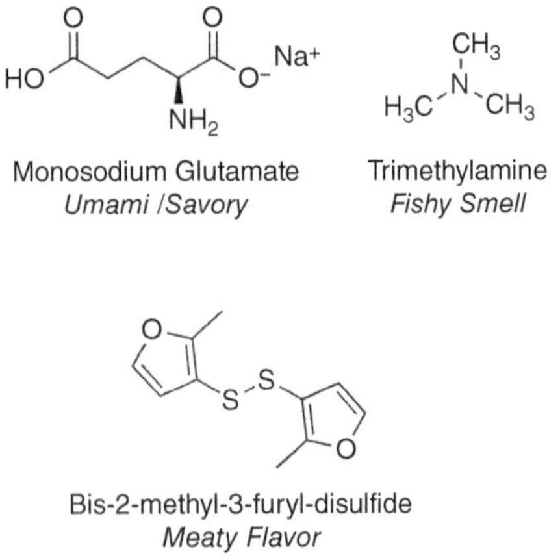

Then, what makes protein and amino acid degradation so distinctive? Why do high-protein foods brown when you cook them at high

Monosodium Glutamate
Umami /Savory

Trimethylamine
Fishy Smell

Bis-2-methyl-3-furyl-disulfide
Meaty Flavor

Figure 15.23   Amino acid products that have odor/flavors.

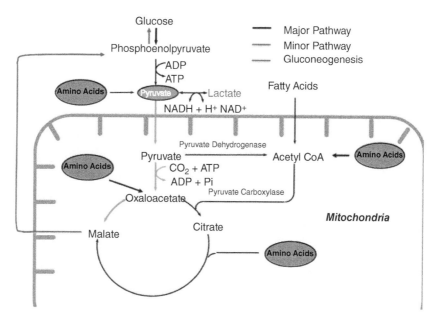

Figure 15.24 Amino acid metabolism. Amino acids feed into several pathways.

temperatures? And why do cooked meats smell differently than raw meats? The molecular feature that distinguishes amino acids from most carbohydrates and fats is the presence of nitrogen, an amino group. Amines have really distinct chemical properties relative to carbon, hydrogen, and oxygen. It naturally follows that you see a difference in the flavors, aromas, and properties of foods that have a high amine or protein content. One example of cooking chemistry that involves proteins is a series of reactions known as the Maillard reactions. When proteins are heated together with carbohydrates (either polysaccharides or smaller mono- or disaccharides) to high temperatures, the two begin to break down into smaller sugars and proteins or amino acids. Under these conditions, a reducing sugar "ring" opens into a linear form, and the resulting aldehyde or carboxylic acid reacts with the amine group (of an amino acid or protein) to produce a new intermediate compound, an Amadori compound (Figure 15.25).

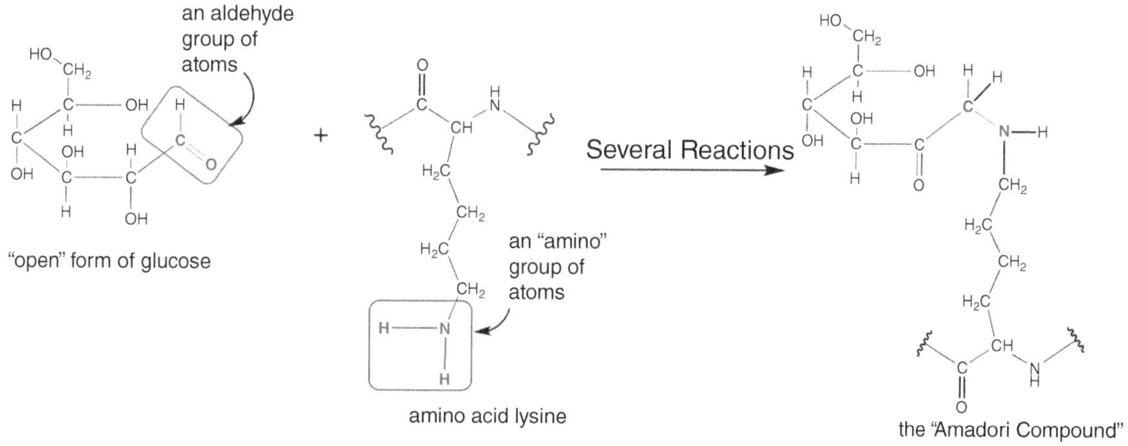

Figure 15.25 The Maillard reaction. See Chapter 3 for a detailed discussion of the Maillard reaction.

The newly formed compound (shown in Figure 15.25) can react over and over again to produce additional compounds. Maillard reactions are very complicated and are not fully understood by food chemists, but some of the known flavorful and aromatic products of this chemistry include the pyrroles, pyridines, pyrazines, thiophenes, thiazoles, and oxazoles (see Chapter 3 for details). These reactions are also referred to as "nonenzymatic browning" reactions due to the color changes in many of the foods when they undergo this chemistry: the browned crust of freshly baked bread, a soft baked pretzel, or a grilled steak.

Box 15.4 Oxidation in food and cooking

Historically, if an element reacted with oxygen to produce an oxide, you would classify the reaction as an oxidation reaction. You know this type of reaction well, as it describes the conversion of elemental iron (Fe) to iron oxide (Fe_2O_3), also known as rust:

$$2\ Fe(s) + 3\ O_2(g) \rightarrow 2\ Fe_2O_3(s)$$

But, as we saw in Section 15.3.2, oxidation has a much broader meaning; oxidation reactions occur when an element, compound, or ion loses electrons (Figure 15.26).

How does a chemist know when an oxidation reaction takes place? An oxidation number is assigned to each atom in a compound, which indicates whether the atom is electron rich, neutral, or electron poor. If the oxidation number for a particular atom changes over the course of a reaction, an oxidation reaction has taken place.

Oxidation reactions occur all around us and are key in food preparation and cooking processes. For example, some of the color differences that you see in meat are due to numerous oxidation reactions that occur in the iron (Fe) that is found in the muscle protein, myoglobin. As the iron changes its oxidation state (between +2 and +3) and its bonding partner (O_2 or H_2O), it changes color, resulting in a color change in the meat (Figure 15.27).

In raw fresh meat, iron has an oxidation state of +2; this means that the iron has two fewer electrons than protons, since protons are positively charged and electrons are negatively charged. Iron in an oxidation state of +2 in myoglobin typically has a red or pink color associated with it. As meat cooks, the iron is oxidized and loses one electron, so its oxidation state increases by one (there are now three more positively charged protons than negatively charged electrons). In other words, the iron is oxidized to Fe^{3+}. Iron in myoglobin with a +3 oxidation state is typically brownish in color, as is observed in well-done beef.

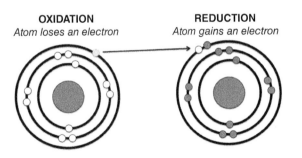

OXIDATION
Atom loses an electron

REDUCTION
Atom gains an electron

Figure 15.26 Oxidation and reduction.

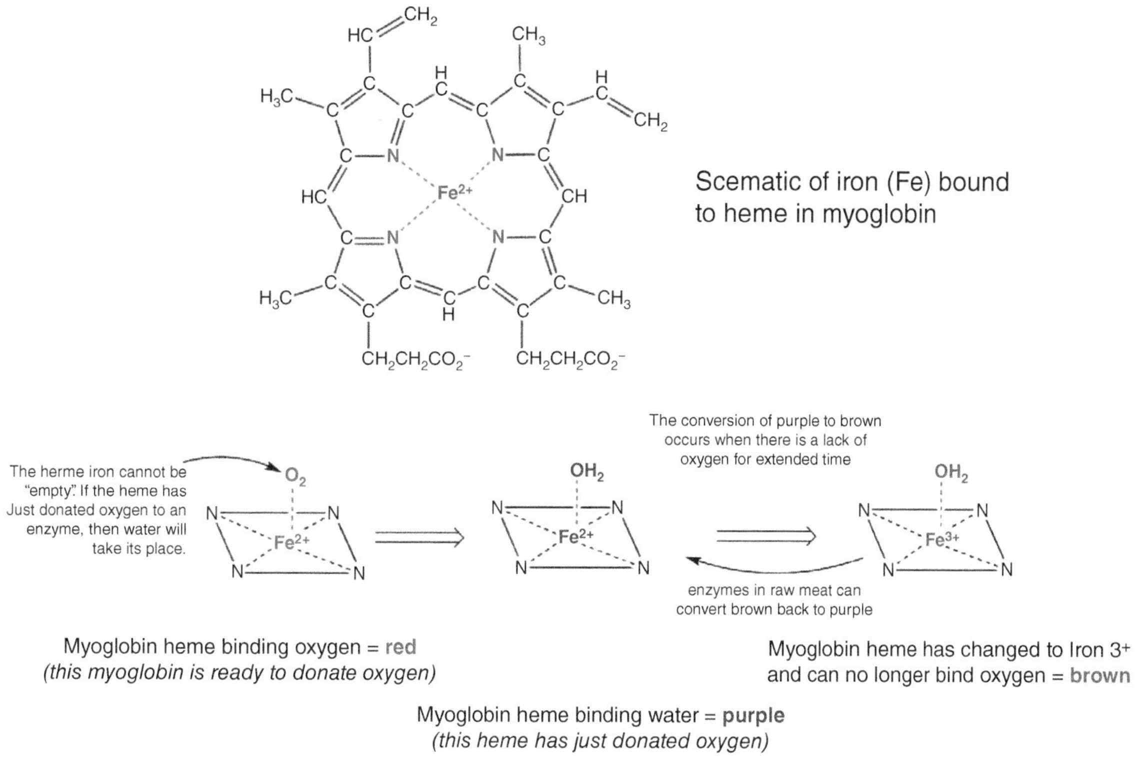

Scematic of iron (Fe) bound to heme in myoglobin

The herme iron cannot be "empty". If the heme has Just donated oxygen to an enzyme, then water will take its place.

The conversion of purple to brown occurs when there is a lack of oxygen for extended time

enzymes in raw meat can convert brown back to purple

Myoglobin heme binding oxygen = **red**
(*this myoglobin is ready to donate oxygen*)

Myoglobin heme binding water = **purple**
(*this heme has just donated oxygen*)

Myoglobin heme has changed to Iron 3+ and can no longer bind oxygen = **brown**

Figure 15.27   Myoglobin and iron. The oxidation state of iron held in place by the heme of myoglobin is shown here.

A different type of oxidation reaction occurs when you watch a cut apple turn brown. A browning apple actually undergoes two different oxidation reactions. In the first reaction, the oxidation reaction is simply identifiable by the addition of oxygen to a molecule, similar to what you see in the elemental iron to rust reaction. Oxygen is added to the monophenol and electrons are lost (which is not obvious in this example) to yield catechol.

In the second oxidation reaction (catechol to *o*-quinone), no additional oxygen is added, but the catechol undergoes a dehydrogenation reaction; it loses hydrogens. This is another way to identify and define oxidation reactions, particularly those in which oxidation numbers are not obvious. Following *o*-quinone production, a variety of enzyme- and nonenzyme-catalyzed reactions take place to yield colored compounds like melanin, which is the color that we observe in our browning apple.

Have you ever noticed that a partially used bottle of wine eventually develops a sour taste? Upon exposure to air (for a long time), the ethanol in the wine oxidizes to acetic acid (Figure 15.28). This process doesn't happen very readily on its own, but occurs in the presence of enzymes that are produced by a bacterium called *Acetobacter aceti*. When vinegar is made, this type of bacteria is purposefully added to wine or other fermented alcoholic beverages to produce the vinegar that you might use in making salad dressings or dyeing Easter eggs.

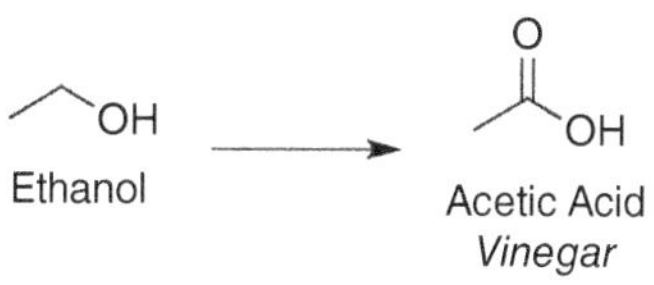

Ethanol

Acetic Acid
Vinegar

Figure 15.28   Ethanol conversion to acetic acid.

(Continued)

Box 15.4 (Continued)

Using the myoglobin, apple browning, and vinegar reactions as examples, you now have the expertise to identify oxidation reactions. If there is a loss of electrons (as in myoglobin), a gain in bonds to oxygen (as in browning and vinegar), or a loss of bonds to hydrogen (as in browning and vinegar), you have an oxidation reaction. To be clear, in all of these reactions, a compound, element, or ion loses electrons. In the case of the browning apple and vinegar, it is harder for a novice to identify the electron loss.

Where do the electrons go? In these examples, it appears that the electrons just get lost in some unknown place in the surroundings. However, this is not the case. As we learned in Section 15.3.2, when oxidation reactions occur, the electrons lost by one molecule must be transferred to or gained by another molecule. The molecule that gains the electrons is reduced. If a molecule loses electrons, another molecule must be available to accept the electrons. In other words, whenever an oxidation occurs, a reduction must also occur. If you think about oxidation/reduction reactions in terms of hydrogen and oxygen, the molecule that is oxidized gains oxygen or loses hydrogen, while the reduced molecule gains hydrogen or loses oxygen (Figure 15.26).

Science for the Chef: Kimchi

Kimchi is an excellent example of a fermented food that showcases the principles of fermentation and metabolism, resulting in a dish that is both flavorful and rich in probiotics, beneficial bacteria that support gut health. This recipe allows students to explore these scientific concepts while enjoying a classic and versatile dish from Korean cuisine. This traditional Korean dish involves the fermentation of vegetables, primarily napa cabbage and daikon radish, through the action of lactic acid bacteria, resulting in a tangy, spicy, and umami-rich food.

Ingredients:
- 1 large napa cabbage
- 1/4 cup sea salt
- 1 daikon radish, julienned
- 4 green onions, chopped
- 1 carrot, julienned
- 1 tablespoon sugar
- 1/4 cup gochugaru (Korean red chili flakes)
- 2 tablespoons fish sauce
- 1 tablespoon grated ginger
- 4 cloves garlic, minced
- 1 tablespoon soy sauce (optional)

Instructions:
Prepare the cabbage:
- Cut the napa cabbage into quarters and remove the core. Chop the cabbage into bite-sized pieces.
- Place the cabbage in a large bowl and sprinkle with sea salt, tossing to ensure even distribution. Let it sit for 2–3 hours, tossing occasionally. The salt will draw out moisture from the cabbage, softening it and beginning the fermentation process.
- After 2–3 hours, rinse the cabbage thoroughly under cold water to remove excess salt. Drain well.

Make the kimchi paste:
- In a separate bowl, combine the gochugaru, fish sauce, grated ginger, minced garlic, sugar, and soy sauce (if using) to make the kimchi paste. Mix until a thick, even paste forms.
- Add the julienned daikon radish, green onions, and carrots to the paste, mixing well to coat the vegetables.

Mix and pack:
- Add the drained cabbage to the kimchi paste and toss until all the cabbage pieces are evenly coated.
- Pack the kimchi tightly into a clean, airtight jar or container, pressing down to remove air bubbles. Leave some space at the top of the jar, as the kimchi will expand during fermentation.
- Seal the jar and let it sit at room temperature for 1–2 days to start the fermentation process. The ideal temperature for fermentation is around 68–72 °F (20–22 °C).

Fermentation:
- After 1–2 days, check the kimchi for fermentation. It should have a tangy, slightly sour taste. If you prefer a stronger flavor, let it ferment for another day or two.
- Once fermented to your liking, store the kimchi in the refrigerator, where it will continue to develop flavor over time.

Science Behind the Recipe

Primary microorganisms involved in kimchi fermentation are lactic acid bacteria, particularly species like *Lactobacillus* and *Leuconostoc*. These bacteria thrive in the anaerobic environment of the packed kimchi and metabolize the sugars present in the cabbage and other vegetables. Through the process of lactic acid fermentation, they convert these sugars into lactic acid, which lowers the pH of the kimchi, giving it a sour taste and preserving the vegetables. The lactic acid bacteria metabolize glucose through glycolysis, producing pyruvate, which is then converted into lactic acid via lactate dehydrogenase. This process is an example of anaerobic respiration, where energy is derived without the use of oxygen, highlighting the fundamental biochemical pathways of microbial metabolism. Salt is crucial in kimchi fermentation for two reasons: it draws out water from the cabbage through osmosis, creating a brine that helps preserve the vegetables, and it inhibits the growth of undesirable microorganisms, allowing lactic acid bacteria to dominate the fermentation process. As the fermentation progresses, the breakdown of proteins in the vegetables by enzymes and bacteria leads to the formation of free amino acids and peptides, which contribute to the rich umami flavor of kimchi. The fish sauce also adds glutamates, enhancing the overall savory taste.

Key Concepts

1. **Metabolism** is the sum of all biochemical reactions occurring within an organism and is essential for energy production, growth, and maintenance of cell function. **Anabolic** pathways or reactions generate the molecules important to the building up of structural and functional units of a cell and require energy. **Catabolic** pathways break down biomolecules and produce energy. **Adenosine triphosphate (ATP)** serves as the primary energy production/use molecule in metabolism.

2. A cell is the basic structural and functional unit of all living organisms. All cells have an exterior barrier that separates their interior from the external environment and requires a way to generate or obtain energy for function and survival. **Prokaryotic** cells lack a true nucleus and membrane-bound organelles. **Eukaryotic** cells have a nucleus and membrane-bound organelles, including mitochondria, which are the site of ATP generation through cellular respiration. Bacteria are an example of prokaryotes. Plants, animals, and yeast (fungi) are examples of eukaryotes.

3. **Oxidation-reduction** reactions involve the transfer of elections, and often, hydrogen and oxygen. The molecule that loses electrons is oxidized; the molecule that gains electrons is reduced. In the oxidized molecule, there is frequently a gain in oxygen and a loss in hydrogen. NAD^+/NADH and FAD/$FADH_2$ act as **coenzymes** in metabolic oxidation-reduction reactions where they act as electron carriers to temporarily hold electrons before transferring them to the final product.

4. **Glycolysis** is a 10-step metabolic process that involves the breakdown of **glucose** into **pyruvate**, yielding two molecules of ATP. The first five steps of glycolysis are called the preparatory phase, where ATP is used to phosphorylate glucose and its intermediates. The last five steps are called the payoff phase, where ATP and NADH are produced, along with two molecules of pyruvate. Glycolysis can occur under **aerobic** (i.e. in the presence of oxygen) or **anerobic** (i.e. in the absence of oxygen) conditions and is thus the primary means that most organisms/cells use to generate energy from glucose or other molecules.

5. **Fermentation** is an anaerobic process that allows cells to produce energy without oxygen. Two common types of fermentation include lactate fermentation, where pyruvate is converted to lactate in a one-step reaction that regenerates NAD^+, common in bacteria and important in cheese and yogurt production. In alcohol fermentation, occurring in yeast and some bacteria, pyruvate is converted to ethanol and CO_2 in a two-step process, regenerating NAD^+. **Alcohol fermentation** is important in the production of beer, wine, and baked goods. Both types of fermentation allow continued ATP production (via glycolysis) due to the regeneration of NAD^+, albeit yielding less ATP than aerobic respiration processes.

6. Following glycolysis, the generated pyruvate is converted to acetyl CoA by **pyruvate dehydrogenase** (i.e. pre-TCA) in order to enter the **tricarboxylic acid (TCA) cycle** also known as the **citric acid cycle** or the **Krebs Cycle**. The TCA cycle is a series of eight **enzyme-catalyzed** reactions, occurring in the mitochondria, where acetyl-CoA is oxidized to CO_2, generating energy in the form of ATP/GTP, NADH, and $FADH_2$. The NADH and $FADH_2$ serve as electron carriers, moving into the electron transport system, which involves a series of protein complexes that generate a proton gradient (due to electron transfer from NADH and $FADH_2$), which drives the formation of ATP and electron transfer to O_2 to form H_2O. A full cycle of aerobic respiration will generate 30–32 molecules of ATP from one glucose molecule, making it a more efficient energy generation cycle than anaerobic respiration.

7. Fat metabolism involves the breakdown of **triglycerides** into fatty acids and glycerol, followed by their oxidation to generate ATP. The glycerol can enter glycolysis, while the fatty acids undergo a process called beta-oxidation, which involves cleavage by two-carbon units to generate acetyl CoA, which can enter the Krebs cycle. The breakdown of a single fatty acid can yield over one hundred molecules of ATP, under conditions of aerobic respiration, making it an effective long-term energy source. This process is crucial for energy production, especially during periods of fasting or prolonged exercise.

8. The metabolism of proteins and amino acids eventually occurs via glycolysis, the pre-TCA, and TCA cycles, as they are composed of carbon and oxygen. However, their molecular differences from carbohydrates and fats lead to some characteristic flavors and aromas of foods, including glutamic acid (also known as MSG), which has a savory or umami taste, sulfur-based amino acids that provide an eggy aroma, and the breakdown of casein proteins in milk. One of the features of amino acids that distinguishes them is the presence of nitrogen. The Maillard reaction involves the reaction of an amine group (nitrogen) with an aldehyde or carboxylic acid of an oxidized sugar under high temperature conditions. The resulting reactions, also known as nonenzymatic browning, yield the brown crust of freshly baked bread or a grilled steak.

References

1 Jencks, W. P. (1976). Physical and chemical data. In: *Handbook of Biochemistry and Molecular Biology*, 3e (ed. G. D. Fasman), Vol. I, 296–304. Cleveland, OH: CRC Press.

2 Volker, M. (2008). *Bacterial Fermentation*. Frankfurt/Main: Goethe Universität.

Additional Readings

Attie, Alan D., Qi-Qun Tang, & Karin E. Bornfeldt. (2021), The insulin centennial—100 years of milestones in biochemistry, *J. Lipid Res.* 62.

Dashty, Monireh. (2013), A quick look at biochemistry: carbohydrate metabolism. *Clin Biochem* 46(15), 1339-1352.

Laurence Cole & Peter R. Kramer. (2016). Human Physiology, Biochemistry and Basic Medicine: Chapter 1.3 - Sugars, Fatty Acids, and Energy Biochemistry (1st ed.). Academic Press.

End of Chapter 15 Questions

1. Consider the reaction of alcohol dehydrogenase shown below:

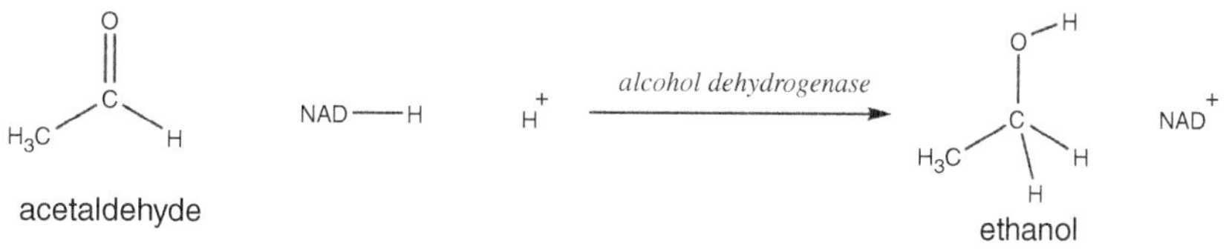

a. In this reaction, what is *being oxidized*? How do you know?
b. In this reaction, what is *being reduced*? How do you know?
c. What is the role of *alcohol dehydrogenase* in this reaction?

2. Consider the process below which occurs in yeast:

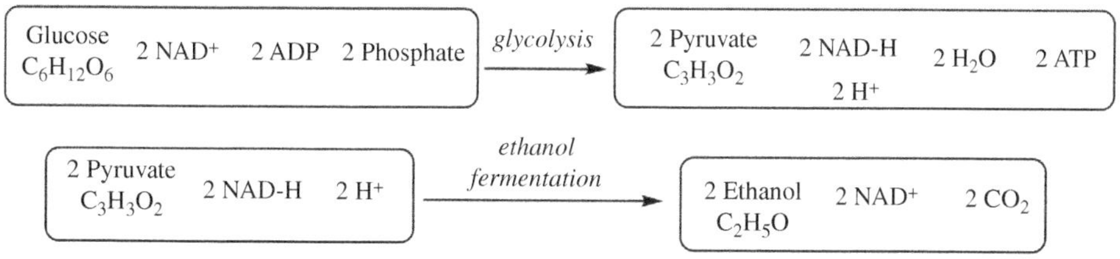

a. Why do yeast perform glycolysis?
b. Why do yeast perform ethanol fermentation?
c. If we purchased some yeast from the grocery store and fed them glucose (i.e. sugar)—how would we know the yeast were performing the reactions shown above? What physical observations would be evident?

3. Which of the following statements are TRUE about this enzyme catalyzed reaction?

a. Acetaldehyde is reduced to ethanol
b. NADH is oxidized to NAD
c. Acetaldehyde is oxidized to ethanol
d. NAD is oxidized to NADH
e. Both oxidation and reduction are occurring
f. NADH reduces acetaldehyde

4. Which of the following statements about glycolysis is/are true? *check any and all correct answers*
   a. Glycolysis is a pathway that contains 10 enzymes
   b. Glycolysis is an enzyme that catalyzes a reaction with 10 steps
   c. Glycolysis is the process (in yeast or in humans) by which sugar is broken down
   d. Glycolysis results in a net consumption of ATP
   e. Glycolysis results in a net consumption of NAD
   f. Glycolysis results in a net production of ATP
   g. Glycolysis results in a net production of NAD

5. Bacteria produce lactate in metabolism of glucose. Why?
   a. To regenerate NAD^+ for continued ATP production
   b. To produce acid to kill other bacteria
   c. To use the ATP produced during glucose metabolism
   d. All of the above

6. Yeast metabolism describes _____
   a. how the yeast creates acid as in sourdough breads
   b. is the disease yeast cause if the wrong yeast infects bread dough
   c. the production of glucose from carbon dioxide and oxygen
   d. all of the above

7. Fermentation is...
   a. Anaerobic
   b. Aerobic
   c. Happens in the absence of oxygen
   d. Happens in the presence of oxygen

8. The complete breakdown of glucose to carbon dioxide and water that occurs in the presence of oxygen is...
   a. Lactic acid fermentation
   b. Ethanol fermentation
   c. Aerobic respiration
   d. Glycolysis

9. In order to catabolize glucose via aerobic respiration, cells must have...
   a. Mitochondria
   b. Oxygen
   c. The Krebs Cycle (i.e. the citric acid or TCA cycle)
   d. A nucleus
   e. Chloroplasts

10. Which of the following pathways for metabolizing glucose yields the *most* ATP for each carbon of glucose?
    a. Aerobic respiration (including glycolysis, the Krebs cycle and the electron transport chain)
    b. Ethanol fermentation
    c. Lactic acid fermentation
    d. Glycolysis only

Appendix A.1

Functional Groups

The structure of a molecule defines how it functions in a cell and how food may taste or react when cooking or baking. Special groups of molecules called functional groups define the behavior of molecules. Functional groups are arrangements of atoms that have specific chemical and biochemical behavior. These groups of atoms are useful to predict and understand the properties of molecules important in food and cooking. Specific functional groups and examples of molecules that are important in food and cooking are shown throughout the book.

Alcohol —OH An alcohol is the simplest of all functional groups. It is an oxygen atom covalently bonded to a hydrogen atom, often designated as —OH (Figure A1.1). Sugars, like fructose, have many alcohol groups. Molecules of ethanol and glycerol both contain alcohol functional groups. The —OH plays a key role in allowing these molecules to interact with and dissolve in water. You likely already know a little about or have experienced the use of fructose (honey) and ethanol in food or drinks. Glycerol is a sweet, sticky, and thick compound that is often added to bread, cookies, and cakes to keep them moist. A glycerol molecule also provides the molecular framework for fat molecules.

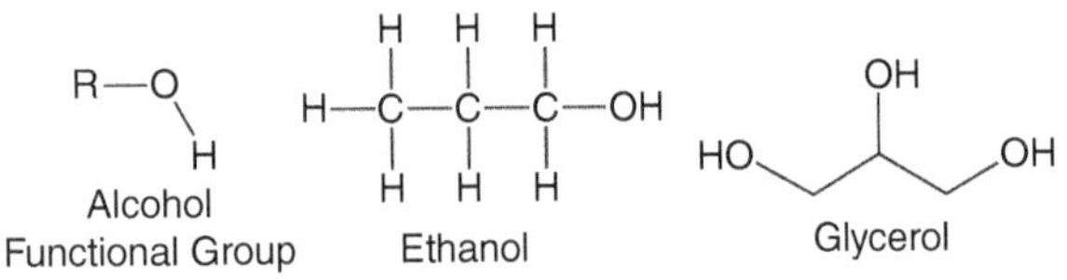

Figure A1.1 Alcohol functional groups. (a) The basic convention for alcohol with R as an undetermined carbon group. (b) A structural drawing of the two-carbon ethanol. (c) Glycerol without the hydrogens and carbons explicitly drawn. In this structure, the intersection of each line is a carbon, and the two hydrogens attached to each of those carbons are implied.

Amino —NH₂ and —NH₃⁺ A group of atoms containing a nitrogen covalently bonded to hydrogen is called an amine or amino group (Figure A1.2). Two or three hydrogen atoms can bond to the nitrogen, creating a neutral ($-NH_2$) or positively charged ($-NH_3^+$) group where the nitrogen atom carries a positive charge. Amino acids, which combine to make proteins, contain an amine functional group.

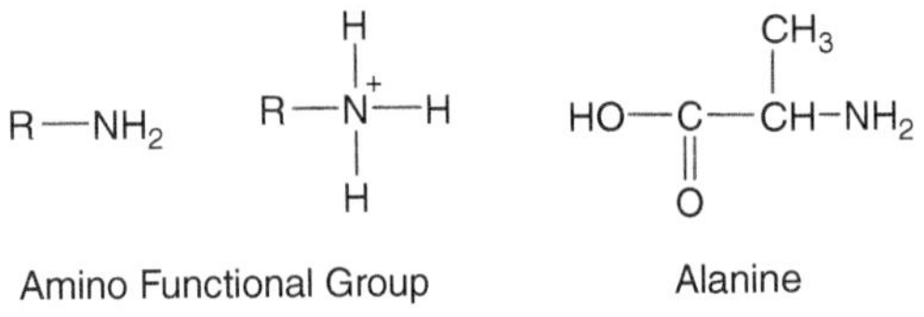

Figure A1.2 Amino functional groups. (a) The basic convention for an amino group with R as an undetermined carbon group. With three hydrogen atoms bonded, the N carries a positive charge. (b) l-Alanine, one of the common 20 amino acids used to make proteins contains an amino group.

The Science of Cooking: Understanding the Biology and Chemistry Behind Food and Cooking, Second Edition. Joseph J. Provost et al.
© 2025 John Wiley & Sons, Inc. Published 2025 by John Wiley & Sons, Inc.
Companion website: www.wiley.com/go/provost/food_science_2e

Carboxylic Acid —COOH   The tangy taste associated with a nice, cool glass of lemonade or a sour citrus hard candy is provided by carboxylic acids (Figure A1.3). This functional group consists of a carbon bound to two oxygen atoms, where one of the oxygen atoms may also be bonded to a hydrogen; it is designated as R—COOH. Sometimes the carboxylic acid group can give up an H+ and become R—COO⁻. Why is the hydrogen atom sometimes absent? Due to oxygen's affinity for electrons, the bond between the hydrogen and oxygen in carboxyl groups is easily broken, yet the oxygen keeps the electron from the previously shared covalent bond. Chemists call this property of oxygen *electronegativity*, but you can just think of oxygen as an electron hog. Oxygen keeps the two electrons from the formerly shared O-H bond, yielding a carboxyl group that lacks a hydrogen ion (H^+) and instead has a negatively charged oxygen (R—COO⁻). Because the R—COOH group can give up an H+ ion, it is a weak organic acid, hence the name carboxylic acid. Carboxylic acids are found throughout food and cooking, most notably in citrus fruits (citric acid) and vinegar (acetic acid). The acid component of these foods stimulates the sour taste receptor on our tongues giving these foods a sour taste. An example is malic acid. Malic acid is an organic acid that is found in unripe fruit like green apples and gives the food a sour green apple flavor.

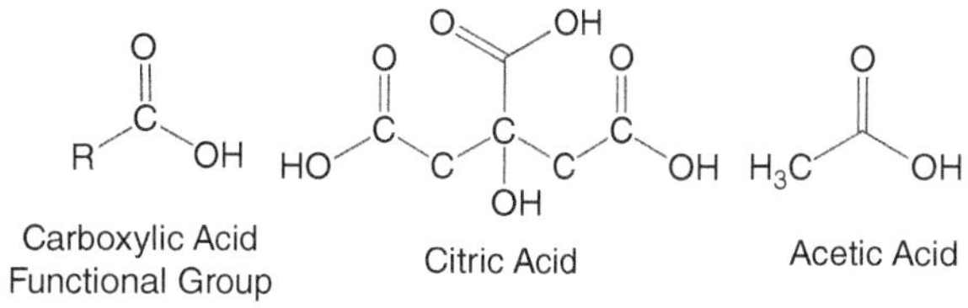

Figure A1.3   Carboxylic acid functional groups. (a) The basic convention for carboxylic acid with R as an undetermined carbon group. (b) The sour tasting weak acid citrate with three carboxylic groups. (c) The molecular structure of acetic acid whose household name is vinegar.

Carboxylic Acid Functional Group

Citric Acid

Acetic Acid

Sulfhydryl (Thiol) Group —SH   Covalently bonded sulfur atoms have very important and diverse roles in cooking and baking. For example, many proteins contain the amino acid cysteine, and cysteine has an —SH group. When sulfur is bonded to a hydrogen atom, we call the functional group a sulfhydryl or thiol group and designate it as —SH (Figure A1.4). Most proteins found in plant and animal tissues have various amounts of cysteine and therefore sulfhydryl groups. However, the sulfur in cysteine does not have to remain bonded to a hydrogen; it can also be bonded to another sulfur atom (often found in a different cysteine amino acid) when a chemical reaction, called an oxidation reaction, occurs, resulting in the formation of a covalent disulfide bond (—S—S—). Proteins often require disulfide bonds to be present to keep the protein folded in a functional, native state and in solution (Figure A1.5). However, because an S—S bond is weaker than a C—C bond, heat can break disulfide bonds. The more disulfide bonds, the more heat that is required to break them and unfold the proteins. Some compounds will change the "oxidation state" of disulfide bonds and will contribute to the denaturing of the protein. In cooking, we visualize this process of protein unfolding when we cook eggs. Eggs have several different kinds of proteins. Those found in egg whites have relatively few disulfide bonds, and low levels of heat cause the proteins to denature. You observe this when the egg whites change from clear to white and "cook" in your warm skillet. In contrast, proteins found in the egg yolk have more disulfides and require a higher temperature to unfold and "cook" these proteins. Disulfides also play an important role in baking and wheat.

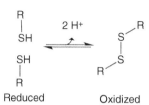

Reduced   Oxidized

Figure A1.4   Thiol functional groups. (a) The basic convention for a reduced sulfhydryl with R as an undetermined carbon group. (b) The change in oxidation state of a sulfhydryl group from reduced (R—SH) to oxidized (R—S—S—R).

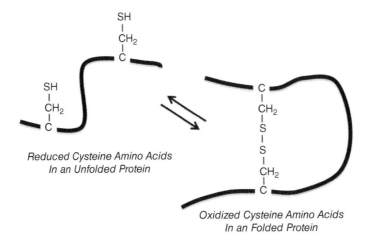

Figure A1.5 The important role of cysteine sulfhydryl functional groups. When proteins are folded, the sulfhydryl groups of two cysteine amino acids are involved in maintaining the shape of the protein. Loss of the bond by reduction will result in the loss or denaturation of the shape of the protein.

Answers

Chapter 1   Atoms, Elements, Compounds, and Molecules

1. **a.** Gas, **b.** Solid, **d.** Liquid
2. **a.** Electrons, **b.** Protons, **c.** Protons and Neutrons, **d.** Neutrons, **e.** Electrons
3. **a.** O, Oxygen, **b.** K, Potassium, **c.** Fe, Iron, **d.** Cl, Chlorine
4. **a** and **d**
5. **b, c,** and **e**
6. **a, d,** and **e**
7. **a.** Hydrogen bonding **b.** Electrostatic **c.** hydrogen bonding **d.** hydrophobic
8. Figure should show polar water molecules interacting with the polar O-H bond of ethanol via the attraction of negative and positive charges. Hydrogen bonds are shown between opposite partial charges as dotted/dashed lines.
9. The salt ions interact with the polar water molecules. The polar water molecules cluster around the negative and positive salt ions disrupting the formation of the ordered hexagonal array needed to make ice.
10. **a.** Water expands when it freezes. The solid water packed into an organized hexagonal array with space between the molecules. The solid takes up more space than the liquid water – so it expands and ruptures the cells.
    b. Smaller ice crystals are less likely to puncture the cells so the fluid doesn't leak out when the fruit/veggie thaws.

Chapter 2   Macromolecules of Food and Cooking

1. **a.** Amino acid
   b. Peptide bonds (covalent) connect amino acids together
   c. 4 amino acids
2. Structure A will have a higher melting point than B. Structure A is a saturated fatty acid. The straight chain of carbons allows the chains to pack together closely for maximum hydrophobic (van der Waals) interactions. Unsaturated fatty acids like Structure B have kinked chains, which prevent the chains from packing tightly together, lowering the melting point. Structure A also has a longer carbon chain than B, increasing the melting point.
3. To prevent the oil from decomposing, select an oil with a smoke point higher than 400°F, such as clarified butter, corn, avocado, peanut, or safflower oil.
4. Both amylose and cellulose are derived from plants, contain glucose monomers, and are linked by glycosidic bonds. However, amylose is a water-soluble storage polysaccharide linked by α-1, 4-glycosidic bonds, while cellulose is a water-insoluble structural polysaccharide linked by β-1, 4-glycosidic bonds. Unlike ruminant animals, humans do not have an enzyme capable of breaking the β-1, 4-glycosidic bonds, making it nutritionally unavailable.

The Science of Cooking: Understanding the Biology and Chemistry Behind Food and Cooking, Second Edition. Joseph J. Provost et al.
© 2025 John Wiley & Sons, Inc. Published 2025 by John Wiley & Sons, Inc.
Companion website: www.wiley.com/go/provost/food_science_2e

5. Denaturation causes the secondary, tertiary, and quaternary structures to be disrupted. When cooking, one can see the denaturation process through color change (transparent to opaque), the protein firms up, or water weeps from the denatured proteins.

6. **a.** Ionic bond.
   b. By adding acid, the H^+ ion will be attracted to the negative charge on the oxygen atom. Once the H^+ ion binds to the oxygen atom, the side chain of aspartate (left) becomes neutral and no longer attracted to the positive charge on lysine.

7.

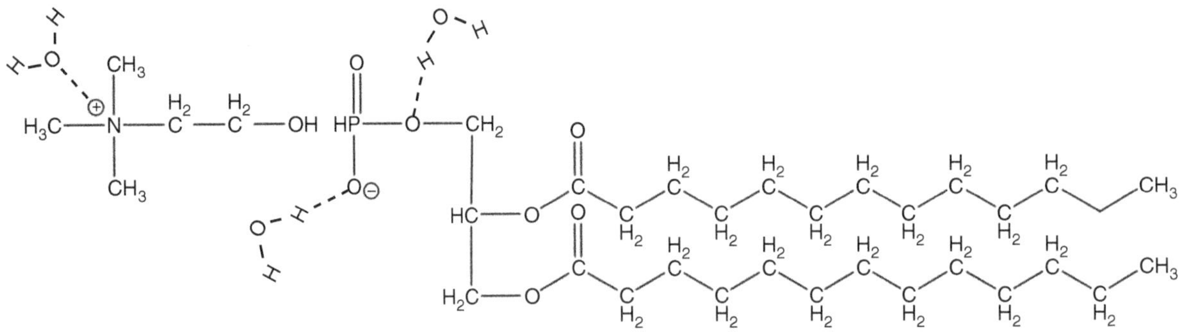

8. The oil (hydrophobic) and vinegar (hydrophilic) would naturally separate into two distinct phases. The egg yolks contain lecithin, a phospholipid emulsifier. Lecithin is able to bind both the hydrophobic oil and hydrophilic vinegar, making a stable emulsion.

9. **a.** 9 Cal/g × 4.5 g = 40.5 Calories are from fat
   b. 8 g Starch: Starch = Total Carbohydrate − Dietary Fiber − Total Sugars
   c. Amylose and amylopectin (plant starch)
   d. No, all sugars are added.
   e. Yes, a %DV close to or greater than 20% is considered a high source (dietary fiber = 18%)

10. **a.** Unsaturated fatty acid (monounsaturated, omega-6)
    b. Emulsifier (phospholipid)
    c. Carbohydrate, sugar, simple carbohydrate, disaccharide
    d. Amino acid
    e. Protein
    f. Carbohydrate, complex carbohydrate, polysaccharide

Chapter 3   Browning Reactions

1. **c**
2. **d**
3. **b**
4. **a.** 3, **b.** 1, **c.** 2, **d.** 4
5. Caramelan, caramelen, and caramelin are responsible for both the texture and brown color of caramel.
6. Proteins and sugars in butter react to form Maillard products. Adding "brown butter" can incorporate flavorful aroma compounds into your cookies that regular butter alone doesn't contain. It deepens the flavor profile of baked goods with nutty, toasty, and roasted aroma compounds.

7. Bread is carbohydrate-rich but low in protein. Therefore, more protein is needed To make appreciable products from the Maillard reaction. Adding an egg wash provides the protein (amino acids) needed to create a Maillard reaction on the surface of the bread loaf.

8. The addition of baking soda increases the pH of the food, which speeds up the Maillard reaction. High pH favors the open ring structure of reducing sugars, which are necessary for the reaction to proceed. A higher pH also ensures the nitrogen in amino acids remains unprotonated.

9. Because plants cannot move away from pests, they rely on chemical reactions to discourage attacks. Once an insect or pest bites into a plant and the vacuole ruptures, the phenols and polyphenol oxidase enzymes mix, resulting in the fruit browning reaction. The brown, bitter products that form prevent further attacks from pests and inhibit the formation of mold and fungus in the wounded plant.

10. **a.** Add acid like lemon or lime juice, lowering the pH to be outside the optimal range of the polyphenol oxidase (PPO) enzyme.

    b. Remove oxygen by placing the guacamole in an air-tight container or adding a layer of water to the top of the guacamole.

    c. Reduce the temperature by placing the leftovers in the refrigerator or freezer. The PPO enzyme is more active at higher temperatures.

Chapter 4   Milk and Ice Cream

1. **d.** Lactose
2. **b.** Convert lactose into glucose and galactose
3. **b.** Lactobacilli
4. **c.** Carbohydrates
5. **c.** Casein and whey
6. **a.** Casein proteins
7. **c.** To support the stability of micelles
8. **d.** To break the milk fat globules into smaller and more uniformly sized particles
9. **b.** It heats milk at temperatures sufficient to kill harmful microorganisms while preserving its nutritive properties.
10. **c.** Denatured proteins
11. **a.** Heavy cream has a higher fat content than skim milk.
12. **c.** Double bonds in their fatty acid chains
13. **a.** It increases the level of saturated fats.
14. **d.** They stabilize the mixture of immiscible ingredients.
15. **d.** Diacetyl
16. **b.** To stabilize the mixture and prevent the separation of fat and water phases
17. **c.** It helps in achieving a smooth and creamy texture.
18. **a.** Air adds body and chewiness to the ice cream

Chapter 5   Cheese

1. **b.** Lactobacteria and lactose
2. **d.** Lactose
3. **b.** Glucose and galactose
4. **b.** Lactobacteria
5. **b.** It becomes more acidic.

6. **b.** To kill harmful bacteria and extend shelf life
7. **c.** Denaturation of whey proteins
8. **d.** Emulsification of milk fat
9. The purpose of adding starter bacteria to pasteurized milk during the cheesemaking process is to initiate lactic acid fermentation. The starter bacteria ferment lactose, the primary sugar in milk, and convert it into lactic acid. This fermentation process lowers the pH of the milk, which is crucial for curd formation and the development of flavor and texture in cheese. Additionally, the starter bacteria outcompete harmful bacteria, ensuring the safety and quality of the cheese.
10. **b.** Whey is the liquid byproduct that is separated from the curd during cheesemaking.
11. **b.** Rennet cleaves casein proteins, specifically κ-casein, to induce milk coagulation and curd formation.
12. Overall, the enzymatic action of rennet, particularly chymosin, is essential for coagulation, curd formation, and the overall quality of cheese during the cheesemaking process. Rennet contains an enzyme called chymosin (or rennin), which plays a pivotal role in the cheesemaking process. Chymosin acts on the milk protein called casein. Specifically, chymosin cleaves the kappa-casein protein, one of the subtypes of casein, at a specific site. This cleavage disrupts the stabilizing bonds within the casein micelles, causing them to aggregate and form a gel-like network. As a result, the milk undergoes coagulation, leading to the separation of solid curds and liquid whey. The enzymatic action of rennet not only initiates coagulation but also influences the texture, flavor, and aging characteristics of the cheese. By breaking down the kappa-casein, chymosin alters the protein matrix and facilitates the expulsion of whey from the curds. This expulsion of whey further concentrates the remaining proteins and solids in the curds, contributing to the desired texture and firmness of the cheese. Additionally, chymosin affects the proteolysis (breakdown of proteins) that occurs during cheese aging, influencing the development of complex flavors and aromas in the final product.
13. Hydrolysis
14. The chemical processes result in the formation of various aromatic and flavorful compounds. For example, the breakdown of fatty acids can produce short-chain fatty acids, such as butyric acid, which imparts a buttery taste. Other compounds, such as diacetyl, ethanal (acetaldehyde), and ethanol, are formed during the breakdown of lactic acid and citric acid produced by bacterial fermentation. These compounds contribute to the overall aroma and flavor profile of the cheese. During cheese ripening, the breakdown of lipids/fats involves several chemical processes. Lipases, which are enzymes that degrade lipids, play a crucial role in this process. Here is an explanation of the chemical processes involved: Hydrolysis: Lipases catalyze the hydrolysis of triacylglycerol fats present in milk fat. This results in the cleavage of the ester bonds in the fats, producing free fatty acids. Oxidation/Reduction: The free fatty acids undergo oxidation/reduction reactions. Beta-oxidation is a key process where the fatty acid is oxidized at the carbon adjacent to the carboxylic acid group. This leads to the release of acetyl-CoA units in a cyclical manner. Decarboxylation: The acetyl-CoA units undergo decarboxylation, reducing the fatty acid chain length by two-carbon units and resulting in the formation of ketones or alcohols. Delta-Oxidation: In delta-oxidation, an oxygen atom is added to the carbon located four units away from the carboxylic acid group. This leads to the formation of δ-hydroxyacids, which can further cyclize to form lactones. Ester Formation: Fatty acids can also be converted into esters through esterification reactions.

Chapter 6   Fruits and Vegetables

1. **c.** Endosperm
2. **a.** It stops
   b. The peach will spoil first. The respiration rate measures how quickly the plant uses up its stores of sugar for energy. The faster a plant respires (without any photosynthesis to restore energy supplies), the sooner it will die (spoil).
   c. Keep it cold (slow down the chemistry). Limit oxygen.

3. **a.** The enzyme (PPO) that causes the browning reactions is inactivated by the heat.
   b. Vitamin C is an antioxidant. Since the reaction to make the brown pigment is an oxidation, adding Vitamin C prevents the reaction from occurring.
   c. The acid in the lemon/lime juice slows the PPO enzyme-catalyzed oxidation down, preventing the browning reaction.
4. **a.** Paper bag traps the hormone ethylene (a gas) around the fruit and accelerates the ripening.
   b. The ripe bananas are emitting ethylene. The green tomatoes are climacteric ripeners so exposure to ethylene causes them to ripen.
   c. Strawberries are non-climacteric ripeners—they do not ripen with exposure to ethylene/removal from the plant.
5. Alkaline—alkaline conditions accelerate the breakdown of hemicelluloses (a component of the cell wall that gives plants their rigidity). The carrots will quickly become mushy

   Neutral—faster than acidic conditions, but slower than alkaline.

   Acidic—hemicelluloses are resistant to breakdown in acidic conditions—so the carrots retain firmness for longer.
6. **a.** This is an Atwater "general factor" calculation:

   19.6 g total carb – 2.1 g dietary fiber = 17.5 g digestible carbohydrate

   17.5 g digestible carb × 4 cal/g carb = 70 calories from carbohydrate

   2.0 g protein × 4 cal/g protein = 8 calories from protein

   0.1 g fat × 9 cal/g fat = 0.9 cal from fat

   70(carb) + 8 (protein) + 0.9 (fat) = 78.9 calories

   Why don't the calories exactly match up? The USDA food data central database (https://fdc.nal.usda.gov/) uses an Atwater "specific factor" calculation according to Merrill and Watt (1973)[1]: 4.03 cal/g for carbohydrate, 8.37 cal/g for fat and 2.78 cal/g for protein coming from starchy roots like potato. Food manufacturers may use either an Atwater general or an Atwater specific calculation.
   b. Cauliflower has more dietary fiber per serving of carbohydrate (compared to potato). Dietary fiber is not digestible by humans—so it does not count toward the total calories.
   c. Sugar

   Dietary fiber

   Starch (amylose and amylopectin)
   d. The cells of the cauliflower have thick cell walls with lots of cellulose, hemicelluloses, etc.—this requires longer time to cook and break down.
7. The starch granules absorb water. The starch molecules move around and form hydrogen bonds between the chains; the intermolecular attractions create a soft solid called a gel.
8. **a.** Volatile molecules—terpenes (primarily linalool and pinene).
   b. The smell/taste would be different because the volatile terpenes would have escaped/evaporated in the drying process.
   c. Both foods have the same flavor molecules (specifically, the terpenes pinene and linalool) creating similar flavor profiles.
9. All of the above
10. **b.** Amylopectin

---

1 Annabel Merrill; Bernice Watt (1973). Energy Values of Food … basis and derivation (PDF). United States Department of Agriculture.

Chapter 7   Meat and Fish

1. **b.** Collagen converts to gelatin during cooking, contributing to tenderness.
2. **d.** Oxidation state and binding of myoglobin
3. **a.** Utilizing plant proteins, fats, and flavor enhancers.
4. **a.** Increased collagen cross-linking and stiffness.
5. **c.** The tissue will tear, giving a mushy feel to the tooth
6. **d.** Trimethylamine oxides found in ocean fish
7. **c.** the need to balance the salt content of their environment, limiting osmotic water loss
8. **b.** Myotomes
9. **c.** Density of muscle fibers
10. **d.** Enhancing flavor and tenderizing the meat
11. **a.** Breaking down proteins into amino acids and peptides
12. Answer: During rigor mortis, the depletion of ATP in muscle fibers has profound effects on the meat's structure and quality. ATP is essential for muscle contraction and relaxation when an animal is alive. However, after death, ATP levels decline, and the ATP-dependent ion pumps fail to maintain the ion balance in the muscle cells. These enzymes start breaking down the muscle fibers' structural proteins, such as myosin and actin, resulting in the tenderization of meat over time. Furthermore, the absence of ATP impedes the cross-bridge detachment between myosin and actin filaments, further contributing to the development of rigor mortis. The muscle fibers become stiff and rigid due to the constant attachment of myosin heads to actin filaments. As rigor mortis progresses, the breakdown of glycogen in the muscles leads to an accumulation of lactic acid, causing the meat's pH to decrease. The decline in pH, in turn, affects various meat properties, including water-holding capacity, color, and flavor.
13. Answer: Plant-based meat products aim to replicate the texture and mouthfeel of animal-based meat, providing a similar sensory experience to consumers. This is achieved through careful selection and processing of plant proteins. Soy, pea, and wheat gluten proteins are often used due to their ability to form a protein matrix that mimics the fibrous structure of animal muscle. These proteins can be modified through various techniques, including extrusion, denaturation, and restructuring, to enhance their texture and create a more meat-like appearance. Another approach involves incorporating additional ingredients, such as fats and binders, to improve the juiciness and binding properties of plant-based meat. Fats, such as coconut oil or sunflower oil, are added to provide a similar mouthfeel and improve the overall flavor profile. Binders like methylcellulose or carrageenan are used to enhance the cohesiveness and firmness of the product. Additionally, texture modifiers, such as cellulose or starch, may be included to mimic the texture of animal-based meat. These modifiers help create a fibrous structure and give plant-based meat a chewy texture.
14. The term GRAS, which stands for Generally Recognized as Safe, has been utilized by Impossible Foods to argue against the need for a GMO label on their plant-based meat products. This decision has generated debate, with varying perspectives regarding its merits and drawbacks.

    Arguments in favor of the decision:

    1. Scientific consensus on safety: Proponents of the use of GRAS argue that extensive scientific research has affirmed the safety of genetically modified organisms (GMOs) used in plant-based meats. They contend that rigorous testing and regulatory assessments have demonstrated that these GMO ingredients pose no significant risks to human health or the environment.
    2. Consumer acceptance and accessibility: Advocates argue that by not labeling their plant-based meat products as GMOs, companies like Impossible Foods can increase consumer acceptance and accessibility. For some individuals, the presence of a GMO label may create unwarranted concerns or misconceptions about the safety or quality of the product. Removing the label can facilitate wider adoption and market penetration.

3. Promoting sustainable alternatives: Supporters assert that by focusing on the sustainability and environmental benefits of plant-based meats, the emphasis should be on encouraging the shift away from conventional meat consumption. They argue that the absence of a GMO label allows consumers to concentrate on the positive ecological impact of plant-based options, such as reduced greenhouse gas emissions and land use.

Arguments against the decision:

1. Right to know and transparency: Critics argue that consumers have the right to know whether their products contain GMO ingredients. They believe that labeling provides transparency and empowers individuals to make informed choices based on their personal preferences or concerns, even if the scientific consensus suggests that GMOs are safe.

2. Ethical and moral considerations: Opponents contend that concealing the presence of GMOs through the GRAS designation undermines the principles of transparency and honesty. They argue that consumers deserve accurate information about the production methods and ingredients used in their products, including GMOs.

3. Potential long-term effects: Skeptics raise concerns about the long-term health and environmental effects of GMOs. They assert that labeling allows for ongoing monitoring and assessment of these effects, enabling researchers and regulatory bodies to identify any unforeseen risks or impacts that may emerge over time.

In summary, the decision by Impossible Foods to argue around the GMO label using the term GRAS is a subject of debate. While some say it promotes consumer acceptance and focuses on sustainability, others contend it compromises transparency and the right to know. Ultimately, perspectives on this matter may vary based on individual beliefs, values, and priorities regarding food safety, consumer awareness, and the environmental impact of food production.

15. Answer: When fish is cooked, especially ocean fish, it can release volatile compounds that contribute to its characteristic fishy odor. One of the primary compounds responsible for this odor is trimethylamine (TMA), which has a chemical formula of $(CH_3)_3N$. TMA is a weak base and has a characteristic aroma of rotten fish. It exists in equilibrium with its protonated form, known as trimethylammonium ion (TMAH), in an acid-base reaction:

$$(CH_3)_3N \rightleftharpoons (CH_3)_3NH^+ + OH^-$$

The odor of TMA is more pronounced when it is in its free base form, $(CH_3)_3N$. The intensity of the fishy smell can be reduced by adding acidic liquids, such as lemon juice, which lowers the pH of the surrounding environment. The pH value of a solution determines the concentration of hydrogen ions (H^+). The lower the pH, the higher the concentration of H^+ ions. By adding an acidic liquid like lemon juice with a low pH, the concentration of H^+ ions increases, favoring the protonation of trimethylamine (TMA) to its less volatile and less odorous protonated form, trimethylammonium ion (TMAH). At a pH below 9.8, the compound will predominantly exist in its protonated (charged) and water soluble form (TMAH), reducing fishy odor. Therefore, when lemon juice or other acidic liquids with a pH below 9.8 are added to fish during cooking, they create an acidic environment that promotes the protonation of trimethylamine, reducing its volatility and fishy odor.

Chapter 8   Eggs, Custards, and Foams

1. **b.** The proteins with less intermolecular forces take less energy (heat) to denature the protein
2. **c.** Often unfolds and denatures proteins
3. Answer: When an egg is laid, it contains carbon dioxide (CO_2) within the egg white. The CO_2 combines with water (H_2O) to form carbonic acid (H_2CO_3) in a reversible reaction:

$$CO_2 + H_2O \rightleftharpoons H_2CO_3 \,(\text{Reaction \#1})$$

The carbonic acid contributes to the initial acidity of the fresh egg.

Eventually, the egg loses moisture and carbon dioxide through the pores in its shell, and the concentrations inside the egg decrease. This causes a shift in equilibrium in Reaction #1, resulting in the breakdown of carbonic acid to produce more water and carbon dioxide gas – ultimately replacing the molecules that escaped through the eggshell.

As the carbonic acid is converted to water and carbon dioxide gas, the acidity decreases, and the pH increases inside the eggshell. Over time, the egg becomes more alkaline/basic, as the concentration of carbonic acid decreases.

   i. Answer: Since Grade B eggs are older than Grade A or AA eggs, more water carbon dioxide has escaped through the pores in the eggshell. This means Grade B eggs have less carbonic acid and, therefore, a higher (more alkaline) pH than Grade A or AA eggs.

  ii. Answer: Because the yolk contains considerably less water than the white, water gradually crosses the membrane from the white into the yolk via osmosis. The increase in water makes the yolk swell, and the yolk membrane weakens. As eggs age, the yolk's membrane weakens, making it more fragile and susceptible to breaking. Therefore, the yolks in older Grade B eggs would break easier than those in Grade A or AA eggs.

 iii. Answer: Grade B eggs have less carbonic acid and, therefore, a higher (more alkaline) pH than Grade A or AA eggs. Cysteine amino acids inside the egg white degrade more readily under alkaline/basic conditions, producing higher amounts of hydrogen sulfide gas (i.e. eggy smell).

4. **b.** Using a copper bowl; e. Adding Acid

5. **b.** Metal bound ovotransferrin requires more energy to denature because it contains more non-covalent interactions between its atoms

  d. The number of covalent bonds in the protein doesn't change. The greater number of non-covalent interactions (due to the organizing effect of the metal) means more energy is needed to break them.

6.   **i.** Answer: Hydrogen sulfide, H_2S.

  ii. Answer: As the egg ages, the egg loses moisture and carbon dioxide through the pores in its shell, and the concentrations inside the egg decrease. Carbonic acid inside the egg is converted to water and carbon dioxide gas to replace the molecules that escaped the eggshell. As carbonic acid is reacted, the acidity of the egg decreases, and the pH increases.

 iii. Answer: Older eggs are more alkaline and cysteine amino acids inside egg whites react faster in alkaline conditions. This causes a higher production of hydrogen sulfide gas.

 iv. Answer: Inside the hardboiled egg, you would see a green-gray ring formed at the interface between the egg yolk and egg white. Hydrogen sulfide forms from the decomposition of cysteine amino acids inside the egg white, which reacts with the iron in the egg yolk. The green-gray color indicates the formation of iron sulfide, FeS.

7. Answer: Due to different intermolecular forces, a salad dressing made of only oil (hydrophobic) and vinegar (hydrophilic) will separate into a heterogeneous mixture. Egg yolks, which contain an amphiphilic phospholipid (i.e. lecithin), are added to dressings like mayonnaise to stabilize the emulsion of oil and vinegar. The hydrophobic region of the phospholipid interacts with the oil, while the hydrophilic part of the phospholipid interacts with the vinegar.

Chapter 9 Breads, Cakes, and Pastry

1. **d.** Germ, endosperm, and bran
2. **d.** Germ
3. **b.** Hard flour has higher protein content than soft flour
4. **c.** Glutenin and gliadin

5. **a.** They contribute to the stability and mechanical rigidity of the gluten network
6. **b.** CO_2 and alcohol
7. **c.** Maillard reactions
8. **c.** Flour
9. **a.** Baking powder
10. Answer: Starch gelatinization is a process that occurs when starch is heated in the presence of water, resulting in the swelling and rupture of starch granules. This process involves the absorption of water by starch granules, causing them to increase in size and form a viscous gel. As the temperature rises, the hydrogen bonds holding the starch molecules together break, leading to an increase in viscosity.

    Retrogradation, on the other hand, refers to the process in which the gelatinized starch undergoes structural changes upon cooling. During retrogradation, the starch molecules realign and recrystallize, forming a network with decreased solubility and increased firmness. This can result in the formation of a gel or the reoccurrence of a granular texture in cooked starches.

    Changing starch sources can have a significant impact on the type of thickening results during cooking. Different starches have varying compositions of amylose and amylopectin, which influence their thickening properties. For example, starches with a higher amylose content, such as cornstarch, tend to form thicker gels with a more cohesive texture. In contrast, starches with a higher amylopectin content, like waxy rice starch, produce a softer gel with less water loss (weeping of liquid).
11. Answer: Gluten is a complex protein network that provides structure, elasticity, and texture to baked goods. It is formed when two wheat proteins, gliadin and glutenin, interact and form gluten strands. These proteins combine with water and undergo mechanical manipulation, such as stirring, mixing, or kneading, which aligns the glutenin molecules and allows them to form cross-linkages with gliadin, resulting in the development of gluten.
12. Answer: The Maillard reactions occur between reducing sugars (such as glucose and fructose) and amino acids, primarily found in proteins, during baking. These reactions involve a series of complex chemical reactions, including sugar degradation, condensation, and rearrangement. The Maillard reactions contribute to the browning of the bread crust, enhancing its color and forming a range of desirable flavors and aromas, such as nutty, toasty, and caramel-like notes.
13. Answer: Celiac disease is an autoimmune disorder in which the ingestion of gluten triggers an immune response, causing damage to the lining of the small intestine. Individuals with celiac disease produce antibodies that mistakenly recognize gluten as a harmful substance. This immune response leads to inflammation and damage to the intestinal villi, impairing nutrient absorption. When baking gluten-free bread, the absence of gluten poses challenges in achieving a desirable texture and structure. Gluten provides elasticity and structure to dough, which is difficult to replicate with alternative gluten-free ingredients. Achieving a similar level of rise, softness, and chewiness in gluten-free bread requires the use of alternative binding agents, such as xanthan gum or guar gum, and a careful balance of other ingredients to compensate for the absence of gluten.

Chapter 10   Seasonings

1. **a.** A compound formed by the reaction of an acid and a base
2. **b.** It prevents protein aggregation and promotes protein solubility.
3. **a.** Salt is an inorganic mineral.
4. **a.** Leaves
5. **d.** Volatile organic compounds
6. **c.** It ensures the growth of vanilla vines in different tropical areas.
7. **a.** Vanillin

8. **a.** Single nucleotide polymorphism (SNP)

9. **b.** Crocin

10. Answer: Curry powder derives its unique flavor and aroma from a complex mixture of chemical compounds present in its constituent spices. The primary compounds responsible for these sensory characteristics include; Turmeric, a key ingredient in curry powder, which contains a group of compounds known as curcuminoids. The most abundant curcuminoid is curcumin, which imparts the characteristic bright yellow color to curry powder. Curcumin also exhibits antioxidant and anti-inflammatory properties, contributing to the potential health benefits associated with consuming curry. Terpenes: Several spices used in curry powder, such as coriander, cumin, and fenugreek, contain terpenes. These organic compounds contribute to the aroma and flavor of curry powder. For example, coriander contains linalool and pinene, which provide a citrusy and pine-like scent, while cumin contains cuminaldehyde, contributing to its warm and earthy aroma. Capsaicinoids: Chili peppers, often included in curry powder, contain capsaicinoids, with capsaicin being the most well-known. Capsaicin activates heat receptors in the mouth, resulting in the characteristic spiciness or heat sensation associated with curry. The concentration of capsaicinoids determines the overall spiciness of the curry powder. Piperine: Black pepper, another common ingredient in curry powder, contains piperine. Piperine enhances the perceived spiciness by interacting with taste receptors and has been shown to improve the bioavailability of certain nutrients.

11. **b.** Carotenoids

12. **b.** By activating pain receptors in the nervous system

13. Answer: The mechanism of action of TRPV receptors in sensory perception holds significant implications for understanding pain and temperature sensations. These receptors play a crucial role in detecting and transmitting sensory signals related to pain and temperature extremes, such as heat or cold. By interpreting the mechanism of action of TRPV receptors, we can gain insights into the intricate workings of our sensory system. TRPV receptors, found on specialized sensory nerves called nociceptors, are activated by specific stimuli. Upon activation, these receptors open ion channels, allowing the influx of calcium and other ions into the nerve cell. This activation triggers a series of intracellular events that result in the transmission of pain signals to the central nervous system.

14. Answer: The claims regarding the health benefits of Himalayan salts and sea salt are often attributed to their mineral content. Both salts contain various minerals that are considered essential for human health, such as sodium, potassium, magnesium, and calcium. Proponents of Himalayan salt argue that its unique pink color is due to the presence of trace minerals, which are believed to offer additional health benefits. These minerals include iron, zinc, copper, and selenium, among others. It is suggested that these trace minerals may contribute to improved hydration, electrolyte balance, and even detoxification processes in the body.

Similarly, sea salt is touted for its mineral-rich composition, obtained from the evaporation of seawater. The minerals present in sea salt may vary depending on the source and processing method. Sea salt is often praised for its higher levels of magnesium and trace elements compared to refined table salt, which is heavily processed and stripped of most minerals.

However, it is important to note that the mineral content in both Himalayan salt and sea salt is relatively minimal compared to the overall dietary intake of minerals from other food sources. The actual contribution of these salts to overall nutrient intake is minimal, and the health benefits associated with their mineral content may be overstated. Furthermore, the claims surrounding the health benefits of Himalayan salt and sea salt have been met with skepticism by some experts. The limited scientific evidence available does not support significant health advantages of these salts over regular table salt in terms of nutritional value or health outcomes. While Himalayan salt and sea salt may contain trace minerals that can contribute to overall mineral intake, the extent of their health benefits is still a subject of debate. The claims made about their superiority or unique health-promoting properties are not well-supported by scientific evidence. It is advisable to focus on a balanced and varied diet to meet the body's nutritional needs rather than relying solely on specific types of salt for health benefits.

15. Answer: The different kinds of capsaicin found in pepper plants include capsaicin, dihydrocapsaicin, nordihydrocapsaicin, homocapsaicin, and homodihydrocapsaicin. Among these, capsaicin is the primary compound responsible for the heat in peppers. Dihydrocapsaicin is another prominent capsaicinoid. The heat level of a pepper is determined by the concentration of capsaicinoids, particularly capsaicin. The more capsaicin present, the hotter the pepper. Individual sensitivity to capsaicinoids can vary, and the Scoville heat unit (SHU) scale is used to measure spiciness based on the concentration of capsaicinoids.

Chapter 11   Beer and Wine

1. **a.** To produce carbon dioxide and ethanol
2. **b.** Isobutanol and isoamyl alcohol
3. **b.** Hydroxyl group
4. **c.** The point where both ethanol and water evaporate at the same rate
5. **b.** Alcohol dehydrogenase (ADH)
6. Alcohol metabolism occurs in a two-step process. First, alcohol dehydrogenase (ADH) in the liver and stomach breaks down ethanol into acetaldehyde. Second, aldehyde dehydrogenase (ALDH) further converts acetaldehyde into acetic acid. The resulting acetic acid is then metabolized into carbon dioxide and water, allowing the body to eliminate alcohol efficiently.
7. Genetic variations in ADH and ALDH enzymes can impact an individual's response to alcohol consumption. Specific alleles may lead to increased or decreased enzyme activity, affecting the rate of ethanol and acetaldehyde metabolism. Some variants may increase the risk of alcohol use disorder (AUD), while others may cause adverse reactions like the "Asian Flush." Additionally, certain genetic profiles may be associated with an increased risk of fetal alcohol syndrome (FAS) in pregnant women who consume alcohol. However, it's important to note that alcohol dependence is a multifactorial condition influenced by both genetic and environmental factors.
8. When alcohol is consumed acutely, it quickly affects the brain's NMDA and GABA(A) receptors. Acute Alcohol Intake: Alcohol enhances GABA(A) receptors, causing relaxation, and inhibits NMDA receptors, reducing brain activity, leading to euphoria. Chronic Alcohol Use: With regular use, tolerance develops, and GABA(A) receptors become less sensitive. Glutamate activity increases, compensating for alcohol's effects, potentially causing brain damage. Alcohol Withdrawal: Without alcohol, the brain becomes hyperexcitable due to reduced GABA(A) activity. Increased glutamate can lead to seizures and psychological symptoms like anxiety and depression. Post-Withdrawal Phase: After withdrawal, the brain slowly adjusts, but some changes may persist, affecting anxiety and stress sensitivity. Seeking professional help during withdrawal is crucial.
9. **c.** Mashing
10. **c.** Maceration
11. **b.** Polyphenols
12. **c.** They metabolize sugar into ethanol and other flavor compounds.
13. **a.** To enhance color extraction during maceration.
14. Different yeast strains used during fermentation can produce varying levels of flavor compounds, such as esters and aromatics, which contribute to the unique character of the wine. Some yeast strains may produce fruity or floral notes, while others can create more complex and savory flavors. The choice of yeast strain can significantly influence the overall flavor development of the wine.
15. Bottom-fermenting yeasts are essential in lager production due to their unique characteristics and the distinct flavor profile they impart to the beer. These yeasts ferment at cooler temperatures, usually between 7 °C and 13 °C, which slows down the fermentation process and contributes to a more refined and clean taste. Unlike top-fermenting yeasts used in ales, bottom-fermenting yeasts settle at the bottom of the fermenter during the

process. This settling contributes to a clearer beer with fewer impurities. Furthermore, bottom-fermenting yeasts produce fewer esters and other byproducts compared to their top-fermenting counterparts, resulting in lagers with a crisp and clear flavor profile, often described as smooth and clean. This particular fermentation method is preferred when aiming to produce lagers that are refreshing and have a subtle flavor, making it ideal for those who appreciate a beer with a smooth finish and less fruity or spicy notes.

Chapter 12   Non-Alcoholic Beverages

1. **c**
2. **d**
3. **b**
4. **d**
5. **a**
6. **b**
7. Consumption is linked to a reduced risk of neurodegenerative disorders due to caffeine's ability to impact dopamine levels, and coffee is rich in antioxidants such as polyphenols and diterpenes, which help protect cells from damage.
8. Grind size affects the extraction rate, with coarser grinds leading to under-extraction and finer grinds extracting faster, potentially leading to an over-extracted, astringent coffee. Water temperature influences the solubility of compounds, with higher temperatures increasing solubility and aiding in the extraction process. Lower temperatures lead to a weak, acidic coffee, while higher temperatures produce a bitter, astringent taste.
9. White tea is the least processed of all tea types, which helps retain the highest level of antioxidants. It is made from dried young leaves and dried buds, preserving their natural chemical composition.
10. Tea leaves are heated to high temperatures during fixation to deactivate polyphenol oxidase (PPO) enzymes. This step is crucial because it prevents the enzymatic browning and oxidation of catechins, preserving the green color and high antioxidant content of teas like green tea.
11. Maillard reactions occur during the fixing or drying steps when L-theanine reacts with reducing sugars in the tea leaves. These reactions contribute to developing complex flavor molecules in the tea, enhancing its taste profile.
12. Cafestol is a hydrophobic compound due to the low percentage of polar bonds (C—O, O-H) and a high percentage of nonpolar bonds (C—C, C—H). Coffee filters are made of cellulose, a hydrophobic, water-insoluble material. Due to similar intermolecular forces, cafestol preferentially interacts with the hydrophobic coffee filter. Therefore, filter papers remove diterpenes in drip-brewed coffee.

Chapter 13   Sweets: Chocolates and Candies

1. **a.** Glucose and fructose
2. **e. a, b,** and **c**
3. **a.** The size of the sugar particles/crystals
4. **d. a** and **b**
5. **b.** Fermentation, roasting, grinding, conching, tempering
6. **f.** All of the above
7. **c.** Cacao beans contain high amounts of fat and sugar, so additional sugar does not need to be added to most chocolates

8. **a.** Hard, butterscotch disk candy

9. Answer: An invert sugar is a sugar mixture that contains sucrose, glucose, and fructose. Invert sugar is produced under acidic conditions when the disaccharide bond of sucrose is broken (by the enzyme invertase), and the free glucose and fructose molecules are released. Because invert sugars contain a mixture of different types of sugars, they don't crystallize easily and are more likely to be in a syrup/viscous liquid state, rather than a solid state. An invert sugar is even more hygroscopic than sucrose because of the additional water molecules that can surround and interact with two sugar molecules (glucose and fructose) relative to one molecule (sucrose). For cookies that contain invert sugars (like brown sugar cookies), this hygroscopic character results in a chewier cookie.

10. Answer: Cocoa butter is added to a chocolate liquor to balance out the added sugar, ensuring that there is enough fat for the cocoa and added sugar to remain suspended. This allows for the chocolate liquor to flow. Without added cocoa butter, the chocolate liquor would be a pasty, grainy solid.

11. Answer: Chocolate bloom occurs when the ideal fat crystals (polymorph IV) found in the original chocolate melt and then recrystallize in a less desirable form due to the temperature conditions of cooling. The desirable form of chocolate has fat particles that crystallize in the 29–32 °C (84.2–89.6 °F) temperature range (depending upon the chocolate type). When heated above this temperature, all of the fat in the chocolate melts. Bloom occurs when the chocolate is cooled down to a temperature that is less than 82 degrees or is cooled in an uncontrolled manner. Under these conditions, the less stable polymorph forms crystallize first, and then the more stable polymorph crystals develop over a period of hours to days. The crystal matrix becomes more compact (more polymorph V), which pushes the cocoa solids and sugar particles to the surface. It is these particles that lead to the whitish/grayish surface appearance of bloomed chocolate. The chocolate can be recovered to its original form by subjecting it to the entire tempering process.

12. Answer: Natural Cocoa powder is quite acidic, which impacts its taste, characteristics, and reactivity. Dutch-processed cocoa powder contains added base (or an alkaline solution) to counter this acidity, giving it a more neutral pH. Dutched cocoa powder has a darker brown appearance than natural cocoa powder due to the neutralization of naturally acidic cocoa powder. In a comparison of recipes, a recipe containing Dutch-processed cocoa powder is more often paired with baking powder, and natural cocoa powder is paired with baking soda. In addition, the taste of natural/regular cocoa powder can be more astringent and harsh, thus use Dutch-processed cocoa powder for a milder chocolate taste.

Chapter 14   The Science of Taste and Smell

1. **A** and **D**
2. Fill in the blank
   - **a.** Sucrose, glucose/dextrose, fructose
   - **b.** Any acid- citric, malic, lactic acid
   - **c.** Monosodium glutamate
3. Bitter alkaloids (i.e. caffeine) > tongue; volatile aromatics > nose
4. **a.** Acid
5. **b.** GPCRs
6. **d.** Salty and Sour
7. **b.** Sour
8. **d.** Amino acids and nucleotides
9. **a.** Pretty bitter
10. All of the above

Chapter 15   The Metabolism of Food

1. Answers
   a. NAD-H, because it "loses" the two electrons and the H
   b. Acetaldehyde. Because it "gains" the two electrons and the H
   c. A catalyst – it speeds up the reaction (instead of catalyst, they could say it speeds up the reaction without being changed by the reaction).
2. Answers
   a. To make ATP – for energy
   b. To recycle NAD – so they can eat more food, to make more energy
   c. The CO_2 bubbles would form a head of foam.
3. a. Acetaldehyde is reduced to ethanol
   b. NADH is oxidized to NAD
   e. Both oxidation and reduction are occurring
   f. NADH reduces acetaldehyde
4. a. Glycolysis is a pathway that contains 10 enzymes
   b. Glycolysis is the process (in yeast or in humans) by which sugar is broken down
   e. Glycolysis results in a net consumption of NAD
   f. Glycolysis results in a net production of ATP
5. a. To regenerate NAD+ for continued ATP production
6. a. The production of carbon dioxide and alcohol from glucose and oxygen
7. a. Anaerobic
   c. Happens in the absence of oxygen
8. Aerobic respiration
9. a. Mitochondria
   b. Oxygen
   c. The Krebs Cycle (i.e. the citric acid or TCA cycle)
10. Aerobic respiration

Index

The Science of Cooking: Understanding the Biology and Chemistry Behind Food and Cooking, Second Edition. Joseph J. Provost et al.
© 2025 John Wiley & Sons, Inc. Published 2025 by John Wiley & Sons, Inc.
Companion website: www.wiley.com/go/provost/food_science_2e